AF751291

Sustainability in Plant and Crop Protection

Volume 1

Series editor
Aurelio Ciancio, Sezione di Bari, Consiglio Nazionale delle Ricerche Istituto per la Protezione delle Piante, Bari, Italy

More information about this series at http://www.springer.com/series/13031

Raquel Campos-Herrera
Editor

Nematode Pathogenesis of Insects and Other Pests

Ecology and Applied Technologies for Sustainable Plant and Crop Protection

Springer

Editor
Raquel Campos-Herrera
University of Neuchâtel
Neuchâtel, Switzerland

Sustainability in Plant and Crop Protection
ISBN 978-3-319-18265-0 ISBN 978-3-319-18266-7 (eBook)
DOI 10.1007/978-3-319-18266-7

Library of Congress Control Number: 2015944708

Springer Cham Heidelberg New York Dordrecht London
© Springer International Publishing Switzerland 2015
Chapters 8, 9 were created within the capacity of an US governmental employment. US copyright protection does not apply.
This work is subject to copyright. All rights are reserved by the Publisher, whether the whole or part of the material is concerned, specifically the rights of translation, reprinting, reuse of illustrations, recitation, broadcasting, reproduction on microfilms or in any other physical way, and transmission or information storage and retrieval, electronic adaptation, computer software, or by similar or dissimilar methodology now known or hereafter developed.
The use of general descriptive names, registered names, trademarks, service marks, etc. in this publication does not imply, even in the absence of a specific statement, that such names are exempt from the relevant protective laws and regulations and therefore free for general use.
The publisher, the authors and the editors are safe to assume that the advice and information in this book are believed to be true and accurate at the date of publication. Neither the publisher nor the authors or the editors give a warranty, express or implied, with respect to the material contained herein or for any errors or omissions that may have been made.

Printed on acid-free paper

Springer International Publishing AG Switzerland is part of Springer Science+Business Media (www.springer.com)

To my father

Series Preface

This is the first Volume of the Series *Sustainability in Plant and Crop Protection*, produced thanks to the enthusiastic dedication and endless efforts of the Volume Editor, Raquel Campos Herrera, and all the contributing authors.

The Series stems from the previous *Integrated Management of Plants Pests and Diseases* Series that I had the pleasure to edit until its completion with the late Prof. K. G. Mukerji. Some considerations underpinning that Series hold for this new endeavour.

SUPP aims at presenting an annual coverage of topics in the field of sustainable agriculture. Each contribution will be dedicated to a specific theme, covering aspects having in common the concept of sustainability. In particular, the Series will target key issues in crop protection, including not only pest and disease control, but also methods applied to manage biotic and abiotic stresses, both at the plant and the whole crop levels.

The keyword "sustainable" is today commonly encountered among agricultural research projects and activities. It has various facets, deriving from the Latin word *sustineo* which means to endure or give support. By itself, the concept is simple and particularly suitable for the actual situation of the world agriculture. Enduring productions require the conservation of all the natural resources involved, to be passed to the new generations. In a parallel with the economy and finance, it is in some way like preserving the capital (or savings), living by the interests.

Are the agricultural "savings" still intact today? In my personal opinion we are already consuming this heritage, and the growing population expected by the next decades will keep this rate increasing. In this perspective, "sustainability" concerns actions aiming at saving the natural *and* agricultural resources and the benefits (food, commodities) they produce. This assumption may appear irrelevant or academic on the rich side of the world, but it assumes a dramatic and concrete meaning if the whole planet is considered. Shortage of food and environment destruction are not only a shame for the technological society we are immersed in, but also represent a potent, although unpredictable, force in history.

There are two ways to increase food production. The first is by increasing the land (or waters) invested. The second is by raising the average productivity per surface unit. The former cannot grow forever, since most fertile soils are already invested in agriculture. Furthermore, the rates of deforestation or oceans and coastal waters exploitation already affect the whole planet ecology. The second option requires an increase in the production of knowledge. In particular, it concerns the substitution of energy-based production cycles (i.e. based on massive use of energy-demanding pesticides or fertilizers) by information-based conservative approaches. Information means all data present *in forma*, inside a given system, including, i.e. data on ecology, species, genes and genomes, methods, management, cultural or anthropological aspects and so on. This is the challenge of these years, since it is the information we produce that will lead us during this century. In fact, we do not know yet how the actual food production and consumption rates will evolve, which kind of changes are coming, nor how the human population will sustain itself on earth, hopefully in a peaceful way.

The Series, hence, aims at providing a forum contribution to present, organize and discuss some aspects of the actual sustainability issues, as described.

This Volume is organized in three sections providing the reader a progressive focus, from basic research approaches to more defined topics, towards final applied aspects of insect control. The amount of information provided is impressive, updated through the research data produced by the laboratory and field work of the authors. The scientists who edited and contributed to the Volume have a high credibility and represent the leading edge of their field, both for their experience and scientific production. I consider this is a fortunate start for the Series and hope it will be followed by equally comprehensive and attractive contributions.

Bari, Italy Aurelio Ciancio

Preface

The ever-increasing scale of human activity, including agriculture, produces pollution and disturbances that have long been recognized as unsustainable. Governments and the public are now generally aware of the need to not only increase crop production levels, but to do it in ways that use fewer resources and are less destructive to natural ecosystems. This series, *Sustainability in Plant and Crop Protection*, is launched with the aim of providing comprehensive, multidisciplinary reviews of information that can be used to move toward more sustainable agricultural systems. In this first volume, we focus on nematodes as nonchemical alternatives for insect pest control, with special emphasis on entomopathogenic nematodes (EPNs). Native and commercial EPN populations are used in every continent for controlling insect pests in target crops. However, their efficacy and success require in-depth knowledge of their biology and ecology in our changing environment. This volume presents new advances on the biodiversity and distribution of native EPNs, discusses some key multitrophic interactions in the soil, and highlights critical factors affecting the survival and activity of EPNs in the environment as well as new production and release systems. To illustrate these fundamental bases, we provide several case studies from different countries and crop systems, to reflect new, rational, and optimal uses of EPNs.

The volume is organized in three sections. The first section covers the diversity, biology, and evolutionary aspects of the nematodes and their bacteria (Chap. 1), advances in improvement using a genetic approach (Chap. 2), behavior and population dynamics (Chap. 3), natural forces that drive EPN activity and survival in the soil (Chap. 4), trophic relationships in agricultural habitats and their implication for biocontrol (Chap. 5), and conservation or classical approaches for using EPNs (Chap. 6). The second section serves to transition between the basic research and those applications required to use EPNs for biological control. This section discusses key aspects of application technology and formulation (Chap. 7), presents for the first time the full details of "insect cadaver" application (Chap. 8), reviews recent advances in EPN application technology (Chap. 9), and presents the regulations and ecological impacts of EPN production and release in a global market (Chap. 10). The final section provides in-depth descriptions of experiences using EPNs to manage critical pests in different countries and systems. First, three long-term research programs developed in the USA illustrate the relevance of linking pest and EPN biology and ecology: alfalfa in New York (Chap. 11), turfgrass in New Jersey (Chap. 12), and orchards in Florida (Chap. 13). Advances in countries in which EPN legacy is more recent require special attention. Two chapters discuss the future and current reality in Cuba (Chap. 14) and Venezuela (Chap. 15), countries with limited resources but profound determination to advance the sustainable use of the resources. Three examples from Europe that include Spain (Chap. 16), the Czech Republic (Chap. 17), and Italy (Chap. 18) show some of the broad and diverse contributions by the "old-continent" to the employment of EPNs in agriculture. Advances in Iran (Chap. 19) and South Africa (Chap. 20) round out this global tour of the use of EPNs. Conspicuously absent from this volume are examples from still unknown and unexplored areas of Africa, Asia, and Oceania. Because this book focuses on the use of nematodes for biological control, we include also the significant advances in understanding and developing uses for the slug parasitic nematode *Phasmarhabditis hermaphrodita* (Chap. 21).

This volume is the product of 57 scientists, experts in each of their respective topics. To all of the authors, I extend my sincere appreciation for their efforts in writing these valuable contributions. It is to be hoped that the information shared here will be used by many others in future endeavors to develop tactics for pest management in sustainable cropping systems. Moreover, I hope this volume serves to inspire and guide a younger generation of scientists who may encounter it as their first contact with these fascinating nematodes that have so much potential importance on our farms and in our landscapes.

Neuchâtel, Switzerland Raquel Campos-Herrera

Contents

Part II Advances on Entomopathogenic Nematodes Production and Release

Part III Entomopathogenic Nematode Exploitation: Case-Studies in Crop Protection in Different Crops and Countries

Contributors

Mary E. Barbercheck Department of Entomology, Pennsylvania State University, University Park, PA, USA

Raquel Campos-Herrera Faculty of Sciences, University of Neuchâtel, Neuchâtel, Switzerland

Mirella Clausi Dipartimento di Scienze Biologiche, Geologiche e Ambientali, Sez. Biologia Animale "M. La Greca", University of Catania, Catania, Italy

Giovanna Curto Servizio Fitosanitario, Regione Emilia–Romagna, Laboratorio di Nematologia, Bologna, Italy

Francesca De Luca Istituto per la Protezione Sostenibile delle Piante, Consiglio Nazionale delle Ricerche, Bari, Italy

Claudia Dolinski Laboratório de Entomologia e Fitopatologia, Universidade Estadual do Norte Fluminense Darcy Ribeiro, Campos dos Goytacazes, RJ, Brazil

Larry W. Duncan Citrus Research and Education Center, University of Florida, (US), Lake Alfred, FL, USA

Lemma Ebssa Department of Entomology, Institute for Health, Health Care Policy, and Aging Research, Rutgers School of Environmental and Biological Sciences, Rutgers University, New Brunswick, NJ, USA

Fahiem E. El-Borai Citrus Research and Education Center, University of Florida, (US), Lake Alfred, FL, USA

Plant Protection Department, Faculty of Agriculture, Zagazig University (Egypt), Ash Sharqia Governorate, Egypt

Fernando García del Pino Department of Animal Biology, Universidad Autónoma de Barcelona, Barcelona, Spain

Itamar Glazer Division of Nematology, Department of Entomology, Institute of Plant Protection, Agricultural Research Organization, The Volcani Center, Rishon LeZion, Israel

Parwinder S. Grewal Entomology and Plant Pathology Department, University of Tennessee, Knoxville, TN, USA

Christine T. Griffin Department of Biology, Maynooth University, Maynooth, County Kildare, Ireland

Baris Gulcu Department of Biology, Faculty of Arts and Science, Duzce University, Duzce, Turkey

Justin L. Hatting South African Agricultural Research Council (ARC), Small Grain Institute, Bethlehem, South Africa

Selcuk Hazir Department of Biology, Faculty of Arts and Sciences, Adnan Menderes University, Aydin, Turkey

Ivan Hiltpold Hawkesbury Institute for the Environment, University of Western Sydney, Penrith, NSW, Australia

Amanda Hodson Department of Entomology and Nematology, University of California, Davis, CA, USA

Javad Karimi Faculty of Agricultura, Department of Plant Protection, College of Agriculture, Ferdowsi University of Mashhad, Mashhad, Iran

Albrecht M. Koppenhöfer Department of Entomology, Rutgers University, New Brunswick, NJ, USA

Olga S. Kostromytska Department of Entomology, Rutgers University, New Brunswick, NJ, USA

Edwin E. Lewis Department of Entomology and Nematology, University of California, Davis, CA, USA

Javier López-Cepero COPLACA, Organización de Productores de Plátanos, Tenerife, Spain

Antoinette P. Malan Department of Conservation Ecology and Entomology, Faculty of AgriSciences, Stellenbosch University, Stellenbosch, South Africa

Benjamin A. McGraw Department of Plant Science, Pennsylvania State University, University Park, PA, USA

Ana Morton Department of Animal Biology, Universidad Autónoma de Barcelona, Barcelona, Spain

Jiří Nermut' Department of Biosystematics and Ecology, Institute of Entomology, Biology Centre, Czech Academy of Sciences, České Budějovice, Czech Republic

Ana Piedra-Buena Instituto Canario de Investigaciones Agrarias, Tenerife, Spain

Vladimír Půža Department of Biosystematics and Ecology, Institute of Entomology, Biology Centre, Czech Academy of Sciences, České Budějovice, Czech Republic

Robert Rae School of Natural Sciences & Psychology, Liverpool John Moores University, Liverpool, UK

Giancarlo Rappazzo Dipartimento di Scienze Biologiche, Geologiche e Ambientali, Sez. Biologia Animale "M. La Greca", University of Catania, Catania, Italy

Mayra G. Rodríguez Hernández Departamento de Proteccion Vegetal, Centro Nacional de Sanidad Agropecuaria (CENSA), San José de las Lajas, Provincia Mayabeque, Cuba

Carolina Rosales Laboratorio de Nematología, Instituto Nacional de Investigaciones Agrícolas, Maracay, Estado Aragua, Venezuela

Elham Salari Faculty of Agricultura, Department of Plant Protection, College of Agriculture, Ferdowsi University of Mashhad, Mashhad, Iran

Ernesto San-Blas Instituto Venezolano de Investigaciones Científicas, Centro de Estudios Botánicos y Agroforestales, Laboratorio de Protección Vegetal, Maracaibo, Estado Zulia, Venezuela

David Shapiro-Ilan Southeastern Fruit and Nut Research Laboratory, USDA–ARS, Byron, GA, USA

Elson J. Shields Department of Entomology, Cornell University, Ithaca, NY, USA

S. Patricia Stock Department of Entomology, University of Arizona, Tucson, AZ, USA

Robin J. Stuart Dundee Biological Control Laboratory, Division of Plant Industry, Florida Department of Agriculture and Consumer Services, Dundee, FL, USA

Eustachio Tarasco Dipartimento di Scienze del Suolo, della Pianta e degli Alimenti (DiSSPA), Sez. Entomologia e Zoologia, University of Bari "Aldo Moro", Bari, Italy

Antonio M. Testa Department of Entomology, Cornell University, Ithaca, NY, USA

Ángel Torres Campo Experimental Bramon, Instituto Nacional de Investigaciones Agrícolas, Bramón, Estado Táchira, Venezuela

Alberto Troccoli Istituto per la Protezione Sostenibile delle Piante, Consiglio Nazionale delle Ricerche, Bari, Italy

Michael Wilson Innovative Farm Systems, AgResearch, Ruakura, New Zealand

Part I
Biological and Environmental Factors Affecting Entomopathogenic Nematodes as Biological Control Agents

Chapter 1
Diversity, Biology and Evolutionary Relationships

S. Patricia Stock

1.1 Introduction

Nematodes are among the most abundant organisms on Earth, as they exist in almost every possible habitat and ecosystem (Bernard, 1992; De Ley, 2006; Ettema, 1998; Powers et al. 2009). Indeed, these organisms can be found in aquatic (marine and fresh water) and terrestrial ecosystems ranging from the tropics to the poles and from the highest to the lowest of elevations. Furthermore, nematodes have exploited a wide range of ecological niches encompassing free–living and parasitic species. Parasites have received the most attention and have been the subject of extensive research because of the damage they cause to crops, livestock, and humans (Anderson, 2000; Norton, 1978; Poinar, 1983; Stirling, Poinar, & Jansson, 1988; Wallace, 1963; Zuckerman & Rhode, 1981). However, several parasitic species are considered beneficial organisms to humans as they can be used as control agents of pests that are of agriculture, forestry or health importance (Bedding, Akhurst, & Kaya, 1993; Gaugler & Kaya, 1990; Grewal, Grewal, & Adams, 2003; Petersen, 1985; Poinar; Stock & Hunt, 2005; Wilson & Gaugler, 2000; Wilson, Glen, & George, 1993).

Among these beneficial groups, there are the so–called entomopathogenic nematodes (EPN), which traditionally comprise two families, Steinernematidae Travassos and Heterorhabditidae Poinar. These nematodes have been used in classical, conservational, and augmentative biological control programs (Grewal, Ehlers, & Shapiro-Ilan, 2005; Kaya & Gaugler, 1993; Lacey & Georgis, 2012). Extensive research has demonstrated both their successes and failures for control of insect pests of crops, ornamental plants, trees, lawn and turf (Georgis et al., 2006;

S.P. Stock (✉)
Department of Entomology, University of Arizona, Tucson, AZ 85721-0036, USA
e-mail: spstock@email.arizona.edu

© Springer International Publishing Switzerland 2015

R. Campos-Herrera (ed.), *Nematode Pathogenesis of Insects and Other Pests*, Sustainability in Plant and Crop Protection 1, DOI 10.1007/978-3-319-18266-7_1

Grewal et al., 2005; Kaya & Gaugler, 1993; Koppenhöfer, 2007; Shapiro-Ilan, Gouge, & Koppenhöfer, 2002). A few studies have also shown that EPN can have direct and/or indirect effects on populations of plant parasitic nematodes and plant pathogens (Grewal et al.; Kaya & Gaugler; 1993; Navarro, McMullen II, & Stock, 2014; Shapiro et al., 2002; Shapiro-Ilan, Han, & Dolinski, 2012).

Over the past decade, realization of the practical use of these nematodes has spurred developments beyond agriculture and pest management (Campos-Herrera, Barbercheck, Hoy, & Stock, 2012; Stock & Goodrich-Blair, 2008a). Indeed, EPN and their bacterial symbionts are now considered a tractable model system that is amenable to study physiological, chemical, structural and developmental aspects of beneficial symbiotic associations (Dillman et al., 2012; Eleftherianos, Joyce, Ffrench-Constant, Clarke, & Reynolds, 2010; Hallem, Rengarajan, Ciche, & Sternberg et al., 2007; Stock & Goodrich-Blair).

Furthermore, it is clear that EPN can also serve as excellent model systems for understanding biological, ecological and evolutionary processes involving other soil organisms (Denno, Gruner, & Kaplan, 2008; El-Borai, Duncan, & Preston, 2005; Enright & Griffin, 2005; Navarro et al, 2014; Shapiro-Ilan etal., 2002; Strong, 2002, 2007; Timper & Kaya, 1992). Yet, this is a research area that remains in its infancy and needs to be further explored (Campos-Herrera et al., 2012).

This chapter summarises the latest information regarding the biological diversity and distribution of EPN, recapitulates hypotheses on their evolutionary origins and discusses their relationship with their symbiotic bacteria. Special focus is placed on the role of EPN as model organisms in pest management and other biological disciplines.

1.2 Life Cycle

Similar to *Caenorhabditis elegans* Maupas (Rhabditida: Rhabditidae), EPN form a dauer or stress–resistant stage known as the infective juvenile (IJ), which is the only free–living stage in the nematodes' life cycle. Unlike normal feeding stages, IJs can live for many months without food (Poinar, 1990). This developmentally arrested stage also plays a key role in the dispersal of the nematodes in the soil as they actively seek and infect suitable insect hosts. Additionally, the IJs are responsible for vectoring entomopathogenic bacteria (Gram–negative γ–Proteobacteria) from one insect host to another (Boemare, 2002; Poinar, 1990). Steinernematids vector *Xenorhabdus*, while heterorhabditids partner with *Photorhabdus*.

Like primary symbionts of insects, *Xenorhabdus* are harbored by *Steinernema* IJs in a specialized structure called the receptacle (Flores-Lara, Renneckar, Forst, Goodrich-Blair, & Stock 2007; Snyder, Stock, Kim, Flores-Lara, & Forst, 2007). This receptacle is a modification of the two most anterior intestinal cells (Kim, Flores-Lara, & Stock, 2012; Snyder et al. 2007) (Fig. 1.1a, b). Unlike steinernematids, *Heterorhabditis* IJs do not have a specialized structure in their intestine to harbor *Photorhabdus* bacteria (Fig. 1.1c, d). In *Heterorhabditis*, the bacteria adhere

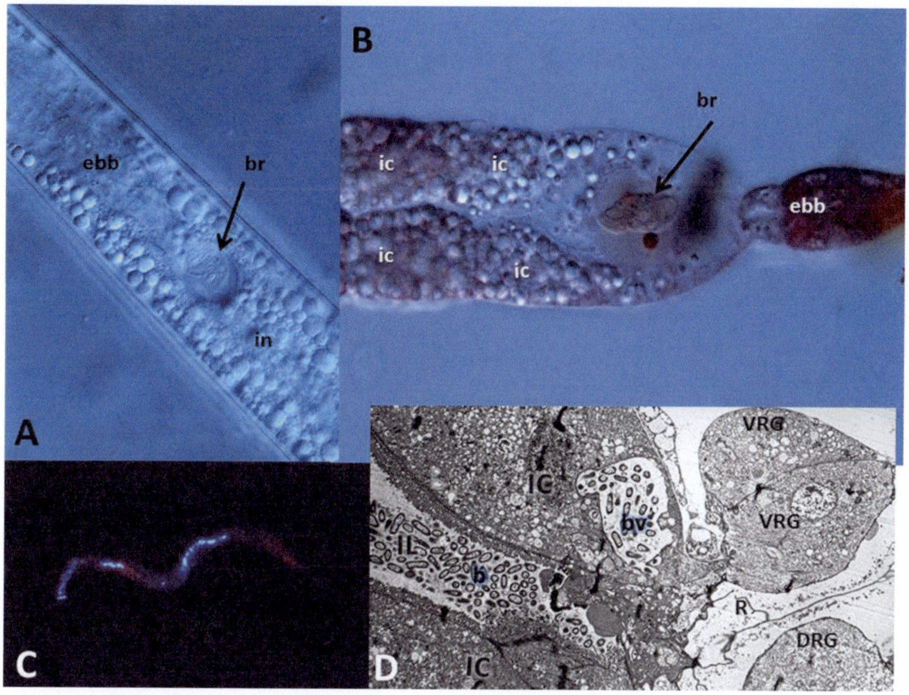

Fig. 1.1 (**a**, **b**) Bacterial receptacle of Steinernema oregonense Liu & Berry (Rhabditida: Steinernematidae) IJ in situ (**a**) and in extruded intestine (**b**); C. Heterorhabditis bacteriophora IJ with GFP–labelled Photorhabdus bacteria in intestinal lumen; D. Transmission electron micrograph of the posterior intestine of H. bacteriophora hermaphrodite showing bacteria in intestinal lumen and inside vesicle in posterior intestinal (Notice rectum and rectal gland cells which are not colonized by bacterial symbionts). References: = bacteria; *br* bacterial receptacle, *bv* bacterial vesicle, *DRG* dorsal rectal gland, *ebb* esophageal basal bulb, *ic* intestinal cell, *il* intestinal lumen, *in* intestine, *r* receptacle, *VRG* ventral rectal gland

first to the esophago–intestinal valve of the IJ and then migrate to the intestine where they proliferate in the intestinal lumen. It has been reported that an average population of 50–150 *Photorhabdus* CFU (colony forming units) may exit per IJ (Ciche & Ensign, 2003) implying the bacteria have access to nutrients in the nematode's intestine (Forst & Clarke 2002).

Upon locating and entering an insect host (Fig. 1.2 [1]), IJs migrate to the hemolymph where they recover from their arrested state of development and release their bacterial symbionts (Fig. 1.2 [2]). The bacteria reproduce, release toxins and kill the insect within 24–48 h. Both *Xenorhabdus* and *Photorhabdus*, are unable to live freely in the soil, thus they benefit from being carried by the IJs from one insect host to another (Forst, Dowds, Boemare, & Stackebrandt, 1997; Froy, 2005). In turn, the nematodes benefit from the symbionts' ability to convert the insect cadaver in a suitable environment for their growth and development (Kaya & Gaugler, 1993; Poinar, 1990). EPN and their bacterial symbionts live inside

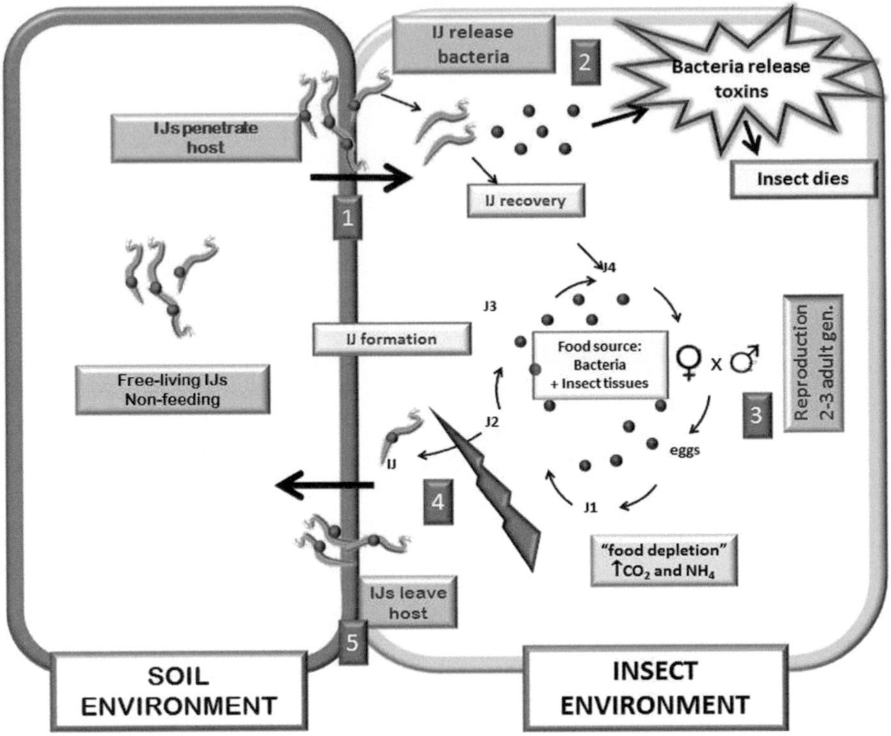

Fig. 1.2 Generalized life cycle of entomopathogenic nematodes and their bacterial symbionts

the host cadaver for a period of one to three nematode generations before the bacteria colonize a new generation of IJs (Fig. 1.2 [3]). The bacteria are known to also perform a protective role for their nematode partners inside the insect cadaver, by creating a near–exclusive environment for themselves and their specific nematode partners producing bacteriocins, antibiotics and antimicrobials (Akhurst, 1982; Chen, Dunphy, & Webster, 1994).

While food resources are available in the insect cadaver, the nematodes undergo normal development: IJs molt twice and become adults (Fig. 1.2 [3]). All *Heterorhabditis* species described to date have a hermaphroditic first adult generation, followed by a gonochoristic second generation where adult males and females are present and copulate. Most *Steinernema* species are gonochoristic as all adult generations have male and females, and copulation is considered essential for propagation. One exception is *Steinernema hermaphroditum* Stock, Griffin & Chaerani (Rhabditida: Steinernematidae), which has a first adult generation represented by self–fertile females (hermaphrodites) which carry sperm in the reproductive tract (Griffin, O'Callaghan, & Dix, 2001). In this species, only 1–6 % of the IJs develop into males in both first and second adult generations and copulation between male and females has not been observed (Stock, Griffin, & Chaerani, 2004). In this respect, it could be hypothesized that first generation males may have lost their ability to copulate or alternatively, hermaphrodites perhaps lost their ability to respond to the males.

When food resources in the insect cadaver become scant, second–stage juveniles (J2) (Fig. 1.2 [4]) develop into IJs as an alternative to the third juvenile stage (J3) (Hirao & Ehlers, 2009; Jensen, Strauch, Wyss, Luttmann, & Ehlers, 2000). Anatomical and physiological changes in this stage include the cessation of feeding and the closing of both mouth and anus. IJs are also characterized by the presence of a double cuticle layer (Poinar, 1990; Sommer & Ogawa, 2011). Before emerging from the insect cadaver (Fig. 1.2 [5]), *Steinernema* IJs retain 1–3 bacterial cells (Snyder et al., 2007) and migrate into the soil environment where they will seek a new host. In *Heterorhabditis*, studies suggest that symbiont–colonized IJs are formed exclusively in hermaphrodites that undergo *endotokia matricida*, a process that involves the intrauterine hatching of juveniles and matricide (Ciche, Kim, Kaufmann–Daszczuk, Nguyen, & Hall, 2008). Juvenile nematodes (J1 and J2) feed on intestinal storage granules and *Photorhabdus* bacteria which become available through the destruction of the hermaphrodite's intestine (Fig. 1.1e) (Johnigk & Ehlers, 1999; Stock, Lee, & Flores-Lara, 2012).

EPN development in the insect cadaver is also dependent on various factors produced individually and/or interactively by the three players in this system: bacteria, nematodes and insect. For example, factors such as accumulation of CO_2 and NH_4 (due to crowding of nematodes) have been shown to trigger IJ formation and to induce the exit of IJs from the insect cadaver (Jensen et al., 2000; Ryder & Griffin, 2002; San-Blas, Gowen, & Pwembroke, 2008; Wang & Bedding, 1996; Wright, 2004). Also, food supply (symbiotic bacteria + insect tissues) (Miranda, Navarro, Davidowitz, Bronstein, & Stock, 2013) and nematode pheromones (Kaplan et al., 2012) seem to play an important role in nematode reproduction and IJ progeny production).

The relationship between *Steinernema* and *Heterorhabditis* nematodes with their symbiotic bacteria is considered mutualistic and it is obligate under natural conditions. However, under laboratory conditions the level of dependence of the nematodes for their bacterial symbionts may vary from obligate to facultative (Stock, pers. comm.). While *Heterorhabditis* spp. have an obligate requirement for their cognate *Photorhabdus* symbionts (they are required for reproduction and proper development) the situation seems to be different for the *Steinernema–Xenorhabdus* mutualism. Nematode–bacterium pairs can be disassociated under experimental conditions and apparently nematode fitness is not affected at least for few generations (Cowles & Goodrich-Blair, 2004, 2008; Sicard et al., 2003, 2004). However, the current body of literature offers contrasting evidence depending on the *Steinernema–Xenorhabdus* pairs considered. For example, Sicard, Ramone, Le Brun, Pagès, and Moulia (2005) showed that non–native *Xenorhabdus* or *Photorhabdus* strains can be pathogenic to *S. carpocapsae*. Contrarily, recent investigations suggest that fitness (i.e., virulence and reproduction) of different *Steinernema* spp. that are hosts of *X. bovienii* (Akhurst) is optimal when they are associate with either their cognate bacteria or a bacterial strain that comes from a closely related *Steinernema* host (Murfin et al., 2015). In this respect, *Xenorhabdus* and *Photorhabdus* can be considered similar to primary endosymbionts of insects, which are often known to perform crucial metabolic roles for their hosts enhancing their fitness (Baumann & Baumann, 2005; Moran, Tran, & Gerardo, 2005; Thao et al., 2000).

1.3 To Be or Not To Be an Entomopathogen

The term "entomopathogenic" has been used to differentiate nematodes in the Steinernematidae and Heterorhabditidae, which vector insect–pathogenic *Xenorhabdus* and *Photorhabdus* bacteria (Gaugler & Kaya, 1990). However, recent evidence suggests entomopathogenesis may not be an exclusive lifestyle of steinernematid and heterorhabditid nematodes inferring this unique type of insect parasitism, may have arisen at least three times during the evolution of Nematoda (Fig. 1.3).

Specifically, a few studies have reported two species of *Oscheius* (Andrassy, 1976), such as *Oscheius caroliniesis* Ye, Torres–Barragan & Cardoza (Rhabditida: Rhabditidae) and *Oscheius chongmingensis* Ye, Torres–Barragan & Cardoza (Rhab-

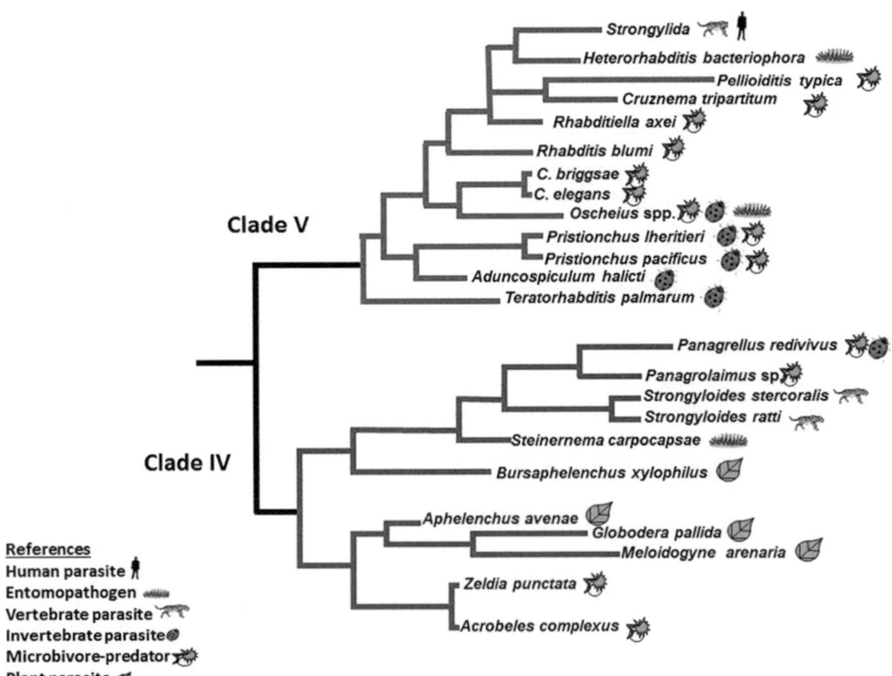

Fig. 1.3 Clades IV and V (Based on Blaxter et al. (1998) phylogenetic framewrork of Nematoda) showing evolution of entomopathogenesis

ditida: Rhabditidae), and one *Caenorhabditis* sp., *Caenorhabditis briggsae* Daugherty & Nigon, (Rhabditida: Rhabditidae), as entomopathogens because of their association with entomopathogenic *Serratia* bacteria (Abebe et al., 2010; Ye, Torres-Barragan, & Cardoza, 2010; Zhang et al., 2008, 2009). These nematode–bacterium associations have also been shown to cause disease in an insect host (Abebe et al., 2010; Torres–Barragan, Suazo, Buhler, & Cardoza, 2011; Ye et al., 2010; Zhang et al., 2009). However, the association of IJs of these species with the bacteria seems to be fortuitous or sporadic. Dillman et al. (2012) recently proposed criteria based on fundamental principles of the EPN lifestyle, for determining if a nematode should be considered entomopathogenic or not. These criteria are: (1) nematodes may have a symbiotic relationship with bacteria to facilitate pathogenesis (their association may not necessarily be obligate, but it should not be transient), and (2) insect death should occur sufficiently rapidly, usually in less than 120 h.

Based on these standards, *O. carolinensis* and *O. chongmingensis* may be considered EPN, but *C. briggsae* should not be, because its virulence is less than that observed for the bacteria alone (Abebe, Bonner, Gray, & Thomas, 2011). However, further research is warranted to resolve the level of dependence these transient entomopathogenic nematode–bacteria associations have on each other and on insect hosts. Additionally, studies on symbiont transmission of their bacterial associates should be explored.

1.4 Historical Overview of Approaches Considered for the Identification of Entomopathogenic Nematodes and Their Bacterial Symbionts

1.4.1 Steinernema and Heterorhabditis

The most recent taxonomic account recognizes two genera in Steinernematidae: (1) *Neosteinernema* which comprises only one species, *Neosteinernema longicurvicauda* Nguyen & Smart (Rhabditida: Steinernematidae), and (2) *Steinernema* (type genus) with more than 70 recognized species. Heterorhabditidae contains a single genus, *Heterorhabditis*, with more than 20 currently recognized species (Stock & Goodrich-Blair, 2012).

Species in both EPN genera are usually identified considering various criteria including phenetic (based on morphology/morphometric analyses), biological (based on cross hybridization assays) and phylogenetic (taxa are delineated based on their evolutionary relationships mostly considering sequence data of various genes) (Adams, Burnell, & Powers, 1998; Hominick, Reid, Bohan, & Briscoe, 1996). These criteria require expertise and extensive time for the proper identification of taxa. Precise and swift identification of EPN is an essential need for their implementation and release in pest management programs. However, given the dearth of expertise in traditional morphological diagnostic methods, supplementary approaches such as molecular methods have been widely adopted to properly characterize and diagnose EPN (Stock, 2009; Stock & Hunt, 2005).

Of all considered and available methods, analysis of nucleotide sequence data has proven to be a useful tool not only for diagnostics at different taxonomic levels, but has also provided valuable data for phylogenetic inference of EPN (Adams et al., 1998; Liu, Berry, & Moldenke, 1997; Nadler, Bolotin, & Stock, 2006; Nguyen & Duncan, 2002; Nguyen, Maruniak, & Adams, 2001; Spiridonov, Reid, Podrucka, Subbotin, & Moens, 2004; Stock, Campbell, & Nadler, 2001). At present two nuclear genes including partial sequence of the 28S rDNA gene (~1Kb including the D2/D3 domain) and the complete Internal Transcribed Spacer (ITS) rDNA region (including ITS1, 5.8 S and ITS 2 genes) are the most widely used markers for diagnostics and assessment of phylogenetic relationships among taxa. Additionally, two mitochondrial genes: COI (cytochrome oxidase I) and 12S are considered informative for diagnostics of *Steinernema* species and have expanded the gene repertoire used in the identification of these taxa (Nadler et al., 2006).

PCR–based diagnostic approaches that consider species–specific primers and quantitative real–time PCR methods have also shown potential in facilitating ecological studies of EPN including the monitoring of their establishment, population dynamics and interactions with organisms of different trophic levels in the soil (Campos-Herrera et al., 2011; Read, Sheppard, Bruford, Glen, & Symondson, 2006; Torr, Spiridonov, Heritage, & Wilson, 2007). For example, Campos-Herrera et al. demonstrated that qPCR assay was a more efficient method for detecting EPN communities in soil and solved some of the limitations (i.e., time for isolating samples from the soil, morphological diagnosis of single species) of the conventional methods such as the *Galleria mellonella* L. (Lepidoptera: Pyralidae) baiting method.

More recently, next–generation sequencing coupled with powerful bioinformatics tools has made possible *de novo* acquisition of many nematode genomes (Schwartz, Antoshechkin, & Sternberg, 2011). Genomes of four *Heterorhabditis* (*H. bacteriophora* Poinar, *H. megidis* Poinar, Jackson & Klein, *H. indica* Poinar, Karunakar & David, and *H. sonorensis* Stock, Rivera-Orduño & Flores-Lara) species and five *Steinernema* (*S. carpocapsae* (Weiseri), *S. scapterisci* (Nguyen & Smart), *S. feltiae*, *S. glaseri* (Steiner) and *S. monticolum*) have been thus far sequenced and assembled (Bai et al., 2007; Dillman et al., review; Sandhu, Jagdale, Hogenhout, & Grewal, 2006; Schwartz et al., 2011). For example, the draft genome of *H. bacteriophora* revealed 1,263 scaffolds totalling 77 Mb after quality trimming and assembly. The overall GC content of this genome is 32.2 %, similar to the free–living nematode *C. elegans*, plant–parasitic nematode *Meloidogyne hapla* Chitwood (Order: Family), and human–parasitic nematode *Brugia malayi* (Brug) (Spirurida: Onchocercidae) (Bai et al.). The genome size for other sequenced *Heterorhabditis* spp. ranges from 64 (*H. indica*) to 69 Mb (*H. megidis*) (Schwartz et al.).

The estimated genome size for *Steinernema* ranges from 79.5 Mb (*S. scapterisci*) to 92.0 Mb (*S. glaseri*), and 28,000 (*S. carpocapsae*) to 36,000 (*S. monticolum*) genes. The overall GC content varies from 42 % for *S. monticolum* to 48 % for *S.*

scapterisci (Dillman et al., review). Protein domain analyses of these genomes have also revealed a striking expansion of numerous putative parasitism genes, including certain protease and protease inhibitor families as well as fatty acid– and retinol–binding proteins. Furthermore, these analyses have also revealed the rapid evolution and expansion of the Hox gene cluster, an important family of genes involved in nematode development (Dillman et al., review).

1.4.2 Symbiotic Bacteria

As previously mentioned, *Xenorhabdus* and *Photorhabdus* are vectored by *Steinernema* and *Heterorhabditis* nematodes, respectively. The two genera are considered phenotypically different from other members of the Enterobacteriaceae in a number of traits including their inability reduce nitrate to nitrite (Boemare, 2002). These two genera can be differentiated from each other by some phenotypic traits. For example, *Xenorhabdus* are catalase negative and non–bioluminescent, while *Photorhabdus* are catalase positive and have bioluminescence (Boemare & Akhurst, 2006; Koppenhöfer, 2007).

More than 20 species of *Xenorhabdus* have been described, but there are only three species of *Photorhabdus* with more than a dozen subspecies (Boemare, 2002; Boemare & Akhurst, 2006; Orozco, Hill, & Stock, 2013; Tailliez, Pagès, Ginibre, & Boemare, 2006; Tailliez et al., 2010). Identification of these bacteria also requires expertise in various fields including biochemistry, serology, physiology and molecular biology. Phenotypic screening is usually considered the first step in the differentiation of bacterial strains as it allows a rough comparison and placement into similar or different group. However, similar to their nematode hosts, the need for fast and accurate detection of species and/or isolates has prompted the consideration of molecular methods. In this respect, it has been widely accepted that molecular biology has revolutionized systematics and classification of all bacteria.

For *Xenorhabdus* and *Photorhabdus*, methods such as DNA reassociation and riboprinting were considered for faster identification of species and strains avoiding tedious phenotypic characterization techniques (Szállás et al., 2001). More recently, researchers adopted sequence data of single and multigene datasets for identification of species and/or strains and also to address evolutionary questions (Lee & Stock, 2010a; Liu, Berry, & Blouin, 1999; Orozco et al., 2013; Tailliez et al., 2006, 2010). In particular, comparative sequence analyses of the 16S rRNA gene and several housekeeping genes have been widely used for the diagnosis of both *Xenorhabdus* and *Photorhabdus* species and to infer phylogenetic relationships of novel bacterial isolates/species (Ferreira, Van Reenen, Endo et al. 2013; Ferreira, Van Reenen, Pagès et al. 2013; Kuwata et al., 2012; Orozco et al.; Tailliez et al.).

As for the nematodes, full genome approaches are currently being explored to study many of these bacterial symbionts. The first genome sequenced was that of *Photorhabdus luminescens* subsp. *laumondi* Fischer–LeSaux, Viallard, Brunnel, Normand & Boemare (Enterobacteriales: Enterobacteriaceae) (isolate

TT01) (Duchaud et al., 2003). The size of this genome is 5.7 Mb and it contains 4,839 predicted protein–coding genes. Other *Photorhabdus* species sequenced include *Photorhabdus asymbiotica asymbiotica* Fischer–LeSaux, Viallard, Brunnel, Normand & Boemare (Enterobacteriales: Enterobacteriaceae) and *Photorhabdus temperata temperata* Fischer–LeSaux, Viallard, Brunnel, Normand & Boemare (Enterobacteriales: Enterobacteriaceae) (Park et al., 2013; Wilkinson et al., 2009). Genome size for these taxa ranges from 5.065 Mb with a G + C content of 42.15 % for *P. a. asymbiotica* to 5.5 Mb with a G + C content of 43.7 %, for *P. t. temperata*.

Within *Xenorhabdus*, several species and strains have been sequenced thus far, including *Xenorhabdus nematophila* Poinar & Thomas (Enterobacteriales: Enterobacteriaceae) (isolate ATCC 19061), *Xenorhabdus zsentirmaii* Lengyel, Lang, Fodor, Szállás, Schumann & Stackebrandt (Enterobacteriales: Enterobacteriaceae) (isolate DSM 16338), *Xenorhabdus poinarii* Lengyel, Lang, Szállás, Schumann & Stackebrandt (Enterobacteriales: Enterobacteriaceae) (isolate G6) and *Xenorhabdus bovienii* (Akhurst) (Enterobacteriales: Enterobacteriaceae) (10 different strains). Genome size of these bacteria varies between 4.5 Mb (*X. nematophila*) and 4.8 Mb (*X. szentirmaii*). The GC content varies between 44.2 % (*X. nematophila*) and 45 % (*X. bovienii*) (Chaston et al., 2011; Gualtieri, Ogier, Pagès, Givaudan, & Gaudriault, 2014; Lanois et al., 2013). Analyses of these genomes has revealed that there is a large number of encoded adhesins, toxins, hemolysins, proteases and lipases, as well as a wide array of antibiotic synthesizing genes (Bode, 2009). These proteins are thought to play an important role in host colonization, invasion and bioconversion of the insect cadaver. Comparison of these genomes with related bacteria has also shown that EPN bacteria have acquired virulence factors by extensive horizontal transfer (Chaston et al., 2011; Duchaud et al., 2003; Gualtieri et al., 2014; Lanois et al., 2013).

1.5 Biological Diversity and Geographic Distribution

Surveys with focus on EPN have been conducted worldwide with the goal of discovering new species and populations adapted to local conditions and insect pests. Studies have shown that in spite that at a local scale they have a patchy distribution, EPN are omnipresent at a global scale (reviewed by Campos-Herrera et al., 2012; Hominick, 2002). These nematodes have been recovered in 49 countries (19 from Europe, 9 from Asian, 3 from North America, 7 from Central America and the Caribbean, 8 from South America and 3 from Africa). However, current knowledge of their geographic distribution is considered an artefact of sampling efforts by researchers whose research interest is on this group of nematodes (San-Blas, 2013; Stock, 2005).

Species such as *S. carpocapsae* and *S. feltiae* have a cosmopolitan distribution and both species have been reported to coexist in the same habitat, and presumably interact with each other (Hominick, 2002). Within Heterorhabditidae, *H. indica* and *H. bacteriophora* have been found on all continents with the exception of Antarctica

(Griffin, Downes, & Block, 1990; Hominick, 2002). The wide geographic distribution of these taxa suggests their dispersal mechanisms may be highly effective and may be the result of a combination of active and passive dissemination mechanisms (Adams et al., 2006).

Europe is thus far the most extensively and intensively sampled continent. Nine named *Steinernema* spp. have been recorded, with *S. feltiae* and *Steinernema affine* (Bovien) (Rhabditida: Rhabditidae) being the most ubiquitous species. Three *Heterorhabditis* species have been reported, with *H. megidis* as the species most widely distributed (Boag, Neilson, & Gordon, 1992; Hominick et al., 1996; Kary, Niknam, Griffin, Mohammadi, & Moghaddam, 2009; Mráček, Bečvář, Kindlmann, & Jersáková, 2005).

In the Australian continent, three *Steinernema* and three *Heterorhabditis* species have been isolated (Akhurst & Bedding, 1986). Apparently, the number of species has not increased since that report (Campos-Herrera et al., 2012).

North America and Asia hold the highest diversity of *Steinernema* species. Approximately 70 % of all described species (i.e., more than 50 nominal species) have been reported in these two continents. Seven *Heterorhabditis* spp. have been isolated thus far. Of all of them, *H. bacteriophora* is the most ubiquitous species, followed followed by *H. indica* and *Heterorhabditis baujardi* Phan, Subbotin, Nguyen & Moens (Rhabditida: Heterorhabditidae).

In Africa, sampling efforts have increased dramatically in the last decade. Many novel and known species have been discovered and are at present being identified (Abu-Shadi, Shamseldean, Abd-Elbary, & Stock, 2011; Çimen, Lee, Hatting, Hazir, & Stock, 2014a, 2014b; Hatting, Hazir, & Stock, 2008; Kanga, Waeyenberge, Hauser, & Moens, 2012; Malan, Knoetze, & Moore, 2011; See also Chap. 20). South Africa is the most extensively surveyed country and many novel species of *Steinernema* and *Heterorhabditis* have been isolated (Çimen et al., 2014a, 2014b; Malan et al., 2011; Stokwe, 2009)

A few studies suggest that factors such as soil and vegetation type as well as the distribution of suitable hosts, are key factors in affecting the distribution of EPN species (See Chap. 4).

1.6 Evolutionary Origins and Phylogenetic Relationships

1.6.1 Steinernema and Heterorhabditis

Steinernematidae and Heterorhabditidae share many common traits including their life histories as well as morphological and ecological features. Poinar (1993) suggested that the similarities observed among these families are the result of convergent evolution. He also speculated that these two groups of EPNs independently developed mutualistic relationships with Gram–negative enteric bacteria (Enterobacteriaceae) about 350 million years ago (i.e., in the mid–Paleozoic). Poinar also speculated that heterorhabditids probably arose from a free–living marine bacterivore precursor.

According to the phylogenetic framework for the Nematoda developed by Blaxter et al. (1998), the Heterorhabditidae belong to clade V and are most closely related to the Strongylida, a group of vertebrate parasites that shares a most recent common ancestor with *Pellioditis*, a free–living rhabditid bacterivore. On the other hand, Steinernematidae are considered members of clade IV and more closely related to the Panagrolaimoidea (free–living and insect associates) and Strongyloididae (vertebrate parasites). Another classification scheme developed by Holterman et al. (2006) depicted Steinernematidae within the suborder Tylenchina (Clade 10), which also includes other insect parasites such as allantonematids and neotylenchids. In this classification, Heterorhabditidae were placed in Clade 9 together Strongyloidea (animal parasites) and other bacterivore groups including members in the Diplogasteridae and Rhabditidae (Holterman et al., 2006). In spite of the differences that exit in the topologies of these two phylogenetic frameworks, they both support the separate origin of Steinernematidae and Heterorhabditidae.

Explicit evolutionary hypotheses for *Steinernema* spp. have mainly been based on analyses of single genes, mostly nuclear ribosomal DNA (rDNA) (Nguyen & Duncan, 2002; Spiridonov et al., 2004; Stock et al., 2001), though Szalanski, Taylor, and Mullin (2000) also explored mitochondrial DNA sequences. Nadler et al. (2006) were the first ones to consider a multigene (28S rDNA and two mitochondrial, *COX*I and 12S genes) approach in combination with morphological traits to assess evolutionary relationships among *Steinernema* spp. Although the combined phylogenetic analysis of the 3–gene dataset in this study yielded well–resolved and highly similar trees, analysis of the morphological dataset yielded poorly resolved trees. These results were expected given the conservation of morphological traits in this genus. Furthermore, mapping of morphological characters on the 3–gene trees provided further evidence in support of the homoplasy of morphological traits (Nadler et al.).

Recently, full genome approaches have been considered to investigate evolutionary relationships in *Steinernema* (see Sect. 1.2). Although taxon sampling is limited to a few taxa, phylogenomic analyses mostly support previous hypotheses of their evolutionary relationships. However, one contrasting difference encountered was that *S. monticolum*, once thought to be a close relative of *S. carpocapsae* and *S. scapterisci* (clade I) (Nadler et al., 2006), was depicted as more closely related to *S. feltiae*, a member of clade III and more distantly related to the clade I where *S. carpocapsae* and *S. scapterisci* belong to.

Evolutionary relationships among *Heterorhabditis* spp. were first inferred by analyses of sequence data of the ITS–1 region of the tandem repeat unit of rDNA (Adams et al., 1998). This study provided good resolution for closely related 'species' such as *H. indica* and *H. hawaiiensis* Gardner, Stock & Kaya (Rhabditida: Heterorhabditidae), *H. bacteriophora* and *H. argentinensis* Stock (Rhabditida: Heterorhabditidae). However, the analyses could not resolve filiation for more distantly related species. The discovery of new species and their inclusion in analyses has provided better resolution for *Heterorhabditis* phylogenetic trees. For example, studies by Phan, Subbotin, Nguyen, and Moens (2003) and Andaló, Nguyen, and Moino (2006) have placed tropical and subtropical *Heterorhabditis* species, i.e., *Heterorhabditis amazonensis* Andaló, Nguyen & Moino (Rhabditida: Heterorhabditidae), *H. indica*, *Heterorhabditis floridensis* Nguyen, Gozel & Koppenhőfer & Adams (Rhabditida: Heterorhabditidae) and *H. baujardi*, into one distinct clade that is separated from other clades that include mostly taxa from temperate regions.

1.6.2 *Xenorhabdus* and *Photorhabdus*

Members in the Enterobacteriaceae have evolved a variety of symbiotic associations with eukaryotic hosts ranging from obligate intracellular to more loose commensalistic gut–associated or pathogenic interactions, and these associations have emerged multiple times in their evolutionary history (Husnik, Chrudimský, & Hypša, 2011). It has been hypothesized *Xenorhabdus* and *Photorhabdus* may have diverged from a common enterobacterial ancestor that was capable of colonizing both *Steinernema* and *Heterorhabditis* hosts (Poinar, 1990). A long–term association with their nematode hosts may have independently given rise to each of these genera. Subsequently, selective pressures may have contributed to the maintenance of their symbiotic association with the nematode hosts, leading both bacterial groups to evolve different mechanisms to converge upon the same lifestyle (Poinar; Boemare, 2002). Recent phylogenetic studies of *Xenorhabdus* and *Photorhabdus* genomes has provided evidence for genomic divergence between *Xenorhabdus* and *Photorhabdus* bacteria. These findings suggest that evolutionary changes, shaped by symbiotic interactions, can follow different routes to achieve similar end points (Chaston et al., 2011).

1.6.3 Entomopathogenic Nematodes–Bacteria Cophylogenetic Studies

With respect to *Steinernema–Xenorhabdus*, Lee and Stock (2010b) were the first to explore their coevolutionary histories. The study considered a multigene approach for both nematodes and their symbionts. Specifically, three *Steinernema* genes: 28S rRNA, *COI* and 12S rRNA; and three bacterial genes: 16S rRNA, *serC* (phosphoserine aminotransferase) and *recA* (recombinase A) were studied. Although no perfect co–cladogenesis was found, the analysis revealed 12 cospeciation events among the 30 *Steinernema–Xenorhabdus* pairs sampled. According to this study, host switches happened at least 17 times. For example, *X. nematophila* a symbiont of two *Steinernema* spp., *Steinernema websteri* Cutler and Stock (Rhabditida: Steinernematidae) and *Steinernema anatoliense* Hazir, Stock & Keskin (Rhabditida: Steinernematidae) members of clade IV, switched to another host, *S. carpocapsae*,

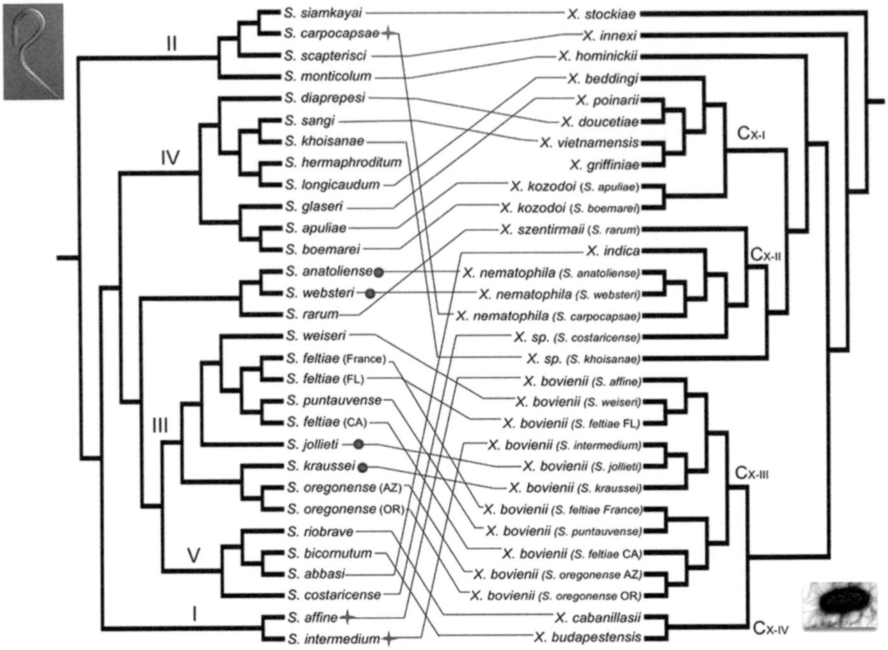

Fig. 1.4 Co–evolutionary hypothesis for the evolution of Steinernema and Xenorhabdus mutualism (Modified from Lee and Stock (2010b)). Tanglegram showing examples of host switches (+) and co–evolutionary (O) events

that belongs to a more distantly related *Steinernema* clade (clade II) (Fig. 1.4). According to this study, most host switches occurred near terminal nodes on the *Steinernema* tree, implying the relatively recent association of the symbiotic partners, without a period of species diversification after the switch (Lee & Stock).

Artificial disassociation and switches of *Xenorhabdus* bacteria and *Steinernema* hosts have been induced under laboratory conditions (McMullen, Lee, and Stock, (2014); Sicard et al. 2004, 2005). But the results obtained have been diverse depending on the *Xenorhabdus–Steinernema* pairs considered. For example, Sicard et al. found that the coupling of *S. scapterisci* with non–native *Xenorhabdus* spp. made no significant difference to the nematodes' fitness compared to cognate associations. However, it was observed that *S. scapterisci* was only able to transmit its natural symbiont, *X. innexi* (Sicard et al.).

In this respect, it has been suggested that phylogenetic distance between native and non–native bacterial symbionts plays a keys role in the resultant levels of fitness in their nematode partners. For example, ongoing studies in the Goodrich-Blair and Stock laboratories have revealed significant differences in nematode fitness when different hosts of *X. bovienii* associate with non–cognate strains of this bacterium. Furthermore, chemotaxis studies have shown that, when given a choice, different hosts of *X. bovienii* prefer either their cognate strain or a strain that comes from a host that is phylogenetically closer to their cognate host (McMullen et al., 2014)

The association between *Heterorhabditis* and *Photorhabdus* was originally thought to be strictly one–to–one in terms of cospeciation. However, with the increasing number of nematode and bacterial isolates, it has come to light that some *Photorhabdus* species such as *P. temperata* (Kuwata, Yoshiga, Yoshida, & Kondo, 2007) may have different *Heterorhabditis* hosts including *H. bacteriophora*, *Heterorhabditis zealandica* Poinar (Rhabditida: Heterorhabditidae), and *Heterorhabditis downesi* Stock, Griffin & Burnell (Rhabditida: Heterorhabditidae) (Adams et al., 2006; An & Grewal, 2010; Boemare, 2002; Tóth & Lakatos, 2008). Recently, Maneesakorn et al. (2011) investigated coevolutionary relationships between *Photorhabdus* and *Heterorhabditis* considering a single gene approach. The ribosomal ITS region was used for *Heterorhabditis* while the housekeeping *gyrase B* (*gyrB*) gene was considered for *Photorhabdus*.

In spite of the low resolution of nematodes and bacteria phylograms, the coevolutionary analysis suggests most *Heterorhabditis–Photorhabdus* pairs shared similar evolutionary trajectories (Maneesakorn et al., 2011). However, host switching events were observed between a few *H. bacteriophora–Photorhabdus* pairs (Fig. 1.5). For example, certain geographically distant *Heterorhabditis* populations of *H. bacteriophora* and *H. georgiana*, showed no detectable phylogenetic divergence, but their respective symbionts apparently speciated into two different *Photorhabdus* species (i.e., *P. luminescens* and *P. temperata*). These results suggest that duplication events may have occurred in the evolutionary trajectories of these *Photorhabdus* species (Maneesakorn et al., 2011). In this respect, Brooks, Leon–Regagnon, McLennan,

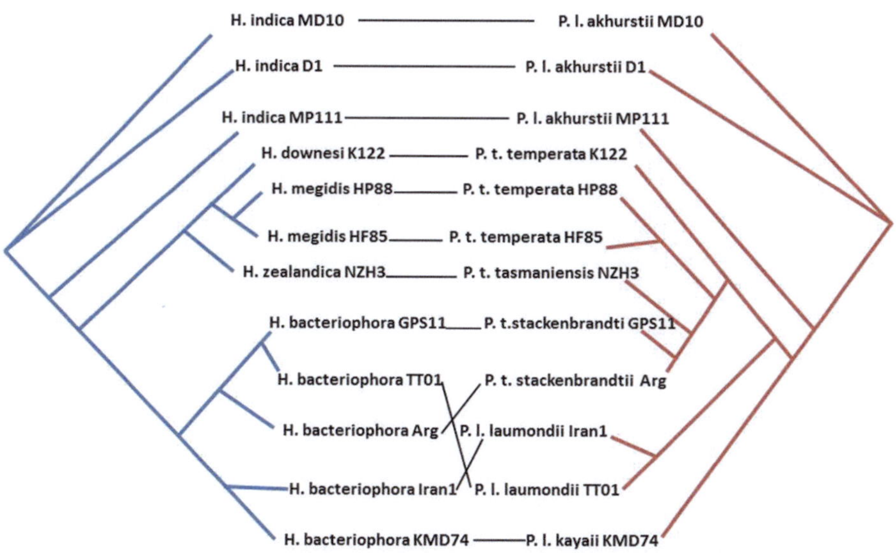

Fig. 1.5 Co–evolutionary hypothesis for the evolution of Heterorhabditis (*left*) and Photorhabdus (*right*) mutualism (Modified from Maneesakorn et al. (2011))

and Zelmer (2006) suggested that host switching is an evolutionary phenomenon that may be related to ecological fitting (i.e., the widespread distribution of phylogenetically conservative traits). Maneesakorn et al. (2011) speculated that this may be a plausible explanation that might explain the shifting of the *Photorhabdus* symbionts to two different *Heterorhabditis* hosts which are morphologically similar and closely related taxa.

1.7 Entomopathogenic Nematodes As Model Organisms

Although model organisms constitute only a small fraction of the biodiversity of this planet, the research that has resulted from their study has provided core knowledge in biology and other disciplines (Hedges, 2002). Until now, the concept of "model organism" has been applied to those species of prokaryotes, protists, fungi, plants and animals that meet specific traits such as: their small size, short generation times and ease for experimental laboratory research. However, modern technologies, such as genome–sequencing and powerful bioinformatics tools have broadened the definition to include what are now considered "genomic models" (Hedges, 2002). The rationale for selecting these models has been driven by their relevance to human health and their agricultural importance. Thus, many nematodes taxa that are parasites of plants, animals and humans have been and are currently being investigated as genomic models (Abad et al., 2008; Ghedin et al., 2007; Kumar, Schiffer, & Blaxter, 2012; Schwartz et al., 2011).

Like *C. elegans* (one of the first eukaryotic model organisms), EPN have many traits that make them excellent genomic models (Table 1.1). In this respect, over the past decades, many researchers have advocated EPNs and their bacterial symbionts as tractable model organisms (Burnell & Stock, 2000; Chaston et al., 2011; Clarke, 2008; Stock, 2005; Stock & Goodrich-Blair, 2008b). Indeed, scientists from diverse disciplines have come to appreciate the potential that these organisms have, embracing them in research fields beyond agriculture or pest management. Furthermore, research networks, such as NemaSym (a USA–based Nematode–Bacterium Symbioses Research Coordination Network, funded by the National Science Foundation), have been established to encourage the intellectual discourse among scientists studying nematode–bacteria associations (including EPN) and promote them as biological model systems both in science and education (Stock & Goodrich-Blair). For example, EPN are now viewed as excellent model systems for advancing research in soil ecology. Specifically, recognition of their role in the diversity and complex interactions with other soil organisms makes them suitable models for understanding ecological interactions involving other soil organisms (Campos-Herrera et al., 2012; El-Borai, Brentu, & Duncan, 2007). EPN and their bacterial symbionts are also being investigated as model systems to advance knowledge on prokaryote–eukaryote interactions. Research has centred in understanding their symbiotic relationships and communication with their bacteria, including virulence mechanisms and bacterial metabolites (Bright & Bulgheresi 2010; Chaston et al., 2011; Kaplan et al., 2012; Noguez et al., 2012).

Table 1.1 Comparative list of traits of EPN and Caenorhabditis elegans as model organisms

	C. elegans	*Steinernema* spp. *Heterorhabditis* spp.
Rearing conditions	Room temperature	Room temperature
	In vitro → agar plates with lawn *E. coli*	*In vitro* → with or without their symbiotic bacteria
		In vivo → various insect hosts
Size and storage	Adults are 1 mm in length	Adults range from 1–6 mm in length, juveniles 0.5–2 mm in length
Morphology	Transparent → easy to directly observe	Transparent → easy to directly observe
	Cellular changes can be seen with/standard microscopy	Cellular changes w/ standard
Life cycle	Short, ~ 3–7 days	Short 8–21 days
Propagation	Simple, hermaphrodites	Simple, hermaphrodites and gonochoristic adults
	Inbreeding depression mostly absent	Inbreeding depression may occur
Cryopreservation	Yes	Yes
Genetic crosses	Possible, when males exist	Yes, males exist
Genome	Available, reference source for other nematodes	Currently 5 *Steinernema* spp., 4 *Heterorhabditis* spp. but more are becoming available

Furthermore, the availability of an increasing number of genomes of both EPN and their bacterial symbionts has opened venues for exploring their potential in medicine and pharmaceutical bioprospecting. In this respect, ongoing genome–sequencing projects have revealed the capacity of *Xenorhabdus* and *Photorhabdus* to produce several different secondary metabolites including peptides, polyketides, and hybrids of both (Bode, 2009). For example, analysis of *P. luminescens* TT01 has revealed that nearly 6 % of its genome is involved in the production of secondary metabolites (Duchaud et al., 2003; Wilkinson et al., 2009).

1.8 Conclusions and Future Directions

There is no doubt that new research strategies and experimental technologies are generating a continuous flow of knowledge and complex data sets that are transforming basic and applied research of all life processes. In this respect, applied fields such as crop protection will become more and more dependent on state–of–the–art knowledge and biotechnology that will have to be integrated into traditional agricultural practices.

With specific reference to EPN and their bacterial symbionts, there is no doubt that realization of their practical use is spurring developments across broader scientific fronts. The path has been set to further advance knowledge of these organisms, not only as unique and intrinsically interesting biological model systems but also for their practical application as effective biological pesticides.

References

Abad, P., Gouzy, J., Aury, J. M., Castagnone-Sereno, P., Danchin, E. G., Deleury, E., et al. (2008). Genome sequence of the metazoan plant–parasitic nematode *Meloidogyne incognita. Nature Biotechnology, 26*, 909–915.

Abebe, E., Bonner, K., Gray, V., & Thomas, W. K. (2011). Response to 'Bugs don't make worms kill'. *Journal of Experimental Biology, 214*, 1053–1054.

Abebe, E., Jumba, M., Bonner, K., Gray, V., Morris, K., & Thomas, W. K. (2010). An entomopathogenic *Caenorhabditis briggsae. Journal of Experimental Biology, 213*, 3223–3229.

Abu-Shadi, N., Shamseldean, M. M., Abd-Elbary, N. A., & Stock, S. P. (2011). Diversity and distribution of entomopathogenic nematodes (Heterorhabditidae and Steinernematidae) in Egypt. In *Society of Nematologists 50th Annual Meeting, Corvallis OR* (p. 45).

Adams, B. J., Burnell, A. M., & Powers, T. O. (1998). A phylogenetic analysis of *Heterorhabditis* (Nemata: Rhabditidae) based on internal transcribed spacer 1 DNA sequence data. *Journal of Nematology, 30*, 22–39.

Adams, B. J., Fodor, A., Koppenhöfer, H. S., Stackebrandt, E., Stock, S. P., & Klein, M. G. (2006). Biodiversity and systematics of nematode–bacterium entomopathogens. *Biological Control, 38*, 4–21.

Akhurst, R. J. (1982). Antibiotic activity of *Xenorhabdus* spp., bacteria symbiotically associated with insect pathogenic nematodes of the families Heterorhabditidae and Steinernematidae. *Journal of General Microbiology, 128*, 3061–3065.

Akhurst, R. J., & Bedding, R. A. (1986). Natural occurrence of insect pathogenic nematodes (Steinernematidae and Heterorhabditidae) in soil in Australia. *Australian Journal of Entomology, 25*, 241–244.

An, R., & Grewal, P. S. (2010). Molecular mechanisms of persistence of mutualistic bacteria *Photorhabdus* in the entomopathogenic nematode host. *PLoS One, 5*, e13154.

Andaló, V., Nguyen, K. B., & Moino, A. (2006). *Heterorhabditis amazonensis* n. sp. (Rhabditida: Heterorhabditidae) from Amazonas, Brazil. *Nematology, 8*, 853–868.

Anderson, R. C. (2000). *Nematode parasites of vertebrates: Their development and transmission*. Wallingford, U.K: CABI Publishing.

Andrassy, I. (1976). *Evolution as a basis for the sytematization of nematodes*. Budapest: Akademiai Kiado. 278 pp.

Bai, X., Grewal, P. S., Hogenhout, S. A., Adams, B. J., Ciche, T. A., Gaugler, R., et al. (2007). Expressed sequence tag analysis of gene representation in insect parasitic nematode *Heterorhabditis bacteriophora. Journal of Parasitology, 93*, 1343–1349.

Baumann, L., & Baumann, P. (2005). Cospeciation between the primary endosymbionts of mealybugs and their hosts. *Current Microbiology, 50*, 84–87.

Bedding, R. A., Akhurst, R. J., & Kaya, H. K. (1993). *Nematodes and the biological control of insect pests*. East Melbourne: CSIRO Publishing.

Bernard, E. C. (1992). Soil nematode biodiversity. *Biology and Fertility of Soils, 14*, 99–103.

Blaxter, M. L., de Ley, P., Garey, J. R., Liu, L. X., Scheldeman, P., Vierstraete, A., et al. (1998). A molecular evolutionary framework for the phylum Nematoda. *Nature, 392*, 71–75.

Boag, B., Neilson, R., & Gordon, S. C. (1992). Distribution and prevalence of the entomopathogenic nematode *Steinernema feltiae* in Scotland. *Annals of Applied Biology, 121*, 355–360.

Bode, H. B. (2009). Entomopathogenic bacteria as a source of secondary metabolites. *Current Opinion in Chemical Biology, 13*, 224–230.

Boemare, N. (2002). Systematics of *Photorhabdus* and *Xenorhabdus*. In R. Gaugler (Ed.), *Entomopathogenic nematology* (pp. 35–56). Wallingford, U.K.: CABI Publishing.

Boemare, N., & Akhurst, R. (2006). The genera *Photorhabdus* and *Xenorhabdus*. In M. Dworkin & S. Falkow (Eds.), *The prokaryotes* (pp. 451–494). New York: Springer.

Bright, M., & Bulgheresi, S. (2010). A complex journey: Transmission of microbial symbionts. *Nature Reviews Microbiology, 8*, 218–230.

Brooks, D. R., Leon-Regagnon, V., McLennan, D. A., & Zelme, D. (2006). Ecological fitting as a determinant of the community structure of platyhelminth parasites of anurans. *Ecology, 87*, 76–85.

Burnell, A. M., & Stock, S. P. (2000). *Heterorhabditis*, *Steinernema* and their bacterial symbionts–lethal pathogens of insects. *Nematology, 2*, 31–42.

Campos-Herrera, R., Barbercheck, M., Hoy, C. W., & Stock, S. P. (2012). Entomopathogenic nematodes as a model system for advancing the frontiers of ecology. *Journal of Nematology, 44*, 162–176.

Campos-Herrera, R., Johnson, E. G., El-Borai, F. E., Stuart, R. J., Graham, J. H., & Duncan, L. W. (2011). Long–term stability of entomopathogenic nematode spatial patterns in soil as measured by sentinel insects and real–time PCR assays. *Annals of Applied Biology, 158*, 55–68.

Chaston, J. M., Suen, G., Tucker, S. L., Andersen, A. W., Bhasin, A., Bode, E., et al. (2011). The entomopathogenic bacterial endosymbionts *Xenorhabdus* and *Photorhabdus*: Convergent lifestyles from divergent genomes. *PloS One, 611*, e27909.

Chen, G., Dunphy, G. B., & Webster, J. M. (1994). Antifungal activity of two *Xenorhabdus* species and *Photorhabdus luminescens*, bacteria associated with the nematodes *Steinernema* species and *Heterorhabditis megidis*. *Biological Control, 42*, 157–162.

Ciche, T. A., & Ensign, J. C. (2003). For the insect pathogen *Photorhabdus luminescens*, which end of a nematode is out? *Applied and Environmental Microbiology, 69*, 1890–1897.

Ciche, T. A., Kim, K. S., Kaufmann-Daszczuk, B., Nguyen, N. C., & Hall, D. H. (2008). Cell invasion and matricide during *Photorhabdus luminescens* transmission by *Heterorhabditis bacteriophora* nematodes. *Applied and Environmental Microbiology, 74*, 2275–2287.

Çimen, H., Lee, M. M., Hatting, J., Hazir, S., & Stock, S. P. (2014a). *Steinernema tophus* sp. n. Nematoda: Steinernematidae, a new entomopathogenic nematode from South Africa. *Zootaxa*, 38, 337–353.

Çimen, H., Lee, M. M., Hatting, J., Hazir, S., & Stock, S. P. (2014b). *Steinernema innovationi* n. sp. Panagrolaimomorpha: Steinernematidae, a new entomopathogenic nematode species from South Africa. *Journal of Helminthology*, CJO2014.

Clarke, D. J. (2008). *Photorhabdus*: A model for the analysis of pathogenicity and mutualism. *Cellular Microbiology, 10*, 2159–2167.

Cowles, C. E., & Goodrich-Blair, H. (2004). Characterization of a lipoprotein, NilC, required by *Xenorhabdus nematophila* for mutualism with its nematode host. *Molecular Microbiology, 54*, 464–477.

Cowles, C. E., & Goodrich-Blair, H. (2008). The *Xenorhabdus nematophila* nilABC genes confer the ability of *Xenorhabdus* spp. to colonize *Steinernema carpocapsae* nematodes. *Journal of Bacteriology, 190*, 4121–4128.

De Ley, P. (2006). A quick tour of nematode diversity and the backbone of nematode phylogeny. *WormBook: The Online Review of C. elegans Biology*. http://www.ncbi.nlm.nih.gov/books/NBK19684.

Denno, R. F., Gruner, D. S., & Kaplan, I. (2008). Potential for entomopathogenic nematodes in biological control: A meta–analytical synthesis and insights from trophic cascade theory. *Journal of Nematology, 40*, 61–72.

Dillman, A. R., Chaston, J. M., Adams, B. J., Ciche, T. A., Goodrich-Blair, H., Stock, S. P., et al. (2012). An entomopathogenic nematode by any other name. *PLoS Pathogens, 83*, e1002527.

Duchaud, E., Rusniok, C., Frangeul, L., Buchrieser, C., Givaudan, A., Taourit, S., et al. (2003). The genome sequence of the entomopathogenic bacterium *Photorhabdus luminescens*. *Nature Biotechnology, 21*, 1307–1313.

El-Borai, F. E., Brentu, C. F., & Duncan, L. W. (2007). Augmenting entomopathogenic nematodes in soil from a Florida citrus orchard: Non–target effects of a trophic cascade. *Journal of Nematology, 39*, 203–210.

El-Borai, F. E., Duncan, L. W., & Preston, J. F. (2005). Bionomics of a phoretic association between *Paenibacillus* sp. and the entomopathogenic nematode *Steinernema diaprepesi. Journal of Nematology, 37*, 18–25.

Eleftherianos, I., Joyce, S., Ffrench-Constant, R. H., Clarke, D. J., & Reynolds, S. E. (2010). Probing the tri–trophic interaction between insects, nematodes and *Photorhabdus. Parasitology, 137*, 1695–1706.

Enright, M. R., & Griffin, C. T. (2005). Effects of *Paenibacillus nematophilus* on the entomopathogenic nematode *Heterorhabditis megidis. Journal of Invertebrate Pathology, 88*, 40–48.

Ettema, C. H. (1998). Soil nematode diversity: Species coexistence and ecosystem function. *Journal of Nematology, 30*, 159–169.

Ferreira, T., Van Reenen, C. A., Endo, A., Spröer, C., Malan, A. P., & Dicks, L. M. (2013). Description of *Xenorhabdus khoisanae* sp. nov., the symbiont of the entomopathogenic nematode *Steinernema khoisanae. International Journal of Systematic and Evolutionary Microbiology, 63*, 3220–3224.

Ferreira, T., Van Reenen, C., Pagès, S., Tailliez, P., Malan, A. P., & Dicks, L. M. (2013). *Photorhabdus luminescens* subsp. *noenieputensis* subsp. nov., a symbiotic bacterium associated with a novel *Heterorhabditis* species related to *Heterorhabditis indica. International Journal of Systematic and Evolutionary Microbiology, 63*, 1853–1858.

Flores-Lara, Y., Renneckar, D., Forst, S., Goodrich-Blair, H., & Stock, S. P. (2007). Influence of nematode age and culture conditions on morphological and physiological parameters in the bacterial vesicle of *Steinernema carpocapsae* (Nematoda: Steinernematidae). *Journal of Invertebrate Pathology, 95*, 110–118.

Forst, S., & Clarke, D. J. (2002). Nematode–bacterium symbiosis. In R. Gaugler (Ed.), *Entomopathogenic nematology* (pp. 57–77). Wallingford, U.K.: CABI Publishing.

Forst, S., Dowds, B., Boemare, N., & Stackebrandt, E. (1997). *Xenorhabdus* and *Photorhabdus* spp.: Bugs that kill bugs. *Annual Review of Microbiology, 51*, 47–72.

Froy, O. (2005). Convergent evolution of invertebrate defensins and nematode antibacterial factors. *Trends in Microbiology, 13*, 314–319.

Gaugler, R., & Kaya, H. K. (1990). *Entomopathogenic nematodes in biological control.* Boca Raton: CRC Press.

Georgis, R., Koppenhöfer, A. M., Lacey, L. A., Bélair, G., Duncan, L. W., Grewal, P. S., et al. (2006). Successes and failures in the use of parasitic nematodes for pest control. *Biological Control, 38*, 103–123.

Ghedin, E., Wang, S., Spiro, D., Caler, E., Zhao, Q., Crabtree, J., et al. (2007). Draft genome of the filarial nematode parasite Brugia malayi. *Science, 317*, 1756–1760.

Grewal, P. S., Ehlers, R. U., & Shapiro-Ilan, D. I. (2005). *Nematodes as biocontrol agents.* Wallingford, UK: CABI Publishing.

Grewal, P. S., Grewal, S. K., & Adams, B. J. (2003). Parasitism of mollusks by nematodes: Types of associations and evolutionary trends. *Journal of Nematology, 35*, 145–156.

Griffin, C. T., O'Callaghan, K. M., & Dix, I. (2001). A self–fertile species of *Steinernema* from Indonesia: Further evidence of convergent evolution amongst entomopathogenic nematodes? *Parasitology, 12202*, 181–186.

Griffin, C. T., Downes, M. J., & Block, W. (1990). Tests of Antarctic soils for insect parasitic nematodes. *Antarctic Science, 2*, 221–222.

Gualtieri, M., Ogier, J.-C., Pagès, S., Givaudan, A., & Gaudriault, S. (2014). Draft genome sequence and annotation of the Entomopathogenic Bacterium *Xenorhabdus szentirmaii* Strain DSM16338. *Genome Announcements, 2*(2), e00190-14.

Hallem, E. A., Rengarajan, M., Ciche, T. A., & Sternberg, P. W. (2007). Nematodes, bacteria, and flies: A tripartite model for nematode parasitism. *Current Biology, 17*, 898–904.

Hatting, J., Hazir, S., & Stock, S. P. (2008). Diversity and distribution of entomopathogenic nematodes Steinernematidae, Heterorhabditidae in South Africa. *Journal of Invertebrate Pathology, 102*, 120–128.

Hedges, S. B. (2002). The origin and evolution of model organisms. *Nature Reviews Genetics, 311*, 838–849.

Hirao, A., & Ehlers, R. U. (2009). Effect of temperature on the development of *Steinernema carpocapsae* and *Steinernema feltiae* Nematoda: Rhabditida in liquid culture. *Applied Microbiology and Biotechnology, 84*, 1061–1067.

Holterman, M., van der Wurff, A., van den Elsen, S., van Megen, H., Bongers, T., Holovachov, O., et al. (2006). Phylum–wide analysis of SSU rDNA reveals deep phylogenetic relationships among nematodes and accelerated evolution toward crown clades. *Molecular Biology and Evolution, 23*, 1792–1800.

Hominick, W. M. (2002). Biogeography. In R. Gaugler (Ed.), *Entomopathhogenic nematology* (pp. 115–143). Wallinford, Oxon, UK: CABI Publishing.

Hominick, W. M., Reid, A. P., Bohan, D. A., & Briscoe, B. R. (1996). Entomopathogenic nematodes: Biodiversity, geographical distribution and the convention on biological diversity. *Biocontrol Science and Technology, 63*, 317–332.

Husník, F., Chrudimský, T., & Hypša, V. (2011). Multiple origins of endosymbiosis within the Enterobacteriaceae γ–Proteobacteria: Convergence of complex phylogenetic approaches. *BMC Biology, 91*, 87.

Jensen, P., Strauch, O., Wyss, U., Luttmann, R., & Ehlers, R. O. (2000). Carbon dioxide triggers recovery from dauer juvenile stage in entomopathogenic nematodes *Heterorhabditis* spp. *Nematology, 2*, 319–324.

Johnigk, S. A., & Ehlers, R. U. (1999). Endotokia matricida in hermaphrodites of *Heterorhabditis* spp. and the effect of the food supply. *Nematology, 17*, 717–726.

Kanga, F. N., Waeyenberge, L., Hauser, S., & Moens, M. (2012). Distribution of entomopathogenic nematodes in Southern Cameroon. *Journal of Invertebrate Pathology, 106*, 41–51.

Kaplan, F., Alborn, H. T., von Reuss, S. H., Ajredini, R., Ali, J. G., Akyazi, F., et al. (2012). Interspecific nematode signals regulate dispersal behavior. *PLoS One, 7*, e38735.

Kary, N. E., Niknam, G., Griffin, C. T., Mohammadi, A. S., & Moghaddam, M. (2009). A survey of entomopathogenic nematodes of the families Steinernematidae and Heterorhabditidae (Nematoda: Rhabditida) in the north–west of Iran. *Nematology, 11*, 107–116.

Kaya, H. K., & Gaugler, R. (1993). Entomopathogenic nematodes. *Annual Revue de Nematology, 38*, 181–206.

Kim, S. K., Flores-Lara, Y., & Stock, S. P. (2012). Morphology and ultrastructure of the bacterial receptacle in *Steinernema* nematodes Nematoda: Steinernematidae. *Journal of Invertebrate Pathology, 110*, 366–374.

Koppenhöfer, H. S. (2007). Bacterial symbionts of *Steinernema* and *Heterorhabditis*. In K. B. Nguyen & D. Hunt (Eds.), *Entomopathogenic nematodes: Systematics, phylogeny and bacterial symbionts* (pp. 735–808), *Nematology monigraphs and perspectives*, Vol. 5. Leiden: Koninklijke Brill NV.

Kumar, S., Schiffer, P. H., & Blaxter, M. (2012). 959 Nematode Genomes: A semantic wiki for coordinating sequencing projects. *Nucleic Acids Research, 40*, 1295–1300.

Kuwata, R., Qiu, L. H., Wang, W., Harada, Y., Yoshida, M., Kondo, E., et al. (2012). *Xenorhabdus ishibashii* sp. nov., a bacterium from the entomopathogenic nematode *Steinernema aciari*. *International Journal of Systematic and Evolutionary Microbiology, 63*, 1690–1695.

Kuwata, R., Yoshiga, T., Yoshida, M., & Kondo, E. (2007). Phylogenetic relationships of japanese *Heterorhabditis* nematodes and their symbiotic *Photorhabdus* bacteria. *Journal of Nematology, 37*, 39–50.

Lacey, L. A., & Georgis, R. (2012). Entomopathogenic nematodes for control of insect pests above and below ground with comments on commercial production. *Journal of Nematology, 44*, 218.

Lanois, A., Ogier, J. C., Gouzy, J., Laroui, C., Rouy, Z., Givaudan, A., et al. (2013). Draft genome sequence and annotation of the entomopathogenic bacterium *Xenorhabdus nematophila* strain F1. *Genome Announcements, 13*, e00342–13.

Lee, M. M., & Stock, S. P. (2010a). A multilocus approach to assessing co–evolutionary relationships between *Steinernema* spp. Nematoda: Steinernematidae and their bacterial symbionts *Xenorhabdus* spp. Gamma–Proteobacteria: Enterobacteriaceae. *Systematic Parasitology, 77*, 1–12.

Lee, M. M., & Stock, S. P. (2010b). A multigene approach for assessing evolutionary relationships of *Xenorhabdus* spp. Gamma–Proteobacteria, the bacterial symbionts of entomopathogenic *Steinernema* nematodes. *Journal of Invertebrate Pathology, 104*, 67–74.

Liu, J., Berry, R. E., & Blouin, M. S. (1999). Molecular differentiation and phylogeny of entomopathogenic nematodes Rhabditida: Heterorhabditidae based on ND4 gene sequences of mitochondrial DNA. *Journal of Parasitology, 85*, 709–715.

Liu, J., Berry, R. E., & Moldenke, A. F. (1997). Phylogenetic relationships of entomopathogenic nematodes Heterorhabditdae and Steinernematidae inferred from partial 18S rRNA gene sequences. *Journal of Invertebrate Pathology, 69*, 246–252.

Malan, A., Knoetze, R., & Moore, S. D. (2011). Isolation and identification of entomopathogenic nematodes from citrus orchards in South Africa and their biocontrol potential against false codling moth. *Journal of Invertebrate Pathology, 108*, 115–125.

Maneesakorn, P., An, R., Daneshvar, H., Taylor, K., Bai, X., Adams, B. J., et al. (2011). Phylogenetic and cophylogenetic relationships of entomopathogenic nematodes *Heterorhabditis* (Rhabditida) and their symbiotic bacteria *Photorhabdus* (Enterobacteriaceae). *Molecular Phylogenetics and Evolution, 59*, 271–280.

McMullen, J. G., Lee, M. M., & Stock, S. P. (2014). *In vivo* and *in vitro* rearing of entomopathogenic nematodes (Steinernematidae and Heterorhabditidae). *Journal of Visualized Experiments, 91*, e52096.

Miranda, V. A., Navarro, P. D., Davidowitz, G., Bronstein, J., & Stock, S. P. (2013). Effect of insect host age and diet on the fitness of the entomopathogenic nematode–bacteria mutualism. *Symbiosis, 61*, 145–153.

Morán, N. A., Tran, P., & Gerardo, N. A. (2005). Symbiosis and insect diversification: an ancient symbiont of sap–feeding insects from the bacterial phylum *Bacteroidetes. Applied and Environmental Microbiology, 71*, 8802–8810.

Mráček, Z., Bečvář, S., Kindlmann, P., & Jersáková, J. (2005). Habitat preference for entomopathogenic nematodes, their insect hosts and new faunistic records for the Czech Republic. *Biological Control, 34*, 27–37.

Murfin, K. E., Lee, M. M., Klassen, J., McDonald, B., Larget, B., Forst, S., Stock, S. P., Currie, C., & Goodrich-Blair, H. (2015). *Xenorhabdus bovienii* bacterial strain diversity impacts coevolution and symbiotic maintenance with *Steinernema* spp. nematode hosts. mBio 015-03-26 14:51.

Nadler, S. A., Bolotin, E., & Stock, S. P. (2006). Phylogenetic relationships of *Steinernema* (Cephalobina, Steinernematidae) based on nuclear, mitochondrial, and morphological data. *Systematic Parasitology, 63*, 159–179.

Navarro, P. D., McMullen, J. G., II, & Stock, S. P. (2014). Interactions between the entomopathogenic nematode *Heterorhabditis sonorensis* (Nematoda: Heterorhabditidae) and the saprobic fungus *Fusarium oxysporum* (Ascomycota: Hypocreales). *Journal of Invertebrate Pathology, 115*, 41–47.

Nguyen, K. B., & Duncan, L. W. (2002). *Steinernema diaprepesi* n. sp. Rhabditida: Steinernematidae, a parasite of the citrus root weevil *Diaprepes abbreviatus* L Coleoptera: Curculionidae. *Journal of Nematology, 34*, 159–170.

Nguyen, K. B., Maruniak, J., & Adams, J. B. (2001). Diagnostic and phylogenetic utility of the rDNA internal transcribed spacer sequences of *Steinernema. Journal of Nematology, 33*, 73–82.

Noguez, J. H., Conner, E. S., Zhou, Y., Ciche, T. A., Ragains, J. R., & Butcher, R. A. (2012). A novel ascaroside controls the parasitic life cycle of the entomopathogenic nematode *Heterorhabditis bacteriophora. ACS Chemical Biology, 7*, 961–966.

Norton, D. C. (1978). *Ecology of plant–parasitic nematodes*. New York, NY: Wiley.

Orozco, R. A., Hill, T., & Stock, S. P. (2013). Characterization and phylogenetic relationship of *Photorhabdus luminescens* subsp. *sonorensis* (γ–Proteobacteria: Enterobacteriaceae) the bacterial symbiont of the entomopathogenic nematode *Heterorhabditis sonorensis* (Nematoda: Heterorhabditidae). *Current Microbiology, 66*, 30–39.

Park, G. S., Khan, A. R., Hong, S. J., Jang, E. K., Ullah, I., Jung, B. K., et al. (2013). Draft genome sequence of entomopathogenic bacterium *Photorhabdus temperata* strain M1021, isolated from nematodes. *Genome Announcements, 15*, e00747–13.

Petersen, J. J. (1985). Nematodes as biological control agents: Part I. Mermithidae. *Advances in Parasitology, 24*, 307–344.

Phan, K. L., Subbotin, S. A., Nguyen, N. C., & Moens, M. (2003). *Heterorhabditis baujardi* sp. n. (Rhabditida: Heterorhabditidae) from Vietnam and morphometric data for *H. indica* populations. *Nematology, 53*, 367–382.

Poinar, G. O., Jr. (1983). *The natural history of nematodes*. Englewood Cliffs, NJ: Prentice-Hall, Inc. 323pp.

Poinar, G. O., Jr. (1990). Entomopathogenic nematodes in biological control. In R. Gaugler & K. H. Kaya (Eds.), *Taxonomy and biology of Steinernematidae and Heterorhabditidae* (pp. 23–74). Boca Raton, Fl: CRC Press, Inc.

Poinar, G. O., Jr. (1993). Origins and phylogenetic relationships of the entomophilic rhabditids, *Heterorhabditis* and *Steinernema. Fundamental and Applied Nematology, 16*, 333–338.

Powers, T. O., Neher, D. A., Mullin, P., Esquivel, A., Giblin-Davis, R. M., Kanzaki, N., et al. (2009). Tropical nematode diversity: Vertical stratification of nematode communities in a Costa Rican humid lowland rainforest. *Molecular Ecology, 185*, 985–996.

Read, D. S., Sheppard, S. K., Bruford, M. W., Glen, D. M., & Symondson, W. O. C. (2006). Molecular detection of predation by soil micro–arthropods on nematodes. *Molecular Ecology, 15*, 1963–1972.

Ryder, J. J., & Griffin, C. T. (2002). Density–dependent fecundity and infective juvenile production in the entomopathogenic nematode, *Heterorhabditis megidis. Parasitology, 125*, 83–92.

San-Blas, E. (2013). Progress on entomopathogenic nematology research: A bibliometric study of the last three decades: 1980–2010. *Biological Control, 66*, 102–124.

San-Blas, E., Gowen, S. R., & Pwembroke, B. (2008). *Steinernema feltiae*: Ammonia triggers the emergence of their infective juveniles. *Experimental Parasitology, 119*, 180–185.

Sandhu, S. K., Jagdale, G. B., Hogenhout, S. A., & Grewal, P. S. (2006). Comparative analysis of the expressed genome of the infective juvenile entomopathogenic nematode, *Heterorhabditis bacteriophora. Molecular and Biochemical Parasitology, 145*, 239–244.

Schwartz, H. T., Antoshechkin, I., & Sternberg, P. W. (2011). Applications of high–throughput sequencing to symbiotic nematodes of the genus *Heterorhabditis. Symbiosis, 55*, 111–118.

Shapiro-Ilan, D., Gouge, D. H., & Koppenhöfer, A. M. (2002). Factors affecting commercial success: case study in cotton, turf and citrus. In R. Gauger (Ed.), *Entomopathogenic nematology* (pp. 333–353). Wallinfoed, UK: CAB International.

Shapiro-Ilan, D. I., Han, R., & Dolinski, C. (2012). Entomopathogenic nematode production and application technology. *Journal of Nematology, 44*, 226–235.

Sicard, M., Ferdy, J. B., Le Brun, N., Godelle, B., Boemare, N., & Moulia, C. (2004). When mutualists are pathogens: An experimental study of the symbioses between *Steinernema* entomopathogenic nematodes and *Xenorhabdus* bacteria. *Journal of Evolutionary Biology, 17*, 985–993.

Sicard, M., Le Brun, N., Pages, S., Godelle, B., Boemare, N., & Moulia, C. (2003). Effect of native *Xenorhabdus* on the fitness of their *Steinernema* hosts: Contrasting types of interaction. *Parasitology Research, 91*, 520–524.

Sicard, M., Ramone, H., Le Brun, N., Pagès, S., & Moulia, C. (2005). Specialization of the entomopathogenic nematode *Steinernema scapterisci* with its mutualistic *Xenorhabdus* symbiont. *Naturwissenschaften, 9210*, 472–476.

Snyder, H., Stock, S. P., Kim, S. K., Flores-Lara, Y., & Forst, S. (2007). New insights into the colonization and release processes of *Xenorhabdus nematophila* and the morphology and ultrastructure of the bacterial receptacle of its nematode host, *Steinernema carpocapsae. Applied and Environmental Microbiology, 73*, 5338–5346.

Sommer, R., & Ogawa, A. (2011). Hormone signaling and phenotypic plasticity review in nematode development and evolution. *Current Biology, 21*, 758–766.

Spiridonov, S. E., Reid, A. P., Podrucka, K., Subbotin, S. A., & Moens, M. (2004). Phylogenetic relationships within the genus *Steinernema* Nematoda: Rhabditida as inferred from analyses of sequences of the ITS1–5.8S–ITS2 region of rDNA and morphological features. *Nematology, 6*, 547–566.

Stirling, G. R., Poinar, G. O., Jr., & Jansson, H. B. (1988). *Biological control of plant–parasitic nematodes*. Wallingford: CRC Press.

Stock, S. P. (2005). Insect–parasitic nematodes: From lab curiosities to model organisms. *Journal of Invertebrate Pathology, 89*, 57–66.

Stock, S. P. (2009). Molecular approaches and the taxonomy of insect–parasitic and pathogenic nematodes. In S. P. Stock, J. Vandenberg, N. E. Boemare, & I. Glazer (Eds.), *Insect pathogens: Molecular approaches and techniques* (pp. 71–100). Wallingford, Oxon, England, UK: CABI Publishing.

Stock, S. P., Campbell, J. F., & Nadler, S. A. (2001). Phylogeny of *Steinernema* Travassos, 1927 (Cephalobina: Steinernematidae) inferred from ribosomal DNA sequences and morphological characters. *Journal of Parasitology, 87*, 877–889.

Stock, S. P., & Goodrich-Blair, H. (2008a). Nematode–bacterium symbioses: Crossing kingdom and disciplinary boundaries. *Symbiosis, 46*, 61–64.

Stock, S. P., & Goodrich-Blair, H. (2008b). Entomopathogenic nematodes and their bacterial symbionts: The inside out of a mutualistic association. *Symbiosis, 46*, 65–75.

Stock, S. P., & Goodrich-Blair, H. (2012). Nematode parasites, pathogens and associated of insects and invertebrates of economic importance. In L. A. Lacey (Ed.), *Manual of techniques in invertebrate pathology* (pp. 373–426). Yakima, WA, U.S.: Elsevier.

Stock, S. P., Griffin, S. P., & Chaerani, R. T. (2004). Morphological and molecular characterisation of *Steinernema hermaphroditum* n. sp. (Nematoda: Steinernematidae), an entomopathogenic nematode from Indonesia, and its phylogenetic relationships with other members of the genus. *Nematology, 63*, 401–412.

Stock, S. P., & Hunt, D. J. (2005). Morphology and systematics of nematodes used in biocontrol. In P. S. Grewal, R. U. Ehlers, & D. Shapiro-Ilan (Eds.), *Nematodes as biocontrol agents*. New York, NY, U.S: CABI Publishing.

Stock, S. P., Lee, M. M., & Flores-Lara, Y. (2012). The rectal glands of *Heterorhabditis bacteriophora* Rhabditida: Heterorhabditidae hermaphrodites and their role in symbiont transmission. *Journal of Invertebrate Pathology, 101*, 135–138.

Stokwe, N. M. (2009). *Entomopathogenic nematodes: characterization of a new species, long–term storage and control of obscure mealybug, Pseudococcus viburni (Hemiptera: Pseudococcidae) under laboratory conditions*. MSc thesis, Department of Conservation Ecology and Entomology, Stellenbosch University, South Africa.

Strong, D. R. (2002). Populations of entomopathogenic nematodes in foodwebs. In R. Gaugler (Ed.), *Entomopathogenic nematology* (pp. 225–240). Wallingford, Oxon, U.K.: CABI Publishing.

Strong, D. R. (2007). Spatial ecology of food webs with entomopathogenic nematodes. *Phytopathology, 97*, S143–S143.

Szalanski, A. L., Taylor, D. B., & Mullin, P. G. (2000). Assessing nuclear and mitochondrial DNA sequence variation within *Steinernema* (Rhabditida: Steinernematidae). *Journal of Nematology, 32*, 229–233.

Szállás, E., Pukall, R., Pamjav, H., Kovacs, G., Buzas, Z., Fodor, A., et al. (2001). Passengers who missed the train: Comparative sequence analysis, PhastSystem PAGE PCR–RFLP and automated RiboPrint phenotypes of *Photorhabdus* strains. In C. T. Gri Yn, A. M. Burnell, M. J. Downes, & R. Mulder (Eds.), *Developments in entomopathogenic nematode/bacterial research* (pp. 36–53). Luxumberg: European Commission Publications.

Tailliez, P., Laroui, C., Ginibre, N., Paule, A., Pagès, S., & Boemare, N. (2010). Phylogeny of *Photorhabdus* and *Xenorhabdus* based on universally conserved protein–coding sequences and implications for the taxonomy of these two genera. Proposal of new taxa: *X. vietnamensis* sp nov., *P. luminescens* subsp. *caribbeanensis* subsp. nov., *P. luminescens* subsp. *hainanensis* subsp nov., *P. temperata* subsp *khanii* subsp. nov., *P. temperata* subsp *tasmaniensis* subsp. nov., and the reclassification of *P. luminescens* subsp. *thracensis* as *P. temperata* subsp. *thracensis* comb. nov. *International Journal of Systematic and Evolutionary Microbiology, 60*, 1921–1937.

Tailliez, P., Pagès, S., Ginibre, N., & Boemare, N. (2006). New insight into diversity in the genus *Xenorhabdus*, including the description of ten novel species. *International Journal of Systematic and Evolutionary Microbiology, 56*, 2805–2818.

Thao, M. L., Moran, N. A., Abbot, P., Brennan, E. B., Burkhardt, D. H., & Baumann, P. (2000). Cospeciation of psyllids and their primary prokaryotic endosymbionts. *Applied and Environmental Microbiology, 66*, 2898–2905.

Timper, P., & Kaya, H. K. (1992). Impact of a nematode–parasitic fungus on the effectiveness of entomogenous nematodes. *Journal of Nematology, 24*, 1–8.

Torr, P., Spiridonov, S. E., Heritage, S., & Wilson, M. J. (2007). Habitat associations of two entomopathogenic nematodes: a quantitative study using real–time quantitative polymerase chain reactions. *Journal of Animal Ecology, 76*, 238–245.

Torres-Barragan, A., Suazo, A., Buhler, W. G., & Cardoza, Y. J. (2011). Studies on the entomopathogenicity and bacterial associates of the nematode *Oscheius carolinensis. Biological Control, 59*, 123–129.

Tóth, T., & Lakatos, T. (2008). *Photorhabdus temperata* subsp. *cinerea* subsp. nov., isolated from Heterorhabditis nematodes. *International Journal of Systematic and Evolutionary Microbiology, 58*, 2579–2581.

Wallace, H. R. (1963). *The biology of plant parasitic nematodes* (p. 280). London: CABI Publishing.

Wang, J., & Bedding, R. A. (1996). Population development of *Heterorhabditis bacteriophora* and *Steinernema carpocapsae* in the larvae of *Galleria mellonella. Fundamental and Applied Nematology, 19*, 363–367.

Wilkinson, P., Waterfield, N. R., Crossman, L., Corton, C., Sanchez-Contreras, M., Vlisidou, I., et al. (2009). Comparative genomics of the emerging human pathogen *Photorhabdus asymbiotica* with the insect pathogen *Photorhabdus luminescens. BMC Genomics, 10*, 302.

Wilson, M. J., & Gaugler, R. (2000). Terrestrial mollusk pests. In L. Lacey & H. K. Kaya (Eds.), *Field manual of techniques in invertebrate pathology* (pp. 787–804). Dordrecht, The Netherlands: Kluwer Academic Publishers.

Wilson, M. J., Glen, D. M., & George, S. K. (1993). The rhabditid nematode *Phasmarhabditis hermaphrodita* as a potential biological control agent for slugs. *Biocontrol Science and Technology, 34*, 503–511.

Wright, D. J. (2004). Osmoregulatory and excretory behaviour. In R. Gaugler & A. L. Bilgrami (Eds.), *Nematode behaviour* (pp. 177–196). Wallingford, U.K.: CAB International.

Ye, W., Torres-Barragan, A., & Cardoza, Y. J. (2010). *Oscheius carolinensis* n. sp. (Nematoda: Rhabditidae), a potential entomopathogenic nematode from vermicompost. *Nematology, 12*, 121–135.

Zhang, C., Liu, J., Xu, M., Sun, J., Yang, S., An, X., et al. (2008). *Heterorhabditidoides chongmingensis* gen. nov., sp. nov. (Rhabditida: Rhabditidae), a novel member of the entomopathogenic nematodes. *Journal of Invertebrate Pathology, 98*, 153–168.

Zhang, C. X., Yang, S. Y., Xu, M. X., Sun, J., Liu, H., L., J. R., Kan, F., Sun, J., Lai, R., Zhang, K. Y. (2009). *Serratia nematodiphila* sp. nov., associated symbiotically with the entomopathogenic nematode *Heterorhabditidoides chongmingensis* (Rhabditida: Rhabditidae). *International Journal of Systematic and Evolutionary Microbiology,* 59, 1603–1608.

Zuckerman, B., & Rhode, R. A. (1981). *Plant parasitic nematodes* (Vol. 3). New York, NY, U.S: Academic.

Chapter 2
Improvement of Entomopathogenic Nematodes: A Genetic Approach

Itamar Glazer

2.1 Introduction

Domestication and improvement of crop plants and animals have been part of agriculture for thousands of years, and many agricultural systems are therefore artificial. Genetic manipulation of other beneficial arthropods, such as silkworms and honeybees, has been conducted for hundreds of years (Hoy, 1990; Yokoyama, 1973). As in crop breeding, four potential genetic–manipulation strategies exist: artificial selection, hybridization (use of heterosis), mutation, and recombinant DNA techniques.

Genetic improvement programs (GIPs) have also provided innovative methods for controlling insect pests (Hoy, 1985c, 1986). Beneficial arthropods have been selected for climate tolerance (White, DeBach, & Garber, 1970; Wilkes, 1942), host–finding ability, host preference (Allen, 1954; Box, 1956), improved sex ratio (Simmonds, 1947; Wilkes, 1947), increased fecundity (Ram & Sharma, 1977; Wilkes, 1947), and resistance to insecticides (Havron, Kenan, & Rosen, 1991; Havron, Rosen, Prag, & Rossler, 1991; Hoy, 1984; Hoy & Cave, 1991; Hoy, Conley, & Robinson, 1988; Pielou & Glasser, 1952; Roush & Hoy, 1981).

Hoy (1985a, 1990) suggested a number of steps that need to be taken for the genetic improvement of biological control agents of arthropods. These include:

1. Identification of the factors limiting the efficacy of the natural enemy and more specifically, identification of the traits that need improving.
2. Genetic variability must be available for artificial selection.

I. Glazer (✉)
Division of Nematology, Department of Entomology, Institute of Plant Protection, Agricultural Research Organization, The Volcani Center, Rishon LeZion, Israel
e-mail: glazerit@vulcani.agri.gov.il

© Springer International Publishing Switzerland 2015
R. Campos-Herrera (ed.), *Nematode Pathogenesis of Insects and Other Pests*, Sustainability in Plant and Crop Protection 1, DOI 10.1007/978-3-319-18266-7_2

3. Genetic improvement by selection, hybridization, mutagenesis or recombinant DNA methods.
4. Evaluation of genetically improved strains in laboratory, greenhouse and field for efficacy, fitness and stability.
5. Cost–benefit evaluation: one must assume that the cost of the project will be justified by the benefits achieved.

For example *Metaseiulus occidentalis* (Nesbitt) (Mesostigmata; Phytoseiida) is an effective predator of spider mites in deciduous orchards and vineyards in western North America (Hoy, 1985b). It acquired resistance to organophosphorus (OP) insecticides through natural selection in apple orchards in Washington State, and this resistance allowed the predator to survive in orchards even though an OP insecticide azinphos–methyl was applied to control codling moth (Hoyt, 1969). A GIP for *M. occidentalis* involving selection for resistance to carbaryl and permethrin was initiated, and multiresistant strains of *M. occidentalis* were obtained through laboratory crosses and additional selections. The laboratory–selected strains were then tested in small–plot trials for 2 years to determine whether they could become established in orchards or vineyards, survive the relevant pesticide applications in the field, spread, multiply, overwinter, and control the spider mites (Hoy). The small–plot trials were then followed by 3 years of research into how to implement the predators in an integrated mite–management program in almonds (Hoy). The economic analysis suggested that almond growers adopting this program would save $60–$110 per hectare. Thus, genetic improvement of *M. occidentalis* has been shown to be efficacious and cost–effective.

Unlike the long history and vast research on the use of beneficial insects for biological control, the use of entomopathogenic nematodes (EPNs) is only in its third decade; consequently, research and development of GIPs for EPNs is in its infancy. In the present chapter, the need for, and advances in the establishment of genetic approaches for trait improvement in steinernematid and heterorhabditid EPNs will be reviewed, using the scheme proposed by Hoy (1990) as a benchmark.

As the use of EPNs for biological control of insect pests becomes practical and commercial, due to improvements in production methods (Grewal, Ehlers, & Shapiro-Ilan, 2005; Shapiro-Ilan, Han, & Dolinksi, 2012), the use of powerful genetic tools to improve their performance has been strongly advocated (see reviews by Burnell & Dowds, 1996; Gaugler, 1987; Fodor, Vecseri, & Farkas, 1990; Segal & Glazer, 1998, 2000). In this chapter, I describe the reported GIPs of EPNs, the experience gained from these attempts, and future possibilities.

2.2 Identification of Traits for Improvement by Genetic Means

To consider a GIP, one must identify the traits that need to be improved. In general, two directions for EPN improvement have been suggested. The first is enhancement of EPN field efficacy by improving their infectivity to certain insects or their ability to overcome environmental factors affecting their performance and biocontrol consistency. The second is enhancement of their commercial suitability, i.e., increased production efficiency and consistency as well as shelf–life stability. These goals need to be "translated" into defined procedures that can be used in a GIP. Knowledge of the genetic, molecular and physiological architecture of particular traits related to the improvement goal must be established.

2.2.1 *Infectivity*

It has been established that nematode infectivity to different insects or developmental stages of a host varies considerably (Caroli, Glazer, & Gaugler, 1996; Ricci, Glazer, & Gaugler, 1996). Moreover, the process of EPN infectivity and virulence is quite complex (see Chap. 1 for more details). Successful infection and establishment of nematodes in the insect rely on their ability to locate and invade the insect, overcome its immune system and successfully release their symbiotic bacteria. Each of these steps, and possibly some secondary ones, should be considered for improvement and must be studied to determine their effect on the overall infectivity process, before a GIP is initiated. Little is known about the EPN's mechanism of infection or the genes involved in this process. Hao, Montiela, Abubuckerb, Mitrevab, and Simoes (2010), in a study of parasitic mechanisms exhibited by *Steinernema carpocapsae* (Weiser) (Rhabditida: Steinernematidae), generated a cDNA library of the induced *S. carpocapsae* parasitic phase. Comparative analysis identified 377 homologs in *Caenorhabditis elegans* Maupas (Rhabditida (Rhabditoidea)), 431 in *Caenorhabditis briggsae* Osche (Rhabditida; Rhabditoidea), and 75 in other nematodes. Classification of the predicted proteins revealed involvement in diverse cellular, metabolic and extracellular functions: 119 clusters were predicted to encode putatively secreted proteins such as proteases, protease inhibitors, lectins, saposin–like proteins, acetylcholinesterase, antioxidants, and heat–shock proteins, which might interact with the host. This dataset provided the basis for genomic studies toward a better understanding of the events that occur in the parasitic process of this EPN, including invasion of the insect hemocoel, adaptation to the insect's innate immunity and stress responses, and production of virulence factors. The identification of key genes in the parasitic process will provide useful tools for the improvement of *S. carpocapsae* infectivity.

2.2.2 *Survival and Persistence*

Once EPNs are released into the field environment, they encounter numerous factors—including physical, chemical and biological components—that affect their survival and activity as biological control agents (see Chaps. 3, 4 and 5 for more details). It is assumed that during the course of evolution, EPNs, like other terrestrial organisms, adopted unique survival mechanisms to resist environmental extremes. Several researchers noted that enhancement of their ability to survive the environmental factors will increased their efficacy by enabling more nematodes to persist and infect longer periods in the field (Gaugler, 1987; Segal & Glazer, 1998, 2000). Thus, genetic improvement has been suggested as means to achieve such enhancement. However, as of yet, there is no direct evidence for this notion with EPNs.

One of the major obstacles to using EPNs in commercial pest control is their limited shelf life (Strauch, Oestergaard, Hollmer, & Ehlers, 2004). The longevity of the nematode infective juveniles (IJs) can be prolonged by inducing a quiescent state in which their metabolic activity is much reduced. In general, this is done by storage at low temperature (Grewal et al., 2005). However, EPNs are often exposed to higher temperatures during transportation (Mukuka, Strauch, & Ehlers, 2010b), which reduces their viability. Selective breeding for enhancement of desiccation and heat tolerance as a means of prolonging their shelf life, and their capacity for storage and transportation has been suggested.

Nematodes, like bacteria, fungi, and plants, can survive unfavorable environmental conditions in a quiescent state, which considerably prolongs their life span and enables them to withstand the rigors of a fluctuating regime (Barrett, 1991; Watanabe, 2006). Unfavorable environmental conditions include lack of water, extreme temperatures, lack of oxygen, and osmotic stress; the types of quiescence induced in organisms by these conditions are termed anhydrobiosis, thermobiosis/cryobiosis, anoxybiosis and osmobiosis, respectively (Barrett). As the importance of EPNs as biological control agents rises, substantial information regarding their survival mechanisms is being published. Two traits in particular are considered to be most important: desiccation and heat tolerance (Glazer, 2002).

Nematodes are aquatic organisms that need to have a film of water surrounding their body to move (Norton, 1978). Dry conditions adversely affect nematode motility and survival. Some nematode species have adopted anhydrobiosis as a means of surviving prolonged dry periods. This quiescent state in usually reached following a slow rate of water loss (Crowe & Madin, 1975). EPNs can persist for 2–3 weeks in dry soil (Kaya, 1990; Kung & Gaugler, 1990). Most studies investigating the steinernematid nematodes' ability to survive desiccation have focused on *S. carpocapsae* (e.g., Glazer, 1992; Ishibashi, Tojo, & Hatate, 1987; Simons & Poinar, 1973; Womersley, 1990). The general finding has been that various strains of *S. carpocapsae* can survive for appreciable lengths of time under slow drying conditions. In addition, all populations of *S. carpocapsae* survived desiccation better than *Steinernema glaseri* (Steiner) (Rhabditida: Steinernematidae) (Kung, Gaugler, & Kaya, 1990) and *Steinernema riobrave* (Cabanillas, Poinar and Raulston) (Rhabditida: Steinernematidae) (Baur, Kaya, & Thurston, 1995). Solomon, Solomon, Paperna, and Glazer (2000) studied the desiccation tolerance of populations IS–6, IS–15 and SF of *Steinernema feltiae* (Filipjev) (Rhabditida: Steinernematidae). Population IS–6 isolated from the desert region of Israel exhibited the highest survival ability, followed by population IS–15 isolated from northern Israel. The poorest tolerance to desiccation was exhibited by population SF, which was isolated in Germany.

As indicated earlier, fewer studies have been devoted to determining the desiccation tolerance of heterorhabditids. Previous studies have indicated that nematodes belonging to this genus are poor anhydrobionts (Menti, Wright, & Perry, 1997; O'Leary, Stack, Chubb, & Burnell, 1998; Surrey & Wharton, 1995). Liu and Glazer (2000) found wide diversity in desiccation tolerance of heterorhabditid populations from Israel. Furthermore, as a basis for genetic selection, Mukuka, Strauch, Al Zainab, and Ehlers (2010) screened the desiccation tolerance of 43 populations of *Heterorhabditis* spp. and 18 hybrid/inbred strains of *H. bacteriophora* showing significant interspecific variation between nematode populations and species. High variation was found among the different populations, which is an essential prerequisite for the initiation of a GIP (Segal & Glazer, 2000).

The genetic and biochemical mechanisms involved in the induction of anhydrobiosis are not fully understood. One biochemical change that has been reported in anhydrobiotic nematodes is the accumulation of polyols and sugars, in many cases trehalose, which are believed to protect the biological membranes and intracellular proteins during dehydration (e.g., Watanabe, 2006; Womersley, 1990). Solomon, Paperna, and Glazer (1999) showed an increase in trehalose levels in *S. feltiae* IJs exposed to slow dehydration conditions; following rehydration, trehalose concentrations decreased by 50 % within 24 h (Solomon et al., 1999). The synthesis and accumulation of proteins during the desiccation process have been characterized in bacteria, fungi, yeast and plant seeds (Close, 1996; Dure, 1993), but little is known about these aspects in nematodes. In this regard, Chen, Gallop, and Glazer (2005) and Chen et al. (2005) identified novel proteins in osmotically stressed IJs of *S. feltiae* using two–dimensional electrophoresis. Ten novel protein spots and ten upregulated protein spots were detected in the osmotically desiccated IJs. Mass spectrometry analysis of seven significant spots indicated that osmotic stress in desiccated IJs is associated with the induction of actin, proteasome regulatory particle (ATPase–like), GroEL chaperonin, and GroES co–chaperonin.

Gene functions and molecular mechanisms involved in desiccation tolerance in EPNs are just at early stage of investigation. Gal, Solomon, Glazer, and Koltai (2001) identified novel genes of *S. feltiae* population IS–6 that exhibit changes in transcript level upon dehydration. These included the gene encoding glycogen synthase (*Sf–gsy–1*), the rate–limiting enzyme in the synthesis of glycogen, the latter likely playing a role in desiccation survival. Solomon et al. (2000) identified a heat–stable, water stress–related protein with a molecular mass of 47 kDa (designated *desc*47) in *S. feltiae* population IS–6. It was characterized as a late embryogenesis abundant (LEA) homolog protein belonging to the LEA group 3–like proteins. The LEA proteins are a diverse group of water stress–related proteins that are expressed in maturing seeds and in water deficit–stressed vegetative tissues of higher plants (Close, 1996), as well as in nematodes (Burnell & Tunnacliffe, 2011; Tyson, Reardon, Browne, & Burnell, 2007).

Using subtraction hybridization and differential display, several classes of *S. feltiae* genes that are induced in IJs in response to desiccation stress were isolated (Gal, Glazer, & Koltai, 2003). These included transcriptional regulators of metabolic enzymes that are involved in the production of osmoregulants and proteinaceous stress protectants. Among the identified stress–related genes were those encoding *S. feltiae* NAP–1 (nucleosome–assembly protein; *Sf–nap–1*) and CK2 (casein kinase 2; *Sf–ck2*) (Gal et al., 2003; Gal, Glazer, & Koltai, 2005; Gal, Glazer, Sherman, & Koltai, 2005). Interaction between the proteins *Sf*–NAP–1 and *Sf*–CK2was indicated using the yeast two–hybrid system.

A functional role for a LEA protein in *C. elegans* (Ce–LEA–1) in the response to stress conditions was demonstrated (Gal, Glazer, & Koltai, 2004). Group 3 LEA proteins, which are prominent components of the stress response in various organisms (Wise & Tunnacliffe, 2004), are thought to be mainly involved in counteracting the irreversible damage caused by the increased ionic strength that develops in the cytosol during desiccation, perhaps through the binding of both anions and cations to the helical region of the protein (Ingram & Bartels, 1996). The steady–state level of *Ce–lea–1* mRNA increased upon dehydration of *C. elegans* dauer larvae. Partial silencing of this mRNA level steady–state, reduced dauer larvae survival under desiccation, osmotic–stress and heat–stress conditions (Gal et al., 2004). Therefore, it was suggested that Ce–LEA–1 is a critical component of the nematodes' strategy for tolerating the water losses associated with dehydration, osmotic and heat stresses (Gal et al.). The common requirement for Ce–LEA–1 for survival during the examined stresses might support the concept of a molecular mechanism in nematodes that is common to several stress responses, making it a key candidate genetic manipulation toward enhancement of stress tolerance in EPNs.

Somvanshi, Koltai, and Glazer (2008) investigated gene expression in nematodes that were tolerant or susceptible to desiccation stress to determine whether enhanced tolerance in these populations results from a 'gene–expression response' to desiccation or if, for enhanced tolerance, no such response is needed, perhaps due to a state of constant 'readiness'. The expressions of four stress representative genes —*aldehyde dehydrogenase*, *nucleosome assembly protein 1*, *glutathione peroxidase* and *heat–shock protein 40* — were characterized during desiccation stress in five EPN species with differing levels of stress tolerance: *S. feltiae* population IS–6, *S. feltiae* Carmiel population, *S. carpocapsae* Mexican population, *S. riobrave*, and *H. bacteriophora* population TTO1. After 24 h of desiccation, an inverse relationship between the expression of the studied genes and phenotypic desiccation–tolerance capability in the nematodes was observed. *H. bacteriophora* TTO1 was most susceptible to desiccation but showed the highest expression of all studied genes under desiccation. *S. carpocapsae* Mexican population and *S. riobrave* showed the lowest expression of these genes but were most tolerant to desiccation. This study showed no induction of gene expression in stress–tolerant nematodes, whereas the stress–susceptible nematodes responded to stress by induced expression of these genes. Since the different levels of gene expression were found to be related to the different stress–tolerance capabilities of the nematodes, such gene–expression ratios can potentially be used as markers of desiccation tolerance in EPNs. Furthermore, these results imply that molecular tools for the reduction of gene expression, such as RNA interference (RNAi) or genome editing (see further on), may be useful for increasing EPNs' tolerance to stress.

2.3 Genetic Improvement of Entomopathogenic Nematodes

As the advances in production methods rendered the use of EPNs for the biological control of insect pests more practical and commercially feasible (Grewal et al., 2005; Shapiro-Ilan et al., 2012), several attempts were made to improve their performance by genetic means. In this section, I list the reported GIPs of EPNs (summarized in Table 2.1), the experience gained from these attempts, and future possibilities.

Prior to any selection for genetic improvement, it is essential to determine the genetic variation of the particular trait (expressed in terms of 'heritability', h^2) that must be present in the population. Glazer, Gaugler and Segal (1991) assessed the genotypic variation among IJs of *H. bacteriophora* population HP88 under heat, desiccation and ultraviolet light by comparing the performance of inbred lines of this nematode in laboratory assays. Considerable variation in all three traits was detected among the different inbred lines. The heritability values for heat and ultraviolet tolerance were high ($h^2 = 0.98$ and 0.66, respectively), indicating that selection should be an efficient way of improving these traits in the population. The results for desiccation tolerance varied considerably within each line. The heritability value was low ($h^2 = 0.11$), indicating that the results were influenced mainly by environmental variation and suggesting that selective breeding for higher desiccation tolerance would be inefficient. Improvement through the induction of mutations might be a better approach.

To improve the production of *H. bacteriophora* in liquid culture, Johnigk, Hollmer, Strauch, Wyss, and Ehlers (2002), determined the heritability of the disposition to recover in 30 homozygous inbred lines which were established by inbreeding over seven generations. The h^2 values of IJ recovery, as well as final yield, were determined in liquid culture, because the proportion of IJs that recover from the infective stage to the developmental reproductive parasitic stage varies considerably in liquid culture, thus affecting the consistency of production. The calculated heritability for IJ recovery was low ($h^2 = 0.38$). No significant genetic variability could be detected for this trait. In contrast, high heritability ($h^2 = 0.90$) was found for the total number of IJs produced in the liquid medium.

Additional heritability values for different traits which have been the subject of genetic improvement in EPNs are given below and listed in Table 2.1. In the future, additional information and tools for genetic improvement are needed. That includes development of genetic markers as well as identification of specific genes and genetically define traits that can be transferred between designated populations.

Table 2.1 List of genetic improvement programs for entomopathogenic nematodes

Nematode species	Improvement goal	Approach	Heritability (h^2)	Methodology	Results	Fitness change	Trait stability	Comments	References
Steinernema feltiae	Host-finding ability	Selection	0.50	13 rounds in host-finding bioassay	20- to 27-fold increase in host finding	Gain in host penetration and reproductive potential, loss in some storage stability	Gradual decrease	Used high-variation foundation strain	Gaugler and Campbelll (1989) Gaugler, Campbell, and McGuire (1990)
S. feltiae	Efficacy against *L. solani*	Selection	ND[a]	33 rounds in host-infection bioassay	4-fold increase in efficacy	NT[b]	NT	In field trial, showed better efficacy than standard EPNs	Tomalak, (1989, 1994a) Grewal, Tomalak, Keil, and Gaugler (1993)
S. feltiae	Desiccation (rapid and slow) tolerance	Selection	ND	20 selection rounds	Significant increase in survival (>85 %)	No change	Gradual decrease	Heterogeneous foundation population	Salame, Glazer, Chubinishvilli, and Chkhu-bianishvili (2010)
S. feltiae	Desiccation tolerance	Gene transformation	ND	Microinjection of trehalose-6-phosphate synthase	Enhancement of osmotic tolerance (in adults)	NT	NT		Vellai et al. (1999)

S. carpocapsae	Desiccation and Heat tolerance + Virulence	Hybridization	ND	Crosses between strains, 'Italian' and DD-136	Increase in all trait in 2 of 3 modified strains	NT	NT		Shapiro-Ilan et al. (2005)
S. carpocapsae	Enhanced dispersal	Selection	0.60	Capturing the fastest and farthest reaching IJs emanating from an infected *G. mellonella* cadaver, in soil.	21–37 fold increase in the percent IJs dispersing from the source cadaver	Reduced reproduction capacity and nictation ability, a		The selected lines comprised more males (72 %) than the foundation population (44 %)	Bal, Michel, and Grewal (2014)
Heterorhabditis bacteriophora	Heat tolerance	Hybridization	0.98	Crosses between strains IS-5 and HP88	Increase in heat tolerance	Loss in low temperature storage	NT	Mutants were used as genetic markers	Shapiro-Ilan, Glazer, and Segal (1997), Koltai, Glazer, and Segal (1994)
H. bacteriophora	Heat tolerance	Selection	0.68	4 selection rounds	Increase from 38.5 to 39.2 °C	NT	NT		Ehlers, Oestergaard, Hollmer, Wingen, and Strauch (2005)
H. bacteriophora	Cold tolerance	Selection	0.38	4 selection rounds	Reduction from 7.3 to 6.1 °C	NT	NT		Ehlers et al. (2005)

(continued)

Table 2.1 (continued)

Nematode species	Improvement goal	Approach	Heritability (h^2)	Methodology	Results	Fitness change	Trait stability	Comments	References
H. bacteriophora	Desiccation tolerance in osmotic solution	Selection	ND	8 selection rounds	Increased tolerance (from 0.89 to 0.81 a_w)	NT	NT	Used inbred and hybrid lines	Strauch, Oestergaard, Hollmer, and Ehlers (2004)
H. bacteriophora	Heat tolerance	Selection	0.68	11 selection rounds	Increase of 3.0-5.5 °C	No change	NT		Mukuka, Strauch, and Ehlers et al. (2010a)
H. bacteriophora	Desiccation tolerance in osmotic solution	Selection	ND	6 selection rounds	Reduced a_w value	No change	NT		Mukuka, Strauch and Ehlers et al. (2010b)
H. bacteriophora	Nematicidal resistance to:	Selection		11 selection rounds	Fast increase			Cross-resistance was detected	Glazer, Salame, and Segal (1997)
	Enamiphos		0.31		9-fold	No change	Decreased		
	Xamyl		0.71		70-fold	"	Stable		
	Vermectin		0.46		8-fold	"	Stable		
H. bacteriophora	Heat tolerance	Gene transformation	ND	Microinjection of heat-shock protein gene *hsp16*	Enhancement of thermotolerance	No change			Hashmi, Hashmi, and Gaugler (1995, 1998)

[a] *ND* not determined
[b] *NT* not tested

2.3.1 Enhancement of Infectivity

One of the first targets for improvement was the host–finding ability of EPNs. Gaugler and Campbell (1989, 1991) showed that sufficient phenotypical differences exist for host finding between geographical isolates of *S. feltiae* to expect a strong response to selection for this trait. Genetic variation was maximized by hybridizing 10 genetically diverse isolates to create a foundation population for selective breeding. Thirteen rounds of selection resulted in a 20- to 27-fold increase in host finding. Moreover, the proportion of IJs initiating positive chemotaxis increased from less than 33 % to more than 80 %. Nematodes failing to migrate out of the inoculation zone declined from 33 to 8 % after six rounds of selection. Relaxing the selection pressure produced a gradual decrease in host finding. This regression, coupled with the high realized heritability for enhanced host finding (0.64), suggested that wild–type populations take a passive approach to host finding.

The selected population for improved host finding (G–13) was compared to two wild–type populations—the 'All' population and the foundation population from which the G–13 population was derived—for changes in fitness (Gaugler et al., 1990). Acquisition of enhanced host–finding abilities did not appear to be correlated with a serious reduction in overall fitness. Selection did not affect pathogenicity, mobility, sex ratio or morphology. However, population G–13 did show a gain of fitness with regard to host penetration and reproductive potential, and a loss of fitness for storage stability.

Hiltpold, Baroni, Toepfer, Kuhlmann, Turlings (2010a) selected the EPN *Heterorhabditis bacteriophora* Poinar (Rhabditia; Heterorhabditidae) higher responsiveness towards (*E*)–β–caryophyllene (EβC), a sesquiterpene that is emitted by maize roots in response to feeding damage by the western corn rootworm (WCR). EβC is normally only weakly attractive to *H. bacteriophora*, which is one of the most infectious nematodes against WCR. By selecting *H. bacteriophora* to move more readily along a EβC gradient they obtained a population that was almost twice more efficient in controlling WCR population in fields planted with an EβC–producing maize variety. Tomalak (1994a) selected *S. feltiae* population PL for improved efficacy against *Lycoriella solani* (Winn.) (Diptera: Sciaridae) by repeated passage through this fly's larvae *in situ* in compost. There was a four-fold improvement in infectivity of the selected population (ScP) to *L. solani* in laboratory tests after 33 rounds of selection. The control potential of the genetically selected population ScP was further evaluated for the management of *Lycoriella mali* (Fitch), (Diptera: Sciaridae) (Grewal et al., 1993). Trials were conducted at two commercial mushroom farms with high and low levels of fly infestation. The efficacy of population ScP was compared with that of *S. feltiae* population SN and the chitin–synthesis inhibitor diflubenzuron. At low densities of *L. mali*, the two populations did not differ in efficacy, both causing 85–94 % reduction in fly populations. At high fly densities, and a mixed infestation with the phorid fly, *Megaselia halterata* (Wood) (Diptera: Phoridae), population ScP caused 56–83 % reduction in *L. mali* populations, whereas population SN caused 51–73 % reduction. Population ScP persisted longer than population SN. Tomalak (1994b) also reported on the genetic improvement of *S. feltiae* for control of the western flower thrips, *Frankliniella occidentalis* (Pergande) (Thysanoptera: Thripidae). In this study, partially inbred lines and recombinant congenic populations were used. They were selected for high infectivity and small body diameter, showing a substantial increase (>40 %) in infectivity and efficacy.

2.3.2 Improvement of Survival

Nematode tolerance and activity under extreme environmental conditions can limit the shelf life, quality and field performance of nematode–based products. As noted above, enhancement of EPN survival is considered a high priority for genetic improvement.

Shapiro-Ilan, Glazer, and Segal (1997) demonstrated the ability to transfer heat tolerance between two populations of *H. bacteriophora.* Transfer of this trait was accomplished by mating the heat–tolerant population IS–5 (Glazer, Kozodoi, Hashmi, & Gaugler, 1996) with the laboratory population HP88. The hybrid nature of the progeny was confirmed with a mutant of the HP88 population (*Hp–dpy–2*) as marker (Koltai et al., 1994) and by back–crossing. Progeny from the cross were screened for heat tolerance by measuring survival after 2 h exposure to 40 °C. After six passages through last–instar larvae of the wax moth *Galleria mellonella* L. (Lepidoptera: Pyralidae), survival of the hybrid nematodes was significantly greater than that of population HP88 and similar to that of population IS–5. At 32 °C, population IS–5 and the hybrid killed *G. mellonella* larvae at a faster rate than population HP88. Both population IS–5 and the hybrid exhibited sensitivity to cold storage at 10 °C. No differences were detected in reproductive potential.

Hybridization as a mean to improve EPN performance was also demonstrated by Shapiro-Ilan, Stuart, and McCoy (2005). They crossed between a highly virulent population (Italian) of *S. carpocapsae* to the pecan weevil, *Curculio caryae* (Horn) (Coleoptera: Curculionidae) and another population (DD–136) which exhibited high levels of heat and desiccation tolerance but poor virulence to that pest. The crosses resulted in enhancement of persistence as well virulence in two out of three populations generated. Heat and desiccation tolerance in all modified populations was more than 2.5-fold greater than the Italian population and not different from the DD–136 population, except one hybrid had lower heat tolerance than DD–136. Mortality of adult *C. caryae* from the modified populations at 2 or 3 days post–treatment was greater than from DD–136 and similar to the Italian population.

Ehlers, Oestergaard, Hollmer, Wingen, and Strauch (2005) attempted selection of *H. bacteriophora* for heat tolerance and cold activity. Analysis of heritability showed a high value ($h^2 = 0.68$) for heat tolerance and a low value ($h^2 = 0.38$) for activity at low temperature. To increase heat tolerance, four rounds of selection were carried out, which increased the mean tolerated temperature from 38.5 to 39.2 °C. The mean temperature at which the IJs of *H. bacteriophora* were active could be reduced from 7.3 to 6.1 °C after five rounds of selection. However, for unknown reasons, the mean temperature of IJ activity rose during five additional rounds of selection to 7.1 °C. Screening of different isolates of the symbiotic bacterium *Photorhabdus luminescens* for growth at low temperature resulted in several cold–adapted populations from North America, which reached considerable cell density at 6 °C.

To extend the shelf life of a commercial population of *H. bacteriophora*, Strauch et al. (2004) determined the genetic variability of its desiccation tolerance and exploited it for enhancement of this trait by breeding. A hybrid population resulting from crosses of eight *H. bacteriophora* isolates from different geographical origins was used. The desiccation stress was induced by hygroscopic polyethylene glycol (PEG 600) solutions, which lowered the water activity (a_w) of this solution, causing removal of water from the IJs. In this study, the influence of an adaptation phase on desiccation tolerance was also investigated. The lowest mean tolerated a_w value (0.85) was achieved with an adaptation phase of 72 h at an a_w value of 0.96. The variance of the desiccation tolerance increased with a reduction in a_w value during adaptation. The heritability of the trait, determined by using homozygous inbred lines, was 0.46 for non–adapted populations (directly exposed to low a_w values), and 0.48 for adapted ones. A negative heterosis effect could be observed for the desiccation tolerance because nearly all of the inbred lines had higher tolerance to desiccation stress than the hybrid strain. Improvement of desiccation tolerance by breeding was only obtained when the adaptation process was included in the selection process, which was related to higher phenotypic variance in the populations after adaptation. A total of eight rounds of selection and breeding were carried out. Without previous adaptation, the mean tolerated a_w value remained almost constant between 0.94 and 0.93. In contrast, when the IJs were adapted prior to exposure to desiccation stress, the tolerated a_w values dropped continuously from 0.89 to 0.81.

In recent years, Mukuka et al. (2010b), Mukuka, Strauch and Ehlers 2010a, Mukuka, Strauch, Al Zainab, and Ehlers (2010), Mukuka, Strauch, Hoppe, and Ehlers (2010), and Mukuka, Strauch, Waeyenberge, Viaene, and Moens (2010) has initiated a GIP for the improvement of heat and desiccation tolerance in *H. bacteriophora*. To enhance heat tolerance, they first characterized the diversity of this trait among 36 populations of *H. bacteriophora* isolated from diverse environments across the globe, as well as 18 hybrid or inbred strains of these bacteria (Mukuka et al., 2010a; Mukuka, Strauch, Waeyenberge et al., 2010). Five populations of *H. indica* and one of *H. megidis* were also included. Nematodes were tested with or without prior adaptation to heat at 35 °C for 3 h. The mean tolerated temperature ranged from 33.3 °C to 40.1 °C for non–adapted populations, and from 34.8 °C to 39.2 °C for adapted ones. *H. indica* was the most tolerant species, followed by *H. bacteriophora* and *H. megidis*. No correlation was found between the assessed tolerance levels with and without adaptation to heat, implying that different genes are involved. Correlation between a population's heat tolerance and the mean annual temperature in its place of origin was weak. A high variability in tolerance among strains and the relatively high heritability ($h^2 = 0.68$) of the heat tolerance recorded for adapted *H. bacteriophora* provide an excellent foundation for future selective breeding for improved heat tolerance in *H. bacteriophora*.

Desiccation tolerance was evaluated in 43 populations of *Heterorhabditis* spp. and 18 hybrid/inbred populations of *H. bacteriophora* (Mukuka et al., 2010a). Dehydration conditions, measured as a_w values, were produced by treating IJs with different concentrations of the non–ionic polymer PEG 600. Significant interspecific variation was recorded between nematode populations and species. The mean tolerated a_w value (MW50) ranged from 0.90 to 0.95 for non–adapted nematode populations and 0.67 to 0.99 for adapted ones. For selective breeding, only the 10 % most tolerant individuals were used. The lowest a_w value tolerated by this 10 % of the population (MW10) ranged from 0.845 to 0.932 for non–adapted nematode populations and 0.603 to 0.950 for adapted ones. Adaptation significantly increased the desiccation tolerance and a weak correlation was recorded for tolerance with and without adaptation. The nematode populations that were most tolerant to heat or desiccation formed the basis for the foundation of a parental stock produced by cross–breeding and subsequent genetic selection for enhanced tolerance (Mukuka et al., 2010b). In this study, *H. bacteriophora* heat tolerance and desiccation tolerance were significantly increased by cross–breeding tolerant parental strains and successive genetic selection. During the latter process, the selection pressure was constantly increased and only the most tolerant 10 % of the nematode populations were propagated for further selection steps. Assessment of tolerance and selection for both traits were performed with and without prior adaptation to the stress conditions. Eleven rounds of selection were performed to increase heat tolerance. A final overall increase in mean heat tolerance of 5.5 °C was achieved when the nematodes were pre–adapted to the heat stress. For non–adapted tolerance, an increase of 3.0 °C (from 40.1 °C to 43.1 °C) was recorded. For comparison, a commercial population had a mean tolerated temperature after adaptation of 38.2 °C and of 36.5 °C without adaptation. To assess the desiccation tolerance, the mean tolerated a_w value of a population was measured. Cross–breeding the most tolerant populations reduced the a_w value from 0.67 to 0.65 after adaptation, and from 0.9 to 0.7 without prior adaptation. A subsequent six rounds of selection could not increase the tolerance, regardless of whether the nematode had been adapted to the stress.

Monitoring beneficial traits such as infectivity is essential in attempts to genetically improve other traits by crossing tolerant populations or using selective breeding. The fitness of the above–described selected heat– and desiccation–tolerant hybrid strains was evaluated following the selection period (Mukuka, Strauch, Al Zainab et al., 2010; Mukuka, Strauch, Hoppe et al., 2010), in terms of virulence, host penetration and reproductive capacity compared to the commercial population EN 01 of *H. bacteriophora*. Only the heat–tolerant strains were superior or similar in fitness to strain EN 01. The strains with increased desiccation tolerance were generally less fit, possibly reflecting a tradeoff effect of selection for desiccation tolerance. Hybrid strains selected for enhanced tolerance to a stress after adaptation to that stress generally ranked better in terms of fitness than those that were not adapted prior to stress exposure. This could be a result of pleiotropy. The commercial population had the highest reproduction rate per mean number of nematodes penetrating the host insect, a result of automatic selection of inbred lines with high reproductive potential during the commercial production process in liquid culture.

The effect of stress exposure on the infectivity of heat– and desiccation–tolerant hybrid strains of *H. bacteriophora* was assessed against last instars of *G. mellonella* (Mukuka, Strauch, Al Zainab et al., 2010). Nematode IJs were exposed to desiccation stress at an a_w value of 0.85 for 24 h or to a temperature treatment at 40 °C or 0 °C for 24 h prior to inoculation of five IJs per insect. Hybrid strains resulting from crosses of the three very best heat– or desiccation–tolerant strains and crosses of heat– with desiccation–tolerant strains were compared with a commercial population of *H. bacteriophora*. Exposure to desiccation stress caused a significant reduction in infectivity of all strains, not surpassing 25 % mortality, except one strain that was not affected and achieved 37.5 % mortality. Infectivity of untreated IJs of desiccation–tolerant hybrids differed significantly, with a mean insect mortality of 54 %, ranging from 33.8 to 89.6 %. The mean mortality from infection with heat–tolerant hybrids was significantly higher (78.2 %). Infectivity of the commercial population and two other hybrids were not affected by the heat treatment. Consequently, the authors concluded that the infectivity of heat–tolerant strains is not necessarily affected by low–temperature stress.

H. bacteriophora was also subjected to selection for enhancement of nematicidal resistance (Glazer et al., 1997). This is because when applied to the soil, the IJs of this nematode may encounter residual nematicides that will hamper their survival and efficacy. The nematicides used were fenamiphos (an organic phosphate), oxamyl (a carbamate) and avermectin (a biological product). Estimates of heritability (h^2 values) for the three nematicides were 0.31, 0.71 and 0.46, respectively. After 11 rounds of selection for resistance to nematicides, resistance increased dramatically. For fenamiphos and avermectin, and eight- to ninefold increase in resistance was recorded, and a 70-fold increase was recorded for oxamyl. When selection was relaxed, resistance to oxamyl and avermectin was stable while a decrease was recorded with fenamiphos. Fitness was retained in all selected populations when evaluated for infectivity. Cross–resistance was displayed for some, but not all of the nematicides tested.

Hashmi et al. (1995) and Hashmi, Hashmi, Glazer, and Gaugler (1998) used genetic engineering as a means of improving heat tolerance of *H. bacteriophora*. They reported the first successful transformation of an EPN. Foreign genes were introduced into *H. bacteriophora* strain HP88 by microinjection using vectors carrying the *C. elegans* genes coding for the roller phenotype and the 16–kDa heat shock protein (*hsp16*). A translational fusion made by inserting *lacZ* in–frame into *hsp16* was expressed in the body musculature, hypodermis, and pharyngeal muscles. Transcription of the *hsp16/lacZ* transgenes resulted in the rapid synthesis of detectable levels of β–galactosidase. In another study (Hashmi et al., 1998), successful transformation of the *hsp70* gene was confirmed by Southern blot hybridization and polymerase chain reaction (PCR). Blot studies showed that the transgenic nematodes contain 5 to 10 copies per genome of the introduced *hsp70*. Transcripts of *hsp70* mRNA were detected in both wild–type and transgenic nematodes. Transcripts increased several fold in transgenic nematodes upon heat shock. IJs of both transgenic and wild–type nematodes were exposed to a sublethal heat treatment (35 °C) for 2 h followed by a normally lethal heat treatment (40 °C) for 1 h. More than 90 % of the transgenic nematodes survived the heat treatment, compared to 2–3 % of the wild–type strain. Overexpression of *hsp70* resulted an enhanced thermotolerance in the transgenic nematodes, which displayed normal growth and development. Furthermore, the transgenic strain was released in turf grass field microplots in the spring, summer, and fall of 1996 (Gaugler, Wilson, & Shearer, 1997), in accordance with the regulatory procedures at the federal, state, university and local levels needed for field release in the USA. As predicted, persistence of the transgenic and wild–type populations did not differ. This risk–assessment study supports the view that the transgenic nematode population is an unlikely environmental threat.

Vellai et al. (1999) reported successful transformation of the yeast desiccation–related gene encoding trehalose–6–phosphate synthase into the nematode *S. feltiae*. The transformed lines were able to survive well in increased concentrations of NaCl in M9 storage solution, while rapid mortality was recorded in the wild–type population. This study demonstrated the ability to transform steinernematids. However, the bioassays verifying the increase in osmotic–pressure tolerance were performed with a population of adult nematodes, not with IJs. The authors also suggested using the LEA–encoding genes (Browne, Tunnacliffe, & Burnell, 2002; Gal et al., 2004) as candidates for transformation, or the gene promoter, to enhance desiccation tolerance (Vellai et al.).

Unlike *Heterorhabditis*, species of *Steinernema* showed less genetic improvement. Selection for cold tolerance of *S. feltiae* together with its bacterial symbiont, *Xenorhabdus bovienii*, had been conducted by Grewal, Gaugler, and Wang (1996) by repeated passage through *G. mellonella* larvae at 15 °C. Nematode virulence (total insect mortality and speed of kill) and establishment (initiation of nematode development following penetration) were evaluated after 6 (=12–24 generations) and 12 (=24–36 generations) passages. Cold selection enhanced nematode virulence at the cooler temperatures. Virulence measured as total insect mortality at 8 °C improved by 5.3- and 6.6-fold after 6 and 12 passages, respectively. Only small improvements (1.2- to 1.5-fold) were observed in speed of kill. Nematode establishment improved at all temperatures after 12 passages; the highest increase, nine-fold, was observed at 8 °C.

Bal et al. (2014) genetically selected the “ambush” foraging *S. carpocapsae* for enhanced dispersal in the absence of hosts by capturing the fastest and farthest reaching IJs emanating from a nematode–infected *G. mellonella* cadaver, in soil. The selected *S. carpocapsae* showed positive response to selection for dispersal with 13–23 and 21–37 fold increase in the percent IJs dispersing to the farthest distance from the source cadaver, after five and ten rounds of selection, respectively. There was also a significant increase in the average displacement of the selected lines (6.85–7.54 cm/day) than the foundation population (5.54 cm/day) maintained by passing through *G. mellonella* larvae in Petri dishes. The overall mean realized heritability for dispersal was 0.60. The farthest reaching IJs of the selected lines comprised more males (72 %) than the foundation population (44 %) at most time points. Trade–offs associated with enhanced dispersal included reduced reproduction capacity and nictation ability, a trait associated with ambush foraging.

Salame et al. (2010) bred a heterogeneous population of the EPN *S. feltiae* for tolerance to both rapid and slow desiccation. The nematodes were selected for tolerance of rapid desiccation by exposing IJs to ambient conditions [22–25 °C; 50–65 % relative humidity (RH)] for 100 min. A survival rate of 80–90 % was reached after 10 selection cycles. To select for tolerance of slow desiccation, the IJs were exposed to 97 % RH for 72 h, followed by exposure to 85 % RH for an additional 72 h. A high survival rate (>85 %) was obtained after 20 selection cycles. No reduction in fitness was detected in the selected populations. Nevertheless, the population selected for slow desiccation was more tolerant of heat stress than the foundation population.

2.4 Trait Stability

Genetic stability of genetically selected lines has been questioned and evaluated from the earliest attempts to improve nematode performance by genetic means. Gaugler and Campbell (1989), who selected *S. feltiae* for host finding, indicated that "relaxation of selection pressure produced a gradual decrease in host–finding". Similarly, in other studies where trait stability was evaluated after relaxation of the selection regime, a certain reduction was reported (see Table 2.1). Nevertheless, genetic deterioration has also been reported in non–selected EPN populations subjected to continuous laboratory or industrial propagation (Gaugler & Campbell, 1991; Stuart & Gaugler, 1996). Shapiro-Ilan, Glazer, and Segal (1996) reported that the heat–tolerance trait in newly isolated *H. bacteriophora* population IS–5 remained stable after 12 passages of the culture in *G. mellonella*. Other fitness measures (infectivity, reproduction and storage at 25 °C) retained their initial levels. In contrast, Wang and Grewal (2002) reported rapid deterioration of environmental tolerance (to heat, desiccation, and UV) and reproductive potential for *H. bacteriophora* population GPS11 during maintenance in the laboratory.

Similar trait changes were observed by Bilgrami, Gaugler, Shapiro-Ilan, and Adams (2006), who studied the stability of traits important for biological control in *H. bacteriophora* and *S. carpocapsae*; they reported that 20 serial passages in *G. mellonella* results in impaired performance of both nematode species. Virulence, heat tolerance and fecundity deteriorated in all experimental *H. bacteriophora* lines, and four out of five experimental lines deteriorated in host–finding ability. All *S. carpocapsae* experimental lines deteriorated in heat tolerance and nictation, and four out of five experimental lines showed decreased reproductive capacity, whereas virulence declined in two experimental lines. They tested whether trait deterioration was due to changes in the nematode, bacterium, or both symbiotic partners by exchanging nematodes or bacteria from control populations with nematodes or bacteria from the most deteriorated experimental lines and assessing trait recovery. The source of deterioration varied according to the trait, but only the bacterial partner played a role in trait reduction for every trait and species, whereas the nematode was the main source only for *S. carpocapsae* nictation.

Hiltpold, Baroni, Toepfer, Kuhlmann, and Turlings (2010a) who enhance *H. bacteriophora* responsiveness towards (*E*)–β–caryophyllene (EβC) (see above) reported that this process resulted in a slight but significant reduction in infectiousness of the the selected population to the target insect the WCR. Yet, this apparent cost was largely compensated for by the higher responsiveness to the root signal. In further study, Hiltpold, Baroni, Toepfer, Kuhlmann, and Turlings (2010b) showed that the selection process had no negative effect on establishment and persistence of field–released EPN.

Bai, Shapiro-Ilan, Gaugler, and Hopper (2005) suggested stabilizing beneficial traits in *H. bacteriophora* through the creation of genetically homozygous inbred lines that can deter beneficial trait decline. Trait stability was evaluated following serial culturing of three inbred lines and the foundation population in *G. mellonella*. Laboratory data indicated that serial culture of the foundation population (16 passages) results in an over 30 % loss in traits deemed beneficial for biological pest suppression, i.e., virulence to an insect host *Diaprepes abbreviates* L. (Coleoptera: Curculionidae), reproductive capacity, heat tolerance (at 38 °C), and host–seeking ability. In contrast, the inbred lines were impervious to declines in all beneficial traits. A greenhouse test targeting *D. abbreviatus* provided additional evidence that the biocontrol efficacy of the inbred lines remains stable during serial culture.

To stabilize the progress made by selective breeding for desiccation tolerance of *H. bacteriophora*, Anbesse, Sumaya, Dörfler, Strauch, and Ehlers (2013a) tested selection during liquid culturing vs. propagation in host insects. After release of the selection pressure, the tolerance was monitored over additional reproductive cycles *in vivo* and *in vitro* to compare the stability of the trait. Furthermore, they tested whether the virulence of the selected populations was impaired. Exposure to desiccation stress prior to propagation, *in vivo* or *in vitro*, resulted in increased desiccation tolerance. When selection pressure was released, the gained tolerance was lost again during *in vivo* production, whereas the tolerance was maintained at a high level in liquid–cultured EPNs. Anbesse, Sumaya, Dörfler, Strauch, and Ehlers (2013b) further evaluated the stability of *H. bacteriophora* populations selected for heat tolerance using an inbred line reared in liquid culture. After release of the selection pressure, the tolerance was monitored for 15 additional reproductive cycles to determine the stability of the trait. Virulence of the selected populations was assessed to check for negative tradeoff effects. Heat tolerance was successfully increased in *H. bacteriophora* propagated *in vivo* (from 39.03 to 40.85 °C) and *in vitro* (from 39 to 40 °C), but could only be maintained in populations which were serially reared in liquid culture. Anbesse, Strauch, and Ehlers (2012) also investigated possible heterosis effects in desiccation and heat tolerance after cross–breeding of homozygous inbred lines of *H. bacteriophora*. Higher desiccation tolerance of the heterozygous progeny compared to the homozygous inbred lines was recorded, indicating that heterosis is a possible means for further improvement of this trait. In contrast, the heat tolerance of the heterozygous offspring was lower than that of the homozygous population.

When *H. bacteriophora* is cultured *in vivo*, reproduction by cross–fertilization is possible. In *in–vitro* culture, males and females cannot mate and reproduction occurs solely by self–fertilizing hermaphrodites resulting in homozygous inbred lines. Therefore, the studies described above suggest that liquid culture could highly improve and stabilize beneficial traits of heterorhabditid EPNs through selective breeding. Selection using liquid culture technology is thus superior to *in–vivo* propagation in sustaining beneficial traits in *H. bacteriophora*, not only for selective breeding but also for mass production.

Adhikari et al. (2009) generated transcriptional profiles of two experimental lines of *H. bacteriophora*, identified the differentially expressed genes (DEGs) and validated their differential expression in the deteriorated line. The expression–profiling study was performed between experimental lines L5M and OHB of *H. bacteriophora* with probes for 15,220 ESTs from the *H. bacteriophora* transcriptome. Microarray analysis showed 1,185 DEGs comprised of of 469 down– and 716 upregulated genes in nematodes with deteriorated traits. Analysis of the DEGs showed that trait deterioration involves massive changes in the transcripts encoding enzymes involved in metabolism, signal transduction, virulence and longevity. We observed a pattern of reduced expression of enzymes related to primary metabolic processes and induced secondary metabolism. Expression of 16 DEGs in deteriorated–trait nematodes was validated by quantitative reverse transcription–PCR, which revealed similar expression kinetics for all of the genes tested, as shown by microarray.

One of the most powerful tools for genetic analysis and improvement is the induction of mutations (Fodor et al., 1990). The first induction and characterization of mutants of *H. bacteriophora* was reported by Zioni, Glazer, and Segal (1992). A homozygous inbred line was used as the base population for mutagenesis and genetic analysis. Mutagenesis was induced by exposing young hermaphrodites to 0.05 M ethyl–methanesulfonate, and a dumpy (dpy) mutant (*Hdpy–1*) was isolated. Morphological analysis revealed distortion of the head region in adults as well as in IJs. Backcrosses with the wild–type population and genetic analysis revealed that the mutation is recessive. Later on, more recessive dpy mutants—*Hdpy–2 and Hdpy–3*—were isolated and characterized (Koltai et al., 1994). Complementation tests indicated that each of the three mutations affects different genes. The *Hdpy–2* mutant was used as a genetic marker to validate crosses between heat–tolerant and heat–sensitive populations (Shapiro-Ilan et al., 1997) as described above.

Tomalak (1994c) described the first morphological and behavioral mutant in *S. feltiae*. The mutation was spontaneous and occurred in a single gene locus designated *Sfdpy–1*. Action of the new allele was only clearly expressed in IJs. The resulting morphology was classified as 'dumpy' due to the significantly reduced ratio of the nematode's body length to maximum diameter. The identified gene was sex–linked and the new mutant allele remained recessive to the wild–type counterpart responsible for normal morphology. Aside from altering the morphology, pleiotropic action of the dpy allele affected nematode movement and infectivity to insect hosts. The mean dispersion distance of mutant juveniles and their infectivity to *G. mellonella* and *L. solani* larvae were significantly reduced compared to those of wild–type nematodes. Revertant individuals that were occasionally isolated from dpy strains regained the ability to move quickly and dispersed even further than the juveniles from the parental strains. They were also more effective at penetrating the hemocoel of *L. solani* larvae. However, the numbers of infected insects did not significantly differ from those observed for wild–type ScP and SN populations. Following this study, Tomalak (1997) described morphological (Tomalak & Mráček, 1998) and genetic analyses of eight additional mutants of *S. feltiae*. These mutations were also found to be recessive and sex–linked.

In recent years, a new tool for the induction of mutations and genetic modifications has been developed, particularly for *C. elegans* research. Ciche and Sternberg (2007) developed the use of RNAi in heterorhabditid nematodes. This approach was further used by Moshayov, Koltai, and Glazer (2013) to silences genes that are presumably related to the recovery process from the infective to parasitic stage. Understanding of the recovery process and its genetic basis may lead to improvement in nematode recovery. This is an important step in nematode production and can increase its efficiency

Recently, a method to edit the *C. elegans* genome using the clustered, regularly interspersed, short palindromic repeats (CRISPR) RNA–guided Cas9 nuclease has been developed (Dickinson, Ward, Reiner, & Goldstein, 2013; Friedland et al., 2013). Cas9 was able to induce DNA double–strand breaks with specificity for targeted sites, and these breaks could be efficiently repaired by homologous recombination. By supplying engineered homologous repair templates, the researchers generated *gfp* knock–ins and targeted mutations. The results outline a flexible methodology to produce essentially any desired modification in the nematode genome quickly and at low cost. This technology is an important addition to the array of available genetic techniques and can be utilized in EPNs.

Genomic and bioinformatic tools are now available for whole–genome analysis. Bai et al. (2007, 2013) compared *H. bacteriophora* GPS11 expressed sequence tags (ESTs) to the ESTs of animal–parasitic, human–parasitic, plant–parasitic, and free–living nematodes: 127 of them were identified as previously undescribed ESTs, of which 119 had homologs in ESTs and 8 had homologs in proteins of free–living nematodes. These ESTs were assigned putative functions in transcription, signal transduction, cell–cycle control, metabolism, information processing, and cellular processes, thereby providing better insight into *H. bacteriophora* metabolism, sex determination, and signal transduction. In addition, 36 *H. bacteriophora* ESTs had significant similarities to ESTs of parasitic nematodes, but not to ESTs or proteins of free–living nematodes species. Among these were ESTs encoding a centrin, an ankyrin–repeat–containing protein, and a nuclear hormone receptor. The analysis also revealed that parasitic nematode–specific ESTs in this *H. bacteriophora* dataset have more homologs in animal–parasitic nematodes than in those parasitizing humans or plants.

Tyson et al. (2012) recently conducted a molecular analysis of desiccation–tolerance mechanisms in the anhydrobiotic nematode *Panagrolaimus superbus* (Rhabditida: Panagrolaimidae) using ESTs, and revealed a series of candidate genes that may have an important role in stress–tolerance mechanisms. Since this nematode is closely related to EPN (same order) this information will be useful for understanding the basis of stress tolerance of these organisms. In addition, transcriptomic analysis of *H. bacteriophora* (Adhikari et al., 2009) and *S. carpocapsae* (Hao, Montiela, Abubuckerb, Mitrevab, & Simoes, 2010) also listed genes with relevance to stress tolerance. In a recent study, (Yaari et al., 2015) analyzed the transcriptomes of various steinernematid species with various tolerance capabilities to desiccation and heat stresses. This accumulated information will provide the basis for improvement of EPNs by genetic engineering.

Unlike many other beneficial organisms which are subjected to genetic improvement like agricultural plants and animals, EPNs are lacking of useful tools. That includes tools for characterization of heterogenicity/homogenicity of a population, markers to follow transfer or enhancement/degradation of traits as well identified beneficial genes which can be transferred between populations by crosses or molecular means.

2.5 Future Prospects of Genetic Improvement in Entomopathogenic Nematodes

Genetic approaches still hold great promise for improvement of EPNs. Studies have clearly shown that the use of classical and advanced genetic techniques can significantly enhance EPN performance (Table 2.1). However, if we look at the scheme established by Hoy (1990), it is evident that most of the research efforts to date fall within the framework of the second step, i.e., "determination of variability and genetic improvement by selection, mutagenesis or gene transfer". Unlike many other beneficial organisms which are subjected to genetic improvement like agricultural plants and animals, EPN are lacking of useful tools. That includes tools for characterization of heterogenicity/homogenicity of a population (RFLP, AFLP, SNPs, SSRs– see general review by Freeland, 2005), markers to follow transfer or enhancement/degradation (Molecular, morphological and QTLs– Freeland) of traits as well identified beneficial genes which can be transferred between populations by crosses or molecular means. Fundamental research into the genetic architecture of key traits, such as infectivity, stress tolerance and reproduction, is needed. The new genomic, proteomic and bioinformatic technologies should be adapted for EPN genetic research (Dillman, Mortazavi, & Sternberg, 2012). Considering that the genome of *H. bacteriophora* (TTO1 population) has been sequenced (Bai et al., 2013), new tools for genome editing (RNAi, CRISPR–Cas9) may be used for research as well as for the development of genetically improved nematodes. The present review shows that the development of genetically improved "products" stops, for the most part, in the laboratory, with only a very few (Gaugler et al., et al., 1997; Grewal et al., 1993) being tested in field trials. The perception is that improvement of a particular trait (infectivity, stress tolerance, etc.) has yet to be proven under natural conditions.

References

Adhikari, B. N., Lin, C., Bai, X., Ciche, T. A., Grewal, P. S., Dillman, A. R., et al. (2009). Transcriptional profiling of trait deterioration in the insect pathogenic nematode *Heterorhabditis bacteriophora. BioMed Central (BMC) Genomics, 10*, 609–616.

Allen, H. W. (1954). Propagation of *Horogenes molestae*, a parasite of the oriental fruit moth, on the potato tuber–worm. *Journal of Economic Entomology, 47*, 278–281.

Anbesse, S., Strauch, O., & Ehlers, R.-U. (2012). Genetic improvement of the biological control nematode *Heterorhabditis bacteriophora* (Rhabditidomorpha: Heterorhabditidae): Heterosis effect enhances desiccation but not heat tolerance. *Biocontrol Science and Technology, 22*, 1035–1045.

Anbesse, S., Sumaya, H. N., Dörfler, V. A., Strauch, O., & Ehlers, R.-U. (2013a). Selective breeding for desiccation tolerance in liquid culture provides genetically stable inbred lines of the entomopathogenic nematode *Heterorhabditis bacteriophora. Applied Microbiology and Biotechnology, 97*, 731–739.

Anbesse, S., Sumaya, H. N., Dörfler, V. A., Strauch, O., & Ehlers, R.-U. (2013b). Stabilization of heat tolerance traits in *Heterorhabditis bacteriophora* through selective breeding and creation of inbred lines in liquid culture. *BioControl, 58*, 85–93.

Bai, C., Shapiro-Ilan, D. I., Gaugler, R., & Hopper, K. R. (2005). Stabilization of beneficial traits in *Heterorhabditis bacteriophora* through creation of inbred lines. *Biological Control, 32*, 220–227.

Bai, X., Grewal, P. S., Hogenhout, S. A., Adams, B. J., Ciche, T. A., Gaugler, R., et al. (2007). Comparative analysis of *Heterorhabditis bacteriophora* expressed sequence tags (ESTs). *Journal of Parasitology, 93*, 1343–1349.

Bai, X., Adams, B. J., Ciche, T., Clifton, A., Gaugler, R., Kim, K. S., et al. (2013). A lover and a fighter: The genome sequence of an entomopathogenic nematode *Heterorhabditis bacteriophora. PLoS ONE, 8*, e69618.

Bal, K. H., Michel, P. A., & Grewal, S. P. (2014). Genetic selection of the ambush foraging entomopathogenic nematode, *Steinernema carpocapsae* for enhanced dispersal and its associated trade–offs. *Evolutionary Ecology, 28*, 923–939.

Barrett, J. (1991). Anhydrobiotic nematodes. *Agricultural Zoology Reviews, 4*, 161–176.

Baur, M. E., Kaya, H. K., & Thurston, G. S. (1995). Factors affecting entomopathogenic nematode infection of *Plutella xylostella* on a leaf surface. *Entomologia Experimentalis et Applicata, 77*, 239–250.

Bilgrami, A. L., Gaugler, R., Shapiro-Ilan, D. I., & Adams, B. J. (2006). Source of trait deterioration in entomopathogenic nematodes *Heterorhabditis bacteriophora* and *Steinernema carpocapsae* during *in vivo* culture. *Nematology, 23*, 397–409.

Box, H. E. (1956). Battle against Venezuela's cane borer. I. Preliminary investigations and the launching of a general campaign. *Sugar (New York), 51*, 25–27.

Browne, J., Tunnacliffe, A., & Burnell, A. (2002). Plant desiccation gene found in nematode. *Nature, 416*, 3.

Burnell, A. M., & Dowds, B. C. A. (1996). The genetic improvement of entomopathogenic nematodes and their symbiotic bacteria: Phenotypic targets, genetic limitations and an assessment of possible hazards. *Biocontrol Science and Technology, 6*, 435–447.

Burnell, A. M., & Tunnacliffe, A. (2011). Gene induction and desiccation stress in nematodes. In R. N. Perry & D. A. Wharton (Eds.), *Molecular and physiological basis of nematode survival* (pp. 126–156). Wallingford, U.K.: CAB International.

Caroli, L., Glazer, I., & Gaugler, R. (1996). Entomopathogenic nematode infectivity assay: Multi variable comparison of penetration into different hosts. *Biocontrol Science and Technology, 6*, 227–233.

Chen, S., Gallop, N., & Glazer, I. (2005). Cross stress tolerance and expression of stress–related proteins in osmotically desiccated entomopathogenic nematode *Steinernema feltiae* IS–6. *Parasitology, 131*, 695–703.

Chen, S., Glazer, I., Gollop, N., Cash, P., Argo, E., Innes, A., et al. (2005). Proteomic analysis of the entomopathogenic nematode *Steinernema feltiae* IS–6 IJs under evaporative and osmotic stresses. *Molecular Biochemical Parasitology, 145*, 195–204.

Ciche, T. A., & Sternberg, P. W. (2007). Postembryonic RNAi in *Heterorhabditis bacteriophora*: a nematode insect parasite and host for insect pathogenic symbionts. *BioMed Central (BMC) Developmental Biology, 7*, 101–111.

Close, T. J. (1996). Dehydrins: Emergence of a biochemical role of a family of plant dehydration proteins. *Physiological Plantarum, 97*, 795–803.

Crowe, J. H., & Madin, K. A. C. (1975). Anhydrobiosis in nematodes: Evaporative water loss and survival. *Journal of Experimental Zoology, 193*, 323–334.

Dickinson, D. J., Ward, J. D., Reiner, J. D., & Goldstein, B. (2013). Engineering the *Caenorhabditis elegans* genome using *Cas9*–triggered homologous recombination. *Nature Methods.*

Dillman, A. R., Mortazavi, A., & Sternberg, W. P. (2012). Incorporating genomics into the toolkit of nematology. *Journal of Nematology, 44*, 191–205.

Dure, L. (1993). Plant response to cellular dehydration during environmental stress: structural motifs in LEA proteins. In T. J. Close & E. A. Bray (Eds.), *Current topics in plant physiology* (Vol. 10, pp. 91–103). Rockville, MD: American Society of Plant Physiologists.

Ehlers, R.-U., Oestergaard, J., Hollmer, S., Wingen, M., & Strauch, O. (2005). Genetic selection for heat tolerance and low temperature activity of the entomopathogenic nematode–bacterium complex *Heterorhabditis bacteriophora–Photorhabdus luminescens. BioControl, 50*, 699–716.

Fodor, A., Vecseri, G., & Farkas, T. (1990). *Caenorhabditis elegans* as a model for the study of entomopathogenic nematodes. In R. Gaugler & H. K. Kaya (Eds.), *Entomopathogenic nematodes in biological control* (pp. 249–269). Wallingford, U.K.: CAB International.

Freeland, R. J. (2005). *Molecular ecology*. West Sussex, England: Wiley.

Friedland, A. E., Tzur, B. Y., Esvelt, M. K., Colaiácovo, P. M., Church, M. G., & Calarco, A. J. (2013). Heritable genome editing in *C. elegans* via a CRISPR–*Cas9* system. *Nature Methods.*

Gal, T., Glazer, I., & Koltai, H. (2003). Differential gene expression during desiccation stress in *Steinernema feltiae* IS6. *Journal of Parasitology, 89*, 761–766.

Gal, T., Glazer, I., & Koltai, H. (2004). An LEA3 family member is involved in survival of *C. elegans* during exposure to dehydration stress. *FEBS Letters, 577*, 21–26.

Gal, T., Glazer, I., & Koltai, H. (2005). Stressed worms: Responding to the post genomic era. *Molecular and Biochemical Parasitology, 143*, 1–5.

Gal T., Glazer I., Sherman A., & Koltai H. (2005). Protein interaction of Nucleosome Assembly Protein–1 and Casein Kinase 2 during desiccation response in the insect–killing nematode *Steinernema feltiae* IS–6. *Journal of Parasitology* 91, 691–693.

Gal, T., Solomon, A., Glazer, I., & Koltai, H. (2001). Alteration in the levels of glycogen and glycogen synthase transcripts during desiccation in the insect–killing, highly desiccation tolerant nematode *Steinernema feltiae* IS–6. *Journal of Parasitology, 87*, 725–732.

Gaugler, R. (1987). Entomogenous nematodes and their prospects for genetic improvement. In K. Maramorosch (Ed.), *Biotechnology in invertebrate pathology and cell culture* (pp. 457–484). San Diego, CA: Academic.

Gaugler, R., & Campbell, J. F. (1989). Selection for host–finding in *Steinernema feltiae. Journal of Invertebrate Pathology, 54*, 363–372.

Gaugler, R., & Campbell, J. F. (1991). Selection for enhanced host–finding of scarab larvae (Coleoptera: Scarabaeidae) in an entomopathogenic nematode. *Environmental Entomology, 20*, 700–706.

Gaugler, R., Campbell, J. F., & McGuire, T. R. (1990). Fitness of a genetically improved entomopathogenic nematode. *Journal of Invertebrate Pathology, 56*, 106–116.

Gaugler, R., Wilson, M., & Shearer, P. (1997). Field release and environmental fate of a transgenic entomopathogenic nematode. *Biological Control, 9*, 75–80.

Glazer, I. (1992). Survival and efficacy of *Steinernema carpocapsae* in an exposed environment. *Biocontrol Science and Technology, 2*, 101–107.

Glazer, I. (2002). Survival biology. In R. Gaugler (Ed.), *Entomopathogenic nematodes* (pp. 169–187). Wallingford, U.K.: CABI Publishing.

Glazer, I., Gaugler, R., & Segal, D. (1991). Genetics of the entomopathogenic nematode *Heterorhabditis bacteriophora* strain HP88: The diversity of beneficial traits. *Journal of Nematology, 23*, 324–333.

Glazer, I., Kozodoi, E., Hashmi, G., & Gaugler, R. (1996). The entomopathogenic nematode *Heterorhabditis bacteriophora* IS–5; a heat tolerant isolate from Israel. *Nematologica, 42*, 481–492.

Glazer, I., Salame, L., & Segal, D. (1997). Genetic enhancement of nematicidal resistance of entomopathogenic nematodes. *Biocontrol Science and Technology, 7*, 499–512.

Grewal, P. S., Ehlers, R.-U., & Shapiro-Ilan, D. I. (2005). *Nematodes a biocontrol agents*. Oxfordshire, U.K.: CABI Publishing.

Grewal, P. S., Tomalak, M., Keil, C. B. O., & Gaugler, R. (1993). Evaluation of a genetically selected strain of *Steinernema feltiae* against the mushroom sciarid *Lycoriella mali*. *Annals of Applied Biology, 123*, 695–702.

Grewal, P. S., Gaugler, R., & Wang, Y. (1996). Enhanced cold tolerance of the entomopathogenic nematode *Steinernema feltiae* through genetic selection. *Annals of Applied Biology, 129*, 335–341.

Hao, Y.-J., Montiela, R., Abubuckerb, S., Mitrevab, M., & Simoes, N. (2010). Transcripts analysis of the entomopathogenic nematode *Steinernema carpocapsae* induced in vitro with insect haemolymph. *Molecular & Biochemical Parasitology, 169*, 79–86.

Hashmi, S., Hashmi, G., & Gaugler, R. (1995). Genetic transformation of an entomopathogenic nematode by microinjection. *Journal of Invertebrate Pathology, 66*, 293–296.

Hashmi, S., Hashmi, G., Glazer, I., & Gaugler, R. (1998). Thermal response of *Heterorhabditis bacteriophora* transformed with the *Caenorhabditis elegans hsp70* encoding gene. *Journal of Experimental Zoology, 281*, 164–170.

Havron, A., Kenan, G., & Rosen, D. (1991). Selection for pesticide resistance in *Aphytis*. II. A. lingnanensis, a parasite of the California red scale. *Entomologia Experimentalis et Applicata, 61*, 221–228.

Havron, A., Rosen, D., Prag, H., & Rossler, Y. (1991). Selection for pesticide resistance in *Aphytis. I. holoxanthas*, a parasite of the Florida red scale. *Entomologia Experimentalis et Applicata, 61*, 229–235.

Hiltpold, I., Baroni, M., Toepfer, S., Kuhlmann, U., & Turlings, T. C. J. (2010a). Selection of entomopathogenic nematodes for enhanced responsiveness to a volatile root signal helps to control a major root pest. *Journal of Experimental Biology, 213*, 2417–2423.

Hiltpold, I., Baroni, M., Toepfer, S., Kuhlmann, U., & Turlings, T. C. J. (2010b). Selective breeding of entomopathogenic nematodes for enhanced attraction to a root signal did not reduce their establishment or persistence after field release. *Plant Signaling and Behavior, 5*, 1450–1452.

Hoy, M. A. (1984). Genetic improvement of a biological control agent: Multiple pesticide resistance and nondiapause in *Metaseiulus occidentalis* (Nesbitt) (Pytoseiidae). In: D. A., Griffiths & C. E. Bowman (Eds.), *Proceedings VI International Congress of Acarology, Acarology VI*, Vol. 2. (pp. 673–679). New York, NY: Ellis Horwood Ltd., Halsted Press.

Hoy, M. A. (1985a). Improving establishment of arthropod naturel enemies. In: M. A. Hoy & D. C. Herzog (Eds.). *Biological control in agricultural IPM systems* (pp. 151–166). Orlando, FL: Academic Press.

Hoy, M. A. (1985b). Integrated mite management for California almond orchards. In: W. Helle & M. W. Sabelis (Eds.), *Spider mites, their biology, natural enemies and control*, Vol. IB (pp. 299–310). Amsterdam, The Netherlands: Elsevier Science Publ.

Hoy, M. A. (1985c). Recent advances in genetics and genetic improvement of the Phytoseiidae. *Annual Review of Entomology, 30*, 345–370.

Hoy, M. A. (1986). Use of genetic improvement in biological control. *Agriculture, Ecosystems and Environment, 15*, 109–119.

Hoy, M. A. (1990). Genetic improvement of arthropod natural enemies: Becoming a conventional tactic? *UCLA Symposia on Molecular and Cellular Biology, 112*, 405–417.

Hoy, M. A., & Cave, F. E. (1991). Genetic improvement of a parasitoid: Response of Trioxys-pallidus to laboratory selection with Azinphosmethyl. *Biocontrol Science and Technology, 1*, 31–41.

Hoy, M. A., Conley, J., & Robinson, W. (1988). Cyhexatin and fenbutatin–oxide resistance in Pacific spider mite (Atari: Tetranychidae): Stability and mode of inheritance. *Journal of Economic Entomology, 81*, 57–64.

Hoyt, S. C. (1969). Integrated chemical control of insects and biological control of spider mites on apple in Washington. *Journal of Economic Entomology, 62*, 74–86.

Ingram, J., & Bartels, D. (1996). The molecular basis of dehydration tolerance in plants. *Annual Review of Plant Physiology and Plant Molecular Biology, 147*, 377–403.

Ishibashi, N., Tojo, S., & Hatate, H. (1987). Desiccation survival of *Steinernema feltiae* str. DD–136 and possible desiccation protectants for foliage application of nematodes. In: N. Ishibashi (Ed.). *Recent advances in biological control of insect pests by entomogenous nematodes in Japan* (pp. 139–144). Japan: Ministry of Education.

Johnigk, S. A., Hollmer, S., Strauch, O., Wyss, U., & Ehlers, R.-U. (2002). Heritability of the liquid culture mass production potential of the entomopathogenic nematode *Heterorhabditis bacteriophora. Biocontrol Science and Technology, 12*, 267–276.

Kaya, H. K. (1990). Soil ecology. In R. Gaugler & H. K. Kaya (Eds.), *Entomopathogenic nematodes in biological control* (pp. 93–115). Boca Raton, FL: CRC Press.

Koltai, H., Glazer, I., & Segal, D. (1994). Phenotypic and genetic analysis of two mutants of *Heterorhabditis bacteriophora* strain HP88. *Journal of Nematology, 26*, 32–39.

Kung, S. P., & Gaugler, R. (1990). Soil type and entomopathogenic nematode persistence. *Journal of Invertebrate Pathology, 55*, 401–406.

Kung, S. P., Gaugler, R., & Kaya, H. K. (1990). Influence of soil pH and oxygen on entomopathogenic nematode persistence. *Journal of Nematology, 22*, 440–445.

Liu, Q.-Z., & Glazer, I. (2000). Factors affecting desiccation survival of the entomopathogenic nematodes, *Heterorhabditis bacteriophora* HP88. *Phytoparasitica, 28*, 331–340.

Menti, H., Wright, D. J., & Perry, R. N. (1997). Desiccation survival of populations of the entomopathogenic nematodes *Steinernema feltiae* and *Heterorhabditis megidis* from Greece and the UK. *Journal of Helminthology, 71*, 41–46.

Moshayov, A., Koltai, H., & Glazer, I. (2013). Molecular characterization of the recovery process in the entomopathogenic nematode *Heterorhabditis bacteriophora. International Journal of Parasitology, 43*, 843–852.

Mukuka, J., Strauch, O., & Ehlers, R.-U. (2010a). Variability in desiccation tolerance among different strains of the entomopathogenic nematode *Heterorhabditis bacteriophora. Nematology, 12*, 711–720.

Mukuka, J., Strauch, O., & Ehlers, R.-U. (2010b). Improvement of heat and desiccation tolerance in *Heterorhabditis bacteriophora* through cross–breeding of tolerant strains and successive genetic selection. *BioControl, 55*, 511–521.

Mukuka, J., Strauch, O., Al Zainab, H. M., & Ehlers, R.–. U. (2010). Effect of temperature and desiccation stress on infectivity of stress tolerant hybrid strains of *Heterorhabditis bacteriophora. Russian Journal of Nematology, 18*, 111–116.

Mukuka, J., Strauch, O., Hoppe, C., & Ehlers, R.-U. (2010). Fitness of heat and desiccation tolerant hybrid strains of *Heterorhabditis bacteriophora* (Rhabditidomorpha: Heterorhabditidae). *Journal of Pest Sciences, 83*, 281–287.

Mukuka, J., Strauch, O., Waeyenberge, L., Viaene, N., Moens, M., & Ehlers, R.-U. (2010). Heat tolerance among different strains of the entomopathogenic nematode *Heterorhabditis bacteriophora. BioControl, 55*, 423–434.

Norton, D. C. (1978). *Ecology of plant–parasitic nematodes*. Westchester, NY: John Wiley & Sons.

O'Leary, S. A., Stack, C. M., Chubb, M. A., & Burnell, A. M. (1998). The effect of day of emergence from the insect cadaver on the behavior and environmental tolerances of infective juveniles of the entomopathogenic nematode *Heterorhabditis megidis* (strain UK211). *Journal of Parasitology, 84*, 665–672.

Pielou, D. P., & Glasser, R. F. (1952). Selection for DDT resistance in a beneficial insect parasite. *Science, 115*, 117–118.

Ram, A., & Sharma, A. K. (1977). Selective breeding for improving the fecundity and sex ratio of *Thrichogramma fasciatum* (Perkins) (Trichogrammatidae: Hymenoptera), an egg parasite of lepidopterous hosts. *Entomology, 2*, 133–137.

Ricci, M., Glazer, I., & Gaugler, R. (1996). Entomopathogenic nematode infectivity assay: Comparison of laboratory bioassays. *Biocontrol Science and Technology, 6*, 235–245.

Roush, R. T., & Hoy, M. A. (1981). Genetic improvement of *Metaseiulus occidentalis*: Selection with methomyl, dimethoate, and carbaryl and genetic analysis of carbaryl resistance. *Journal of Economic Entomology, 74*, 138–141.

Salame, L., Glazer, I., Chubinishvilli, M. T., & Chkhubianishvili, T. (2010). Genetic improvement of the desiccation tolerance and host-seeking ability of the entomopathogenic nematode *Steinernema feltiae. Phytoparasitica, 38*, 359–368.

Segal, D., & Glazer, I. (1998). Genetic approaches for enhancing beneficial traits in entomopathogenic nematodes. *Japanese Journal of Nematology, 28*, 101–107.

Segal, D., & Glazer, I. (2000). Genetics for improving biological control agents: The case of entomopathogenic nematodes. *Crop Protection, 19*, 685–689.

Shapiro-Ilan, D. I., Glazer, I., & Segal, D. (1996). Trait stability and fitness of the heat tolerant entomopathogenic nematode *Heterorhabditis bacteriophora* IS5 strain. *Biological Control, 6*, 238–244.

Shapiro-Ilan, D. I., Glazer, I., & Segal, D. (1997). Genetic improvement of heat tolerance in *Heterorhabditis bacteriophora* through hybridization. *Biological Control, 8*, 153–159.

Shapiro-Ilan, D. I., Stuart, J. R., & McCoy, W. C. (2005). Targeted improvement of *Steinernema carpocapsae* for control of the pecan weevil, *Curculio caryae* (Horn) (Coleoptera: Curculionidae) through hybridization and bacterial transfer. *Biological Control, 34*, 215–221.

Shapiro-Ilan, D. I., Han, R., & Dolinksi, C. (2012). Entomopathogenic nematode production and application technology. *Journal of Nematology, 44*, 206–217.

Simmonds, F. J. (1947). Improvement of the sex ratio of a parasite by selection. *Canadian Entomology, 79*, 41–44.

Simons, W. R., & Poinar, G. O., Jr. (1973). The ability of *Neoaplectana carpocapsae* (Steinernematidae: Nematodea) to survive extended periods of desiccation. *Journal of Invertebrate Pathology, 22*, 228–230.

Solomon, A., Paperna, I., & Glazer, I. (1999). Desiccation survival of the entomopathogenic nematode *Steinernema feltiae*: Induction of anhydrobiosis. In: I. Glazer, P. Richardson, N. Boemare, & F. Coudert (Eds.), *Survival strategies of entomopathogenic nematodes* (pp. 83–98). Luxemburg: EUR 18855 EN Report.

Solomon, A., Solomon, R., Paperna, I., & Glazer, I. (2000). Desiccation stress of entomopathogenic nematodes induces the accumulation of a novel heat-stable-protein. *Parasitology, 121*, 409–416.

Somvanshi, V. S., Koltai, H., & Glazer, I. (2008). Expression of different desiccation tolerance related genes in various species of entomopathogenic nematodes. *Molecular Biochemical Parasitology, 158*, 65–71.

Strauch, O., Oestergaard, J., Hollmer, S., & Ehlers, R.-U. (2004). Genetic improvement of the desiccation tolerance of the entomopathogenic nematode *Heterorhabditis bacteriophora* through selective breeding. *Biological Control, 31*, 218–226.

Stuart, R. J., & Gaugler, R. (1996). Genetic adaptation and founder effect in laboratory populations of the entomopathogenic nematode *Steinernema glaseri. Canadian Journal of Zoology, 74*, 164–170.

Surrey, M. R., & Wharton, D. A. (1995). Desiccation survival of the infective larvae of the insect parasitic nematode, *Heterorhabditis zealandica* Poinar. *International Journal of Parasitology, 25*, 749–752.

Tomalak, M. (1989). Improvement of the host-search capability as an objective of selective breeding of *Steinernema bibionis* (p. 36). *Proceedings of Symposium on Biopesticides, Theory and Practice*. Ceske Budejovice, Czech Republic.

Tomalak, M. (1994a). Selective breeding of *Steinernema feltiae* (Filipjev) (Nematoda: Steinernematidae) for improved efficacy in control of a mushroom fly, *Lycoriella solani* Winnertz (Diptera: Sciaridae). *Biocontrol Science and Technology, 4*, 187–198.
Tomalak, M. (1994b). Genetic improvement of *Steinernema feltiae* for integrated control of the Western Flower Thrips, *Frankliniella occidentalis. IOBC/WPRS Bulletin, 17*, 17–20.
Tomalak, M. (1994c). Phenotypic and genetic characterization of dumpy infective juvenile mutant in *Steinernema feltiae* (Rhabditida: Steinernematidae). *Fundamental and Applied Nematology, 17*, 485–495.
Tomalak, M. (1997). New morphological variants of infective juveniles associated with mutations in four sex–linked genes of *Steinernema feltiae* (Filipjev) (Nematoda: Steinernematidae). *Fundamental and Applied Nematology, 20*, 541–550.
Tomalak, M., & Mráček, Z. (1998). Scanning electron microscope study on morphological modifications of lateral fields in infective juveniles of mutant *Steinernema feltiae* (Filipjev) (Nematoda: Steinernematidae). *Fundamental and Applied Nematology, 21*, 89–94.
Tyson, T., Reardon, W., Browne, J. A., & Burnell, A. M. (2007). Gene induction by desiccation stress in the entomopathogenic nematode *Steinernema carpocapsae* reveals parallels with drought tolerance mechanisms in plants. *International Journal of Parasitology, 37*, 763–776.
Tyson, T., O'Mahony–Zamora, Z., Wong, S., Skelton, M., Daly, B., Jones, T. J., et al. (2012). A molecular analysis of desiccation tolerance mechanisms in the anhydrobiotic nematode *Panagrolaimus superbus* using expressed sequenced tags. *BioMed Central (BMC) Research Notes, 5*, 68–92.
Vellai, T., Molnár, A., Lakatos, L., Bánfalvi, Z., Fodor, A., & Sáringer, G. (1999). Transgenic nematodes carrying a cloned stress resistant gene from yeast. In I. Glazer, P. Richardson, N. Boemare, & F. Coudert (Eds.), *Survival of entomopathogenic nematodes* (pp. 105–119). Luxemburg: European Commission Publications.
Wang, X., & Grewal, P. S. (2002). Rapid genetic deterioration of environmental tolerance and reproductive potential of an entomopathogenic nematode during laboratory maintenance. *Biological Control, 23*, 71–78.
Watanabe, M. (2006). Anhydrobiosis in invertebrates. *Applied Entomology and Zoology, 41*, 15–31.
White, E. B., DeBach, P., & Garber, M. J. (1970). Artificial selection for genetic adaptation to temperature extremes in *Aphytis lingnanensis* Compere (Hymenoptera: Aphelinidae). *Hilgardia, 40*, 161–192.
Wilkes, A. (1942). The influence of selection on the preferendum of a chalcid (*Microplectron fuscipennis* Zett.) and its significance in the biological control of an insect pest. *Proceedings of the Royal Society of London Series B, 130*, 400–415.
Wilkes, A. (1947). The effects of selective breeding on the laboratory propagation of insect parasites. *Proceedings of the Royal Society of London Series B, 134*, 227–245.
Wise, M. J., & Tunnacliffe, A. (2004). POPP the question: What do LEA proteins do? *Trends in Plant Science, 9*, 13–17.
Womersley, C. Z. (1990). Dehydration survival and anhydrobiotic potential. In R. Gaugler & H. K. Kaya (Eds.), *Entomopathogenic nematodes in biological control* (pp. 117–137). Boca Raton, FL: CRC Press.
Yokoyama, T. (1973). The history of sericultural science in relation to industry. In R. F. Smith, T. E. Mittler, & C. N. Smith (Eds.), *History of entomology* (pp. 267–285). Palo Alto, CA: Annual Reviews, Inc.
Zioni (Cohen–Nissan), S., Glazer, I., & Segal, D. (1992). Phenotypic and genetic analysis of a mutant of *Heterorhabditis bacteriophora* strain HP88. *Journal of Nematology, 24*, 359–364.

Chapter 3
Behaviour and Population Dynamics of Entomopathogenic Nematodes Following Application

Christine T. Griffin

3.1 Introduction

Entomopathogenic nematodes (EPN) of the genera *Steinernema* and *Heterorhabditis* are widely used in inundative biological pest control programmes. It has long been recognised that increased understanding of the ecology of EPN is important for better predictions of field performance and environmental risk (Ehlers & Hokkanen, 1996; Gaugler, Lewis, & Stuart, 1997). Increasingly, EPN are also finding a place as model organisms for fundamental studies in behavioural ecology and evolutionary biology (Campos-Herrera, Barbercheck, Hoy, & Stock, 2012). In this chapter, I consider the fate of EPN used in biocontrol, focussing largely on inundative application to soil. The aim is to provide an overview of the transformation of a biotechnological product to an ecological entity, rather than a review of this rather broad topic. There are already several extensive reviews relevant to the subject, including EPN behaviour and their fate in soil (e.g. Griffin, 2012; Kaya, 2002; Lewis, Campbell, Griffin, Kaya, & Peters, 2006; Stuart, Barbercheck, Grewal, Taylor, & Hoy, 2006; see also Chap. 4). It should be noted that, while the concept of this chapter is to follow the fate of commercially produced EPN when applied to soil, many of the laboratory studies cited have used nematodes produced in insects rather than taken from commercial formulations.

In considering the fate of EPN we can focus on the population or the individual. Smits (1996) proposed a useful model for considering the fate of the applied population, with an initial period of rapid decline, a more gradual decrease in numbers, followed by maintenance at a low level through periodic recycling (Fig. 3.1). Later studies support these general trends. Different factors are likely to

C.T. Griffin (✉)
Department of Biology, Maynooth University, Maynooth, County Kildare, Ireland
e-mail: Christine.griffin@nuim.ie

© Springer International Publishing Switzerland 2015

R. Campos-Herrera (ed.), *Nematode Pathogenesis of Insects and Other Pests*, Sustainability in Plant and Crop Protection 1, DOI 10.1007/978-3-319-18266-7_3

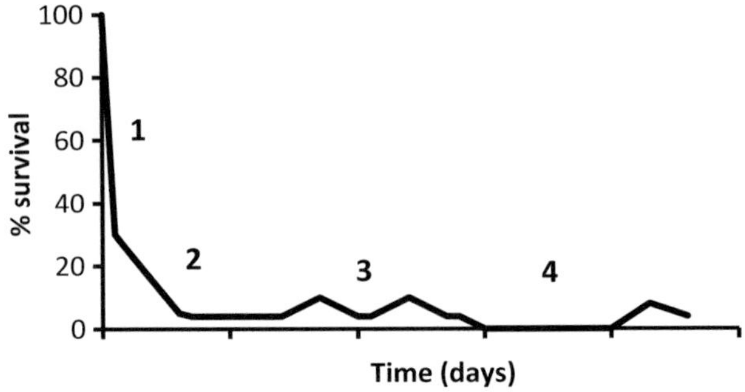

Fig. 3.1 Post–application persistence of entomopathogenic nematode populations, after Smits (1996), showing a rapid decline immediately after application (*1*) followed by a more gradual decline of the applied infective juveniles (*2*) and eventual maintenance of the population through recycling (*3*). Populations may become undetectable (*4*) due to low numbers of infective juveniles free in soil and recover as new hosts are infected or juveniles emerge from previously infected hosts

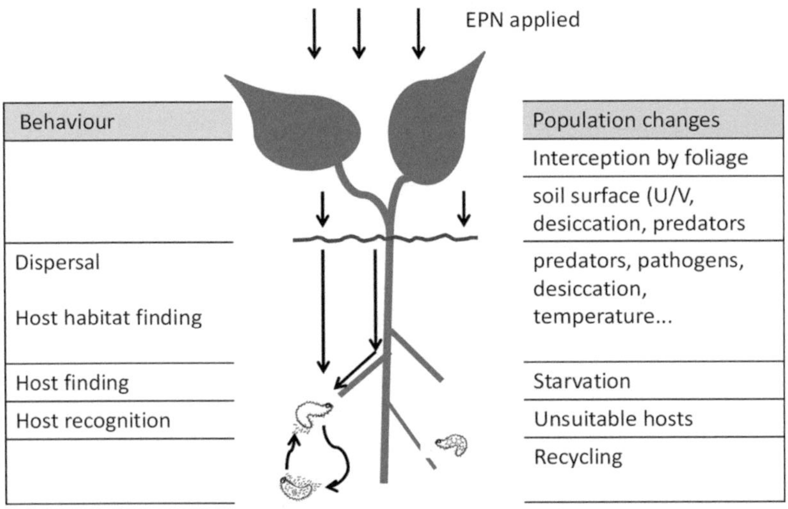

Fig. 3.2 Schematic of behaviour and fate of entomopathogenic nematodes following application to soil for biocontrol

be important at each phase – for example, acute mortality factors such as ultraviolet light, desiccation or predation may be important in the first phase, with starvation, pathogens or invasion into hosts resulting in disappearance of IJs later, while availability of suitable hosts will be critical for longer term population persistence (Fig. 3.2, right panel). While Smits' scheme charts the fate of the population, a fuller appreciation of population dynamics and pest suppression can be obtained by focussing on the individual infective juveniles (IJs) that make up the population. At the individual level, the host–finding behaviour of a parasite can be considered as a hierarchical series of steps including host habitat finding, host finding and host selection (Fig. 3.2, left panel). This scheme, developed for trematodes (Wright, 1959), also fits EPN (Campbell & Lewis, 2002) and will be used here.

3.2 Fate of the Inoculum: Death or Dispersal

For biocontrol purposes, nematodes are typically applied inundatively in high numbers (at least 2.5×10^9 IJs/hectare, Shapiro-Ilan, Han, & Dolinski, 2012). Their fate will depend on a multitude of interacting factors, including soil conditions, crop type and the quality of the applied inoculum. While the scale varies depending on conditions at the application site and the species or population of EPN, in general the scheme represented in Fig. 3.1 describes the fate of the population, with an initial dramatic decline immediately after application (Phase 1 in Fig. 3.1). This rapid decline, with losses varying between 40 and 90 % within hours or days of application, has been attributed to inactivation of IJs by ultraviolet light and desiccation at the soil surface (Smits, 1996), but predation by collembolans and mites may also be important at this stage (Wilson & Gaugler, 2004). Those nematodes that move into soil will be protected from UV, but still vulnerable to abiotic stressors such as desiccation and temperature extremes as well predators and pathogens such as nematode trapping fungi (reviewed by Kaya, 2002 and Chap. 4 in this volume). These mortality factors will contribute to the more gradual decline over succeeding weeks (Phase 2 in Fig. 3.1), but during this phase, starvation will become an additional mortality factor. Infective juveniles do not feed, relying on energy reserves of lipid (mainly triglycerides) and glycogen (Fitters, Patel, Griffin, & Wright, 1999; Patel, Stolinski, & Wright, 1997; Patel & Wright, 1997). Under ideal conditions, individual IJs can survive for months but become visibly lighter as lipid reserves are used up, and eventually die of starvation (Fitters & Griffin, 2006; Hass, Downes, & Griffin, 2002; Patel et al., 1997). Physical soil parameters, especially temperature, moisture and texture influence survival of IJs that have reached the soil (Kaya, Gaugler, & Kung, 1990; Kung & Gaugler, 1991; Molyneux, 1985). Soil factors interact – for example, Pilz et al. (2014) point out that light sandy soils will only favour persistence as long as moisture is not a limiting factor, but in drier regions sandy soils will be subject to desiccation which is inimical to EPN survival. Extreme high temperatures are lethal (Shapiro, Glazer, & Segal, 1996; Somasekhar, Grewal, & Klein, 2002) but permissive temperatures also impact on survival by influencing respiration and motility and hence the rate at which energy reserves are depleted (Andalo, Moino, Maximiniano, Campos, & Mendonca, 2011). Most studies of post–application persistence of EPN do not distinguish between the survival of applied IJs and replenishment of the population by recycling, but Preisser, Dugaw, Dennis, and Strong (2005) found that IJs of *Heterorhabditis marelatus* Liu and Berry (Rhabditida: Heterorhabditidae) survived in the field for at least a year in the absence of hosts.

Survival of applied IJs will be influenced by the quality of the nematodes at time of application – a product of culture and storage conditions – as well as the treatment of the IJs during application (Grewal, 2002; Grewal & Peters, 2005). Genetic quality of the master stock, as well as chemical and physical conditions during production, harvesting, formulation and storage all impact on the quality of the applied inoculum (reviewed by Grewal & Peters). They affect the proportion of IJs retaining bacteria and the number of bacteria per IJ, as well as the quality of the IJs' energy reserves; these in turn influence virulence and the potential for survival of the IJs. Lipid reserves may be depleted during transport and storage, particularly if temperature deviates from the species–specific survival optimum (Grewal). At time of application, IJs may be damaged by exposure to high temperature or UV or by shear forces in the application equipment (Wright, Peters, Schroer, & Fife, 2005). Stresses encountered before application may weaken the IJs and contribute to the initial decline in numbers.

Crop type influences both the number of IJs reaching the soil and their fate in the soil. The percentage of *Heterorhabditis bacteriophora* Poinar (Rhabditida: Heterorhabditidae) IJs reaching soil level immediately after spraying ranged from 5 to 6 % in dense canopy crops (oilseed rape and lupine) to 77–78 % in pasture and potatoes (Susurluk & Ehlers, 2008). As this was measured by placing Petri dishes on the soil it probably overestimates the numbers actually reaching the soil in a pasture with dense thatch, which can form a significant barrier to EPN dispersal (Zimmerman & Cranshaw, 1991). The number of IJs reaching the soil had no impact on short term establishment as detected by baiting one month post–application (Susurluk & Ehlers), and the authors suggested that additional IJs may have been washed from the plant canopy later. While IJs typically survive only short periods on exposed foliage (Schroer, Yi, & Ehlers, 2005; Williams & Macdonald, 1995), high humidity within a canopy and availability of water pooled in nodes of cereals and grasses may enable longer survival in certain circumstances (Fallon, 1998). In field experiments in turfgrass, whether a dramatic decline in recovery of EPN was observed or not varied depending on soil type, turfgrass management regime and time of year that the nematodes were applied, as well as EPN species (Ebssa & Koppenhöfer, 2011). While recovery of all four species tested decreased with time, *Steinernema carpocapsae* (Weiser) Wouts, Mráček, Gerdin & Bedding, (Rhabditida: Steinernematidae) was most likely to undergo a steep drop in the 4 days post–application. The authors attributed the rapid decline of *S. carpocapsae* to its tendency to remain near the soil surface where it would experience more extreme conditions. It should be noted that the cause of a decrease in numbers of EPN in the upper soil layer may be difficult to distinguish between downward migration and attrition of IJs (Elmowitz, Ebssa, & Koppenhöfer, 2014). Crop factors that facilitate larger numbers of EPN reaching soil level may militate against longer term survival; thus, for example, the longer foliage of a golf course "rough" may mean fewer EPN reach the soil than on a closely mown green area, but the longer foliage may provide better protection from UV, high temperature and desiccation (Ebssa & Koppenhöfer).

Persistence of EPN in the field is usually monitored by baiting soil with insects and reporting the proportion of bait insects killed, though methods using qPCR have also been developed (Campos-Herrera et al., 2013; Duncan, Stuart et al., 2013). Even adaptations of the bait method to allow quantification (e.g. Elmowitz et al., 2014; Koppenhöfer, Campbell, Kaya, & Gaugler, 1998) only detect IJs that are infective, and not necessarily total numbers of IJs surviving. Before eventual death by starvation, there is a decline in motility and ability to infect (Fitters & Griffin, 2006; Hass, Griffin, & Downes, 1999; Patel et al., 1997). In a laboratory study, Hass et al. showed that the baiting method (with nematodes quantified following dissection of bait insects) recovered only 20 % of *Heterorhabditis megidis* Poinar, Jackson & Klein (Rhabditida: Heterorhabditidae) from soil immediately after application compared to 56 % recovered by Baermann funnel (a method that relies on activity of the IJs) and 82 % by centrifugal floatation, a mechanical method that does not depend on nematode activity. The Baermann and baiting methods became even less efficient relative to mechanical extraction over the next 28 days, presumably due to declining activity of the IJs (Hass et al.). Detection by baiting gives an indication of the "killing power" of the soil, which is what matters in biocontrol; it can be argued that this is what matters in an ecological context also, as IJs that are not infective cannot reproduce. However, baiting may underestimate the persistence of ecologically relevant IJs if part of the population is temporarily non–infective (Bohan & Hominick, 1996, 1997; Griffin, 1996).

Within–population heterogeneity in survivorship may have important consequences in determining extinction or persistence of a population (Bolnick et al., 2003; Dugaw & Ram, 2011). Population numbers may drop dramatically, but a few individuals that survive (e.g. harsh conditions or periods without hosts) may be responsible for its recovery. Using a modelling approach, Dugaw and Ram showed that a population of *H. marelatus* IJs with individual variation in mortality rates had a good chance of surviving the necessary 5 months in soil until hosts became available, while a population of homogeneous individuals would face almost certain extinction. Demonstrated sources of variation in survivorship include variation in starting lipid reserves, and in the rate at which these lipids are depleted (Fitters & Griffin, 2006; Patel et al., 1997). This heterogeneity in lipid utilisation, where a few individuals remain visibly dark and rich in reserves when others of the population are completely transparent and close to death by starvation, may be indicative of a "bet–hedging" strategy, where parents spread the risk so that at least some offspring survive (Fenton & Hudson, 2002). Apart from genetic variation (Ehlers, Oestergaard, Hollmer, Wingen, & Strauch, 2005; Shapiro, Glazer, & Segal, 1997; Wang, Jung, Son, & Choo, 2013), differences between individuals may arise due to varied conditions experienced during development. IJs emerging from a host at different times differ in size, infectivity and other behaviours (Lewis & Gaugler, 1994; Nguyen & Smart, 1995; Ryder & Griffin, 2003), presumably due to differing conditions of nutrition experienced. Intrinsic (biotic) variation in the population will interact with micro–site variation in soil conditions such as moisture.

3.3 Foraging in Soil and the Root Zone

While thousands of IJs emerge from an insect and miles of millons are applied to control pests, each acts as an individual in its search for a host. In the classic scheme of parasite host–finding developed for schistosomes (MacInnis, 1976; Wright, 1959), there is an initial dispersal phase when parasites move away from the natal host. This dispersal phase is characterised by random movement, though the parasite may also be responsive to signals from the environment that serve to bring it to the host habitat. Once in the host's habitat, the parasite may again move randomly until it encounters the host's "active space" (area of the habitat modified by the presence of the host – gradients of CO_2, other chemicals, temperature) after which more directed host–searching along gradients brings the parasite to the host surface (McInnis). For EPN, the host's active space will frequently be chemical in nature (Dillman et al., 2012; Lewis et al., 2006), though vibrations (Torr, Heritage, & Wilson, 2004) and fine–scale temperature gradients (Burman & Pye, 1980; Byers & Poinar, 1982) may also be effective components of the insect host active space.

As part of their transmission strategy, parasites may modify their behaviour spontaneously depending on their age, coinciding with the various stages of host–finding (Haas, 2003). This is best documented for trematode miracidia, which for the first few hours after hatching are unresponsive to their snail host while moving rapidly in straight lines (Sukhdeo & Sukhdeo, 2004). There is evidence of a similar phasing of activities in certain heterorhabditids (Dempsey & Griffin, 2002; Griffin, 1996). The distance *H. megidis* IJs migrated in sand declined with age, while infectivity (measured as the proportion of IJs entering an insect) increased, suggesting that IJs are initially in a dispersive phase of high mobility and low interest in infecting hosts that should serve to take them away from competitors, and that they subsequently become more motivated to infect (Dempsey & Griffin; Griffin). The behaviour of parasite infective stages is shaped by natural selection to increase the probability of encountering a host, and changes in infectivity and of motility are part of this "optimal transmission strategy" for EPN. Movement is essential for host location, but brings starvation closer and also increases the risk of encountering pathogens. The IJ is thus faced with a classic trade–off situation (McNamara & Houston, 1991), where the optimum strategy for the IJ depends on the characteristic abundance of both hosts and pathogens. In habitats where host abundance is seasonal, the IJ may do best by becoming inactive on reaching a critical starvation level (indicating a failure to find a host) to conserve energy until hosts are again available. We expect the behavioural strategy of native EPN to be adapted to local conditions, while that of applied IJs may not be such a good fit.

A distinction can be made between EPN species based on modes of foraging (Campbell & Gaugler, 1997; Grewal, Lewis, Gaugler, & Campbell, 1994; Lewis, Grewal, & Gaugler, 1995; reviewed by Lewis et al., 2006). While *Heterorhabditis* spp. tend to adopt cruise foraging, *Steinernema* species vary along a continuum from cruise to ambush, with species such as *Steinernema feltiae* (Filipjev) Wouts, Mráček, Gerdin & Bedding (Rhabditida: Steinernematidae) said to employ an intermediate foraging strategy (Campbell & Gaugler). Cruise foragers move actively through soil, and use distant volatile cues to assist in host–finding. In ambush foragers, notably *S. carpocapsae*, and *Steinernema scapterisci* Nguyen & Smart (Rhabditida: Steinernematidae) most IJs remain near the soil surface (Georgis & Poinar, 1983) where they lift their body into the air, facilitating attachment to passing insects (Campbell & Gaugler, 1993). These IJs exhibit jumping behaviour, which is also believed to facilitate attachment to hosts (Campbell & Kaya, 1999; Campbell, Lewis, Stock, Nadler, & Kaya, 2003). For ambushers, volatiles are said to be relatively unimportant in host–finding at a distance (Grewal et al.; Lewis et al.), though ambush species are attracted to both CO_2 and more specific host odours (Dillman et al., 2012). Cruise foragers are expected to infect less mobile hosts underground while ambushers are considered to be more successful at infecting mobile, surface dwelling hosts (Gaugler et al., 1997). While there are definite differences in behaviour between EPN species traditionally classified as ambushers and those classified as cruisers (e.g. nictation and jumping are expressed by "ambush" species), it is becoming increasingly clear that *S. carpocapsae*, an ambusher, can find and infect relatively immobile insects at considerable distances from the point of application (de Altube, Strauch, de Castro, & Pena, 2008; Dembilio, Llacer, de Altube, & Jacas, 2010; Dillon, Ward, Downes, & Griffin, 2006), prompting some to question the usefulness of the classification (Wilson, Ehlers, & Glazer, 2012).

Even true sit–and–wait foragers must disperse, and dispersal is an essential phase preceding and possibly interspersed with bouts of host finding. *S. carpocapsae* IJs move both vertically, reaching depths of 15–20 cm in soil (Ferguson, Schroeder, & Shields, 1995) and laterally: about 4 % of *S. carpocapsae* IJs ("sprinters") dispersed faster than the fastest *H. bacteriophora* (Bal, Taylor, & Grewal, 2014). Dispersal by *S. carpocapsae* IJs appears to be strongly influenced by substrate, being much greater in pure peat than in pure sand (Kruitbos, Heritage, Hapca, & Wilson, 2010). In nature, IJs emerge in their thousands from the depleted natal host and should have experienced strong selection to move away from these overcrowded conditions.

3.3.1 Dispersal and Host Finding in Soil

Active dispersal by IJs after inundative application is usually a few centimetres per day and limited to a scale of meters overall (Downes & Griffin, 1996; Poinar & Hom, 1986). As for survival, abiotic factors strongly influencing EPN dispersal and host–finding include soil texture, moisture and temperature. In general, light–textured (sandy) soils favour nematode movement (Georgis & Poinar, 1983; Koppenhöfer & Fuzy, 2006). Early experiments with plant parasitic nematodes (e.g. Wallace, 1968) illustrate how nematodes move through the water film coating soil particles, and emphasised that what determines the suitability for nematode movement is not the proportions of different sized particles *per se*, but the size of pores relative to the nematodes, and that it is the matric potential (not total water content) that is important, as this is related to the surface tension and water films within the soil. The bulk density or degree of compaction is also important, as it affects the soil pores. Portillo–Aguilar, Villani, Tauber, Tauber, and Nyrop (1999) varied both texture and bulk density and found that rates of movement and infection by three EPN species were strongly correlated with the amount of soil pore space with dimensions similar to or greater than the diameter of the nematodes. The size of particles and their packing determine the channels open to nematode movement, as well as the soil moisture profile and the diffusion of oxygen. While moisture is essential for nematode movement and survival, in saturated soils microbial activity results in anaerobic conditions which may render nematodes quiescent. In fertile soils, the particles are aggregated together in the form of crumbs, which increase the total pore space in the soil, allowing good aeration and drainage (Wallwork, 1970). Burr and Robinson (2004) suggest that nematodes with mean lengths of around 1,000 μm (the typical length for EPN IJs) may be adapted to use channels provided by roots and insects, rather than the soil interstices that are better suited to smaller nematodes of around 400 μm.

Consideration of the effects of soil type on EPN usually focusses on the mineral component – proportions of sand, silt and clay particles. Movement in organic media has received less attention. Many potting media, and peat soils such as those used for coniferous forestry in northern temperate regions, are composed largely of organic matter. EPN can disperse and find hosts in these highly organic media (Ansari & Butt, 2011; Nielsen & Lewis, 2011). Kruitbos et al. (2010) compared pure sand and pure peat as media for EPN dispersal and host–finding, and found contrasting responses for two species; the ambush forager *S. carpocapsae* displayed host–finding behaviour in peat but not in sand, while the reverse was true for the cruise forager *H. megidis*. *S. carpocapsae* also dispersed better in peat than in sand. The authors suggested that the poor performance of *H. megidis* in peat was due to adsorption onto the organic matter of the host volatiles that are used by cruiser species to locate their host. However, in organic media including peat, *H. megidis* and two other heterorhabditids showed superior host finding compared to *S. carpocapsae* (Ansari & Butt). In the field, both *S. carpocapsae* and the cruiser *Heterorhabditis downesi* Stock, Burnell & Griffin (Rhabditida: Heterorhabditidae) performed better in peat than in mineral soils in field trials against pine weevil (Williams, Dillon, Girling, & Griffin, 2013). Movement and host finding of EPN in organic media is worthy of more attention.

Random movement may be important in bringing a parasite into the zone in which signals (from host or host habitat) can be effective for directing movement (MacInnis, 1976). IJs will encounter stimuli that initiate directed search either in response to the host itself, or roots as indicator of likely host habitat. The importance of directed movement in bringing IJs to their hosts should not be overestimated. In laboratory assays, where single stimuli are presented in simplified media such as sterile sand, it is easy to demonstrate that EPN follow gradients or accumulate at stimulus source. In nematodes, directed movement may be superimposed on a large random movement component, and directed movement may be only partially substituted for random movement even in a gradient (Hunt, Wall, DeCrappeo, & Brenner, 2001). Carbon dioxide is one of the main attractants identified for EPN but biologically active soil will be full of sources of it, making it unreliable for finding insects at a distance in such soils. However, following a CO_2 gradient is likely to bring nematodes to plant roots (Dusenbery, 1987), where potential hosts might then be found by moving along the roots randomly and/or in response to further directional signals operating over a shorter scale. CO_2 may be seen as a response–activator that alerts EPN to the presence of living organisms and may enhance responsiveness to other more specific cues (Turlings, Hiltpold, & Rasmann, 2012) that are discussed in more detail below.

3.3.2 Dispersal and Host–Finding in the Root Zone

Many of the hosts naturally utilized by EPN or targeted by their application are root–feeding insects and hence much of IJ behaviour takes place in the context of the rhizosphere. Plant roots may affect the dispersal and host–finding of EPN in a number of ways: by attracting nematodes into the root zone, where hosts are located, by effects on host–finding within the root zone itself, and by influencing conditions for survival. Roots play a major role in shaping the soil environment, influencing the physical structure, pH, water and oxygen availability, as well as gradients of information–rich chemicals (Dini-Andreote & van Elsas, 2013; Hinsinger, Bengough, Vetterlein, & Young, 2009). While the traditional view is that roots dry the surrounding soil by the uptake of water, the effects of plants on the availability of water in soil are complex (Carminati et al., 2011). Redistribution of soil water from moist deep layers to drier surface layers (hydraulic lift) by citrus roots enhanced survival of *Steinernema diaprepesi* Nguyen & Duncan (Rhabditida: Steinernematidae) (Duncan & McCoy, 2001). Similarly, the maintenance of a moist microclimate by the taproot of bush lupine facilitated survival of *H. marelatus* during dry conditions (Preisser, Dugaw, Dennis, & Strong, 2006). Bal et al. (2014) reported that the presence of vegetation enhanced lateral dispersal of both *S. carpocapsae* and *H. bacteriophora*. The presence of roots in soil increased the rate of infection by EPN of non–feeding trap insects (wax moths), but only at low root density (Choo & Kaya, 1991), while high density of roots interfered with host finding (Choo, Kaya, Burlando, & Gaugler, 1989). Similar effects were reported by Cutler and Webster (2003).

Roots alone are attractive to EPN (Bird & Bird, 1986; Hui & Webster, 2000; Kanagy & Kaya, 1996), but especially when wounded or fed on by insects (Boff, van Tol, & Smits, 2002; Rasmann & Turlings, 2008; van Tol et al., 2001; Wang & Gaugler, 1998). Rasmann et al. (2005) reported that maize plants released (E)–β–caryophyllene in response to feeding by larvae of corn rootworm *Diabrotica virgifera virgifera* Le Conte (Coleoptera: Chrysomelidae), and that this volatile was highly attractive to *H. megidis*. Other plant species also release volatiles from their roots in response to insect feeding, that attract EPN, though not all EPN respond similarly (Ali, Alborn, & Stelinski, 2011; Hiltpold, Baroni, Toepfer, Kuhlmann, & Turlings, 2010; Rasmann & Turlings). Species with all categories of foraging strategy respond (Ali et al., 2011). Indeed, the same signal also attracted free–living bacterial feeding nematodes that might compete with EPN for the cadaver as a resource (Ali, Campos-Herrera, Alborn, Duncan, & Stelinski, 2013). That nematode–host specialization may be more important than foraging strategy is indicated by the fact that the response of *S. diaprepesi* to citrus roots damaged by its host *Diaprepes abbreviatus* L. (Coleoptera: Curculionidae) was stronger than that of other EPN species (Ali et al.).

If herbivore–induced volatile signals are a reliable indicator of host presence, then they could be considered to represent host "active space", while the root zone itself could be considered host habitat. Chemical enlargement of host active space can significantly increase the transmission success of parasites, putting hosts themselves under pressure to suppress the emission of attractants. Use by EPN of plant–derived signals would be particularly important in finding hosts that are otherwise unattractive, such as vine weevil *Otiorhynchus sulcatus* Fabricius (Coleoptera: Curculionidae) larvae (Boff, Zoon, & Smits, 2001). However, the usefulness of the volatiles will vary depending on soil type and conditions; for example, being less effective in soil types with high levels of chemical activity (Turlings et al., 2012). Soil texture and moisture also affect diffusion of volatiles. For example, (E)–β–caryophyllene diffused readily through sand but diffusion was limited in more complex soils (Hiltpold & Turlings, 2008). Although (E)–β–caryophyllene appears to diffuse in the gaseous phase, soil moisture facilitated diffusion probably by preventing its adsorption onto soil particles and/or loss by vertical diffusion, with higher water content required in soil than in pure sand (Hiltpold & Turlings, 2008; Chiriboga, Jaffual, Campos-Herrera, Roëder, & Turlings, 2014).

In addition to providing a source of attractant and/or confusing volatiles, roots of trees or other plants may provide EPN with a physical "routeway", facilitating their penetration deep into soil. The presence of plant material (twigs) significantly increased *S. carpocapsae*–induced mortality of large pine weevils *Hylobius abietis* L. (Coleoptera: Curculionidae) buried close to the base of the twigs (Ennis, Dillon, & Griffin, 2010). Such routeways may be particularly important for ambush strategists such as *S. carpocapsae* by stimulating ranging search (Lewis, Gaugler, & Harrison, 1993). Migration along roots provides a plausible explanation for the success of the ambush species *S. carpocapsae* against root–feeding insects (de Altube et al., 2008; Dillon et al., 2006; Jansson, Lecrone, & Gaugler, 1993), despite its reputation for remaining at the soil surface and not responding directionally to volatiles. For example, *S. carpocapsae* can parasitize larvae and pupae of *H. abietis* under the bark of tree roots at depths of up to 40 cm in the soil (Dillon et al.) and has been used as part of an integrated population suppression strategy for this pest (Torr, Wilson, & Heritage, 2005). Similarly, *S. carpocapsae* has provided up to 95 % control of flat–headed root borer *Capnodis tenebrionis* L. (Coleoptera: Buprestidae) in roots of apricot trees (de Altube et al.). Even for those species that actively disperse through a soil matrix, searching along a root may be a particularly efficient way of arriving at root–feeding insects. Host–finding by the cruise–forager

H. megidis was facilitated by a simple unbranched artificial root, but the effect was reduced as branching levels increased (Demarta, Hibbard, Bohn, & Hiltpold, 2014). However, the addition of (E)–β–caryophyllene dramatically changed the results, with more EPN finding hosts where the most complex root models were present (Demarta et al.). Comparative studies on the relative importance for ambush and cruise foragers of roots in facilitating penetration into soil would be instructive, as would further investigations into the impact of root architecture and surface properties on EPN migration. Roots actively release exudates including mucins and secondary metabolites which may facilitate or inhibit movement of nematodes in intimate association with the root surface (Dini-Andreote & van Elsas, 2013; Wuyts, Maung, Swennen, & De Waele, 2006), but this has received little attention for nematodes of any kind.

Insect feeding on plants may result both in the release of volatiles and the propagation of vibrations. For cruise foragers such as heterorhabditids, attraction to roots on which insects are feeding may be largely due to allelochemicals released from the damaged plant, but ambushers such as *S. carpocapsae* are reported to respond poorly to distant volatile signals while searching (Lewis et al., 1995). Torr et al. (2004) showed that artificially–produced vibrations transmitted through peat were attractive to EPN including *S. carpocapsae*, and since acoustic stimuli produced by insects can be transmitted for up to 30 cm through soil (Mankin et al., 2000) they are potentially an important source of information for EPN.

3.3.3 Host Recognition and Acceptance

We assume that the behaviour of nematodes, as other animals, has been shaped by evolution so as to yield the highest number of surviving offspring. The critical choice of which host to invade, which determines both the quality and quantity of resources, and the availability of both mating partners and of competitors, is made by the IJ. Since the IJ will rarely be presented with a simultaneous choice of hosts, the decision can be thought of as a series of binary "invade/do not invade" decisions. Hosts of diverse species and developmental stage may be utilized, as well as already–infected and even dead hosts (Peters, 1996; San-Blas & Gowen, 2008). Having arrived at an insect, the IJ must decide whether to attempt to enter or not (host recognition). For a nematode, entry into a host in which it cannot develop is a dead end, though this may fulfil the requirement for pest population reduction. Following attraction by volatile cues, host recognition based on contact insect cues may involve insect excretory products, cuticle or gut contents (Grewal, Gaugler, & Lewis, 1993; Grewal, Gaugler, & Selvan, 1993; Lewis, Ricci, & Gaugler, 1996). Host recognition behaviour may predict the suitability of invertebrates as hosts: the behavioural recognition response of *S. carpocapsae* to various insect species was correlated with the level of nematode reproduction supported by the species, while non–insect arthropods stimulated no recognition response and were not susceptible to nematode infection (Lewis et al.). However, for a generalist species with a broad host range, such a matching between host acceptance and suitability for reproduction may be less than perfect unless there are some general cues that are reliable predictors of suitability or unsuitability. Unless unsuitable hosts are commonly encountered, selection for recognition will be weak. Maintenance of a weak recognition filter is of advantage in the long term, allowing new hosts to be added to the range, even if some mortality of individuals in unsuitable hosts results in the short term (Combes, 1991).

For any EPN–host interaction, there is an optimal infection rate. For most species of *Steinernema*, which reproduce by amphimixis, a minimum of two IJs must enter and develop in order for reproduction to be possible, while for *Heterorhabditis* a single IJ can colonise a host, as all develop into self–fertile hermaphrodites; a similar situation exists for *Steinernema hermaphroditum* Stock Griffin & Chaerani (Rhabditida: Steinernematidae) (Griffin, O'Callaghan, & Dix, 2001). However, it may take more than one or two IJs to overcome the host defences (Peters & Ehlers, 1994). With increasing numbers, intraspecific competition for host resources results in lower reproductive output per invading founding adult (discussed later). Therefore, once the number of nematodes necessary for reproduction and/or host–killing has entered, it is in the interests of the residents to deter further invasion. Glazer (1997) found that hosts injected with IJs of three *Steinernema* spp. became less attractive for invasion by conspecifics about 6–9 h later, and data indicated that the initial infection induced the release of a substance which reduced the subsequent invasion (Glazer). However, in several other studies IJs were attracted to and entered hosts that were already occupied by conspecifics, even to the point of overcrowding (Christen, Campbell, Lewis, Shapiro-Ilan, & Ramaswamy, 2007; Lewis et al., 2006; Ramos-Rodriguez et al., 2007). Indeed, some species may even prefer to invade an already–killed host – for example, *Steinernema riobrave* Cabanillas, Poinar & Raulston (Rhabditida: Steinernematidae) preferred wax moth larvae infected 24 h previously over an uninfected wax moth (Christen et al.). For an IJ with a short lifespan, limited locomotory ability and only one chance at infection, it may be better to invade a suboptimal host than to reject it and fail to find a better one, on the basis that some reproduction is better than none. Alternatively, a suboptimal host may be accepted because IJs do not recognise it as such, due either to lack of meaningful cues from the host or limited sensory abilities of the IJs.

In some species at least, it appears that the tendency to infect changes with time since emergence from the source cadaver – before the eventual decline in infectivity associated with ageing. There are two models of "phased infectivity": Bohan and Hominick (1996, 1997) described fluctuations in infectivity of *S. feltiae* and attributed it to a proportion of IJs switching between a non–infective and an infective state. Griffin (1996) described an increase in the infectivity of *H. megidis* in the initial weeks after emerging from the natal host and attributed it to a gradual change in infection tendency of individual IJs, rather than a switch between states (Dempsey & Griffin, 2002; Griffin, 1996; Ryder & Griffin, 2003). This is discussed more fully in Lewis et al. (2006). From the individual nematode's perspective, IJs that delay infectivity for some time during which they (or their competitors) migrate away from the natal host from which they have emerged en masse benefit by avoiding the overcrowding that otherwise might be expected in adjacent hosts (Dempsey & Griffin), while for a parent, producing offspring that differ in their infection strategies may be an important adaptation to uncertain conditions such as host availability – a strategy of bet–hedging (Fenton & Hudson, 2002). At a population level, where individuals are not all maximally infective at the same time there is greater probability of successful recycling of EPN (Shields, Testa, Miller, & Flanders, 1999).

3.4 Infection and Reproduction: Recycling in Targets and Non–target Hosts

Following the initial decline in numbers of applied IJs, nematode populations may be boosted and maintained by recycling in target and non–target hosts (Phase 3, Fig. 3.1). According to Smits (1996) the population maintains a fairly stable level of "perhaps 10,000–40,000 nematodes/m^2". Most species of EPN can utilise a fairly broad range of hosts, and as more than 100,000 IJs can be produced from a single large host cadaver (Dutky, Thompson, & Cantwell, 1964; Lindegren, Valero, & Mackey, 1993; Shapiro-Ilan, Gaugler, Tedders, Brown, & Lewis, 2002), a large population of susceptible insects can contribute significantly to EPN population numbers. A higher yield of IJs per available host also contributes to recycling success of applied EPN (Kim & Alston, 2008).

The time scale over which recycling is detected varies from just 1 month (McGraw, Vittum, Cowles, & Koppenhöfer, 2010) to several years (Dillon, Rolston, Meade, Downes, & Griffin, 2008; Ferguson et al., 1995; Shields et al., 1999). Some agronomic systems are more conducive to nematode recycling than others. In field crops, the longest persistence by *H. bacteriophora* of 23 months was following application to beans followed in rotation by wheat with red clover as cover crop (Susurluk & Ehlers, 2008). This was presumed to be due to reproduction in larvae of the bean weevil *Sitona lineatus* L. (Coloptera: Curculionidae) which were abundant in the bean crop and may have persisted in the clover (Susurluk & Ehlers). Similarly, nematode incidence (percentage of soil cores with EPN) increased from spring to autumn in crops with high densities of potential hosts – *S. lineatus* in pea and *Delia radicum* L. (Diptera: Anthomyiidae) in cabbage (Nielsen & Philipsen, 2004b). Availability of suitable insects and low disturbance by ploughing, harrowing or other soil movements were factors that favoured longer term EPN persistence (Susurluk & Ehlers). Incidence of several species of EPN remained high for 2 years after application to tree stumps, only declining by year three, as the stumps became unsuitable for *H. abietis* pine weevils, the target pest (Dillon, Rolston et al.). Infected pine weevils collected shortly after treatment yielded up to 98,000 IJs per insect (Dillon et al., 2006); estimations based on an average of 140 weevils per stump and 50 % infection rate (Dillon et al., 2006), with a conservative 50,000 IJs produced per insect, indicate that recycling could theoretically replace the 3.6 million IJs applied per stump. Similarly, Taylor, Szalanski, Adams, and Peterson (1998) estimated that house fly maggots *Musca domestica* L. (Diptera: Muscidae) found at high density in cattle feedlots could sustain nematode populations at the LC_{99} level, even if only 1.5 % of the maggots were infected. Stable ecosystems such as turfgrass or alfalfa are likely to favour longer term persistence, and applied EPN can persist for several years in these systems (Koppenhöfer & Fuzy, 2009; Shields et al.); indeed, *S. scapterisci* became established in turf grass and pasture following its introduction into Florida, as was intended (Parkman & Smart, 1996).

Non–target insects may also contribute to the persistence of EPN populations. For example, larvae, pupae and adults of the non–target longhorn beetle *Rhagium bifasciatum* Fabricius (Coleoptera: Cerambycidae) all supported reproduction of EPN applied against pine weevil, producing up to 140,000 IJs per insect (Harvey, Alameen, & Griffin, 2012). In a field study, persistence of applied *S. carpocapsae* was positively correlated with abundance of tenebrionid beetles, indicating possible use of these beetles for recycling (Hodson, Siegel, & Lewis, 2012). Non–target insects may be important as reservoir hosts for maintenance of EPN populations over periods where target host susceptible stages are absent.

The suitability of insects encountered in the field to support EPN reproduction will vary depending on factors such as the insect's diet and disease levels (Barbercheck, Wang, & Hirsh, 1995; Randall, Cable, Guschina, Harwood, & Lello, 2013). For example, about ten times as many *S. carpocapsae* IJs were produced from corn rootworm *Diabrotica undecimpunctata* L. (Coleoptera: Chrysomelidae) fed on corn than from those fed on bitter squash (Barbercheck et al.). The effect of host insect diet was less severe for *H. bacteriophora*, and Barbercheck et al. proposed that inhibition of *Xenorhabdus* symbiont by cucurbitacins, secondary plant compounds derived from bitter squash, was a factor in the reduced progeny production by *S. carpocapsae* in squash–fed rootworms. Cockroaches harbouring endemic low–virulence parasites (a gregarine) produced 50 % fewer *S. carpocapsae* IJs, and this was attributed to a reduction in host lipid levels of up to 69 % (Randall et al.). Since low virulence endemic parasites such as protozoa are extremely common in nature (Randall et al.), their prevalence may impact on the recycling potential of applied EPN.

EPN may also enter and reproduce in hosts that have been killed by other causes (Půža & Mráček, 2010a; San-Blas & Gowen, 2008), and San-Blas and Gowen suggest that EPN should be considered facultative scavengers rather than as obligate parasites. The extent to which freeze–killed insects could support nematode reproduction varied between species, from *Heterorhabditis indica* Poinar, Karunakar & David (Rhabditida: Heterorhabditidae), which utilised hosts that had been dead for no more than 3 days, to *Steinernema glaseri* (Steiner) Wouts, Mráček, Gerdin & Bedding (Rhabditida: Steinernematidae) which utilised hosts that had been dead for 10 days (San-Blas & Gowen). The poorer scavenging potential of the two *Heterorhabditis* species than of the steinernemamids tested by San-Blas and Gowen may be due at least in part to the greater reliance of heterorhabditids on their symbiont (Han & Ehlers, 2000). Insects such as wireworms (Elateridae) that are resistant to EPN when alive, may be readily utilised for reproduction when killed by other causes, and since wireworms can reach high densities of 400 individuals/m^2, dead individuals may represent a profitable resource for EPN (Půža & Mráček 2010b), though in the field there will be competition with other scavengers and saprotrophs. EPN can also develop in hosts killed by several insecticides (Hara & Kaya, 1983; Koppenhöfer, Cowles, Cowles, Fuzy, & Kaya, 2003) and by parasitoids (Atwa, Hegazi, Khafagi, & Abd El-Aziz, 2013; Mbata & Shapiro-Ilan, 2010), and in moribund hosts infected by granulosis virus (Kaya & Burlando, 1989). However, where the integrity of the host cuticle is compromised, development may fail. Insects infected with nucleopolyhedrosis virus have a fragile cuticle and when this ruptures, developing EPN may desiccate and die before reproduction (Kaya, 2002). Hosts that have been killed by other pathogens including *Bacillus thuringiensis* Berliner (Bacillales: Bacillaceae) and the entomopathogenic fungi *Beauveria bassiana* (Balsamo) Vuillemin (Ascomycota: Hypocreales) do not support nematode reproduction (Barbercheck & Kaya, 1990; Kaya & Burlando), and may be avoided by IJs (Barbercheck & Kaya, 1991). Although application of EPN together with an entomopathogenic fungus such as *Metarhizium* or *Beauveria* may result in enhanced mortality of target pests in the short term (Anbesse, Adge, & Gebru, 2008; Ansari, Shah, & Butt, 2008; Shapiro-Ilan, Jackson, Reilly, & Hotchkiss, 2004), a strategy of joint application has implications for the recycling potential of both the EPN and the fungus in the pest environment. Both nematode and fungus compete for the host; which of the agents is successful depends to large extent on the time difference in colonisation (Acevedo, Samuels, Machado, & Dolinski, 2007; Barbercheck & Kaya, 1990).

The host cadaver provides a protected environment for nematodes, and IJs may remain inside during adverse conditions such as desiccation and cold (Koppenhöfer et al., 1997; Serwe-Rodriguez, Sonnenberg, Appleman, & Bornstein-Forst, 2004; Spence et al., 2011). When cadavers infected by each of four EPN species were incubated in dry soil for various periods of time and then rehydrated, IJs survived from 27 to 111 days, depending on species (Koppenhöfer et al.). Amongst the *Steinernema* species, those adapted to infect insects near the soil surface (*S. carpocapsae*) or from semiarid regions (*S. riobrave*) survived longer periods of desiccation than the temperate cruise forager *S. glaseri*. Koppenhöfer et al. hypothesised that the outer layer of the insect cuticle dried out first, and the desiccated layers reduced further drying of the cadaver. As free–living IJs are not well adapted to survival in dry soil, the cadaver may be important in allowing nematode populations to persist through dry periods (Půža & Mráček, 2007), as also in survival of freezing. Cadavers frozen when adult *S. carpocapsae* or *H. bacteriophora* were present went on to produce IJs when returned to permissive conditions (Lewis & Shapiro-Ilan, 2002). While EPN IJs are freeze tolerant (Brown & Gaugler, 1996), the cadaver may provide a safer overwintering environment by providing protection not only against freezing but other abiotic and biotic dangers.

During dry or cold periods that are not conducive to IJ dispersal and host–finding, the pool of free IJs in soil will decline without replenishment from additional cadavers. At such times, a large proportion of the EPN population may be contained in infected insects, and would not be detected by standard methods of baiting or extraction of soil–dwelling nematodes (Phase 4 in Fig. 3.1).

Nematodes emerging from hosts in which they recycled may differ significantly from the applied, mass produced nematodes in several ways (physiology, size, behaviour, rate and location of arrival in soil). Firstly, nematodes produced in insects may differ in quality from those produced in fermenters, though the nature of the difference may vary between EPN species (Dillon et al., 2006; Ebssa & Koppenhöfer, 2012; Gaugler & Georgis, 1991; Grewal, Converse, & Georgis, 1999). Moreover, the species of insect in which they develop may influence the lipid content, virulence or reproductive capacity of EPN (Abu Hatab & Gaugler, 1999; Abu Hatab, Gaugler, & Ehlers, 1998; Shapiro-Ilan, Dutcher, & Hatab, 2005). For example, *S. glaseri* and *H. bacteriophora* developing in Japanese beetle *Popillia japonica* Newman (Coleoptera: Scarabaeidae) and *S. carpocapsae* developing in pecan weevil *Curculio caryae* Horn (Coleoptera: Curculionidae) had higher lipid content than those developing in wax moths (Abu Hatab et al.; Abu Hatab & Gaugler; Shapiro-Ilan et al.). Passage through pecan weevil did not affect virulence of steinernematids, but did reduce their subsequent reproductive capacity in that host, leading to the conclusion that the recycling potential of nematodes in that host would diminish over time (Shapiro Ilan et al.). In contrast, the pathogenicity of *S. carpocapsae* increased more than two-fold after two passages through gypsy moth *Lymantria dispar* L. (Lepidoptera: Lymantriidae) larvae and there was no reduction in progeny production in that host (Shapiro, Poinar, & Lindegren, 1985). Small hosts may result in smaller IJs (Gouge & Hague, 1995; Nielsen & Philipsen, 2004a), which could be an advantage in infecting hosts that have small natural openings (Scheepmaker, Geels, Griensven, Van, & Smits, 1998).

Even when all IJs were produced in the same conditions (e.g. wax moth hosts) the behaviour of EPN emerging from cadavers differs from that of IJs applied in aqueous suspension, with enhanced dispersal and infectivity reported for the former (Shapiro & Glazer, 1996; Shapiro & Lewis, 1999). This may either be due to physiological status of recently emerged IJs and/or the presence of host stimuli such as ammonia or pheromones stimulating dispersal (Kaplan et al., 2012; San-Blas, Gowen, & Pembroke, 2008; Shapiro & Glazer). At application time, nematodes are released all at once, while emergence from an insect cadaver can take place over days or weeks (Stuart, Lewis, & Gaugler, 1996; Ryder & Griffin, 2003), essentially a “slow release”. IJs continued emerging from long horn beetle *R. bifasciatum* for at least 8 weeks (Harvey et al., 2012), which would provide increased chances of at least some of the IJs emerging at a time when suitable hosts were available. Cadavers from which IJs emerge will be located in the same area as other insects of the host species, giving them an advantage over surface–sprayed inoculum. The more cryptic the host, the greater the difference in search path between applied and recycled nematodes. For example, IJs emerging from infected pine weevil larvae will already be under the bark of tree stumps and roots, and at depths of up to 50 cm in soil (Dillon et al., 2006) and thus well placed for infecting any remaining live weevils.

3.5 Competition and Cooperation: Effects on Native Entomopathogenic Nematodes and Parasitioids

Entomopathogenic nematodes applied inundatively arrive into the soil together with vast numbers of competitors with which they are applied, and are also faced with an array of resident competitors including native EPN, parasitoids, or pathogens. During their evolutionary history, EPN typically emerge from hosts in groups, many of which are close relatives, and there is thus scope for cooperative behaviour to have been selected for.

3.5.1 Aggregation: Cooperative Behaviour?

Immediately after application, nematodes are expected to have a fairly uniform horizontal spatial distribution in soil (assuming a uniform application); however, with time this may tend towards the patchy or aggregated distribution more typical of natural populations (Campbell, Orza, Yoder, Lewis, & Gaugler, 1998). A number of phenomena may contribute to this, including aggregation in preferred conditions (and extinction in more risky regions), or recycling through hosts resulting in patches of newly emerging IJs (Spiridonov, Moens, & Wilson, 2007). Since natural soil is not uniform, interconnecting spaces may provide opportunity for IJs to be washed with irrigation or rain water into foci, or may form physically easier routes for IJs to traverse, resulting in aggregations. In addition to these processes, which do not require IJs to respond to each other, there is some evidence of aggregative behaviour, which does require animals to respond to each others' presence, in several species of EPN (El-Borai, Campos-Herrera, Stuart, & Duncan, 2011; Shapiro-Ilan, Lewis, & Schliekelman, 2014). This aggregative behaviour ("shoaling" or "herding") was exhibited both by IJs applied to sand in aqueous suspension as well as those emerging naturally from cadavers (El-Borai et al.; Shapiro-Ilan et al.). It is unclear to what extent this shoaling simply results from physical forces acting on the IJs, or involves integration by the nervous system. Little is known of the mechanism of collective movement in nematodes (Gart, Vella, & Jung, 2011; Yuan, Raizen, & Bau, 2014). Studying the movement of *Panagrellus redivivus* Goodey (Rhabditida: Panagrolaimidae), Gart et al. showed that nematodes in a thin layer of fluid come into contact spontaneously. They suggest that the initial aggregation is driven by random collisions between nematodes and continued collective motion is due to an attractive force arising from the surface tension of the water film (Gart et al.). Other physical forces may also contribute to aggregation, and systems of self–propelled particles are known for their tendency to aggregate and display swarming behaviour (Yang, Marceau, & Gompper, 2010). For example, hydrodynamic interactions contribute to synchronisation and attraction of sperm (Yang, Elgeti, & Gompper, 2008). On the other hand, social behaviour in feeding *Caenorhabditis elegans* Maupas (Rhabditida. Rhabditidae) is clearly under neural control (Boender, Roubos, & van der Velde, 2011; Rogers, Persson, Cheung, & de Bono, 2006). The mechanism involved in EPN aggregation is as yet unclear. However, even if it results from physical forces and is not a product of natural selection, aggregation may nevertheless be beneficial to the IJs. For example, there may be a requirement for a critical mass of IJs in order to kill certain insects (Peters & Ehlers, 1994), and aggregation provides protection against natural enemies through dilution and shielding effects (Hamilton, 1971). Under desiccating conditions, IJs in a mass survive much better than isolated individuals, by providing a smaller surface area over which water is lost (O'Leary, Power, Stack, & Burnell, 2001). However, there are also disadvantages to migrating in a group, including competition for host resources and inbreeding (Downes & Griffin, 1996).

3.5.2 Competition

Infective juveniles that survive and infect an insect are not necessarily assured of reproductive success. Firstly, a lone *Steinernema* individual may kill a host (thereby satisfying the requirement of the biocontrol practitioner) but, being amphimictic, cannot reproduce. However, lone male and female *S. feltiae* survived up to 6 weeks within killed wax moth larvae (Rolston, Griffin, & Downes, 2006), which may give them a chance of future mating opportunities. Secondly, they will compete with each other: overcrowding results in lower reproductive output per invading nematode (Koppenhöfer & Kaya, 1995; Ryder & Griffin, 2002; Selvan, Campbell, & Gaugler, 1993; Zervos, Johnson, & Webster, 1991). Above a certain inoculum level there is also a reduced output from the host, therefore lower recycling potential in the environment. For example, wax moth larvae inoculated with 500 *H. bacteriophora* produced no IJs (Zervos et al.), presumably due to overcrowding. The highest yield per host was obtained with an inoculum of 100 IJs, though the optimum inoculum level for the nematodes (yield per IJ of inoculum, rather than per host) was much lower (Zervos et al.). Thus, patterns of host invasion and utilisation that favour EPN population survival may conflict with those that maximise fitness of the individual IJ. As well as competing for host resources with each other, inundatively applied nematodes also compete with endemic pathogens (as discussed in Sect. 3.3) and parasites including native EPN and parasitoids.

Several studies have investigated the outcome of co–infections by two species of EPN in the laboratory. Laboratory studies found that *Steinernema* and *Heterorhabditis* could not co–exist in the same insect; *Steinernema* excluded *Heterorhabditis*, even if they infected up to 6 h later, probably due to bacteria–mediated interference competition (Alatorre-Rosas & Kaya, 1990, 1991). However, in a recent study on native EPN populations, *Galleria* larvae used to bait field soils regularly contained progeny of both a *Heterorhabditis* and a *Steinernema* within the same cadaver (Campos-Herrera et al., 2015). Where two *Steinernema* species co–infect the one host individual, normally one species predominates in the emerging progeny (Bashey, Hawlena, & Lively, 2012; Bashey, Reynolds, Sarin, & Young, 2011; Kondo, 1989; Koppenhöfer, Kaya, Shanmugam, & Wood, 1995; Půža & Mráček 2009, 2010b; Sicard et al., 2006). Thus for example, *S. feltiae* produced scarcely any progeny in co–infections with either *S. carpocapsae* or *S. glaseri* (Kondo). Both exploitation and interference competition are implicated in the dominance of one *Steinernema* species over another, with the bacterial symbiont playing a role here also. Each species of *Steinernema* associates with a single species of *Xenorhabdus*, though one species of *Xenorhabdus* may associate with more than one *Steinernema* species (Adams et al., 2006). Although some species of *Steinernema* may feed on *Xenorhabdus* other than their natural associate, or even develop without symbiont, other *Xenorhabdus* less closely related to the natural associate may be detrimental to reproduction (Sicard et al., 2004; Sicard, Ramone, Le Brun, Pages, & Moulia, 2005). Koppenhöfer et al. proposed that the superiority of *S. glaseri* over *S. carpocapsae* in co–infected hosts was due to both the faster development rate of *S. glaseri* and to its less specific relationship with its bacterial symbiont which allowed it to develop on the symbiont carried by its competitor. The relative numbers of bacteria carried by IJs of each species and the ability of the symbionts to produce bacteriocins (toxins that suppress other related strains of bacteria) (Bashey et al., 2012; Hawlena, Bashey, Mendes-Soares, & Lively, 2010) may affect the outcome of the interaction between nematode species by favouring one symbiont over the other. Recently, a novel form of interference competition has been demonstrated in *Steinernema*, in which males physically injure and kill competitors, both male and female, of other *Steinernema* species (O'Callaghan, Zenner, Hartley, & Griffin, 2014; Zenner, O'Callaghan, & Griffin, 2014). A male wraps its tail end around the body of its competitor and squeezes, with the spicule pointing to the victim (Fig. 3.3). This may result in almost immediate paralysis, followed by death (Zenner et al.). The means by which this is achieved are unclear, but physical injuries including ruptured cuticle and damaged internal organs have been seen (Zenner et al.).

These laboratory studies indicate what may happen when two species find themselves in the same host, but EPN applied at high density may also impact on native populations less directly, through scramble competition for available hosts. Impacts on native EPN at the population level have been detected in the field. An introduced exotic species, *S. riobrave*, suppressed native *H. bacteriophora*, but not *S. carpocapsae* in a North Carolina cornfield (Millar & Barbercheck, 2001). Similarly, Duncan, Graham et al. (2003) detected suppression of native EPN following application of exotic *S. riobrave* in Florida citrus. Since native EPN in Florida citrus are involved in regulating the target pest, *D. abbreviatus* citrus root weevil, suppression of these native EPN in plots treated with *S. riobrave*, combined with inferior persistence by the introduced species, reduced the net efficacy of *S. riobrave* against the weevils (Duncan, Graham et al.). While there may be a risk of competitive displacement of native EPN on a temporary basis, an international

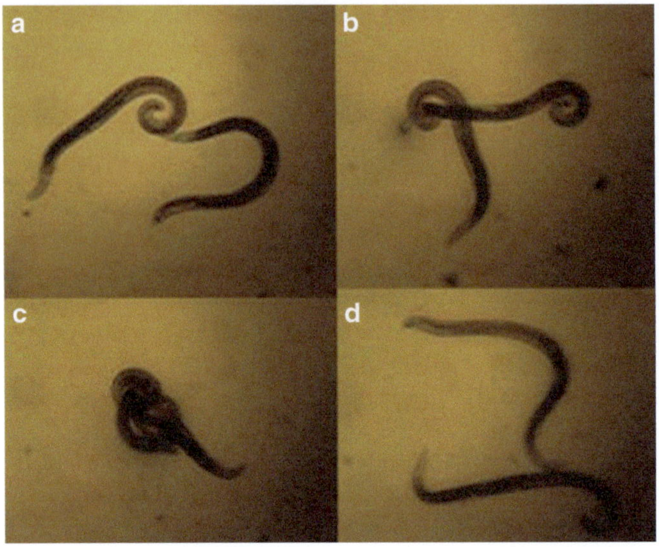

Fig. 3.3 Competition for resources (hosts and mates) may be intense in populations of entomopathogenic nematodes. Here, two *Steinernema* males fight in a drop of haemolymph. (**a**) One male wraps its tail around the other male's head; (**b**, **c**) the grip tightens; (**d**) the victim (lower male in the image) is immobilised within minutes of the encounter

panel of experts considered that there was no risk of permanent displacement (Ehlers & Hokkanen, 1996). Even if there is short–term extinction at an application site, the native population may be re–established from neighbouring areas (Ram, Preisser, Gruner, & Strong, 2008). Where the same species is both applied and indigenous, there is the possibility of hybridisation between the two, if co–infection of hosts occurs. Evidence of introgression was found for *S. feltiae* applied to tree stumps harbouring *H. abietis* pine weevils (Dillon, Rolston, et al., 2008). Genome–wide molecular analysis (Amplified Fragment Length Polymorphism, AFLP) of *S. feltiae* isolates recovered 4 years later suggested possible hybridisation between the persisting and locally colonising strains (Dillon, Rolston et al.).

Top down control can be strengthened where natural enemies complement each other, or dampened by negative interactions (Letourneau, Jedlicka, Bothwell, & Moreno, 2009). Parasitoids are widely introduced as biological control agents, and natural populations may exert considerable mortality of pest populations (Hawkins, Cornell, & Hochberg, 1997), therefore interactions between EPN and parasitoids are of concern. EPN compete with parasitoids for hosts or attack and kill susceptible stages of parasitoid, an example of intraguild predation (Rosenheim, Kaya, Ehler, Marois, & Jaffee, 1995). Ectoparasitoids are susceptible to nematode infection throughout larval development, but frequently become inaccessible at cocoon stage (Everard, Griffin, & Dillon, 2009; Lacey, Unruh, & Headrick, 2003), while endoparasitoids are susceptible for a shorter period, between emerging from the host and completing the cocoon (Kaya, 1978; Kaya & Hotchkin, 1981; Shannag &

Capinera, 2000). In addition, parasitoid death due to premature nematode–induced host death has been reported in several laboratory studies. This is particularly clear in cases where the parasitoid itself is not infected by the nematodes (Head, Palmer, & Walters, 2003; Kaya, 1978; Mráček & Spitzer, 1983). In addition to killing individual parasitoids, nematodes might negatively impact on parasitoid populations if the female parasitoid lays her eggs on nematode–infected weevils where they are unable to complete their development. However, female parasitoids tend to avoid hosts that have been infected by EPN (Everard et al., 2009; Harvey & Griffin, 2012; Lacey et al., 2003; Sher, Parrella, & Kaya, 2000). Females of two ichneumonid species detected and avoided codling moth larvae as little as 12 h after treatment of the host with IJs (Lacey et al.). Such avoidance of oviposition on insect hosts infected with EPN is adaptive for the parasitoid and enhances the complementary effect of EPN for pest suppression (Lacey et al.), since parasitoids may "mop up" weevils that are not hit by nematodes.

Despite the negative effects of EPN on parasitoids demonstrated in the laboratory, the two types of agent may be compatible in the field, resulting in additive or even synergistic effects (Dillon, Moore, Downes, & Griffin, 2008; Mbata & Shapiro-Ilan, 2010). A critical feature is the timing of EPN application relative to peak times of susceptible parasitoid stages. In a field test of EPN against *Cephalcia arvensis* Panzer (Hymenoptera: Pamphiliidae) in Italy, one species of ichneumonid parasitoid was negatively impacted by EPN, while another was not (Battisti, 1994); it was suggested that the difference between species was due to the fact that most of the individuals of the unaffected species were diapausing within cocoons at the time of nematode application.

Free–living bactivorous nematodes (FLBN) may also colonise insect cadavers and represent another class of competitor (Duncan, Graham, et al., 2003; Duncan, Dunn, Bague, & Nguyen, 2003). These nematodes such as *Pelliodítis* were unable to kill insects themselves, but opportunistically invade EPN – killed cadavers and may significantly reduce the number of emerging IJs and hence the recycling capacity of applied nematodes (Duncan, Dunn et al.). If populations of FLBN increase in response to EPN density as suggested by field data (Campos-Herrera, El-Borai, & Duncan, 2012; Somasekhar et al., 2002), FLBN may be a potential regulator of EPN populations (Duncan, Dunn et al.).

3.6 Persistence and Spread of Populations

Application of EPN is not generally aimed at long–term establishment (see Chap. 6). However, release of a large number of propagules may result in establishment and persistence of a population. Several factors conducive to persistence of EPN populations have already been identified, including stability of the ecosystem, availability of suitable hosts and heterogeneity in the nematode population in terms of survival potential and infectivity. The period during which hosts must be available will vary depending on the survival characteristics of the EPN population – the longer the IJs can survive, the shorter the period during which hosts must be present. To establish, the applied population must either be adapted to local climatic and edaphic conditions, or be capable of sufficiently rapid adaptation. Indeed, recent studies suggest that the success of a species in establishing in a new environment may depend more heavily on its ability to respond to natural selection than on having broad physiological tolerance (Lee, 2002; Prentis, Wilson, Dormontt, Richardson, & Lowe, 2008). Both *Heterorhabditis* and *Steinernema* species have responded to artificial selection (e.g. Ehlers et al., 2005; Gaugler & Campbell, 1991) and the short lifecycle of EPN means that adaptation may be rapid. Traditional views of ecological communities assume that they are full or saturated with species, but this may be less general than was previously thought, and even species–rich communities can still accept new–comers, resulting in increased species diversity (Sax et al., 2007). Thus, given time, applied EPN may establish even if not well adapted to local conditions, and despite competition.

Whether or not the site at which EPN are inundatively applied is suitable for long–term establishment of a population, it may provide a jumping off point or beach–head for colonisation of a more suitable habitat. This is of particular interest where using a nematode species that is not native in the region of its use. While active dispersal by nematodes results in local displacement in the order of centimetres, passive dispersal by wind, water or animals may result in translocation to greater distances. Phoresis or other external contamination of animals is the most widely considered explanation for rapid short–range dispersal (Jabbour & Barbercheck, 2008) or long–range dispersal over several hundred meters or kilometres (Barratt, Blossey, & Hokkanen, 2006). Several types of soil invertebrates have potential to act as phoretic hosts for EPN, including earthworms (Campos-Herrera, Trigo, & Gutierrez, 2006; Shapiro, Tylka, Berry, & Lewis, 1995), isopods (Eng, Preisser, & Strong, 2005), predatory carabid beetles (Mertz, Agudelo, Sales, Rohde, & Moino, 2014) and termites (Zadji, Baimey, Afouda, Moens, & Decraemer, 2014). However, only insects capable of flight will result in significant displacement of EPN from the site of application. Lacey, Kaya, and Bettencourt (1995) showed that Japanese beetles *P. japonica* infected in the laboratory were capable of dispersing EPN by flight for at least 50 m, either in their haemocoel or externally. Many of the infected beetles contained enough nematodes to allow reproduction (Lacey et al.). Similarly, infected adult beet armyworm *Spodoptera exigua* Hubner (Lepidoptera: Noctuidae) transported EPN up to 11 m and nematode progeny from the dead moths moved into the soil where they infected larvae of the same species (Timper, Kaya, & Gaugler, 1988). Adult pine weevil *H. abietis*, which are capable of flight, transported EPN on their elytra (Kruitbos, Heritage, & Wilson, 2009). Since they are also susceptible to EPN infection and survive for several days post–infection (Girling, Ennis, Dillon, & Griffin, 2010), internal transport in these weevils is also possible. Following application of *S. scapterisci* to control mole crickets, infected insects were collected as far as 23 km from the nearest site of application, and this method of dissemination was important in establishing the species as part of a strategy of mole cricket suppression in Florida (Parkman & Smart, 1996). Strong et al. (1996) suggest that dispersal of EPN may occur when moist soil particles adhere to fossorial insects and mammals. Dispersal by larger animals including unintentional dispersal by humans is also possible. Human assisted movement of plant parasitic nematodes in soil associated with machinery, vehicles and human footwear as well as in growing media accompanying plants is well documented (reviewed by Singh, Hodda, Ash, and Banks 2013). As well as effecting local dispersal, humans can also be responsible for global dispersal of nematodes. For example, live EPN, both *Heterorhabditis* and *Steinernema* were recovered from soil on footwear from aircraft passengers' baggage (McNeill et al., 2011).

It is likely that EPN can be dispersed by wind and water also. In a theoretical analysis, Carroll and Viglierchio (1981) considered that wind transport of nematode juveniles up to 5 km should be fairly common, with rarer redeposition events up to 40 km from their original location. Wind dissemination of nematodes, including bacterial feeders has been experimentally demonstrated in arid regions including sub–saharan Africa and Antarctica (Baujard & Martiny, 1994; Nkem et al., 2006). According to Nkem et al. the ability to enter anhydrobiosis may be important for wind transport over longer distances, as dry organisms, being lighter, will be carried further (Nathan et al., 2002) and will also survive dry conditions during transport. Although EPN are not capable of true anhydrobiosis they can enter a state of quiescent anhydrobiosis in response to slow drying (Grewal, Bornstein-Forst, Burnell, Glazer, & Jagdale, 2006). Runoff water was shown to be an important transport mechanism for plant parasites at the field level (Chabrier & Queneherve, 2008), while in coastal environments, transport in sea water is an additional possibility (de la Peña, Vandegehuchte, Bonte, & Moens, 2011). Sand dunes are subject to periodic erosion and redeposition by storms (Pye, 1983), providing ample opportunity for redistribution of organisms. Short distance dispersal trapped in mucilaginous foam at the surface (Thornton, 1999) or longer distance dispersal in vegetation rafts (Fuller, Schwarz, & Tierney, 2005) are theoretically possible. IJs of three *Heterorhabditis* species survived prolonged immersion in sea water, and remained infective for 19 weeks, making this a plausible means of dispersal for EPN in coastal locations (Griffin, Finnegan, & Downes, 1994).

Since nematodes at or near the surface are more likely to be picked up by wind (Nkem et al., 2006) and transported by surface water and soil erosion events (Baxter, Rowan, McKenzie, & Neilson, 2013), EPN species that adopt an ambush foraging strategy may be more susceptible to long–distance transport by these means. In line with its surface location, *S. carpocapsae* is generally the most desiccation tolerant species of EPN (Shapiro-Ilan, Brown, & Lewis, 2014), an advantage in wind dispersal. EPN at the soil surface are also more likely to be transported by phoresis. Indeed, the nictation and jumping behaviours that are primarily seen as adaptations to host–finding (Campbell & Kaya, 1999, 2002) may also serve to attach to phoretic hosts (either susceptible or not) resulting in displacement of the nematodes. Dauer juveniles of free–living nematodes such as *C. elegans* also nictate, and for this species nictation primarily serves as a dispersal behaviour (Lee et al., 2012). Nictation may have played a part in the evolution of parasitism from free–living nematodes through necromeny to entomopathogeny (Brown, D'Anna, & Sommer, 2011; Sudhaus, 2008).

While the means by which EPN might disperse can be identified, the extent to which dispersal takes place can best be addressed by molecular population studies, as has been done for other nematodes (e.g. Andersen et al., 2012; Morgan, McGaughran, Ganeshan, Herrmann, & Sommer, 2014). Limited studies that have been done to date with EPN show no correlation between genetic similarity of populations and their geographical proximity, indicating a high level of gene flow (Rolston, Meade, Boyle, Kakouli-Duarte, & Downes, 2009; Wang et al., 2013) and suggesting that long–distance dispersal events of the kind discussed here are relatively common.

Little is known of establishment probabilities of nematodes following dispersal events (McNeill et al., 2011). As for establishment at the site where EPN are applied, establishment following dispersal from such sites will depend on adequate numbers of individuals, their adaptation to local conditions, and the range of available hosts and competitors (Giblin-Davis, Kanzaki, & Davies, 2013; Singh et al., 2013). Hermaphroditic species such as *Heterorhabditis* and *S. hermaphroditum* have an advantage, as each IJ is a potential colonist. Dispersal has important consequences for gene flow, the local and global persistence of species, and the evolution of life–history traits (de la Peña et al., 2011; Ronce, 2007). While even rare dispersal events may allow an applied EPN to establish outside the area of its application, or gradually extend the range of a species, frequent local dispersal events may also facilitate long–term persistence of a species (Simberloff, 2009). Dispersal affects the distribution of genetic diversity contained within populations, and can help mitigate the effect of genetic drift in small populations, decrease mutation load and thereby reduce the risk of extinction (Ronce).

Much can be learnt about factors influencing longer term persistence of applied EPN from natural systems. Natural populations persist for years, and the patchy nature of the distribution within and between sites is consistent with the metapopulation concept (Stuart et al., 2006). An endemic population of *H. marelatus* persisted at high incidence at some but not all sites at Bodega Bay, California, and factors affecting the probability of persistence included local variation in abiotic conditions and metapopulation dynamics (Ram, Gruner, McLaughlin, Preisser, & Strong, 2008; Ram, Preisser, et al.,, 2008). Over a 13 year study period, colonization rates were highly correlated with long–term persistence. Sites with highest long–term persistence experienced the highest rate of colonization, extinction and turnover, leading to the conclusion that *H. marelatus* at Bodega Bay is a dynamic metapopulation (Ram, Preisser et al.).

3.7 Conclusions and Future Directions

EPN have been shaped by millions of years of evolution; like the early wild grasses from which cereals were bred, EPN behavioural and survival strategies may be far from ideal from the human perspective. Studies on their behaviour are broadening beyond foraging strategies and responses to insects reviewed by

Lewis et al. (2006). For example, the demonstration by Rasmann et al. (2005) that EPN respond to herbivore–induced plant volatiles spurred further research in this area, while the recent documentation of a large class of ascaroside signalling molecules in nematodes including EPN opens further exciting horizons (Choe et al., 2012; Kaplan et al., 2012). Research aimed toward the practical goal of genetic improvement of EPN (see Chap. 2) is providing insights into the genetic basis of important traits related to survival and infectivity, while exciting comparative studies encompassing EPN along with other parasites and the intensively studied *C. elegans* holds prospects for understanding both the mechanisms and evolution of behaviours including dispersal and host finding (Hallem et al., 2011; Kaplan et al.).

Although applied in their miles of millons, it is important to remember that IJs are individuals, each of which is attempting to achieve the best reproductive opportunities it can. Even in inbred domesticated strains produced under standardised conditions, not all individuals are the same; genetic variation and the varied conditions experienced by IJs from fermenter to soil will result in physiological and behavioural diversity. The importance of heterogeneity for long–term persistence of the population has been noted. Exploring individual differences in traits like infectivity or host finding is more difficult than for survival, and unravelling the causes of heterogeneity is challenging, especially when each individual is changing over time.

While continued presence of applied EPN may be desirable for sustained suppression of pest populations, establishment and persistence of populations, especially of exotic nematodes, is of environmental concern. Molecular ecology techniques have potential for studying population events including dispersal, introgression, and genetic adaptation, but have not been widely applied. Studies at local level of recent events following inundative application could also contribute to understanding EPN biogeography, since the current natural distribution is a product of earlier dispersal events, some of which might be quite recent.

The diversity of EPN species and experimental questions addressed can sometimes make it difficult to discern patterns from amongst the data. In depth multi–year programmes of interlinked field and laboratory studies, are important in understanding the behaviour and fate of EPN individuals and populations, whether natural (Strong, 2002) or applied (see Chap. 13).

References

Abu Hatab, M., Gaugler, R., & Ehlers, R. U. (1998). Influence of culture method on *Steinernema glaseri* lipids. *Journal of Parasitology, 84*, 215–221.

Abu Hatab, M. A., & Gaugler, R. (1999). Lipids of in vivo and in vitro cultured *Heterorhabditis bacteriophora*. *Biological Control, 15*, 113–118.

Acevedo, J. P. M., Samuels, R. I., Machado, I. R., & Dolinski, C. (2007). Interactions between isolates of the entomopathogenic fungus *Metarhizium anisopliae* and the entomopathogenic nematode *Heterorhabditis bacteriophora* JPM4 during infection of the sugar cane borer *Diatraea saccharalis* (Lepidoptera: Pyralidae). *Journal of Invertebrate Pathology, 96*, 187–192.

Adams, B. J., Fodor, A., Koppenhöfer, H. S., Stackebrandt, E., Stock, S. P., & Klein, M. G. (2006). Biodiversity and systematics of nematode–bacterium entomopathogens. *Biological Control, 37*, 32–49.

Alatorre-Rosas, R., & Kaya, H. K. (1990). Interspecific competition between entomopathogenic nematodes in the genera *Heterorhabditis* and *Steinernema* for an insect host in sand. *Journal of Invertebrate Pathology, 55*, 179–188.

Alatorre-Rosas, R., & Kaya, H. K. (1991). Interaction between two entomopathogenic nematode species in the same host. *Journal of Invertebrate Pathology, 57*, 1–6.

Ali, J. G., Alborn, H. T., & Stelinski, L. L. (2011). Constitutive and induced subterranean plant volatiles attract both entomopathogenic and plant parasitic nematodes. *Journal of Ecology, 99*, 26–35.

Ali, J. G., Campos-Herrera, R., Alborn, H. T., Duncan, L. W., & Stelinski, L. L. (2013). Sending mixed messages: A trophic cascade produced by a belowground herbivore–induced cue. *Journal of Chemical Ecology, 39*, 1140–1147.

Anbesse, S. A., Adge, B. J., & Gebru, W. M. (2008). Laboratory screening for virulent entomopathogenic nematodes (*Heterorhabditis bacteriophora* and *Steinernema yirgalemense*) and fungi (*Metarhizium anisophae* and *Beauveria bassiana*) and assessment of possible synergistic effects of combined use against grubs of the barley chafer *Coptognathus curtipennis. Nematology, 10*, 701–709.

Andalo, V., Moino, A., Maximiniano, C., Campos, V. P., & Mendonca, L. A. (2011). Influence of temperature and duration of storage on the lipid reserves of entomopathogenic nematodes. *Revista Colombiana De Entomologia, 37*, 203–209.

Andersen, E. C., Gerke, J. P., Shapiro, J. A., Crissman, J. R., Ghosh, R., Bloom, J. S., et al. (2012). Chromosome–scale selective sweeps shape *Caenorhabditis elegans* genomic diversity. *Nature Genetics, 44*, 285–290.

Ansari, M. A., & Butt, T. M. (2011). Effect of potting media on the efficacy and dispersal of entomopathogenic nematodes for the control of black vine weevil, *Otiorhynchus sulcatus* (Coleoptera: Curculionidae). *Biological Control, 58*, 310–318.

Ansari, M. A., Shah, F. A., & Butt, T. M. (2008). Combined use of entomopathogenic nematodes and *Metarhizium anisopliae* as a new approach for black vine weevil, *Otiorhynchus sulcatus*, control. *Entomologia Experimentalis et Applicata, 129*, 340–347.

Atwa, A. A., Hegazi, E. M., Khafagi, W. E., & Abd El-Aziz, G. M. (2013). Interaction of the koinobiont parasitoid *Microplitis rufiventris* of the cotton leafworm, *Spodoptera littoralis*, with two entomopathogenic rhabditids, *Heterorhabditis bacteriophora* and *Steinernema carpocapsae. Journal of Insect Science, 13*, 84.

Bal, H. K., Taylor, R. A. J., & Grewal, P. S. (2014). Ambush foraging entomopathogenic nematodes employ 'sprinting emigrants' for long distance dispersal in the absence of hosts. *Journal of Nematology, 44*, 450–451.

Barbercheck, M. E., & Kaya, H. K. (1990). Interactions between *Beauveria bassiana* and the entomogenous nematodes, *Steinernema feltiae* and *Heterorhabditis heliothidis. Journal of Invertebrate Pathology, 55*, 225–234.

Barbercheck, M. E., & Kaya, H. K. (1991). Competitive interactions between entomopathogenic nematodes and *Beauveria bassiana* (Deuteromycotina: Hyphomycetes) in soilborne larvae of *Spodotera exigua* (Lepidoptera: Noctuidae). *Environmental Entomology, 20*, 707–712.

Barbercheck, M. E., Wang, J., & Hirsh, I. S. (1995). Host plant effects on entomopathogenic nematodes. *Journal of Invertebrate Pathology, 66*, 169–177.

Barratt, B. I. P., Blossey, B., & Hokkanen, H. M. T. (2006). Post–release evaluation of non–target effects of biological control agents. In F. Bigler, D. Babendreier, & U. Kuhlmann (Eds.), *Environmental impact of invertebrates for biological control of arthropods: Methods and risk assessment* (pp. 166–186). Wallingford, UK: CABI.

Bashey, F., Hawlena, H., & Lively, C. M. (2012). Alternative paths to success in a parasite community: Within–host competition can favor higher virulence or direct interference. *Evolution, 67*, 900–907.

Bashey, F., Reynolds, C., Sarin, T., & Young, S. K. (2011). Virulence and competitive ability in an obligately killing parasite. *Oikos, 120*, 1539–1545.

Battisti, A. (1994). Effects of entomopathogenic nematodes on the spruce web–spinning sawfly *Cephalcia arvensis* Panzer and its parasitoids in the field. *Biocontrol Science and Technology, 4*, 95–102.

Baujard, P., & Martiny, B. (1994). Transport of nematodes by wind in the peanut cropping area of Senegal, West–Africa. *Fundamental and Applied Nematology, 17*, 543–550.

Baxter, C., Rowan, J. S., McKenzie, B. M., & Neilson, R. (2013). Understanding soil erosion impacts in temperate agroecosystems: Bridging the gap between geomorphology and soil ecology using nematodes as a model organism. *Biogeosciences, 10*, 7133–7145.

Bird, A. F., & Bird, J. (1986). Observations on the use of insect parasitic nematodes as a means of biological control of root–knot nematodes. *International Journal for Parasitology, 16*, 511–516.

Boender, A. J., Roubos, E. W., & van der Velde, G. (2011). Together or alone? Foraging strategies in *Caenorhabditis elegans. Biological Reviews, 86*, 853–862.

Boff, M. I. C., van Tol, R., & Smits, P. H. (2002). Behavioural response of *Heterorhabditis megidis* towards plant roots and insect larvae. *Biocontrol, 47*, 67–83.

Boff, M. I. C., Zoon, F. C., & Smits, P. H. (2001). Orientation of *Heterorhabditis megidis* to insect hosts and plant roots in a Y–tube sand olfactometer. *Entomologia Experimentalis et Applicata, 98*, 329–337.

Bohan, D. A., & Hominick, W. M. (1996). Investigations on the presence of an infectious proportion amongst populations of *Steinernema feltiae* (Site 76 strain) infective stages. *Parasitology, 112*, 113–118.

Bohan, D. A., & Hominick, W. M. (1997). Long term dynamics of infectiousness within the infective stage pool of the entomopathogenic nematode *Steinernema feltiae* (Site 76 strain) Filipjev. *Parasitology, 114*, 301–308.

Bolnick, D. I., Svanback, R., Fordyce, J. A., Yang, L. H., Davis, J. M., Hulsey, C. D., et al. (2003). The ecology of individuals: Incidence and implications of individual specialization. *American Naturalist, 161*, 1–28.

Brown, F. D., D'Anna, I., & Sommer, R. J. (2011). Host–finding behaviour in the nematode *Pristionchus pacificus. Proceedings of the Royal Society B–Biological Sciences, 278*, 3260–3269.

Brown, I. M., & Gaugler, R. (1996). Cold tolerance of steinernematid and heterorhabitid nematodes. *Journal of Thermal Biology, 21*, 115–121.

Burman, M., & Pye, A. E. (1980). *Neoaplectana carpocapsae* – movements of nematode populations on a thermal gradient. *Experimental Parasitology, 49*, 258–265.

Burr, A. H. J., & Robinson, A. F. (2004). Locomotion behaviour. In R. Gaugler & A. L. Bilgrami (Eds.), *Nematode behaviour* (pp. 25–62). Wallingford, UK: CABI.

Byers, J. A., & Poinar, G. O. (1982). Location of insect hosts by the nematode *Neoaplectana carpocapsae*, in response to temperature. *Behaviour, 79*, 1–10.

Campbell, J., & Gaugler, R. (1993). Nictation behaviour and its ecological implications in the host search strategies of entomopathogenic nematodes (Heterorhabditidae and Steinernematidae). *Behaviour, 126*, 154–169.

Campbell, J. F., & Gaugler, R. (1997). Inter–specific variation in entomopathogenic nematode. Foraging strategy: Dichotomy or variation along a continuum? *Fundamental and Applied Nematology, 20*, 393–398.

Campbell, J. F., & Kaya, H. K. (1999). How and why a parasitic nematode jumps. *Nature, 397*, 485–486.

Campbell, J. F., & Kaya, H. K. (2002). Variation in entomopathogenic nematode (Steinernematidae and Heterorhabditidae) infective–stage jumping behaviour. *Nematology, 4*, 471–482.

Campbell, J. F., & Lewis, E. E. (2002). Entomopathogenic nematode host–search strategy. In E. E. Lewis, J. F. Campbell, & M. V. K. Sukhdeo (Eds.), *The behavioural ecology of parasites* (pp. 13–38). Wallingford, UK: CABI.

Campbell, J. F., Lewis, E. E., Stock, S. P., Nadler, S., & Kaya, H. K. (2003). Evolution of host search strategies in entomopathogenic nematodes. *Journal of Nematology, 35*, 142–145.

Campbell, J. F., Orza, G., Yoder, F., Lewis, E., & Gaugler, R. (1998). Spatial and temporal distribution of endemic and released entomopathogenic nematode populations in turfgrass. *Entomologia Experimentalis et Applicata, 86*, 1–11.

Campos-Herrera, R., Barbercheck, M., Hoy, C. W., & Stock, S. P. (2012). Entomopathogenic nematodes as a model system for advancing the frontiers of ecology. *Journal of Nematology, 44*, 162–176.

Campos-Herrera, R., El-Borai, F. E., & Duncan, L. W. (2012). Wide interguild relationships among entomopathogenic and free–living nematodes in soil as measured by real time qPCR. *Journal of Invertebrate Pathology, 111*, 126–135.

Campos-Herrera, R., Jaffuel, G., Chiriboga, X., Blanco-Perez, R., Fesselet, M., Půža, V., et al. (2015). Traditional and molecular detection methods reveal intense interguild competition and other multitrophic interactions associated with native entomopathogenic nematodes in Swiss tillage soils. *Plant and Soil, 389*, 237–255.

Campos-Herrera, R., Pathak, E., El-Borai, F. E., Schumann, A., Abd-Elgawad, M. M. M., & Duncan, L. W. (2013). New citriculture system suppresses native and augmented entomopathogenic nematodes. *Biological Control, 66*, 183–194.

Campos-Herrera, R., Trigo, D., & Gutierrez, C. (2006). Phoresy of the entomopathogenic nematode *Steinernema feltiae* by the earthworm *Eisenia fetida*. *Journal of Invertebrate Pathology, 92*, 50–54.

Carminati, A., Schneider, C. L., Moradi, A. B., Zarebanadkouki, M., Vetterlein, D., Vogel, H. J., et al. (2011). How the rhizosphere may favor water availability to roots. *Vadose Zone Journal, 10*, 988–998.

Carroll, J. J., & Viglierchio, D. R. (1981). On the transport of nematodes by the wind. *Journal of Nematology, 13*, 476–483.

Chabrier, C., & Queneherve, P. (2008). Preventing nematodes from spreading: A case study with *Radopholus similis* (Cobb) Thorne in a banana field. *Crop Protection, 27*, 1237–1243.

Chiriboga, X., Jaffuel, G., Campos–Herrera, R., Roëder, G., & Turlings, T. C. J. (2014, August 17–22). *(E)–b–Caryophyllene diffuses differently regarding soil texture type*. 15th International Symposium on Insect–Plant Relationships, University of Neuchâtel, Switzerland.

Choe, A., von Reuss, S. H., Kogan, D., Gasser, R. B., Platzer, E. G., Schroeder, F. C., et al. (2012). Ascaroside signaling is widely conserved among nematodes. *Current Biology, 22*, 772–780.

Choo, H. Y., & Kaya, H. K. (1991). Influence of soil texture and presence of roots on host finding by *Heterorhabditis bacteriophora*. *Journal of Invertebrate Pathology, 58*, 279–280.

Choo, H. Y., Kaya, H. K., Burlando, T. M., & Gaugler, R. (1989). Entomopathogenic nematodes: Host–finding ability in the presence of plant roots. *Environmental Entomology, 18*, 1136–1140.

Christen, J. M., Campbell, J. F., Lewis, E. E., Shapiro-Ilan, D. I., & Ramaswamy, S. B. (2007). Responses of the entomopathogenic nematode, *Steinernema riobrave* to its insect hosts, *Galleria mellonella* and *Tenebrio molitor*. *Parasitology, 134*, 889–898.

Combes, C. (1991). Evolution of parasite life cycles. In C. A. Tofts, A. Aeschlimann, & C. Bolis (Eds.), *Parasite host associations: Coexistence or conflict?* (pp. 63–82). Oxford, UK: Oxford University Press.

Cutler, C. G., & Webster, J. M. (2003). Host–finding ability of three entomopathogenic nematode isolates in the presence of plant roots. *Nematology, 5*, 601–608.

de Altube, M. D. M., Strauch, O., de Castro, G. F., & Pena, A. M. (2008). Control of the flat–headed root borer *Capnodis tenebrionis* (Linne) (Coleoptera: Buprestidae) with the entomopathogenic nematode *Steinernema carpocapsae* (Weiser) (Nematoda: Steinernematidae) in a chitosan formulation in apricot orchards. *BioControl, 53*, 531–539.

de la Peña, E., Vandegehuchte, M. L., Bonte, D., & Moens, M. (2011). Nematodes surfing the waves: Long–distance dispersal of soil–borne microfauna via sea swept rhizomes. *Oikos, 120*, 1649–1656.

Demarta, L., Hibbard, B. E., Bohn, M. O., & Hiltpold, I. (2014). The role of root architecture in foraging behavior of entomopathogenic nematodes. *Journal of Invertebrate Pathology, 122*, 32–39.

Dembilio, O., Llacer, E., de Altube, M. D. M., & Jacas, J. A. (2010). Field efficacy of imidacloprid and *Steinernema carpocapsae* in a chitosan formulation against the red palm weevil *Rhynchophorus ferrugineus* (Coleoptera: Curculionidae) in *Phoenix canariensis*. *Pest Management Science, 66*, 365–370.

Dempsey, C. M., & Griffin, C. T. (2002). Phased activity in *Heterorhabditis megidis*. *Parasitology, 124*, 605–613.

Dillman, A. R., Guillermin, M. L., Lee, J. H., Kim, B., Sternberg, P. W., & Hallem, E. A. (2012). Olfaction shapes host–parasite interactions in parasitic nematodes. *Proceedings of the National Academy of Sciences of the United States of America, 109*, E2324–E2333.

Dillon, A. B., Moore, C. P., Downes, M. J., & Griffin, C. T. (2008). Evict or infect? Managing populations of the large pine weevil, *Hylobius abietis*, using a bottom–up and top–down approach. *Forest Ecology and Management, 255*, 2634–2642.

Dillon, A. B., Rolston, A. N., Meade, C. V., Downes, M. J., & Griffin, C. T. (2008). Establishment, persistence, and introgression of entomopathogenic nematodes in a forest ecosystem. *Ecological Applications, 18*, 735–747.

Dillon, A. B., Ward, D., Downes, M. J., & Griffin, C. T. (2006). Suppression of the large pine weevil *Hylobius abietis* (L.) (Coleoptera: Curculionidae) in pine stumps by entomopathogenic nematodes with different foraging strategies. *Biological Control, 38*, 217–226.

Dini-Andreote, F., & van Elsas, J. D. (2013). Back to the basics: The need for ecophysiological insights to enhance our understanding of microbial behaviour in the rhizosphere. *Plant and Soil, 373*, 1–15.

Downes, M. J., & Griffin, C. T. (1996). Dispersal behaviour and transmission strategies of the entomopathogenic nematodes *Heterorhabditis* and *Steinernema*. *Biocontrol Science and Technology, 6*, 347–356.

Dugaw, C. J., & Ram, K. (2011). Individual heterogeneity in mortality mediates long–term persistence of a seasonal microparasite. *Oecologia, 166*, 317–325.

Duncan, L., & McCoy, C. (2001). Hydraulic lift increases herbivory by *Diaprepes abbreviatus* larvae and persistence of *Steinernema riobrave* in dry soil. *Nematology, 33*, 142–146.

Duncan, L. W., Dunn, D. C., Bague, G., & Nguyen, K. (2003). Competition between entomopathogenic and free–living bactivorous nematodes in larvae of the weevil *Diaprepes abbreviatus*. *Journal of Nematology, 35*, 187–193.

Duncan, L. W., Graham, J. H., Dunn, D. C., Zellers, J., McCoy, C. W., & Nguyen, K. (2003). Incidence of endemic entomopathogenic nematodes following application of *Steinerema riobrave* for control of *Diaprepes abbreviatus*. *Journal of Nematology, 35*, 178–186.

Duncan, L. W., Stuart, R. J., El-Borai, F. E., Campos-Herrera, R., Pathak, E., Giurcanu, M., et al. (2013). Modifying orchard planting sites conserves entomopathogenic nematodes, reduces weevil herbivory and increases citrus tree growth, survival and fruit yield. *Biological Control, 64*, 26–36.

Dusenbery, D. B. (1987). Theoretical range over which bacteria and nematodes locate plant roots using carbon dioxide. *Journal of Chemical Ecology, 13*, 1617–1624.

Dutky, S. R., Thompson, J. V., & Cantwell, G. E. (1964). A technique for the mass propagation of the DD–136 nematode. *Journal of Insect Pathology, 6*, 417–422.

Ebssa, L., & Koppenhöfer, A. M. (2011). Efficacy and persistence of entomopathogenic nematodes for black cutworm control in turfgrass. *Biocontrol Science and Technology, 21*, 779–796.

Ebssa, L., & Koppenhöfer, A. M. (2012). Entomopathogenic nematodes for the management of *Agrotis ipsilon*: Effect of instar, nematode species and nematode production method. *Pest Management Science, 68*, 947–957.

Ehlers, R. U., & Hokkanen, H. M. T. (1996). Insect biocontrol with non–endemic entomopathogenic nematodes (*Steinernema* and *Heterorhabditis* spp): Conclusions and recommendations of a combined OECD and COST Workshop on Scientific and Regulatory Policy Issues. *Biocontrol Science and Technology, 6*, 295–302.

Ehlers, R. U., Oestergaard, J., Hollmer, S., Wingen, M., & Strauch, O. (2005). Genetic selection for heat tolerance and low temperature activity of the entomopathogenic nematode–bacterium complex *Heterorhabditis bacteriophora–Photorhabdus luminescens*. *BioControl, 50*, 699–716.

El-Borai, F. E., Campos-Herrera, R., Stuart, R. J., & Duncan, L. W. (2011). Substrate modulation, group effects and the behavioral responses of entomopathogenic nematodes to nematophagous fungi. *Journal of Invertebrate Pathology, 106*, 347–356.

Elmowitz, D. E., Ebssa, L., & Koppenhöfer, A. M. (2014). Overwintering behavior of the entomopathogenic nematodes *Steinernema scarabaei* and *Heterorhabditis bacteriophora* and their white grub hosts. *Entomologia Experimentalis et Applicata, 148*, 246–258.

Eng, M. S., Preisser, E. L., & Strong, D. R. (2005). Phoresy of the entomopathogenic nematode *Heterorhabditis marelatus* by a non–host organism, the isopod *Porcellio scaber. Journal of Invertebrate Pathology, 88*, 173–176.

Ennis, D. E., Dillon, A. B., & Griffin, C. T. (2010). Simulated roots and host feeding enhance infection of subterranean insects by the entomopathogenic nematode *Steinernema carpocapsae. Journal of Invertebrate Pathology, 103*, 140–143.

Everard, A., Griffin, C. T., & Dillon, A. B. (2009). Competition and intraguild predation between the braconid parasitoid *Bracon hylobii* and the entomopathogenic nematode *Heterorhabditis downesi*, natural enemies of the large pine weevil, *Hylobius abietis. Bulletin of Entomological Research, 99*, 151–161.

Fallon, D. J. (1998). *The use of indigenous entomopathogenic nematodes* (Heterorhabditis indica *and* Steinernema *spp.) to control rice stem borer in West Java, Indonesia.* Ph.D. Thesis, National University of Ireland, Maynooth.

Fenton, A., & Hudson, P. J. (2002). Optimal infection strategies: Should macroparasites hedge their bets? *Oikos, 96*, 92–101.

Ferguson, C. S., Schroeder, P. C., & Shields, E. J. (1995). Vertical distribution, persistence, and activity of entomopathogenic nematodes (Nematoda: Heterorhabditidae and Steinernematidae) in alfalfa snout beetle (Coleoptera: Curculionidae) infested fields. *Environmental Entomology, 24*, 149–158.

Fitters, P. F. L., & Griffin, C. T. (2006). Survival, starvation, and activity in *Heterorhabditis megidis* (Nematoda: Heterorhabditidae). *Biological Control, 37*, 82–88.

Fitters, P. F. L., Patel, M. N., Griffin, C. T., & Wright, D. J. (1999). Fatty acid composition of *Heterorhabditis* sp during storage. *Comparative Biochemistry and Physiology B–Biochemistry & Molecular Biology, 124*, 81–88.

Fuller, S., Schwarz, M., & Tierney, S. (2005). Phylogenetics of the allodapine bee genus *Braunsapis*: Historical biogeography and long–range dispersal over water. *Journal of Biogeography, 32*, 2135–2144.

Gart, S., Vella, D., & Jung, S. (2011). The collective motion of nematodes in a thin liquid layer. *Soft Matter, 7*, 2444–2448.

Gaugler, R., & Campbell, J. F. (1991). Selection for enhanced host–finding of scarab larvae (Coleoptera: Scarabaeidae) in an entomopathogenic nematode. *Environmental Entomology, 20*, 700–706.

Gaugler, R., & Georgis, R. (1991). Culture method and efficacy of entomopathogenic nematodes (Rhabditida: Steinernematidae and Heterorhabditidae). *Biological Control, 1*, 269–274.

Gaugler, R., Lewis, E., & Stuart, R. J. (1997). Ecology in the service of biological control: The case of entomopathogenic nematodes. *Oecologia, 109*, 483–489.

Georgis, R., & Poinar, G. O. (1983). Effect of soil texture on the distribution and infectivity of *Neoaplectana carpocapasae* (Nematoda: Steinernematidae). *Journal of Nematology, 15*, 308–311.

Giblin-Davis, R. M., Kanzaki, N., & Davies, K. A. (2013). Nematodes that ride insects: Unforeseen consequences of arriving species. *Florida Entomologist, 96*, 770–780.

Girling, R. D., Ennis, D., Dillon, A. B., & Griffin, C. T. (2010). The lethal and sub–lethal consequences of entomopathogenic nematode infestation and exposure for adult pine weevils, *Hylobius abietis* (Coleoptera: Curculionidae). *Journal of Invertebrate Pathology, 104*, 195–202.

Glazer, I. (1997). Effects of infected insects on secondary invasion of steinernematid entomopathogenic nematodes. *Parasitology, 114*, 597–604.

Gouge, D. H., & Hague, N. G. M. (1995). The development of *Steinernema feltiae* (Nematoda: Steinernematidae) in the sciarid fly *Bradysia paupera* (Diptera: Sciaridae). *Annals of Applied Biology, 126*, 395–401.

Grewal, P. S. (2002). Formulation and application technology. In R. Gaugler (Ed.), *Entomopathogenic nematology* (pp. 265–287). Wallingford, UK: CAB International.

Grewal, P. S., Bornstein-Forst, S., Burnell, A. M., Glazer, I., & Jagdale, G. B. (2006). Physiological, genetic, and molecular mechanisms of chemoreception, thermobiosis, and anhydrobiosis in entomopathogenic nematodes. *Biological Control, 38*, 54–65.

Grewal, P. S., Converse, V., & Georgis, R. (1999). Influence of production and bioassay methods on infectivity of two ambush foragers (Nematoda: Steinernematidae). *Journal of Invertebrate Pathology, 73*, 40–44.

Grewal, P. S., Gaugler, R., & Lewis, E. E. (1993). Host recognition by entomopathogenic nematodes during contact with insect gut contents. *Journal of Parasitology, 79*, 495–503.

Grewal, P. S., Gaugler, R., & Selvan, S. (1993). Host recognition by entomopathogenic nematodes: Behavioral response to contact with host feces. *Journal of Chemical Ecology, 19*, 1219–1231.

Grewal, P. S., Lewis, E. E., Gaugler, R., & Campbell, J. F. (1994). Host finding behaviour as a predictor of foraging strategy in entomopathogenic nematodes. *Parasitology, 108*, 207–215.

Grewal, P. S., & Peters, A. (2005). Formulation and quality. In P. S. Grewal, R.U. Ehlers, & D. Shapiro-Ilan (Eds.), *Nematodes as Biocontrol Agents*, (pp. 79–90). Wallingford, UK: CAB International.

Griffin, C. T. (1996). Effects of prior storage conditions on the infectivity of *Heterorhabditis* sp. (Nematoda: Heterorhabditidae). *Fundamental and Applied Nematology, 19*, 95–102.

Griffin, C. T. (2012). Perspectives on the behavior of entomopathogenic nematodes from dispersal to reproduction: traits contributing to nematode fitness and biocontrol efficacy. *Journal of Nematology, 44*, 177–184.

Griffin, C. T., Finnegan, M. M., & Downes, M. J. (1994). Environmental tolerances and the dispersal of *Heterorhabditis*: Survival and infectivity of european *Heterorhabditis* following prolonged immersion in seawater. *Fundamental and Applied Nematology, 17*, 415–421.

Griffin, C. T., O'Callaghan, K., & Dix, I. (2001). A self–fertile species of *Steinernema* from Indonesia: Further evidence of convergent evolution amongst entomopathogenic nematodes? *Parasitology, 122*, 181–186.

Haas, W. (2003). Parasitic worms: Strategies of host finding, recognition and invasion. *Zoology, 106*, 349–364.

Hallem, E. A., Dillman, A. R., Hong, A. V., Zhang, Y. J., Yano, J. M., DeMarco, S. F., et al. (2011). A sensory code for host seeking in parasitic nematodes. *Current Biology, 21*, 377–383.

Hamilton, W. D. (1971). Geometry for the selfish herd. *Journal of Theoretical Biology, 31*, 295–311.

Han, R. C., & Ehlers, R. U. (2000). Pathogenicity, development, and reproduction of *Heterorhabditis bacteriophora* and *Steinernema carpocapsae* under axenic in vivo conditions. *Journal of Invertebrate Pathology, 75*, 55–58.

Hara, A. H., & Kaya, H. K. (1983). Development of the entomogenous nematode, *Neoaplectana carpocapsae* (Rhabditida: Steinernematidae), in insecticide killed beet armyworm (Lepidoptera: Noctuidae). *Journal of Economic Entomology, 76*, 423–426.

Harvey, C. D., Alameen, K. M., & Griffin, C. T. (2012). The impact of entomopathogenic nematodes on a non–target, service–providing longhorn beetle is limited by targeted application when controlling forestry pest *Hylobius abietis*. *Biological Control, 62*, 173–182.

Harvey, C. D., & Griffin, C. T. (2012). Host activity and wasp experience affect parasitoid wasp foraging behaviour and oviposition on nematode–infected larvae of the forestry pest *Hylobius abietis*. *Ecological Entomology, 37*, 269–282.

Hass, B., Downes, M. J., & Griffin, C. T. (2002). Persistence of four *Heterorhabditis* spp. isolates in soil: Role of lipid reserves. *Journal of Nematology, 34*, 151–158.

Hass, B., Griffin, C. T., & Downes, M. J. (1999). Persistence of *Heterorhabditis* infective juveniles in soil: Comparison of extraction and infectivity measurements. *Journal of Nematology, 31*, 508–516.
Hawkins, B. A., Cornell, H. V., & Hochberg, M. E. (1997). Predators, parasitoids, and pathogens as mortality agents in phytophagous insect populations. *Ecology, 78*, 2145–2152.
Hawlena, H., Bashey, F., Mendes-Soares, H., & Lively, C. M. (2010). Spiteful interactions in a natural population of the bacterium *Xenorhabdus bovienii*. *American Naturalist, 175*, 374–381.
Head, J., Palmer, L. F., & Walters, K. E. (2003). The compatibility of control agents used for the control of the South American leafminer, *Liriomyza huidobrensis*. *Biocontrol Science and Technology, 13*, 77–86.
Hiltpold, I., Baroni, M., Toepfer, S., Kuhlmann, U., & Turlings, T. C. J. (2010). Selection of entomopathogenic nematodes for enhanced responsiveness to a volatile root signal helps to control a major root pest. *Journal of Experimental Biology, 213*, 2417–2423.
Hiltpold, I., & Turlings, T. C. J. (2008). Belowground chemical signaling in maize: When simplicity rhymes with efficiency. *Journal of Chemical Ecology, 34*, 628–635.
Hinsinger, P., Bengough, A. G., Vetterlein, D., & Young, I. M. (2009). Rhizosphere: Biophysics, biogeochemistry and ecological relevance. *Plant and Soil, 321*, 117–152.
Hodson, A. K., Siegel, J. P., & Lewis, E. E. (2012). Ecological influence of the entomopathogenic nematode, *Steinernema carpocapsae*, on pistachio orchard soil arthropods. *Pedobiologia, 55*, 51–58.
Hui, E., & Webster, J. M. (2000). Influence of insect larvae and seedling roots on the host–finding ability of *Steinernema feltiae* (Nematoda: Steinernematidae). *Journal of Invertebrate Pathology, 75*, 152–162.
Hunt, H. W., Wall, D. H., DeCrappeo, N. M., & Brenner, J. S. (2001). A model for nematode locomotion in soil. *Nematology, 3*, 705–716.
Jabbour, R., & Barbercheck, M. E. (2008). Soil and habitat complexity effects on movement of the entomopathogenic nematode *Steinernema carpocapsae* in maize. *Biological Control, 47*, 235–243.
Jansson, R. K., Lecrone, S. H., & Gaugler, R. (1993). Field efficacy and persistance of entomopathogenic nematodes (Rhabditida, Steinernematidae, Heterorhabditidae) for control of sweet–potato weevil (Coleoptera, Apionidae) in southern Florida. *Journal of Economic Entomology, 86*, 1055–1063.
Kanagy, J. M. N., & Kaya, H. K. (1996). The possible role of marigold roots and alpha terthienyl in mediating host–finding by steinernematid nematodes. *Nematologica, 42*, 220–231.
Kaplan, F., Alborn, H. T., von Reuss, S. H., Ajredini, R., Ali, J. G., Akyazi, F., et al. (2012). Interspecific nematode signals regulate dispersal behavior. *Plos One, 7*, e38735.
Kaya, H. K. (1978). Interaction between *Neoaplectana carpocapsae* (Nematoda: Steinernematidae) and *Apanteles militaris* (Hymenoptera: Bracondiae), a parasitoid of the armyworm, *Pseudaletia unipuncta*. *Journal of Invertebrate Pathology, 31*, 358–364.
Kaya, H. K. (2002). Natural enemies and other antagonists. In R. Gaugler (Ed.), *Entomopathogenic nematology* (pp. 189–203). Wallingford, UK: CAB International.
Kaya, H. K., & Burlando, T. M. (1989). Development of *Steinernema feltiae* (Rhabditidae: Steinernematidae) in diseased insect hosts. *Journal of Invertebrate Pathology, 53*, 164–168.
Kaya, H. K., Gaugler, R., & Kung, S. P. (1990). Soil type and entomopathogenic nematode persistence. *Journal of Invertebrate Pathology, 55*, 401–406.
Kaya, H. K., & Hotchkin, P. G. (1981). The nematode *Neoaplectana carpocapsae* Weiser and its effect on selected ichneumonid and braconid parasites. *Environmental Entomology, 10*, 474–478.
Kim, H. G., & Alston, D. G. (2008). Potential of two entomopathogenic nematodes for suppression of plum curculio (*Conotrachelus nenuphar*, Coleoptera: Curculionidae) life stages in northern climates. *Environmental Entomology, 37*, 1272–1279.

Kondo, E. (1989). Studies on the infectivity and propogation of entomogenous nematodes, *Steinernema* spp. (Rhabditida: Steinernematidae), in the common cutworm *Spodoptera litura* (Lepidoptera: Noctuidae). *Bulletin of the Faculty of Agriculture, Saga University, 67*, 1–88.

Koppenhöfer, A., Baur, M., Stock, P., Choo, H., Chinnasri, B., & Kaya, H. (1997). Survival of entomopathogenic nematodes within host cadavers in dry soil. *Applied Soil Ecology, 6*, 231–240.

Koppenhöfer, A. M., Campbell, J. F., Kaya, H. K., & Gaugler, R. (1998). Estimation of entomopathogenic nematodes population density in soil by correlation between bait inset mortality and nematode penetration. *Fundamental and Applied Nematology, 21*, 95–102.

Koppenhöfer, A. M., Cowles, R. S., Cowles, E. A., Fuzy, E. M., & Kaya, H. K. (2003). Effect of neonicotinoid synergists on entomopathogenic nematode fitness. *Entomologia Experimentalis et Applicata, 106*, 7–18.

Koppenhöfer, A. M., & Fuzy, E. M. (2006). Effect of soil type on infectivity and persistence of the entomopathogenic nematodes *Steinernema scarabaei, Steinernema glaseri, Heterorhabditis zealandica*, and *Heterorhabditis bacteriophora. Journal of Invertebrate Pathology, 92*, 11–22.

Koppenhöfer, A. M., & Fuzy, E. M. (2009). Long–term effects and persistence of *Steinernema scarabaei* applied for suppression of *Anomala orientalis* (Coleoptera: Scarabaeidae). *Biological Control, 48*, 63–72.

Koppenhöfer, A. M., & Kaya, H. K. (1995). Density–dependent effects on *Steinernema glaseri* (Nematoda: Steinernematidae) within an insect host. *Journal of Parasitology, 81*, 797–799.

Koppenhöfer, A. M., Kaya, H. K., Shanmugam, S., & Wood, G. L. (1995). Interspecific competition between steinernematid nematodes within an insect host. *Journal of Invertebrate Pathology, 66*, 99–103.

Kruitbos, L. M., Heritage, S., Hapca, S., & Wilson, M. J. (2010). The influence of habitat quality on the foraging strategies of the entomopathogenic nematodes *Steinernema carpocapsae* and *Heterorhabditis megidis. Parasitology, 137*, 303–309.

Kruitbos, L. M., Heritage, S., & Wilson, M. J. (2009). Phoretic dispersal of entomopathogenic nematodes by *Hylobius abietis. Nematology, 11*, 419–427.

Kung, S. P., & Gaugler, R. (1991). Effects of soil temperature, moisture, and relative humidity on entomopathogenic nematode persistence. *Journal of Invertebrate Pathology, 57*, 242–249.

Lacey, L. A., Kaya, H. K., & Bettencourt, R. (1995). Dispersal of *Steinernema glaseri* (Nematoda: Steinernematidae) in adult Japanese beetles, *Popillia japonica* (Coleoptera: Scarabaeidae). *Biocontrol Science and Technology, 5*, 121–130.

Lacey, L. A., Unruh, T. R., & Headrick, H. L. (2003). Interactions of two idiobiont parasitoids (Hymenoptera: Ichneumonidae) of codling moth (Lepidoptera: Tortricidae) with the entomopathogenic nematode *Steinernema carpocapsae* (Rhabditida: Steinernematidae). *Journal of Invertebrate Pathology, 83*, 230–239.

Lee, C. E. (2002). Evolutionary genetics of invasive species. *Trends in Ecology & Evolution, 17*, 386–391.

Lee, H., Choi, M. K., Lee, D., Kim, H. S., Hwang, H., Kim, H., et al. (2012). Nictation, a dispersal behavior of the nematode *Caenorhabditis elegans*, is regulated by IL2 neurons. *Nature Neuroscience, 15*, 107–112.

Letourneau, D. K., Jedlicka, J. A., Bothwell, S. G., & Moreno, C. R. (2009). Effects of natural enemy biodiversity on the suppression of arthropod herbivores in terrestrial ecosystems. *Annual Review of Ecology, Evolution, and Systematics, 40*, 573–592.

Lewis, E. E., Campbell, J., Griffin, C., Kaya, H., & Peters, A. (2006). Behavioral ecology of entomopathogenic nematodes. *Biological Control, 38*, 66–79.

Lewis, E. E., & Gaugler, R. (1994). Entomopathogenic nematode (Rhabdita: Steinernemat idae) sex ratio relates to foraging strategy. *Journal of Invertebrate Pathology, 64*, 238–242.

Lewis, E. E., Gaugler, R., & Harrison, R. (1993). Response of cruiser and ambusher entomopathogenic nematodes (Steinernematidae) to host volatile cues. *Canadian Journal of Zoology–Revue Canadienne De Zoologie, 71*, 765–769.

Lewis, E. E., Grewal, P. S., & Gaugler, R. (1995). Hierarchical order of host cues in parasite foraging strategies. *Parasitology, 119*, 207–213.

Lewis, E. E., Ricci, M., & Gaugler, R. (1996). Host recognition behaviour predicts host suitability in the entomopathogenic nematode *Steinernema carpocapsae* (Rhabditida: Steinernematidae). *Parasitology, 113*, 573–579.
Lewis, E. E., & Shapiro-Ilan, D. I. (2002). Host cadavers protect entomopathogenic nematodes during freezing. *Journal of Invertebrate Pathology, 81*, 25–32.
Lindegren, J. E., Valero, K. A., & Mackey, B. E. (1993). Simple in vivo production and storage methods for *Steinernema carpocapsae* infective juveniles. *Journal of Nematology, 25*, 193–197.
MacInnis, A. J. (1976). How parasites find hosts: Some thoughts on the inception of host–parasite integration. In C. R. Kennedy (Ed.), *Ecological aspects of parasitology* (pp. 3–20). Amsterdam: North–Holland.
Mankin, R. W., Brandhorst-Hubbard, J., Flanders, K. L., Zhang, M., Crocker, R. L., Lapointe, S. L., et al. (2000). Eavesdropping on insects hidden in soil and interior structures of plants. *Journal of Economic Entomology, 93*, 1173–1182.
Mbata, G. N., & Shapiro-Ilan, D. I. (2010). Compatibility of *Heterorhabditis indica* (Rhabditida: Heterorhabditidae) and *Habrobracon hebetor* (Hymenoptera: Braconidae) for biological control of *Plodia interpunctella* (Lepidoptera: Pyralidae). *Biological Control, 54*, 75–82.
McGraw, B. A., Vittum, P. J., Cowles, R. S., & Koppenhöfer, A. M. (2010). Field evaluation of entomopathogenic nematodes for the biological control of the annual bluegrass weevil, *Listronotus maculicollis* (Coleoptera: Curculionidae), in golf course turfgrass. *Biocontrol Science and Technology, 20*, 149–163.
McNamara, J. M., & Houston, A. I. (1991). Starvation and predation in a patchy environment. In B. Shorrocks & I. R. Swingland (Eds.), *Living in a patchy environment* (pp. 23–43). Oxford: Oxford University Press.
McNeill, M., Phillips, C., Young, S., Shah, F., Aalders, L., Bell, N., et al. (2011). Transportation of nonindigenous species via soil on international aircraft passengers' footwear. *Biological Invasions, 13*, 2799–2815.
Mertz, N. R., Agudelo, E. J. G., Sales, F. S., Rohde, C., & Moino, A. (2014). Phoretic dispersal of the entomopathogenic nematode *Heterorhabditis amazonensis* by the beetle *Calosoma granulatum. Phytoparasitica, 42*, 179–187.
Millar, L. C., & Barbercheck, M. E. (2001). Interaction between endemic and introduced entomopathogenic nematodes in conventional–till and no–till corn. *Biological Control, 22*, 235–245.
Molyneux, A. S. (1985). Survival of infective juveniles of *Heterorhabditis* spp., and *Steinernema* spp. (Nematoda: Rhabditida) at various temperatures and their subsequent infectivity for insects. *Revue de Nematologie, 8*, 165–170.
Morgan, K., McGaughran, A., Ganeshan, S., Herrmann, M., & Sommer, R. J. (2014). Landscape and oceanic barriers shape dispersal and population structure in the island nematode *Pristionchus pacificus. Biological Journal of the Linnean Society, 112*, 1–15.
Mráček, Z., & Spitzer, K. (1983). Interaction of the predators and parasitoids of the sawfly, *Cephalcia abietis* (Pamphilidae: Hymenoptera) with its nematode *Steinernema kraussei. Journal of Invertebrate Pathology, 42*, 397–399.
Nathan, R., Katul, G. G., Horn, H. S., Thomas, S. M., Oren, R., Avissar, R., et al. (2002). Mechanisms of long–distance dispersal of seeds by wind. *Nature, 418*, 409–413.
Nguyen, K. B., & Smart, G. C. J. (1995). Morphometrics of infective juveniles of *Steinernema* spp. and *Heterorhabditis bacteriophora* (Nemata: Rhabditida). *Journal of Nematology, 27*, 206–212.
Nielsen, A. L., & Lewis, E. E. (2011). Designing the ideal habitat for entomopathogen use in nursery production. *Pest Management Science, 68*, 1053–1061.
Nielsen, O., & Philipsen, H. (2004a). Recycling of entomopathogenic nematodes in *Delia radicum* and in other insects from cruciferous crops. *BioControl, 49*, 285–294.
Nielsen, O., & Philipsen, H. (2004b). Seasonal population dynamics of inoculated and indigenous steinernematid nematodes in an organic cropping system. *Nematology, 6*, 901–909.

Nkem, J. N., Wall, D. H., Virginia, R. A., Barrett, J. E., Broos, E. J., Porazinska, D. L., et al. (2006). Wind dispersal of soil invertebrates in the McMurdo Dry valleys, Antarctica. *Polar Biology, 29*, 346–352.

O'Callaghan, K. M., Zenner, A. N. R. L., Hartley, C. J., & Griffin, C. T. (2014). Interference competition in entomopathogenic nematodes: Male *Steinernema* kill members of their own and other species. *International Journal for Parasitology, 44*, 1009–1017.

O'Leary, S., Power, P., Stack, C., & Burnell, A. (2001). Behavioural and physiological responses of infective juveniles of the entomopathogenic nematode *Heterorhabditis* to desiccation. *BioControl, 46*, 345–362.

Parkman, J. P., & Smart, G. C. (1996). Entomopathogenic nematodes, a case study: Introduction of *Steinernema scapterisci* in Florida. *Biocontrol Science and Technology, 6*, 413–419.

Patel, M. N., Stolinski, M., & Wright, D. J. (1997). Neutral lipids and the assessment of infectivity in entomopathogenic nematodes: Observations on four *Steinernema* species. *Parasitology, 114*, 489–496.

Patel, M. N., & Wright, D. J. (1997). Glycogen: Its importance in the infectivity of aged juveniles of *Steinernema carpocapsae. Parasitology, 114*, 591–596.

Peters, A. (1996). The natural host range of *Steinernema* and *Heterorhabditis* spp. and their impact on insect populations. *Biocontrol Science and Technology, 6*, 389–402.

Peters, A., & Ehlers, R. U. (1994). Susceptibility of leatherjackets (*Tipula paludosa* and *Tipula oleracea*; Tipulidae: Nematocera) to the entomopathogenic nematode *Steinernema feltiae. Journal of Invertebrate Pathology, 63*, 163–171.

Pilz, C., Toepfer, S., Knuth, P., Strimitzer, T., Heimbach, U., & Grabenweger, G. (2014). Persistence of the entomoparasitic nematode *Heterorhabditis bacteriophora* in maize fields. *Journal of Applied Entomology, 138*, 202–212.

Poinar, G. O., & Hom, A. (1986). Survival and horizontal movement of infective stage *Neoaplectana carpocapsae* in the field. *Journal of Nematology, 18*, 34–36.

Portillo-Aguilar, C., Villani, M. G., Tauber, M. J., Tauber, C. A., & Nyrop, J. P. (1999). Entomopathogenic nematode (Rhabditida: Heterorhabditidae and Steinernematidae) response to soil texture and bulk density. *Environmental Entomology, 28*, 1021–1035.

Preisser, E. L., Dugaw, C. J., Dennis, B., & Strong, D. R. (2005). Long–term survival of the entomopathogenic nematode *Heterorhabditis marelatus. Environmental Entomology, 34*, 1501–1506.

Preisser, E. L., Dugaw, C. J., Dennis, B., & Strong, D. R. (2006). Plant facilitation of a belowground predator. *Ecology, 87*, 1116–1123.

Prentis, P. J., Wilson, J. R. U., Dormontt, E. E., Richardson, D. M., & Lowe, A. J. (2008). Adaptive evolution in invasive species. *Trends in Plant Science, 13*, 288–294.

Půža, V., & Mráček, Z. (2007). Natural population dynamics of entomopathogenic nematode *Steinernema affine* (Steinernematidae) under dry conditions: Possible nematode persistence within host cadavers? *Journal of Invertebrate Pathology, 96*, 89–92.

Půža, V., & Mráček, Z. (2009). Mixed infection of *Galleria mellonella* with two entomopathogenic nematode (Nematoda: Rhabditida) species: *Steinernema affine* benefits from the presence of *Steinernema kraussei. Journal of Invertebrate Pathology, 102*, 40–43.

Půža, V., & Mráček, Z. (2010a). Does scavenging extend the host range of entomopathogenic nematodes (Nematoda: Steinernematidae)? *Journal of Invertebrate Pathology, 104*, 1–3.

Půža, V., & Mráček, Z. (2010b). Mechanisms of coexistence of two sympatric entomopathogenic nematodes, *Steinernema affine* and *S. krausse*i (Nematoda: Steinernematidae), in a central European oak woodland soil. *Applied Soil Ecology, 45*, 65–70.

Pye, K. (1983). Coastal dunes. *Progress in Physical Geography, 7*, 531–557.

Ram, K., Gruner, D. S., McLaughlin, J. P., Preisser, E. L., & Strong, D. R. (2008). Dynamics of a subterranean trophic cascade in space and time. *Journal of Nematology, 40*, 85–92.

Ram, K., Preisser, E. L., Gruner, D. S., & Strong, D. R. (2008). Metapopulation dynamics override local limits on long–term parasite persistence. *Ecology, 89*, 3290–3297.

Ramos-Rodriguez, O., Campbell, J. F., Christen, J. M., Shapiro-Ilan, D. I., Lewis, E. E., & Ramaswamy, S. B. (2007). Attraction behaviour of three entomopathogenic nematode species towards infected and uninfected hosts. *Parasitology, 134*, 729–738.

Randall, J., Cable, J., Guschina, I. A., Harwood, J. L., & Lello, J. (2013). Endemic infection reduces transmission potential of an epidemic parasite during co–infection. *Proceedings of the Royal Society B–Biological Sciences, 280*, 20131500.

Rasmann, S., Kollner, T. G., Degenhardt, J., Hiltpold, I., Toepfer, S., Kuhlmann, U., et al. (2005). Recruitment of entomopathogenic nematodes by insect–damaged maize roots. *Nature, 434*, 732–737.

Rasmann, S., & Turlings, T. C. J. (2008). First insights into specificity of belowground tritrophic interactions. *Oikos, 117*, 362–369.

Rogers, C., Persson, A., Cheung, B., & de Bono, M. (2006). Behavioral motifs and neural pathways coordinating O_2 responses and aggregation in *C. elegans*. *Current Biology, 16*, 649–659.

Rolston, A., Meade, C., Boyle, S., Kakouli-Duarte, T., & Downes, M. (2009). Intraspecific variation among isolates of the entomopathogenic nematode *Steinernema feltiae* from Bull Island, Ireland. *Nematology, 11*, 439–451.

Rolston, A. N., Griffin, C. T., & Downes, M. J. (2006). Emergence and dispersal patterns of two isolates of the entomopathogenic nematode *Steinernema feltiae*. *Journal of Nematology, 38*, 221–228.

Ronce, O. (2007). How does it feel to be like a rolling stone? Ten questions about dispersal evolution. *Annual Review of Ecology, Evolution, and Systematics, 38*, 231–253.

Rosenheim, J. A., Kaya, H. K., Ehler, L. E., Marois, J. J., & Jaffee, B. A. (1995). Intraguild predation among biological–control agents: Theory and evidence. *Biological Control, 5*, 303–335.

Ryder, J. J., & Griffin, C. T. (2002). Density dependent fecundity and infective juvenile production in the entomopathogenic nematode, *Heterorhabditis megidis*. *Parasitology, 125*, 83–92.

Ryder, J. J., & Griffin, C. T. (2003). Phased infectivity in *Heterorhabditis megidis*: The effects of infection density in the parental host and filial generation. *International Journal for Parasitology, 33*, 1013–1018.

San-Blas, E., & Gowen, S. R. (2008). Facultative scavenging as a survival strategy of entomopathogenic nematodes. *International Journal for Parasitology, 38*, 85–91.

San-Blas, E., Gowen, S. R., & Pembroke, B. (2008). *Steinernema feltiae*: Ammonia triggers the emergence of their infective juveniles. *Experimental Parasitology, 119*, 180–185.

Sax, D. F., Stachowicz, J. J., Brown, J. H., Bruno, J. F., Dawson, M. N., Gaines, S. D., et al. (2007). Ecological and evolutionary insights from species invasions. *Trends in Ecology & Evolution, 22*, 465–471.

Scheepmaker, J. W. A., Geels, F. P., Griensven, L. J. L., Van, D., & Smits, P. H. (1998). Susceptibility of larvae of the mushroom fly *Megaselia halterata* to the entomopathogenic nematode *Steinernema feltiae* in bioassays. *BioControl, 43*, 201–214.

Schroer, S., Yi, X. L., & Ehlers, R. U. (2005). Evaluation of adjuvants for foliar application of *Steinernema carpocapsae* against larvae of the diamondback moth (*Plutella xylostella*). *Nematology, 7*, 37–44.

Selvan, S., Campbell, J. F., & Gaugler, R. (1993). Density–dependent effects on entomopathogenic nematodes (Heterorhabditidae and Steinernematidae) within an insect host. *Journal of Invertebrate Pathology, 62*, 278–284.

Serwe-Rodriguez, J., Sonnenberg, K., Appleman, B., & Bornstein-Forst, S. (2004). Effects of host desiccation on development, survival, and infectivity of entomopathogenic nematode *Steinernema carpocapsae*. *Journal of Invertebrate Pathology, 85*, 175–181.

Shannag, H. K., & Capinera, J. L. (2000). Interference of *Steinernema carpocapsae* (Nematoda: Steinernematidae) with *Cardiochiles diaphaniae* (Hymenoptera: Braconidae), a parasitoid of melonworm and pickleworm (Lepidoptera: Pyralidae). *Environmental Entomology, 29*, 612–617.

Shapiro, D. I., & Glazer, I. (1996). Comparison of entomopathogenic nematode dispersal from infected hosts versus aqueous suspension. *Environmental Entomology, 25*, 1455–1461.

Shapiro, D. I., Glazer, I., & Segal, D. (1996). Trait stability and fitness of the heat tolerant entomopathogenic nematode *Heterorhabditis bacteriophora* IS5 strain. *Biological Control, 6*, 238–244.

Shapiro, D. I., Glazer, I., & Segal, D. (1997). Genetic diversity in wild and laboratory populations of *Heterorhabditis bacteriophora* as determined by RAPD–PCR analysis. *Fundamental and Applied Nematology, 20*, 581–585.

Shapiro, D. I., & Lewis, E. E. (1999). Comparison of entomopathogenic nematode infectivity from infected hosts versus aqueous suspension. *Environmental Entomology, 28*, 907–911.

Shapiro, D. I., Tylka, G. L., Berry, E. C., & Lewis, L. C. (1995). Effects of earthworms on the dispersal of *Steinernema* spp. *Parasitology, 21*, 21–28.

Shapiro, M., Poinar, G. O., & Lindegren, J. E. (1985). Suitability of *Lymantria dispar* (Lepidoptera: Lymantriidae) as a host for the entomogeneous nematode, *Steinernema feltiae* (Rhabditida: Steinernematidae). *Journal of Economic Entomology, 78*, 342–345.

Shapiro-Ilan, D. I., Brown, I., & Lewis, E. E. (2014). Freezing and desiccation tolerance in entomopathogenic nematodes: Diversity and correlation of traits. *Journal of Nematology, 46*, 27–34.

Shapiro-Ilan, D. I., Dutcher, J. D., & Hatab, M. (2005). Recycling potential and fitness of steinernematid nematodes cultured in *Curculio caryae* and *Galleria mellonella. Journal of Nematology, 37*, 12–17.

Shapiro-Ilan, D. I., Gaugler, R., Tedders, W. L., Brown, I., & Lewis, E. E. (2002). Optimization of inoculation for in vivo production of entomopathogenic nematodes. *Journal of Nematology, 34*, 343–350.

Shapiro-Ilan, D. I., Han, R., & Dolinski, C. (2012). Entomopathogenic nematode production and application technology. *Journal of Nematology, 44*, 226–235.

Shapiro-Ilan, D. I., Jackson, M., Reilly, C. C., & Hotchkiss, M. W. (2004). Effects of combining an entomopathogenic fungi or bacterium with entomopathogenic nematodes on mortality of *Curculio caryae* (Coleoptera: Curculionidae). *Biological Control, 30*, 119–126.

Shapiro-Ilan, D. I., Lewis, E. E., & Schliekelman, P. (2014). Aggregative group behavior in insect parasitic nematode dispersal. *International Journal for Parasitology, 44*, 49–54.

Sher, R. B., Parrella, M. P., & Kaya, H. K. (2000). Biological control of the Leafminer *Liriomyza trifolii* (Burgess): Implications for intraguild predation between *Diglyphus begini* Ashmead and *Steinernema carpocapsae* (Weiser). *Biological Control, 17*, 155–163.

Shields, E. J., Testa, A., Miller, J. M., & Flanders, K. L. (1999). Field efficacy and persistence of the entomopathogenic nematodes *Heterorhabditis bacteroiophora* 'Oswego' and *H. bacteriophora* "NC" on alfalfa snout beetle larvae (Coleoptera: Curculionidae). *Environmental Entomology, 28*, 128–136.

Sicard, M., Ferdy, J. B., Pages, S., Le Brun, N., Godelle, B., Boemare, N., et al. (2004). When mutualists are pathogens: An experimental study of the symbioses between *Steinernema* (entomopathogenic nematodes) and *Xenorhabdus* (bacteria). *Journal of Evolutionary Biology, 17*, 985–993.

Sicard, M., Hinsinger, J., Le Brun, N., Pages, S., Boemare, N., & Moulia, C. (2006). Interspecific competition between entomopathogenic nematodes (*Steinernema*) is modified by their bacterial symbionts (*Xenorhabdus*). *BMC Evolutionary Biology, 6*, 68.

Sicard, M., Ramone, H., Le Brun, N., Pages, S., & Moulia, C. (2005). Specialization of the entomopathogenic nematode *Steinernema scapterisci* with its mutualistic *Xenorhabdus* symbiont. *Naturwissenschaften, 92*, 472–476.

Simberloff, D. (2009). The role of propagule pressure in biological invasions. *Annual Review of Ecology, Evolution, and Systematics, 40*, 81–102.

Singh, S. K., Hodda, M., Ash, G. J., & Banks, N. C. (2013). Plant–parasitic nematodes as invasive species: Characteristics, uncertainty and biosecurity implications. *Annals of Applied Biology, 163*, 323–350.

Smits, P. H. (1996). Post–application persistence of entomopathogenic nematodes. *Biocontrol Science and Technology, 6*, 379–387.

Somasekhar, N., Grewal, P. S., & Klein, M. G. (2002). Genetic variability in stress tolerance and fitness among natural populations of *Steinernema carpocapsae*. *Biological Control, 23*, 303–310.

Spence, K. O., Stevens, G. N., Arimoto, H., Ruiz-Vega, J., Kaya, H. K., & Lewis, E. E. (2011). Effect of insect cadaver desiccation and soil water potential during rehydration on entomopathogenic nematode (Rhabditida: Steinernematidae and Heterorhabditidae) production and virulence. *Journal of Invertebrate Pathology, 106*, 268–273.

Spiridonov, S. E., Moens, M., & Wilson, M. J. (2007). Fine scale spatial distributions of two entomopathogenic nematodes in a grassland soil. *Applied Soil Ecology, 37*, 192–201.

Strong, D. R. (2002). Populations of entomopathogenic nematodes in foodwebs. In R. Gaugler (Ed.), *Entomopathogenic nematology* (pp. 225–240). Wallingford, UK: CAB International.

Strong, D. R., Kaya, H. K., Whipple, A. V., Child, A. L., Kraig, S., Bondonno, M., et al. (1996). Entomopathogenic nematodes: Natural enemies of root–feeding caterpillars on bush lupine. *Oecologia, 108*, 167–173.

Stuart, R. J., Barbercheck, M. E., Grewal, P. S., Taylor, R. A. J., & Hoy, C. W. (2006). Population biology of entomopathogenic nematodes: Concepts, issues, and models. *Biological Control, 38*, 80–102.

Stuart, R. J., Lewis, E. E., & Gaugler, R. (1996). Selection alters the pattern of emergence from the host cadaver in the entomopathogenic neatode, *Steinernema glaseri*. *Parasitology, 113*, 183–189.

Sudhaus, W. (2008). Evolution of insect parasitism in rhabditid and diplogastrid nematodes. Papers dedicated to Prof. Dr Bozidar Curcic. In S. E. Markarov & R. N. Dimittrijevic (Eds.), *Advances in arachnology and developmental biology* (pp. 143–161). Belgrade, Serbia: Institute of Zoology, Faculty of Biology, University of Belgrade.

Sukhdeo, M. V. K., & Sukhdeo, S. C. (2004). Trematode behaviours and the perceptual worlds of parasites. *Canadian Journal of Zoology–Revue Canadienne De Zoologie, 82*, 292–315.

Susurluk, A., & Ehlers, R. U. (2008). Field persistence of the entomopathogenic nematode *Heterorhabditis bacteriophora* in different crops. *BioControl, 53*, 627–641.

Taylor, D. B., Szalanski, A. L., Adams, B. J., & Peterson, R. D. (1998). Susceptibility of house fly (Diptera: Muscidae) larvae to entomopathogenic nematodes (Rhabditida: Heterorhabditidae, Steinernematidae). *Environmental Entomology, 27*, 1514–1519.

Thornton, D. C. O. (1999). Phytoplankton mucilage production in coastal waters: A dispersal mechanism in a front dominated system? *Ethology Ecology & Evolution, 11*, 179–185.

Timper, P., Kaya, H. K., & Gaugler, R. (1988). Dispersal of the entomogenous nematode *Steinernema feltiae* (Rhabditida: Steinernematidae) by infected adult insects. *Environmental Entomology, 17*, 546–550.

Torr, P., Heritage, S., & Wilson, M. J. (2004). Vibrations as a novel signal for host location by parasitic nematodes. *International Journal for Parasitology, 34*, 997–999.

Torr, P., Wilson, M. J., & Heritage, S. (2005). Forestry applications. In P. S. Grewal, R.-U. Ehlers, & D. I. Shapiro-Ilan (Eds.), *Nematodes as biocontrol agents* (pp. 281–293). Wallingford, UK: CAB International.

Turlings, T. C. J., Hiltpold, I., & Rasmann, S. (2012). The importance of root–produced volatiles as foraging cues for entomopathogenic nematodes. *Plant and Soil, 358*, 47–56.

van Tol, R., van der Sommen, A. T. C., Boff, M. I. C., van Bezooijen, J., Sabelis, M. W., & Smits, P. H. (2001). Plants protect their roots by alerting the enemies of grubs. *Ecology Letters, 4*, 292–294.

Wallace, H. R. (1968). Dynamics of nematode movement. *Annual Review of Phytopathology, 6*, 91–114.

Wallwork, J. A. (1970). *Ecology of soil animals*. London: McGraw Hill.

Wang, H., Jung, Y. H., Son, D., & Choo, H. Y. (2013). High level of genetic diversity among *Steinernema monticolum* in Korea revealed by single–enzyme amplified fragment length polymorphism. *Journal of Invertebrate Pathology, 113*, 146–151.

Wang, Y., & Gaugler, R. (1998). Host and penetration site location by entomopathogenic nematodes against Japanese beetle larvae. *Journal of Invertebrate Pathology, 72*, 313–318.

Williams, C. D., Dillon, A. B., Girling, R. D., & Griffin, C. T. (2013). Organic soils promote the efficacy of entomopathogenic nematodes, with different foraging strategies, in the control of a major forest pest: A meta–analysis of field trial data. *Biological Control, 65*, 357–364.

Williams, E. C., & Macdonald, O. C. (1995). Critical factors required by the nematode *Steinernema feltiae* for the control of the leafminers *Liriomyza huidobrensis*, *Liriomyza bryoniae* and *Chromatomyia syngenesiae. Annals of Applied Biology, 127*, 329–341.

Wilson, M., & Gaugler, R. (2004). Factors limiting short–term persistence of entomopathogenic nematodes. *Journal of Applied Entomology, 128*, 250–253.

Wilson, M. J., Ehlers, R.-U., & Glazer, I. (2012). Entomopathogenic nematode foraging strategies – is *Steinernema carpocapsae* really an ambush forager? *Nematology, 14*, 389–394.

Wright, C. A. (1959). Host location by trematode miracidia. *Annals of Tropical Medicine and Parasitology, 53*, 288–292.

Wright, D. J., Peters, A., Schroer, S., & Fife, J. P. (2005). Application technology. In P. S. Grewal, R. U. Ehlers, & D. Shapiro-Ilan (Eds.), *Nematodes as biocontrol agents* (pp. 91–106). Wallingford, UK: CAB International.

Wuyts, N., Maung, Z. T. Z., Swennen, R., & De Waele, D. (2006). Banana rhizodeposition: Characterization of root border cell production and effects on chemotaxis and motility of the parasitic nematode *Radopholus similis. Plant and Soil, 283*, 217–228.

Yang, Y. Z., Elgeti, J., & Gompper, G. (2008). Cooperation of sperm in two dimensions: Synchronization, attraction, and aggregation through hydrodynamic interactions. *Physical Review E, 78*, 061903.

Yang, Y. Z., Marceau, V., & Gompper, G. (2010). Swarm behavior of self–propelled rods and swimming flagella. *Physical Review E, 82*, 031904.

Yuan, J. Z., Raizen, D. M., & Bau, H. H. (2014). Gait synchronization in *Caenorhabditis elegans. Proceedings of the National Academy of Sciences of the United States of America, 111*, 6865–6870.

Zadji, L., Baimey, H., Afouda, L., Moens, M., & Decraemer, W. (2014). Comparative susceptibility of *Macrotermes bellicosus* and *Trinervitermes occidentalis* (Isoptera: Termitidae) to entomopathogenic nematodes from Benin. *Nematology, 16*, 719–727.

Zenner, A. N. R. L., O'Callaghan, K., & Griffin, C. T. (2014). Lethal fighting in nematodes is dependent on developmental pathway: Male–male fighting in the entomopathogenic nematode *Steinernema longicaudum. PloS One, 9*, e89385.

Zervos, S., Johnson, S. C., & Webster, J. M. (1991). Effect of temperature and inoculum size on reproduction and development of *Heterorhabditis heliothidis* and *Steinerema glaseri* (Nematoda: Rhabditoidea) in *Galleria mellonella. Canadian Journal of Zoology, 69*, 1261–1264.

Zimmerman, R. J., & Cranshaw, W. S. (1991). Short–term movement of *Neoaplectana* spp (Rhabditida, Steinernematidae) and *Heterorhabditis* HP–88 Strain (Rhabditida, Heterorhabditidae) through turfgrass thatch. *Journal of Economic Entomology, 84*, 875–878.

Chapter 4
Entomopathogenic Nematodes in the Soil Environment: Distributions, Interactions and the Influence of Biotic and Abiotic Factors

Robin J. Stuart, Mary E. Barbercheck, and Parwinder S. Grewal

4.1 Introduction

Entomopathogenic nematodes (EPNs) in the families Steinernematidae and Heterorhabditidae are important agents for the biological control of soil insect pests in natural and managed ecosystems (Denno, Gruner, & Kaplan, 2008; Grewal, Ehlers, & Shapiro-Ilan, 2005; Lacey & Georgis, 2012). However, like most soil organisms, our knowledge of their activities is relatively limited compared to above ground organisms. Indeed, research on soil biota has long been a challenging aspect of modern ecology because of the inherent difficulties of sampling, manipulating, and otherwise investigating below ground processes (Brown & Gange, 1990; Fierer, Strickland, Liptzin, Bradford, & Cleveland, 2009). Progress is being made with EPNs but we are still a long way from the comprehensive understanding of their soil biology that is required if they are to fulfill their rich potential as manageable biological control agents in cultivated ecosystems. Twenty–five years ago Hominick and Reid (1990) stated: "We are almost completely ignorant of the population biology of entomopathogenic nematodes, yet such information is fundamental to understanding their persistence, distribution, effect on insect populations, and to the development of predictive models for control programs." Subsequently, researchers

R.J. Stuart (✉)
Dundee Biological Control Laboratory, Division of Plant Industry, Florida Department of Agriculture and Consumer Services, Dundee, FL, USA
e-mail: Robin.Stuart@FreshFromFlorida.com

M.E. Barbercheck
Department of Entomology, Pennsylvania State University, University Park, PA, USA
e-mail: meb34@psu.edu

P.S. Grewal
Entomology and Plant Pathology Department, University of Tennessee, Knoxville, TN, USA
e-mail: pgrewal@utk.edu

© Springer International Publishing Switzerland 2015

R. Campos-Herrera (ed.), *Nematode Pathogenesis of Insects and Other Pests*, Sustainability in Plant and Crop Protection 1, DOI 10.1007/978-3-319-18266-7_4

have been chipping away at this problem, more intensive field studies have been conducted, models have been developed for various processes, and molecular techniques have begun providing new ways of exploring fundamental issues (Bai, Adams, Ciche, Clifton, Gaugler, et al., 2013; Campos-Herrera, Barbercheck, Hoy, & Stock, 2012; Campos-Herrera, Pathak, El-Borai, Stuart, Gutiérrez, et al., 2013; Stuart, Barbercheck, Grewal, Taylor, & Hoy, 2006) but much remains to be done. This paper reviews some aspects of the distribution of EPNs in the soil environment, what we know about their interactions, and the various biotic and abiotic factors that influence them.

EPNs have an unusual life history that places important constraints on the structure and dynamics of their populations (Stuart et al., 2006). The only free–living stage is the third stage infective juvenile (IJ), a non–feeding, environmentally resistant "dauer" larva that occurs in the soil and seeks out and penetrates insect hosts (Kaya, Bedding, & Akhurst, 1993; Kaya & Gaugler, 1993). Once inside the insect, the IJ releases symbiotic bacteria (Enterobacteriaceae) from its alimentary tract. The bacteria associated with the Steinernematidae are in the genus *Xenorhabdus* whereas those with the Heterorhabditidae are *Photorhabdus*; and both nematode and bacteria contribute to overwhelming the insect's immune system and killing the host (Boemare, 2002; Dillman, Chaston, Adams, Ciche, Goodrich-Blair, et al., 2012; Forst & Clarke, 2002; Sugar, Murfin, Chaston, Andersen, Richards, et al., 2012). The bacteria proliferate rapidly and soon dominate the insect cadaver. The nematodes feed on the symbiont biomass and insect tissues, develop, mate, and reproduce, often for multiple generations, before producing another generation of IJs that emerge into the soil. Thus, many aspects of the life history and population dynamics of EPNs take place within the host cadaver; and characteristics of the symbiotic bacteria are critical to the outcome of infections (Bashey, Young, Hawlena, & Lively, 2012; Sicard, Hinsinger, Le Brun, Pages, Boemare, et al., 2006). Moreover, because the IJ is the only free–living stage for EPNs, the production, dispersal, persistence, and infection potential of cohorts of IJs derived from individual cadavers provide critical but perhaps tenuous links for the survival, structure and proliferation of local populations (Stuart et al., 2006). The possibility that EPNs might often function as facultative scavengers rather than obligatory lethal parasites (Griffin, 2012; Půža & Mráček, 2010b; San-Blas & Gowen, 2008; San-Blas, Gowen, & Pembroke, 2008) provides a potentially important alternative life history strategy that could profoundly influence population structure, persistence, and competitive interactions. How these possibilities play out in diverse natural and manipulated soil environments remains obscure and provides considerable scope for future research.

Much of the scientific interest and research advances concerning various aspects of the biology of EPNs can be attributed to their great potential as safe, effective and practical agents for augmentative, conservation and classical biological control programs in agriculture and other managed environments (Grewal, 2012; Kaya et al., 1993; Kaya & Gaugler, 1993; Lacey & Georgis, 2012; Shapiro-Ilan, Gouge, & Koppenhöfer, 2002). EPNs are efficacious against numerous insect pests in a variety of habitats, and mass–produced nematodes have achieved commercial success in various niche markets and high value crops around the world (Dolinski, Choo, & Duncan, 2012; Grewal, 2012; Grewal et al., 2005; Lacey & Georgis, 2012; Shapiro-Ilan et al., 2002). Consequently, a small international industry has developed around the production and merchandising of nematode products. However, the commercial production and use of EPNs raises numerous questions regarding population dynamics (Stuart et al., 2006). How does the augmentation of natural populations with commercially produced native or exotic EPNs influence local EPN populations? What is the best strategy for augmentation to maximize control of particular insect pests in particular environments? When and where is the conservation of endemic EPNs a viable alternative to augmentation for pest management and how is it best achieved? These kinds of questions provide much fertile ground for population research.

4.2 Distribution

4.2.1 Patchiness

Entomopathogenic nematodes occur on all continents except Antarctica and in a broad range of habitats and soil types (Adams, Fodor, Koppenhöfer, Stackenbrandt, Stock, et al., 2006; Hominick, 2002). However, there is considerable variability across seasons, habitats and geographic regions; and factors such as soil texture, moisture content, temperature, and availability of hosts are important in determining local distributions (Akhurst & Brooks, 1984; Campos-Herrera, Escuer, Labrador, Robertson, Barrios, et al., 2007; Campos-Herrera, Pathak, El-Borai, Stuart, et al., 2013; Ehlers, Deseö, & Stackebrandt, 1991; Hominick & Briscoe, 1990a, 1990b; Stuart & Gaugler, 1994). Unfortunately, EPN surveys typically only assess occurrence, involve a relatively small number of samples, and provide little quantitative information on relative abundance or distribution at various scales of measurement or across different kinds of sites or habitats. More intensive studies that address these issues have found that populations tend to be extremely patchy, both spatially and temporally, and highly variable in diversity and distribution among sites and habitats (Cabanillas & Raulston, 1994; Campbell, Orza, Yoder, Lewis, & Gaugler, 1998; Campos-Herrera et al., 2007; Campos-Herrera, Pathak, El-Borai, Stuart, et al., 2013; Efron, Nestel, & Glazer, 2001; Garcia Del Pino & Palomo, 1996; Glazer, Kozodoi, Salame, & Nestel, 1996; Koppenhöfer & Kaya, 1996a; Lawrence, Hoy, & Grewal, 2006; Spiridonov, Moens, & Wilson, 2007; Strong, Kaya, Whipple, Child, Kraig, et al., 1996; Stuart & Gaugler, 1994; Taylor, 1999).

In general, populations of organisms can exhibit uniform, random, or patchy distributions but the pattern observed depends upon the scale over which it is measured (Dutilleul & Legendre, 1993; Ettema & Wardle, 2002). Patchy distributions are often an apparent consequence of the distribution of resources or of interactions among conspecifics or heterospecifics. However, whatever the cause, patchy distributions can have important ramifications at the population and community levels by influencing gene flow and altering the dynamics of competition, predation, and parasitism (Ettema & Wardle, 2002; Harrison & Hastings, 1996; McCauley, 1991, 1995).

EPNs are likely to exhibit patchy distributions within and among sites for various reasons including variability in the distribution and abundance of suitable habitat and susceptible hosts, the large number of IJs that emerge from individual hosts (e.g., 30,000–400,000 IJs, (Stuart, Lewis, & Gaugler, 1996)), the limited dispersal capabilities of IJs, and variability in founding, establishment, and persistence ability under different circumstances (Efron et al., 2001; Kaya, 1990; Kaya & Gaugler, 1993; Preisser, Dugaw, Dennis, & Strong, 2006; Strong, 2002; Stuart & Gaugler, 1994). Local extinctions and reintroductions are probably important aspects of the distribution of these species, and populations could often be extremely transitory in space and time (Hominick & Briscoe, 1990a; Ram, Gruner, McLaughlin, Preisser, & Strong, 2008; Ram, Preisser, Gruner, & Strong, 2008). Passive dispersal by water or phoretic dispersal within or upon hosts (Lacey, Kaya, & Bettencourt, 1995; Timper, Kaya, & Gaugler, 1988) or other organisms (e.g., mites, (Epsky, Walter, & Capinera, 1988); earthworms, (Shapiro, Berry, & Lewis, 1993); isopods, (Eng, Preisser, & Strong, 2005)) likely play important roles. The field distribution of *Steinernema feltiae* (Filipjev) (Rhabditida: Steinernematidae) IJs in a grassland habitat appears to consist of a low–level Poisson distributed background of "old" IJs combined with discrete narrow peaks consisting of several dozen "young" IJs resulting from recent insect emergence (Spiridonov & Voronov, 1995; Spiridonov et al., 2007). The characteristic dimension of these IJ peaks along linear transects was 15–20 cm.

All EPN species probably exhibit patchy distributions but the degree of patchiness could be characteristic for particular species and sets of conditions. In a study of spatial distribution in turfgrass in New Jersey, Campbell, Lewis, Yoder, and Gaugler (1995, 1996) recovered *Steinernema carpocapsae* (Weiser) (Rhabditida: Steinernematidae) from a larger proportion of sections along transects than *Heterorhabditis bacteriophora* Poinar (Rhabditida: Heterorhabditidae), and concluded that *S. carpocapsae* populations might generally be more contiguous than those of *H. bacteriophora*. Moreover, *S. carpocapsae* was recovered primarily near the soil surface whereas *H. bacteriophora* was recovered uniformly throughout the soil profile. This result probably relates to differing foraging strategies: *S. carpocapsae* typically appears to function as an "ambush" forager and attacks mobile insects on the soil surface whereas *H. bacteriophora* is a "cruise" forager and attacks more sedentary insects deeper in the soil matrix (Gaugler, Lewis, & Stuart, 1997; Grewal, Lewis, Gaugler, & Campbell, 1994; Lewis, Campbell, Griffin, Kaya, & Peters, 2006; but see Griffin, 2012; Wilson, Ehlers, Wilson, & Glazer, 2012). Differences in foraging strategy and host usage patterns might be the underlying cause for the difference in patchiness for these two species. However, the distribution of a potentially important sedentary host, the Japanese beetle, *Popillia japonica* Newman (Coleoptera: Scarabaeidae), was not related to the distribution of either nematode species (Campbell et al., 1998). Interestingly, even when *H. bacteriophora* is released in a uniform distribution in a turfgrass habitat, it quickly returns to the typical aggregated pattern of natural populations (Campbell et al., 1998; Wilson, Lewis, Yoder, & Gaugler, 2003). Other studies indicate that the natural distributions of EPNs within sites can be remarkably stable over time (Campos-Herrera, Johnson, El-Borai, Stuart, Graham, et al., 2011; Ram, Gruner et al., 2008; Ram, Preisser, et al., 2008).

4.2.2 Metapopulations

Many organisms exhibit complex population structures in which arrays of local populations are interconnected to varying degrees by limited dispersal and gene flow to form what are referred to as metapopulations (Hanski, 1999a, 1999b; Harrison & Taylor, 1997; McCauley, 1995). The study of metapopulation structure and dynamics has become an important theme in population ecology with natural populations being viewed as a series of transient ephemeral local populations with average lifespans that are much shorter than that of the whole network. Rates of birth, death, immigration, and emigration, influence local populations but the persistence of a metapopulation results from a balance between recurrent colonization and extinction events and could involve a high turnover of local populations. The extinction rate generally decreases with increasing patch size, and the colonization rate decreases with increasing distance between patches (Hanski, 1998, 2001; Hanski & Simberloff, 1997). Given the inherently patchy distribution of EPNs (see above), concepts and models developed for metapopulations could have important applications for understanding the population dynamics of these species (Blouin, Liu, & Berry, 1999; Dugaw, Hastings, Preisser, & Strong, 2004; Fenton, Norman, Fairbairn, & Hudson, 2000; Grewal, Wang, & Taylor, 2002; Hominick, Hunt, Reid, Briscoe, & Bohan, 1999; Ram, Gruner, et al., 2008; Ram, Preisser, et al., 2008).

4.2.3 Genetic Diversity

The patchy distribution of EPNs within and among sites is consistent with a metapopulation concept and could have ramifications for the genetic diversity of populations (Harrison & Hastings, 1996; McCauley, 1991, 1995). A fragmented and dynamic population structure might explain the apparent rarity of mixed restriction fragment length polymorphism (RFLP) types from individual collection sites but provide for sufficient gene flow among populations or subpopulations to prevent extensive intraspecific genetic differentiation (Reid & Hominick, 1992; Stuart & Gaugler, 1994). An apparent lack of genetic differentiation within species of EPNs is consistent with an apparent overall lack of adaptive radiation within the group as evidenced by the relatively small number of described species, many of which have broad distributions (Hominick, 2002).

However, for at least some EPN species under certain environmental conditions, genetic diversity among patches within sites can be considerable. Grewal, Wang, and Taylor (2002) found that IJ longevity and tolerance to major environmental stresses including heat, ultraviolet radiation, hypoxia, and desiccation differed significantly among isolates of *H. bacteriophora* taken from an apparently uniform turfgrass site of 200 m^2. The isolates also had different isozyme patterns for several metabolic enzymes (Jagdale, Saeb, Somasekhar, & Grewal, 2006). Similarly, Stuart, Shapiro-Ilan, James, Nguyen, and McCoy (2004) found differences in virulence to larvae of the root weevil, *Diaprepes abbreviatus* (L.) (Coleoptera: Curculionidae), for a series of isolates of *Steinernema riobrave* Cabanillas, Poinar, & Raulston (Rhabditida: Steinernematidae) from a Texas pecan orchard. Other studies also indicate substantial genetic variation within and among EPN populations (Grewal, Grewal, Malik, & Klein, 2002; Rosa & Simões, 2004; Saeb & Grewal, 2008; Somasekhar, Grewal, & Klein, 2002; Wang, Jung, Son, & Choo, 2013).

In contrast, within localized patches at a given site, nematodes might often exhibit very little genetic variation. If the number of individual IJs colonizing an insect host tends to be small, then all of the IJs produced by that cadaver would be very closely related genetically and, consequently, effective population sizes might often be relatively small. Heterorhabditids might be especially well adapted for this eventuality because the first generation inside the host is hermaphroditic. Thus, a study of mitochondrial DNA sequence data for *Heterorhabditis marelatus* Liu & Berry (Rhabditida: Heterorhabditidae), indicated relatively low genetic diversity within and among populations (Blouin et al., 1999). Nonetheless, it is unclear how general this pattern might be because the genetic diversity within a laboratory population of *H. bacteriophora* (HP88 population) appears to be considerable, even though this population was isolated from a single cadaver taken from the field (Glazer, Gaugler, & Segal, 1991).

4.3 Biotic Factors

Various components of the soil biota are likely to influence the distribution and abundance of EPNs. Soils contain rich and diverse communities of flora and fauna that are interconnected in complex trophic webs (Hawksworth, 1991; Neher & Barbercheck, 1999; Strong, Whipple, Child, & Dennis, 1999; Wall & Moore, 1999). From an EPN perspective, the soil contains a broad array of organisms that release various kinds of stimuli that might be beneficially approached, avoided or ignored. These include various plants (Ali, Campos-Herrera, Alborn, Duncan, & Stelinski, 2013; Rasmann, Kollner, Degenhardt, Hiltpold, Toepfer, et al., 2005; Turlings, Hiltpold, & Rasmann, 2012) host and non–host arthropods (Dillman, Guillermin, Lee, Kim, Sternberg, et al., 2012), competitors (Barbercheck & Kaya, 1990, 1991b; Kaya, 2002), predators (Baur, Kaya, & Strong, 1998; Sayre & Walter, 1991), parasites and pathogens (Bellows, 1999; Ishibashi & Kondo, 1987; Kaya; Stirling, 1991; Timper & Kaya, 1992; Timper, Kaya, & Jaffee, 1991). Laboratory evidence clearly indicates that various organisms have the potential to influence the survival and reproduction of EPNs but few field studies have examined the relative importance of different factors in naturally diverse and complex soil environments (Strong et al., 1999). Moreover, omnivory is rampant in soil communities, and trophic webs based on detritus and primary production are linked in various ways that might often produce indirect and diffuse impacts on EPNs (Walter, 1987a, 1987b, 1988a; Walter, Moore, & Loring, 1989). Determining the relative importance of a broad range of biotic factors for the spatial and temporal distribution of EPNs in their appropriate ecological context is a daunting task but the increased use of various molecular, statistical and modeling techniques is contributing to recent progress (Campos-Herrera, El-Borai, & Duncan, 2012; Campos-Herrera, Pathak, El-Borai, Stuart, et al., 2013).

4.3.1 Natural Enemies of Entomopathogenic Nematodes

The primary biotic factor influencing the occurrence and persistence of EPNs at a particular site is probably the presence of suitable hosts (Mráček, Becvár, & Kindlmann, 1999; Mráček, Becvár, Kindlmann, & Jersakova, 2005; Mráček & Webster, 1993; Peters, 1996). However, when hosts are abundant, predators, parasites, and pathogens could regulate populations. The widespread occurrence of nematophagous fungi, bacteria, protozoa, nematodes, mites, collembolans and other microarthropods in soil, and the high rates of predation observed in the laboratory suggest that these organisms might have considerable impact on EPNs in nature (Duncan, Graham, Zellers, Bright, Dunn, et al., 2007; Greenwood, Barbercheck, & Brownie, 2011; Kaya, 2002; Pathak, El-Borai, Campos-Herrera, Johnson, Stuart, et al., 2012; Stirling, 1991). Even specialist nematophagous invertebrates will attack a variety of nematode prey (Small, 1987; Walter, Hunt, & Elliot, 1987).

The potential impact of natural enemies on EPNs has generally been assessed in simplified observation chambers or sterilized soil but activity in these simple systems might not be well correlated with effects in the field. For example, Gilmore and Raffensperger (1970) observed that collembolans consumed large numbers of plant–parasitic nematodes in charcoal–plaster of Paris observation arenas, but predatory activity was substantially reduced when a soil–vermiculite mix was added to the arenas. Similarly, the assay system and host insect used can modify the apparent impact of predation on EPNs. Predatory microarthropods reduced the efficacy of EPNs against a very susceptible but not natural host (wax moth larvae, *Galleria mellonella* (L.) (Lepidoptera: Pyralidae) in soilless assay arenas but had no effect against a natural host (Japanese beetle grubs) in an assay arena containing turf (Epsky et al., 1988; Gilmore & Potter, 1993). Real–time PCR can be useful for assessing the association of EPNs and their natural enemies under field conditions (Campos-Herrera, El-Borai, & Duncan, 2012; Campos-Herrera, Jaffuel, Chiriboga, Blanco-Perez, Fesselet, et al., 2015; Campos-Herrera, Pathak, El-Borai, Stuart, et al., 2013; Pathak et al., 2012).

The effectiveness of a natural enemy can depend on many factors including voracity, specificity, survival at low prey/host densities, dispersal and distribution in relation to the prey/host, and reproductive potential (Pianka, 1999). Under laboratory conditions, omnivorous and nematophagous predators can be voracious feeders. In assays with raw field soil, the presence of astigmatid mites in the genus *Sancassania* (Sarcoptiformes: Acaridae) greatly reduced IJ production by *S. carpocapsae*, *S. riobrave*, and *H. bacteriophora* in *G. mellonella* (Greenwood et al., 2011; also see Cakmak, Hazir, Ulug, & Karagoz, 2013; Ekmen, Hazir, Cakmak, Ozer, Karagoz, et al., 2010; Karagoz, Gulcu, Cakmak, Kaya, & Hazir, 2007). Many nematophagous species have rapid development and high reproductive rates, exhibit some degree of specificity towards nematodes, and are capable of reproducing rapidly by parthenogenesis (e.g., mesostigmatid mites). Mites had faster development times, lower mortality and higher egg–laying rates when feeding on nematodes than when feeding on arthropods (Walter, 1988a, 1988b; Walter et al., 1987). Predatory nematodes typically exhibit high consumption rates with little indication of satiation (Bilgrami & Jairajpuri, 1989a, 1989b).

Insect cadavers that have been infected by EPNs and their bacterial symbionts are subject to predation by various scavengers (Kaya, 2002). However, the cadavers of some EPNs/bacteria are at least somewhat repellent to particular ants and certain other potential scavengers (Baur et al., 1998; Fenton, Magoolagan, Kennedy, & Spencer, 2011; Griffin, 2012; Gulcu, Hazir, & Kaya, 2012; Ulug, Hazir, Kaya, & Lewis, 2014; Zhou, Kaya, Heungens, & Goodrich-Blair, 2002). The benefit of such protection for the developing nematodes is clear but the extent and effectiveness of this repellency for various EPN/bacterial species remains to be explored.

4.3.2 Competition and Displacement

The relative importance of competition in determining the characteristics of organisms, populations, and communities has long been a major issue in ecology (Chase, Abrams, Grover, Diehl, Chesson, et al., 2002; Pianka, 1999; Wootton, 1994), and there are practical reasons to examine the role of competition in the biology of EPNs (Stuart et al., 2006). A greater understanding of competitive abilities could aid evaluation of the suitability of particular EPN species for control programs because these abilities could impact the establishment, persistence, and population dynamics of introduced EPNs. More importantly, when EPN applications are made, what are the risks of displacing non–target natural enemies, including endemic EPNs? What are the ecological consequences of augmentation?

Competition is defined as any mutually negative interaction that does not directly involve predation or parasitism (Pianka, 1999; Wootton, 1994). Competition is most obvious and dramatic when it occurs between species but also occurs and can have important consequences within species. Competition theory predicts that coexisting species that share limited resources will compete, and that competing species must diverge in resource use and reduce niche overlap for competitive exclusion to be avoided and coexistence to continue. The resulting competitive (or character) displacement tends to produce a regular segregation of coexisting species (and their characteristics) in resource space. Variation in resource availability can influence the dynamics of this phenomenon as coexisting species respond opportunistically to superabundant resources, specialize when resources are more limiting, and converge when resources are scarce.

There has been little field research on inter– and intraspecific competition among EPNs, and we can only speculate on whether observed phenomena are due to competitive interactions (i.e., "the ghost of competition past" (Connell, 1980)). However, numerous aspects of EPN ecology and behavior could have evolved in this context and might enable the coexistence of certain species. To conclusively demonstrate the coevolutionary divergence of competitors, one must demonstrate that changes occurred, that the changes have a genetic basis, and that competition was responsible (Pianka, 1999). These requirements have yet to be met for EPNs.

Surveys show that multiple species of EPNs can coexist with as many as five species being reported from a single site (Akhurst & Brooks, 1984; Campos-Herrera et al., 2011; Duncan, Graham, Dunn, Zellers, McCoy, et al., 2003; Stuart & Gaugler, 1994). Coexistence would be expected when behavioral differences and variability in environmental factors enable strong niche separation and avoidance of competition. In laboratory and greenhouse studies, differences in foraging behaviors apparently reduce competition among some EPN species and permit coexistence (Koppenhöfer & Kaya, 1996a, 1996b). The foraging strategies of EPNs are thought to vary along a continuum (Gaugler et al., 1997; Gaugler, Wang, & Campbell, 1994; Lewis et al., 2006) with the extremes represented by "ambushers", which tend to remain relatively sedentary at or near the soil surface, exhibit standing, body waving and jumping behaviors, and attack mobile insects (e.g., *S. carpocapsae*); and "cruisers", which move longer distances and actively seek out sedentary hosts deeper in the soil profile (e.g., *H. bacteriophora*). Intermediate species have characteristics of both (e.g., *S. riobrave*). In studies in North Carolina cornfields, these three species coexisted with each moving to a different location in the soil profile (Millar & Barbercheck, 2001). The native *H. bacteriophora* was found deepest in the soil, introduced *S. riobrave* occurred at intermediate depths, and the native *S. carpocapsae* remained near the surface. Thus, differences in foraging behavior might explain the ability of these species to coexist. Similar factors might also explain the coexistence of *Steinernema affine* (Bovien) (Rhabditida: Steinernematidae) and *Steinernema kraussei* (Steiner) (Rhabditida: Steinernematidae) in central European oak woodland habitats (Půža & Mráček, 2010a). Interestingly, under certain soil conditions or in the presence of roots, the foraging behavior of *S. carpocapsae* apparently departs from the usual ambushing paradigm: they travel greater distances, move more deeply into the soil profile, use roots as "routeways", penetrate into the roots themselves, and effectively attack root–feeding larvae (Griffin, 2012; Wilson et al., 2012). Indeed, some *S. carpocapsae* IJs, so–called "sprinters", move faster and farther than IJs of the cruiser *H. bacteriophora* (Bal, Michael, & Grewal, 2014; Bal, Taylor, & Grewal, 2014). This complexity suggests that certain EPN species might exhibit differential foraging strategies in differing environments and provides further scope for studies of species interactions, competition and coexistence in diverse habitats.

Habitat heterogeneity can also facilitate coexistence and avoidance of competition. Habitat patches exist in a matrix within which the numbers, arrangement and size of patches influence the movements of organisms, food web interactions, and the persistence of populations (Polis, Anderson, & Holt, 1997; Wiens, Schooley, & Weeks, 1997; With, Pavuk, Worchuck, Oates, & Fisher, 2002). Habitat heterogeneity and complexity can contribute to population persistence by presenting microhabitats in a mosaic that spatially and temporally separate competitors, predators and prey (Ettema, 1998). These processes might help explain the highly aggregated distribution of soil–dwelling species (Adl, 2003; Coleman & Crossley, 1996) and the coexistence of multiple species of EPNs within sites (Campos-Herrera et al., 2011; Duncan, Graham, et al., 2003; Koppenhöfer & Kaya, 1996a; Stuart & Gaugler, 1994).

4.3.3 Types of Competition and Other Interactions

Interspecific competition can be direct or indirect. Direct competition involves direct interference between two species, whereas indirect competition applies to a wide range of effects that are mediated by the presence of one or more other species or by a change in the chemical or physical environment (Wootton, 1994; Pianka, 1999). Indirect effects in ecological communities can be either positive or negative, with only the negative effects being considered competitive. Negative indirect effects promote traits that minimize the indirect effects, reduce competition, and facilitate coexistence whereas positive indirect effects tend to move species toward increased sympatry and the maximization of the indirect effects in mutualistic or commensalistic interactions. Indirect effects require strong interactions but high levels of environmental variation, stress or disturbance might keep populations at such low levels that the species do not interact strongly and effects do not occur. Because EPNs are associated with symbiotic bacteria, both direct competition among the nematodes and indirect competition mediated by the bacteria could be common within host cadavers (Bashey, Hawlena, & Lively, 2013; Bashey et al., 2012; Půža & Mráček, 2009; Sicard et al., 2006).

At least five types of indirect effects (both positive and negative) have been demonstrated in ecological communities (Wootton, 1994) and include: (1) exploitative competition, (2) trophic cascades, (3) apparent competition, (4) indirect mutualism and commensalism, and (5) higher order interactions. In *exploitative competition* one species indirectly reduces a second species by reducing the abundance of a shared resource. For example, an insect is rarely infected by more than one species of EPN (Ehlers et al., 1991), and heterorhabditids and steinernematids apparently cannot develop on each other's symbiotic bacteria (Alatorre-Rosas & Kaya, 1990, 1991). However, two steinernematids, *S. carpocapsae* and *Steinernema glaseri* (Steiner) (Rhabditida: Steinernematidae), can coinfect and produce progeny from a single *G. mellonella* (Koppenhöfer, Kaya, & Taormino, 1995). In this case, *S. glaseri* is less negatively affected by the mixed infection than *S. carpocapsae*, perhaps because of its faster development and ability to use the symbiont of *S. carpocapsae*. Similar competition experiments have produced various outcomes and underline the importance of species and population variability for EPNs and their symbiotic bacteria during these kinds of interactions (Bashey et al., 2013; Půža & Mráček, 2009; Sicard et al., 2006).

A *trophic cascade* is an indirect effect mediated through a series of consumer–resource interactions. In general, successful biological control manipulations induce trophic cascades in which certain natural enemies effectively reduce the abundance of particular pest organisms and thereby provide enhanced protection for crops or other organisms of benefit to man (Stuart, El-Borai, & Duncan, 2008). In coastal California, endemic *H. marelatus* are dynamically linked with populations of root–feeding larvae of a hepialid moth, *Hepialus californicus* (Boisduval) (Lepidoptera: Hepialidae), and its bush lupine host plant, *Lupinus arboreus* (Sims) (Fabales: Fabaceae) (Ram, Gruner, et al., 2008; Ram, Preisser, et al., 2008; Strong, 2002; Strong et al., 1995, 1996, 1999). Hepialid larvae inflict heavy root damage and can kill bush lupines but *H. marelatus* causes high mortality of hepialid larvae, and the spatial distribution of *H. marelatus* is positively correlated with long–term fluctuations in the local distribution of lupines. Any other organisms that influence the abundance of this resource, either positively or negatively, are likely to impact *H. marelatus* through this trophic cascade. Moreover, by protecting bush lupine, a nitrogen–fixer, *H. marelatus* probably mediates additional community effects (Preisser, 2003).

Apparent competition occurs when two prey species share a common natural enemy, with an increase in one prey resulting in an increase in the natural enemy and a decline in the second prey. For example, when one EPN occurs naturally at a site and another EPN is applied then the resulting increase in the overall abundance of EPNs could cause a numerical response in predatory mites and a subsequent reduction of both EPN species. Some soil mites exhibit a numerical response when fed EPNs in laboratory studies (Walter, Hudgens, & Freckman, 1986).

Indirect mutualism and commensalism are positive effects and typically involve a consumer–resource interaction linked to either exploitative (indirect) or interference (direct) competition. For example, certain ants might preferentially prey on steinernematid rather than heterorhabditid infected cadavers (Alatorre-Rosas & Kaya, 1990, 1991; Baur et al., 1998; Gulcu et al., 2012; Koppenhöfer, Kaya, Shanmugam, & Wood, 1995). Preferential predation on a competitive dominant could allow an otherwise inferior sympatric competitor to increase through indirect commensalism.

Higher order interactions are non–additive effects among groups of species or individuals. The interactions do not meet the assumption that the combined effect of several species on a species of interest can be represented by adding up all the pair–wise effects. For example, characteristics of certain food plants could modify the susceptibility of an herbivorous insect to one EPN species but not to another (Barbercheck, Wang, & Hirsh, 1995).

4.3.4 *Interspecific Competition Among Entomopathogenic Nematodes*

The dynamics and ramifications of interspecific competition among EPN species are of special interest because of the potential effects that biological control applications of either exotic or native nematodes might have on native nematode communities. In simple laboratory assays, when larvae were exposed to various concentrations of IJs of two EPN species, *S. carpocapsae* successfully infected and reproduced in a greater number of cadavers than *H. bacteriophora* at all concentrations of IJs tested and also displayed a competitive advantage when directly inoculated into the hemocoel (Alatorre-Rosas & Kaya, 1991). The authors concluded that the result was caused by interference competition between the nematodes within the host cadaver, a direct effect. However, because of the symbiotic bacteria, the result for the nematodes might better be interpreted as exploitative competition, an indirect effect in which one species indirectly reduces a second species by reducing the abundance of a shared resource. Nonetheless, since *Xenorhabdus* species are known to produce bacteriocins that kill *Photorhabdus* species (Boemare, 2002), the interaction between the bacteria could be direct interference competition. In other assays, *Spodoptera litura* (F.) (Lepidoptera: Noctuidae) co–infected with *S. glaseri* and *S. feltiae* produced mixed progeny (Kondo, 1989) but *G. mellonella* co–infected with *S. glaseri* and *S. carpocapsae* depressed *S. carpocapsae* IJ production (Koppenhöfer, Kaya, Shanmugam, et al., 1995). As mentioned above, the competitive advantage of *S. glaseri* over *S. carpocapsae* might be due to its faster development and less specific association with its symbiotic bacteria (Koppenhöfer & Kaya, 1996b; also see Bashey et al., 2013; Sicard et al., 2006; Půža & Mráček, 2009). Moreover, males of at least some *Steinernema* species can also engage in interference competition within host cadavers by physically injuring or killing males of their own species, and both males and females of other species (O'Callaghan, Zenner, Hartley, & Griffin, 2014; Zenner, O'Callaghan, & Griffin, 2014).

Competition between EPN species might often be mediated by multiple factors and reflect complex interactions. Millar and Barbercheck (2001) applied an exotic EPN, *S. riobrave,* to a corn field in North Carolina that contained endemic *H. bacteriophora* and *S. carpocapsae*, and monitored the outcome by baiting soil samples with *G. mellonella* larvae. One week after application, *S. riobrave* was detected in less than half of the samples. Subsequently, the distributions of the three species rarely overlapped, and multiple species were rarely found in the same soil sample. The lack of overlap was further indicated by the absence of insects coinfected by multiple species even though coinfections had been demonstrated in laboratory tests. Overall, the detection of *H. bacteriophora* was significantly reduced in the presence of *S. riobrave* but was not completely displaced 2 years after the introduction. Detection of *S. carpocapsae* and *S. riobrave* was not affected by the presence of each other, and detection of *S. riobrave* was not affected by the presence of *H. bacteriophora. H. bacteriophora* had the strongest tendency to be detected deeper in the soil profile, followed by *S. riobrave*, and then *S. carpocapsae*.

In this case, differences in environmental tolerance, foraging behavior, host usage, vertical distribution, and patchiness probably contributed to coexistence.

Augmentation with an exotic EPN could lead to unexpected interactions with native species. In a Florida citrus grove, twice yearly applications of an exotic EPN, *S. riobrave*, to control the root weevil, *D. abbreviatus*, suppressed native EPNs and provided levels of weevil control higher than those caused by native EPNs in untreated plots only during months of treatment while providing less control during non–treatment months (Duncan, Graham, et al., 2003). The native EPNs included *Steinernema diaprepesi* Nguyen & Duncan (Rhabditida: Steinernematidae), *H. bacteriophora*, *Heterorhabditis indica* Poinar, Karunakar & David (Rhabditida: Heterorhabditidae), and *Heterorhabditis zealandica* Poinar (Rhabditida: Heterorhabditidae), and the abundance of adult weevils was directly correlated with the proportion of sentinel weevil larvae infected by the endemic EPNs but was inversely correlated with the proportion of larvae infected by *S. riobrave*. Apparently, *S. riobrave* partially displaced the native EPNs but reproduced and persisted poorly, partly because of competition for cadavers with native bacterial–feeding nematodes (Campos-Herrera, El-Borai & Duncan, 2012; Duncan, Dunn, Bague, & Nguyen, 2003; Duncan et al., 2007, 2013; see below).

4.3.5 Intraspecific Competition and Cooperation Among Entomopathogenic Nematodes

Intraspecific competition could influence various aspects of the biology of EPNs including emergence patterns, foraging strategies and the dynamics of host invasion, establishment and reproduction. The emergence of IJs from the host cadaver often begins abruptly, peaks during the first few days, and involves tens or hundreds of thousands of IJs (e.g., Stuart et al., 1996; Rolston, Griffin, & Downes, 2006). Certain traits of IJs vary predictably as the emergence progresses: early emerging IJs are typically larger and, in steinernematids, have either a more male–biased or female–biased sex ratio compared to later emerging IJs (Alsaiyah, Ebssa, Zenner, O'Callaghan, & Griffin, 2009; Lewis & Gaugler, 1994; Nguyen & Smart, 1995; Stuart et al., 1996; Therese & Bashey, 2012). The pattern of IJ emergence and certain characteristics of IJs could be associated with the population dynamics of the nematodes and their bacteria within the host cadaver, the availability and utilization of resources, and other conditions and cues that trigger the formation and release of IJs; and much of this could be associated with intense intraspecific competition. Moreover, the emergence pattern sets the stage for dispersal, host finding, and host colonization, and could influence potential competition and cooperation among IJs during the infection process.

The adaptive value of a particular emergence pattern might reflect the relative reproductive success of IJs emerging at different times (Stuart et al., 1996). For cruise foragers that exploit sedentary hosts, early emerging IJs from a particular cadaver are likely to have the best opportunity to locate, infect, and reproduce within nearby hosts. Those emerging just a few days later might be required to disperse farther before encountering additional uninfected hosts or might suffer negative fitness consequences because of their late arrival within already infected hosts where other IJs have a head start in development, mating, and reproduction. Consequently, later emerging IJs might often have lower reproductive success than early emerging IJs. This difference might not apply to ambush foragers that exploit mobile hosts. However, EPNs occur in various habitats and use a broad range of hosts (Kaya & Gaugler, 1993), and the temporal and spatial distribution of hosts might vary considerably across habitats or times of the year. Variability in the adaptive value of particular emergence patterns could maintain genetic variability for this trait in the general population. Other traits of the nematodes and their symbiotic bacteria that are correlated with the pattern of emergence (see above) might also impose trade–offs or constraints on adaptive modifications in the emergence pattern.

Competitive and cooperative interactions among IJs could form the basis for the evolution of alternative infection strategies by early and late emerging IJs (Stuart et al., 1996). When a host is colonized by EPNs, a certain number of IJs are necessary to overcome host defenses (Gaugler et al., 1994; Wang, Gaugler, & Cui, 1994) and to guarantee mating for steinernematids but too many IJs might impede development, survival, and reproduction (Selvan, Campbell, & Gaugler, 1993; Zervos, Johnson, & Webster, 1991; but see Therese & Bashey, 2012). Given the large number of IJs that emerge from a single cadaver, various strategies could have evolved to regulate dispersal and infectivity. Such strategies might be especially likely if the IJs emerging from a cadaver are often close relatives since kin selection could be involved (Maynard Smith, 1989). The repellency of cadavers with EPNs (Glazer, 1997; Grewal, Lewis, & Gaugler, 1997) and staggered patterns of infectivity (Bohan & Hominick, 1995, 1996, 1997b; Griffin, 2012; Hominick & Reid, 1990; Kaya & Koppenhöfer, 1996; Yodder, Grewal, & Taylor, 2004; but see Campbell, Koppenhöfer, Kaya, & Chinnasri, 1999) might have evolved in this context. Size differences among IJs probably correlate with lipid reserves and longevity (Lewis & Gaugler, 1994; Selvan, Gaugler, & Grewal, 1993; Selvan, Gaugler, & Lewis, 1993) but, since early emerging IJs are larger than later emerging IJs (Lewis & Gaugler, 1994; Stuart et al., 1996; Therese & Bashey, 2012), this is not indicative of a greater potential for delayed infectivity by the latter. However, later emerging IJs are often more mobile and less responsive to host cues than early emerging IJs (Lewis & Gaugler, 1994), and these traits would facilitate dispersal.

The first IJs to successfully invade a host and develop into adults are likely to have reproductive advantages but early host colonization is probably risky since early invaders could suffer high mortality from host defenses (Gaugler et al., 1994; Peters & Ehlers, 1994; Wang et al., 1994). Nonetheless, if IJs emerging from a cadaver and arriving at a new host are often close relatives, then kin selection might confer fitness benefits on IJs that contribute to subduing a host but die in the process if their relatives are thereby able to reproduce (Maynard Smith, 1989; Stuart, Abu Hatab, & Gaugler, 1998). Apparent group movement or aggregation of IJs dispersing through the soil (El-Borai, Campos-Herrera, Stuart, & Duncan, 2011; Shapiro-Ilan, Lewis,

& Schliekelman, 2014) could set the stage for this phenomenon. Thus, the optimal times for IJs to invade a host might be a function of numerous factors including host–induced mortality rates, development times, reproductive competition, and genetic relatedness.

Experiments indicate that optimal invasion times might exist for EPNs invading hosts. In the short term, initial infections facilitate subsequent infections (Grewal, Lewis & Gaugler, 1997; Grewal, Selvan, Lewis, & Gaugler, 1993; Hay & Fenlon, 1997), but ongoing infections eventually cause the release of a chemical that deters further infection (Fairbairn, Fenton, Norman, & Hudson, 2000; Glazer, 1997; Grewal, Lewis, & Gaugler, 1997). Nonetheless, it is unclear how constrained optimal invasion times might be. Glazer (1997) found that invasion into insects injected with IJs of certain steinernematid species was reduced 6–9 h after injection whereas Stuart et al. (1998) found that IJs of *S. glaseri* invade *G. mellonella* larvae up to at least 14 h after the first IJ has entered. Optimal invasion intervals could be quite plastic and depend on the rate of invasion and dynamics of the interaction between particular EPN species, bacterial symbionts, and hosts.

Research suggests that either males (Grewal et al., 1993) or females (Bohan & Hominick, 1997a) might show a bias toward early host colonization, results that might indicate fundamental differences in reproductive competition for the species involved. However, Stuart et al. (1998) found no sex bias in host colonization by *S. glaseri* even though this species exhibits the various behavioral differences between male and female IJs that suggest that there could be an early male colonization bias (Grewal et al., 1993). Similarly, Alsaiyah et al. (2009) found no sex bias for host colonization in an array of EPN species.

Intraspecific competition could also be a factor in commercial production of EPNs because it might influence optimal inoculation rates and conditions for in vivo and in vitro production systems. Indeed, artificial rearing conditions themselves could have an important influence on the development and reproduction of EPNs, alter the dynamics involved, and select for an array of different traits from those that are important in nature. This kind of inadvertent selection has been documented for laboratory cultures of EPNs (Bilgrami, Gaugler, Shapiro-Ilan, & Adams, 2006; Stuart & Gaugler, 1996; Wang & Grewal, 2002) and could influence the establishment and persistence abilities of mass–reared nematodes when applied in the field.

4.3.6 Competition with Non–entomopathogenic Nematodes

Free–living bactivorous, fungivorous, predatory and omnivorous nematodes constitute important components of decomposition and nutrient cycling food webs in the soil. Duncan et al. (2003) examined interactions between introduced *S. riobrave*, native *S. diaprepesi*, and a native free–living bacterial feeding nematode, *Pellioditis* sp. (Rhabditida: Rhabditidae), with respect to mortality of *D. abbreviatus* larvae in Florida citrus groves. The presence of *S. riobrave* increased the number of *Pellioditis* that developed in insect cadavers, and the presence of *Pellioditis* suppressed the number of *S. riobrave* that developed. However, there was no interaction observed between *Pellioditis* and *S. diaprepesi.* Similarly, addition of *S. carpocapsae* or *S. glaseri* to soil resulted in a temporary increase in predatory and free–living rhabditid nematodes (Ishibashi & Kondo, 1986, 1987). In contrast, Grewal, Martin, Miller, & Lewis (1997) found no effects of application of *S. carpocapsae* or *S. glaseri* on free–living nematodes. EPNs appear to interact negatively with certain plant parasitic nematodes, and can reduce their populations and associated plant damage (Grewal, Lewis, & Venkatachari, 1999; Ishibashi & Kondo, 1986; Jagdale, Kamoun, & Grewal, 2009; Jagdale, Somasekhar, Grewal, & Klein, 2002; Lewis, Grewal, & Sardanelli, 2001; Somasekhar, Grewal, DeNardo, & Stinner, 2002).

Competition for insect resources can also occur between EPNs and other microbial insect pathogens. EPNs will infect certain virus–infected insects but the insect cadavers have a fragile integument that can break open and reduce IJ production (Kaya & Brayton, 1978; Kaya & Burlando, 1989). *S. carpocapsae* and *Bacillus thuringiensis* (Berliner) (Bacillales: Bacillaceae) (Bt) can develop simultaneously in co–infected hosts, but the development of the EPNs is abnormal and the resulting IJs are smaller and have less food reserves than do IJs produced from insects that are not infected with Bt (Kaya & Burlando, 1989). Nonetheless, combinations of EPNs and Bt can additively or synergistically increase levels of mortality of scarab grubs for certain combinations of EPN and grub species (Koppenhöfer, Choo, Kaya, Lee, & Gelernter, 1999; Koppenhöfer & Kaya, 1997); and the combined use of EPNs and *Metarhizium anisopliae* (Metchnikoff) Sorokin (Hypocreales: Clavicipitaceae) has been suggested for black vine weevil control (Ansari, Shah, & Butt, 2008).

Environmental conditions can influence the outcome of competitive interactions. When EPNs compete with other insect pathogens for a host insect, the host usually dies but EPN progeny may not be produced from the co–infected hosts (Barbercheck & Kaya, 1990, 1991b). When insects are co–infected with the fungus, *Beauveria bassiana* (Balsamo) Vuillemin, and EPNs, the EPNs usually out–compete the fungus but this result is influenced by temperature and the time of infection (Barbercheck & Kaya, 1990, 1991b). If *B. bassiana* is given a head start of 3–4 days at 30 °C, 1–2 days at 22 °C and 1 day at 15 °C, then the fungus will develop to the exclusion of the EPNs.

4.3.7 Habitats, Plants and Hosts

Various studies indicate that EPN species show preferences for certain kinds of habitats, a finding that could reflect evolved relationships and the general suitability of available hosts and other habitat characteristics for particular EPN species. Heterorhabditids appear to prefer sandy coastal soils, with some species apparently preferring calcareous soils (*H. bacteriophora*) whereas others prefer more acidic soils (*H. marelatus*); and some species range beyond coastal regions and are broadly distributed in turf and weedy habitats (*H. bacteriophora, H. megidis*) (Constant,

Marchay, Fischer-Le Saux, Briand-Panoma, & Mauleon, 1998; Stuart & Gaugler, 1994; Stock, Strong, & Gardner, 1996). In contrast, steinernematids are usually most prevalent in woodlands (Hominick, Reid, Bohan, & Briscoe, 1996), with some species exhibiting somewhat broader or narrower preferences. *S. feltiae* occurs in grasslands and woodlands (Hominick, 2002), and *S. feltiae* and *S. affine* are virtually the only steinernematids found in arable soils in Germany (Sturhan, 1999). *S. kraussei* and *Steinernema intermedium* Poinar (Mamiya) (Rhabditida: Steinernematidae) are mainly forest and woodland specialists, with *S. kraussei* occurring in coniferous forests in Europe and North America (Sturhan, 1999; Sturhan & Lisková, 1999) and in high altitude woodland in Spain (Campos-Herrera et al., 2007). In Scotland, *S. kraussei* is equally abundant in grassland and woodland (Torr, Spiridonov, Heritage, & Wilson, 2007); and in the Czech Republic EPNs are generally more abundant in tree habitats and light soils (Mráček et al., 2005). Further research is necessary to assess the broader significance of these kinds of associations and the factors that drive them.

Within habitats, particular plants and their characteristics can play an important role in EPN interactions in the soil environment. The presence, density and architecture of plant roots influence dispersal and host finding by EPNs (Bal, Michael & Grewal, 2014; Choo & Kaya, 1991; Demarta, Hibbard, Bohn, & Hiltpold, 2014); and insect damaged maize and citrus roots have been shown to release HIPVs (herbivore–induced plant volatiles) that attract EPNs that subsequently attack the insect pests (Ali, Alborn, & Stelinski, 2010; Rasmann et al., 2005). Moreover, potential host insects feeding on certain plants must contend with endophytes and plant secondary chemicals, which can influence the insects' growth and resistance to EPN attack, and subsequently impact reproduction of the EPNs in complex multitrophic interactions (Barbercheck, 1993; Barbercheck et al., 1995; Gassmann, Stock, Tabashnik, & Singer, 2010; Grewal, Grewal, & Gaugler, 1995; Kunkel & Grewal, 2003; Kunkel, Grewal, & Quigley, 2004; Rasmann et al., 2005; Richmond, Kunkel, Somasekhar, & Grewal, 2004).

EPNs have likely been selected to respond in various ways and with various degrees of specificity to stimuli emanating from hosts, habitats and other organisms and, consequently, these responses will influence their distribution and abundance. For example, EPNs respond to temperature (Burman & Pye, 1980; Byers & Poinar, 1982), CO_2 (Lewis, 2002; Lewis, Gaugler, & Harrison, 1993), and various chemical compounds, some of which can be narrowly host specific (Dillman, Guillermin et al., 2012; Pye & Burman, 1981; Shapiro, McCoy, Fares, Obreza, & Dou, 2000). These responses presumably enhance their chances of finding a host but might also enable them to avoid adverse environmental conditions (e.g., temperature extremes; (Ishibashi & Kondo, 1990)). In terms of host finding, certain attributes of soil such as organic matter could inhibit the transfer of chemical signals and render this modality of little use except at very short range (Torr, Heritage, & Wilson, 2004). In these and other soils, vibrational cues might be especially useful indications of host movement or feeding behavior, and EPNs have been shown to respond to such cues (Torr et al., 2004). EPNs have also been shown to respond differentially to electrical fields, responses that could facilitate attraction to roots (Shapiro-Ilan, Campbell, Lewis, Elkon, & Kim-Shapiro, 2009; Shapiro-Ilan, Lewis, Campbell, & Kim-Shapiro, 2012).

4.4 Abiotic Factors

Many abiotic factors can affect the occurrence and persistence of EPNs. These include natural physical or chemical factors (e.g., climate, soil pH, soil texture, soil structure) as well as those resulting from human activities (e.g., physical or chemical disturbance). The effects of abiotic factors on EPNs have been widely studied under simplified laboratory conditions with soils or artificial substrates treated to reduce interactions with other abiotic and biotic factors (Barbercheck, 1992; Glazer, 2002). However, in nature, complex interactions are common and extrapolation from simple laboratory studies to ecosystems is problematic. Nonetheless, this laboratory research does indicate the likely importance of various factors, and recent field studies are providing real–world tests of how these factors interact and their relative importance.

4.4.1 Soil Texture and Structure

Most studies of soil effects on EPNs have focused on soil texture (i.e., the composition of soil solids by particle size range) rather than on soil structure (i.e., the arrangement of soil particles into aggregates of varying size, geometry and porosity) (Hillel, 1982). Structural pore space is determined largely by size and arrangement of aggregates, and affects the movement of water, air and organisms in soil. In laboratory studies, nematodes are differentially affected by soil texture and structure (Barbercheck, 1993; Barbercheck & Kaya, 1991a; Kung, Gaugler, & Kaya, 1990a). Movement is more restricted in soils with restrictive pore space (heavy or poorly structured soils) than in soils with a more porous structure. In the laboratory, survival and movement of *H. bacteriophora*, *S. carpocapsae*, and *S. glaseri* varied with soil texture and bulk density (Portillo-Aguilar, Villani, Tauber, Tauber, & Nyrop, 1999). All three species moved more in sandy loam than in loam or silty clay loam, and movement generally decreased as bulk density increased. However, the degree to which soils of high bulk density reduced movement differed among species and soil textures: *H. bacteriophora* was least restricted, whereas *S. carpocapsae* was most restricted. Survival of *S. glaseri* was positively correlated with bulk density but survival of *H. bacteriophora* was negatively correlated, and survival of *S. carpocapsae* was unaffected. The infection rate of *G. mellonella* by *H. bacteriophora* and *S. glaseri* did not vary with bulk density but the infection rate for *S. carpocapsae* increased with bulk density. In general, rates of movement and infection were strongly correlated with the amount of soil pore space having dimensions similar to or greater than the diameter of the EPNs.

In natural ecosystems, soil type might have more influence on heterorhabditids than on steinernematids (Hominick, 2002). However, for *H. bacteriophora* in turfgrass, edaphic factors were relatively uniform along transects and only weakly correlated with EPN recovery (Campbell et al., 1998). In no–till and conventional–till maize fields in North Carolina, no significant relationships were detected between the occurrence of endemic *S. carpocapsae* or *H. bacteriophora* and soil organic matter, pH or soil texture (Millar & Barbercheck, 2002). In Florida citrus groves, soil type was not correlated with infection of root weevils by *S. carpocapsae* (Beavers, McCoy, & Kaplan, 1983) but suppression of root weevils by *S. riobrave* was greater in coarse, sandy soils than in fine textured soils (Duncan, Genta, Zellers, Fares, & Stansly, 2001; Shapiro et al., 2000). Indeed, across a broad range of Florida citrus groves, soil texture (and associated variables that influence soil water potential; see below) appears to strongly influence the distribution of EPN species and the natural regulation of root weevils (Campos-Herrera, Pathak, El-Borai, Schumann, Abd-Elgawad, et al., 2013; Campos-Herrera, Pathak, El-Borai, Stuart, et al., 2013; Duncan, Stuart, El-Borai, Campos-Herrera, Pathak, et al., 2013; El-Borai, Stuart, Campos-Herrera, Pathak, & Duncan, 2012; Stuart et al., 2008).

4.4.2 Soil Moisture

Moisture is arguably the most critical abiotic factor affecting soil nematodes (Nickle, 1984). Terrestrial nematodes require water films of sufficient thickness and continuity to allow movement. In very wet or saturated soils, oxygen may be limiting and nematode movement can be restricted due to lack of surface tension forces (Wallace, 1971). Numerous laboratory studies have examined the effect of soil moisture on the efficacy and survival of EPNs (Gaugler & Kaya, 1990; Glazer, 2002; Kaya & Gaugler, 1993; Shapiro-Ilan et al., 2002). In the laboratory, virulence of *H. bacteriophora*, *S. glaseri*, *S. feltiae*, and *S. carpocapsae* increased with soil moisture content in sandy loam soils ranging in moisture content from below the permanent wilting point to near saturation (Grant & Villani, 2003). Hudson and Nguyen (1989) tested the infectivity of *Steinernema scapterisci* Nguyen & Smart (Rhabditida: Steinernematidae) to the mole crickets, *Scapteriscus vicinus* Scudder and *Scapteriscus acletus* Rehn & Hebard (Orthoptera: Gryllotalpidae), under a variety of conditions in the laboratory and found that soil moisture that varied from 5 to 15 % had no effect on infection. In a survey of Spanish soils for EPNs, Garcia Del Pino and Palomo (1996) concluded that soil moisture and temperature regimes are more important than other factors in determining the prevalence of EPNs in cold moist soils. In conventional–till and no–till maize in North Carolina there was a quadratic relationship between soil moisture content and numbers of sentinel *G. mellonella* infected by *S. carpocapsae* but not by *S. riobrave* or *H. bacteriophora* (Millar & Barbercheck, 2002). Many nematodes have physiological or behavioral adaptations that allow resumption of activity after quiescence induced by moisture limiting conditions (Glazer, 2002). Reduced virulence of EPNs in low moisture conditions can be increased by rehydrating the soil to simulate rainfall or irrigation (Grant & Villani, 2003). Moreover, the survival of *S. riobrave* is apparently enhanced following quiescence induced by moisture deficits (Duncan, Dunn, & McCoy, 1996).

Nematode activity and survival are reduced in waterlogged soils (e.g., low oxygen conditions) and can be influenced by the relative humidity of the soil atmosphere (Kung, Gaugler, & Kaya, 1990b; Qiu & Bedding, 1999). Under normal field conditions where moisture levels are high enough to support plant growth, the soil atmosphere is nearly always vapor saturated. Survival and pathogenicity of *S. carpocapsae* and *S. glaseri* decreased as relative humidity decreased from 100 to 25 % over a 32–day test period (Kung et al., 1990b). Brown and Gaugler (1997) found that IJs could survive adverse environmental conditions by remaining in host cadavers for up to 50 days. Survival varied among species and was dependent upon environmental conditions. *S. carpocapsae*, an ambush forager, might be especially well adapted to survive in cadavers in dry soil because of its tendency to infect insects near the soil surface (Koppenhöfer, Baur, Stock, Choo, Chinnasri, et al., 1997); and *S. riobrave* might have similar adaptations because of the subtropical, semiarid climate of its area of origin in southern Texas (Koppenhöfer et al., 1997; Koppenhöfer, Kaya, Shanmugam, et al., 1995).

In Florida citrus groves, the efficacy of EPN applications for management of the root weevil, *D. abbreviatus*, appears to depend largely on soil type (Stuart et al., 2008). These soils tend to be extremely sandy (94 % and higher) and contain only small quantities of silt, clay and organic matter but applications are very effective in the well–drained coarse sands of the Central Ridge but often much less so in the poorly–drained finer–textured soils of the coastal and inland Flatwoods. Interestingly, groves on the Central Ridge often harbor rich communities of native EPNs (especially *S. diaprepesi, H. zealandica*, and *H. indica*) that often appear to suppress weevil populations below economic thresholds. In contrast, Flatwoods groves contain similar numbers of EPNs to Central Ridge groves but are generally dominated by a single EPN species (*H. indica*) and have frequent weevil problems (Campos-Herrera, Pathak, El-Borai, Stuart, et al., 2013; Stuart et al., 2008). In a redundancy analysis based on a survey of 53 citrus groves, four variables that affect soil water potential (ground water depth, available water capacity, clay and organic matter) significantly contributed to explain variability in soil communities (Campos-Herrera, Pathak, El-Borai, Stuart et al., 2013). Greenhouse and field manipulation studies support the idea that EPNs are more effective at suppressing root weevils and protecting citrus root systems in coarse Central Ridge soils than in fine Flatwoods soils (El-Borai et al., 2012; Duncan et al., 2013), and variables that modulate soil water potential might provide the key to how these agricultural soils might be modified to better conserve EPNs and increase levels of natural biological control (Campos-Herrera, Pathak, El-Borai, Stuart et al., 2013). Moisture also appears to play an important role in pest suppression by EPNs in a coastal prairie in California (Preisser & Strong, 2004); and might be a major factor impacting EPN efficacy after applications in peaty versus mineral soils in Ireland (Williams, Dillon, Girling, & Griffin, 2013).

4.4.3 Soil Temperature

Soil temperature can be an important factor for the survival of EPNs and for rates of biological processes with temperatures above 32 °C generally being problematic and above 37 °C generally being fatal (Ghally, 1995; Glazer, 2002; Grewal, Selvan, & Gaugler, 1994; Griffin, 1993; Hudson & Nguyen, 1989; Kung, Gaugler, & Kaya, 1991; Townsend, Johnson, & Steinkraus, 1998; Zervos et al., 1991). However, EPNs have been isolated from a broad range of locations from the sub–arctic to the tropics (Glazer, 2002; Hominick et al., 1996; Khatri-Chhetri, Waeyenberge, Manandhar, & Moens, 2010; Poinar, 1990), and different EPN species and populations exhibit various temperature tolerances, which can be modified by selection (Grewal, Gaugler, & Shupe, 1996; Grewal, Gaugler, & Wang, 1996; Grewal, Selvan, & Gaugler, 1994; Jagdale & Gordon, 1998; Mason & Hominick, 1995; Westerman, 1998). Heterorhabditid isolates from Israel (Glazer et al., 1991), Egypt (Shamseldean & Abd-Elgawad, 1994) and Sri Lanka (Amarasinghe, Hominick, Briscoe, & Reid, 1994) have been shown to be especially heat tolerant. This heat tolerance trait was successfully passed on and maintained in a hybrid *H. bacteriophora* laboratory population that resulted from mating a heat tolerant population from Israeli (IS–5) to the HP88 population (Shapiro, Glazer, & Segal, 1997).

EPNs from cold regions can withstand freezing, and this also varies among species and laboratory populations (Glazer, 2002). Brown and Gaugler (1997) demonstrated that *S. feltiae*, *S. anomali* (= *S. arenarium* (Artyukhovsky)) and *H. bacteriophora* were freezing tolerant with lower lethal temperatures of −22, −14 and −19 °C, respectively. Wharton and Surrey (1994) found that *H. zealandica* is freeze avoiding in that the sheath of *H. zealandica* prevents innoculative freezing and allows extensive supercooling to −32 °C whereas exsheathed IJs freeze and die at temperatures above −6 °C. IJs can also overwinter in living insects following infection when temperatures are too cold for bacteria to proliferate and kill the host (Brown, Lovett, Grewal, & Gaugler, 2002).

4.4.4 Soil Chemistry

The chemistry and pH of the soil solution affect EPNs but nematodes tolerate a wide range of soil pH. Kung et al. (1990b) found reduced survival of steinernematid nematodes at pH 10 but no differences from pH 4 to 8. Mortality of the cotton leafworm, *Spodoptera littoralis* (Boisduval) (Lepidoptera: Noctuidae), from *H. bacteriophora* and *S. carpocapsae* was higher and more rapid at pH 6.9 and 8.0 than at pH 5.6 (Ghally, 1995). Acid deposition may be a limiting factor in some areas (Sharpe & Drohan, 1999) but no studies appear to have documented such effects on EPNs. In laboratory experiments, acid pH reduced the efficacy of *S. carpocapsae*, *S. feltiae* and *H. bacteriophora* against diapausing larvae of *Cephalcia abietis* (L.) (Hymenoptera, Pamphiliidae), and it was suggested that application of lime or magnesium fertilizers that raise soil pH might induce EPN epizootics by increasing the activity of EPNs (Jaworska, 1993). At high concentrations, NaCl, KCl, and $CaCl_2$ inhibited the ability of *S. glaseri* to move through a soil column and to locate and infect a susceptible host (Thurston, Ni, & Kaya, 1994). Calcium chloride and KCl had no effect on *H. bacteriophora* survival, infection efficiency, or movement through a soil column, but moderate concentrations of these salts enhanced *H. bacteriophora* virulence. NaCl at high salinities ($>$16 dS/m) adversely affected all of these parameters (Thurston et al., 1994). Nematode activity and survival are reduced by low oxygen conditions (e.g., waterlogged soils; (Kung et al. 1990b; Qiu & Bedding, 1999)). A field survey in an Ohio vegetable production area indicated that both biotic and abiotic factors were associated with EPN abundance, and these factors included increased enrichment and food web structure as well as lower P, higher K, and a lower C:N ratio (Hoy, Grewal, Lawrence, Jagdale, & Acosta, 2008). These studies provide some possibilities for how soils might be manipulated to better conserve EPNs and enhance natural biological control.

4.5 Managed Ecosystems and Conservation Biological Control

EPNs have generally been used for short–term inundative or augmentative biological control but longer–term strategies of conservation biological control might ultimately be more practical and cost effective (Lewis, Campbell, & Gaugler, 1998; Stuart et al., 2006). In the context of conservation biological control, various aspects of agricultural and other managed ecosystems can influence populations of insects and their natural enemies. Two distinct components of biodiversity, planned and associated, exist in managed ecosystems (Vandermeer & Perfecto, 1995). Planned biodiversity is associated with crops or animals intentionally included by the farmer or land manager, and varies depending on management system and practices in space and time. Associated biodiversity includes the flora and fauna that colonize the ecosystem from surrounding habitats and establish and persist depending on management and structure. The microenvironment in a field can be altered significantly by crop species and practices such as irrigation, planting density, variety selection, tillage regime, fertility inputs, pesticide use and various other factors. These modifications can affect the abundance and diversity of pests and their natural enemies or enhance host plant resistance to herbivores (Cook & Baker, 1983).

A goal of conservation biological control is to identify the type of biodiversity that is needed to maintain or enhance biological control. Conservation of naturally occurring EPNs through choice of production practices could improve the persistence and efficacy of native EPNs as insect control agents (Lewis et al., 1998). However, it is difficult to assess mechanisms or causal effects of production practices on EPNs or on biological control because of the interaction of direct and indirect biotic and abiotic effects. For example, tillage can have far reaching consequences on community composition either directly by killing pests and beneficial organisms or indirectly by changing soil temperature, moisture, and structure. Biotic interactions and their mediation by physical factors could be critical for conservation biological control with EPNs but practices that favor EPNs and soil biodiversity in general might also favor the natural enemies of EPNs (Bellows, 1999; Jabbour, Crowder, Aultman, & Snyder, 2011; Sayre & Walter, 1991; Stirling, 1991). In laboratory and greenhouse experiments, EPNs that give effective control of pests in depauperate planting media often show lower efficacy in native soil with more complex soil communities (Ishibashi & Kondo, 1986; Timper & Kaya, 1992; Timper et al., 1991).

The success of natural enemies of above ground herbivorous insects can often be related to plant species or variety (Barbosa & Benrey, 1998). Similarly, crop varieties directly affect the soil abiotic environment (e.g., soil temperature and moisture) through shading and water uptake, and the biotic environment through the provision of particular insect hosts associated with the crop. Root density and architecture can affect the ability of EPNs to find a host insect (Choo & Kaya, 1991; Demarta et al., 2014) and hydraulic lift associated with plant roots can create favorable conditions for EPNs and their insect hosts in otherwise dry surface soils (Duncan & McCoy, 2001). The efficacy of natural enemies of herbivorous insects can often be related to plant secondary chemistry, and this has been demonstrated for several pathogen groups, including EPNs (Barbercheck, 1993; Barbercheck et al., 1995; Epsky & Capinera, 1994; Grewal et al., 1995; Kunkel & Grewal, 2003; Kunkel et al., 2004; Richmond et al., 2004).

In agriculture, tillage is especially disruptive to the soil environment and can influence the survival and persistence of EPNs. Soil faunal biomass often drops with increased agricultural usage, especially where conventional tillage is practiced (Stinner, McCartney, & Van Doren, 1988). Diversity and abundance of predators are greater under no tillage than under conventional tillage, and natural control of pest insects in soil may be enhanced in conservation tillage systems (Brust, 1991; Letourneau, 1998; Stinner & House, 1990). The greater complexity of the soil environment associated with relatively high levels of crop residue in conservation tillage regimes might influence the abundance of EPNs through provision of a greater number and diversity of hosts. Under a conventional tillage regime, the soil surface tends to have greater fluctuations in temperature and moisture than under no–till or reduced tillage, and EPNs are often more frequently detected in reduced tillage regimes (Hsiao & All, 1998; Hummel, Walgenbach, Barbercheck, Kennedy, Hoyt, et al., 2002; Millar & Barbercheck, 2002; Shapiro, Obrycki, Lewis, & Jackson, 1999; but see Campos-Herrera et al., 2015). However, the effects of tillage can also depend on EPN species (Millar & Barbercheck, 2002). When non–native *S. riobrave* were applied to conservation tillage and conventional till cornfields containing native *H. bacteriophora* and *S. carpocapsae*, both *H. bacteriophora* and *S. riobrave* were favored by tillage whereas *S. carpocapsae* was favored by the conservation tillage regime. This result might be explained by differences in EPN foraging strategies (Gaugler et al., 1997)

The application of fertilizers to soil represents a nutrient disturbance that can have profound direct and indirect effects on the abundance and community composition of soil biota (Campos-Herrera, Gómez-Ros, Escuer, Cuadra, Barrios, et al., 2008; Campos-Herrera, Pathak, El-Borai, Schumann, et al., 2013; Neher & Barbercheck, 1999). High concentrations of mineral or manure–based fertilizers can be detrimental to soil biota because of toxicity (e.g., anhydrous ammonia) or high osmotic pressure from salts (Andrén & Lagerlöf, 1983). In the laboratory, 10—20–day exposure to high inorganic fertilizer concentrations inhibited EPN infectivity and reproduction, whereas 1–day exposures increased infectivity (Bednarek & Gaugler, 1997). *Heterorhabditis bacteriophora* was more sensitive to the adverse effects of fertilizer than were two *Steinernema* species.

Additions of organic matter effectively change the soil environment and can increase the diversity of organisms. Organic materials can improve the physical properties of the soil that directly and indirectly affect EPNs (e.g., bulk density, porosity, moisture–holding capacity), and enhance plant growth and health. Organic amendments can be highly variable and have been used successfully to create phytopathogen–suppressive soils, but almost no documentation exists on the effects of these amendments on populations of EPNs. The strategy for increasing the suppressiveness of soils for phytopathogens is based on stimulation of high levels of biological diversity (Windels, 1997). In the creation or restoration of disease–suppressive soils, it is rare that any single biotic or abiotic factor accounts for suppression of disease (Hoitink & Fahy, 1986).

In field studies, organic manure used as fertilizer has either increased or decreased the establishment and recycling of EPNs. Bednarek and Gaugler (1997) found that the application of organic manure resulted in increased densities of native *S. feltiae*, whereas NPK fertilizer suppressed densities. The authors concluded that inorganic fertilizers are likely to be compatible with EPNs in tank mixes and should not reduce the effectiveness of EPNs applied for short–term control as biological insecticides, but might interfere with the use of EPNs as inoculative agents for long–term control. Shapiro, Lewis, Obrycki, and Abbas (1999) found that applications of *S. carpocapsae* reduced damage to seedling corn by the black cutworm, *Agrotis ipsilon* (Hufnagel) (Lepidoptera: Noctuidae), in soil amended with fresh cow manure, composted manure, or urea except at the higher rate of fresh manure. Black cutworm damage in EPN–treated plots was greater in plots with fresh manure than in plots without fertilizer. Amendments of urea or composted manure did not have a detrimental effect on suppression of the black cutworm by *S. carpocapsae*. In field and laboratory testing, pathogenicity of *S. carpocapsae* was reduced by poultry, swine, and beef cattle manure (Hsiao & All, 1997).

In Florida citrus groves, the addition of composted animal manure mulch is a potentially useful technique for conserving and enhancing EPN populations for control of Diaprepes root weevils (Duncan et al., 2007; Stuart et al., 2008).

However, another habitat manipulation, one that was initiated by growers for other purposes, might be even more effective (Stuart et al., 2008; Duncan et al., 2013). As mentioned above, EPN applications for management of the root weevil, *Diaprepes abbreviatus*, in Florida citrus groves are much more effective in the well–drained coarse sands of the Central Ridge than the poorly–drained finer–textured soils of the coastal and inland Flatwoods; and groves on the Central Ridge often harbor richer communities of endemic EPNs that often appear to suppress weevil populations below economic thresholds compared to the species poor but numerically equivalent EPN populations in Flatwoods groves (Stuart et al., 2008; Campos-Herrera, Pathak, El-Borai, Stuart, et al., 2013). Recently, citrus growers in the Flatwoods began planting new trees in oversized planting holes filled with coarse sand of the kind found on the Central Ridge to improve drainage, reduce infection by plant pathogens, and promote tree growth. Incidentally, this practice also appears to provide an improved habitat for EPNs and biological control around the root system of the young tree. Field studies in which such coarse sand islands were inoculated with EPNs from the Central Ridge showed that they promoted tree growth and greater mortality of sentinel weevil larvae compared to native soil plots (Stuart et al., 2008; Duncan et al., 2013). A conservation biological control program based on this kind of habitat manipulation could serve as an important model for similar programs in diverse agroecosystems worldwide.

4.6 Conclusions and Future Directions

Entomopathogenic nematodes are wide spread and important natural enemies of insects in soils throughout the world, and likely play a fundamental role in the regulation of insect populations in various habitats. The inoculation, augmentation and conservation of EPN populations for biological control purposes in agricultural and other managed habitats provides unique opportunities for the effective and environmentally benign control of soil insect pests. However, the extent to which such manipulations impact soil ecosystems are largely unknown and we are only beginning to comprehend the full range and complexity of the trophic webs that are involved. Understanding the distribution, abundance, and dynamics of EPN populations within their broader ecological context presents interesting and complex challenges for future research. We know relatively little about the spatial and genetic structure of EPN populations, how these populations vary over time and among sites, their various interactions with the biotic and abiotic environment, and how fundamental EPN characteristics influencing these interactions vary among species, populations and habitats. Research to date has laid an important foundation upon which to build but provides little more than a glimpse of the rich complexity, diversity, and variability that could be present. Future studies will benefit from the increased use of molecular techniques, multivariate analyses, and modeling strategies to identify, quantify and determine the relative importance of key components of soil food webs and unravel how EPNs are associated and depend upon a broad range of biotic and abiotic factors. Meanwhile, the economic and social realities of modern agriculture combined with continuing environmental degradation and human population growth in developed and developing countries provide a critical backdrop for these studies and assure that EPNs will continue to be important subjects for basic and applied ecological research.

References

Adams, B. J., Fodor, A., Koppenhöfer, H. S., Stackenbrandt, E., Stock, S. P., & Klein, M. G. (2006). Biodiversity and systematics of nematode–bacterium entomopathogens. *Biological Control, 38*, 4–21.

Adl, S. M. (2003). *The ecology of soil decomposition*. Wallingford, UK: CABI.

Akhurst, R. J., & Brooks, W. M. (1984). The distribution of entomophilic nematodes (Heterorhabditidae and Steinernematidae) in North Carolina. *Journal of Invertebrate Pathology, 44*, 140–145.

Alatorre-Rosas, R., & Kaya, H. K. (1990). Interspecific competition between entomopathogenic nematodes in the genera *Heterorhabditis* and *Steinernema* for an insect host in sand. *Journal of Invertebrate Pathology, 55*, 179–188.

Alatorre-Rosas, R., & Kaya, H. K. (1991). Interactions between two entomopathogenic nematode species in the same host. *Journal of Invertebrate Pathology, 57*, 1–6.

Ali, J. G., Alborn, H. T., & Stelinski, L. L. (2010). Subterranean herbivore–induced volatiles released by citrus roots upon feeding by *Diaprepes abbreviatus* recruit entomopathogenic nematodes. *Journal of Chemical Ecology, 36*, 361–368.

Ali, J. G., Campos-Herrera, R., Alborn, H. T., Duncan, L. W., & Stelinski, L. L. (2013). Sending mixed messages: A trophic cascade produced by a belowground herbivore–induced cue. *Journal of Chemical Ecology, 39*(8), 1140–1147.

Alsaiyah, M. A. M., Ebssa, L., Zenner, A., O'Callaghan, K. M., & Griffin, C. T. (2009). Sex ratios and sex–biased infection behaviour in the entomopathogenic nematode genus *Steinernema*. *International Journal for Parasitology, 39*, 725–734.

Amarasinghe, L. D., Hominick, W. M., Briscoe, B. R., & Reid, A. P. (1994). Occurrence and distribution of entomopathogenic nematodes in Sri Lanka. *Journal of Helminthology, 68*(4), 277–286.

Andrén, O., & Lagerlöf, J. (1983). Soil fauna (microarthropods, enchytraeids, nematodes) in Swedish agricultural cropping systems. *Acta Agriculturae Scandinavica, 33*, 33–52.

Ansari, M. A., Shah, F. A., & Butt, T. M. (2008). Combined use of entomopathogenic nematodes and *Metarhizium anisopliae* as a new approach for black vine weevil, *Otiorhynchus sulcatus*, control. *Entomologia Experimentalis et Applicata, 129*(3), 340–347.

Bai, X., Adams, B. J., Ciche, T. A., Clifton, S., Gaugler, R., Kim, K., et al. (2013). A lover and a fighter: The genome sequence of an entomopathogenic nematode. *PloS One, 8*(7), e69618.

Bal, H. K., Michael, A., & Grewal, P. S. (2014). Genetic selection of the ambush foraging entomopathogenic nematode *Steinernema carpocapsae* for enhanced dispersal and its associated trade–offs. *Evolutionary Ecology, 28*, 923–939.

Bal, H. K., Taylor, R. A. J., & Grewal, P. S. (2014). Ambush foraging entomopathogenic nematodes employ sprinting emigrants for long distance dispersal in the absence of hosts. *Parasitology, 100*, 422–432.

Barbercheck, M. E. (1992). Effect of soil physical factors on biological control agents of soil insect pests. *Florida Entomologist, 75*, 539–548.

Barbercheck, M. E. (1993). Tritrophic level effects on entomopathogenic nematodes. *Environmental Entomology, 22*, 1166–1171.

Barbercheck, M. E., & Kaya, H. K. (1990). Interactions between *Beauveria bassiana* and the entomogenous nematodes, *Steinernema feltiae* and *Heterorhabditis heliothidis*. *Journal of Invertebrate Pathology, 55*, 225–234.

Barbercheck, M. E., & Kaya, H. K. (1991a). Effect of host condition and soil texture on host finding by the entomogenous nematodes *Heterorhabditis bacteriophora* (Rhabditida: Heterorhabditidae) and *Steinernema carpocapsae* (Rhabditida: Steinernematidae). *Environmental Entomology, 20*, 582–589.

Barbercheck, M. E., & Kaya, H. K. (1991b). Competitive interactions between entomopathogenic nematodes and *Beauveria bassiana* (Deuteromycotina: Hyphomycetes) in soil borne larvae of *Spodoptera exigua* (Lepidoptera: Noctuidae). *Environmental Entomology, 20*, 707–712.

Barbercheck, M. E., Wang, J., & Hirsh, I. S. (1995). Host plant effects on entomopathogenic nematodes. *Journal of Invertebrate Pathology, 66*, 169–177.

Barbosa, P., & Benrey, B. (1998). The influence of plants on insect parasitoids: Implications for conservation biological control. In P. Barbosa (Ed.), *Conservation biological control* (pp. 235–254). San Diego, CA: Academic.

Bashey, F., Hawlena, H., & Lively, C. M. (2013). Alternative paths to success in a parasite community: Within–host competition can favor higher virulence or direct interference. *Evolution, 67*(3), 900–907.

Bashey, F., Young, K., Hawlena, H., & Lively, C. M. (2012). Spiteful interactions between sympatric natural isolates of *Xenorhabdus bovienii* benefit kin and reduce virulence. *Journal of Evolutionary Biology, 25*(3), 431–437.

Baur, M. E., Kaya, H. K., & Strong, D. R. (1998). Foraging ants as scavengers on entomopathogenic nematode–killed insects. *Biological Control, 12*, 231–236.

Beavers, J. B., McCoy, C. W., & Kaplan, D. T. (1983). Natural enemies of subterranean *Diaprepes abbreviatus* (Coleoptera: Curculionidae) larvae in Florida. *Environmental Entomology, 12*, 840–843.

Bednarek, A., & Gaugler, R. (1997). Compatibility of soil amendments with entomopathogenic nematodes. *Journal of Nematology, 29*, 220–227.

Bellows, T. S. (1999). Controlling soil–borne plant pathogens. In T. S. Bellows & T. W. Fisher (Eds.), *Handbook of biological control* (pp. 699–711). San Diego, CA: Academic.

Bilgrami, A. L., Gaugler, R., Shapiro-Ilan, D. I., & Adams, B. J. (2006). Source of trait deterioration in entomopathogenic nematodes *Heterorhabditis bacteriophora* and *Steinernema carpocapsae* during *in vivo* culture. *Nematology, 8*(3), 397–410.

Bilgrami, A. L., & Jairajpuri, M. S. (1989a). Predatory abilities of *Mononchoides longicaudatus* and *M. fortidens* (Nematoda: Diplogasterida) and factors influencing predation. *Nematologica, 35*, 475–488.

Bilgrami, A. L., & Jairajpuri, M. S. (1989b). Resistance of prey to predation and strike rate of the predators *Mononchoides longicaudatus* and *M. fortidans* (Nematoda: Diplogasterida). *Revue de Nematologie, 12*, 45–49.

Blouin, M. S., Liu, J., & Berry, R. E. (1999). Life cycle variation and the genetic structure of nematode populations. *Heredity, 8*, 253–259.

Boemare, N. (2002). Biology, taxonomy, and systematics of *Photorhabdus* and *Xenorhabdus*. In R. Gaugler (Ed.), *Entomopathogenic nematology* (pp. 35–56). New York: CABI.

Bohan, D. A., & Hominick, W. M. (1995). Intra–population infectious structure and temporal variation in *Steinernema feltiae*. In C. T. Griffin, R. L. Gwynn, & J. P. Masson (Eds.), *Ecology and transmission strategies of entomopathogenic nematodes* (pp. 83–94). Luxembourg: European Commission.

Bohan, D. A., & Hominick, W. M. (1996). Investigations on the presence of an infectious proportion amongst populations of *Steinernema feltiae* (site 76 strain) infective stages. *Parasitology, 112*, 113–118.

Bohan, D. A., & Hominick, W. M. (1997a). Sex and the dynamics of infection in the entomopathogenic nematode *Steinernema feltiae*. *Journal of Helminthology, 71*, 197–201.

Bohan, D. A., & Hominick, W. M. (1997b). Long–term dynamics of infectiousness within the infective–state pool of the entomopathogenic nematode *Steinernema feltiae* (site 76 strain) Filipjev. *Parasitology, 114*, 301–308.
Brown, I. M., & Gaugler, R. (1997). Temperature and humidity influence emergence and survival of entomopathogenic nematodes. *Nematologica, 43*, 363–375.
Brown, I., Lovett, B., Grewal, P. S., & Gaugler, R. (2002). Latent infection: A low temperature survival strategy in steinernematid nematodes. *Journal of Thermal Biology, 27*, 531–539.
Brown, V. K., & Gange, A. C. (1990). Insect herbivory below ground. *Advances in Ecological Research, 20*, 1–58.
Brust, G. E. (1991). Augmentation of an endemic entomogenous nematode by agroecosystem manipulation for the control of a soil pest. *Agriculture, Ecosystems & Environment, 36*, 175–184.
Burman, M., & Pye, A. E. (1980). *Neoaplectana carpocapsae*: Movements of nematode populations on a thermal gradient. *Experimental Parasitology, 49*(2), 258–265.
Byers, J. A., & Poinar, G. O., Jr. (1982). Location of insect hosts by the nematode, *Neoaplectana carpocapsae*, in response to temperature. *Behaviour, 79*(1), 1–10.
Cabanillas, H. E., & Raulston, J. R. (1994). Evaluation of the spatial pattern of *Steinernema riobravis* in corn plots. *Journal of Nematology, 26*(1), 25–31.
Cakmak, I., Hazir, S., Ulug, D., & Karagoz, M. (2013). Olfactory response of *Sancassania polyphyllae* (Acari: Acaridae) to its phoretic host larva killed by the entomopathogenic nematode, *Steinernema glaseri* (Rhabditida: Steinernematidae). *Biological Control, 65*, 212–217.
Campbell, J. F., Koppenhöfer, A. M., Kaya, H. K., & Chinnasri, B. (1999). Are there temporarily non–infectious dauer stages in entomopathogenic nematode populations: A test of the phased infectivity hypothesis. *Parasitology, 118*, 499–508.
Campbell, J. F., Lewis, E., Yoder, F., & Gaugler, R. (1995). Entomopathogenic nematode (Heterorhabditidae and Steinernematidae) seasonal population dynamics and impact on insect populations in turfgrass. *Biological Control, 5*, 598–606.
Campbell, J. F., Lewis, E., Yoder, F., & Gaugler, R. (1996). Entomopathogenic nematode (Heterorhabditidae and Steinernematidae) spatial distribution in turfgrass. *Parasitology, 113*, 473–482.
Campbell, J. F., Orza, G., Yoder, F., Lewis, E., & Gaugler, R. (1998). Spatial and temporal distribution of endemic and released entomopathogenic nematode populations in turfgrass. *Entomologia Experimentalis et Applicata, 86*, 1–11.
Campos-Herrera, R., Barbercheck, M., Hoy, C. W., & Stock, S. P. (2012). Entomopathogenic nematodes as a model system for advancing the frontiers of ecology. *Journal of Nematology, 44*(2), 162–176.
Campos-Herrera, R., El-Borai, F. E., & Duncan, L. W. (2012). Wide interguild relationships among entomopathogenic and free–living nematodes in soil as measured by real time qPCR. *Journal of Invertebrate Pathology, 111*, 126–135.
Campos-Herrera, R., Escuer, M., Labrador, S., Robertson, L., Barrios, L., & Gutiérrez, C. (2007). Distribution of the entomopathogenic nematodes from La Rioja (Northern Spain). *Journal of Invertebrate Pathology, 95*, 125–139.
Campos-Herrera, R., Gómez-Ros, J. M., Escuer, M., Cuadra, L., Barrios, L., & Gutiérrez, C. (2008). Diversity, occurrence, and life characteristics of natural entomopathogenic nematode populations from La Rioja (Northern Spain) under different agricultural management and their relationships with soil factors. *Soil Biology and Biochemistry, 40*(6), 1474–1484.
Campos-Herrera, R., Jaffuel, G., Chiriboga, X., Blanco-Perez, R., Fesselet, M., Puza, V., et al. (2015). Traditional and molecular detection methods reveal intense interguild competition and other multitrophic interactions associated with native entomopathogenic nematodes in Swiss tillage soils. *Plant and Soil, 389*, 237–255.
Campos-Herrera, R., Johnson, E. G., El-Borai, F. E., Stuart, R. J., Graham, J. H., & Duncan, L. W. (2011). Long–term stability of entomopathogenic nematode spatial patterns in soil as measured by sentinel insects and real–time PCR assays. *Annals of Applied Biology, 158*, 55–68.

Campos-Herrera, R., Pathak, E., El-Borai, F. E., Schumann, A., Abd-Elgawad, M. M. M., & Duncan, L. W. (2013). New citriculture system suppresses native and augmented entomopathogenic nematodes. *Biological Control, 66*, 183–194.

Campos-Herrera, R., Pathak, E., El-Borai, F. E., Stuart, R. J., Gutiérrez, C., Rodríguez-Martín, J. A., et al. (2013). Geospatial patterns of soil properties and the biological control potential of entomopathogenic nematodes in Florida citrus groves. *Soil Biology and Biochemistry, 66*, 163–174.

Chase, J. M., Abrams, P. A., Grover, J. P., Diehl, S., Chesson, P., Holt, R. D., et al. (2002). The interaction between predation and competition: a review and synthesis. *Ecology Letters, 5*(2), 302–315.

Choo, H. Y., & Kaya, H. K. (1991). Influence of soil texture and presence of roots on host finding by *Heterorhabditis bacteriophora. Journal of Invertebrate Pathology, 58*, 279–280.

Coleman, D. C., & Crossley, D. A., Jr. (1996). *Fundamentals of soil ecology*. San Diego, CA: Academic.

Connell, J. (1980). Diversity and the coevolution of competitors, or the ghost of competition past. *Oikos, 35*, 131–138.

Constant, P., Marchay, L., Fischer-Le Saux, M., Briand-Panoma, S., & Mauleon, H. (1998). Natural occurrence of entomopathogenic nematodes (Rhabditida: Steinernematidae and Heterorhabditidae) in Guadalupe islands. *Fundamental and Applied Nematology, 21*, 667–672.

Cook, R. J., & Baker, K. F. (1983). *The nature and practice of biological control of plant pathogens*. St. Paul, MN: American Phytopathological Society.

Demarta, D., Hibbard, B. E., Bohn, M. O., & Hiltpold, I. (2014). The role of root architecture in foraging behavior of entomopathogenic nematodes. *Journal of Invertebrate Pathology, 122*, 32–39.

Denno, R. F., Gruner, D. S., & Kaplan, I. (2008). Potential for entomopathogenic nematodes in biological control: A meta–analytical synthesis and insights from trophic cascade theory. *Journal of Nematology, 40*(2), 61–72.

Dillman, A. R., Chaston, J. M., Adams, B. J., Ciche, T. A., Goodrich-Blair, H., Stock, S. P., et al. (2012). An entomopathogenic nematode by any other name. *PLoS Pathogens, 8*, e1002527.

Dillman, A. R., Guillermin, M. L., Lee, J. H., Kim, B., Sternberg, P. W., & Hallem, E. A. (2012). Olfaction shapes host–parasite interactions in parasitic nematodes. *Proceedings of the National Academy of Sciences, 109*(35), E2324–E2333.

Dolinski, C., Choo, H. Y., & Duncan, L. W. (2012). Grower acceptance of entomopathogenic nematodes: Case studies on three continents. *Journal of Nematology, 44*, 226–235.

Dugaw, C. J., Hastings, A., Preisser, E. L., & Strong, D. R. (2004). Seasonally limited host supply generates microparasite population cycles. *Bulletin of Mathematical Biology, 66*, 583–594.

Duncan, L. W., Dunn, D. C., Bague, G., & Nguyen, K. (2003). Competition between entomopathogenic and free–living bactivorous nematodes in larvae of the weevil *Diaprepes abbreviatus. Journal of Nematology, 35*, 187–193.

Duncan, L. W., Dunn, D. C., & McCoy, C. W. (1996). Spatial patterns of entomopathogenic nematodes in microcosms: Implications for laboratory experiments. *Journal of Nematology, 28*, 252–258.

Duncan, L. W., Genta, J. G., Zellers, J., Fares, A., & Stansly, P. A. (2001). Efficacy of *Steinernema riobrave* against larvae of *Diaprepes abbreviatus* in Florida soils of different texture. *Nematropica, 31*, 130.

Duncan, L. W., Graham, J. H., Dunn, D. C., Zellers, J., McCoy, C. W., & Nguyen, K. (2003). Incidence of endemic entomopathogenic nematodes following application of *Steinernema riobrave* for control of *Diaprepes abbreviatus. Journal of Nematology, 35*, 178–186.

Duncan, L. W., Graham, J. H., Zellers, J., Bright, D., Dunn, D. C., El-Borai, F. E., et al. (2007). Food web responses to augmenting the entomopathogenic nematodes in bare and animal manure–mulched soil. *Journal of Nematology, 39*(2), 176–189.

Duncan, L. W., & McCoy, C. W. (2001). Hydraulic lift increases herbivory by *Diaprepes abbreviatus* larvae and persistence of *Steinernema riobrave* in dry soil. *Journal of Nematology, 33*, 142–146.
Duncan, L. W., Stuart, R. J., El-Borai, F. E., Campos-Herrera, R., Pathak, E., Giurcanu, M., et al. (2013). Modifying orchard planting sites conserves entomopathogenic nematodes, reduces weevil herbivory and increases citrus tree growth, survival and fruit yield. *Biological Control, 64*, 26–36.
Dutilleul, P., & Legendre, P. (1993). Spatial heterogeneity against heteroscedasticity: An ecological paradigm versus a statistical concept. *Oikos, 66*, 152–171.
Efron, D., Nestel, D., & Glazer, I. (2001). Spatial analysis of entomopathogenic nematodes and insect hosts in a citrus grove in a semi–arid region in Israel. *Environmental Entomology, 30*, 254–261.
Ehlers, R.–. U., Deseö, K. V., & Stackebrandt, E. (1991). Identification of *Steinernema* spp. (Nematoda) and their symbiotic bacteria *Xenorhabdus* spp. from Italian and German soils. *Nematologia, 37*, 360–366.
Ekmen, Z. I., Hazir, S., Cakmak, I., Ozer, N., Karagoz, M., & Kaya, H. K. (2010). Potential negative effects on biological control by *Sancassania polyphyllae* (Acari: Acaridae) on an entomopathogenic nematode species. *Biological Control, 54*, 166–171.
El-Borai, F. E., Campos-Herrera, R., Stuart, R. J., & Duncan, L. W. (2011). Substrate modulation, group effects and the behavioral responses of entomopathogenic nematodes to nematophagous fungi. *Journal of Invertebrate Pathology, 106*(3), 347–356.
El-Borai, F. E., Stuart, R. J., Campos-Herrera, R., Pathak, E., & Duncan, L. W. (2012). Entomopathogenic nematodes, root weevil larvae, and dynamic interactions among soil texture, plant growth, herbivory, and predation. *Journal of Invertebrate Pathology, 109*(1), 134–142.
Eng, M. S., Preisser, E. L., & Strong, D. R. (2005). Phoresy of the entomopathogenic nematode *Heterorhabditis marelatus* by a non–host organism, the isopod *Porcellio scaber. Journal of Invertebrate Pathology, 88*, 173–176.
Epsky, N. D., & Capinera, J. L. (1994). Influence of herbivore diet on the pathogenesis of *Steinernema carpocapsae* (Nematoda: Steinernematidae). *Environmental Entomology, 23*, 487–491.
Epsky, N. D., Walter, D. E., & Capinera, J. L. (1988). Potential role of nematophagous microarthropods as biotic mortality factors of entomopathogenic nematodes (Rhabditid, Steinernematidae and Heterorhabditidae). *Journal of Economic Entomology, 81*, 821–825.
Ettema, C. H. (1998). Soil nematode diversity: Species coexistence and ecosystem function. *Journal of Nematology, 30*, 159–169.
Ettema, C. H., & Wardle, D. A. (2002). Spatial soil ecology. *Trends in Ecology & Evolution, 17*(4), 177–183.
Fairbairn, J. P., Fenton, A., Norman, R., & Hudson, P. J. (2000). Re–assessing the infection strategies of the entomopathogenic nematode *Steinernema feltiae* (Rhabditidae; Steinernematidae). *Parasitology, 121*, 211–216.
Fenton, A., Magoolagan, L., Kennedy, Z., & Spencer, K. A. (2011). Parasite–induced warning coloration: A novel form of host manipulation. *Animal Behaviour, 81*(2), 417–422.
Fenton, A., Norman, R., Fairbairn, J. P., & Hudson, P. J. (2000). Modeling the efficacy of entomopathogenic nematodes in the regulation of invertebrate pests in glasshouse crops. *Journal of Applied Ecology, 37*, 309–320.
Fierer, N., Strickland, M. S., Liptzin, D., Bradford, M. A., & Cleveland, C. C. (2009). Global patterns in belowground communities. *Ecology Letters, 12*, 1–12.
Forst, S., & Clarke, D. (2002). Bacteria–nematode symbiosis. In R. Gaugler (Ed.), *Entomopathogenic nematology* (pp. 57–77). New York: CABI.
Garcia Del Pino, F., & Palomo, A. (1996). Natural occurrence of entomopathogenic nematodes (Rhabditida: Steinernematidae and Heterorhabditidae) in Spanish soils. *Journal of Invertebrate Pathology, 68*, 84–90.

Gassmann, A. J., Stock, S. P., Tabashnik, B. E., & Singer, M. S. (2010). Tritrophic effects of host plants on an herbivore–pathogen interaction. *Annals of the Entomological Society of America, 103*, 371–378.

Gaugler, R., & Kaya, H. K. (1990). *Entomopathogenic nematodes in biological control*. Boca Raton, FL: CRC.

Gaugler, R., Lewis, E. E., & Stuart, R. J. (1997). Ecology in the service of biological control: The case of entomopathogenic nematodes. *Oecologia, 109*, 483–489.

Gaugler, R., Wang, Y., & Campbell, J. F. (1994). Aggressive and evasive behaviors in *Popillia japonica* (Coleoptera: Scarabaeidae) larvae: Defences against entomopathogenic nematode attack. *Journal of Invertebrate Pathology, 64*, 193–199.

Ghally, S. E. (1995). Some factors affecting the activity and pathogenicity of *Heterorhabditis heliothidis* and *Steinernema carpocapsae* nematodes. *Journal of the Egyptian Society of Parasitology, 25*, 125–135.

Gilmore, S. K., & Potter, D. A. (1993). Potential role of collembola as biotic mortality agents for entomopathogenic nematodes. *Pedobiologia, 37*, 30–38.

Gilmore, S. K., & Raffensperger, E. M. (1970). Foods ingested by *Tomocerus* spp. (Collembola, Entomobryidae), in relation to habitat. *Pedobiologia, 10*, 135–140.

Glazer, I. (1997). Effects of infected insects on secondary invasion of steinernematid entomopathogenic nematodes. *Parasitology, 114*, 597–604.

Glazer, I. (2002). Survival biology. In R. Gaugler (Ed.), *Entomopathogenic nematology* (pp. 169–187). Wallingford, UK: CABI.

Glazer, I., Gaugler, R., & Segal, D. (1991). Genetics of the nematode *Heterorhabditis bacteriophora* HP88 strain: The diversity of beneficial traits. *Journal of Nematology, 23*, 324–333.

Glazer, I., Kozodoi, E., Salame, L., & Nestel, D. (1996). Spatial and temporal occurrence of natural populations of *Heterorhabditis* spp. (Nematoda: Rhabditida) in a semiarid region. *Biological Control, 6*, 130–136.

Grant, J. A., & Villani, M. G. (2003). Soil moisture effects on entomopathogenic nematodes. *Environmental Entomology, 32*, 80–87.

Greenwood, C. M., Barbercheck, M. E., & Brownie, C. (2011). Short term response of soil microinvertebrates to application of entomopathogenic nematode–infected insects in two tillage systems. *Pedobiologia, 54*, 177–186.

Grewal, P. S. (2012). Entomopathogenic nematodes as tools in integrated pest management. In D. P. Abrol & U. Shankar (Eds.), *Integrated pest management: Principles and practice* (pp. 162–236). Wallingford, UK: CABI.

Grewal, P. S., Ehlers, R.–. U., & Shapiro-Ilan, D. I. (Eds.). (2005). *Nematodes as biocontrol agents*. Wallingford, UK: CABI.

Grewal, P. S., Gaugler, R., & Shupe, C. (1996). Rapid changes in thermal sensitivity of entomopathogenic nematodes in response to selection at temperature extremes. *Journal of Invertebrate Pathology, 68*, 65–73.

Grewal, P. S., Gaugler, R., & Wang, Y. (1996). Enhanced cold tolerance of the entomopathogenic nematode *Steinernema feltiae* through genetic selection. *Annals of Applied Biology, 129*, 335–341.

Grewal, P. S., Grewal, S. K., Malik, V. S., & Klein, M. G. (2002). Differences in susceptibility of introduced and native white grub species to entomopathogenic nematodes from various geographic localities. *Biological Control, 24*, 230–237.

Grewal, P. S., Lewis, E. E., Gaugler, R., & Campbell, J. F. (1994). Host finding behavior as a predictor of foraging strategy of entomopathogenic nematodes. *Parasitology, 108*, 207–215.

Grewal, P. S., Lewis, E. E., & Gaugler, R. (1997). Response of infective stage parasites (Nematoda: Steinernematidae) to volatile cues from infected hosts. *Journal of Chemical Ecology, 23*, 503–515.

Grewal, P. S., Lewis, E. E., & Venkatachari, S. (1999). Allelopathy: A possible mechanism of suppression of plant–parasitic nematodes by entomopathogenic nematodes. *Nematology, 1*, 735–743.

Grewal, P. S., Martin, W. R., Miller, R. W., & Lewis, E. E. (1997). Suppression of plant–parasitic nematode populations in turfgrass by application of entomopathogenic nematodes. *Biocontrol Science and Technology, 7*, 393–399.

Grewal, P. S., Selvan, S., Lewis, E. E., & Gaugler, R. (1993). Male insect–parasitic nematodes: A colonizing sex. *Experientia, 49*, 605–608.

Grewal, P. S., Selvan, S., & Gaugler, R. (1994). Thermal adaptation of entomopathogenic nematodes: Niche breadth for infection, establishment, and reproduction. *Journal of Thermal Biology, 19*, 245–253.

Grewal, P. S., Wang, X., & Taylor, R. A. J. (2002). Dauer juvenile longevity and stress tolerance in natural populations of an entomopathogenic nematode: Is there a relationship? *International Journal of Parasitology, 32*, 717–725.

Grewal, S. K., Grewal, P. S., & Gaugler, R. (1995). Endophytes of fescue grasses enhance susceptibility of *Popillia japonica* larvae to an entomopathogenic nematode. *Entomologia Experimentalis et Applicata, 74*, 219–224.

Griffin, C. T. (1993). Temperature responses of entomopathogenic nematodes: Implications for the success of biological control programmes. In R. Bedding, R. Akhurst, & H. K. Kaya (Eds.), *Nematodes and the biological control of insects* (pp. 115–126). East Melbourne, Australia: CSIRO.

Griffin, C. T. (2012). Perspectives on the behavior of entomopathogenic nematodes from dispersal to reproduction: Traits contributing to nematode fitness and biocontrol efficacy. *Journal of Nematology, 44*(2), 177–184.

Gulcu, B., Hazir, S., & Kaya, H. K. (2012). Scavenger deterrent factor (SDF) from symbiotic bacteria of entomopathogenic nematodes. *Journal of Invertebrate Pathology, 110*, 326–333.

Hanski, I. (1998). Metapopulation dynamics. *Nature, 396*, 41–49.

Hanski, I. (1999a). Habitat connectivity, habitat continuity, and metapopulations in dynamic landscapes. *Oikos, 87*, 209–219.

Hanski, I. (1999b). *Metapopulation ecology*. Oxford, UK: Oxford University Press.

Hanski, I. (2001). Spatially realistic theory of metapopulation ecology. *Naturwissenschaften, 88*, 372–381.

Hanski, I., & Simberloff, D. (1997). The metapopulation approach, its history, conceptual domain and application to conservation. In I. Hanski & M. E. Gilpin (Eds.), *Metapopulation biology: Ecology, genetics, and evolution* (pp. 5–26). London: Academic.

Harrison, S., & Hastings, A. (1996). Genetic and evolutionary consequences of metapopulation structure. *Trends in Ecology and Evolution, 11*, 180–183.

Harrison, S., & Taylor, A. D. (1997). Empirical evidence for metapopulation dynamics. In I. Hanski & M. E. Gilpin (Eds.), *Metapopulation biology: Ecology, genetics, and evolution* (pp. 27–42). London: Academic.

Hawksworth, D. L. (1991). *The biodiversity of microorganisms and invertebrates: Its role in sustainable agriculture*. Wallingford, UK: CABI.

Hay, D. B., & Fenlon, J. S. (1997). A modified binomial model that describes the infection dynamics of the entomopathogenic nematode *Steinernema feltiae* (Steinernematidae; Nematoda). *Parasitology, 111*, 627–633.

Hillel, P. M. (1982). *Fundamentals of soil physics*. New York: Academic.

Hoitink, H. A. J., & Fahy, P. C. (1986). Basis for the control of soilborne plant pathogens with composts. *Annual Review of Phytopathology, 24*, 93–114.

Hominick, W. M. (2002). Biogeography. In R. Gaugler (Ed.), *Entomopathogenic nematology* (pp. 115–143). Wallingford, UK: CABI.

Hominick, W. M., & Briscoe, B. R. (1990a). Survey of 15 sites over 28 months for entomopathogenic nematodes (Rhabditida: Steinernematidae). *Parasitology, 100*, 289–294.

Hominick, W. M., & Briscoe, B. R. (1990b). Occurrence of entomopathogenic nematodes (Rhabditida: Steinernematidae and Heterorhabditidae) in British soils. *Parasitology, 100*, 295–302.

Hominick, W. M., Hunt, D. J., Reid, A. P., Briscoe, B. R., & Bohan, D. A. (1999). Biosystematics, phylogeny and population genetics of entomopathogenic nematodes. In N. Boemare, P.

Richardson, & F. Coudert (Eds.), *COST 819. Taxonomy, phylogeny and gnotobiological studies of entomopathogenic nematodes bacterium complexes* (pp. 45–53). Brussels, Belgium: European Commission.

Hominick, W. M., & Reid, A. P. (1990). Perspectives on entomopathogenic nematology. In R. Gaugler & H. K. Kaya (Eds.), *Entomopathogenic nematodes in biological control* (pp. 327–345). Boca Raton, FL: CRC Press.

Hominick, W. M., Reid, A. P., Bohan, D. A., & Briscoe, B. R. (1996). Entomopathogenic nematodes: biodiversity, geographical distribution and the convention on biological diversity. *Biocontrol Science and Technology, 6*, 317–331.

Hoy, C. W., Grewal, P. S., Lawrence, J. L., Jagdale, G., & Acosta, N. (2008). Canonical correspondence analysis demonstrates unique soil conditions for entomopathogenic nematode species compared with other free–living nematode species. *Biological Control, 46*, 371–379.

Hsiao, W. F., & All, J. N. (1997). Effect of animal manure on the survival and pathogenicity of the entomopathogenic nematode, *Steinernema carpocapsae. Zhonghua Kunchong, 17*, 53–65.

Hsiao, W. F., & All, J. N. (1998). Survey of the entomopathogenic nematode *Steinernema carpocapsae* (Rhabditida: Steinernematidae) natural populations and its dispersal in the field. *Chinese Journal of Entomology, 18*, 39–49.

Hudson, W. G., & Nguyen, K. B. (1989). Effects of soil moisture, exposure time, nematode age, and nematode density on laboratory infection of *Scapteriscus vicinus* and *Scapteriscus acletus* (Orthoptera: Gryllotalpidae) by *Neoaplectana* sp. (Rhabditida: Steinernematidae). *Environmental Entomology, 18*, 719–722.

Hummel, R., Walgenbach, J. F., Barbercheck, M. E., Kennedy, G. G., Hoyt, G. D., & Arellano, C. (2002). Effects of production practices on soil–borne entomopathogens in western North Carolina vegetable systems. *Environmental Entomology, 31*, 84–91.

Ishibashi, N., & Kondo, E. (1986). *Steinernema feltiae* (DD–136) and *S. glaseri*: persistence in soil and bark compost and their influence on native nematodes. *Journal of Nematology, 18*, 310–316.

Ishibashi, N., & Kondo, E. (1987). Dynamics of the entomogenous nematode *Steinernema feltiae* applied to soil with and without nematicide treatment. *Journal of Nematology, 19*, 404–412.

Ishibashi, N., & Kondo, E. (1990). Behaviour of infective juveniles. In R. Gaugler & K. K. Kaya (Eds.), *Entomopathogenic nematodes in biological control* (pp. 139–150). Boca Raton, FL: CRC Press.

Jabbour, R., Crowder, D. W., Aultman, E. A., & Snyder, W. E. (2011). Entomopathogen biodiversity increases host mortality. *Biological Control, 59*, 277–283.

Jagdale, G. B., & Gordon, R. (1998). Effect of propagation temperatures on temperature tolerances of entomopathogenic nematodes. *Fundamental and Applied Nematology, 21*, 177–183.

Jagdale, G. B., Kamoun, S., & Grewal, P. S. (2009). Entomopathogenic nematodes induce components of systemic resistance in plants: Biochemical and molecular evidence. *Biological Control, 51*, 102–109.

Jagdale, G. B., Saeb, A. T. M., Somasekhar, N., & Grewal, P. S. (2006). Genetic variation and relationships between isolates and species of the entompathogenic nematode genus *Heterorhabditis* deciphered through isozyme profiles. *Journal of Parasitology, 92*, 509–516.

Jagdale, G. B., Somasekhar, N., Grewal, P. S., & Klein, M. G. (2002). Suppression of plant–parasitic nematodes by application of live and dead infective juveniles of an entomopathogenic nematode, *Steinernema carpocapsae*, on boxwood (*Buxus* spp.). *Biological Control, 24*, 42–49.

Jaworska, M. (1993). Investigations on the possibility of using entomophilic nematodes in reduction of *Cephalcia abietis* (L.) (Hym. Pamphiliidae) population. *Polskie Pismo Entomologiczne, 62*, 201–213.

Karagoz, M., Gulcu, B., Cakmak, I., Kaya, H. K., & Hazir, S. (2007). Predation of entomopathogenic nematodes by *Sancassania* sp. (Acari: Acaridae). *Experimental and Applied Acarology, 43*, 85–95.

Kaya, H. K. (1990). Soil ecology. In R. Gaugler & H. K. Kaya (Eds.), *Entomopathogenic nematodes in biological control* (pp. 93–115). Boca Raton, FL: CRC Press.

Kaya, H. K. (2002). Natural enemies and other antagonists. In R. Gaugler (Ed.), *Entomopathogenic nematology* (pp. 189–203). Wallingford, UK: CABI.

Kaya, H. K., Bedding, R. A., & Akhurst, R. J. (1993). An overview of insect–parasitic and entomopathogenic nematodes. In R. Bedding, R. Akhurst, & H. K. Kaya (Eds.), *Nematodes and the biological control of insect pests* (pp. 1–10). East Melbourne, Australia: CSIRO Publications.

Kaya, H. K., & Brayton, M. A. (1978). Interaction between *Neoaplectana carpocapsae* and a granulosis virus of the armyworm *Pseudaletia unipuncta*. *Journal of Nematology, 10*, 350–354.

Kaya, H. K., & Burlando, T. M. (1989). Development of *Steinernema feltiae* (Rhabditida: Steinernematidae) in diseased insect hosts. *Journal of Invertebrate Pathology, 53*, 164–168.

Kaya, H. K., & Gaugler, R. (1993). Entomopathogenic nematodes. *Annual Review of Entomology, 38*, 181–206.

Kaya, H. K., & Koppenhöfer, A. M. (1996). Effects of microbial and other antagonistic organisms and competition on entomopathogenic nematodes. *Biocontrol Science and Technology, 6*, 357–371.

Khatri-Chhetri, H. B., Waeyenberge, L., Manandhar, H. K., & Moens, M. (2010). Natural occurrence and distribution of entomopathogenic nematodes (Steinernematidae and Heterorhabditidae) in Nepal. *Journal of Invertebrate Pathology, 103*(1), 74–78.

Kondo, E. (1989). Studies on the infectivity and propagation of entomogenous nematodes, *Steinernema* spp. (Rhabdititida: Steinernematidae) in the common cutworm, *Spodoptera litura* (Lepidoptera: Noctuidae). *Bulletin of the Faculty of Agriculture. Saga University, 67*, 1–87.

Koppenhöfer, A. M., Baur, M. E., Stock, S. P., Choo, H. Y., Chinnasri, B., & Kaya, H. K. (1997). Survival of entomopathogenic nematodes within host cadavers in dry soil. *Applied Soil Ecology, 6*, 231–240.

Koppenhöfer, A. M., Choo, H. Y., Kaya, H. K., Lee, D. W., & Gelernter, W. D. (1999). Increased field and greenhouse efficacy against scarab grubs with a combination of an entomopathogenic nematode and *Bacillus thuringiensis*. *Biological Control, 14*, 37–44.

Koppenhöfer, A. M., & Kaya, H. K. (1996a). Coexistence of two steinernematid nematode species (Rhabditida: Steinernematidae) in the presence of two host species. *Applied Soil Ecology, 4*, 221–230.

Koppenhöfer, A. M., & Kaya, H. K. (1996b). Coexistence of entomopathogenic nematode species (Steinernematidae and Heterorhabditidae) with different foraging behavior. *Fundamental and Applied Nematology, 19*, 175–183.

Koppenhöfer, A. M., & Kaya, H. K. (1997). Additive and synergistic interaction between entomopathogenic nematodes and *Bacillus thuringiensis* for scarab grub control. *Biological Control, 8*, 131–137.

Koppenhöfer, A. M., Kaya, H. K., Shanmugam, S., & Wood, G. L. (1995). Interspecific competition between steinernematid nematodes within an insect host. *Journal of Invertebrate Pathology, 66*, 99–103.

Koppenhöfer, A. M., Kaya, H. K., & Taormino, S. (1995). Infectivity of entomopathogenic nematodes (Rhabditida: Steinernematidae) at different soil depths and moistures. *Journal of Invertebrate Pathology, 65*, 193–199.

Kung, S. P., Gaugler, R., & Kaya, H. K. (1990a). Soil type and entomopathogenic nematode persistence. *Journal of Invertebrate Pathology, 55*, 401–406.

Kung, S. P., Gaugler, R., & Kaya, H. K. (1990b). Influence of soil pH and oxygen on persistence of *Steinernema* spp. *Journal of Nematology, 22*, 440–445.

Kung, S. P., Gaugler, R., & Kaya, H. K. (1991). Effects of soil temperature, moisture, and relative humidity on entomopathogenic nematode persistence. *Journal of Invertebrate Pathology, 57*, 242–249.

Kunkel, B., & Grewal, P. S. (2003). Endophytic infection in perennial ryegrass reduces the susceptibility of *Agrotis ipsilon* (Lepidoptera) to an entomopathogenic nematode. *Entomologia Experimentalis et Applicata, 107*, 95–104.

Kunkel, B., Grewal, P. S., & Quigley, M. F. (2004). A mechanism of acquired resistance by the black cutworm *Agrotis ipsilon* (Lepidoptera) to an entomopathogenic nematode. *Biological Control, 29*, 100–108.

Lacey, L. A., & Georgis, R. (2012). Entomopathogenic nematodes for control of insect pests above and below ground with comments on commercial production. *Journal of Nematology, 44*(2), 218–225.

Lacey, L. A., Kaya, H. K., & Bettencourt, R. (1995). Dispersal of *Steinernema glaseri* (Nematoda: Steinernematidae) in adult Japanese beetles, *Popillia japonica* (Coleoptera: Scarabaeidae). *Biocontrol Science and Technology, 5*, 121–130.

Lawrence, J. L., Hoy, C. W., & Grewal, P. S. (2006). Spatial and temporal distribution of endemic entomopathogenic nematodes in a heterogeneous vegetable production landscape. *Biological Control, 37*, 247–255.

Letourneau, D. K. (1998). Conservation biological control: Lessons for conserving natural enemies. In P. Barbosa (Ed.), *Conservation biological control* (pp. 9–38). New York: Academic.

Lewis, E. E. (2002). Behavioural ecology. In R. Gaugler (Ed.), *Entomopathogenic nematology* (pp. 205–223). New York: CABI.

Lewis, E. E., Campbell, J. F., Griffin, C., Kaya, H., & Peters, A. (2006). Behavioral ecology of entomopathogenic nematodes. *Biological Control, 38*(1), 66–79.

Lewis, E. E., Campbell, J. F., & Gaugler, R. (1998). A conservation approach to using entomopathogenic nematodes in turf and landscapes. In P. Barbosa (Ed.), *Conservation biological control* (pp. 235–254). New York: Academic.

Lewis, E. E., & Gaugler, R. (1994). Entomopathogenic nematode (Rhabdita: Steinernematidae) sex ratio relates to foraging strategy. *Journal of Invertebrate Pathology, 64*, 238–242.

Lewis, E. E., Gaugler, R., & Harrison, R. (1993). Response of cruiser and ambusher entomopathogenic nematodes (Steinernematidae) to host volatile cues. *Canadian Journal of Zoology, 71*(4), 765–769.

Lewis, E. E., Grewal, P. S., & Sardanelli, S. (2001). Interactions between *Steinernema feltiae–Xenorhabdus bovienii* insect pathogen complex and root–knot nematode *Meloidogyne incognita. Biological Control, 21*, 55–62.

Mason, J. M., & Hominick, W. M. (1995). The effect of temperature on infection, development and reproduction of heterorhabditids. *Journal of Helminthology, 69*, 337–345.

Maynard Smith, J. (1989). *Evolutionary genetics*. New York: Oxford University Press.

McCauley, D. E. (1991). Genetic consequences of local population extinction and recolonization. *Trends in Ecology and Evolution, 6*, 5–8.

McCauley, D. E. (1995). Effects of population dynamics on genetics in mosaic landscapes. In L. Hansson, L. Fahrig, & G. Merriam (Eds.), *Mosaic landscapes and ecological processes* (pp. 178–198). London: Chapman and Hall.

Millar, L. C., & Barbercheck, M. E. (2001). Interaction between endemic and introduced entomopathogenic nematodes in conventional–till and no–till corn. *Biological Control, 22*, 235–245.

Millar, L. C., & Barbercheck, M. E. (2002). Effects of tillage practices on entomopathogenic nematodes in a corn agroecosystem. *Biological Control, 25*, 1–11.

Mráček, Z., Becvár, S., Kindlmann, P., & Jersakova, J. (2005). Habitat preference for entomopathogenic nematodes, their insect hosts and new faunistic records for the Czech Republic. *Biological Control, 34*, 27–37.

Mráček, Z., Becvár, S., & Kindlmann, P. (1999). Survey of entomopathogenic nematodes from the families Steinernematidae and Heterorhabditidae (Nematoda: Rhabditida) in the Czech Republic. *Folia Parasitologica, 46*, 145–148.

Mráček, Z., & Webster, J. M. (1993). Survey of Heterorhabditidae and Steinernematidae (Rhabditida: Nematoda) in western Canada. *Journal of Nematology, 25*, 710–717.

Neher, D., & Barbercheck, M. E. (1999). Diversity and function of soil mesofauna. In W. Collins & C. O. Qualset (Eds.), *The biodiversity of agroecosystems* (pp. 27–47). Boca Raton, FL: CRC Press.

Nguyen, K. B., & Smart, G. C., Jr. (1995). Morphometrics of infective juveniles of *Steinernema* spp. and *Heterorhabditis bacteriophora* (Nematoda: Rhabditida). *Journal of Nematology, 27*, 206–212.

Nickle, W. R. (1984). *Plant and insect nematodes*. New York: Marcel Decker.

O'Callaghan, K. M., Zenner, A. N., Hartley, C. J., & Griffin, C. T. (2014). Interference competition in entomopathogenic nematodes: Male *Steinernema* kill members of their own and other species. *International Journal for Parasitology, 44*(13), 1009–1017.

Pathak, E., El-Borai, F. E., Campos-Herrera, R., Johnson, E. G., Stuart, R. J., Graham, J. H., et al. (2012). Use of real–time PCR to discriminate predatory and saprophagous behavior by nematophagous fungi. *Fungal Biology, 116*, 563–573.

Peters, A. (1996). The natural host range of *Steinernema* and *Heterorhabditis* spp. and their impact on insect populations. *Biological Control Science and Technology, 6*, 389–402.

Peters, A., & Ehlers, R.–. U. (1994). Susceptibility of leather jackets (*Tipula paludosa* and *Tipula oleracea*; Tipulidae; Nematocera) to the entomopathogenic nematode *Steinernema feltiae*. *Journal of Invertebrate Pathology, 63*, 163–171.

Pianka, E. P. (1999). *Evolutionary ecology* (6th ed.). New York: Harper and Row.

Poinar, G. O. (1990). Biology and taxonomy of Steinernematidae and Heterorhabditidae. In R. Gaugler & H. K. Kaya (Eds.), *Entomopathogenic nematodes in biological control* (pp. 23–61). Boca Raton, FL: CRC Press.

Polis, G. A., Anderson, W. B., & Holt, R. D. (1997). Toward an integration of landscape and food web ecology: The dynamics of spatially subsidized food webs. *Annual Review of Ecology and Systematics, 28*, 289–316.

Portillo-Aguilar, C., Villani, M. G., Tauber, M. J., Tauber, C. A., & Nyrop, J. P. (1999). Entomopathogenic nematode (Rhabditida: Heterorhabditidae and Steinernematidae) response to soil texture and bulk density. *Environmental Entomology, 28*, 1021–1035.

Preisser, E. L. (2003). Field evidence for a rapidly cascading underground food web. *Ecology, 84*, 869–874.

Preisser, E. L., Dugaw, C. J., Dennis, B., & Strong, D. R. (2006). Plant facilitation of a belowground predator. *Ecology, 87*, 1116–1123.

Preisser, E. L., & Strong, D. R. (2004). Climate affects predator control of an herbivore outbreak. *American Naturalist, 163*, 754–762.

Půža, V., & Mráček, Z. (2009). Mixed infection of *Galleria mellonella* with two entomopathogenic nematodes (Nematoda: Rhabditida) species: *Steinernema affine* benefits from the presence of *Steinernema kraussei*. *Journal of Invertebrate Pathology, 102*, 40–43.

Půža, V., & Mráček, Z. (2010a). Does scavenging extend the host range of entomopathogenic nematodes (Nematoda: Steinernematidae)? *Journal of Invertebrate Pathology, 104*(1), 1–3.

Půža, V., & Mráček, Z. (2010b). Mechanisms of coexistence of two sympatric entomopathogenic nematodes, *Steinernema affine* and *S. kraussei* (Nematoda: Steinernematidae), in a central European oak woodland soil. *Applied Soil Ecology, 45*, 65–70.

Pye, A. E., & Burman, M. (1981). *Neoaplectana carpocapsae*: Nematode accumulations on chemical and bacterial gradients. *Experimental Parasitology, 51*(1), 13–20.

Qiu, L., & Bedding, R. (1999). The relationship between energy metabolism and survival of the infective juveniles of *Steinernema carpocapsae* under unstressed–aerobic and anaerobic conditions. In I. Glazer, P. Richardson, M. E. Boemare, & F. Coudert (Eds.), *Survival strategies of entomopathogenic nematodes* (pp. 149–156). Brussels, Belgium: European Commission.

Ram, K., Gruner, D. S., McLaughlin, J. P., Preisser, E. L., & Strong, D. R. (2008). Dynamics of a subterranean trophic cascade in space and time. *Journal of Nematology, 40*(2), 85–92.

Ram, K., Preisser, E. L., Gruner, D. S., & Strong, D. R. (2008). Metapopulation dynamics override local limits on long–term parasite persistence. *Ecology, 89*(12), 3290–3297.

Rasmann, S., Kollner, T. G., Degenhardt, J., Hiltpold, I., Toepfer, S., Kuhlmann, U., et al. (2005). Recruitment of entomopathogenic nematodes by insect–damaged maize roots. *Nature, 434*, 732–737.

Reid, A. P., & Hominick, W. M. (1992). Restriction fragment length polymorphisms within the ribosomal DNA repeat unit of British entomopathogenic nematodes (Rhabditida: Steinernematidae). *Parasitology, 105*, 317–323.

Richmond, D., Kunkel, B. A., Somasekhar, N., & Grewal, P. S. (2004). Top–down and bottom–up regulation of herbivores: *Spodoptera frugiperda* turns tables on endophyte–mediated plant defense and virulence of an entomopathogenic nematode. *Ecological Entomology, 29*, 353–360.

Rolston, A. N., Griffin, C. T., & Downes, M. J. (2006). Emergence and dispersal patterns of two isolates of the entomopathogenic nematode *Steinernema feltiae. Journal of Nematology, 38*, 221–228.

Rosa, J. S., & Simões, N. (2004). Evaluation of twenty–eight strains of *Heterorhabditis bacteriophora* isolated in Azores for biocontrol of the armyworm, *Pseudaletia unipuncta* (Lepidoptera: Noctuidae). *Biological Control, 29*(3), 409–417.

Saeb, A. T. M., & Grewal, P. S. (2008). Genetic variation and sub–species structure of entomopathogenic nematode *Heterorhabditis bacteriophora* based on the major sperm protein gene. *International Journal of Nematology, 17*, 187–198.

San-Blas, E., & Gowen, S. R. (2008). Facultative scavenging as a survival strategy of entomopathogenic nematodes. *International Journal for Parasitology, 38*(1), 85–91.

San-Blas, E., Gowen, S. R., & Pembroke, B. (2008). Scavenging or infection? Possible host choosing by entomopathogenic nematodes. *Nematology, 10*(2), 251–259.

Sayre, R. M., & Walter, D. E. (1991). Factors affecting the efficacy of natural enemies of nematodes. *Annual Review of Phytopathology, 29*, 149–166.

Selvan, S., Campbell, J. F., & Gaugler, R. (1993). Density–dependent effects on entomopathogenic nematodes (Heterorhabditidae and Steinernematidae) within an insect host. *Journal of Invertebrate Pathology, 62*, 278–284.

Selvan, S., Gaugler, R., & Grewal, P. S. (1993). Water content and fatty acid composition of infective juvenile entomopathogenic nematodes during storage. *Journal of Parasitology, 79*, 510–516.

Selvan, S., Gaugler, R., & Lewis, E. E. (1993). Biochemical energy reserves of entomopathogenic nematodes. *Journal of Parasitology, 79*, 167–172.

Shamseldean, M. M., & Abd-Elgawad, M. M. (1994). Natural occurrence of insect pathogenic nematodes (Rhabditida: Heterorhabditidae) in Egyptian soils. *Afro–Asian Journal of Nematology, 4*, 151–154.

Shapiro, D. I., Berry, E. C., & Lewis, L. C. (1993). Interactions between nematodes and earthworms: Enhanced dispersal of *Steinernema carpocapsae. Journal of Nematology, 25*, 189–192.

Shapiro, D. I., Glazer, I., & Segal, D. (1997). Genetic improvement of heat tolerance in *Heterorhabditis bacteriophora* through hybridization. *Biological Control, 8*(2), 153–159.

Shapiro, D. I., Lewis, L. C., Obrycki, J. J., & Abbas, M. (1999). Effects of fertilizers on suppression of black cutworm (*Agrotis ipsilon*) damage with *Steinernema carpocapsae. Journal of Nematology, 31*, 690–693.

Shapiro, D. I., McCoy, C. W., Fares, A., Obreza, T., & Dou, H. (2000). Effects of soil type on virulence and persistence of entomopathogenic nematodes in relation to control of *Diaprepes abbreviatus. Environmental Entomology, 29*, 1083–1087.

Shapiro, D. I., Obrycki, J. J., Lewis, L. C., & Jackson, J. J. (1999). Effects of crop residue on the persistence of *Steinernema carpocapsae. Journal of Nematology, 31*, 517–519.

Shapiro-Ilan, D. I., Campbell, J. F., Lewis, E. E., Elkon, J. M., & Kim-Shapiro, D. B. (2009). Directional movement of steinernematid nematodes in response to electrical current. *Journal of Invertebrate Pathology, 100*, 134–137.

Shapiro-Ilan, D. I., Gouge, D. H., & Koppenhöfer, A. M. (2002). Factors affecting commercial success: Case studies in cotton, turf and citrus. In R. Gaugler (Ed.), *Entomopathogenic nematology* (pp. 333–355). Wallingford, UK: CABI.

Shapiro-Ilan, D. I., Lewis, E. E., Campbell, J. F., & Kim-Shapiro, D. B. (2012). Directional movement of entomopathogenic nematodes in response to electrical fields: Effects of species, magnitude of voltage, and infective juvenile age. *Journal of Invertebrate Pathology, 109*, 34–40.

Shapiro-Ilan, D. I., Lewis, E. E., & Schliekelman, P. (2014). Aggregative group behavior in insect parasitic nematode dispersal. *International Journal for Parasitology, 44*(1), 49–54.

Sharpe, W. E., & Drohan, J. R. (1999). *The effects of acid deposition on Pennsylvania's forests*. University Park, PA: Environmental Resources Research Institute.

Sicard, M., Hinsinger, J., Le Brun, N., Pages, S., Boemare, N., & Moulia, C. (2006). Interspecific competition between entomopathogenic nematodes (*Steinernema*) is modified by their bacterial symbionts (*Xenorhabdus*). *BMC Evolutionary Biology, 6*(1), 68.

Small, R. W. (1987). A review of the prey of predatory soil nematodes. *Pedobiologia, 30*, 179–206.

Somasekhar, N., Grewal, P. S., DeNardo, E. A. B., & Stinner, B. R. (2002). Non–target effects of entomopathogenic nematodes on the soil nematode community. *Journal of Applied Ecology, 39*, 735–744.

Somasekhar, N., Grewal, P. S., & Klein, M. G. (2002). Genetic variability in stress tolerance and fitness among natural populations of *Steinernema carpocapsae*. *Biological Control, 23*, 303–310.

Spiridonov, S. E., Moens, M., & Wilson, M. J. (2007). Fine scale spatial distributions of two entomopathogenic nematodes in a grassland soil. *Applied Soil Ecology, 37*, 192–201.

Spiridonov, S. E., & Voronov, D. A. (1995). Small scale distribution of *Steinernema feltiae* juveniles in cultivated soil. In C. T. Griffin, R. L. Gwynn, & J. P. Masson (Eds.), *Ecology and transmission strategies of entomopathogenic nematodes* (pp. 36–41). Luxembourg, Luxembourg: European Commission.

Stinner, B. R., & House, G. J. (1990). Arthropods and other invertebrates in conservation–tillage agriculture. *Annual Review of Entomology, 35*, 299–318.

Stinner, B. R., McCartney, D. A., & Van Doren, D. M., Jr. (1988). Soil and foliage arthropod communities in conventional, reduced and no–tillage corn (maize, *Zea mays* L.) systems: A comparison after 20 years of continuous cropping. *Soil and Tillage Research, 11*, 147–158.

Stirling, G. R. (1991). *Biological control of plant parasitic nematodes*. Wallingford, UK: CABI.

Stock, S. P., Strong, D., & Gardner, S. L. (1996). Identification of *Heterorhabditis* (Nematoda: Heterorhabditidae) from California with a new species isolated from the larvae of the ghost moth *Hepialis californicus* (Lepidoptera: Hepialidae) from the Bodega Bay Natural Reserve. *Fundamental and Applied Nematology, 19*, 585–595.

Strong, D. R. (2002). Populations of entomopathogenic nematodes in foodwebs. In R. Gaugler (Ed.), *Entomopathogenic nematology* (pp. 225–240). Wallingford, UK: CABI.

Strong, D. R., Kaya, H. K., Whipple, A. V., Child, A. L., Kraig, S., Bondonno, M., et al. (1996). Entomopathogenic nematodes: Natural enemies of root–feeding caterpillars on bush lupine. *Oecologia, 108*, 167–173.

Strong, D. R., Maron, J. L., Connors, P. G., Whipple, A., Harrison, S., & Jeffries, R. L. (1995). High mortality, fluctuation in numbers, and heavy subterranean insect herbivory in bush lupine, *Lupinus arboreus*. *Oecologia, 104*, 85–92.

Strong, D. R., Whipple, A. V., Child, A. L., & Dennis, B. (1999). Model selection for a subterranean trophic cascade: Root–feeding caterpillars and entomopathogenic nematodes. *Ecology, 80*, 2750–2761.

Stuart, R. J., Abu Hatab, M., & Gaugler, R. (1998). Sex ratio and the infection process in entomopathogenic nematodes: Are males the colonizing sex? *Journal of Invertebrate Pathology, 72*, 288–295.

Stuart, R. J., Barbercheck, M. E., Grewal, P. S., Taylor, R. A. J., & Hoy, C. W. (2006). Population biology of entomopathogenic nematodes: Concepts, issues, and models. *Biological Control, 38*, 80–102.

Stuart, R. J., El-Borai, F. E., & Duncan, L. W. (2008). From augmentation to conservation of entomopathogenic nematodes: Trophic cascades, habitat manipulation and enhanced biological control of *Diaprepes abbreviatus* root weevils in Florida citrus groves. *Journal of Nematology, 40*, 73–84.

Stuart, R. J., & Gaugler, R. (1994). Patchiness in populations of entomopathogenic nematodes. *Journal of Invertebrate Pathology, 64*, 39–45.
Stuart, R. J., & Gaugler, R. (1996). Genetic adaptation and founder effect in laboratory populations of the entomopathogenic nematode, *Steinernema glaseri. Canadian Journal of Zoology, 74*, 164–170.
Stuart, R. J., Lewis, E. E., & Gaugler, R. (1996). Selection alters the pattern of emergence from the host cadaver in the entomopathogenic nematode, *Steinernema glaseri. Parasitology, 113*, 183–189.
Stuart, R. J., Shapiro-Ilan, D. I., James, R. R., Nguyen, K. B., & McCoy, C. W. (2004). Virulence of new and mixed strains of the entomopathogenic nematode *Steinernema riobrave* to larvae of the citrus root weevil, *Diaprepes abbreviatus. Biological Control, 24*, 199–206.
Sturhan, D. (1999). Prevalence and habitat specificity of entomopathogenic nematodes in Germany. In R. L. Gwynn, P. H. Smits, C. Griffin, R.–. U. Ehlers, N. E. Boemare, & J.–. P. Masson (Eds.), *Entomopathogenic nematodes: Application and persistence of entomopathogenic nematodes* (pp. 123–132). Brussels, Belgium: European Commission.
Sturhan, D., & Lisková, M. (1999). Occurrence and distribution of entomopathogenic nematodes in the Slovak Republic. *Nematology, 1*, 273–277.
Sugar, D. R., Murfin, K. E., Chaston, J. M., Andersen, A. W., Richards, G. R., de Léon, L., et al. (2012). Phenotypic variation and host interactions of *Xenorhabdus bovienii* SS-2004, the entomopathogenic symbiont of *Steinernema jollieti* nematodes. *Environmental Microbiology, 14*(4), 924–939.
Taylor, R. A. J. (1999). Sampling entomopathogenic nematodes and measuring their spatial distribution. In R. L. Gwynn, P. H. Smits, C. Griffin, R.–. U. Ehlers, N. E. Boemare, & J.–. P. Masson (Eds.), *Application and persistence of entomopathogenic nematodes* (pp. 43–60). Brussels, Belgium: European Commission.
Therese, M. O., & Bashey, F. (2012). Natal–host environmental effects on juvenile size, transmission success, and operational sex ratio in the entomopathogenic nematode *Steinernema carpocapsae. The Journal of Parasitology, 98*(6), 1095–1100.
Thurston, G. S., Ni, Y., & Kaya, H. K. (1994). Influence of salinity on survival and infectivity of entomopathogenic nematodes. *Journal of Nematology, 26*, 345–351.
Timper, P., & Kaya, H. K. (1992). Impact of a nematode–parasitic fungus on the effectiveness of entomogenous nematodes. *Journal of Nematology, 24*, 1–8.
Timper, P., Kaya, H. K., & Gaugler, R. (1988). Dispersal of the entomogenous nematode *Steinernema feltiae* (Rhabditida: Steinernematidae) by infected adult insects. *Environmental Entomology, 17*, 546–550.
Timper, P., Kaya, H. K., & Jaffee, B. A. (1991). Survival of entomogenous nematodes in soil infested with the nematode–parasitic fungus *Hirsutella rhossiliensis* (Deuteromycotina: Hyphomycetes). *Biological Control, 1*, 42–50.
Torr, P., Heritage, S., & Wilson, M. J. (2004). Vibrations as a novel signal for host location by parasitic nematodes. *International Journal for Parasitology, 34*(9), 997–999.
Torr, P., Spiridonov, S. E., Heritage, S., & Wilson, M. J. (2007). Habitat associations of two entomopathogenic nematodes: A quantitative study using real-time quantitative polymerase chain reactions. *Journal of Animal Ecology, 76*(2), 238–245.
Townsend, M. L., Johnson, D. T., & Steinkraus, D. C. (1998). Laboratory studies of the interactions of environmental conditions on the susceptibility of green June beetle (Coleoptera: Scarabaeidae) grubs to entomopathogenic nematodes. *Journal of Entomological Science, 33*, 40–48.
Turlings, T., Hiltpold, I., & Rasmann, S. (2012). The importance of root produced volatiles as foraging cues for entomopathogenic nematodes. *Plant and Soil, 358*, 51–60.
Ulug, D., Hazir, S., Kaya, H. K., & Lewis, E. (2014). Natural enemies of natural enemies: The potential top-down impact of predators on entomopathogenic nematode populations. *Ecological Entomology, 39*(4), 462–469.
Vandermeer, J., & Perfecto, I. (1995). *Breakfast of biodiversity: The truth about rain forest destruction*. Oakland, CA: Food First Books.

Wall, D., & Moore, J. C. (1999). Interactions underground: Soil biodiversity, mutualism, and ecosystem processes. *BioScience, 49*, 109–117.
Wallace, H. R. (1971). Abiotic influences in the soil environment. In B. M. Zuckerman, W. F. Mai, & R. A. Rohde (Eds.), *Plant parasitic nematodes* (Vol. 1, pp. 257–280). New York: Academic.
Walter, D. E. (1987a). Trophic behavior of 'mycophagous' microarthropods. *Ecology, 68*, 226–229.
Walter, D. E. (1987b). Life history, trophic behaviour and description of *Gamasellodes vermivorax* n. sp. (Mesostigmatoa: Ascidae), a predator of nematodes and arthropods in semiarid grassland soils. *Canadian Journal of Zoology, 65*, 1689–1695.
Walter, D. E. (1988a). Predation and mycophagy by endostigmatid mites (Acariformes: Prostigmata). *Experimental and Applied Acarology, 4*, 159–166.
Walter, D. E. (1988b). Nematophagy by soil arthropods from the shortgrass steppe, Chihauahuan Desert and Rocky Mountains of the central United States. *Agriculture, Ecosystems & Environment, 24*, 307–316.
Walter, D. E., Hudgens, R. A., & Freckman, D. W. (1986). Consumption of nematodes by fungivorous mites, *Tyrophagus* spp. (Acarina: Astigmata: Acaridae). *Oecologia, 70*, 357–361.
Walter, D. E., Hunt, H. W., & Elliot, E. T. (1987). The influence of prey type on the development and reproduction of some predatory soil mites. *Pedobiologia, 30*, 419–424.
Walter, D. E., Moore, J. C., & Loring, S. J. (1989). *Symphylella* sp. (Symphyla: Scolopendrellidae) predators of arthropods and nematodes in grassland soils. *Pedobiologia, 33*, 113–116.
Wang, H., Jung, Y. H., Son, D., & Choo, H. Y. (2013). High level of genetic diversity among *Steinernema monticolum* in Korea revealed by single–enzyme amplified fragment length polymorphism. *Journal of Invertebrate Pathology, 113*(2), 146–151.
Wang, X., & Grewal, P. S. (2002). Rapid genetic deterioration of environmental tolerance and reproductive potential of an entomopathogenic nematode during laboratory maintenance. *Biological Control, 23*, 71–78.
Wang, Y., Gaugler, R., & Cui, L. (1994). Variation in immune response of *Popillia japonica* and *Acheta domesticus* to *Heterorhabditis bacteriophora* and *Steinernema* species. *Journal of Nematology, 26*, 11–18.
Westerman, P. R. (1998). Penetration of the entomopathogenic nematode *Heterorhabditis* spp. into host insects at 9 and 20 degrees C. *Journal of Invertebrate Pathology, 72*, 197–205.
Wharton, D. A., & Surrey, M. R. (1994). Cold tolerance mechanisms of the infective larvae of the insect parasitic nematode, *Heterorhabditis zealandica* Poinar. *Cryo Letters, 15*, 353–360.
Wiens, J. A., Schooley, R. L., & Weeks, R. D., Jr. (1997). Patchy landscapes and animal movements: Do beetles percolate? *Oikos, 78*, 257–264.
Williams, C. D., Dillon, A. B., Girling, R. D., & Griffin, C. T. (2013). Organic soils promote the efficacy of entomopathogenic nematodes, with different foraging strategies, in the control of a major forest pest: A meta–analysis of field trial data. *Biological Control, 65*(3), 357–364.
Wilson, M. J., Ehlers, R.-U., Wilson, M. J., & Glazer, I. (2012). Entomopathogenic nematode foraging strategies – is *Steinernema carpocapsae* really an ambush forager? *Nematology, 14*, 389–394.
Wilson, M. J., Lewis, E. E., Yoder, F., & Gaugler, R. (2003). Application pattern and persistence of the entomopathogenic nematode *Heterorhabditis bacteriophora. Biological Control, 26*, 180–188.
Windels, C. E. (1997). Altering community balance: Organic amendments, selection pressures, and biological control. In D. A. Andow, D. W. Ragsdale, & R. F. Nyvall (Eds.), *Ecological interactions in biological control* (pp. 282–300). Boulder, CO: Westview Press.
With, K. A., Pavuk, D. M., Worchuck, J. L., Oates, R. K., & Fisher, J. L. (2002). Threshold effects of landscape structure on biological control in agroecosystems. *Ecological Applications, 12*, 52–65.
Wootton, J. T. (1994). The nature and consequences of indirect effects in ecological communities. *Annual Review of Ecology and Systematics, 25*, 443–466.
Yodder, C. A., Grewal, P. S., & Taylor, R. A. J. (2004). Rapid age–related changes in infection behavior of entomopathogenic nematodes. *Journal of Parasitology, 90*, 1229–1234.

Zenner, A. N. R. L., O'Callaghan, K. M., & Griffin, C. T. (2014). Lethal fighting in nematodes is dependent on developmental pathway: Male–male fighting in the entomopathogenic nematode *Steinernema longicaudum*. *PloS One, 9*(2), e89385.

Zervos, S., Johnson, S. C., & Webster, J. M. (1991). Effect of temperature and inoculum size on reproduction and development of *Heterorhabditis heliothidis* and *Steinernema glaseri* (Nematoda: Rhabditoidea) in *Galleria mellonella*. *Canadian Journal of Zoology, 69*, 1261–1264.

Zhou, X., Kaya, H. K., Heungens, K., & Goodrich-Blair, H. (2002). Response of ants to a deterrent factor(s) produced by the symbiotic bacteria of entomopathogenic nematodes. *Applied and Environmental Microbiology, 68*, 6202–6209.

Chapter 5
Trophic Relationships of Entomopathogenic Nematodes in Agricultural Habitats

Edwin E. Lewis, Selcuk Hazir, Amanda Hodson, and Baris Gulcu

5.1 Introduction

5.1.1 What is the Role of Entomopathogenic Nematodes in Soil?

Entomopathogenic nematodes (EPNs) play several roles in the soil ecosystem. While EPNs are generally thought of in the context of reducing the density of pest populations when they are applied, they are also natural components of soil food webs and exert considerable influence on the population dynamics of many players in the system in addition to the intended targets of biological control efforts (Hodson, Siegel, & Lewis, 2012). They are lethal parasites of insects, but not all of the species they infect are the targets for which they are applied. They are also prey and hosts to a variety of other soil organisms. Here, we attempt explain the fate of EPNs after they are applied to soil in the context of the complicated interactions among members of soil food webs.

E.E. Lewis (✉) • A. Hodson
Department of Entomology and Nematology, University of California, Davis, CA, USA
e-mail: eelewis@ucdavis.edu

S. Hazir
Department of Biology, Faculty of Arts and Sciences, Adnan Menderes University, Aydin, Turkey

B. Gulcu
Department of Biology, Faculty of Arts and Sciences, Duzce University, Duzce, Turkey

© Springer International Publishing Switzerland 2015
R. Campos-Herrera (ed.), *Nematode Pathogenesis of Insects and Other Pests*, Sustainability in Plant and Crop Protection 1, DOI 10.1007/978-3-319-18266-7_5

5.1.2 Trophic Cascades and Biological Control

Trophic cascades occur when one trophic level (predators) reduces the density of a lower trophic level (herbivores), which in turn benefits the next lower trophic level (plants). In terms of biological control, we attempt to manipulate trophic cascades in ways that protect a resource of value, like crops. However, a challenge to the success of biological control is that trophic cascades are rarely this simple. Thus, biological control agents are released into functioning food webs that will affect their survival, the likelihood of long–term establishment, and their efficacy. For every beneficial organism released to accomplish biological control, there is a potential host of natural enemies, competitors and alternate prey/host species. Predictability has always been challenging in biological control (Georgis & Gaugler, 1991), and these interactions are, to some extent, why. These interactions have been studied extensively in above–ground systems (see Rosenheim, Kaya, Ehler, Marois, & Jaffee, 2002), but similar work in below–ground systems lags. Our objective is to explore and categorize the interactions between EPNs and other soil organisms and relate these to the challenges of biological control. First, we categorized the types of relationships as "direct" or "indirect". Then, we describe what drives both positive and negative outcomes of these interactions.

5.2 Direct Trophic Relationships

5.2.1 Interactions Between Entomopathogenic Nematodes and Target Insects

All living organisms have evolved mechanisms to overcome adverse conditions. The soil–dwelling EPNs have evolved various strategies to overcome defenses of their perspective hosts. To survive in the soil, EPNs have a free–living, third–stage infective juvenile (IJ) stage that searches for their insect host (Kaya & Gaugler, 1993). Moreover, EPNs have evolved a mutualistic relationship with a group of pathogenic bacteria, *Xenorhabdus* spp. (steinernematids) or *Photorhabdus* spp. (heterorhabditids), where the IJs have the bacterial cells sequestered in their intestine (Boemare, 2002). In turn, insects display a great variety of defenses against invasion and infection by EPNs. It has been estimated that about 90 % of insect species have one or more life stages in the soil, and hence are exposed regularly to EPNs. Yet, not all stages or individuals of insects are susceptible to EPNs or other natural enemies, indicating that they must have developed behavioral, morphological and/or physiological defense mechanisms to overcome these enemies (Castillo, Reynolds, & Eleftherianos, 2011).

The pathogenicity of the EPN–bacterium complex is regulated by the interaction of three biological phenomena: (1) defense reactions of the insect host, (2) the pathogenic properties of the nematode and (3) the pathogenic action of the mutualistic bacteria, respectively (Boemare, Givaudan, Brehelin, & Laumond, 1997). The interaction of EPNs with potential hosts at this level is extremely important in determining EPN host range.

5.2.1.1 Behavioral Interactions Between Entomopathogenic Nematodes and Hosts

EPNs must breach the morphological barriers of insects to reach the hemocoel. They generally penetrate into the host via natural openings (anus, mouth and spiracle) (Poinar, 1990). In some insects, the usual routes of entry may be inaccessible because the mouth may be obstructed by oral filters (wireworms) or may be too narrow (insects with sucking/piercing mouthparts or small insects with chewing mouthparts). The anus may be constricted by muscles or other structures (wireworms), and the spiracles may be covered with septa (wireworms) or sieve plates (scarab grubs) (Forschler & Gardner, 1991) or simply be too narrow for nematode entry (some dipterans and lepidopterans, especially during early instars) (Bastidas, Edgar, & San Blas, 2014). Insects reduce the probability of nematode infection by several actions: they obstruct their anus with a large amount of fecal material (scarab grubs), minimize their CO_2 output or release CO_2 in intermittent bursts that minimize the presence of chemical cues (lepidopterous pupae and scarab grubs), and generate impenetrable cocoons or soil cells before pupation that serve as physical barriers (many lepidopterans, scarabs, and weevils) (Hazir, Kaya, Stock, & Keskin, 2003). Heterorhabditid IJs have an anterior tooth that can be used to assist in penetrating directly through soft, thin cuticle (Bedding & Molyneux, 1983) and steinernematids have also been reported to penetrate through the soft cuticle (Koppenhöfer, Grewal, & Fuzy, 2007; Peters & Ehlers, 1994). Accordingly, a thick or hard cuticle also serves as a physical barrier to EPN infection.

Insects also have behavioral responses to EPN contact that can reduce infection rates. For example, grooming by larval scarabs has been shown to remove IJs from their bodies, but stressed scarab larvae show a reduced grooming resulting in higher EPN mortality (Gaugler, Wang, & Campbell, 1994; Koppenhöfer, Grewal, & Kaya, 2000). Scarab larvae are also known to have evasive behavior by moving away from areas infested with EPN IJs (Gaugler et al., 1994). Worker termites also groom to remove IJs, whereas soldier termites do not have the mouthparts for grooming behavior and are more susceptible to EPN infection (Mankowski, Kaya, Grace, & Sipes, 2005). Moreover, social insects such as termites and ants are known to wall off or remove EPN–infected individuals to avoid or reduce contamination to other individuals in a nest (termites) (Epsky & Capinera, 1988; Yu, Gouge, & Baker, 2006).

When EPN IJs penetrate into the insect's hemocoel, they cause physical damage to various insect tissues and organs, such as the gut, trachea, and fat body; they also release their symbiotic bacteria into the hemolymph, which leads to septicemia (Castillo et al., 2011). Although insects do not have the same type of antibody and antigen response as in mammals, infection and injury to insects can activate their immune system with potent cellular and humoral responses analogous to the mammalian immune system. Upon infection, the physical presence of the nematodes and bacteria, along with any tissue damage, elicits the release of proteins that recognize and bind to surface sugar moieties and trigger further immune responses, e.g. C–type lectins, hemolin, peptidoglycan recognition proteins and β–1,3–glucan recognition proteins (Hinchliffe, Hares, Dowling, & ffrench-Constant, 2010). In response to these signals, the cellular immune responses include phagocytosis by hemocytes and the formation of hemocyte aggregates (nodules) (Hinchliffe et al., 2010; Lavine & Strand, 2002), whereas the humoral immune response secretes a variety of antimicrobial peptides and proteins including lysozymes, cecropins and attacin, and the prophenol–oxidase system (Hoffmann, 2003; Kanost, Jiang, & Yu, 2004; Li, Cowles, Cowles, Gaugler, & Cox-Foster, 2007). Nodules rapidly darken due to the synthesis of insoluble melanin related to the production of reactive oxygen species, causing the bacteria and/or nematodes to become immobile and isolated from nutrients (Cox-Foster & Stehr, 1994; Hinchliffe et al.; Nappi & Vass, 2001). Humoral immune responses involve the induced transcriptional activation of a large number of effector genes that leads to the synthesis of antimicrobial peptides (AMPs), predominantly in the fat body and their secretion into the hemolymph. Insect AMPs are small, cationic, membrane–active molecules that accumulate in the hemolymph, reaching high concentrations in response to infection; they exhibit a broad range of activities against different classes of pathogens (Castillo et al.).

5.2.1.2 Entomopathogenic Nematode Infectivity/Host Immunity

The Role of Nematodes and the Host Response EPNs avoid recognition by the host immune system in one of two ways; immune evasion or immune suppression. Both strategies involve the nematode surface (Politz & Philipp, 1992). Li et al. (2007) and Li, Cowles, Cowles, Gaugler, & Cox-Foster (2009) showed that evasion of the immune reaction of hosts is species–specific. Hemocytes from the same host species reacted to nematodes differently depending on nematode species or populations. For example, *Manduca sexta* L. (Lepidoptera: Sphingidae) can resist infection by *Heterorhabditis bacteriophora* Poinar (Rhabditida: Heterorhabditidae) because the parasite is quickly recognized as "non–self" and hemocytes encapsulate the nematodes (>99 % recognized). In the semi–susceptible *M. sexta* host, *Steinernema glaseri* (Steiner) Wouts, Mráček, Gerdin & Bedding (Rhabditida: Steinernematidae) IJs were not encapsulated as well (28 % recognized). *Steinernema glaseri* and *H. bacteriophora* induced strong melanization in the house cricket (>94 % of both nematode species were recognized) and were killed at 24 h. In suppression response strategies, EPNs interfere, disrupt or manipulate immune defenses. Like parasitic nematodes of mammals, EPNs use surface coat proteins (SCPs) and other surface components to suppress the host's immune response and destroy host hemocytes (Li et al.; Maizels et al., 2004). *Steinernema glaseri* avoids host melanotic encapsulation in larvae of the *Popillia japonica* Newman (Coleoptera: Scarabaeidae) by using the SCPs that not only destroys host hemocytes, but also protects unrelated nematode species from being detected and eliminated (Li et al.). Brivio, Pagani, and Restelli (2002) reported that the cuticle of *Steinernema feltiae* (Filipjev) Wouts, Mráček, Gerdin & Bedding (Rhabditida: Steinernematidae) plays an important role in inactivation of the pro–phenoloxidase pathway, an enzyme involved in the melanization process. *Steinernema feltiae* has also been shown to protect its associated bacteria by sequestering opsonization factors from the insect hemolymph that resulted in reduced phagocytosis by host hemocytes in *Galleria mellonella* L. (Lepidoptera: Pyralidae) (Brivio, Mastore, & Nappi, 2010). The soluble SCPs of *S. glaseri* NC populations were extracted and tested if these could protect *H. bacteriophora* from being melanized and encapsulated in oriental beetles *Exomala orientalis* Waterhouse (Coleoptera: Scarabaeidae) and *M. sexta* larvae. *Steinernema glaseri* NC SCPs suppressed melanization and encapsulation of *H. bacteriophora* in *E. orientalis* but not in *M. sexta*. Melanization of *H. bacteriophora* decreased from 92 % in the untreated check to 1 ± 3 % after treatment with SCPs from *S. glaseri*. The comparison of SCPs from different populations and species of nematodes suggests that the differences in SCPs may be correlated with their ability to infect the insect hosts. For example, 23 % of IJs of *S. glaseri* FL populations remained free from encapsulation in *M. sexta* whereas all were encapsulated in *P. japonica* larvae. However, more than 70 and 40 % of *S. glaseri* NC nematodes were free–moving in *M. sexta* and *P. japonica*, respectively. This differential response indicates that the two populations of the same nematode species elicited different intensities of immune responses in the same species of host. Most *H. bacteriophora* and *S. glaseri* IJs were melanized in the masked chafer, *Cyclocephala borealis* Arrow (Coleoptera: Scarabaeidae) larvae, which may explain why these species of EPNs were unable to reproduce in this insect (Li et al. 2007, 2009).

In some instances, neither evasion nor suppression of the immune system happens. After entering the insect's hemocoel, not all species of EPNs release their symbiotic bacteria immediately. For example, *S. glaseri* IJs release their symbiotic bacteria, *Xenorhabdus poinarii* (Akhurst) Akhurst & Boemare (Enterobacteriales: Enterobacteriaceae), 4–6 h after entry into the host hemocoel, whereas *H. bacteriophora* IJs release *Photorhabdus luminescens* (Thomas & Poinar) Boemare, Akhurst, & Mourant (Enterobacteriales: Enterobacteriaceae) 30 min after entry (Bowen & Ensign, 1998; Wang, Gaugler, & Cui, 1994). Thus, adult house crickets, *Acheta domesticus* L. (Orthoptera: Gryllidae) melanize and encapsulate *H. bacteriophora*, *S. glaseri* and *Steinernema carpocapsae* (Weiser) Wouts, Mráček, Gerdin & Bedding (Rhabditida: Steinernematidae) before their symbionts can be released (Wang et al., 1994), reducing the likelihood of a successful infection. In cases where bacteria release is delayed, EPNs must evade or suppress the host immune responses to ensure the release of their symbionts (Brivio et al., 2002). The nematodes resistant to insect antimicrobial peptides (AMPs) are able to use their cuticular lipids to recruit host hemolymph proteins. They use these hemolymph proteins to coat themselves which allows them to avoid opsonization and encapsulation by insect hemocytes (Castillo et al., 2011). In *G. mellonella*, *H. bacteriophora* and its symbiotic bacterium *P. luminescens* were found to produce a specific extracellular proteinase that is secreted into the insect and inhibits the AMP cecropin (Jarosz 1998). *Heterorhabditis* sp. avoids encapsulation in tipulid larvae by exsheathing from the second–stage cuticle during host penetration (Peters, Gouge, Ehlers, Hague, 1997). Nematodes may also produce paralyzing exotoxins and cytotoxic and proteolytic extracellular enzymes. The above reactions are dependent on the insect host and nematode/bacterium complex (Dowds & Peters, 2002) and contribute to the variable efficacy of EPNs against different insect species. *S. carpocapsae* is effective against webworms but ineffective for mushroom flies yet *S. feltiae* is an excellent match against these flies (Koppenhöfer, 2000).

5.2.1.3 Role of Symbiotic Bacteria in the Infection Process

Successful infection of the insect host is only achieved by overcoming the multitude of anti–microbial defenses that form the insect immune system. The symbiotic bacteria of EPNs (*Photorhabdus* and *Xenorhabdus*) produce several toxins and virulence factors that are expressed upon infection and are capable of disarming insect humoral (degradation of AMPs), cellular (apoptosis of hemocytes) and melanization (inactivation of the phenoloxidase cascade) responses (Forst, Dowds, Boemare, & Stackebrandt, 1997; Owuama, 2001). The humoral response centers on the production of AMPs/proteins. *Xenorhabdus* species specifically inhibit the host expression of AMPs such as lysozyme and cecropins (Dickinson, Russell, & Dunn, 1988).

Bacterial proteases which target the AMPs have also been implicated in evasion of the insect immune response. In order to prevent the melanization response, both *Photorhabdus* and *Xenorhabdus* appear to specifically inhibit phospolipase A2 (PLA2) resulting in inhibition of the eicosanoid pathway which controls hemocyte aggregation and nodulation by activation of the prophenoloxidase cascade (Kim, Ji, Cho, & Park, 2005; Park, Kim, Tunaz, & Stanley, 2004). After supressing the insect immune system, while in the log–phase, *Photorhabdus* and *Xenorhabdus* bacteria produce a large variety of antimicrobial compounds to prevent microbial contamination, mainly from the insect intestinal microflora (Boemare & Akhurst, 2006).

Several small molecule antibiotics by *Photorhabdus* and *Xenorhabdus* spp. are secreted and protect the cadaver from contaminating organisms during its bioconversion. Not only does the cadaver need protecting from bacterial and fungal opportunists, but also from foraging insects. Both *Photorhabdus* and *Xenorhabdus* have been shown to be able to repel scavengers including ants (Baur, Kaya, & Strong, 1998; Zhou, Kaya, Heungens, & Goodrich-Blair, 2002), crickets, cockroaches, springtails, wasps (Gulcu, Hazir, & Kaya, 2012; Ulug, Hazir, Kaya, & Lewis, 2014), predator insects (Foltan & Půža, 2009) and even birds (Fenton, Magoolagan, Kennedy, & Spencer, 2011) by the production of Scavenger Deterrent Factor (SDF) which has been not unidentified chemically.

5.2.2 *Interactions Between Entomopathogenic Nematodes and Non–targets*

5.2.2.1 Other Susceptible Arthropods in the System Are Infected

Entomopathogenic nematodes may infect susceptible arthropods in the ecosystems to which they are applied, although long term negative effects are rare. For example, Bathon (1996) reported that *S. feltiae* and *Heterorhabditis megidis* Poinar, Jackson & Klein (Rhabditida: Heterorhabditidae), when applied separately in a range of agricultural and natural habitats, generally had little impact on non–target arthropods, but some minimal impacts occurred on non–pest species of Coleoptera and Diptera. In a large scale multi–year study in citrus, microarthropods and enchytraeid worm densities were reduced in *Steinernema riobrave* Cabanillas, Poinar, & Raulston (Rhabditida: Steinernematidae) treated mesocosms but populations quickly returned to baseline levels (Duncan et al., 2007). However, these and other studies finding no non target effects on resident arthropod communities often only identified insects to the family or order level (Georgis, Kaya, & Gaugler, 1991) perhaps hiding effects on individual species. Studies that have measured changes in individual species abundance have found that brief reductions can occur in some populations after EPN application. For example, in *H. megidis* treated plots, the densities of one species of Curculionidae (*Barypeithes* spp.) was reduced and in *S. feltiae* treated plots the densities of four non–target species of chrysomelid and carabid beetles were reduced (Buck & Bathon 1993; Koch & Bathon 1993). In large scale applications of *S. carpocapsae* in pistachios, populations of the earwig, *Forficula auricularia* L. (Dermaptera: Forficulidae), and the tenebrionid beetle, *Blapstinus discolor* Horn (Coleoptera: Tenebrionidae), were temporarily reduced (Hodson et al. 2012) and the susceptibility of earwigs confirmed in the laboratory (Hodson, Friedman, Wu, & Lewis, 2011).

5.2.2.2 Predators of Entomopathogenic Nematodes

High losses of IJs occur within the first few days after their application because of biotic (e.g., host availability, plant metabolites, and natural enemies) and abiotic factors (e.g., low moisture, temperature extremes, UV light, dehydration, and soil type) (Baur & Kaya, 2001; Kaya, 2002; Smits, 1996). Once the IJs disperse into the soil away from UV light, temperature extremes, and desiccating conditions, natural enemies such as bacteria, protozoa, nematophagous fungi and invertebrate predators may cause additional mortality (Kaya, 2002). A number of invertebrate predators including collembolans, oligochaetes, tardigrades, turbellarians, predatory nematodes, mites and insects will feed on EPN IJs (Hazir et al., 2003; Kaya, 2002; Small, 1987). Ulug et al. (2014) suggest that predators of EPNs can have significant top–down population regulatory impact on applications of EPNs.

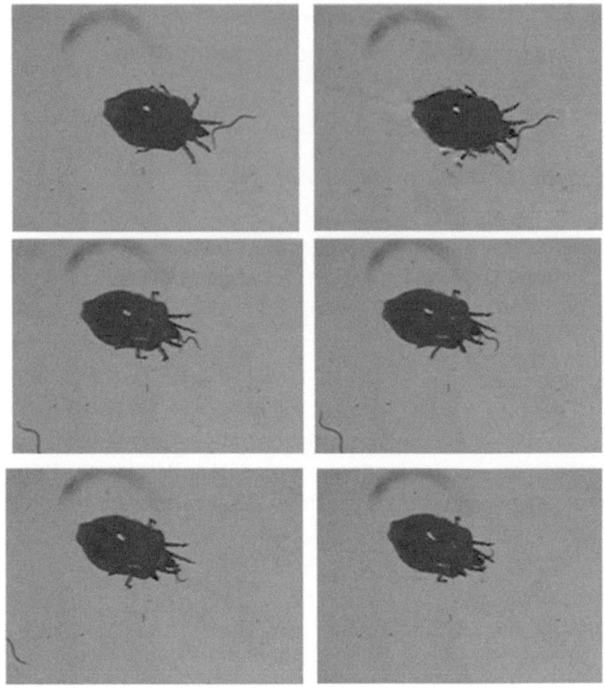

Fig. 5.1 *Predation of a* Steinernema feltiae *infective juvenile by a female* Sancassania polyphyllae. (*Sarcoptiformes*: *Acaridae*)

The most studied EPN predators are nematophagous mites. Poinar (1979) reported that mites in the genus *Macrocheles* preyed on *S. feltiae* IJs. Subsequently, Ishibashi, Young, Nakashima, Abiru, and Haraguchi (1987) found that the mesostigmatid mite, *Eugamasus* sp., fed on *S. carpocapsae* IJs. Thirteen nematophagous mite species from different groups were tested against *S. carpocapsae* IJs and 12 out of 13 mite species fed on this nematode species (Epsky, Walter, & Capinera, 1988). In a field study, Wilson & Gaugler (2004) correlated the decline of applied IJs to the rise in mites and collembolans but did not observe them feeding on the nematodes or identify the arthropod species. Finally, Karagoz, Gulcu, Cakmak, Kaya, and Hazir (2007) demonstrated using a video recorder that *Sancassania* sp. [subsequently identified as *Sancassania polyphyllae* (Zachvatkin) (Acari: Acaridae)] mites consumed *S. feltiae* IJs (Fig. 5.1) and *H. bacteriophora* IJs. In a laboratory study they found that (1) two adult females of *S. polyphyllae* consumed more than 80 of 100 *S. feltiae* IJs on an agar substrate within 24 h, (2) the mites fed more on *S. feltiae* than of *H. bacteriophora* on an agar substrate and (3) they fed on more *S. feltiae* IJs in sandy than in loamy soil. The reason for the greater consumption of *S. feltiae* over *H. bacteriophora* by *S. polyphyllae* is not known, but Karagoz et al. (2007) hypothesized that IJ size, differences in foraging strategies, or retention of the second–stage cuticle by *H. bacteriophora* may be involved. More detailed research with *S. polyphyllae* showed that this mite could potentially interfere with biological pest control by feeding on purposely released EPN IJs, cadavers containing EPNs, or IJs emerging from the cadavers (Cakmak, Ekmen, Karagoz, Hazir, & Kaya, 2010; Cakmak, Karagoz, Ekmen, Hazir, & Kaya, 2011, Ekmen, Cakmak et al. 2010; Ekmen, Hazir et al. 2010; Karagoz et al. 2007). Furthermore, Cakmak et al. showed that *S. polyphyllae* successfully completed its development and reproduced when feeding on *S. feltiae* and *H. bacteriophora* IJs. Ekmen, Hazir et al. (2010) reported that significantly more *S. polyphyllae* gathered near or on nematode–killed larvae of the medfly, *Ceratitis capitata* (Wiedemann) (Diptera: Tephritidae), compared to freeze–killed larvae or bamboo pieces used to mimic medfly larvae and consumed 96 % of IJs that emerged from a cadaver. Ekmen, Hazir et al. (2010) hypothesized that a chemical or an odor from the nematode–killed larvae attracted the mites. Thus, in soil containing a nematode–killed insect, the average number of *S. feltiae* IJs recovered was <30 when mites were present, whereas the average number of IJs recovered was >375 when mites were absent. When the IJs alone were placed at different depths in relation to mites in the soil column for 4 and 10 days, *S. polyphyllae* was not as efficient at finding the IJs when they were separated from each other in the soil. Recently, Cakmak et al. (2013) used a Y–tube olfactometer and showed that *S. polyphyllae* preferred odors from tissues of its phoretic host, *Polyphylla fullo* L. (Coleoptera: Scarabaeidae), followed by *S. feltiae* IJs, *G. mellonella* and *H. bacteriophora* IJs, respectively.

Because many mite species are not specialists on nematodes, perhaps the ability to use alternative foods such as nematodes allows for survival or even population growth during periods of low density populations of more typical prey (McMurtry, 1984). As a result of applying a huge amount of EPNs into the soil environment during field applications, IJs may serve as food for native soil arthropods. Forschler & Gardner (1991) found increases in predatory mites (family Rodararidae) 1–4 weeks after field–application of EPN s and poor persistence of EPNs has been positively correlated with numbers of total mites and collembolans. Hodson et al. (2012) investigated the effects of EPN applications on soil arthropods in two large orchards. They found significantly more isotomid collembolans, predatory anystid mites and gnaphosid spiders where nematodes were applied under trees, indicating either direct predation or indirect trophic effects.

5.3 Indirect Trophic Relationships

5.3.1 Repellent Effect of Bacteria

The genera *Xenorhabdus* and *Photorhabdus* are motile, gram–negative bacteria in the family Enterobacteriaceae which belongs to the γ–subclass of the Protobacteria (Boemare & Akhurst, 2006; Hinchliffe et. al., 2010). During their life cycle, bacteria produce a variety of small molecules and compounds many of which not only protect the cadaver from microbial contaminants but also repel scavenging arthropods or nematodes (Boemare & Akhurst, 2006; Hinchliffe et. al.). These chemicals, such as insecticidal toxins or antibiotics produced by either *Photorhabdus* spp. or *Xenorhabdus* spp., have been categorized in Table 5.1. Data showing antibiotic activity of either *Photorhabdus* or *Xenorhabdus* highlight the importance of the role these materials play in protecting the cadaver from microbial invasion during nematode development and maintaining pure cultures of *Photorhabdus* and *Xenorhabdus* (Boemare & Akhurst, 2006; Derzelle, Duchaud, Kunst, Danchin, & Bertin, 2002; Hinchliffe et. al.).

5.3.1.1 Protecting the Cadaver from Invasion by Microbial Contaminants

In their complex life cycle, *Photorhabdus* and *Xenorhabdus* not only kill the host but also protect the cadaver from competing microorganisms. They secrete different chemicals such as antibacterial, antifungal, nematicidal, insecticidal or repellent. In this section we classify them as antibacterial and antifungal.

Anti–bacterial Activity Six different compounds for *Photorhabdus* and ten compounds for *Xenorhabdus* have been reported as antibacterial (Table 5.1). One class of compounds, bacteriocins, includes bactericidal and antibiotic–like substances (Daw & Falkiner, 1996) that are produced by almost all bacteria (Hawlena, Bashey, & Lively, 2012). Two types of bacteriocins, lumicin (Sharma et al., 2002) and xenorhabdicin (Thaler, Baghdiguian & Boemare, 1995) were identified from *Photorhabdus* and *Xenorhabdus*, respectively. When Sharma et al. isolated four new bacteriocins, they defined them as lumicins by referring to "lum" (related to luminescens) from *Photorhabdus luminescens* subsp. *akhurstii* Fischer-Le Saux, Viallard, Brunel, Normand, & Boemare (Enterobacteriales: Enterobacteriaceae) strain W14. In addition to identifying these compounds, they showed the capability of killing other *Photorhabdus* strains and *Escherichia coli* Escherich (Enterobacteriales: Enterobacteriaceae). They also hypothesized that lumicins remove the intestinal flora of the insects (Sharma et al.). Two bacteriocins have been identified from *Xenorhabdus*, xenorhabdicin (Thaler et. al., 1995) and xenoxin (Singh, 2012). Boemare and Akhurst (2006) mentioned that xenorhabdicin from *Xenorhabdus nematophila* Thomas & Poinar (Enterobacteriales: Enterobacteriaceae) is active against *P. luminescens* and *Xenorhabdus beddingii* Akhurst & Boemare (Enterobacteriales: Enterobacteriaceae) and inhibit the more distantly related *Morganella morganii* Fulton (Enterobacteriales: Enterobacteriaceae) and *Proteus vulgaris* Hauser (Enterobacteriales: Enterobacteriaceae). On the other hand, it has been shown that different strains of *X. nematophila* are unaffected by xenorhabdicin (Boemare & Akhurst, 2006; Boemare, Boyer-Giglio, Thaler, Akhurst, & Brehelin, 1992). Recently, Hawlena et al. (2012) observed bacteriocin–mediated interactions within and between the strains of two bacteria, *Xenorhabdus bovienii* Akhurst & Boemare (Enterobacteriales: Enterobacteriaceae) and *Xenorhabdus koppenhoeferi* Tailliez, Pagès, Ginibre, & Boemare (Enterobacteriales: Enterobacteriaceae) species isolated a few meters apart from each other. They suggested that bacteriocins mediate mostly intraspecific competition.

Table 5.1 *Compounds and activities of* Photorhabdus *and* Xenorhabdus *bacteria*

Bacteria	Antibacterial	Antifungal	Nematicidal	Insecticidal	Repellent	
Photorhabdus	Bacteriocin (lumicin) β-lactam carbapenem Stilbenes 1. Ethylstilbene 2. Epoxy-stilbene 3. Hydroxy-stilbenes (Isopropyl-stilbene) Macrolides (madumycin II) Nucleosides (puromycin) Siderophores (photobactin)	Hydroxy-stilbenes (isopropylstilbene) Ethylstilbene Nucleosides (puromycin)	Hydroxy-stilbenes (isopropyl-stilbene)	Ethylstilbene Toxincomplexis (Tca, Tcb, Tcc, Tcd, PirA/PirB)	Nematode repellent	Insect repellent
					Hydroxy-stilbenes (isopropyl-stilbene)	Anthraquinones (poly-ketides)
Xenorhabdus	Bacteriocin (xenorhabdicin, xenoxin) Benzylideneacetone Bicornutin A, B, C Indoles 1. Indole, 2. indole derivative, 3. nematophin, 4. nematophin derivative Isoflavonoids (genistein) Xenematide Xenocoumacins 1, 2 Xenomin Xenorhabdins I, II, III, IV, V Xenorxide I, II	Bicornutin A, B, C Nematophin Xenocoumacins 1 Xenorhabdins I, II, III, IV, V Xenorxide I, II	Indole	Toxincomplexis (XaxA/XaxB, Tcc) Xenematide Xenorhabdin II Xenortide A, B Xenorxide I, II	Indole	

Stilbenes are common, simple and normally plant–related compounds. They include pharmacologically active chemicals like resveratrol in grape or wine. However *Photorhabdus* spp. is reportedly the only group of bacteria that produce stilbenes (Bode, 2009). Moreover stilbenes are multipotent compounds which have different activities such as antibiotic against gram–positive bacteria, nematicidal effects and they inhibit the phenoloxidase cascade in the insect immune system (Bode, 2009; Lewis & Clarke, 2012). Joyce et al. (2008) also showed their role as a signal molecule essential to nematode development and called 3′,5′–dihydroxy–4–isopropylstilbene. So far, three different stilbenes, ethylstilbene (Hu, Li, Li, Webster, & Chen, 2006), epoxystilbene (Hu et al., 2006), and hydroxy–stilbenes (isopropylstilbene), have shown antibacterial activity.

There are three more antibacterial compounds identified from *Photorhabdus*, macrolides (madumycin II) and nucleosides (puromycin) (Webster, Chen, Hu, & Li, 2002) and siderophores (photobactin) (Ciche, Blackburn, Carney, & Ensign, 2003). Ciche et al. showed that siderophores support *P. luminescens* growth and the reproduction of the nematode symbiont. Two secondary metabolites, carbapenem and antraquinon pigments, have been reported as antibacterial by Boemare and Akhurst, (2006) β–lactam carbapenem has been shown to be effective against *E. coli*, *Klebsiella pneumonia* (Schroeter) Trevisan (Enterobacteriales: Enterobacteriaceae), and *Enterobacter cloacae* (Jordan) Hormaeche & Edwards (Enterobacteriales: Enterobacteriaceae) (Brachmann et al., 2007; Derzelle et. al., 2002).

Different antibacterial compounds have been reported from *Xenorhabdus* spp. One of them benzylideneacetone (trans–4–phenyl–3–buten–2–one), first found by Ji et al. (2004), was extracted from *X. nematophila* and exhibited a significant antibiotic activity against five plant–pathogenic bacterial strains, *Agrobacterium vitis* (Ophel & Kerr) (Rhizobiales: Rhizobiaceae), *Pectobacterium carotovorum* subsp. *Atrosepticum* (Enterobacteriales: Enterobacteriaceae), *P. carotovorum* subsp. *Carotovorum* (Enterobacteriales: Enterobacteriaceae), *Pseudomonas syringae* pv. *tabaci*, Pseudomonadales: Pseudomonadaceae) and *Ralstonia solanacearum* (Smith) (Burkholderiales: Ralstoniaceae). Three bicornutins, (A, B and C), have also been detected (Böszörményi et al., 2009; Fodor et al., 2012; Hinchliffe et. al., 2010). One of them, bicornutin A, was isolated from *Xenorhabdus budapestensis* Lengyel, Lang, Fodor, Szállás, Schumann & Stackenbrandt (Enterobacteriales: Enterobacteriaceae) and shows strong antibiotic activity against the fire blight bacterium *Erwinia amylovora* (Burril) (Enterobacterales: Enterobacteriaceae) (Böszörményi et al.). Indole derivatives (Furgani et al., 2008), xenorhabdins (McInerney et al., 1991), xenorxides (oxidized xenorhabdins) (Li, Hu, Webster, 1998), xenocoumacins (McInerney, Taylor, Lacey, Akhurst, & Gregson, 1991), nematophin (Li, Chen, & Webster, 1997), xenematide (Lang, Kalvelage, Peters, Wiese, & Imhoff, 2008) xenomins, and isoflavonoids (reviewed by Webster et al., 2002) also have antibiotic activity.

Anti–fungal Activity Some compounds produced by *Photorhabdus* and *Xenorhabdus* spp. not only antibacterial activity but also show antifungal effects. Li, Chen, Wu, and Webster (1995) reported some antifungal effects of hydroxy–stilbenes (isopropylstilbene) on some fungal pathogens, *Aspergillus flavus* Link (Eurotiales: Trichocomaceae), *Aspergillus fumigates* Fresenius (Eurotiales: Trichocomaceae), *Botrytis cinerea* Pers (Helotiales: Sclerotiniaceae), *Candida tropicalis* Berkhout (Saccharomycetales: Saccharomycetaceae), and *Cryptococcus neoformans* (San Felice) Vuill (Tremellales: Tremellaceae). It has also reported that trans–stilbene from *Photorhabdus luminescens* subsp. *luminescens* and *Photorhabdus temperate* Fisher-Le Saux, Villard, Brunel, Normand & Boemare (Enterobacterales: Enterobacteraceae) strains have an antifungal effects (Boemare & Akhurst, 2006; Brachmann, Forst, Furgani, Fodor, & Bode, 2006). More over ethylstilbene and nucleoside–puromycin from *Photorhabdus* spp. have been reviewed by Bode (2009) and Webster et al. (2002), respectively.

Xenorhabdus spp. produce a number of different compounds which are also antifungal. One of these antifungal metabolites, bicornutin A, has been identified by Böszörményi et al. (2009), who found that chemicals isolated from *Xenorhabdus budapestensis* and *Xenorhabdus szentirmaii* Lengyel, Lang, Fodor, Szállás, Schumann & Stackenbrandt (Enterobacteriales: Enterobacteriaceae) cultures inhibit colony formation and mycelial growth of *Phytophthora nicotianae* Breda de Haan (Peronosporales: Pythiaceae). More metabolites such as nematophin, xenocaumacin I and xenorhabdin I, II, III, IV, V, nematophin, xenorxides (oxidized xenorhabdins) have been indicated as antifungal molecules by Bode (2009) and Boemare and Akhurst (2006).

5.3.1.2 Protecting the Cadaver from Invading Nematodes

Protection from Other EPNs The symbionts of EPNs produce a narrow spectrum of antimicrobial compounds called bacteriocins. These compounds are important for the EPNs, as it has been hypothesized that the bacteriocins are important for survival of *Steinernema–Xenorhabdus* complexes if there is any co–infection with other *Steinernema* or *Heterorhabditis* nematodes (Morales-Soto, Synder, & Forst, 2009) within the same host. Sicard et al., (2004) showed that when aposymbiotic (axenic) *S. carpocapsae* and non–symbiont *Xenorhabdus* sp., e.g. *Xenorhabdus innexi* Lengyel, Lang, Fodor, Szállás, Schumann & Stackenbrandt (Enterobacterales: Enterobacteraceae) or *X. bovienii*, are co–injected to *G. mellonella* larvae, the non–symbiont bacteria prevent nematode reproduction. But it also has been reported by Sicard et al. (2006) that if *X. nematophila* and non–symbiont *X. bovineii* and *S. carpocapsae* are injected into the *G. mellonella* larvae, *S. carpocapsae* can reproduce because the R–type bacteriocins produced by *X. nematophila* have a negative effect on *X. bovineii*. Sicard, Tabart, Boemare, Thaler & Moulia (2005) observed that *X. innexi* has an antagonistic effect on nematode reproduction when *X. nematophila* and *X. innexi* and *S. carpocapsae* are injected together because *X. innexi* is insensitive to the R–type bacteriocins produced by *X. nematophila*. A small number of *Xenorhabdus* spp. and *Photorhabdus* spp. have been tested for *in vitro* activity of bacteriocins, and Boemare et al. (1992) and Sicard et al. reported that bacteriocins were active against *X. beddingii* and *X. bovienii* (symbiont of *S. feltiae*) and *Photorhabdus luminescens* and were inactive against *X. cabanillasii* and *X. innexi*.

Protection from Other Nematodes Nematicidal effects of the secondary metabolites, stilbene (3,5–dihydroxy–4–isopropylstilbene) and indole, were reported by Hu, Li, & Webster (1999). They observed that stilbene killed almost 100 % of J4 and adults of *Aphelenchoides rhytium* Massey (Tylenchida: Aphelenchoididae), *Bursaphelenchus* spp. and *Caenorhabditis elegans* Maupas (Rhabditida: Rhabditidae) but did not affect J2 of *Meloidogyne incognita* (Kofoid & White) (Tylenchida: Heteroderidae) or IJs of *H. megidis*. On the other hand, they showed that indole has a lethal effect or inhibition on *Bursaphelenchus* spp. (J4 and adults), *M. incognita* (J2) and *Heterorhabditis* spp. (IJ). Also egg hatch of *M. incognita* was negatively affected by stilbene and indole. Moreover stilbene repels IJs of some *Steinernema* species and indole repels IJs of some species of both *Steinernema* and *Heterorhabditis*. Lewis, Campbell, Griffin, Kaya, & Peters (2006) reported that more than one *Steinernema* species can coexist in one insect host, but that *Heterorhabditis* and *Steinernema* cannot develop together. For example *S. carpocapsae* can eliminate the *Heterorhabditis* when the *S. carpocapsae* infection started at least 6 h before the exposure. These observations are consistent with data Hu et al. (1999).

5.3.1.3 Protecting the Cadaver from Scavengers

Omnivores and scavengers have an important role on population dynamics of EPNs. There are several reports about the interaction between the nematode–bacteria complex and scavengers. The first one was reported by Poinar (1979), who observed mites in the genus *Macroheles* eat *S. carpocapsae* IJs. Since then, the mesostigmata mite, *Eugamasus* sp., the tardigrade, *Macrobiotus richtersi* Murray (Parachela: Macrobiotidae), the nematodes, *Clarkus* sp., and *Actinolaimus* sp. (Ishibashi et al., 1987), the collembolan, *Folsomia candida* Willem (Collembola: Isotomidae) and *Sinella caeca* (Schott) (Collembola: Entomobryidae) (Gilmore & Potter, 1993) have all been reported to feed on *S. carpocapsae* IJs.

Baur et al. (1998) and Zhou et al. (2002) documented that different ant species respond to EPN–bacteria infected cadavers. Baur et al. reported that Argentine ants *Linepithema humile* (Mayr) (Hymenoptera: Formicidae) preferred Steinernema–*Xenorhabdus* infected cadavers over *Heterorhabditis–Photorhabdus* infected ones. In addition, several different ant species *Veromessor andrei* (Mayr) (Hymenoptera: Formicidae), *Pheidole vistana* Forel (Hymenoptera: Formicidae), *Formica pacifica* Francoeur (Hymenoptera: Formicidae), and *Monomorium ergatogyna* Wheeler (Hymenoptera: Formicidae) scavenged *Steinernema* killed cadavers (Baur et al.). Zhou et al. also indicated that a factor produced by *Photorhabdus* spp. and *X. nematophila* bacteria deterred the ants *L. humile*, *Lasius alienus* (Foerster) (Hymenoptera: Formicidae), and *Formica subsericea* (Say) (Hymenoptera: Formicidae) from feeding on cadavers. Gulcu et al. (2012) reported similar behavior of the ant, *Lepisiota frauenfeldi* (Mayr) (Hymenoptera: Formicidae). The workers of *L. frauenfeldi* only consumed 1– day old *H. bacteriophora* and *S. feltiae* infected cadavers, but avoided cadavers more than 2 days after infection. In another study, ants would feed on control (5 % sucrose solution) and 1– to 3– day old cultures of *P. luminescens* containing 5 % sucrose equally, but avoided older cultures of *P. luminescens*. Neither of the ants *Tetramorium chefketi* Forel (Hymenoptera: Formicidae) nor *Pheidole pallidula* (Nylander) (Hymenoptera: Formicidae) consumed *H. bacteriophora* killed cadavers not more than 2– day old (Ulug et al., 2014).

Recent studies showed that the chemicals produced by symbionts of EPNs deter different scavengers such as crickets, *Gryllus bimaculatus* De Geer (Orthoptera: Gryllidae) (Gulcu et al., 2012; Ulug et al., 2014), American cockroaches, *Peripleneta americana* L. (Blattodea: Blattidae) (Ulug et. al.), wasps *Vespa orientalis* L. (Hymenoptera: Vespidae) and *Paravespula* sp. and the calliphorid fly *Chrysomya albiceps* Wiedemann (Diptera: Calliphoridae) (Gulcu et al.). Feeding and oviposition behaviours were also tested in these studies. The observations and data about the experiments have been summarized in Table 5.2.

5.3.2 Interactions Among Entomopathogenic Nematode Species

One concern of applying introduced species of EPNs is that they will disrupt native EPN communities and the pest suppression services they provide. For example, when the exotic *S. riobrave* was applied, detection of the endemic species *H. bacteriophora* decreased, with possible effects on long term pest suppression (Millar & Barbercheck, 2001). Indirect interactions between EPN species and their natural enemies is also possible. In laboratory studies using raw soil from citrus orchards, pretreatment with *S. riobrave* decreased subsequent survival of applications of *S. riobrave* or *S. diaprepesi*, although this effect was not evident in air dried soil, leading the authors to suspect that nematode trapping fungi populations increased after applications, creating a hostile environment for nematodes applied later (El Borai, Brentu, & Duncan, 2007).

5.3.2.1 Competing for Hosts

EPNs compete for hosts by repelling other IJs from the cadaver after infection. Since the decision of whether to infect a host is irreversible, IJs use cues to determine host suitability, including cues which indicate whether a host is already infected (Grewal, Lewis, Gauger, & Campbell, 1993; Lewis, Ricci, & Gaugler, 1996). For example, the number of infecting *S. riobrave* IJs declines over time in conspecific infected hosts (Christen, Campbell, Lewis, Shapiro-Ilan, & Ramaswamy, 2007) and

Table 5.2 Response of various insects to Scavenger Deterrent Factor

Groups	Type of experiment groups	*Gryllus bimaculatus* Gulcu et al. (2012), Ulug et al. (2014)	*Periplenata americana* Ulug et al. (2014)	*Paravespula* sp. Gulcu et al. (2012)	*Vespa orientalis* Gulcu et al. (2012)	*Chrysomya albiceps* Gulcu et al. (2012)
Hb-Pl infected *Galleria*	1–5 day-old	1-day-old consumed				
	1–6 day-old			1-day-old consumed		
	1–3 day-old	1-day-old consumed	1-day-old consumed			
Sf-Xb infected *Galleria*	1–5 day old	1-day-old consumed				
	1–3 day-old	1-day-old consumed	1–2-day-old consumed			
aSf infected *Galleria*	1–7 day-old	All consumed				
Sm infected *Galleria*	1–5 day-old	1–4–5-day-old consumed				
Pl infected *Galleria*	1–5 day-old	1-day-old consumed				
Pl supernatant	Soaked meat			Fed on only interior part	Fed on only interior part	Eggs not laid
Ec supernatant or water	Soaked meat			Totally consumed	Totally consumed	Laid eggs

Hb-PI: *Heterorhabditis bacteriophora-Photorhabdus luminescens*
Sf-Xb: *Steinernema feltiae-Xenorhabdus bovienii*
aSf: Axenic *Steinernema feltiae*
Sm: *Serratia marcescens*
PI: *Photorhabdus luminescens*
Ec: *Escherichia coli*

other *Steinernema* spp. are inhibited from entering hosts previously injected with conspecifics 6–9 h earlier (Glazer, 1997). But IJs of some conspecific infections will continue entering hosts past the point at which it is developmentally worthwhile (Lewis et al., 2006), reducing their reproductive potential (Koppenhöfer & Kaya, 1995; Ryder & Griffin, 2002). The decision to invade may depend on the availability of hosts in the environment and in these cases, an overcrowded host may be better than no host at all (Christen et al., 2007).

Once inside the host, other competitive interactions may emerge, where one invading species wins and the other loses (see Půža & Mráček, 2009). One potential explanation for these outcomes is male–male fighting in the species *Steinernema longicaudum* Shen & Wang (Rhabditida: Rhabditidae), which coil around each other, causing injury and frequent death. Observed in drops of insect haemolymph, males from newly colonized hosts (that had passed through an IJ stage) had more aggressive combats, with higher probability of death than those males that developed later in the hosts and did not pass through and IJ stage (Zenner, O'Callaghan, & Griffin 2014). Killing competing males early in an infection may secure access to females and host resources for the male's offspring. In contrast, the males of later generations in a host are often closely related, making fighting less worthwhile (Zenner et al., 2014).

5.3.2.2 Maintaining Reproductive Isolation

Host Range The natural host range for many species of EPNs remains unclear, partly due to their ease of isolation from the environment. Based on laboratory assays, some species infect a broad range of hosts, such as *H. bacteriophora* and *S. carpocapsae* (Poinar 1979), while other species have more narrow ranges, such as the *S. scapterisci*, which is only know to naturally infect mole crickets (Nguyen & Smart, 1990, 1991) or *S. diaprepsci*, isolated from the weevil pest *Diaprepes abbreviates* L. (Coleoptera: Curculionidae) (Nguyen & Duncan, 2002). A factor which complicates measures of host range is the observation that many EPN species can infect a larger variety of hosts than they would naturally when applied in extremely high numbers (de Doucet, Bertolotti, Giayetto, & Miranda, 1999; Samish & Glazer 1992). For any species, host range will be determined partly by the nematode's ability to find, recognize, and infect hosts that are able to support its development (Kaya & Gaugler, 1993; Lewis et al., 2006).

Foraging Strategy Entomopathogenic nematodes use strategies to find hosts that vary in a continuum from ambush to cruise foraging (Campbell & Gaugler, 1997). To ambush prey, some *Steinernema* species nictate, or raise their bodies off the soil surface, so they are better poised to attach to passing insects (Campbell & Gaugler, 1993). Many *Steinernema* are able to jump by forming a loop with their bodies that creates stored energy which, when released, propels them through the air (Campbell & Kaya, 2000). Other species rarely nictate and instead roam through the soil searching for potential hosts. These foraging strategies influence which hosts nematodes infect. For example, ambush predators such as *S. carpocapsae* generally infect more insects on the surface, while cruising predators like *H. bacteriophora* infect insects that live deep in the soil (Campbell & Gaugler, 1993). However, these designations are not static, and the foraging strategies of EPNs can vary with environmental cues (Wilson, Ehlers, & Glazer, 2012). For example, *H. megidis* forage more effectively for hosts in environments with complex root architectures, likely using the roots as pathways to find hosts (Demarta, Hibbard, Bohn, & Hiltpold, 2014). *Steinernema carpocapase* foraging behavior is also enhanced by high organic matter content, leading Kruitbos et al. (2010) to suggest that laboratory experiments using pure sand may have mischaracterized this species as an ambush forager.

5.4 Conclusions and Future Directions

A persistent and challenging problem in using EPNs for biological control is the seeming instability of their spatial structure after application. When applied in a homogenous blanket at rates up to several miles of millons per hectare, EPNs are known to be distributed in patches within days of application (Wilson & Gaugler, 2004). The relationship they have with hosts is partly responsible; they are attracted to hosts which are patchily distributed. Further, once a host is infected, it becomes more attractive than nearby uninfected hosts (Fushing, Zhu, Shapiro-Ilan, Campbell & Lewis, 2008). Abiotic conditions can also contribute to the formation of "nematode deserts" due to soils with high clay content, high salt content or a lack of moisture (Kaspi et al., 2010; see Chap. 4). But perhaps the most difficult to predict impacts on EPN distribution stem from their interactions with natural enemies and competitors. As these relationships are further studied and understanding of their impacts improves, it is important to recognize that none of these interactions happens in a vacuum.

The trophic relationships described have both positive and negative impacts on EPNs in the context of biological control, though the negatives outweigh the positives. The presence of alternate hosts in an agricultural ecosystem could be considered positive, since this may favor the long–term establishment of an EPN application. However, this may also reduce the efficacy of the same application in reducing pest populations. From the negative point of view, there are many members of the soil food web that can impact EPN applications. Arthropod predators and scavengers reduce EPN numbers directly. Pathogens of nematodes infect EPNs. Other EPN species that occur naturally compete for hosts. Even the seemingly simple relationship between the EPN and the target hosts is complicated by the immune response and other defensive strategies employed by many soil insect species. While we have a reasonably good understanding of these interactions on individual bases, the impact of all of these interactions working in concert presents an enormous challenge to developing predictable biological control programs. As we move to the era of "big data" analysis, perhaps understanding complicated systems will not remain insurmountable. Indeed, "ecoinformatics" is a new area of study that attempts to accomplish exactly this task. The parts are available, putting them together remains to be accomplished.

References

Bastidas, B., Edgar, P., & San Blas, E. (2014). Size does matter: The life cycle of *Steinernema* spp. in micro–insect hosts. *Journal of Invertebrate Pathology, 121*, 46–55.

Bathon, H. (1996). Impact of entomopathogenic nematodes on non–target hosts. *Biocontrol Science and Technology, 6*, 421–434.

Baur, M. E. & Kaya, H. K. (2001). Persistence of entomopathogenic nematodes. In Baur, M. E. & Fuxa J. R. (Eds.), *Southern Cooperative Series Bulletin* (26 pp.). http://www.agctr.lsu.edu/s265/Baur.

Baur, M. E., Kaya, H. K., & Strong, D. R. (1998). Foraging ants as scavengers on entomopathogenic nematode–killed insects. *Biological Control* 12, 231–236.

Bedding, R. A., & Molyneux, A. S. (1983). Penetration of insect cuticle by infective juvenile of *Heterorhabditis* spp. (Heterorhabditidae: Nematoda). *Nematologica, 28*, 354–359.

Bode, H. B. (2009). Entomopathogenic bacteria as a source of secondary metabolites. *Current Opinion in Chemical Biology, 13*, 1–7.

Boemare, N. (2002). Biology, taxonomy, and systematics of *Photorhabdus* and *Xenorhabdus*. In R. Gaugler (Ed.), *Entomopathogenic nematology* (pp. 35–56). New York: CABI.

Boemare, N., & Akhurst, R. (2006). The genera *Photorhabdus* and *Xenorhabdus*. In M. Dworkin, S. Falkow, E. Rosenberg, K.-H. Schleifer, & E. Stackebrandt (Eds.), *The prokaryotes 6* (pp. 451–494). New York: Springer Science + Business Media, Inc.

Boemare, N. E., Boyer-Giglio, M.–. H., Thaler, J.–. O., Akhurst, R. J., & Brehelin, M. (1992). Lysogeny and bacteriocinogeny in *Xenorhabdus nematophilus* and other *Xenorhabdus* spp. *Applied and Environmental Microbiology, 58*, 3032–3037.

Boemare, N. E., Givaudan, A., Brehelin, M., & Laumond, C. (1997). Symbiosis and pathogenicity of nematode–bacterium complexes. *Symbiosis, 22*, 21–45.

Böszörményi, E., Ersek, T., Fodor, A., Fodor, A. M., Földes, L. S., Hevesi, M., et al. (2009). Isolation and activity of *Xenorhabdus* antimicrobial compounds against the plant pathogens *Erwinia amylovora* and *Phytophthora nicotianae. Journal of Applied Microbiology, 107*, 746–759.

Bowen, D. J., & Ensign, J. C. (1998). Purification and characterization of a high molecular weight insecticidal protein complex produced by the entomopathogenic bacterium *Photorhabdus luminescens. Applied Environmental Microbiology, 64*, 3029–3035.

Brachmann, A. O., Bode, H. B., Joyce, S. A., Jenke-Kodama, H., Schwär, G., & Clarke, D. J. (2007). A type II polyketide synthase is responsible for anthraquinone biosynthesis in *Photorhabdus luminescens. ChemBioChem, 24*, 1721–1728.

Brachmann, A. O., Forst, S., Furgani, G. M., Fodor, A., & Bode, H. B. (2006). Xenofuranones A and B: Phenylpyruvate dimers from *Xenorhabdus szentirmaii. Journal of Natural Products, 69*, 1830–1832.

Brivio, M. F., Mastore, M., & Nappi, A. J. (2010). A pathogenic parasite interferes with phagocytosis of insect immunocompetent cells. *Developmental and Comparative Immunology, 34*, 991–998.

Brivio, M. F., Pagani, M., & Restelli, S. (2002). Immune suppression of *Galleria mellonella* (Insecta, Lepidoptera) humoral defenses induced by *Steinernema feltiae* (Nematoda: Rhabditida): Involvement of the parasite cuticle. *Experimental Parasitology, 101*, 149–156.

Buck, M., & Bathon, H. (1993). Effects of a field application of entomopathogenic nematodes (*Heterorhabditis* sp.) on the non–target fauna, Part 2: Diptera. *Anzeiger Fur Schadlingskunde Pflanzenschutz Umweltschultz, 66*, 84–88.

Cakmak, I., Ekmen, Z. I., Karagoz, M., Hazir, S., & Kaya, H. K. (2010). Development and reproduction of *Sancassania polyphyllae* (Acari: Acaridae) feeding on entomopathogenic nematodes and tissues of insect larvae. *Pedobiologia, 53*, 235–240.

Cakmak, I., Karagoz, M., Ekmen, Z. I., Hazir, S., & Kaya, H. K. (2011). Life history of *Sancassania polyphyllae* (Acari: Acaridae) feeding on dissected tissues of its phoretic host, *Polyphylla fullo* (Coleoptera: Scarabaeidae): Temperature effects. *Experimental and Applied Acarology, 53*, 41–49.

Cakmak, I., Hazir, S., Ulug, D., & Karagoz, M. (2013). Olfactory response of *Sancassania polyphyllae* (Acari: Acaridae) to its phoretic host larva killed by the entomopathogenic nematode, *Steinernema glaseri* (Rhabditida: Steinernematidae). *Biological Control, 65*, 212–217.

Campbell, J. F., & Gaugler, R. (1993). Nictation behavior and its ecological implications in the host search strategies of entomopathogenic nematodes (Heterorhabditidae and Steinernematodidae). *Behaviour, 126*, 155–169.

Campbell, J. F., & Gaugler, R. (1997). Inter–specific variation in entomopathogenic nematode. Foraging strategy: Dichotomy or variation along a continuum? *Fundamental and Applied Nematology, 20*, 393–398.

Campbell, J. F., & Kaya, H. K. (2000). Influence of insect–associated cues on the jumping behavior of entomopathogenic nematodes (*Steinernema* spp.). *Behaviour, 137*(5), 591–609.

Castillo, J. C., Reynolds, S. E., & Eleftherianos, I. (2011). Insect immune responses to nematode parasites. *Trends in Parasitology, 27*, 537–547.

Christen, J. M., Campbell, J. F., Lewis, E. E., Shapiro-Ilan, D. I., & Ramaswamy, S. B. (2007). Responses of the entomopathogenic nematode, *Steinernema riobrave* to its insect hosts, *Galleria mellonella* and *Tenebrio molitor. Parasitology, 134*, 889–898.

Ciche, T. A., Blackburn, M., Carney, J. R., & Ensign, J. C. (2003). Photobactin: A catechol siderophore produced by *Photorhabdus luminescens*, an Entomopathogen Mutually Associated with *Heterorhabditis bacteriophora* NC1 Nematodes. *Applied Environmental Microbiology, 69*, 4706–4713.

Cox-Foster, D. L., & Stehr, J. E. (1994). Induction and localization of FAD–Glucose Dehydrogenase (GLD) during encapsulation of abiotic implants in *Manduca sexta* larvae. *Journal of Insect Physiology, 40*(3), 235–250.

Daw, M. A., & Falkiner, F. R. (1996). Acteriocins: Nature, function and structure. *Micron, 27*, 467–479.

de Doucet, M. M., Bertolotti, M. A., Giayetto, A. L., & Miranda, M. B. (1999). Host range, specificity, and virulence of *Steinernema feltiae*, *Steinernema rarum*, and *Heterorhabditis bacteriophora* (Steinernematidae and Heterorhabditidae) from Argentina. *Journal of Invertebrate Pathology, 73*, 237–242.

Demarta, L., Hibbard, B. E., Bohn, M. O., & Hiltpold, I. (2014). The role of root architecture in foraging behavior of entomopathogenic nematodes. *Journal of Invertebrate Pathology, 122*, 32–39.

Derzelle, S., Duchaud, E., Kunst, F., Danchin, A., & Bertin, P. (2002). Identification, characterization, and regulation of a cluster of genes involved in carbapenem biosynthesis in *Photorhabdus luminescens. Applied Environmental Microbiology, 68*, 3780–3789.

Dickinson, L., Russell, V. W., & Dunn, P. E. (1988). A family of bacteria–regulated cecropin D–like peptides from *Manduca sexta. Journal of Biological Chemistry, 263*, 19424–19429.

Dowds, B. C. A., & Peters, A. (2002). Virulence mechanisms. In R. Guagler (Ed.), *Entomopathhogenic nematology* (pp. 79–98). Wallingford, U.K.: CABI Publishing.

Duncan, L. W., Graham, J. H., Zellers, J., Bright, D., Dunn, D. C., El-Borai, F. E., et al. (2007). Food web responses to augmenting the entomopathogenic nematodes in bare and animal manure–mulched soil. *Journal of Nematology, 39*, 176–189.

Ekmen, Z. I., Cakmak, I., Karagoz, M., Hazir, S., Ozer, N., & Kaya, H. K. (2010). Food preference of *Sancassania polyphyllae* (Acari: Acaridae): Living entomopathogenic nematodes or insect tissues? *Biocontrol Science and Technology, 20*, 553–566.

Ekmen, Z. I., Hazir, S., Cakmak, I., Ozer, N., Karagoz, M., & Kaya, H. K. (2010). Potential negative effects on biological control by *Sancassania polyphyllae* (Acari: Acaridae) on an entomopathogenic nematode species. *Biological Control, 54*, 166–171.

El-Borai, F. E., Brentu, C. F., & Duncan, L. W. (2007). Augmenting entomopathogenic nematodes in soil from a Florida citrus orchard: Non–target effects of a trophic cascade. *Journal of Nematology, 39*, 203–210.

Epsky, N., & Capinera, J. L. (1988). Efficacy of the entomogenous nematode *Steinernema feltiae* against a subterranean termite, *Reticulitermes tibialis* (Isoptera: Rhinotermitidae). *Journal of Economic Entomology, 81*, 1313–1317.

Epsky, N. D., Walter, D. E., & Capinera, J. L. (1988). Potential role of nematophagous microarthropods as biotic mortality factors of entomogenous nematodes (Rhabditida: Steinernematidae and Heterorhabditidae). *Journal of Economic Entomology, 81*, 821–825.

Fenton, A., Magoolagan, L., Kennedy, Z., & Spencer, K. A. (2011). Parasite–induced warning coloration: A novel form of host manipulation. *Animal Behavior, 81*, 417–422.

Fodor, A., Varga, I., Hevesi, M., Máthé-Fodor, A., Racsko, J., & Hogan, J. A. (2012). Anti–microbial peptides of *Xenorhabdus* origin against multidrug resistant plant pathogens. In V. Bobbarala (Ed.), *A search for antibacterial agents* (pp. 147–195). Rijeka: InTech Press.

Foltan, P., & Půža, V. (2009). To complete their life cycle, pathogenic nematode–bacteria complexes deter scavengers from feeding on their host cadaver. *Behavioural Processes, 80*, 76–79.

Forschler, B. T., & Gardner, W. A. (1991). Parasitism of *Phyllophaga hirticula* (Coleoptera: Scarabaeidae) by *Heterorhabditis heliothidis* and *Steinernema carpocapsae. Journal of Invertebrate Pathology, 58*, 386–407.

Forst, S., Dowds, B., Boemare, N., & Stackebrandt, E. (1997). *Xenorhabdus* and *Photorhabdus* spp.: Bugs that kill bugs. *Annual Review of Microbiology, 51*, 47–72.

Furgani, G., Böszörményi, E., Fodor, A., Fodor, A. M., Forst, S., Hogan, J., et al. (2008). *Xenorhabdus* antibiotics: A comparative analysis and potential utility for controlling mastitis caused by bacteria. *Journal of Applied Microbiology, 104*, 745–758.

Fushing, H., Zhu, L., Shapiro-Ilan, D., Campbell, J., & Lewis, E. (2008). State–space based mass event–history model I: Many decision–making agents with one target. *The Annals of Applied Statistics, 2*(4), 1503–1522.

Gaugler, R., Wang, Y., & Campbell, J. F. (1994). Aggressive and evasive behaviors in *Popillia japonica* (Coleoptera: Scarabaeidae) larvae: Defenses against entomopathogenic nematode attack. *Journal of Invertebrate Pathology, 64*, 193–199.

Georgis, R., & Gaugler, R. (1991). Predictability in biological control using entomopathogenic nematodes. *Journal of Economic Entomology, 84*, 713–720.

Georgis, R., Kaya, H. K., & Gaugler, R. (1991). Effect of Steinernematid and Heterorhabditid nematodes (Rhabditida: Steinternematidae and Heterorhabditidae) on nontarget arthropods. *Environmental Entomology, 20*(3), 815–822.

Gilmore, S. K., & Potter, D. A. (1993). Potential role of collembola as biotic mortality agents for entomopathogenic nematodes. *Pedobiologia, 37*, 30–38.

Glazer, I. (1997). Effects of infected insects on secondary invasion of steinernematid entomopathogenic nematodes. *Parasitology, 114*, 597–604.

Grewal, P. S., Lewis, E. E., Gauger, R., & Campbell, J. F. (1993). Host finding behavior as a predictor of foraging strategy in entomopathogenic nematodes. *Parasitology, 108*(4), 207–215.

Gulcu, B., Hazir, S., & Kaya, H. K. (2012). Scavenger deterrent factor (SDF) from symbiotic bacteria of entomopathogenic nematodes. *Journal of Invertebrate Pathology, 110*, 326–333.

Hawlena, H., Bashey, F., & Lively, C. M. (2012). Bacteriocin–mediated interactions within and between coexisting species. *Ecology and Evolution, 2*(10), 2516–2521.

Hazir, S., Kaya, H. K., Stock, S. P., & Keskin, N. (2003). Entomopathogenic nematodes (Steinernematidae and Heterorhabditidae) for biological control of soil pests. *Turkish Journal of Biology, 27*, 181–202.

Hinchliffe, S. J., Hares, M. C., Dowling, A. J., & Ffrench-Constant, R. H. (2010). Insecticidal toxins from the *Photorhabdus* and *Xenorhabdus* bacteria. *The Open Toxinology Journal, 3*, 101–118.

Hodson, A. K., Friedman, M. L., Wu, L. N., & Lewis, E. E. (2011). European earwig (*Forficula auricularia*) as a novel host for the entomopathogenic nematode *Steinernema carpocapsae. Journal of Invertebrate Pathology, 107*, 60–64.

Hodson, A. K., Siegel, J. P., & Lewis, E. E. (2012). Ecological influence of the entomopathogenic nematode, *Steinernema carpocapsae*, on pistachio orchard soil arthropods. *Pedobiologia, 55*, 51–58.

Hoffmann, J. A. (2003). The immune response of *Drosophila. Nature, 426*, 33–38.

Hu, K., Li, J., & Webster, J. M. (1999). Nematicidal metabolites produced by *Photorhabdus luminescens* (Enterobacteriaceae), bacterial symbiont of entomopathogenic nematodes. *Nematology, 1*(5), 457–469.

Hu, K. J., Li, J. X., Li, B., Webster, J. M., & Chen, G. H. (2006). A novel antimicrobial epoxide isolated from larval *Galleria mellonella* infected by the nematode symbiont, *Photorhabdus luminescens* (Enterobacteriaceae). *Bioorganic and Medical Chemistry, 14*, 4677–4681.

Ishibashi, N., Young, F. Z., Nakashima, M., Abiru, C., & Haraguchi, N. (1987). Effects of application of DD–136 on silkworm, *Bombyx mori* predatory insect, *Agriosphodorus dohrni*, parasitoid, *Trichomalus apanteloctenus*, soil mites and other non–target soil arthropods with brief notes on feeding behaviour and predatory pressure of soil mites, tardigrates and predatory nematodes on DD–136 nematodes. In N. Ishibashi (Ed.). *Recent advances in biological control of insect pests by entomogenous nematodes in Japan* (pp. 158–164). Ministry of Education, Japan, Grant No. 59860005, (In Japanese, English abstract).

Jarosz, J. (1998). Active resistance of entomophagous rhabditid *Heterorhabditis bacteriophora* to insect immunity. *Parasitology, 117*, 201–208.

Ji, D., Yi, Y., Kang, G. H., Choi, Y. H., Kim, P., Baek, N. I., et al. (2004). Identification of an antibacterial compound, benzylideneacetone, from *Xenorhabdus nematophila* against major plant–pathogenic bacteria. *FEMS Microbiol Letters, 239*, 241–248.

Joyce, S. A., Brachmann, A. O., Glazer, I., Lango, L., Schwr, G., Clarke, D. J., et al. (2008). Bacterial biosynthesis of a multipotent stilbene. *Angewandte Chemie International Edition, 47*, 1942–1945.

Kanost, M. R., Jiang, H., & Yu, X.-Q. (2004). Innate immune responses of a lepidopteran insect, *Manduca sexta. Immunological Reviews, 198*(1), 97–105.

Karagoz, M., Gulcu, B., Cakmak, I., Kaya, H. K., & Hazir, S. (2007). Predation of entomopathogenic nematodes by *Sancassania* sp. (Acari: Acaridae). *Experimental and Applied Acarology, 43*, 85–95.

Kaspi, R., Ross, A., Hodson, A., Stevens, G., Kaya, H., & Lewis, E. (2010). Foraging efficacy of the entomopathogenic nematode *Steinernema riobrave* in different soil types from California citrus groves. *Applied Soil Ecology, 45*(3), 243–253.

Kaya, H. K. (2002). Natural enemies and other antagonists. In R. Gaugler (Ed.), *Entomopathogenic nematology* (pp. 189–202). New York: CABI Publishing.

Kaya, H. K., & Gaugler, R. (1993). Entomopathogenic nematodes. *Annual Review of Entomology, 38*, 181–206.

Kim, Y., Ji, D., Cho, S., & Park, Y. (2005). Two groups of entomopathogenic bacteria, *Photorhabdus* and *Xenorhabdus*, share an inhibitory action against phospholipase A2 to induce host immunodepression. *Journal of Invertebrate Pathology, 89*, 258–264.

Koch, U., & Bathon, H. (1993). Auswirkungen des einsatzes entomopathogener nematoden im freiland auf die nicht–zielfauna. I. Teil: Coleoptera. *Anzeiger fur Schadlingskunde. Pflanzenschutz Umweltschutz, 66*, 65–68.

Koppenhöfer, A. M. (2000). Nematodes. In L. A. Lacey & H. K. Kaya (Eds.), *Field manual of techniques in invertebrate pathology* (pp. 249–264). Dordrecht, The Netherlands: Kluwer.

Koppenhöfer, A. M., Grewal, P. S., & Fuzy, E. M. (2007). Differences in penetration routes and establishment rates of four entomopathogenic nematode species into four white grub species. *Journal of Invertebrate Pathology, 94*, 184–195.

Koppenhöfer, A. M., Grewal, P. S., & Kaya, H. K. (2000). Synergism of imidacloprid and entomopathogenic nematodes against white grubs: The mechanism. *Entomologia Experimentalis et Applicata, 94*, 283–293.

Koppenhöfer, A. M., & Kaya, H. K. (1995). Density–dependent effects on *Steinernema glaseri* (Nematoda: Steinernematidae) within an insect host. *Journal of Parasitology, 81*, 797–799.

Kruitbos, L. M., Heritage, S., Hapca, S., & Wilson, M. J. (2010). The influence of habitat quality on the foraging strategies of the entomopathogenic nematodes *Steinernema carpocapsae* and *Heterorhabditis megidis. Parasitology, 137*, 303–309.

Lang, G., Kalvelage, T., Peters, A., Wiese, J., & Imhoff, J. F. (2008). Linear and cyclic peptides from the entomopathogenic bacterium *Xenorhabdus nematophilus. Journal of Natural Products, 71*, 1074–1077.

Lavine, M. D., & Strand, M. R. (2002). Insect haemocytes and their role in immunity. *Insect Biochemistry and Molecular Biology, 32*, 1295–1309.

Lewis, E. E., Campbell, J., Griffin, C., Kaya, K. H., & Peters, A. (2006). Behavioral ecology of entomopathogenic nematodes. *Biological Control, 38*, 66–79.

Lewis, E. E., & Clarke, D. J. (2012). Nematode parasites and entomopathogens. In E. V. Fernando & H. K. Kaya (Eds.), *Insect pathology* (2nd ed., pp. 395–424). London, UK: Elsevier Inc.

Lewis, E. E., Ricci, M., & Gaugler, R. (1996). Host recognition behaviour predicts host suitability in the entomopathogenic nematode *Steinernema carpocapsae* (Rhabditida: Steinernematidae). *Parasitology, 113*, 573–579.

Li, J., Chen, G. H., & Webster, J. M. (1997). Nematophin, a novel antimicrobial substance produced by *Xenorhabdus nematophilus* (Enterobacteriaceae). *Canadian Journal of Microbiology, 43*, 770–773.

Li, J., Hu, K., & Webster, J. M. (1998). Antibiotics from *Xenorhabdus* spp. and *Photorhabdus* spp. (Enterobacteriaceae). *Chemistry of Heterocyclic Compounds, 34*, 1331–1339.

Li, J. X., Chen, G. H., Wu, H. M., & Webster, J. M. (1995). Identification of two pigments and a hydroxystilbene antibiotic from *Photorhabdus luminescens. Applied Environmental Microbiology, 61*, 4329–4333.

Li, X. Y., Cowles, E. A., Cowles, R. S., Gaugler, R., & Cox-Foster, D. L. (2009). Characterization of immunosuppressive surface coat proteins from *Steinernema glaseri* that selectively kill blood cells in susceptible hosts. *Molecular & Biochemical Parasitology, 165*, 162–169.

Li, X. Y., Cowles, R. S., Cowles, E. A., Gaugler, R., & Cox-Foster, D. L. (2007). Relationship between the successful infection by entomopathogenic nematodes and the host immune response. *International Journal for Parasitology, 37*, 365–374.

Maizels, R. M., Balic, A., Gomez-Escobar, N., Nair, M., Taylor, M., & Allen, J. E. (2004). Helminth parasites: Masters of regulation. *Immunological Reviews, 201*, 89–116.

Mankowski, M. E., Kaya, H. K., Grace, J. K., & Sipes, B. S. (2005). Differential susceptibility of subterranean termite castes to entomopathogenic nematodes. *Biocontrol Science and Technology, 15*, 367–377.

McInerney, B. V., Gregson, R. P., Lacey, M. J., Akhurst, R. J., Lyons, G. R., Rhodes, S. H., et al. (1991). Biologically active metabolites from *Xenorhabdus* spp. Part 1: Dithiolopyrrolone derivatives with antibiotic activity. *Journal of Natural Products, 54*, 774–784.

McInerney, B. V., Taylor, W. C., Lacey, M. J., Akhurst, R. J., & Gregson, R. P. (1991). Biologically active metabolites from *Xenorhabdus* spp. Part 2: Benzopyran–1–one derivatives with gastroprotective activity. *Journal of Natural Products, 54*, 785–795.

McMurtry, J. A. (1984). A consideration of the role of predators in the control of acarine pests. In D. A. Griffiths & C. E. Bowman (Eds.), *Acarology VI* (Vol. 1, pp. 109–121). New York: Ellis Horwood Ltd.

Millar, L. C., & Barbercheck, M. E. (2001). Interaction between endemic and introduced entomopathogenic nematodes in conventional–till and no–till corn. *Biological Control, 22*, 235–245.

Morales-Soto, N., Synder, H., & Forst, S. (2009). Interspecies competition in a bacteria–nematode mutualism. In J. F. White Jr. & M. S. Torres (Eds.), *Defensive mutualism in microbial symbiosis* (pp. 117–128). Boca Raton, FL: CRC press, Taylor & Francis Group.

Nappi, A. J., & Vass, E. (2001). Cytotoxic reactions associated with insect immunity. *Advances in Experimental Medicine and Biology, 484*, 329–348.

Nguyen, K. B., & Duncan, L. W. (2002). *Steinernema diaprepesi* n. sp. (Rhabditida: Steinernematidae), a parasite of the citrus root weevil *Diaprepes abbreviatus* (L) (Coleoptera: Curculionidae). *Journal of Nematology, 34*, 159–170.

Nguyen, K. B., & Smart, G. C. (1990). *Steinernema scapterisci* n. sp. (Rhabditida: Steinernematidae). *Journal of Nematology, 22*, 187–199.

Nguyen, K. B., & Smart, G. C. (1991). Pathogenicity of *Steinernema scapterisci* to selected invertebrates. *Journal of Nematology, 23*, 7–11.

Owuama, C. I. (2001). Entomopathogenic symbiotic bacteria, *Xenorhabdus* 687 and *Photorhabdus* of nematodes. *World Journal of Microbiology and Biotechnology, 17*(688), 505–515.

Park, Y., Kim, Y., Tunaz, H., & Stanley, D. (2004). An entomopathogenic bacterium *Xenorhabdus nematophila* inhibits hemocytic phospholipase A2 (PLA2) in tobacco hornworms. *Journal of Invertebrate Pathology, 86*, 65–71.

Peters, A., & Ehlers, R.–. U. (1994). Susceptibility of leatherjackets (*Tipula paludosa* and *T. oleracea*; Tipulidae: Nematocera) to the entomopathogenic nematode *Steinernema feltiae. Journal of Invertebrate Pathology, 63*, 163–171.

Peters, A., Gouge, D. H., Ehlers, R.-U., & Hague, N. G. M. (1997). Avoidance of encapsulation by *Heterorhabditis* spp. infecting laravae of *Tipula oleracea. Journal of Invertebrate Pathology, 70*, 161–164.

Poinar, G. O., Jr. (1979). *Nematodes for biological control of insects*. Boca Raton, FL: CRC Press Inc.

Poinar, G. O. (1990). Biology and taxonomy of Steinernematidae and Heterorhabditidae. In R. Gaugler & H. K. Kaya (Eds.), *Entomopathogenic nematodes in biological control* (pp. 23–61). Boca Raton, FL: CRC Press.

Politz, S. M., & Philipp, M. (1992). *Caenorhabditis elegans* as a model for parasitic nematodes—a focus on the cuticle. *Parasitology Today, 8*, 6–12.

Půža, V., & Mráček, Z. (2009). Mixed infection of *Galleria mellonella* with two entomopathogenic nematodes (Nematoda: Rhabditida) species: *Steinernema affine* benefits from the presence of *Steinernema kraussei. Journal of Invertebrate Pathology, 102*, 40–43.

Rosenheim, J. A., Kaya, H. K., Ehler, L. E., Marois, J. J., & Jaffee, B. A. (2002). Intraguild predation among biological control agents: Theory and evidence. *Biological Control, 5*, 303–335.

Ryder, J. J., & Griffin, C. T. (2002). Density dependent fecundity and infective juvenile production in the entomopathogenic nematode, *Heterorhabditis megidis. Parasitology, 125*, 83–92.

Samish, M., & Glazer, I. (1992). Infectivity of entomopathogenic nematodes (Steinernematidae and Heterorhabditidae) to female ticks of *Boophilus annulatus* (Arachnida: Ixodidae). *Journal of Medical Entomology, 29*, 614–618.

Sharma, S., Waterfield, N., Bowen, D., Thomas, R., Holland, L., James, R., et al. (2002). The lumicins: Novel bacteriocins from *Photorhabdus luminescens* with similarity to the uropathogenic–specific protein (USP) from uropathogenic *Escherichia coli. FEMS Microbiology Letters, 214*, 241–249.

Sicard, M., Ferdy, J. B., Pages, S., Le Brun, N., Godelle, B., Boemare, N., et al. (2004). When mutualists are pathogens: An experimental study of the symbioses between *Steinernema* (Entomopathogenic nematodes) and *Xenorhabdus* (bacteria). *Journal of Evolutionary Biology, 17*, 985–993.

Sicard, M., Hissinger, J., Le Brun, N., Pages, S., Boemare, N., & Moulia, C. (2006). Interspecific competition between entomopathogenic nematodes (*Steinernema*) is modified by their bacterial symbionts (*Xenorhabdus*). *BMC Evolutionary Biolology, 6*, 68.

Sicard, M., Tabart, J., Boemare, N. E., Thaler, O., & Moulia, C. (2005). Effect of phenotypic variation in *Xenorhabdus nematophila* on its mutualistic relationship with the entomopathogenic nematode *Steinernema carpocapsae. Parasitology, 131*, 687–694.

Singh, J. (2012). Structural and functional interferences from a molecular structural model of xenocin toxin from *Xenorhabdus nematophila. American Journal of Bioinformatics Research, 2*(4), 55–60.

Small, R. W. (1987). A review of the prey of predatory soil nematodes. *Pedobiologia, 30*, 179–206.

Smits, H. P. (1996). Post–application persistence of entomopathogenic nematodes. *Biocontrol Science and Technology, 6*, 379–387.

Thaler, J. O., Baghdiguian, S., & Boemare, N. (1995). Purification and characterization of xenorhabdicin, a phage tail–like bacteriocin, from the lysogenic strain F1 of *Xenorhabdus nematophila. Applied Environmental Microbiology, 61*, 2049–2052.

Ulug, D., Hazir, S., Kaya, H. K., & Lewis, E. E. (2014). Natural enemies of natural enemies: The potential top–down impact of predators on entomopathogenic nematode populations. *Ecological Entomology, 39*(4), 462–469.

Wang, Y., Gaugler, R., & Cui, L. (1994). Variations in immune response of *Popillia japonica* and *Acheta domesticus* to *Heterorhabditis bacteriophora* and *Steinernema* species. *Journal of Nematology, 26*, 11–18.

Webster, J. M., Chen, G., Hu, K., & Li, J. (2002). Bacterial metabolites. In R. Gaugler (Ed.), *Entomopathogenic nematology* (pp. 99–114). New York: CABI Publishing.

Wilson, M., & Gaugler, R. (2004). Factors limiting short–term persistence of entomopathogenic nematodes. *Journal of Applied Entomology, 128*(4), 250–253.

Wilson, M. J., Ehlers, R. U., & Glazer, I. (2012). Entomopathogenic nematode foraging strategies – is *Steinernema carpocapsae* an ambush forager? *Nematology, 14*, 389–394.

Yu, H., Gouge, D. H., & Baker, B. (2006). Parasitism of subterranean termites (Isoptera: Rhinotermitidae: Termitidae) by entomopathogenic nematodes (Rhabditida: Steinernematidae; Heterorhabditidae). *Journal of Economic Entomology, 99*, 1112–1119.

Zenner, A. N. R. L., O'Callaghan, K. M., & Griffin, C. T. (2014). Lethal fighting in nematodes is dependent on developmental pathway: Male–male fighting in the entomopathogenic nematode *Steinernema longicaudum. PLoS ONE, 9*(2), e89385.

Zhou, X. S., Kaya, H. K., Heungens, K., & Goodrich-Blair, H. (2002). Response of ants to a deterrent factor(s) produced by the symbiotic bacteria of entomopathogenic nematodes. *Applied and Environmental Microbiology, 68*, 6202–6209.

Chapter 6
Utilizing Persistent Entomopathogenic Nematodes in a Conservation or a More Classical Biological Control Approach

Elson J. Shields

6.1 Introduction

Entomopathogenic nematodes (EPNs) have been the focus of a significant amount of research since the early 1980s due to their many favorable attributes as biological control agents of potential pest insects (Kaya & Gaugler, 1993). As a group, EPNs have a broad host range, variation in foraging strategies and host associations which suggest the potential to control pest species with diverse life histories (e.g. Campbell & Gaugler, 1997; Grewal, Lewis, Gaugler, & Campbell, 1994, Lewis, Gaugler, & Harrison, 1992, 1993; Lewis, Ricci, & Gaugler, 1996; Wilson, Ehlers, & Glazer, 2012). The vast majority of the EPN research has been focused on important components required to utilize EPNs in an inundative release strategy or a biopesticide. These components include mass rearing techniques, isolating/propagating populations with the highest laboratory efficacy with little regard to ecological adaptation to the release environment, species/populations with the longest storage "shelf" life, and application technology. EPNs are used almost exclusively as a biological insecticide, typically at high densities to the soil (2.5×10^9 per ha, 25 IJ/cm^2) with little concern for the fate of applied EPNs. Evaluation of EPN population is usually focused on pest reduction and is limited to a few days or weeks after application (Lewis, Campbell, & Gaugler, 1998). In these releases, little emphasis is placed on the long term establishment of the EPN in the soil profile for long-term pest suppression through pest recycling and the selection of an EPN population which retains the genetic coding for long-term persistence under low host density and unfavorable environmental conditions.

E.J. Shields (✉)
Department of Entomology, Cornell University, Ithaca, NY, USA
e-mail: es28@cornell.edu

© Springer International Publishing Switzerland 2015

R. Campos-Herrera (ed.), *Nematode Pathogenesis of Insects and Other Pests*, Sustainability in Plant and Crop Protection 1, DOI 10.1007/978-3-319-18266-7_6

In contrast, little research has been focused on the utilization of naturally occurring or adapted EPNs in the long-term suppression of pest outbreaks in managed systems, ranging from turf to agricultural fields. EPNs have been isolated from every inhabited continent, in virtually every type of soil habitat where concentrated effort has been made to find them (Adams et al., 2006; Hominick, 2002). Isolation records demonstrate the great diversity of habitats exploited by EPNs (Hara, Gaugler, Kaya, & Lebeck, 1991; Kaya & Gaugler, 1993). Natural populations are extremely common, though poorly understood and range from >1 to 100 % of soil samples collected (e.g. Campos-Herrera et al., et al., 2013; Gaugler, Campbell, Selvan, & Lewis, 1992; Mráček & Becvár, 2000). Native populations of *Steinernema carpocapsae* (Weiser) Wouts, Mráček, Gerdin & Bedding, (Rhabditida: Steinernematidae), *Steinernema feltiae* (Filipjev) Wouts, Mráček, Gerdin & Bedding (Rhabditida: Steinernematidae) and *Heterorhabditis bacteriophora* Poinar (Rhabditida: Heterorhabditidae) have been isolated from numerous samples of NY agricultural soil samples ranging in levels from 1 to 15 % (Shields, unpublished; see also Chap. 11).

The use of EPNs has been primarily focused on inundative releases, yet the origins and future use of EPNs is inoculative and conservation biological control (Lewis et al., 1998). After the discovery of *Steinernema glaseri* (Steiner) Wouts, Mráček, Gerdin & Bedding (Rhabditida: Steinernematidae) from scarab larvae in 1929, small plot field successes and the development of in vitro rearing, 10^9 of IJs were reared and released in an inoculative control program between 1939 and 1942, across New Jersey, focused on Japanese beetle, *Popillia japonica* Newman (Coleoptera: Scarabeidae), an introduced pest (Glaser, 1932, Glaser & Farrell, 1935, McCoy & Glaser, 1936). This program was largely unsuccessful due to the elimination of the bacterial symbionts by the use of antimicrobials in the artificial rearing media and *S. glaseri* was only re-isolated from southern New Jersey (Gaugler et al., 1992).

Long-term multigenerational survival and recycling through hosts is required for inoculative release programs. Kaya (1990) suggests that for an inoculative release program to be successful, three conditions are required: (1) moderately susceptible pests need to be present throughout most of the year, (2) pests should have a high economic threshold level, and (3) soil conditions should be favorable for nematode survival. These suggestions may be more appropriate for the tropical and subtropical regions where temperature conditions are favorable for continuous insect activity and nematode attack/recycling. In the more northern temperate regions, native EPNs must survive long periods of adverse condition including frozen soils and cold temperatures. In these areas, EPNs must have the ability to conserve their resources and pass significant time between host availability (Shields, Testa, Miller, & Flanders, 1999).

In the laboratory under optimum conditions only 30–40 % of the IJs infect the hosts present (Fan & Hominick, 1991) and these results are independent of species and population. Multiple authors have reported that EPNs utilize phased infectivity to bridge periods of time of environmental stress and lack of host availability (Griffin, 2012). The loss of field persistence in some commercial populations

(Ferguson, Schroeder, & Shields, 1995, Shields et al., 1999), supports the idea that these survival mechanisms are genetically encoded and are easily lost under conditions of continuous rearing (Griffin 2012).

Sampling for native EPN populations in both agricultural fields and non-cultivated areas indicates that the population is clumped in distribution and highly variable in frequency (e.g. Cabanillas & Raulston, 1994; Campbell, Lewis, Yoder, & Gaugler, 1995; Campbell, Lewis, Yoder, & Gaugler, 1997). In natural areas, the distribution of the flora with its associated fauna are believed to contribute to the clumped EPN distribution as the nematodes attack and recycle in susceptible hosts feeding on the local flora. The lifecycle of the insects feeding on the local flora and their duration of soil contact along with EPN susceptibility also influences the population and distribution of the native EPN population. In addition, the lack of soil movement and the relative stability of the native areas also contribute to the observed clumped distribution (e.g. Campos-Herrera et al., 2008; Stock & Gress, 2006; Stock, Pryor, & Kaya, 1999). Over time as the flora and susceptible fauna change within a habitat in response to climatic cycles and pest outbreaks, the associated native EPN species, abundance and distribution also might change in response to the availability of susceptible hosts.

When these natural habitats become agricultural production areas, major changes in the historically susceptible fauna usually occurs, directly impacting the abundance and distribution of the native EPN species mix and populations. In the absence of subsequent EPN introductions, the EPN fauna in a geographical area is a historical remnant of the pre-agricultural native ecosystem where the EPN species complex evolved and maintained its population under the pre-agriculture succession of flora and fauna. A shift from natural areas to agricultural fields disrupts the historical relationship between the native EPN population and the availability of the array of their historical hosts. This disruption may force the native EPN to ecological extinction or reduce the population to the observed low levels in most agricultural fields (Campos-Herrera et al., 2008).

The intensity and type of agriculture following the conversion from native habitat has potentially a controlling impact on the native EPN population distribution, and size. For example, the establishment of semi-permanent crops (i.e. orchards, vineyards) with intertwining areas of grassy mixed species would be less detrimental to the native EPN population than a frequently rotated set of monoculture crops. These intertwining grassy mixed species areas are attractive to an array of susceptible EPN hosts, many of which may be native and historically served as susceptible hosts (Campos-Herrera et al., 2008; Duncan et al., 2013).

In the highly disturbed agricultural system, native EPN populations may remain clumped and may be relatively rare in abundance due to a lack of susceptible hosts or an ecological mismatch between the soil inhabiting herbivores and the EPN species present. However, soil tillage redistributes the resident EPN population. This EPN movement by tillage is illustrated in Fig. 6.1. In this experiment, two 10 m × 200 m strips of an established alfalfa field were treated with persistent native EPNs (*S. carpocapsae* 'NY 001' and *S. feltiae* 'NY 04') at a rate of 1.25×10^8 per species per ha (total $= 2.5 \times 10^8$ IJs per ha, 2.5 IJ/cm^2) in 2009. Pre-sampling the field indicated

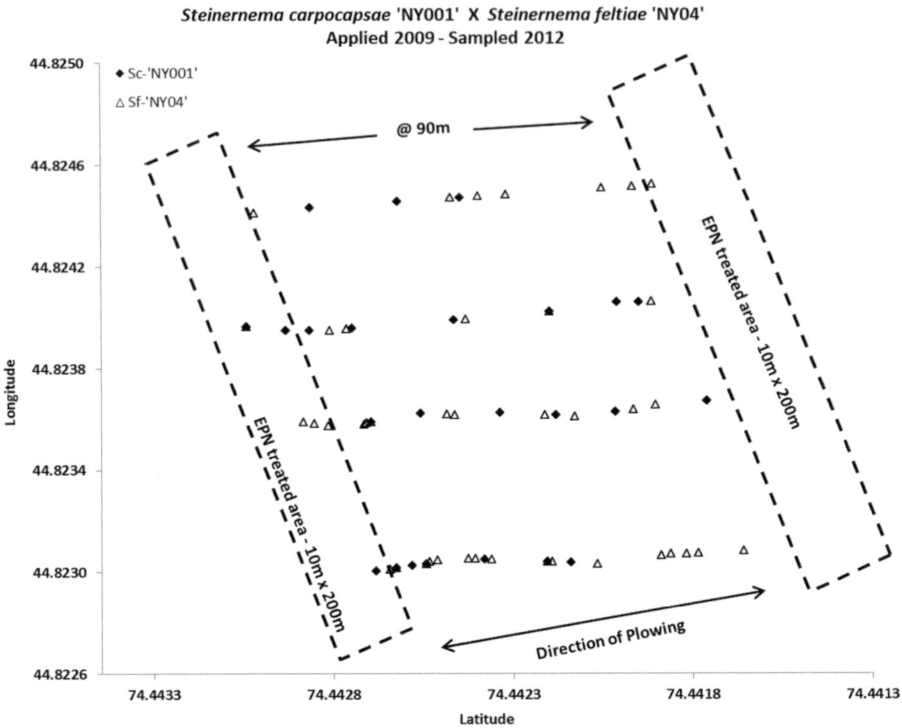

Fig. 6.1 Areas within the treated areas were inoculated with *Steinernema carpocapsae* 'NY 001' and *Steinernema feltiae* 'NY 04' during July 2009 at 1.25×10^8 IJs per ha (1.25 IJ/cm^2) per species (total 2.5×10^8 IJs per ha, 2.5 IJ/cm^2) when the field was an alfalfa/grass mixture. The field was plowed and rotated to corn in 2010–2012 when it was sampled for EPN movement in June 2012. Both EPN species had been moved a minimum of 45 m between applications and subsequent

no detectable natural populations of EPNs. Orientation of the EPN treated strips was perpendicular to the direction of field tillage. The field was rotated to corn in 2010 and remained in corn through 2012. Each year during the corn rotation, the field was aggressively tilled before planting. In 2012, soil cores were collected in four transects between the two treated strips and the GPS coordinates of each soil core site was recorded. Each soil core was divided into the top 5 cm and the lower 15 cm and placed into individual containers. The soil cores were returned to the laboratory and bioassayed for the presence of EPNs using *Galleria mellonella* L. (Lepidoptera: Pyralidae) larvae as bait. After 7 days of incubation at 22 °C, the presence of EPNs and the species of EPN was recorded for each soil core. Figure 6.1 clearly shows that the EPNs applied in the two treated strips in 2009 were redistributed a minimum of 45 m into the untreated area between 2010 and 2012 as a result of three aggressive tillage operations. Similar results were recorded at seven other locations were a similar experimental area was established.

A classic example of this ecological mismatch between the native EPN and potential hosts is the ambush nematode, *S. carpocapsae* and its preference to the top 5 cm of the soil profile. Susceptible insects are exposed to attack until they move below the top 5 cm within the soil profile and significantly reduce the probability of attack (Ferguson et al., 1995). Under some conditions, *S. carpocapsae* utilizes plant roots as physical routeways and conduits to move deeper in the soil in response to feeding-associated stimuli and attack host deeper in the soil (Ennis, Dillon, & Griffin 2010). Given this soil ecological niche, *S. carpocapsae* is usually effective against susceptible adult insects hiding at the duff/soil interface and larvae moving across the soil surface and within the top 5 cm of soil. The cutworm complex (Noctuidae) would be easily attacked while the deeper soil insects like western corn rootworm (WCR) larvae *Diabrotica virgifera virgifera* LeConte (Coleoptera: Chrysomelidae) would be out of the zone of attack by this nematode species. In contrast, if the native nematode species was *S. feltiae*, its soil niche extends down from the surface to 20 cm and the deeper penetrating soil insects would be susceptible to attack until the insect moved below 20 cm (Ferguson et al. 1995).

Annual crop rotation often changes or disrupts the flow of susceptible hosts into fields, limiting the buildup of native EPN populations. In addition, EPN population buildup may be significantly limited by the mismatch of the EPN species and its preferred ecological niche in the soil profile with the movement in the soil by potential insect hosts.

This historical ecological perspective raises the following questions:

1. If native EPN species with the following characteristics were reintroduced into the agricultural system, would the EPN population maintain itself at a sufficient level to provide a significant impact on soil herbivores attacking the array of crops?
   - Persist across unfavorable conditions
   - Efficacious against the array of potential hosts in the agricultural system
   - Overlapped in soil profile preference with the potential host complex.
2. With the level of residual native EPN populations ranging from 1 to 50 % of individual soil cores testing positive for the presence of EPNs (Neumann & Shields, 2004), the question arises about the population level required to maintain EPNs in an agricultural situation where they can persist for long periods of time, yet be numerous enough to respond to insect invasion and reduce the economic losses from those invaders.

The reintroduction of a significant EPN population persisting across years into a cropping system would have at least two distinct benefits. The most obvious benefit would be a reduction in the frequency of economic outbreaks by the herbivores attacking the crop. In the case where herbivore populations build over several generations before the populations become large enough to cause economic damage, the addition of an additional mortality factor (residual EPNs) may have enough impact to prevent or at least delay the herbivore population increase and the subsequent economic damage. A perfect example would be in fields with rotations from alfalfa/grass to a row crop (corn, soybeans). Native white grubs (*Phyllophaga sp.*) and wire worms (Elateridae) with multi-year lifecycles are attracted to the grass within the alfalfa fields and the populations continue to build over the 4+ years while the field is in alfalfa/grass. When the field is rotated to a row crop, the white grubs/wireworms in the soil then begin attacking the corn or soybean plants, requiring an insecticide seed treatment to prevent economic losses. If native persistent EPNs with a level of efficacy and overlapping soil profile preferences were reintroduced into the agricultural field, the EPNs would actively reduce the white grub/wireworm populations during the years in alfalfa/grass and thereby reduce or eliminate problems when the field was rotated to a row crop.

Four examples of this EPN interaction with a crop rotation are illustrated in Fig. 6.2. For this study, a subset of 51 fields was selected from 87 fields treated with EPNs. In each of the 51 fields, the field was inoculated a single time with a combination of EPNs species (*S. carpocapsae* 'NY 001' and *S. feltiae* 'NY 04') at a rate of 1.25×10^8 per species per ha (total $= 2.5 \times 10^8$ IJs per ha, 2.5 IJ/cm^2) in 2008 or 2009. Each field was sampled each year a single time during the growing season at approximately the same time of year by collecting 100 soil cores in a transect across the treated area of each field. The location of each soil core was recorded using GPS. Each soil core was divided into the top 5 cm and the lower 15 cm and placed into individual containers. The soil cores were returned to the laboratory and bioassayed for the presence of EPNs using *Galleria* larvae as bait. After 7 days of incubation at 22 °C, the presence of EPNs and the species of EPN was recorded for each soil core. The four fields illustrated in Fig. 6.2 are representative of EPN response across all the fields rotated to corn during the 6 years of the study. These data suggests that the persistent EPN field population is very dynamic with its response to insect invasion and residual EPN populations in the 10–15 % range (positive soil cores) is capable of a significant response to host invasion, even when the field is rotated to a different crop with a completely different array of insect herbivores.

The second but less obvious benefit would be the introduction of a second mortality factor for soil insects attacking crops with Plant Incorporated Protectants (PIP) (i.e. BT-corn). Herbivores attacking plants with incorporated toxins often experience a longer larval stage, increasing their vulnerability to attack by other mortality factors. A second mortality factor working with the PIP could have a significant impact on reducing the risk and speed of resistance development to the PIP (Pertzold-Maxwell, Jaronski, & Gassmann, 2012). The EPN response in Fig. 6.2 during the corn rotation years was probably due to EPN recycling in corn rootworm (*D. virgifera*). However, is it realistic to expect native EPN populations to persist at a high enough level in our intensive agricultural systems to have a significant impact on invading insects with part of their lifecycle at the soil interface or within the soil profile? In the case of a highly migratory and invasive insect such as common army worm, *Pseudaletia unipuncta* (Haworth) (Lepidoptera: Noctuidae), residual EPNs would have minimal impact because of the large number of invaders and the relatively slow EPN recycling time.

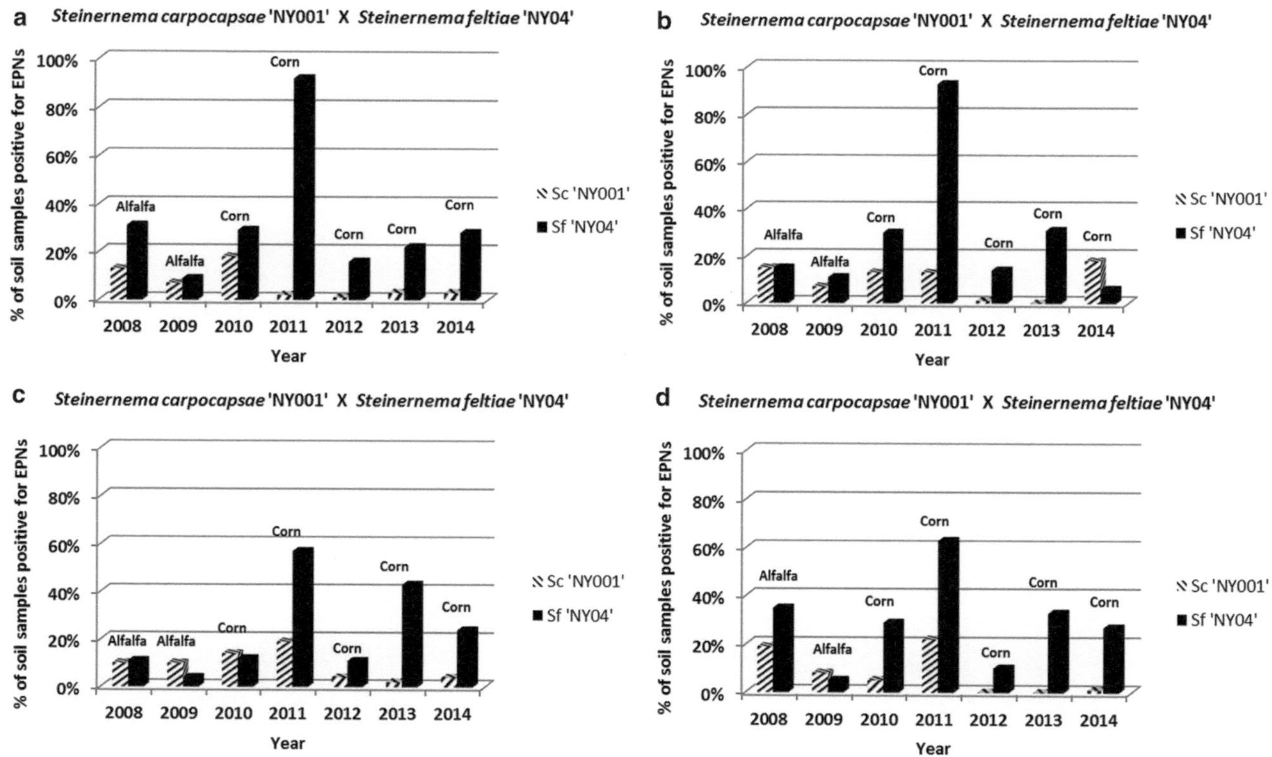

Fig. 6.2 Four different fields in Northern NY where native EPNs were applied once in 2008, inoculated with *Steinernema carpocapsae* 'NY 001' and *Steinernema feltiae* 'NY 04' during July 2009 at 1.25×10^8 IJs per ha (1.25 IJ/cm^2) per species (total 2.5×10^8 IJs per ha, 2.5 IJ/cm^2) when the field was an alfalfa/grass mixture. Entomopathogenic nematodes (EPNs) population frequency was measured once per year during the growing season (200 samples/field/year). Each column represents percent of the 200 samples positive for EPNs in a *Galleria*-based laboratory bioassay. EPN recycling on western corn rootworm (WCR), *Diabrotica virgifera virgifera* larvae during the corn years is inferred based on insect biology but not directly measured.

6.2 Matching Entomopathogenic Nematodes to the Cropping System and Pests

In our own research using native persistent EPNs in the alfalfa/grass (4 years) rotated to corn (4 years) system, fields inoculated once at a rate of approximately 1.25×10^8 per species per ha (total $= 2.5 \times 10^8$ IJs per ha, 2.5 IJ/cm^2) had stable EPN populations become established and remain in the field for multiple years across a corn rotation. EPN numbers maintained themselves at 10–25 % of the soil cores positive for EPNs and responded to insect invasion with the percentage of soil cores testing positive for EPNs increasing from 40 to 80 % (Figs. 6.2 and 6.3).

Since native EPN populations in agricultural soils are a remnant of the pre agricultural ecosystem and agriculture is usually comprised of non-native plants with its associated non-native insect pests, an ecological or efficacy mismatch between the native EPNs and the available non-native hosts frequently exists. This mismatch prevents the residual native EPN population from suppressing the non-native pest insect complex below damaging levels. This mismatch is illustrated by the examples presented below.

Alfalfa snout beetle (ASB), *Otiorhynchus ligustici* (L.) (Coleoptera: Curculionidae) larvae move through the top 45–60 cm of soil during their root feeding stage. Within the infested regions of New York, *S. carpocapsae* can frequently be found in the alfalfa fields infested with snout beetle and at relatively high numbers. In spite of the high *S. carpocapsae* numbers, enough ASB larvae survive to completely destroy the alfalfa stand within 1–2 growing seasons. In regions of Hungary where ASB is native and at very low population levels, native populations of *S. carpocapsae* are found to co-exist with native populations of *S. feltiae*. In these Hungarian alfalfa fields, ASB is a non-economic pest, alfalfa roots show few ASB feeding scars and alfalfa stands last for more than 4 years (Neumann & Shields, 2004). Within the New York ASB infested area, fields inoculated with a combination of native NY *S. carpocapsae* 'NY001' with native NY *S. feltiae* 'NY 04' demonstrate the same impact on ASB larval populations as observed in Hungary (Neumann & Shields, 2004). The explanation to this example is suggested in Ferguson et al. (1995) when the authors show the different EPN species have different preferred, sometimes overlapping soil niches. Interactions between nematode species when coexisting was demonstrated in the laboratory using sand columns where *S. carpocapsae* dominated the upper 5 cm, pushing both *S. feltiae* and *H. bacteriophora* to the lower portions of the sand column (Neumann & Shields, 2006).

A second insect which illustrates this potential mismatch with native nematodes is *D. v. virgifera* (WCR) which feeds on corn roots in the upper 20–30 cm of the soil profile. When only *S. carpocapsae* is present, this nematode usually restricts itself to the top 15 cm and often to the top 1–2 cm (Georgis & Poinar, 1983; Moyle & Kaya, 1981; Schroeder & Beavers, 1987) when surface applied. A significant amount of CRW larval feeding occurs under the "carpocapsae layer" and a significant number of the larvae are not attacked when below the "carpocapsae layer". Although, some authors have reported *S. carpocapsae* utilizing root channels to move deeper in the soil following host cues (Ennis et al., 2010). However, adults laying eggs on the soil surface and within the upper few cm of soil and newly hatched larvae would be vulnerable and attacked by *S. carpocapsae* when present. The addition of a second nematode species like *S. feltiae* which ranges deeper in the soil profile adds additional mortality pressure on CRW larvae deeper in the soil as shown by Fig. 6.2 and indicated by the increase in the frequency of *S. feltiae* compared to *S. carpocapsae*.

Our research suggests that an effective EPN biocontrol program works best with a two species EPN mix to cover the typical profile inhabited by most potential hosts. Depending on the cropping system and array of potential insect hosts, any two species combination of *S. carpocapsae*, *S. feltiae* or *H. bacteriophora* persist well together and do not drive the other species to extinction. When all three species are applied together, *S. feltiae* is squeezed between *S. carpocapsae* in the upper layer and *H. bacteriophora* and forced to extinction (Newmann & Shields 2006, 2008, 2011). When *S. feltiae* or *H. bacteriophora* is matched with *S. carpocapsae*, the second species occupies the area of the soil profile below the top 5 cm. Interestingly, when *S. feltiae* is matched with *H. bacteriophora*, *S. feltiae* is moved to the upper layers of the soil profile whereas *H. bacteriophora* occupies the soil profile below 10 cm (Neumann & Shields, 2006, 2008, 2011).

6.3 Techniques for Preserving Persistence Genetics in Culture

Organisms reared in the laboratory adapt to those laboratory conditions and lose traits which help them survive under field conditions. This laboratory adaptation also occurs in EPNs when cultured continuously in the laboratory. As reported earlier, Fan and Hominick (1991) found that even under optimum laboratory conditions only 30–40 % of the IJs infect the host. In many cases, the continuous culture of EPNs in the laboratory or commercial production facility utilizes only this 30–40 % of the population, the IJs immediately infective upon emergence from the host cadaver in their rearing. Continuous culture utilizing only these immediately infective IJs narrows the gene pool to those individuals who do not carry the coding for any delayed infectivity and its associated benefit of the ability to persist across unfavorable conditions. This loss of field persistence is reported between a commercial and native population of *H. bacteriophora* (Ferguson et al., 1995, Shields et al., 1999) and can be demonstrated in the laboratory in less than 15 generations in native populations of *S. carpocapsae* 'NY 001' and *H. bacteriophora* 'Oswego' (Shields, unpublished).

Since the inoculative approach requires the EPNs used to retain their adaptation for field survival, several approaches can be utilized to help to retain these genes in laboratory culture.

6.3.1 Re-isolation from the Field

We re-isolate the native NY populations from a number of areas in Northern NY every 2 years and these isolations are used to completely restart our cultures. We have been successful in retaining a high level of field adaptation and persistence in our cultures across all three native NY species that we have worked with over the past two decades by restarting our cultures every 2 years. We have noticed that the initial few cycles of these new cultures in the laboratory produce reduced yields of IJs until some adaptation to laboratory conditions and rearing host are made. Mixing of new genetic material from re-isolated individuals with ongoing laboratory cultures only delays the loss of persistence.

6.3.2 Establishing "Wild Populations" in Easily Accessible Areas

A second strategy we currently use as a backup plan to maintain EPN populations adapted to NY conditions is to utilize areas around the Cornell University campus as "natural storage areas". Every university and many commercial buildings are surrounded by ornamental plants and significant grassy areas, which are invaded by insect hosts. We have selected individual areas which are isolated and inoculated these areas with individual populations of NY native species of EPN. Subsequent sampling has documented persistence of the inoculated populations/species in these "wild" areas.

6.3.3 Laboratory Culturing to Retain Persistence

A third strategy we have recently initiated is to change our laboratory culturing methodology to help preserve the persistence genetics. We inoculate 200 g of moistened autoclaved loam soil placed in a plastic container with a fitting lid with 1,000 IJs of a species/population and store it at room temperature (20–22 °C) on a dark shelf in a cabinet. A new container is established each month in a similar manner until containers representing at least a 12 month period are available. To maintain the laboratory culture, 1–2 wax moth larvae are placed in each container spanning the entire time period (12–24 months). After death, cadavers are removed from each container and grouped together for IJ emergence, creating a mixture of IJs from across the stored soil time period. Soil containers are discarded once there are no more deaths from EPNs, usually 12–18 months after initial inoculation. Since a new soil container is added each month, there is a continuous array of "aging" soil samples with IJs becoming infective at any point in time. This technique has been successful in retaining the delayed infectivity trait but does little to prevent adaptation to rearing under moderate temperatures.

6.4 Entomopathogenic Nematode Application Timing and Rate in the Inoculative Approach in NY State Program

Since the focus of the inoculative approach is to introduce or reintroduce persistent adapted EPNs into the agricultural system, the timing of the application is not very critical. Most native EPN species which have not lost their ability to persist in culture will maintain their presence in the soil profile for a minimum of a year without hosts to recycle. Figure 6.3 illustrates the results of a recent laboratory persistent study where two species of native NY EPNs retaining their ability to persist were inoculated into small cups containing moistened soil and stored on a darkened shelf at room temperatures without host available. After >300 days in storage, there is

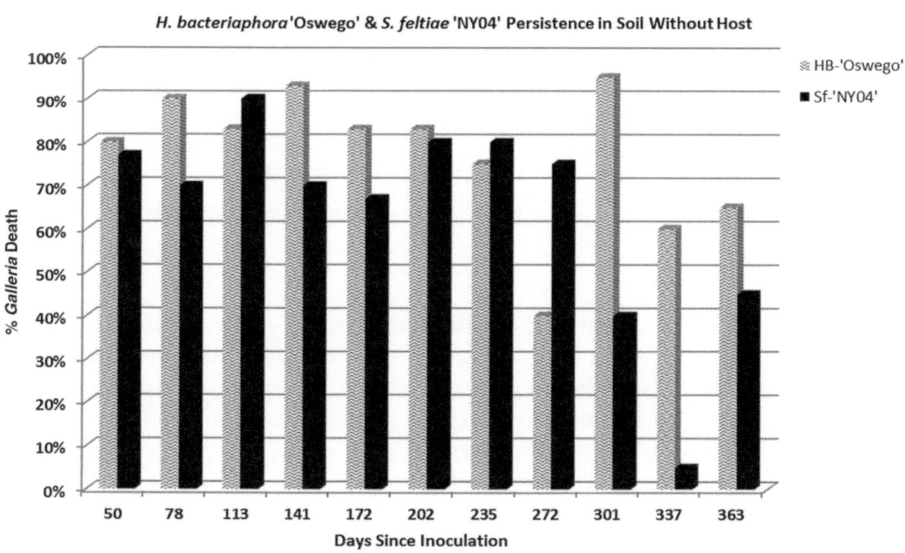

Fig. 6.3 Persistence of two New York native nematode populations at laboratory temperature without hosts. Samples were removed from storage at room temperature (22–24 °C) on the date tested, bait larvae were introduced and mortality was recorded 7 days later. The test samples were then discarded. A new set of samples were then evaluated on the next test date. There was no recycling in the sample cups. No error bars are shown because there is no variation between reps. Decline over time is suggested to be reduction in viable infective juveniles (IJs) and/or a reduction in IJs activated to be infective.

little difference in EPN infectivity between soil samples tested after 30 days or 300 days. Field applications also support this conclusion. Native NY EPNs in our research have been applied late fall, early spring and throughout the summer with no significant differences in establishment. This application strategy is a conceptual difference to EPN inundative releases of commercial nematode populations which have poor long-term persistence in the soil and require accurate timing of releases to match susceptible host presence.

The required EPN rates for the inoculative approach with persistent EPNs is significantly lower than required for the inundative approach with non-persisting EPNs. With the latter approach, a high enough dose must be applied to effectively kill the majority of the host population within a short window while the applied EPNs remain viable. Typically, an inundative application consists of 2.5×10^9 IJs per ha (25 IJ/cm^2). In contrast, the inoculative approach with persistent EPNs relies on successful colonizing on host present in the soil, recycling in those hosts and persisting in the soil profile between periods of host availability. As illustrated in the previous section, EPNs populations, retaining the genetics for persistence, remain infective for more than 300 days at room temperature without host to recycle in. In the field where soil temperatures fall below EPN activity thresholds, the cold temperatures assist the adapted EPN with retaining its infectivity across periods of host unavailability. Our research has indicated that effective inoculation of the soil can be accomplished with 2.5×10^8 IJs per ha (2.5 IJ/cm^2), a ten-fold lower rate than the inundative rate. Perhaps future research will indicate an even lower rate may be also effective for uniform EPN establishment.

6.5 Entomopathogenic Nematodes Movement and Application Strategies in Inoculative Approach in NY State Program

The ability of EPNs to move in the soil by themselves, move in infected insects before the insect dies or be redistributed in the field from soil movement with agricultural practices directly impacts application strategies in the inoculative approach. If movement is limited, then EPNs require a uniform application to the soil surface for effective soil inoculation and host suppression. However, if EPN movement is significant, application does not need to be completely uniform and areas between zones of application will naturally fill in with EPNs. As a result, areas of inoculation can be reduced, resulting in a less expensive application.

In the greenhouse, *H. bacteriophora* 'Oswego', moved 26 cm within 7 days after application (Schroeder, Villani, Ferguson, Nyrop, & Shields, 1993). Shanks and Agudelo-Silva (1990) reported movement of the nematode *Heterorhabditis heliothidis* (Khan, Brooks & Hirschmann) (Rhabditida: Heterorhabditidae) (syn. of

H. bacteriophora, Adams & Nguyen, 2002) into neighboring untreated plots within 3 weeks. This question has also been addressed in the NY alfalfa system indirectly and directly in several studies. Neuman and Shields (2011) reported a minimum of 3 m movement for all three native NY species detected by contamination of neighboring plots where the EPN species was not inoculated. Movement of the EPNs was detected after 12 months, suggesting the movement could have been the result of the movement of infected adult ASB before death as well as the physical movement of EPN infective juveniles (IJs) in the soil. EPN movement has also been reported in when *Steinernema scapterisci* Nguyen & Smart (Rhabditida: Steinernematidae) was used as a biocontrol agent against mole cricket (*Scapteriscus spp.*) in Florida. Parkman, Frank, Nguyen, and Smart (1993) report the nematode dispersed at least 150 m from the release point in three instances and the mean dispersal was 60 m within 21 months. Suggestions about the possible large scale movement of EPNs with the movement of soil during farming operations was reported by Shields, Testa, Neumann, Flanders, and Schroeder (2009). Two subsequent studies have documented unassisted EPN movement (all three species) of 1–1.5 m in a single growing season in the field and a longer distance movement of 25–45 m within soil when the field is plowed (Shields, unpublished) (Fig. 6.1).

This level of EPN movement and mechanical redistribution raises an interesting question regarding the interaction between application rate and the uniformity of the application. Would a better strategy be to inoculate a smaller area at a higher EPN rate to insure establishment and then have the EPNs move into the unapplied areas via natural movement or mechanical redistribution? Or, would it be a better strategy to treat the entire field with a reduced rate of EPNs risking poor establishment while gaining a uniform initial distribution? Research reported by Parkman et al. (1993) and Parkman, Frank, Nguyen, and Smart (1994) suggests that inoculating small areas with the persistent EPN and allow the EPN to spread naturally in infected insects is an inexpensive and viable way to introduce persistent EPNs into the desired host and ecosystem.

Our application/establishment research has been focused on the first option, thinking that EPN establishment is a restricting/limiting factor in the alfalfa ecosystem. With the previous documentation of a rather robust EPN movement in field situations, a completely uniform application of IJs to the soil surface is less important using an inoculative approach because the applied EPNs are persistent for many months without recycling, and will fill in the areas with low EPN density during efforts to locate hosts.

In addition, the focus of the inoculative approach is to establish a long-term EPN presence in the soil profile with the objective of long-term pest suppression. Active and passive EPN movement can then be utilized to reduce the cost of EPN application and still result in the effective inoculation of the soil profile. In semi-permanent agricultural ecosystems with minimum soil tillage and soil movement, EPN natural movement would be slower and a more uniform application would more quickly fill in. In these situations, the use of fertilizer stream nozzles (nozzle type – 0008) on a commercial sprayer with the screens and filters removed would effectively apply the EPNs to the soil surface in continuous strips separated by ca. 55 cm. The area between the application strips would fill in with EPN within 30 days under field conditions. If the nozzles were separated by 1 m, EPN fill in would take a longer time period but EPNs would be found within the untreated areas within a growing season. In a more disturbed agricultural ecosystem where soil tillage occurs at least annually, EPNs will be redistributed with the movement of soil during soil tillage. For example, EPNs were applied in our initial large scale field research in concentrated zones at EPN rate of 2.5×10^8 IJs per ha (2.5 IJ/cm^2), perpendicular to the direction of soil movement during tillage operations. EPN sampling after field tillage indicated a movement-by-tillage of 45 m (Fig. 6.1).

As innovative farmers, participating in an area wide EPN-root weevil pilot program, using *S. carpocapsae* 'NY 001' with *S. feltiae* 'NY04', listened and thought about EPN movement, they derived a more effective application strategy. In their commercial sprayers, nozzles are separated by 55 cm. By fitting the nozzle bodies with fertilizer stream nozzles (nozzle type – 0006–0015) and blocking some of the nozzles, only a portion of the field is treated with EPNs even though the sprayer is driven over the entire field. For example, if two out every three nozzles are blocked, only 33 % of the field is treated in higher concentrated EPN strips separated by 165 cm. EPNs moving inward from each application stream would need to move 82 cm for complete fill-in of the untreated area. Depending on the EPN species and their mobility, these untreated zones would be occupied with EPNs within 60–90 days and the cost of the EPNs for the application would be reduced by 66 %. Some farmers have taken the reduced application strategy a step further with only apply EPNs out of every 6th nozzle, blocking the 5 in between nozzles, and applying to only 17 % of the field while driving over the entire field, reducing the nematode cost by 83 %. With EPN application bands every 330 cm, EPN need to move (or be moved with soil movement) 165 cm for complete fill-in of the untreated area. Complete fill in of these untreated would be accomplished with 1–2 growing seasons or during a single major tillage operation. In these EPN inoculated fields, ASB damage declined to sub economic levels within 3–5 years depending on the application strategy and the intensity of the ASB population.

6.6 Nematode Persistence Results and Impact: Future Research

The utilization of these concepts have been successful in the development of an effective EPN-based biological control program against alfalfa snout beetle, *O. ligustici* and the full detail of the project is discussed under the case study section of this book (Chap. 11). Research projects are currently focused on the application of these concepts to other agricultural systems. Projects have been initiated in three other systems with semi-permanent ecosystems (Fig. 6.4).

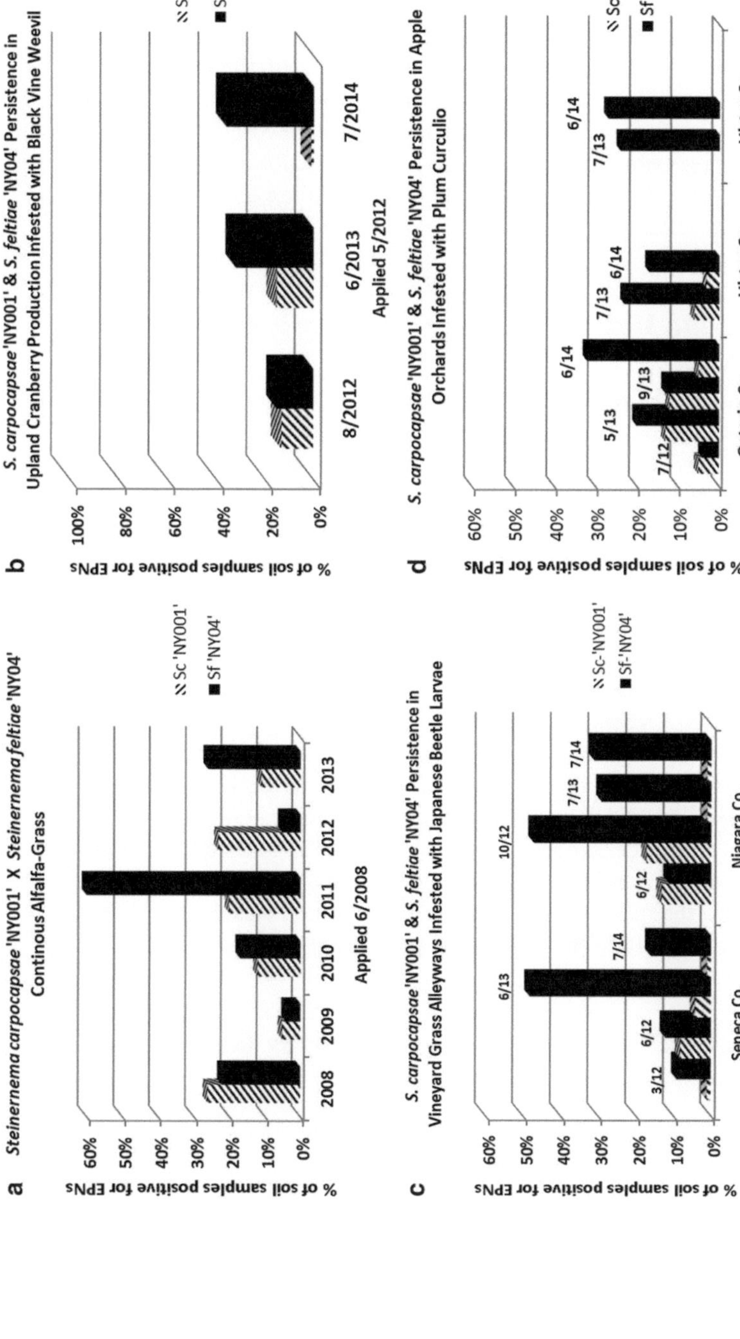

Fig. 6.4 Four different cropping systems in NY where native EPNs were applied once. Entomopathogenic nematode (EPN) population frequency was measured one-two times per year during the growing season. See EPN rates in text. EPN recycling and persistence evident in each cropping system. No error bars because each column is a single value representing the percent of 200 soil cores positive for EPNs using a Galleria-based laboratory bioassay. **(a)** Continous alfalfa-grass. **(b)** Upland cranberry production. **(c)** Grape vineyard. **(d)** Apple orchards.

6.6.1 Alfalfa-Grass

As baseline data for this discussion, the continuous alfalfa-grass field is considered a stable environment with a wide array of potential EPN hosts feeding on the roots of the plants. EPNs were applied in 2008 at 5.0×10^8 IJs per ha (5 IJ/cm^2) and the EPN population was monitored once per growing season (Fig. 6.4a). Several interesting points can be concluded from this graph. Native EPNs persisted and responded to insect invasion in a classical predator-prey response and using the “static” soil core bioassay sampling method, 15–20 % of the cores positive with EPNs appears to be a stable population capable of responding to insect invasion. Research is currently underway which indicates the soil core bioassay sampling technique significantly underestimates the actual potential for EPN host attack in the field.

6.6.2 Black Vine Weevil – Cranberries/Strawberries

Large areas within 80 ha of cranberries grown in a non-flooding culture in central NY were being killed out and large numbers of Black vine weevil (BVW), *Otiorhynchus sulcatus* (Fabricius) (Coleoptera: Curculionidae) were present in the areas surrounding the killed out plants. Since larvae of BVW are soil inhabiting, have a ca. 10 months of soil contact, and are reported in the literature to be susceptible to EPNs, native NY populations of *S. carpocapsae* ‘NY 001’ and *S. feltiae* ‘NY 04’ were applied to all 80 ha (Fig. 6.4b). EPNs were reared on the farm by the producer and self-applied at 2.5×10^8 IJs per ha (2.5 IJ/cm^2) per species (total 5.0×10^8 IJs per ha, 5 IJ/cm^2).

Within a single growing season, BVW was reduced to non-economic levels and EPN have persisted 2 years from a single application in May 2012. It appears that *S. feltiae* is becoming the dominate EPN in a sandy soil environment which requires frequent irrigation. In 2014, the grower noticed several small areas of BVW damage but subsequent EPN sampling of those areas indicated that the EPNs were responding to the insect presence. A similar project was initiated in response to a similar call regarding a severe economic outbreak of BVW in strawberries. Native EPNs were applied the fall 2013 and the insect population was declining during the 2014 growing season. Research is continuing with this system.

6.6.3 Japanese Beetle – Grape Vineyard Grass Roadways

Japanese beetle adults, *Popillia japonica* Newman (Coleoptera: Scarabaeidae) causes leaf feeding damage on grape vines used for wine production in NY during the mid-summer months, requiring multiple applications of broad spectrum insecticides to minimize damage. Eggs are laid in grassy areas and the larvae feed on grass roots starting in late summer through the following spring and early summer.

In many vineyards, alleyways between the vine rows are frequently planted to a grass mixture and frequently, the entire vineyard is surrounded by grassy habitat. At five locations, the grassy areas between the vine rows and surrounding the vineyard were inoculated with native NY populations of *S. carpocapsae* and *S. feltiae*, targeting the beetle larvae. The focus was to determine that if the larvae population was suppressed within the vineyard, would there be a reduction in the adult population feeding on the grape foliage. Four of the five sites had very low Japanese beetle populations during the 3 years of the study and little adult feeding was observed in the check plot. However, in the fifth site, the area of the vineyard treated with EPNs recorded a 27 % reduction in adult Japanese beetle feeding on the grape foliage in 2014. Across all locations, the EPN persisted with the EPN population responding to insect invasion and two of the locations are illustrated in Fig. 6.4c.

6.6.4 Plum Curculio – Organic Apple Production

Plum curculio, *Conotrachelus nenuphar* Herbst (Coleoptera: Curculionidae) is the single most limiting insect pest of organic apple production, causing fruit damage up to 100 %. In conventional production apple orchards, 1–3 broad spectrum insecticide applications are targeted toward controlling this pest. Adults overwinter either in the orchard floor or outside of the orchard in nearby sheltered areas. Adults enter the orchard after fruit set and lay eggs in the small developing fruit. If the apples remain on the tree, the eggs-young larvae is crushed by the growing apple leaving only the disfiguring oviposition scar(s). Many apples after oviposition abort and fall to the orchard floor where the larvae continue to development in the apple. Mature larvae leave dropped fruit on the orchard floor and burrow into the soil for pupation. This allows about a 30 day period which the mature larvae and pupa are susceptible to EPN attack. A project was initiated in 2012 and persistent native NY populations of *S. carpocapsae* and *S. feltiae* were applied to the orchard floor of several organic apple production orchards. EPNs are persisting in all locations at >20 % (Fig. 6.4d) within the grassy habitat on the orchard floor and the incidence of fruit damage has fallen within the treated areas 70–90 %. This research is currently in progress.

References

Adams, B. J., Fodor, A., Koppenhöfer, H. S., Stackebrandt, E., Stock, S. P., & Klein, M. G. (2006). Biodiversity and systematics of nematode-bacterium entomopathogens. *Biological Control, 37*, 32–49.

Adams, B. J., & Nguyen, K. B. (2002). Taxonomy and systematics. In R. Gaugler (Ed.), *Entomopathogenic nematology* (pp. 1–33). Wallinford, UK: CABI Publishing.

Cabanillas, H. E., & Raulston, J. R. (1994). Evaluation of the spatial pattern of *Steinernema riobravis* in corn plots. *Journal of Nematology, 26*, 5–31.
Campbell, J. F., & Gaugler, R. (1997). Inter-specific variation in entomopathogenic nematode foraging strategy: Dichotomy or variation along a continuum? *Fundamental and Applied Nematology, 20*, 393–398.
Campbell, J. F., Lewis, E. E., Yoder, F., & Gaugler, R. (1995). Entomopathogenic nematode (Heterorhabditidae and Steinernematidae) seasonal population dynamics and impact on insect populations in turfgrass. *Biological Control, 5*, 598–606.
Campbell, J. F., Lewis, E. E., Yoder, F., & Gaugler, R. (1997). Entomopathogenic nematode (Heterorhabditidae and Steinernematidae) spatial distribution in turf grass. *Parasitology, 113*, 473–482.
Campos-Herrera, R., Gomez-Ros, J. M., Escuer, M., Cuadra, L., Barrios, L., & Gutierrez, C. (2008). Diversity, occurrence and life characteristics of natural entomopathogenic nematode populations from La Rioja (Northern Spain) under different agricultural management and their relationships with soil factors. *Soil Biology and Biochemistry, 40*, 1474–1484.
Campos-Herrera, R., Pathak, E., El-Borai, F. E., Stuart, R. J., Gutierrez, C., Rodriguez-Martin, J. A., et al. (2013). Geospatial patterns of soil properties and the biological control potential of entomopathogenic nematodes in Florida citrus groves. *Soil Biology and Biochemistry, 66*, 163–174.
Duncan, L. W., Stuart, R. J., El-Borai, F. E., Campos-Herrera, R., Pathak, E., Giurcanu, M., et al. (2013). Modifying orchard planting sites conserves entomopathgenic nematodes, reduced weevil herbivory and increases citrus growth, survival and fruit yield. *Biological Control, 64*, 26–36.
Ennis, D. E., Dillon, A. B., & Griffin, C. T. (2010). Simulated roots and host feeding enhance infection of subterranean insects by the entomopathogenic nematode *Steinernema carpocapsae*. *Journal of Invertebrate Pathology, 103*, 140–143.
Fan, X., & Hominick, W. M. (1991). Efficiency of the *Galleria* (wax moth) baiting technique for recovering infective stages of entomopathogenic rhabditids (Heterorhabditidae and Steinernematidae) from sand and soil. *Revue de Nematologie, 14*, 381–387.
Ferguson, C. S., Schroeder, P. C., & Shields, E. J. (1995). Vertical distribution, persistence and activity of entomopathogenic nematodes (Nematoda: Heterorhabditidae and Steinernematidae) in alfalfa snout beetle (Coleoptera: Curculionidae) infested fields. *Environmental Entomology, 24*, 149–158.
Gaugler, R., Campbell, J. F., Selvan, S., & Lewis, E. E. (1992). Large scale inoculative releases of the entomopathogen, *Steinernema glasseri*: Assessment 50 years later. *Biological Control, 2*, 181–187.
Georgis, R., & Poinar, G. O., Jr. (1983). Effects of soil texture on the distribution and infectivity of *Neoaplectana carpocapsae* (Nematoda: Steinernematidae). *Journal of Nematology, 15*, 308–311.
Glaser, R. W. (1932). *Studies on* Neoaplectana glasseri*, a nematode parasite of the Japanese beetle (Popillia japonica)*. NJ Department of Agriculture, Circ. No. 211.
Glaser, R. W., & Farrell, C. C. (1935). Field experiments with the Japanese beetle and its nematode parasite. *Journal of the New York Entomology Society, 43*, 345–371.
Grewal, P. S., Lewis, E. E., Gaugler, R., & Campbell, J. F. (1994). Host finding behavior as a predictor of foraging strategy in entomopathogenic nematodes. *Parasitology, 108*, 207–215.
Griffin, C. T. (2012). Perspectives on the behavior of entomopathogenic nematodes from dispersal to reproduction: Traits contributing to nematode fitness and biocontrol efficacy. *Journal of Nematology, 44*, 177–184.
Hara, A. H. R., Gaugler, R., Kaya, H. K., & Lebeck, L. M. (1991). Natural populations of the entomopathogenic nematodes (Rhabditida: Heterorhabditidae, Steinernematidae) from the Hawaiian Islands. *Environmental Entomology, 20*, 211–216.
Hominick, W. M. (2002). Biogeography. In R. Gaugler (Ed.), *Entomopathogenic nematology* (pp. 115–143). Wallingford, U.K.: CABI.

Kaya, H. K. (1990). Soil ecology. In R. Gaugler & H. K. Kaya (Eds.), *Entomopathogenic nematodes in biological control* (pp. 93–116). Boca Raton, FL: CRC Press.
Kaya, H. K., & Gaugler, R. (1993). Entomopathogenic nematodes. *Annual Review Entomology, 38*, 181–206.
Lewis, E. E., Campbell, J. F., & Gaugler, R. (1998). A conservation approach to using entomopathogenic nematodes in turf and landscapes. In P. Barbosa (Ed.), *Conservation biological control* (pp. 235–254). San Diego, CA: Academic Press.
Lewis, E. E., Gaugler, R., & Harrison, R. (1992). Entomopathogenic nematode host finding: Response to contact cues by cruise and ambush foragers. *Parasitology, 105*, 309–315.
Lewis, E. E., Gaugler, R., & Harrison, R. (1993). Response of cruiser and ambusher entomopathogenic nematodes (Steinernematidae) to host volatile cues. *Canadian Journal of Zoology, 71*, 765–769.
Lewis, E. E., Ricci, M., & Gaugler, R. (1996). Host recognition behavior reflects host suitability for the entomopathogenic nematode, *Steinernema carpocapsae. Parasitology, 113*, 573–579.
McCoy, E. E., & Glaser, R. W. (1936). *Nematode culture for Japanese beetle control.* NJ Department of Agriculture, Circular No. 265.
Moyle, P. L., & Kaya, H. K. (1981). Dispersal and infectivity of the entomogenous nematode, *Neoaplectana carpocapsae* Weiser (Rhabditida: Steinernematidae) in sand. *Journal of Nematology, 13*, 295–300.
Mráček, Z., & Becvár, S. (2000). Insect aggregations and entompathogenic nematode occurrence. *Nematology, 2*, 297–301.
Neumann, G., & Shields, E. J. (2004). Survey for entomopathogenic nematodes and entomopathogenic fungi in alfalfa snout beetle, *Otiorhynchus ligustici* (L.) (Coleoptera: Curculionidae) in Hungary and New York. *Great Lakes Entomologist, 37*, 152–158.
Neumann, G., & Shields, E. J. (2006). Interspecific interactions among three entomopathogenic nematodes, *Steinernema carpocapsae*, *S. feltiae*, and *Heterorhabditis bacteriophora*, with different foraging strategies for hosts in multi-piece sand columns. *Environmental Entomology, 35*, 1578–1583.
Neumann, G., & Shields, E. J. (2008). Multiple-species natural enemy approach for the biological control of the alfalfa snout beetle, *Otiorhynchus ligustici* L. (Coleoptera: Curculionidae), using entomopathogenic nematodes. *Journal of Economic Entomology, 101*, 1533–1539.
Neumann, G., & Shields, E. J. (2011). Long term field persistence and coexistence in three entomopathogenic nematodes, *Steinernema carpocapsae* Weiser, *S. feltiae* Filipjev, and *Heterorhabditis bacteriophora* Poinar. *The Great Lakes Entomologist, 44*, 42–52.
Parkman, J. P., Frank, J. H., Nguyen, K. B., & Smart, G. C., Jr. (1993). Dispersal of *Steinernema scapterisci* (Rhabditida: Steinernematidae) after inoculative applications for mole cricket (Orthoptera: Gryllotalpidae) control in pastures. *Biological Control, 3*, 226–232.
Parkman, J. P., Frank, J. H., Nguyen, K. B., & Smart, G. C., Jr. (1994). Inoculative release of *Steinernema scapterisci* (Rhabditida: Steinernematidae) to suppress pest mole crickets (Orthoptera: Gryllotalpidae) on golf courses. *Environmental Entomology, 23*, 1331–1337.
Pertzold-Maxwell, J. L., Jaronski, S. T., & Gassmann, A. J. (2012). Tritrophic interactions among Bt maize, and insect pest and entomopathogens: Effects on development and survival of western corn rootworm. *Annals of Appied Biology, 160*, 43–55.
Schroeder, P. C., Villani, M. G., Ferguson, C. S., Nyrop, J. P., & Shields, E. J. (1993). Behavioral interactions between Japanese beetle (Coleoptera: Scarabaeidae) grubs and an entomopathogenic nematode (Nematode: Heterorhabditidae) within turf microcosms. *Environmetal Entomology, 22*, 595–600.
Schroeder, W. J., & Beavers, J. B. (1987). Movement of the entomogenous nematodes of the families Heterorhabditidae and Steinernematidae in soil. *Journal of Nematology, 19*, 257–259.
Shanks, C. H., & Agudelo-Silva, F. (1990). Field pathogenicity and persistence of heterorhabditid and steinernematid nematodes infecting black vine weevil larvae (Coleoptera: Curculionidae) in cranberry bogs. *Journal of Economic Entomology, 83*, 107–110.

Shields, E. J., Testa, A., Miller, J. M., & Flanders, K. L. (1999). Field efficacy and persistence of the entomopathogenic nematodes *Heterorhabditis bacteriophora* 'Oswego' and *H. bacteriophora* 'NC' on alfalfa snout beetle larvae (Coleoptera: Curculionidae). *Environmental Entomology, 28*, 128–136.

Shields, E. J., Testa, A., Neumann, G., Flanders, K. L., & Schroeder, P. C. (2009). Biological control of alfalfa snout beetle with a multi-species application of locally-adapted persistent entomopathogenic nematodes: The first success. *American Entomologist, 55*, 250–257.

Stock, S. P., & Gress, J. C. (2006). Diversity and phylogenetic relationships of entomopathogenic nematodes (Steinernematidae and Heterorhabditidae) from the Sky Islands of southern Arizona. *Journal of Invertebrate Pathology, 92*, 66–72.

Stock, S. P., Pryor, B. M., & Kaya, H. K. (1999). Distribution of entomopathogenic nematodes (Steinernematidae and Heterorhabditidae) in natural habitats in California, USA. *Biodiversity and Conservation, 8*, 535–549.

Wilson, M. J., Ehlers, R.-U., & Glazer, I. (2012). Entomopathogenic nematode foraging strategies – is *Steinernema carpocapsae* really an ambush forager? *Nematology, 14*, 389–394.

Part II
Advances on Entomopathogenic Nematodes Production and Release

Chapter 7
Prospects in the Application Technology and Formulation of Entomopathogenic Nematodes for Biological Control of Insect Pests

Ivan Hiltpold

7.1 Introduction

Efficient application of the most suitable entomopathogenic nematode (EPN) species and/or population is pivotal to successful biological control strategies. Prior to the release of EPNs, particular attention should therefore be paid to the selection of the appropriate species or population of EPN to optimize their efficacy under given conditions. Additionally, appropriate timing of EPN release and the use of optimal equipment will result in successful biological control programs. Both biotic and abiotic factors influencing EPN success are then to be taken into account when planning to implement EPN pest control programs (also discussed in Chap. 9 of this volume). This chapter reviews the latest knowledge and ideas in EPN release methods and the ways of improving their effectiveness in controlling target insect pests. Prospects for integrated approaches are also discussed.

7.2 General Considerations for the Selection and Release of Entomopathogenic Nematodes

Prior to their release, particular attention should be paid to the selection of the EPN species or population. Among the two genera commercially available (*Heterorhabditis* and *Steinernema*, Lewis & Clake, 2012), several species and

I. Hiltpold (✉)
Hawkesbury Institute for the Environment, University of Western Sydney,
Penrith, NSW, Australia
e-mail: i.hiltpold@uws.edu.au

© Springer International Publishing Switzerland 2015
R. Campos-Herrera (ed.), *Nematode Pathogenesis of Insects and Other Pests*,
Sustainability in Plant and Crop Protection 1, DOI 10.1007/978-3-319-18266-7_7

populations can be selected in regards to their various behavioral traits and resistance to abiotic factors. Laboratory screening is the usual approach to select the most suitable species and/or population among EPN candidates. Well–established assays, as described for instance in Grunder (2005), allow quick selection of the appropriate species and/or populations to specifically match with the target insect pest, and, to some extent, with the biotic and abiotic conditions where the EPNs will be released. This laboratory–based approach has been helpful in isolating EPNs that later proved successful in the field to control several insect pest such as the western corn rootworm *Diabrotica virgifera virgifera* LeConte (Coleoptera: Chrysomelidae) (Kurtz, Hiltpold, Turlings, Kuhlmann, & Toepfer, 2009; Toepfer, Gueldenzoph, Ehlers, & Kuhlmann, 2005; Toepfer, Peters, Ehlers, & Kuhlmann, 2008) or *Diaprepes abbreviatus* L. (Coleoptera: Curculionidae) (Duncan & McCoy, 1996; Shapiro, Cate, Pena, Hunsberger, & McCoy, 1999).

Despite these successful implementations of EPNs in biological control campaigns, the importance of comparative analyses of laboratory EPN efficacy experiments with data from field trials cannot be underestimated. A large spectrum of biotic and abiotic factors in the natural environment can render virulent EPN populations selected under highly controlled laboratory conditions poorly effective, and lead to a failure in the control of the insect pest (Shapiro-Ilan, Bruck, & Lacey, 2012). For instance, both *Steinernema feltiae* (Filipjev) Wouts, Mráček, Gerdin & Bedding (Heterorhabditida: Steinernematidae) and *Steinernema riobravae* Cabanillas Poinar & Raulston (Heterorhabditida: Steinernematidae) showed comparable virulence to the plum curculio *Conotrachelus nenuphar* (Herbst) (Coleoptera: Curculionidae) in laboratory assays (Shapiro-Ilan, Mizell, & Campbell, 2002). Despite this, only *S. riobravae* was effective against this weevil in Georgian peach orchards, probably because of differences in the suitability of certain abiotic factors between the two EPN species (Shapiro-Ilan, Mizell, Cottrell, & Horton, 2004).

Laboratory screening does not often take account of the importance of the foraging strategies adopted by the tested EPN, which are considered to vary along a continuum between ambushers (mostly inactive, waiting for a host to pass by) and cruisers (mostly active, searching for host) (Campbell & Gaugler, 1997; Lewis & Clarke, 2012). The foraging behavior possibly influences the ability of a particular nematode to control a specific insect pest; ambushers are likely to infect highly active insects whereas cruisers will be more effective at killing rather sessile insects (Campbell, Lewis, Stock, Nadler, & Kaya, 2003; Gaugler, 1988). However, recent work suggests that EPN foraging behavior not only depends on their taxonomy but also on the substratum into which they are released (Kruitbos, Heritage, Hapca, & Wilson, 2009, 2010). Further, laboratory tests are often performed on agar or on homogenized soil matrices, and are therefore only suggestive of EPN efficacy in the field. In addition to these species/populations specific factors, a large array of abiotic conditions such as UV radiation, temperature, and/or moisture also affect EPN performance (detailed in Chap. 9 of this volume). Consequently, discrepancies between laboratory and field conditions make laboratory screenings poor sole predictors of the field efficacy and ultimate success of laboratory-selected species or populations. Therefore, EPN field performance should be assessed before the transfer of a particular species or population from the laboratory to the field for successful implementation of a biological control plan.

Once the efficient EPN species/population are selected and tested in the field, release methods and technologies are to be considered. These aspects are reviewed in detail by e.g. Bateman, Matthew, and Hall (2007) and also discussed in Chap. 9 of this volume. Briefly, release methods can be categorized in two broad classes; application of EPNs onto or into the soil. Both application approaches present various advantages and drawbacks. For instance, as UV radiation and soil moisture are two major factors affecting EPN survival and effectiveness (Koppenhöfer & Fuzy, 2007), the application of EPNs directly into the ground (i.e., EPNs applied at sowing with a soil fluid insecticide injector) reduces exposure to solar radiation and desiccating conditions compared to the other delivery method. Direct application of EPNs into the soil also dramatically reduces water consumption as compared to their application onto the ground (Hiltpold, Hibbard, French, & Turlings, 2012). Despite these advantages, EPN application into the ground can only be carried out during sowing and thus likely requires a higher number of nematodes to ensure prolonged control of soil–dwelling pests throughout the growing season (Kurtz, Toepfer, Ehlers, & Kuhlmann, 2007). In contrast, applying EPNs by spraying over the soil surface can be done later in the season when the insect pest is about to prevail. However, the latter approach exposes EPNs to harsher conditions than the former, reducing the likelihood of successful biological control using EPNs. In both cases, spatio–temporal occurrence of the pest and the EPNs has to be considered, as EPNs tend to eventually have a patchy distribution even if applied uniformly (Wilson, Lewis, Yoder, & Gaugler, 2003).

In addition to the temporal aspect of EPN release, the ecology of the target insect pests should be considered when choosing whether EPNs are to be applied onto or into the ground. Application of EPNs onto the ground potentially favors the control of litter–dwelling pests such as ticks (Alekseev, Glazer, & Samish, 2006) or lesser mealworms (Szalanski, Palmer, McKay, & Steelman, 2004), whereas EPNs released below ground are more effective in the control of soil–dwelling pests such as *D. v. virgifera* (Hiltpold, Toepfer, Kuhlmann, & Turlings, 2010; Pilz, Keller, Kuhlmann, & Toepfer, 2009) or the citrus root weevil *D. abbreviatus* (Downing, Erickson, & Kraus, 1991; Schroeder, 1994).

7.3 Enhancement of Release Methods and Intrinsic Entomopathogenic Nematode Traits

Based on the factors impairing EPN efficacy described above and in Chap. 10 of this volume (also see Bateman et al., 2007; Shapiro-Ilan, Bruck, & Lacey, 2012), EPN effectiveness can be improved by (1) using/selecting for a more competitive species or population; (2) developing more adequate release methods/formulations and timing; and (3) manipulating the environment in which the EPNs will be released. The following sections illustrate each approach and give detailed examples where the enhancement of EPN application has been successful. Taken from highly collaborative work between various fields of science, these examples underpin the importance of integrative research to ensure the integration of EPNs into pest control programs.

7.3.1 Use/Selection for a More Competitive Species or Population

Using superior EPN populations will enhance the chances of a successful control of insect pests (e.g. Gaugler & Campbell, 1991; Gaugler, Campbell, & McGuire, 1989; Hiltpold et al., 2010a; Hiltpold, Toepfer, et al., 2010). The favored method of obtaining such a population is to survey EPNs at the sites of future applications to isolate and identify new EPN candidates, as indigenous EPNs are predicted to be most effective against the local insect pests. Isolated EPNs are then screened to compare performance against other EPN species/populations (e.g., Campos-Herrera et al., 2008; Shapiro-Ilan et al., 2003). Indigenous EPNs isolated this way should therefore represent very potent alternatives to commercially available populations to solve local pest issues.

Despite superior performances, native and adapted EPN populations may still fail in controlling pest populations under harsh conditions in their natural environment. Previous studies have demonstrated that genetic selection of EPNs can overcome certain physiological limitations by improving various traits of particular interest for biological control such as host finding (e.g., (Gaugler & Campbell, 1991; Gaugler et al., 1989) or other particular behaviors (Bal, Michel, & Grewal, 2014), virulence (e.g., Peters & Ehlers, 1998; Tomalak, 1994), tolerance to extreme temperatures (e.g., Ehlers, Oestergaard, Hollmer, Wingen, & Strauch, 2005; Shapiro, Glazer, & Segal, 1997; Susurluk, Ulu, & Kongu, 2013) and desiccation (e.g., Anbesse, Sumaya, Dörfler, Strauch, & Ehlers, 2013; Strauch, Oestergaard, Hollmer, & Ehlers, 2004). While selective breeding offers an efficient and relatively quick method to obtain EPN populations with desired or improved traits (discussed in Chap. 2 of this volume), it occasionally results in the inadvertent selection of inferior traits such as lower storage stability (Gaugler et al.) or the size of the infective juveniles emerging from the insect host (Stuart, Lewis, & Gaugler, 1996). Therefore, even if such traits are not the purpose of the selection, important attributes for biological control (i.e., shelf–life, persistence, host finding, virulence) have to be tested over the breeding process to ensure selection of a superior EPN population. Recent genome sequencing of EPN species (Bai et al., 2013) will certainly facilitate the process of genetic selection.

Beside conspicuous traits such as virulence and persistence, there are other characteristics specifically related to host finding behavior that are of particular interest when improving EPN efficacy. Over the last decade, the increasing interest in EPN signaling has opened new avenues for the selection of superior populations in biological control. Boff, Zoon, and Smits (2001), and van Tol et al. (2001) first showed that the EPN *Heterorhabditis megidis* Poinar, Jackson & Klein (Rhabditida: Heterorhabditidae) was attracted towards the roots of strawberry and thuja when they were both damaged by *Otiorhynchus sulcatus* Fabricius (Coleoptera: Curculionidae). The authors suggested that plant roots were emitting chemical cues that subsequently attracted EPNs (Boff et al., 2001; van Tol et al., 2001). Later on, Rasmann et al. (2005) showed that damage by *D. v. virgifera* on maize roots induces indirect plant defenses and the subsequent emission of volatile organic compounds, attracting *H. megidis*. The attracted EPNs successfully reduced *D. v. virgifera* population in the field and consequently significantly reduced maize root damage (Hiltpold, Toepfer, et al., 2010; Rasmann et al., 2005). In this particular example, the volatile compound attracting the EPNs was identified as (*E*)–β–caryophylene (Rasmann et al.), a sesquiterpene strongly produced and emitted after the induction of the root defenses by *D. v. virgifera* (Hiltpold, Erb, Robert, & Turlings, 2011; Rasmann et al., 2005). The emission of this sesquiterpene varies among maize cultivars as some lack the ability to produce (*E*)–β–caryophylene and, therefore, do not recruit EPNs when attacked by insect pests (Hiltpold, Toepfer, et al., 2010; Köllner et al., 2008; Rasmann et al., 2005).

Yet, not all nematode species or populations respond to this root alarm signal (Anbesse & Ehlers, 2013; Hiltpold, Toepfer, et al., 2010; Laznik & Trdan, 2013). While *Heterorhabditis bacteriophora* Poinar (Rhabditida: Heterorhabditidae) exhibits a high level of virulence against *D. v. virgifera* (Kurtz et al., 2009), it does not strongly respond to the emission of (*E*)–β–caryophylene (Hiltpold et al., 2010a; Hiltpold, Toepfer, et al., 2010). In the laboratory, a population of this EPN was selected for an enhanced response to this root volatile (Hiltpold et al., 2010a). After six generations of selective breeding, the selected population of *H. bacteriophora* responded 4-fold more quickly and strongly to (*E*)–β–caryophylene than the original one (Hiltpold et al., 2010a). As the selection did not impair the infectiousness of the EPNs (Hiltpold et al., 2010b), both populations were tested in the field and were exposed to two maize cultivars either emitting or lacking the ability to emit (*E*)–β–caryophylene (Hiltpold et al., 2010a; Rasmann et al., 2005). The application of EPNs significantly reduced the survival of *D. v. virgifera* irrespective of maize cultivars. When applied to the maize cultivar emitting (*E*)–β–caryophylene, the selected EPN population performed significantly better than the original (Hiltpold et al., 2010a, 2010b).

The example above demonstrates the potential of selective breeding in the frame of nematode signalling to improve pest control with EPNs. Chemotaxis is repeatedly used by soil–dwelling nematodes (Rasmann, Ali, Helder, & van der Putten, 2012). Several other root–mediated interactions between EPNs and insects and the underpinning volatile emissions have been described since the report of Rasmann et al. (2005) (e.g., Ali, Alborn, & Stelinski, 2010; Rasmann, Erwin, Halitschke, & Agrawal, 2011; Turlings, Hiltpold, & Rasmann, 2012), opening new avenues for the improvement of EPN efficacy in controlling root–damaging pests. EPNs also respond to volatiles emitted from insect hosts (Dillman et al., 2012; Hallem et al., 2011), thereby, enlarging the range of compounds (and blends) to which they can be selected.

7.3.2 Development of Release Methods and Formulations

To date, most of the equipment used to apply EPNs has been adapted from commonly used farming machinery in order to reduce costs and ease handling (Toepfer, Hatala-Zseller, Ehlers, Peters, & Kuhlmann, 2010). EPNs can be applied with virtually any commercially available sprayers (Bateman et al., 2007). Yet, the effect of particular parts of equipment on the post–application EPN efficacy cannot be overstressed. For instance, pressure (Fife, Derksen, Ozkan, & Grewal, 2003; Shapiro-Ilan, Gouge, Piggott, & Fife, 2006), nozzle shape (Fife, Ozkan, Derksen, Grewal, & Krause, 2005) or pump type (Fife, Ozkan, Derksen, & Grewal, 2007) can reduce the post–application effectiveness of EPNs and have to be carefully selected (Bateman et al., 2007). It would therefore be interesting to develop specific hardware to increase compatibility with EPNs as demonstrated in several studies (e.g., Beck et al., 2013, 2014; Brusselman et al., 2012).

Beyond the hardware, using appropriate formulation facilitates application of EPNs. There has been much recent progression in EPN formulation for aboveground application, specifically tackling the issue of UV radiation and risk of desiccation. As a recent example, de Waal, Malan, and Addison (2013) reported improved control of *Cydia pomonella* L. (Lepidoptera: Tortricidae) in the laboratory with *Heterorhabditis zealandica* Poinar (Rhabditida: Heterorhabditidae) when EPNs were applied with a superabsorbent hydrogel based on maize starch. The field application of the same species of EPN in this particular polymer enhanced survival of nematodes and their infectiousness when targeting cryptic niches in pear tree orchards (de Waal et al., 2013). Appropriate formulation of *S. carpocapsae* significantly controlled the population of the diamond–back moth, *Plutella xylostella* L. (Lepidoptera: Plutellidae) on cabbage foliage through prevention of EPN sedimentation in the tankers (Schroer, Ziermann, & Ehlers, 2005) and providing improved environmental conditions, supporting better nematode invasion into the insect host (Schroer & Ehlers, 2005). Greenhouse tests on cotton showed that EPN populations formulated in an alginate gel significantly controlled two lepidopteran pests: *Spodoptera littoralis* Boisduval (Lepidoptera: Noctuidae) and *Helicoverpa armigera* Hübner (Lepidoptera: Noctuidae), whereas when applied in water, the EPNs failed to control the pests, resulting in plant defoliation (Navon, Nagalakshmi, Levski, Salame, & Glazer, 2002).

Alginate was also used in attempts to develop pioneering EPN delivery methods. This polysaccharide isolated from brown algae polymerizes in contact with the right complexing ions (i.e., Ca^{2+} in a solution of $CaCl_2$). This controlled polymerization allows the formation of capsules or beads dependent on the production method (Fig. 7.1) (e.g., Vemmer & Patel, 2013). Exploiting these properties, Kaya and Nelsen (1985) encapsulated *S. feltiae* and *Heterorhabditis heliothidis* Kan, Brooks & Hirschmann (Rhabditida: Heterorhabditidae). Encapsulated EPNs were still able to infect *Spodoptera exigua* Hübner (Lepidoptera: Noctuidae). There was no detectable reduction in survival nor in infectivity during the encapsulation process (Kaya & Nelsen, 1985). In a similar attempt to improve biological control, beads

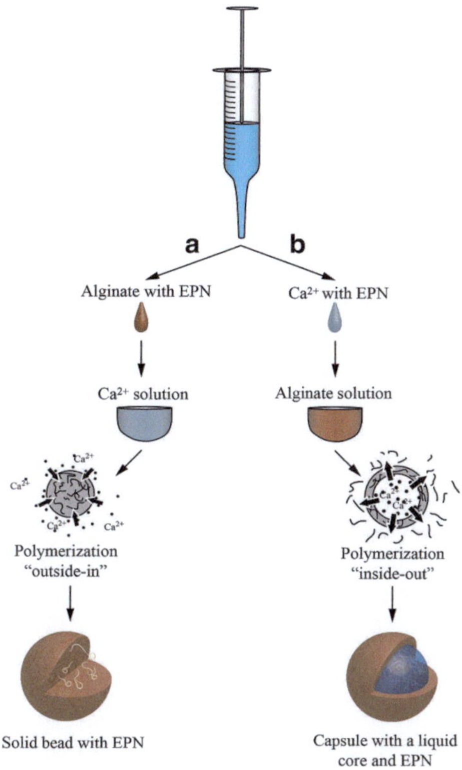

Fig. 7.1 Formation of beads and capsules with alginate. (**a**) A droplets of alginate containing EPNs is dropped in a solution of Ca^{2+} (i.e., $CaCl_2$). The surface of the droplet solidifies by ionotropic gelation while Ca^{2+} reacts with alginate chains to form a three dimensional solid polymer. The ions are diffusing "outside–in", forming a bead with a solidified core. (**b**) A droplet of Ca^{2+} solution containing EPNs is dropped in an alginate solution. The surface of the droplet solidifies by ionotropic gelation while Ca^{2+} reacts with alginate chains to form a three dimensional solid polymer. The ions are diffusing "inside–out", forming a capsule with a liquid core. Both techniques can be used to release EPNs in the field. Signaling chemicals can be added to the alginate solution to interfere with the insect pest behavior and use the alginate structure as a Trojan Horse (Hiltpold et al., 2012) or possibly boost the EPN effectiveness (Hiltpold, Jaffuel, & Turlings, 2015) (Drawing: I. Hiltpold)

of alginate containing hyphae of the endoparasitic nematophagous fungi *Hirsutella rhossiliensis* Minter & Brady (Deuteromycotina: Hyphomycetes) were produced to successfully control plant–parasitic nematodes in microcosms (Lackey, Muldoon, & Jaffee, 1993). Recently, *H. bacteriophora* was encapsulated in alginate capsules (Hiltpold et al., 2012). Nematodes breaking through the capsule shell were still infectious and the polymerization period (in other words, the thickness of the shell) influenced the time needed for the *H. bacteriophora* to escape the capsule (Hiltpold et al., 2012). Following up the idea hypothesized in Kaya and Nelsen (1985), the authors created a blend mimicking the plume of maize roots and blended it with the shell of the capsules to bait the larvae of *D. v. virgifera* (Hiltpold et al., 2012). In the laboratory, the insect larvae were equally attracted towards the capsules and maize seedlings over a distance of ca. 20 cm (Hiltpold et al., 2012). Feeding stimulants were also included in the shell but did not influence the behavior of the insect larvae (Hiltpold et al., 2012). Under laboratory conditions, such an approach had been successful with *S. litorallis* feeding on alginate capsules coated with yeast (Navon, Keren, Salame, & Glazer, 1998). Bringing the capsules to a maize field, Hiltpold et al. (2012) improved the level of control of *D. v. virgifera* as compared to the level reached when nematodes were sprayed in water onto the ground. Though the addition of artificial attractants to the shell did not improve the control of the insect pest when tested in the field, a significant amount of water could be saved; ca. 0.5 L was used to produce the capsules, whereas ca. 2,000 L were needed to treat the plots where nematodes were sprayed (Hiltpold et al., 2012). While promising, the capsules developed by Hiltpold et al. (2012) were soft and did not retain EPNs over an extended period of time, therefore limiting long term storage of capsules containing nematodes. In this context, the temperature at which the polymerization occurs appears to be crucial. The formation of alginate capsules at 4 °C resulted in thinner shelled, yet harder capsules than when the polymerization was performed at 24 °C (Kim, Jaffuel, & Turlings, 2015). The lowest temperature probably resulted in a finer alignment of the alginate polymer, yet prolonged polymerization tended to weaken the capsule, supposedly because of the limited availability of Ca^{2+} (Kim et al., 2015). Post–treatment of the capsules with additional Ca^{2+} noticeably enhanced the hardness of the capsules (Kim et al., 2015). Hardened capsules retained EPN significantly better (Kim et al., 2015) yet post–treatment with Ca^{2+} surprisingly exerted an adverse effect of EPN retainment (Kim et al., 2015). Limiting the EPN activity or slowing down their metabolism, without impairing their infectiousness, can also improve their retainment in capsules.

Alginate gels have been used to reduce EPN mobility (Georgis, 1990), whereas EPN metabolism was slowed down when forcing them into partial anhydrobiosis through slow desiccation (Kondo & Ishibashi, 1989). Examples of such formulations include anhydrous polyacrylamide gel (Bedding & Butler, 1994), powders (Bedding, 1988), and granules (Connick, Nickle, & Vinyard, 1993). However, none of these approaches could support the survival of the nematodes longer than 4 months at room temperature. For instance, at room temperature, *S. feltiae* survived ca. 10 days in vermiculite, 1 month in alginate gels, 2 months in dispersible granules, and 3 months in wet table powder (Grewal, 2002). In the attempt to slow down EPN metabolism, Chen and Glazer (2005) used a combination of the approaches described above and developed beads of alginate (Fig. 7.1) with EPN suspended in glycerol. As glycerol induced a state of dormancy in *S. feltiae*, the EPN survived prolonged exposure to room temperature up to 180 days without noticeable reduction of their survival (Chen & Glazer, 2005). Such dormancy has also been observed when the EPNs are in contact with certain root exudates (Hiltpold et al., 2015). In addition to the improvement of shelf life at room temperature, the EPN infectiousness was enhanced once the state of dormancy was broken (Hiltpold et al., 2015). Encapsulation of EPNs still needs improvements, yet, this approach has some very good potential as it simplifies nematode transport, storage, handling and application, and saves substantial amounts of water, a critical resource in agriculture.

The application of EPNs inside the carcases of their infected hosts recently gained attention (Chapter 8 of this volume is dedicated to this topic). Following the principles discussed above, a similar "Trojan Horse" approach could be imagined if insect feeding stimulants and attractants are added to the coating materials used to harden insect–host cadavers (Shapiro-Ilan, Lewis, Behle, & McGuire, 2001).

7.3.3 Manipulating the Environment in Which the Entomopathogenic Nematodes Will Be Released

Manipulating the environment to favor the biological control agent can be achieved in certain circumstances. In the greenhouse, the dispersal of seven EPN species and their efficacy against the black vine weevil *O. sulcatus* depended on the potting media the nematodes were applied in (Ansari & Butt, 2011). The influence of the environment on the behavior of EPNs has also been suggested in Kruitbos et al. (2009). These examples suggest that using the right EPN species in the appropriate milieu can improve its efficacy in controlling a given target pest.

Manipulating the environment to favor earthworms prior to the application of EPNs, or application of EPNs alongside earthworms may enhance the biological control capacity of the nematodes. Various soil–dwelling organisms, such as isopods or mites, are known to serve as phoretic hosts to EPNs. Phoretic associations were described between the earthworm species *Eisenia fetida* Savigny (Oligocheata: Lumbricidae) and *Rhabditis maupasi* Seurat (Rhabditida: Rhabditidae) (Poinar, 1978) and virulent individuals of the EPN *Pellioditis pelio* Schneider (Rhabditida: Rhabditidae) were found in the excretory system of the earthworm species *Aporrectodea trapezoides* Dugés (Oligocheata: Lumbricidae) (Poinar & Thomas, 1975). Similar phoretic relationships have been described between *S. carpocapsae* and *A. trapezoides* (Shapiro, Berry, & Lewis, 1993) and *Steinernema scapterisci* Nguyen & Smart (Rhabditida: Steinernematidae) and the earthworm species *Allolobophora caliginosa* Sauvigny (Oligochaeta: Lumbricidae) (Nguyen & Smart, 1991). Campos-Herrera, Trigo, and Gutiérrez (2006) have shown that phoresy with *E. fetida* severely impairs *S. feltiae* mobility and virulence. Yet, dispersal of *Steinernema* spp. increased in the presence of earthworms (Shapiro et al., 1993; Shapiro, Tylka, Berry, & Lewis, 1995), suggesting actual phoretic associations as earthworm burrows alone did not enhance EPN dispersal (Shapiro et al., 1995). The earthworm species *Lombriscus terrestris* L. (Oligocheata: Lumbricidae) has enhanced the biological control capacity of *S. carpocapsae* and served as phoretic hosts to entomopathogenic fungi (Shapiro-Ilan & Brown, 2013). As phoresy between *L. terrestris* and EPNs does not seem to be harmful to the nematodes (Shapiro et al., 1995), Shapiro-Ilan and Brown (2013) suggested that phoretic dispersal by earthworms could assist in the regulation of insect pests.

In the field, cover crop, crop residues or mulch are also options to manipulate the environment where EPNs are released. There are numerous examples of the use of cover crops to enhance the control of plant parasitic nematodes (e.g., Lazzeri, Curto, Leoni, & Dallavalle, 2004; Lazzeri, D'Avino, & Gies, 2010; Matthiessen & Kirkegaard, 2006; Oliveira et al., 2011; Potter, Davies, & Rathjen, 1998). In regard to the improvement of EPN efficacy, cover crops have the potential to support alternative hosts and to minimize variation in soil moisture and temperature, thereby increasing persistence of EPNs (Susurluk & Ehlers, 2008) (benefits and challenges of EPN persistence are detailed in Chaps. 4, 5 and 6 of this volume). In addition, cover crops can also favor other antagonists of the target insect pests (e.g., Shapiro-Ilan, Gardner, Wells, & Wood, 2012), thereby offering cumulative or synergistic controlling benefits. Similarly, the use of crop residue, protecting EPNs from abiotic extremes, improved the persistence of *S. carpocapsae* (Shapiro, Obrycki, Lewis, & Jackson, 1999), and the addition of wood–chip mulch enhanced the control of overwintering *C. pomonella* with steinernematid EPNs (Lacey, Granatstein, Arthurs, Headrick, & Fritts, 2006). In citrus orchards, horse manure mulch did not increase populations of free living bacterivorous nematodes and EPNs whereas both populations increased when chicken manure was applied (Duncan et al., 2007). Therefore, case–by–case studies might be necessary to guarantee a positive effect of cover crops on EPN populations.

In an innovative and dramatic attempt to manipulate the habitat of EPNs to favor the biological control of *D. abbreviatus*, Duncan et al. (2013) planted citrus trees in sandy soil, being more favorable to EPNs than the removed native soil. This in–depth manipulation of the rhizospheric environment aimed at improving the health of the citrus tree, as they had previously shown higher insect damage in fine–textured soil (El-Borai, Stuart, Campos-Herrera, Pathak, & Duncan, 2012) than in coarse sandy soil (Futch, Duncan, & Zekri, 2005), where EPN species richness and diversity is greater (Campos-Herrera, Pathak, El-Borai, Stuart, et al., 2013; Campos-Herrera, Stuart, El-Borai, Gutierrez, & Duncan, 2010). Although the characterization of all the underpinning mechanisms still has to be elucidated, the use of sand in place of native soil in citrus orchards increased EPN richness and diversity and reduced the number of emerging weevils, thereby increasing the survival, growth and fruit production of young citrus trees (Duncan et al.). However, the manipulations of the aboveground environment of citrus trees to control a bacterial disease had severe detrimental effects on belowground food webs and EPN population (Campos-Herrera, El-Borai, Ebert, Schumann, & Duncan, 2014; Campos-Herrera, Pathak, El-Borai, Schumann, et al., 2013), underpinning the need of caution while adopting such approach.

Plant secondary metabolites demonstrate strong potential in the manipulation of the EPN environment to enhance biological control of soil–dwelling insect pests. As described above, roots damaged by insects release exudates in the rhizosphere and these have a dramatic effect on the behavior of particular EPN species (Ali et al., 2010; Rasmann et al., 2011; Rasmann et al., 2005; Turlings et al., 2012), often subsequently increasing nematode effectiveness (Ali et al., 2012; Degenhardt et al., 2009; Hiltpold et al., 2010a). Plants can be genetically engineered to cnhance the production of certain volatiles, thereby favoring insect pest control with EPNs (Degenhardt et al., 2009). Particular root exudates have been shown to have a dual effect in the rhizosphere, subduing plant pathogenic nematodes while invigorating EPNs (Hiltpold et al., 2015). While concentrated root–tip exudates of green pea, *Pisum sativum* L. (Fabales: Fabaceae) induced quiescence in several EPN species, lower concentrations, likely to be encountered in the rhizosphere, did not induce quiescence in EPNs but significantly increased their activity and infectiousness (Hiltpold et al., 2015). Incorporating such chemicals into the EPN formulations under development, including alginate capsules (see description above), has the potential to improve EPN shelf life due to reduced metabolic rates in quiescent EPNs (more lipids were measured in EPNs stored in root exudate than in EPNs stored in water). In addition, root exudate additives could boost EPN efficacy in controlling soil–dwelling insect pests due to increased nematode activity when woken up from quiescence (Hiltpold et al., 2015).

Root and rhizosphere chemical ecology is an emerging field of research and, to date, only its surface has been scratched (see Hiltpold et al., 2013; Hiltpold & Turlings, 2012, 2013; Rasmann, Hiltpold, & Ali, 2012). Despite the benefits from attracting EPNs towards sites of insect feeding, it must be noted that root–emitted volatiles are not only affecting the behavior of these beneficial organisms but also attracting other rhizospheric microorganisms, some of which will be detrimental to the plant or the EPNs themselves. For instance, the plant parasitic nematode *Tylenchulus semipenetrans* Cobb (Tylenchida: Tylenchulidae) was significantly more attracted towards citrus roots induced by *D. abbreviatus* than towards undamaged tree roots (Ali, Alborn, & Stelinski, 2011). Still in the rhizosphere of citrus trees, nematophagous fungi tended to be more sampled in areas where volatiles typically emitted by insect–induced roots were applied than in sites where only solvent was used. Free living nematodes, potentially competing with EPNs, were also more abundant in the presence of the root cues (Ali, Campos-Herrera, Alborn, Duncan, & Stelinski, 2013). It is therefore clear that the impact of belowground cues is complex, yet these new avenues of research will still provide scientists and biotechnologists with environmentally-friendly molecules with various applications (Hiltpold & Turlings, 2012). Manipulation of the EPN habitat for better pest control is tightly related to a better and more comprehensive understanding of EPNs (and insect pests) signaling and behavior in response to plant secondary metabolites. Each step further in this field of research will allow us to better fine-tune current pest management approaches as well as avoid detrimental impacts on soil communities.

7.4 Conclusions and Future Directions

The successful use of EPNs in biological control is punctuated with failures (Georgis et al., 2006), unfortunately making this technology unreliable in the eyes of those who may use EPNs. However, recent successes and developments of new approaches are likely to bring attention back to EPNs and to their high potential in biological control or integrated pest management.

Recent advances in the mass production of EPNs (Ehlers, 2001, 2007; Ehlers & Shapiro-Ilan, 2005; McMullen & Stock, 2014) make EPNs cost competitive compared to other pest management tactics, provided that their application is adequate and effective. The usage of equipment routinely used for application of chemical pesticides has enabled easier implementation of EPNs as biological control agents. However, their susceptibility to biotic and abiotic factors cannot be overstressed and hardware adapted to specific EPN species and cropping systems should be eventually developed to ensure ideal application techniques.

Innovations in EPN formulation technologies have been essential for unraveling new fields of applications that were previously thought beyond the potential of EPNs, such as targeting pests in cryptic belowround niches. Development of new formulations should not only aim at improving EPN effectiveness but also be oriented toward easing handling and application procedures, improving their storage and transport, and reducing overall costs of use.

Most advances in the improvement of EPN application will depend on the advancement of our knowledge of the interactions between the nematodes and their surrounding environment. EPNs respond to numerous biotic and abiotic factors, which influence their success in controlling a specific pest in a particular environment and crop. A more comprehensive understanding of their ecology and behavior will possibly increase their competitiveness, optimize their use, and reduce the costs involved. Such holistic knowledge will result from collaborative research and expertise in nematology, entomology, plant physiology and biochemistry, chemistry, and physics. Recent breakthroughs in EPN biology open new research avenues, which will possibly lead to wider EPN use in pest management schemes.

References

Alekseev, E., Glazer, I., & Samish, M. (2006). Effect of soil texture and moisture on the activity of entomopathogenic nematodes against female *Boophilus annulatus* ticks. *BioControl, 51*(4), 507–518.

Ali, J. G., Alborn, H. T., Campos-Herrera, R., Kaplan, F., Duncan, L. W., Rodriguez-Saona, C., et al. (2012). Subterranean, herbivore–induced plant volatile increases biological control activity of multiple beneficial nematode species in distinct habitats. *Plos One, 7*(6), e38146.

Ali, J. G., Alborn, H. T., & Stelinski, L. L. (2010). Subterranean herbivore–induced volatiles released by citrus roots upon feeding by *Diaprepes abbreviatus* recruit entomopathogenic nematodes. *Journal of Chemical Ecology, 36*(4), 361–368.

Ali, J. G., Alborn, H. T., & Stelinski, L. L. (2011). Constitutive and induced subterranean plant volatiles attract both entomopathogenic and plant parasitic nematodes. *Journal of Ecology, 99*(1), 26–35.

Ali, J. G., Campos-Herrera, R., Alborn, H. T., Duncan, L. W., & Stelinski, L. L. (2013). Sending mixed messages: A trophic cascade produced by a belowground herbivore–induced cue. *Journal of Chemical Ecology, 39*(8), 1140–1147.

Anbesse, S., & Ehlers, R. U. (2013). *Heterorhabditis* sp. not attracted to synthetic (E)–B–caryophyllene, a volatile emitted by roots upon feeding by corn rootworm. *Journal of Applied Entomology, 137*(1–2), 88–96.

Anbesse, S., Sumaya, N. H., Dörfler, A. V., Strauch, O., & Ehlers, R. U. (2013). Selective breeding for desiccation tolerance in liquid culture provides genetically stable inbred lines of the entomopathogenic nematode *Heterorhabditis bacteriophora. Applied Microbiology and Biotechnology, 97*(2), 731–739.

Ansari, M. A., & Butt, T. M. (2011). Effect of potting media on the efficacy and dispersal of entomopathogenic nematodes for the control of black vine weevil, *Otiorhynchus sulcatus* (Coleoptera: Curculionidae). *Biological Control, 58*(3), 310–318.

Bai, X., Adams, B. J., Ciche, T. A., Clifton, S., Gaugler, R., Kim, K., et al. (2013). A lover and a fighter: The genome sequence of an entomopathogenic nematode. *PLoS ONE, 8*(7), e69618.

Bal, H. K., Michel, A. P., & Grewal, P. S. (2014). Genetic selection of the ambush foraging entomopathogenic nematode, *Steinernema carpocapsae* for enhanced dispersal and its associated trade–offs. *Evolutionary Ecology, 28*(5), 923–939.

Bateman, R. P., Matthew, G. A., & Hall, F. R. (2007). Ground–based application equipment. In L. A. Lacey & H. K. Kaya (Eds.), *Field manual of techniques in invertebrate pathology* (2nd ed., pp. 73–98). Dordrecht, The Netherlands: Springer.

Beck, B., Brusselman, E., Nuyttens, D., Moens, M., Pollet, S., Temmerman, F., et al. (2013). Improving foliar applications of entomopathogenic nematodes by selecting adjuvants and spray nozzles. *Biocontrol Science and Technology, 23*(5), 507–520.

Beck, B., Brusselman, E., Nuyttens, D., Moens, M., Temmerman, F., Pollet, S., et al. (2014). Improving the biocontrol potential of entomopathogenic nematodes against *Mamestra brassicae*: effect of spray application technique, adjuvants and an attractant. *Pest Management Science, 70*(1), 103–112.

Bedding, R. A. (1988). *Storage of insecticidal nematodes*. World patent No. WO 88/08668.

Bedding, R. A., & Butler, K. L. (1994). *Method for storage of insecticidal nematodes*. World patent No. WO 94/05150.

Boff, M. I. C., Zoon, F. C., & Smits, P. H. (2001). Orientation of *Heterorhabditis megidis* to insect hosts and plant roots in a Y–tube sand olfactometer. *Entomologia Experimentalis et Applicata, 98*(3), 329–337.

Brusselman, E., Beck, B., Pollet, S., Temmerman, F., Spanoghe, P., Moens, M., et al. (2012). Effect of the spray application technique on the deposition of entomopathogenic nematodes in vegetables. *Pest Management Science, 68*(3), 444–453.

Campbell, J. F., & Gaugler, R. R. (1997). Inter–specific variation in entomopathogenic nematode foraging strategy: Dichotomy or variation along a continuum? *Fundamental and Applied Nematology, 20*, 393–398.

Campbell, J. F., Lewis, E. E., Stock, S. P., Nadler, S., & Kaya, H. K. (2003). Evolution of host search strategies in entomopathogenic nematodes. *Journal of Nematology, 35*(2), 142–145.

Campos-Herrera, R., El-Borai, F. E., Ebert, T. E., Schumann, A., & Duncan, L. W. (2014). Management to control citrus greening alters the soil food web and severity of a pest–disease complex. *Biological Control, 76*, 41–51.

Campos-Herrera, R., Gómez-Ros, J. M., Escuer, M., Cuadra, L., Barrios, L., & Gutiérrez, C. (2008). Diversity, occurrence, and life characteristics of natural entomopathogenic nematode populations from La Rioja (Northern Spain) under different agricultural management and their relationships with soil factors. *Soil Biology and Biochemistry, 40*(6), 1474–1484.

Campos-Herrera, R., Pathak, E., El-Borai, F. E., Schumann, A., Abd-Elgawad, M. M. M., & Duncan, L. W. (2013). New citriculture system suppresses native and augmented entomopathogenic nematodes. *Biological Control, 66*(3), 183–194.

Campos-Herrera, R., Pathak, E., El-Borai, F. E., Stuart, R. J., Gutiérrez, C., Rodríguez-Martín, J. A., et al. (2013). Geospatial patterns of soil properties and the biological control potential of entomopathogenic nematodes in Florida citrus groves. *Soil Biology and Biochemistry, 66*, 163–174.

Campos-Herrera, R., Stuart, R., El-Borai, F., Gutierrez, C., & Duncan, L. (2010). Entomopathogenic nematode ecology and biological control in Florida citrus orchards. In A. Ciancio & K. G. Mukerji (Eds.), *Integrated management of arthropod pests and insect borne diseases* (Vol. 5, pp. 101–130). Dordrecht, The Netherlands: Springer.

Campos-Herrera, R., Trigo, D., & Gutiérrez, C. (2006). Phoresy of the entomopathogenic nematode *Steinernema feltiae* by the earthworm *Eisenia fetida*. *Journal of Invertebrate Pathology, 92*(1), 50–54.

Chen, S. B., & Glazer, I. (2005). A novel method for long–term storage of the entomopathogenic nematode *Steinernema feltiae* at room temperature. *Biological Control, 32*(1), 104–110.

Connick, W. J., Nickle, W. R., & Vinyard, B. T. (1993). PESTA – New granular formulations for *Steinernema carpocapsae*. *Journal of Nematology, 25*(2), 198–203.

de Waal, J. Y., Malan, A. P., & Addison, M. F. (2013). Effect of humidity and a superabsorbent polymer formulation on the efficacy of *Heterorhabditis zealandica* (Rhabditida: Heterorhabditidae) to control codling moth, *Cydia pomonella* (L.) (Lepidoptera: Tortricidae). *Biocontrol Science and Technology, 23*(1), 62–78.

Degenhardt, J., Hiltpold, I., Kollner, T. G., Frey, M., Gierl, A., Gershenzon, J., et al. (2009). Restoring a maize root signal that attracts insect–killing nematodes to control a major pest. *Proceedings of the National Academy of Sciences of the United States of America, 106*(32), 13213–13218.

Dillman, A. R., Guillermin, M. L., Lee, J., Kim, B., Sternberg, P. W., & Hallem, E. A. (2012). Olfaction shapes host–parasite interactions in parasitic nematodes. *Proceedings of the National Academy of Sciences of the United States of America, 109*(35), E2324–E2333.

Downing, A. S., Erickson, C. G., & Kraus, M. J. (1991). Filed–evaluation of entomopathogenic nematodes against citrus root weevils (Coleptera, Curculionidae) in Florida citrus. *Florida Entomologist, 74*(4), 584–586.

Duncan, L. W., Graham, J. H., Zellers, J., Bright, D., Dunn, D. C., El-Borai, F. E., et al. (2007). Food web responses to augmenting the entomopathogenic nematodes in bare and animal manure–mulched soil. *Journal of Nematology, 39*(2), 176–189.

Duncan, L. W., & McCoy, C. W. (1996). Vertical distribution in soil, persistence, and efficacy against citrus root weevil (Coleoptera: Curculionidae) of two species of entomogenous nematodes (Rhabditida: Steinernematidae; Heterorhabditidae). *Environmental Entomology, 25*(1), 174–178.

Duncan, L. W., Stuart, R. J., El-Borai, F. E., Campos-Herrera, R., Pathak, E., Giurcanu, M., et al. (2013). Modifying orchard planting sites conserves entomopathogenic nematodes, reduces weevil herbivory and increases citrus tree growth, survival and fruit yield. *Biological Control, 64*(1), 26–36.

Ehlers, R. U. (2001). Mass production of entomopathogenic nematodes for plant protection. *Applied Microbiology and Biotechnology, 56*, 623–633.

Ehlers, R. U., Oestergaard, J., Hollmer, S., Wingen, M., & Strauch, O. (2005). Genetic selection for heat tolerance and low temperature activity of the entomopathogenic nematode–bacterium complex *Heterorhabditis bacteriophora–Photorhabdus luminescens*. *Biocontrol, 50*(5), 699–716.

Ehlers, R. U., & Shapiro-Ilan, D. I. (2005). Mass production. In P. S. Grewal, R. U. Ehlers, & D. I. Shapiro-Ilan (Eds.), *Nematodes as biocontrol agents* (pp. 65–78). Oxfordshire, UK: CABI Publishing.

Ehlers, R.-U. (2007). Entomopathogenic nematodes: From science to commercial use. In C. Vincent, M. S. Toettel, & G. Lazarovits (Eds.), *Biological control: A global perspective* (pp. 136–151). Oxfordshire, UK: CABI Publishing.

El-Borai, F. E., Stuart, R. J., Campos-Herrera, R., Pathak, E., & Duncan, L. W. (2012). Entomopathogenic nematodes, root weevil larvae, and dynamic interactions among soil texture, plant growth, herbivory, and predation. *Journal of Invertebrate Pathology, 109*(1), 134–142.

Fife, J. P., Derksen, R. C., Ozkan, H. E., & Grewal, P. S. (2003). Effects of pressure differentials on the viability and infectivity of entomopathogenic nematodes. *Biological Control, 27*(1), 65–72.

Fife, J. P., Ozkan, H. E., Derksen, R. C., & Grewal, P. S. (2007). Effects of pumping on entomopathogenic nematodes and temperature increase within a spray system. *Applied Engineering in Agriculture, 23*(4), 405–412.

Fife, J. P., Ozkan, H. E., Derksen, R. C., Grewal, P. S., & Krause, C. R. (2005). Viability of a biological pest control agent through hydraulic nozzles. *Transactions of the Asae, 48*(1), 45–54.

Futch, S. H., Duncan, L. W., & Zekri, M. (2005). Validation of an area–wide extension program to estimate the seasonal abundance of adult citrus root weevils with unbaited pyramidal traps. *Proceedings of the Florida State Horticultural Society, 117*, 143–147.

Gaugler, R. (1988). Ecological considerations in the biological control of soil–inhabiting insects with entomopathogenic nematodes. *Agriculture, Ecosystems and Environment, 24*(1–3), 351–360.

Gaugler, R., & Campbell, J. F. (1991). Selection for enhanced host–finding of scarab larvae (Coleoptera, Scarabaeidae) in an entomopathogenic nematode. *Environmental Entomology, 20*(2), 700–706.

Gaugler, R., Campbell, J. F., & McGuire, T. R. (1989). Selection for host–finding in *Steinernema feltiae. Journal of Invertebrate Pathology, 54*(3), 363–372.

Georgis, R. (1990). Formulation and application technology. In R. Gaugler & H. K. Kaya (Eds.), *Entomopathogenic nematodes in biological control* (pp. 173–191). Boca Raton, FL: CRC Press.

Georgis, R., Koppenhöfer, A. M., Lacey, L. A., Bélair, G., Duncan, L. W., Grewal, P. S., et al. (2006). Successes and failures in the use of parasitic nematodes for pest control. *Biological Control, 38*, 103–123.

Grewal, P. (2002). Formulation and application technology. In R. Gaugler (Ed.), *Entomopathogenic nematology* (pp. 265–287). Wallingford, UK/Oxfordshire, UK: CABI Publishing.

Grunder, J. M. (2005). *Quality control of entomopathogenic nematodes*. Wädenswil, CH: Agroscope FAW.

Hallem, E. A., Dillman, A. R., Hong, A. V., Zhang, Y. J., Yano, J. M., DeMarco, S. F., et al. (2011). A sensory code for host seeking in parasitic nematodes. *Current Biology, 21*(5), 377–383.

Hiltpold, I., Baroni, M., Toepfer, S., Kuhlmann, U., & Turlings, T. C. J. (2010a). Selection of entomopathogenic nematodes for enhanced responsiveness to a volatile root signal helps to control a major root pest. *Journal of Experimental Biology, 213*(14), 2417–2423.

Hiltpold, I., Baroni, M., Toepfer, S., Kuhlmann, U., & Turlings, T. C. J. (2010b). Selective breeding of entomopathogenic nematodes for enhanced attraction to a root signal did not reduce their establishment or persistence after field release. *Plant Signaling and Behavior, 5*(11), 1450–1452.

Hiltpold, I., Bernklau, E., Bjostad, L. B., Alvarez, N., Miller-Struttmann, N. E., Lundgren, J. G., et al. (2013). Nature, evolution and characterisation of rhizospheric chemical exudates affecting root herbivores. In S. N. Johnson, I. Hiltpold, & T. C. J. Turlings (Eds.), *Advances in insect physiology* (Vol. 45, pp. 97–157). Oxford, UK: Academic.

Hiltpold, I., Erb, M., Robert, C. A. M., & Turlings, T. C. J. (2011). Systemic root signalling in a belowground, volatile–mediated tritrophic interaction. *Plant, Cell and Environment, 34*(8), 1267–1275.

Hiltpold, I., Hibbard, B. E., French, D. W., & Turlings, T. C. J. (2012). Capsules containing entomopathogenic nematodes as a Trojan horse approach to control the western corn rootworm. *Plant and Soil, 358*, 11–25.

Hiltpold, I., Jaffuel, G., & Turlings, T. C. J. (2015). The dual effects of root–cap exudates on nematodes: From quiescence in plant–parasitic nematodes to frenzy in entomopathogenic nematodes. *Journal of Experimental Botany, 66*(2), 603–611.

Hiltpold, I., Toepfer, S., Kuhlmann, U., & Turlings, T. C. J. (2010). How maize root volatiles influence the efficacy of entomopathogenic nematodes against the western corn rootworm? *Chemoecology, 20*(2), 155–162.

Hiltpold, I., & Turlings, T. C. J. (2012). Manipulation of chemically mediated interactions in agricultural soils to enhance the control of crop pests and to improve crop yield. *Journal of Chemical Ecology, 38*(6), 641–650.

Hiltpold, I., & Turlings, T. C. J. (2013). Manipulation of below–ground scents for the benefit of agriculture. In: K. V. Wurms, A. J. Popay, N. L. Bell, & M. V. Jaspers (Eds.), *Utilising plant defence for pest control* (pp. 39–56). Christchurch, New Zealand: The New Zealand Plant Protection Society (Incorporated).

Kaya, H. K., & Nelsen, C. E. (1985). Encapsulation of steinernematid and heterorhabditid nematodes with calcium alginate: A new approach for insect control and other applications. *Environmental Entomology, 14*(5), 572–574.

Kim, J., Jaffuel, G., & Turlings, T. C. J. (2015). Enhanced alginate capsule properties as a formulation of entomopathogenic nematodes. *BioControl*. In press.

Köllner, T. G., Held, M., Lenk, C., Hiltpold, I., Turlings, T. C. J., Gershenzon, J., et al. (2008). A maize (*E*)–β–caryophyllene synthase implicated in indirect defense responses against herbivores is not expressed in most American maize varieties. *The Plant Cell, 20*(2), 482–494.

Kondo, E., & Ishibashi, N. (1989). Ultrastructural characteristics of the infective juveniles of *Steinernema* spp (Rhabditida, Steinernematidae) with reference to their motility and survival. *Applied Entomology and Zoology, 24*, 103–111.

Koppenhöfer, A. M., & Fuzy, E. M. (2007). Soil moisture effects on infectivity and persistence of the entomopathogenic nematodes *Steinernema scarabaei*, *S. glaseri*, *Heterorhabditis zealandica*, and *H. bacteriophora*. *Applied Soil Ecology, 35*(1), 128–139.

Kruitbos, L. M., Heritage, S., Hapca, S., & Wilson, M. J. (2009). Influence of substrate on the body–waving behaviour of nematodes. *Nematology, 11*(6), 917–925.

Kruitbos, L. M., Heritage, S., Hapca, S., & Wilson, M. J. (2010). The influence of habitat quality on the foraging strategies of the entomopathogenic nematodes *Steinernema carpocapsae* and *Heterorhabditis megidis*. *Parasitology, 137*(2), 303–309.

Kurtz, B., Hiltpold, I., Turlings, T. C. J., Kuhlmann, U., & Toepfer, S. (2009). Comparative susceptibility of larval instars and pupae of the western corn rootworm to infection by three entomopathogenic nematodes. *BioControl, 54*(2), 255–262.

Kurtz, B., Toepfer, S., Ehlers, R. U., & Kuhlmann, U. (2007). Assessment of establishment and persistence of entomopathogenic nematodes for biological control of western corn rootworm. *Journal of Applied Entomology, 131*(6), 420–425.

Lacey, L. A., Granatstein, D., Arthurs, S. P., Headrick, H., & Fritts, R. (2006). Use of entomopathogenic nematodes (Steinernematidae) in conjunction with mulches for control of overwintering codling moth (Lepidoptera: Tortricidae). *Journal of Entomological Science, 41*, 107–119.

Lackey, B. A., Muldoon, A. E., & Jaffee, B. A. (1993). Alginate pellet formulation of *Hirsutella rhossiliensis* for biological–control of plant–parsitic nematodes. *Biological Control, 3*(2), 155–160.

Laznik, Ž., & Trdan, S. (2013). An investigation on the chemotactic responses of different entomopathogenic nematode strains to mechanically damaged maize root volatile compounds. *Experimental Parasitology, 134*, 349–355.

Lazzeri, L., Curto, G., Leoni, O., & Dallavalle, E. (2004). Effects of glucosinolates and their enzymatic hydrolysis products via myrosinase on the root–knot nematode *Meloidogyne incognita* (Kofoid et White). *Journal of Agricultural and Food Chemistry, 52*(22), 6703–6707.

Lazzeri, L., D'Avino, L., & Gies, D. (2010). Additional benefits of the efficacy in containing soilborne pest and pathogens with biofumigant plants and materials. *Acta Horticulturae, 883*, 323–339.

Lewis, E. E., & Clarke, D. J. (2012). Nematode parasites and entomopathogens. In E. Vega & H. K. Kaya (Eds.), *Insect pathology* (2nd ed., pp. 395–424). Amsterdam: Elsevier.

Matthiessen, J., & Kirkegaard, J. (2006). Biofumigation and enhanced biodegradation: Opportunity and challenge in soilborne pest and disease management. *Critical Reviews in Plant Sciences, 25*(3), 235–265.

McMullen, J. G., II, & Stock, S. P. (2014). *In vivo* and *in vitro*rearing of entomopathogenic nematodes (Steinernematidae and Heterorhabditidae). *Journal of Visualized Experiments: JoVE, 91*, e52096.

Navon, A., Keren, S., Salame, L., & Glazer, I. (1998). An edible–to–insects calcium alginate gel as a carrier for entomopathogenic nematodes. *Biocontrol Science and Technology, 8*(3), 429–437.
Navon, A., Nagalakshmi, V. K., Levski, S., Salame, L., & Glazer, I. (2002). Effectiveness of entomopathogenic nematodes in an alginate gel formulation against lepidopterous pests. *Biocontrol Science and Technology, 12*(6), 737–746.
Nguyen, K. B., & Smart, G. C. (1991). Pathogenicity of *Steinerneam scapterisci* to selected invertebrates. *Journal of Nematology, 23*, 7–11.
Oliveira, R. D. L., Dhingra, O. D., Lima, A. O., Jham, G. N., Berhow, M. A., Holloway, R. K., et al. (2011). Glucosinolate content and nematicidal activity of Brazilian wild mustard tissues against *Meloidogyne incognita* in tomato. *Plant and Soil, 341*(1–2), 155–164.
Peters, A., & Ehlers, R. U. (1998). Evaluation and selection for enhanced nematode pathogenicity against *Tipula* spp. In N. Simoes, N. Boemare, & R. U. Ehlers (Eds.), *Pathogenicity of entomopathogenic nematodes versus insect defense mechanisms: Impact on selection of virulent strains* (pp. 225–241): Brussels, Belgium: European Commission Publication COST819.
Pilz, C., Keller, S., Kuhlmann, U., & Toepfer, S. (2009). Comparative efficacy assessment of fungi, nematodes and insecticides to control western corn rootworm larvae in maize. *BioControl, 54*(5), 671–684.
Poinar, G. O. (1978). Associations between nematodes (Nematoda) and oligocheates (Annelida). *Proceedings of the Helminthological Society of Washington, 45*, 202–210.
Poinar, G. O., & Thomas, G. M. (1975). *Rhabditis–Pellio* Schneider (Nematoda) from earthworm, *Aporrectodae tapezoides* Duges (Annelida). *Journal of Nematology, 7*, 374–379.
Potter, M. J., Davies, K., & Rathjen, A. J. (1998). Suppressive impact of glucosinolates in Brassica vegetative tissues on root lesion nematode *Pratylenchus neglectus. Journal of Chemical Ecology, 24*(1), 67–80.
Rasmann, S., Ali, J. G., Helder, J., & van der Putten, W. H. (2012). Ecology and evolution of soil nematode chemotaxis. *Journal of Chemical Ecology, 38*(6), 615–628.
Rasmann, S., Erwin, A. C., Halitschke, R., & Agrawal, A. A. (2011). Direct and indirect root defences of milkweed (*Asclepias syriaca*): Trophic cascades, trade–offs and novel methods for studying subterranean herbivory. *Journal of Ecology, 99*(1), 16–25.
Rasmann, S., Hiltpold, I., & Ali, J. G. (2012). The role of root–produced volatile secondary metabolites in mediating soil interactions. In G. Montanaro & B. Dichio (Eds.), *Advances in selected plant physiology aspects* (pp. 269–290). Rijeka, Croatia: InTech.
Rasmann, S., Kollner, T. G., Degenhardt, J., Hiltpold, I., Toepfer, S., Kuhlmann, U., et al. (2005). Recruitment of entomopathogenic nematodes by insect–damaged maize roots. *Nature, 434*, 732–737.
Schroeder, W. J. (1994). Comparison of 2 Steinernematid species for control fo the root weevil *Diaprepes abbreviatus. Journal of Nematology, 26*(3), 360–362.
Schroer, S., & Ehlers, R. U. (2005). Foliar application of the entomopathogenic nematode *Steinernema carpocapsae* for biological control of diamondback moth larvae (*Plutella xylostella*). *Biological Control, 33*(1), 81–86.
Schroer, S., Ziermann, D., & Ehlers, R. U. (2005). Mode of action of a surfactant–polymer formulation to support performance of the entomopathogenic nematode *Steinernema carpocapsae* for control of diamondback moth larvae (*Plutella xylostella*). *Biocontrol Science and Technology, 15*(6), 601–613.
Shapiro, D. I., Berry, E. C., & Lewis, L. C. (1993). Interactions between nematodes and earthworms – Enhanced dispersal of *Steinernema carpocapsae. Journal of Nematology, 25*, 189–192.
Shapiro, D. I., Cate, J. R., Pena, J., Hunsberger, A., & McCoy, C. W. (1999). Effects of temperature and host age on suppression of *Diaprepes abbreviatus* (Coleoptera: Curculionidae) by entomopathogenic nematodes. *Journal of Economic Entomology, 92*(5), 1086–1092.
Shapiro, D. I., Glazer, I., & Segal, D. (1997). Genetic improvement of heat tolerance in *Heterorhabditis bacteriophora* through hybridization. *Biological Control, 8*(2), 153–159.
Shapiro, D. I., Obrycki, J. J., Lewis, L. C., & Jackson, J. J. (1999). Effects of crop residue on the persistence of *Steinernema carpocapsae. Journal of Nematology, 31*, 517–519.

Shapiro, D. I., Tylka, G. L., Berry, E. C., & Lewis, L. C. (1995). Effect of earthworms on the dispersal of *Steinernema* spp. *Journal of Nematology, 27*, 21–28.

Shapiro-Ilan, D. I., & Brown, I. (2013). Earthworms as phoretic hosts for *Steinernema carpocapsae* and *Beauveria bassiana*: Implications for enhanced biological control. *Biological Control, 66*(1), 41–48.

Shapiro-Ilan, D. I., Bruck, D. J., & Lacey, L. A. (2012). Principles of epizootiology and microbial control. In E. Vega & H. K. Kaya (Eds.), *Insect pathology* (2nd ed., pp. 29–72). Amsterdam: Elsevier.

Shapiro-Ilan, D. I., Gardner, W. A., Fuxa, J. R., Wood, B. W., Nguyen, K. B., Adams, B. J., et al. (2003). Survey of entomopathogenic nematodes and fungi endemic to pecan orchards of the Southeastern United States and their virulence to the pecan weevil (Coleoptera: Curculionidae). *Environmental Entomology, 32*(1), 187–195.

Shapiro-Ilan, D. I., Gardner, W. A., Wells, L., & Wood, B. W. (2012). Cumulative impact of a clover cover crop on the persistence and efficacy of *Beauveria bassiana* in suppressing the pecan weevil (Coleoptera: Curculionidae). *Environmental Entomology, 41*, 298–307.

Shapiro-Ilan, D. I., Gouge, D. H., Piggott, S. J., & Fife, J. P. (2006). Application technology and environmental considerations for use of entomopathogenic nematodes in biological control. *Biological Control, 38*(1), 124–133.

Shapiro-Ilan, D. I., Lewis, E. E., Behle, R. W., & McGuire, M. R. (2001). Formulation of entomopathogenic nematode–infected cadavers. *Journal of Invertebrate Pathology, 78*(1), 17–23.

Shapiro-Ilan, D. I., Mizell, R. F., & Campbell, J. F. (2002). Susceptibility of the plum curculio, *Conotrachelus nenuphar,* to entomopathogenic nematodes. *Journal of Nematology, 34*(3), 246–249.

Shapiro-Ilan, D. I., Mizell, R. F., Cottrell, T. E., & Horton, D. L. (2004). Measuring field efficacy of *Steinernema feltiae* and *Steinernema riobrave* for suppression of plum curculio, *Conotrachelus nenuphar*, larvae. *Biological Control, 30*(2), 496–503.

Strauch, O., Oestergaard, J., Hollmer, S., & Ehlers, R. U. (2004). Genetic improvement of the desiccation tolerance of the entomopathogenic nematode *Heterorhabditis bacteriophora* through selective breeding. *Biological Control, 31*(2), 218–226.

Stuart, R. J., Lewis, E. E., & Gaugler, R. (1996). Selection alters the pattern of emergence from the host cadaver in the entomopathogenic nematode, *Steinernema glaseri. Parasitology, 113*, 183–189.

Susurluk, A., & Ehlers, R. U. (2008). Field persistence of the entomopathogenic nematode *Heterorhabditis bacteriophora* in different crops. *BioControl, 53*(4), 627–641.

Susurluk, I. A., Ulu, T. C., & Kongu, Y. (2013). Tolerances of hybridized entomopathogenic nematode *Heterorhabditis bacteriophora* (Rhabditida: Heterorhabditidae) strains to heat and desiccation. *Turkiye Entomoloji Dergisi–Turkish Journal of Entomology, 37*(2), 221–228.

Szalanski, A. L., Palmer, T. W., McKay, T., & Steelman, C. D. (2004). Infectivity of *Steinernema* spp. (Nematoda: Steinernematidae) to adult litter beetles, *Alphitobius diaperinus* (Coleoptera: Tenebrionidae) in the laboratory. *Biocontrol Science and Technology, 14*(1), 81–85.

Toepfer, S., Gueldenzoph, C., Ehlers, R. U., & Kuhlmann, U. (2005). Screening of entomopathogenic nematodes for virulence against the invasive western corn rootworm, *Diabrotica virgifera virgifera* (Coleoptera: Chrysomelidae) in Europe. *Bulletin of Entomological Research, 95*(1–10), 473–482.

Toepfer, S., Hatala-Zseller, I., Ehlers, R. U., Peters, A., & Kuhlmann, U. (2010). The effect of application techniques on field–scale efficacy: Can the use of entomopathogenic nematodes reduce damage by western corn rootworm larvae? *Agricultural and Forest Entomology, 12*, 389–402.

Toepfer, S., Peters, A., Ehlers, R. U., & Kuhlmann, U. (2008). Comparative assessment of the efficacy of entomopathogenic nematode species at reducing western corn rootworm larvae and root damage in maize. *Journal of Applied Entomology, 132*(5), 337–348.

Tomalak, M. (1994). Selective breeding of *Steinernema feltiae* Filipjev (Nematoda, Steinernematidae) for improved efficacy in control of a mushroom fly, *Lycoriella solani* Winnertz (Diptera, Sciaridae). *Biocontrol Science and Technology, 4*(2), 187–198.

Turlings, T. C. J., Hiltpold, I., & Rasmann, S. (2012). The importance of root–produced volatiles as foraging cues for entomopathogenic nematodes. *Plant and Soil, 359*, 51–60.

van Tol, R. W. H. M., van der Sommen, A. T. C., Boff, M. I. C., Van Bezooijen, J., Sabelis, M. W., & Smits, P. H. (2001). Plants protect their roots by alerting the enemies of grubs. *Ecology Letters, 4*(4), 292–294.

Vemmer, M., & Patel, A. V. (2013). Review of encapsulation methods suitable for microbial biological control agents. *Biological Control, 67*(3), 380–389.

Wilson, M. J., Lewis, E. E., Yoder, F., & Gaugler, R. (2003). Application pattern and persistence of the entomopathogenic nematode *Heterorhabditis bacteriophora. Biological Control, 26*(2), 180–188.

Chapter 8
Insect Cadaver Applications: Pros and Cons

Claudia Dolinski, David Shapiro-Ilan, and Edwin E. Lewis

8.1 Introduction

Application of entomopathogenic nematodes (EPNs) formulated as insect cadavers has become an alternative to aqueous application for the control of agricultural pests. In this approach, the infected insect host cadaver is applied directly to the target site and pest suppression is achieved by the infective juveniles (IJs) that emerge from the host cadavers. This type of technology could be especially effective for small– and medium–sized growers, with planted areas up to 10 ha, or for use in flower pots and home orchards, etc. The cost of production for cadaver–based formulations is low because it eliminates the need to capture and concentrate infective juveniles (IJs) and reduces storage costs required in production systems that involve aqueous suspension of IJs. Also, the insect cadaver represents a shelter from environmental extremes such as freezing. However, the insect cadaver approach has a number of downsides that demand further study and adaptation. In this chapter we review the development and application of the "cadaver approach" and discuss advantages and disadvantages as well as avenues for future research.

C. Dolinski (✉)
Laboratório de Entomologia e Fitopatologia, Universidade Estadual do Norte Fluminense Darcy Ribeiro, Campos dos Goytacazes, RJ, Brazil
e-mail: Claudia.dolinski@censanet.com.br

D. Shapiro-Ilan
Southeastern Fruit and Nut Research Laboratory, USDA–ARS, Byron, GA, USA
e-mail: david.shapiro@ars.usda.gov

E.E. Lewis
Department of Entomology and Nematology, University of California, Davis, CA, USA
e-mail: eelewis@ucdavis.edu

© Springer International Publishing Switzerland (outside the USA) 2015
R. Campos-Herrera (ed.), *Nematode Pathogenesis of Insects and Other Pests*, Sustainability in Plant and Crop Protection 1, DOI 10.1007/978-3-319-18266-7_8

8.2 Production Techniques

Techniques associated with the production of nematode–infected insect cadavers generally follow the same basic approaches required for laboratory–based or commercial–based *in vivo* production of EPNs. Further details on methods of *in vivo* EPN production may be found in Kaya and Stock (1997), Shapiro-Ilan, Han, and Dolinski (2012) and Shapiro-Ilan, Han, and Qiu (2014). *In vivo* EPN production, for purposes of aqueous application or storage of IJs, generally consists of four steps: inoculation, incubation, harvest and concentration. A notable and key difference in insect cadaver production is that the harvest and concentration steps are eliminated because the IJs emerge from the insect cadaver at the application site. Thus, production of insect cadavers simply consists of inoculation and temporary storage. In an additional step, formulation of cadavers to protect them from rupturing or sticking together upon application has also been suggested.

For simple laboratory experiments or small experimental trials, hosts are usually exposed to IJs in Petri dishes lined with an absorbent substrate such as filter paper (Kaya & Stock, 1997). Other substrates that may be used include plaster of Paris, peat, or any other absorbent and inert material. It is important that the substrate is moist upon introduction of IJs, but standing water is detrimental. The rate of application per insect will vary among host and EPN species. For larger–scale applications such as in the greenhouse or especially field applications and other commercial ventures, more efficient procedures and a large–scale process are needed. Foremost, rather than using Petri dishes, larger containers or a system of trays can be used to accommodate numbers of hosts in the tens–of–thousands. Furthermore, the inoculation method can be automated by dunking large numbers of insect hosts in into a suspension of IJs, then draining and transferring them onto a large tray lined with a suitable substrate (such as absorbent paper) (Shapiro-Ilan & Gaugler, 2002; Shapiro-Ilan, Han, & Qiu, 2014). Efficiency can be greatly enhanced by automating each step of the process from production of insect hosts to preparation for packaging and shipping; in that vein significant advancements were made in automating the system for production and inoculation of *Tenebrio molitor* L. (Coleoptera: Tenebrionidae) (Morales-Ramos, Rojas, Shapiro-Ilan, & Tedders, 2011; Morales-Ramos, Tedders, Dean, Shapiro-Ilan, & Rojas, 2013).

To obtain maximum yields per cadaver, it is important to know the host density in the inoculation arena and the inoculation rate that is specific to each particular nematode species (Shapiro-Ilan, Gaugler, Tedders, Brown, & Lewis, 2002; Shapiro-Ilan, Han, & Qiu, 2014). Generally, infectivity increases with nematode concentration and decreases with host density per unit area (Shapiro-Ilan et al., 2002). An inoculation rate that is too low results in low host mortality, whereas a rate that is too high could result in failed infections due to intra specific competition (Woodring & Kaya, 1988). Thus, yields are likely to be maximized using intermediate concentrations of IJs (Boff, Wiegers, Gerritsen, & Smits, 2000). For example, rates of approximately 25–200 IJs per insect are generally sufficient (depending on nematode species and method of inoculation) to infect *Galleria mellonella* (L.) (Lepidoptera: Pyralidae), whereas a higher rate is necessary to infect *T. molitor* (e.g. 100–600 IJs per insect). Crowding of hosts can cause oxygen deprivation or build-up of ammonia, which suppresses nematode yield (Shapiro, Lewis, Paramasivam, & McCoy, 2000; Shapiro-Ilan et al., 2002).

Throughout the production process environmental conditions conducive to nematode development must be maintained. Once the nematode inoculum is applied to the insect hosts, the relative humidity should remain high (around 90 %) and the substrate should remain moist throughout the incubation period. Additionally, an optimum temperature for inoculation and nematode development should be implemented. Temperature tolerances vary in nematode species, e.g., some species such as *Heterorhabditis indica* Poinar, Karunakar, & David (Rhabditida: Heterorhabditidae), and *Steinernema riobrave* Cabanillas, Poinar, & Raulston (Rhabditida: Steinernematidae) are relatively heat tolerant while other species, such as *Heterorhabditis megidis* Poinar, Jackson, & Klein (Rhabditida: Heterorhabditidae), and *Steinernema feltiae* (Filipjev) Wouts, Mráček, Gerdin & Bedding (Rhabditida: Steinernematidae), are generally more tolerant to cooler temperatures (Grewal, Selvan, & Gaugler, 1994; Shapiro & McCoy, 2000). Adequate aeration must also be maintained throughout the production cycle. An optimum balance between humidity and aeration (thus avoiding build up ammonia or other harmful gases) should be achieved, such as by using high efficiency particulate air filters and a humidifying system.

8.3 Application Techniques

The simplest and most straight forward method to apply infected host cadavers is by inserting them into the target site by hand. Small holes, around five cm deep, are made in the soil or in pots, where the insect cadavers will be placed and then covered with soil (Fig. 8.1). The number of insect cadavers will be related to the area to be covered by the infective juveniles (IJs). Many characteristics related to EPN biology and physiology applied as insect cadavers are already known. Among these characteristics are the dispersal consequences of emergent IJs, their infectivity and survival.

Fig. 8.1 Grower applying insect cadaver infected with *Heterorhabditis baujardi* Phan, Subbotin, Nguyen & Moens (Rhabditida: Heterorhabditidae) LPP7 in a commercial guava orchard in Cachoreiras de Macacu, RJ, Brazil

For larger scale insect cadaver applications, a mechanized approach would be beneficial. Toward that end, Zhu, Grewal, and Reding (2011) developed a delivery system for nematode–infected hosts based on a modified crop seed planter. The machine applies partially desiccated cadavers at a constant rate into slits in the soil that it creates and then covers up. In laboratory and field experiments, the delivery system performed effectively in distributing *G. mellonella* cadavers infected with *Heterorhabditis bacteriophora* Poinar (Rhabditida: Heterorhabditidae) (Zhu et al., 2011). Additional research is needed to develop mechanized equipment and distribution systems for insect cadavers in systems not amendable to the above–described method.

8.3.1 Dispersal Behaviour

The distribution can be caused by several factors in varying combinations, such as the host–seeking strategy, infectivity, density of IJs, dispersal capacity, IJs' remaining energy reserves, desiccation and temperature tolerance. These interact with extrinsic factors that are biotic (i.e. predation, competition, phoretic relationships, synergism and plant roots) and abiotic (i.e., soil moisture, radiation, temperature, aeration, soil characteristics and others). Each nematode species has its own dispersal capacity and foraging strategies, making it important to treat them individually when making application decisions.

As for host seeking strategy, some EPNs have been classified as ambushers, [i.e. *Steinernema carpocapsae* (Weiser) (Rhabditida: Steinernematidea)], while others are considered cruisers (i.e. *H. bacteriophora*) (Campbell & Gaugler, 1993; Lewis, Gaugler, & Harrison, 1993; Lewis, Grewal, & Gaugler, 1995). Finally, some EPN species are classified as "intermediate foragers", such as *S. feltiae*, which neither stand on their tails like ambushers, nor respond to long–range host volatile cues like cruisers (Grewal, Lewis, & Gaugler, 1994). More details in Chap. 3.

Shapiro and Glazer (1996) presented evidence that *S. carpocapsae* and *H. bacteriophora* emerging from insect cadavers have greater dispersal capacity than those applied in aqueous suspensions. The authors speculated that treatment effects were based on physiological or behavioral differences associated with the IJs from the cadaver *versus* aqueous suspension. Further, the enhanced dispersal was deemed to be advantageous in terms of biocontrol potential.

Density of IJs may also interfere in their dispersion. In the case of *H. baujardi* LPP7, a cruiser species, IJs showed a wider horizontal distribution when the insect cadavers were applied in larger numbers (15 insect cadavers applied in the same spot), compared to one insect cadaver application. Also, the high concentration of IJs defined in space and time (e.g., 90 cm from the point of application and 5th week post–application) suggests that they move *en masse*, which can result in a patchy dispersion (Del Valle, Dolinski, & Souza, 2008). Several researchers have described this patchy spatial and temporal distribution in many different EPNs species (Cabanillas & Raulston, 1994; Efron, Nestel, & Glazer, 2001; Glazer,

Kozodoi, Salame, & Nestel, 1996; Stuart & Gaugler, 1994). Shapiro-Ilan, Lewis, and Schliekelman (2014), studying six EPN species (*H. bacteriophora*, *H. indica*, *S. carpocapsae* (Weiser), *S. feltiae*, *Steinernema glaseri* (Weiser) (Rhabditida: Steinernematidae) and *S. riobrave* applied in aqueous suspension via filter paper discs or in infected insect host cadavers observed that nematode dispersal resulted in an aggregated dispersion pattern rather than a random or uniform distribution. These findings have implications for nematode spatial distribution and suggest that group behaviour is involved in nematode foraging. This type of behavior is akin to "follow–the–leader" or the herding paradigm, where risk–tolerant nematodes infect the host, followed by nematodes that are more risk averse.

This "follow–the–leader" behavior was also confirmed for the ambusher nematode *S. carpocapsae*, comparing it with the cruiser, *H. bacteriophora*, using infected insect cadavers in autoclaved, silt–loam soil in large microcosms (0.05–1.5 m^2) with or without vegetation in the absence of hosts. Bal, Taylor, and Grewal (2014) observed that the majority of the *S. carpocapsae* population stayed closed to the source cadaver (<3.8 cm), whereas a majority of the *H. bacteriophora* population dispersed between 7 and 12 cm away from the cadaver. The most interesting behavior was observed when about 4 % of the *S. carpocapsae* population dispersed faster than the fastest *H. bacteriophora*, reaching 30–61 cm, compared to only 2 % of the *H. bacteriophora* population dispersing this far. They called those "faster" nematodes 'sprinters' for long–distance dispersal and concluded that the presence of sprinters may represent an adaptive dispersal strategy for the ambush forager *S. carpocapsae* in the absence of hosts.

Recently, liquid chromatography–mass spectrometry analysis of *Caenorhabditis elegans* Maupas (Rhabditida: Rhabditidae) dauer conditioned media, which show strong dispersing activity, revealed four known ascarosides (ascr#2, ascr#3, ascr#8, icas#9). A synthetic blend of these ascarosides at physiologically relevant concentrations dispersed *C. elegans* dauer stages in the presence of food and also caused dispersal of IJs of *S. feltiae* and J2s of plant parasitic *Meloidogyne* spp. Assay–guided fractionation revealed structural analogues as major active components of the *S. feltiae* (ascr#9) and *C. elegans* (ascr#2) dispersal blends. Further analysis revealed ascr#9 in all *Steinernema* spp. and *Heterorhabditis* spp. infected insect cadavers (Kaplan et al., 2012). These findings indicate that compounds produced in the infected host induce nematode dispersal, thus providing an explanation for the earlier observations of enhanced nematode dispersal from cadavers by Shapiro and Glazer (1996). More *in vitro* tests are necessary.

8.3.2 Infectivity

Shapiro and Lewis (1999) compared infectivity between IJs emerging directly from infected insect cadavers into sand with IJs that were applied to sand in aqueous suspensions after collection in a White trap. Infectivity of *H. bacteriophora* was significantly greater when nematodes emerged directly into sand compared with the aqueous application. After 24 h, $\simeq$11 % of *H. bacteriophora* from cadaver treatments had penetrated into a host, whereas infectivity of aqueous treatments was 4–10 folder lower. A similar trend was observed after 48 h. The infectivity of *S. carpocapsae* was $\simeq$24 % in both treatments. An additional experiment was conducted for *H. bacteriophora*, which consisted of three treatments: (1) nematodes directly from insect cadavers, (2) nematodes applied in aqueous suspension, and (3) nematodes in aqueous suspension that were exposed to an extract from macerated host cadavers. Higher infectivity was observed in the amended aqueous suspension compared with aqueous nematodes without the extract. This study points out that differences in fitness and behavior must exist between nematodes reared under standard laboratory procedures (they are captured in water as IJs) and those in nature (see also Chap. 6). Furthermore, the increased infectivity in the treatment containing macerated cadaver extract indicates that factors in the host cadaver induce nematodes to infect.

8.3.3 Survival

For a biological control agent to be effective against agricultural soil pests, they must remain infective in soil for at least 2 weeks (Shapiro-Ilan, Gouge, Piggott, & Fife, 2006). Abiotic and biotic factors play a major role in that, but it will be discussed latter. The prolonged emergence of IJs from the cadaver could be beneficial. Infective juveniles of *H. baujardi* LPP7 emerged from *G. mellonella* insect cadavers remained infective during 6 weeks (Del Valle, Dolinski, & Souza, 2008). The survival in the field maybe affected by the concentration of insect cadavers and/or IJs. Infective juveniles persisted longer when 15 insect cadavers were applied in one spot, compared to application of only one insect cadaver (Del Valle, Dolinski, & Souza, 2008). The heterorhabditid's habit of sticking together might help their survival by avoiding desiccation (Shapiro-Ilan, Lewis, & Schliekelman, 2014). Perez, Lewis, and Shaprio-Ilan (2003) reported that three nematode species, *S. carpocapsae*, *S. riobrave*, and *H. bacteriophora*, survived longer when applied via insect cadaver method relative to aqueous application.

8.4 Factors Affecting Efficiency

8.4.1 Biotic Factors

Ants are the most apparent invertebrate scavengers observed foraging on EPN–insect cadavers. Baur, Kaya, and Strong (1998) observed that workers of the Argentine ant, *Linepithema humile* (Mayr) (Hymenoptera: Formicidae) scavenged insect cadavers on the surface and those buried 2 cm below the surface. Ant workers scavenged significantly more steinernematid–killed (60–85 %) than heterorhabditid–

killed (10–20 %) insect cadavers. None of the insects were scavenged when insects were killed by *Photorhabdus luminescens* (Thomas & Poinar) (Enterobacteriales: Enterobacteriaceae) (from *H. bacteriophora*), 70 % were scavenged, when killed by *Xenorhabdus nematophilus* (Poinar & Thomas) (Enterobacteriales: Enterobacteriaceae) (from *S. carpocapsae*), and 90 % of the insects killed by *Bacillus thuringiensis* Berliner (Bacilliales: Bacilliaceae) were scavenged by the Argentine ant. Therefore, the authors suggested that chemicals associated with *P. luminescens* might be responsible for preventing ants from foraging on heterorhabditid–killed hosts. Also, Zhou, Kaya, Heungens, and Goodrich-Blair (2002) suggested that the *Photorhabdus* bacteria within the insect cadaver might be responsible for this apparent ant repellence.

Other ant species, including *Veromessorandrei* (Mayr) (Hymenoptera: Formicidae), *Pheidolevistana* Forel (Hymenoptera: Formicidae), *Formica pacifica* Francoeur (Hymenoptera: Formicidae), and *Monomorium mergatogyna* Wheeler (Hymenoptera: Formicidae), are also insect cadaver scavengers. They either removed or destroyed steinernematid–killed insects. These results suggest that survival of steinernematid nematodes may be more significantly impacted by invertebrate scavengers, especially ants, than that of heterorhabditid nematodes, so placement of steinernematid–killed insects in the field for biological control may be an ineffective release strategy (Baur et al., 1998).

When the behavior of two *Ectatomma* species were observed, the percentages of heterorhabtid–insect cadavers removed by *Ectatomma* sp. 1 ranged between 73 and 80 %, while for *Ectatomma* sp. 2 ranged between 40 and 67 %. In this case, the ants did not destroy the insect cadavers, but just moved them away from their nest (Del Valle, Dolinski, Barreto, & Souza, 2009). However, when the insect cadavers were covered by gelatine capsules, neither *Ectatomma* species responded to their presence. These results indicate that both ant species are attracted by the insect cadavers, and remove them from the nest opening. The authors observed that the ants transported insect cadavers to a distance greater than 1 m away from the nest, reaching up to 4 m. *Ectatomma* sp. 1 presented more aggressive response in the presence of insect cadavers, since these ants removed them in larger quantities and more quickly (Del Valle et al., 2009).

Gülcü, Hazir, and Kaya (2012) investigated the response of several scavenger insects to nematode infected host. Scavengers included the ant species *Lepisiota frauenfeldi* Mayr (Hymenoptera: Formicidae), the cricket *Gryllus bimaculatus* (DeGeer) (Orthoptera: Gryllidae), and wasps *Vespa orientalis* L. and *Paravespula* sp., (Hymenoptera: Vespidae). The scavengers were repelled by the infected hosts. Specifically, these insects (ants, crickets and wasps) did not feed on nematode–killed insects containing the nematode/bacterium complex that were 2 days old and older but fed on 1-day–old nematode–killed as well as freeze–killed insects. Additionally, the calliphorid fly *Chrysomya albiceps* Wiedemann (Diptera: Calliphoridae) did not oviposit on meat treated with *P. luminescens* supernatant but did oviposit on untreated meat. Thus, the authors suggested that a broad "scavenger deterrent factor" is contained in nematode infected hosts (Gülcü et al., 2012) (see more details in Chap. 5).

8.4.2 Abiotic Factors

8.4.2.1 Temperature

The effectiveness of controlling insect pests with EPNs depends on the habitat temperature when they are released in the soil (Grewal, Lewis, & Gaugler, 1994; Kaya, 1990). Temperature influences many processes, including the proportion of food reserves (i.e., lipids, proteins and carbohydrates) that are used by nematodes for their motility, survival, infectivity, development and reproduction (Kaya). The virulence and reproduction of *Heterorhabditis* spp. are reduced at temperatures above 30 °C (Grewal, Lewis, & Gaugler, 1994). In general, the insecticidal activity of EPNs is more effective in the temperature range of 18–28 °C (Akhurst & Boemare, 1990). When *H. baujardi* LPP7 was applied as insect cadavers in a guava orchard, it was observed that the temperature might have adversely affected the emergence of IJs from insect cadavers in the experiment's first week. The ideal temperature for this stran is 26 °C, but the average soil temperature during that time was 18 °C (Del Valle, Dolinski, Souza, & Samuels, 2005; Del Valle, Dolinski, & Souza, 2008). Freezing temperatures (e.g. during overwintering) are detrimental to EPNs; however, Lewis and Shapiro-Ilan (2002) discovered that EPNs survive freezing temperatures better in infected hosts relative to IJs outside of the hosts.

8.4.2.2 Desiccation and Soil Moisture

Several authors have shown that desiccation significantly reduces the number of IJs produced, their infectivity and survival in insect cadavers (Koppenhöfer et al., 1997; Perez et al., 2003). Similar to freezing conditions (Lewis & Shapiro-Ilan, 2002), Koppenhöfer et al. (1997) discovered that the infected host could serve as a survival refuge for nematodes under desiccating conditions. Spence et al. (2011) pointed out that the number of IJs produced depends on the nematode species, desiccation period, and soil moisture during rehydration. They suggested that while desiccation generally has a negative effect on EPN infection success and IJ production, at least some individuals of certain EPN species are capable of surviving within 30 % desiccated insect cadavers (30 % of the original moisture), and they suggested this method as a low–tech strategy for production of insect cadavers that would not require subsequent formulations. It must be pointed out that the reduced yield in desiccated cadavers creates an additional barrier to achieving the rates required for wide–scale field applications. However, there are particular environments where application of desiccated cadavers may not only be effective, but even advantageous, for example in plant nurseries. In pots the insect cadavers can provide a long–term, slowly emerging source of IJs that can infect root–feeding larvae and limit the spread of insect pests. In addition, the reduced technological requirements of this *in vivo* production method can be particularly attractive for farmer cooperatives in developing countries where manual labor is not costly. They also mentioned another advantage of this method, the insect cadaver texture, since the cadaver can be compared to a raisin and can easily be hand–planted, as mentioned and shown before (Fig. 8.1).

Molyneux and Bedding (1984) demonstrated that the EPN activity is influenced by the thickness of a water film surrounding soil particles. This water film is a key component that drives IJ emergence from insect cadavers, causing little or no emergence of IJs at lower ranges of soil moisture (Spence et al., 2011). However, high water content of soil may also constrain movement of IJs emerging from the host cadavers. During a field experiment, after many rainy days, Del Valle, Dolinski, and Souza (2008) observed no IJ movement in the soil and they inferred that IJs may have emerged from the insect cadavers, but remained in the same place until the soil water content reached a level that facilitated their locomotion.

8.4.3 Insect Cadavers' Preparation

8.4.3.1 Coatings

Application of nematode–infected insect cadavers has been suggested as a suitable biological control technique in a range of agricultural systems. Unfortunately, such insect cadavers can be easily ruptured, stick together, or present other difficulties that complicate their utilization in the field due to their physical characteristics. To produce insect cadavers that have better handling, packaging and application properties, they can be formulated with protective coverings. These coverings can give the body durability without affecting the nematodes inside the host. Thus, infected larvae can be packed, transported, and distributed to the target site without the risk of breaking or sticking to each other (Shapiro-Ilan, Lewis, Behle, & McGuire, 2001). In addition, coatings can reduce the stress generated by adverse environmental conditions (Hussaini, Nagesh, Rajeshwari, & Dar, 2004). For example, gelatin capsules can effectively isolate insect cadavers from the soil, even preventing damage caused by scavenging ants (Del Valle et al., 2009).

Different coverings have been attempted. Shapiro-Ilan et al. (2001) tested clay, gluten, starch and lignin in *G. mellonella* larvae infected with *H. bacteriophora* Hb. All formulations provided greater IJ survival and tolerance to desiccation for insect cadavers formulated 4 days after infection compared with those formulated after 8 days. Del Valle et al. (2009) tried commercial calcitic calcareum, formulated as a powder and as an aqueous suspension; talc (containing sulphur) formulated as a powder and as an aqueous suspension; and gelatin capsules. These formulations did not negatively impact IJ production except the calcareum formulation applied 8 days after infection. Perhaps this was a consequence of the shorter period between formulation and the beginning of IJ emergence, since the IJs started emerging only 2 days after formulation. Lower calcareum concentration around cadavers formulated with the aqueous suspension produced more IJs, compared to the cadavers formulated with calcareum only. Apparently the osmotic potential is

Fig. 8.2 Different coverings over *Galleria mellonella* cadavers. *From left to right*: calcitic calcareum formulated as a powder and as an aqueous suspension; gelatin capsule, talc formulated as a powder and as an aqueous suspension

reduced by dilution and time. Therefore, in field applications the physical–chemical characteristics of the soil (Andrén & Lagerlöf, 1983; Barbercheck, 1992) must be considered, since these can influence the covering used on cadavers (Fig. 8.2).

Ansari, Hussain, and Moens (2009) demonstrated that a kaolin–starch coating provided protection and preservation of insect cadavers, since nematode reproductive potential and virulence were not affected. They compared emergence from *G. mellonella* cadavers infected by *H. bacteriophora* CLO51 in pots, greenhouse and in the field and found that cadavers formulated in kaolin was significantly higher than from non–coated cadavers. Under greenhouse conditions, efficacy of freshly formulated (8 days post–infection) cadavers of *H. bacteriophora* CLO51 provided significantly higher control of *Hoplia philanthus* Füessly (Coleoptera: Scarabaeidae) (62 %) than 3-month–old cadavers (31 %) or aqueous applications of the same EPN (39 and 43 %). Similarly, under field conditions, significantly higher *H. philanthus* control was achieved with freshly formulated cadavers (39 %) than with 3-month–old cadavers (21 %) or with aqueous applications (24 and 28 %) of *H. bacteriophora* CLO51 2 weeks after application. Additionally, after 1 year, cadaver applications provided >90 % *H. philanthus* control, while aqueous applications of *H. bacteriophora* CLO51 gave only 55 % control. This work confirms that this technique can prolong the persistence of the IJs in the soil.

The insect cadaver firmness can vary with nematode species and life cycle within the host. Insect cadavers infected with *H. bacteriophora* Hb have levels of physical damage above 80 % 9 days after infection (Shapiro-Ilan et al., 2001), whereas insect cadavers infected with *H. baujardi* LPP7 still have physical integrity 8 days after infection (Del Valle et al., 2009). As mentioned above, hard–bodied insects such as *T. molitor* have natural resistance to rupturing or sticking together and thus coating formulations are not critical (Shapiro-Ilan, Tedders, & Lewis, 2008). Nonetheless, a formulation for protecting *T. molitor* infected cadavers using tape (where the nematodes emerge from the sides of the tape) was developed; the tape packaging approach further enhances handling and was facilitated with an automated machine that places the cadavers on a moving belt (Morales-Ramos et al., 2013; Shapiro-Ilan, Morales-Ramos, Rojas, & Tedders, 2010).

8.4.3.2 Number of Infective Juveniles

The application of insect cadavers in the field presents some challenges in terms of quantifying nematode application rates. Monteiro, Matos, Araújo, Perinotto, et al. (2014) tested different nematode species against engorged females of *Boophilus microplus* (Canestrini) (Acari: Ixodidae). The percentage of control was higher than 99 % for all groups treated with *H. bacteriophora* HP88 and *H. indica* LPP1 (2, 4 or 6 insect cadavers). Regarding the groups treated with *S. carpocapsae* ALL and *S. feltiae* SN, the highest level of control was found in the treatments with 2 and 4 cadavers, respectively, but an increase in the number of cadavers in subsequent treatments resulted in a decrease in effectiveness. The addition of cadavers in the treatment certainly caused a considerable increase in the number of nematodes, since one *G. mellonella*larvae produces on average approximately 100,000–300,000 IJs (Dolinski, Del Valle, Burla, & Machado, 2007; Poinar, 1990). In many cases there is a positive correlation between the number of IJs applied and increased efficacy (Monteiro, Prata, et al., 2010). Therefore, it is advisable to know how many IJs come out of the insect cadavers and over what sort of time period, being all of that in a species–specific basis.

8.4.3.3 Insect Choice

The choice of insect host can impact yield and efficacy of cadaver applications. Cadavers of *G. mellonella* and *T. molitor* larvae are the most commonly used host species to introduce IJs to the soil. In general, most EPN species have higher rates of production from *G. mellonella* compared with *T. molitor* (Molina, Moino, & Cavalcanti, 2004; Monteiro, Matos, Araújo, Campos, et al., 2014; Shapiro-Ilan et al., 2002).

Despite lower IJ yield, the formulation of EPNs in larvae of *T. molitor*, especially of the genus *Heterorhabditis*, persists because the cost of producing *T. molitor* larvae is 4–5 times lower than the production cost of *G. mellonella*. Shapiro-Ilan et al. (2002) reported that the current cost on a per–individual basis in the United States is approximately $0.0120 and $0.0025 for *G. mellonella* and *T. molitor*, respectively. Thus, the inferior performance can be offset by the possibility of using a larger number of *T. molitor* cadavers per area still with a lower production cost. Another aspect that can favor the use of *T. molitor* larvae, is the existence of several companies that breed this coleopteran for various purposes such as feed for birds and fish. In other countries commercial mass rearing of other hosts such as *G. mellonella* is also in place and could facilitate production of cadavers using these insects. Another positive point that can offset the lower performance of the formulation in *T. molitor* is that larvae present a tougher texture after EPN infection, making it easier to preserve the integrity of the body during storage, handling and application of cadavers (Shapiro-Ilan et al., 2008).

Various other hosts have been investigated for *in vivo* nematode production, including the navel orange worm, *Amyelois transitella* (Walker) (Lepidoptera: Pyralidae), tobacco bud worm, *Heliothis virescens* (Fabricus) (Lepidoptera: Noctuidae), cabbage looper, *Trichoplusi ani* (Hübner) (Lepidoptera: Noctuidae), pink bollworm, *Pectinophora gossypiella* (Saunders) (Lepidoptera: Gelechiidae), beet armyworm, *Spodoptera exigua* (Hübner) (Lepidoptera: Noctuidae), corn earworm, *Helicoverpa zea* (Boddie) (Lepidoptera: Noctuidae), house cricket, *Acheta domesticus* (L.) (Orthoptera: Gryllidae), and various beetles (Coleoptera) (Blinova & Ivanova, 1987; Cabanillas & Raulston, 1994; Elawad, Gowen, & Hague, 2001; Grewal, Converse, & Georgis, 1999; Lindegren, Hoffman, Collier, & Fries, 1979). Generally, nematode yield is proportional to insect size (Blinova & Ivanova, 1987; Flanders, Miller, & Shields, 1996). However, IJ yield per unit mass of insect within host species, and susceptibility to infection, is often inversely proportional to host size or age (Blinova & Ivanova, 1987; Shapiro, Cate, Pena, Hunsberger, & McCoy, 1999; Shapiro-Ilan, Han, & Qiu, 2014). In addition to nematode yield, other factors that play into choosing the host species include ease of insect culture and susceptibility to IJs (Blinova & Ivanova, 1987; Shapiro-Ilan & Gaugler, 2002). Susceptibility of the host can be enhanced through improved host diets that increase both nematode virulence and insect production efficiency (Shapiro-Ilan, Rojas, Morales-Ramos, & Tedders, 2012). Ultimately, the choice of host species and nematode for *in vivo* production should depend on nematode yield per cost of insect and the suitability of the nematode to the pests that are being targeted (Blinova & Ivanova, 1987; Shapiro-Ilan et al., 2002).

8.4.4 Interval Between Nematode Infection and Soil Application

Understanding the consequences on IJ emergence at different intervals between infection and soil application is crucial to establish an ideal timeframe for application. Therefore, it is necessary to evaluate the influence of different time periods between the infection and soil application of insect cadavers, infected with different nematode species on the emergence of the IJs nematodes.

In the case of *H. baujardi* LPP7, *G. mellonella* larvae cadavers were placed individually in plastic cups filled with soil from a commercial guava orchard after 6, 8, 10 and 12 days after infection. The number of emerging IJs was counted after 2 weeks. The time period of 6–10 days between infection and application in soil resulted in higher emergence of IJs, so this method was recommended when insect cadavers are chosen as the application method. On the other hand, handling and manipulating the insect cadavers before the development of hermaphrodites in the insect cadaver are not recommended (Dolinski et al., 2007).

8.5 Practical Application Examples

A number of studies have indicated efficacy of the insect cadaver approach. For instance, in an early example of the approach, Jansson, Lecrone, and Gaugler (1993) demonstrated field efficacy of *G. mellonella* infected with *H. bacteriophora* for control of the sweet potato weevil, *Cylasformicarius elegantulus* (Summers) (Coleoptera: Curculionidae). In a later example, Dillon, Downes, Ward, and Griffin (2007) investigated control of *Hylobius abietis* (L.) (Coleoptera: Curculionidae) in pine stumps using *G. mellonella* infected with *H. bacteriophora*; higher *H. abietis* control was observed with formulated cadavers (90 % control) compared with aqueous application of EPNs 1 year post–treatment (55 % control). In the sections below we review some case study examples in more detail.

8.5.1 Greenhouse and Field Studies

In a greenhouse study, Shapiro-Ilan, Lewis, and Tedders (2003) compared the efficacy of EPNs applied in aqueous suspension with application in infected *G. mellonella* cadavers. Two species of EPN were tested against two pest species; *Diaprepes* root weevil, *Diaprepes abbreviatus* (L.) (Coleoptera: Curculionidae) with *H. indica* Hom1, and the black vine weevil, *Otiorhynchus sulcatus* (F.) (Coleoptera: Curculionidae) with *H. bacteriophora* Oswego. They observed that the mortalities of *D. abbreviatus* and *O. sulcatus* were higher in the field infected insect cadaver treatment than the aqueous applications. This study indicated that EPN application in infected insect cadavers tends to be more effective than applied in aqueous solution. The increased efficacy observed in the insect cadaver applications may have been due to additional physiological stress in the aqueous application (during temporary storage in water or upon application). Superior efficacy in the cadaver application might also have been due to compounds in the infected host cadaver that can enhance nematode infectivity or dispersal.

In another study conducted in small pots filled with soil, the efficacy of EPNs emerging from *T. molitor* infected cadavers was tested against *D. abbreviatus* and the small hive beetle, *Aethina tumida* (Murray) (Coleoptera: Nitidulidae) (Shapiro-Ilan et al., 2010). The insect cadavers were enclosed in a masking tape package (see text above on coatings and formulations) and were infected with *H. indica*. Insect cadavers without tape packaging were also applied. A greenhouse experiment was also conducted in a similar manner measuring survival of *D. abbreviatus*. In all experiments, both the tape and no–tape treatments caused significant reductions in insect survival relative to the control, with no differences between the nematode treatments. Fifteen days post–application, the infected host treatments caused up to 78 % control in *A. tumida*, and 75–91 % control of *D. abbreviatus*.

However, application of insect cadavers infected with *Heterorhabditis* sp. CCA and *Heterorhabditis* sp. JPM3 against *Dysmicoccus texensis* (Tinsley) (Hemiptera: Pseudococcidae) resulted in unsatisfactory control levels both in the greenhouse and field application (Alves, Moino, Santa-Cecilia, Rohde, & Silva, 2006). Although specific reasons for that are not mentioned by the authors, we can hypothesize that the environmental conditions were not favorable for IJs to exit from the insect cadaver or for their movement in the soil.

8.5.2 The Guava Weevil

Different pests attack fruits, leaves and trunks of the guava plant, causing more or less damage depending on the region or country. The main pest to the fruit itself is the guava weevil, *Conotrachelus psidii* Marshall (Coleoptera: Curculionidae) because adults and larvae directly affect fruit quality (Souza, Haga, & Souza, 2003). In Brazil, adults are present in orchards during the summer, appearing in September–October and remaining until March. Females lay eggs in immature fruit (3–4 cm diameter) and larvae progress through four instars as the fruit develops. Infestation leads to accelerated fruit maturation and fruit drop on ripening. Subsequently, larvae crawl into the soil where they develop into pre-pupae. Individuals may remain in this stage for up to 6 months before pupation and development into adults (Bailez, Viana-Bailez, Lima, & Moreira, 2003; Boscán de Martinez & Cásares, 1982) (Fig. 8.3a–d). The usual control methods involve weekly applications of insecticides to suppress adults, but most of those currently in use for guava weevil control will be discontinued (Agência Nacional de Vigilância Sanitária – ANVISA, 2004; Souza et al., 2003). Without chemical control, the percentage of damaged fruit in heavily infested orchards can reach 100 % (Boscán de Martinez, & Cásares, 1980). The amount of fruit attacked has been increasing over the past 3 years, possibly due to the development of insecticide resistance Dolinski (unpublished). Poorly timed chemical applications and the tendency for adult weevils to hide in the litter around trees and avoid contact with the chemicals could also be involved (Denholm & Rolland, 1992).

The virulence of nine species/populations of EPNs to 4th–instar weevils was assessed in the laboratory. Larval mortality in Petri dish assays with sterile sand at 100 IJs/larva, ranged from 33.5 to 84.5 %, with the heterorhabditids being the most virulent (Fig. 8.3e). For *H. baujardi* LPP7, the LT_{50} and LT_{90} for 100 IJs were 6.3 and 9.9 days, whereas the LC_{50} and LC_{90} over 7 days were 52 and 122.2 IJs (Dolinski, Del Valle, & Stuart, 2006). In a greenhouse study with guava trees in 20-L pots (10 weevil larvae/pot), and doses of 500, 1,000 or 2,000 IJs/pot (0.17, 0.35 or 0.7 IJ/cm^2), *H. baujardi* LPP7 caused 30 and 58 % mortality at the two highest doses (Dolinski et al., 2006). It is important to point out that *H. baujardi* LPP7 was the only indigenous nematode tested, and it was isolated from the tropical forest of Rondônia, in the north of Brazil (Dolinski, Kamitani, Machado, & Winter, 2008).

Fig. 8.3 (**a**) Healthy guava fruit. (**b**) Guava fruit damaged by the guava weevil. (**c**) 4th instar larvae leaving the fruit going to the soil. (**d**) Adult guava weevil. (**e**) 4th instar larva infected with *Heterorhabditis baujardi* LPP7. (**f**) Guava growers from GOIACAM

Del Valle, Dolinski, Barreto, Souza, and Samuels (2008) assessed the susceptibility of the guava weevil to *H. baujardi* LPP7 IJs in the greenhouse and under field conditions, and applied nematodes though infected insect cadavers. Field persistence of these nematodes in the soil was evaluated through *G. mellonella*–baiting. Insect cadaver concentrations of 2, 4, and 6 applied in pots in the greenhouse experiment caused significantly greater mortality than the control. Significant differences were observed in the field between the control and treatments only when

6 cadavers/0.25 m^2 were applied. Infective juveniles from the cadavers persisted 6 weeks after application in the field, but decreased sharply thereafter (Del Valle, Dolinski, Barreto, et al., 2008).

In 2008, the Universidade Estadual do Norte Fluminense Darcy Ribeiro, lead by C. Dolinski started working with a small group of farmers with the objective of establishing an integrated pest management (IPM) program in their orchards. Approximately 20 farmers organized in an association named GOIACAM (Associação do Produtores de Goiabas de Cachoeiras de Macacu) in the state of Rio de Janeiro, Brazil, (Fig. 8.3f) paid for EPN registration and were trained to rear *G. mellonella* themselves. The *G. mellonella* larvae were taken to the lab, infected with *H. baujardi* LPP7 and then returned to the farms as infected–cadavers for application. The first results showed a decrease of 40–70 % in adult weevil emergence, in plots of 9 m^2 where 20 cadavers were used, compared to control trees with no nematode application. When neem cake was also applied below the canopy for larval control, an additive effect occurred, with the reduction in emergence reaching almost 80 %. The farmers also initiated culture control by removing all damaged fruit from the orchards, which reduced the pest inoculum for the following year. Between rows, *Crotalaria* sp. and other Leguminosae were planted to increase the soil fertility and serve as refugia for natural enemies. Because insecticides are not being used in those experimental areas, beneficial arthropods such as coccinellids and chrysopids could be seen more often within the orchards where nematodes were applied. Using these strategies reduced the production costs by 40 %. Some growers use sprinkler irrigation, so Lara, Dolinski, Sousa, and Daher (2008) evaluated the influence of irrigation application of EPNs on the viability, infectivity and host search capability of *H. baujardi* LPP7 IJs. The results demonstrated that the irrigation system did not adversely affect any of the factors described previously. Today, growers can choose from these different application methods, based on their situation.

The guava weevil population in the areas where the IPM was implemented in 2008 was very low compared to what it was when we first applied nematodes. We routinely recovered 20–30 adults from each tree prior to initiating GOIACAM, compared to just 2 or 3 adults in 2014. This population reduction reflects directly on the fruit damage rate, which is now typically 1 % of the fruits produced. The weevil is still present, but there is an understanding that it is at a new lower equilibrium in the orchards. Currently there are at least 10 ha of guava trees in the IPM program.

In 2012, a new project was started aiming to convert orchards under both IPM and conventional management to organic management systems. The consumption and demand for organically grown fruit is increasing. Because guava planted in these areas is exclusively for direct consumption, there is a customer desire for less pesticide use. One grower showed interest, and the results have been very promising. In the organic area, the guava weevil has been controlled by *H. indica* LPP30, a population isolated from a guava orchard (Minas, 2012), and other insect pests are being controlled by neem oil. Cattle manure was used instead of chemical fertilization. The fruit production wasanalyzed in 2013 from organic and conventional areas and the results were similar, although the guavas from the organic areas were smaller (less total weight). There is a need for continuing evaluation of insect presence and fruit quality instead of only the quantity to make sure the organic system can be economically beneficial.

8.5.3 Cattle Ticks

The cattle tick *Rhipicephalus microplus* Canestrini (Acari: Ixodidae) causes extensive economic damage to milk and meat production systems in various regions of the world. The application of chemical acaricides remains the main method of control. However, the systematic and often improper use of these chemicals has resulted in selection of resistant ticks (Furlong, Martins, & Prata, 2007).

Studies under laboratory conditions have shown that EPNs represent a promising alternative to control *R. microplus* (Monteiro, Furlong, et al., 2010; Monteiro, Prata, et al., 2010, 2012; Silva et al., 2012; Vasconcelos et al., 2004). Also, previous simulation studies have shown that application of different EPNs to the soil is effective against *Rhipicephalus annulatus* (Say) (Acari: Ixodidae) (Alekseev, Glazer, & Samish, 2006; Samish & Glazer, 2001) and *Dermacentor nitens* Neumann (Acari: Ixodidae) (Monteiro, Matos, Araújo, Perinotto, et al., 2014). The biological method using EPNs focusing on control of the non–parasitic phase of ticks can be effective, since engorged females at the time of oviposition seek environments with high moisture which are protected from solar radiation, a condition that also favors the survival of EPNs.

Insect cadaver application has been tested with different EPNs using two different insect host species against *R. microplus* placed in Petri dishes with sand. The percentage of control was above 95 % in all groups treated with *H. bacteriophora* HP88 and *H. indica* LPP1 and in the treatment with four insect cadavers infected with *S. feltiae* SN. In the groups treated with *S. carpocapsae* All, higher efficacy was observed in the treatment with two insect cadavers, with a control percentage of 80 %. When *G. mellonella* and *T. molitor* larvae were tested, each infected by the two most virulent EPNs, there were differences between hosts. *H. bacteriophora* HP88 and *H. indica* LPP1 in *T. molitor* caused control levels of 82.4 and 84.9 %, respectively, but 99.9 % was reached when the EPNs were in *G. mellonella* larvae (Monteiro, 2014).

Under semi–natural conditions (pots with sand in a greenhouse), *H. bacteriophora* HP88, *H. baujardi* LPP7, *H. indica* LPP1 and *H. bacteriophora* LPP30 were tested against engorged females of *R. microplus* using *G. mellonella* as the insect cadaver. The mortality was 78 % in the groups treated with *H. bacteriophora* LPP30 and *H. indica* LPP1, and reached 100 and 98 % for the treatments with *H. bacteriophora* HP88 and *H. baujardi* LPP7, respectively. Population LPP30 was virulent under laboratory conditions, but it was not the most effective in semi–natural conditions. This experiment also demonstrated the IJs' persistence, because

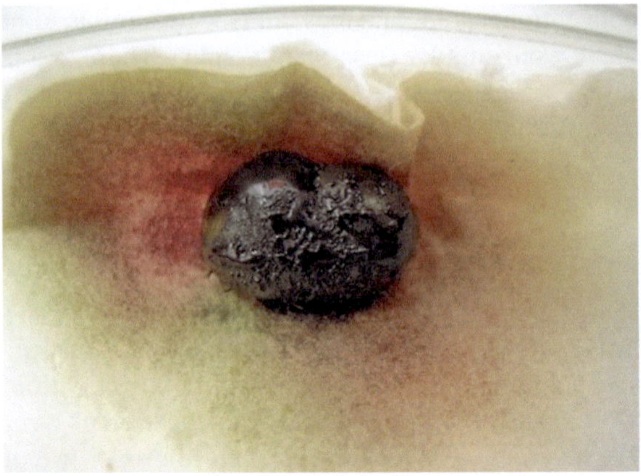

Fig. 8.4 Engorged female cattle tick infected with *Heterorhabditis bacteriophora* HP88. Nematodes living in the tick cadaver. Reddish hemolymphin the filter paper caused by the symbiotic bacteria

nematodes remained active and virulent for over 65 days after application in the soil, and were still able to infect engorged female cattle ticks at the end of the period (Fig. 8.4) (Monteiro, 2014).

8.6 Conclusions and Future Directions

Pest control using EPNs formulated as insect cadavers is an attractive approach for several reasons. Infective juveniles emerged from cadavers are more infective and have a higher dispersal capacity, and prolonged longevity (as observed under laboratory conditions) compared with IJs applied in aqueous suspension. Furthermore, the cadaver itself appears to serve as protection against harmful environmental extremes such as freezing and desiccation. For *in vivo* production, the insect cadaver approach is economically advantageous because harvesting and IJ concentration is not necessary (thereby reducing labor costs). The use of insect cadavers also avoids the costs of storage of IJs and avoids the energy expenditure associated with the metabolism of IJs. This "optimization" in the use of energy reserves of IJs favors the search for hosts and increases tolerance against environmental stresses. Moreover, IJs emerge at the same place where target hosts are, potentially facilitating nematode–host contact. Additionally, the effectiveness of the application can be enhanced through the use of coatings on the cadavers. The implementation of this technology is simple and inexpensive, and may be used by farmers in various technological situations.

Despite the advantages listed above, and the evidence that cadaver application can be highly efficacious in the field, additional research is needed before the approach is likely to be used widely. Like many biocontrol approaches, use of the EPN cadaver can be limited by competition with other pest control tactics. It remains to be seen whether the cadaver approach can be economically competitive with chemical or other biorational insecticides as well as *in vitro* produced EPN products.

The cost of the cadaver approach can be reduced if *in vivo* production methods to generate the infected host are further streamlined and automated; thus research toward improved production is warranted. Additional research on mechanized distribution of cadavers and optimization of rates is also needed. Finally, fundamental research is required to elucidate aspects of the cadaver approach, e.g., to determine the chemicals associated with the cadaver that trigger infection or dispersal, and investigate the ecological impact of cadaver application on EPN populations and the soil community post–application.

References

Agência Nacional de Vigilância Sanitária – ANVISA. (2004). www.anvisa.gov.br. Revised 15 Jan 2014.

Akhurst, R. J., & Boemare, N. E. (1990). Biology and taxonomy of *Xenorhabdus*. In R. Gaugler & H. K. Kaya (Eds.), *Entomopathogenic Nematodes in Biological Control* (pp. 75–90). Boca Raton, FL: CRC Press.

Alekseev, E., Glazer, I., & Samish, M. (2006). Effect of soil texture and moisture on the activity of entomopathogenic nematodes against female *Boophilus annulatus* ticks. *BioControl, 51*, 507–518.

Alves, V. S., Moino, A., Jr., Santa-Cecilia, L. V. C., Rohde, C., & Silva, M. A. T. (2006). Testes em condições para o controle de *Dysmicoccus texensis* (Tinsley) (Hemiptera, Pseudococcidae) em cafeeiro com nematóides entomopatogênicos do gênero *Heterorhabditis* (Rhabditida, Heterorhabditidae). *Revista Brasileira de Entomologia, 53*(1), 139–143.

Andrén, O., & Lagerlöf, J. (1983). Soil fauna (microarthropods, enchytraeids, nematodes) in Swedish agricultural cropping systems. *Acta Agriculturae Scandinavica, 33*, 33–52.

Ansari, M. A., Hussain, M. A., & Moens, M. (2009). Formulation and application of entomopathogenic nematode–infected cadavers for control of *Hopliaphilanthus* in turfgrass. *Pest Management Science, 65*, 367–374.

Bailez, O. E., Viana-Bailez, A. M., Lima, J. O. G., & Moreira, D. D. O. (2003). Life–history of the guava weevil, *Conotracheluspsidii* Marshall (Coleoptera: Curculionidae), under laboratory conditions. *Neotropical Entomology, 32*, 203–207.

Bal, H. K., Taylor, R. A. J., & Grewal, P. S. (2014). Ambush foraging entomopathogenic nematodes employ "sprinters" for long–distance dispersal in the absence of hosts. *Journal of Parasitology, 100*, 422–432.

Barbercheck, M. E. (1992). Effect of soil physical factors on biological control agents of soil pests. *Florida Entomology, 75*, 539–548.

Baur, M. E., Kaya, H. K., & Strong, D. R. (1998). Foraging ants as scavengers of entomopathogenic nematode–killed insects. *Biological Control, 12*, 231–236.

Blinova, S. L., & Ivanova, E. S. (1987). Culturing the nematode–bacterial complex of *Neoaplectana carpocapsae* in insects. In M. D. Sonin (Ed.), *Helminths of Insects* (pp. 22–26). New Delhi, India: American Publishing Co.

Boff, M. I. C., Wiegers, G. L., Gerritsen, L. J. M., & Smits, P. H. (2000). Development of the entomopathogenic nematode *Heterorhabditis megidis* population NLH–E87.3 in *Galleria mellonella. Nematology, 2*, 303–308.

Boscán de Martinez, N., & Cásares, R. (1980). El gorgojo de la guayaba *Conotrachelus psiddi* Marshal (Coleptera: Curculionidae). I. Evaluacion de daños. *Agronomia Tropical, 30*, 77–83.

Boscán de Martinez, N., & Cásares, R. (1982). Distribución en el tiempo de las fases del gorgojo de la guayaba *Conotrachelus psidii* Marshall (Coleoptera: Curculionidae) en el campo. *Agronomia Tropical, 31*, 123–130.

Cabanillas, H. E., & Raulston, J. R. (1994). Evaluation of the spatial pattern of *Steinernema riobravis* in corn plots. *Journal of Nematology, 26*, 25–31.

Campbell, J., & Gaugler, R. (1993). Nictation behaviour and its ecological implications in the host search strategies of entomopathogenic nematodes (Heterorhabditidae and Steinernematidae). *Behaviour, 126*, 154–169.

Del Valle, E. E., Dolinski, C., Barreto, E. L. S., Souza, R. M., & Samuels, R. I. (2008). Efficacy of *Heterorhabditis baujardi* LPP7 (Nematoda: Rhabditida) applied in *Galleria mellonella* (Lepidoptera: Pyralidae) insect cadavers to *Conotrachelus psidii* (Coleoptera: Curculionidae) larvae. *Biocontrol Science and Technology, 18*, 33–41.

Del Valle, E. E., Dolinski, C., Barreto, E. L. S., & Souza, R. M. (2009). Effect of cadaver coatings on emergence and infectivity of the entomopathogenic nematode *Heterorhabditis baujardi* LPP7 (Rhabditida: Heterorhabditidae) and the removal of cadavers by ants. *Biological Control, 50*, 21–24.

Del Valle, E. E., Dolinski, C., Souza, R. M., & Samuels, R. I. (2005). Performance de *Heterorhabditis baujardi* LPP7 (28) (Nematoda: Rhabditida), selecionada para tolerância a elevadastemperaturas, no controle de *Conotrachelus psidii* (Coleoptera: Curculionidae). *Nematologia Brasileira, 29*, 199–205.

Del Valle, E. E., Dolinski, C. M., & Souza, R. M. (2008). Dispersal of *Heterorhabditis baujardi* LPP7 (Nematoda: Rhabditida) applied to the soil as infected host cadavers. *International Journal of Pest Management, 54*(2), 115–122.

Denholm, I., & Rolland, M. W. (1992). Tactics for managing pesticide resistance in arthropods: Theory and practice. *Annual Review Entomology, 37*, 92–112.

Dillon, A. B., Downes, M. J., Ward, D., & Griffin, C. T. (2007). Optimizing application of entomopathogenic nematodes to manage large pine weevil, *Hylobius abietis* L. (Coleoptera: Curculionidae) populations developing in pine stumps, *Pinus sylvestris*. *Biological Control, 40*, 253–263.

Dolinski, C., Del Valle, E. E., Burla, R. S., & Machado, I. R. (2007). Biological traits of two native Brazilian entomopathogenic nematodes (Rhabditida: Heterorhabditidae). *Nematologia Brasileira, 31*(3), 180–185.

Dolinski, C., Del Valle, E. E., & Stuart, R. J. (2006). Virulence of entomopathogenic nematodes to larvae of guava weevil, *Conotrachelus psidii* (Coleoptera: Curculionidae), in laboratory and green house experiments. *Biological Control, 38*, 422–427.

Dolinski, C., Kamitani, F. L., Machado, I. R., & Winter, C. E. (2008). Molecular and morphological characterization of heterorhabditid entomopathogenic nematodes from the tropical rainforest in Brazil. *Memórias do Instituto Oswaldo Cruz, 103*, 150–159.

Efron, D., Nestel, D., & Glazer, I. (2001). Spatial analysis of entomopathogenic nematodes and insect hosts in a citrus grove in a semi–arid region in Israel. *Environmental Entomology, 30*(2), 254–261.

Elawad, S. A., Gowen, S. R., & Hague, N. G. M. (2001). Progeny production of *Steinernema abbasi* in lepidopterous larvae. *International Journal of Pest Management, 47*, 17–21.

Flanders, K. L., Miller, J. M., & Shields, E. J. (1996). *In vivo* production of *Heterorhabditis bacteriophora* 'Oswego' (Rhabditida: Heterorhabditidae), a potential biological control agent for soil inhabiting insects in temperate regions. *Journal of Economic Entomology, 89*, 373–380.

Furlong, J., Martins, J. R. S., & Prata, M. C. A. (2007). O carrapato dos bovinos e a resistência: temos o que comemorar? Controle estratégico do carrapato dos bovinos. *A Hora Veterinária, 27*, 53–56.

Glazer, I., Kozodoi, E., Salame, L., & Nestel, D. (1996). Spatial and temporal occurrence of natural populations of *Heterorhabditis* spp. (Nematoda: Rhabditida) in a semiarid region. *Biological Control, 6*, 130–136.

Grewal, P. S., Converse, V., & Georgis, R. (1999). Influence of production and bioassay methods on infectivity of two ambush foragers (Nematoda: Steinernematidae). *Journal of Invertebrate Pathology, 73*, 40–44.

Grewal, P. S., Lewis, E. E., & Gaugler, R. (1994). Host finding behavior as a predictor of foraging strategy in entomopathogenic nematodes. *Parasitology, 108*, 207–215.

Grewal, P. S., Selvan, S., & Gaugler, R. (1994). Thermal adaptation of entomopathogenic nematodes: Niche breadth for infection, establishment and reproduction. *Journal of Thermal Biology, 19*, 245–253.
Gülcü, B., Hazir, S., & Kaya, H. K. (2012). Scavenger deterrent factor (SDF) from symbiotic bacteria of entomopathogenic nematodes. *Journal of Invertebrate Pathology, 110*, 326–333.
Hussaini, S. S., Nagesh, M., Rajeshwari, R., & Dar, M. H. (2004). Formulation of host cadavers infected with indigenous *Heterorhabditis* spp. (Nematoda: Heterorhabditidae: Rhabditida) isolates. *Entomon Journal, 29*(4), 339–344.
Jansson, R. K., Lecrone, S. H., & Gaugler, R. (1993). Field efficacy and persistence of entomopathogenic nematodes (Rhabditida: Steinernematidae, Heterorhabditidae) for control of sweet potato weevil (Coleoptera: Apionidae) in southern Florida. *Journal of Economic Entomology, 86*, 1055–1063.
Kaplan, F., Alborn, H. T., von Reuss, S. H., Ajredini, R., Ali, J. G., Akyazi, F., et al. (2012). Interspecific nematode signals regulate dispersal behavior. *PLoS ONE, 7*(6), e38735.
Kaya, H. K. (1990). Soil ecology. In R. Gaugler & H. K. Kaya (Eds.), *Entomopathogenic nematodes in biological control* (pp. 93–115). Boca Raton, FL: CRC Press.
Kaya, H. K., & Stock, P. (1997). Techniques in insect nematology. In L. A. Lacey (Ed.), *Manual of techniques in insect pathology* (pp. 281–324). San Diego, CA: Academic.
Koppenhöfer, A. M., Baur, M. E., Stock, S. P., Choo, H. Y., Chinnasri, B., & Kaya, H. K. (1997). Survival of entomopathogenic nematodes within host cadavers in dry soil. *Applied Soil Ecology, 6*, 231–240.
Lara, J. C., Dolinski, C., Sousa, E. F., & Daher, R. F. (2008). Effect of mini–sprinkler irrigation system on *Heterorhabditis baujardi* LPP7 (Nematoda: Heterorhabditidae) infective juvenile. *Scientia Agricola, 65*, 433–437.
Lewis, E. E., Gaugler, R., & Harrison, R. (1993). Response of cruiser and ambusher entomopathonic nematodes (Steinernematidae) to host volatile cues. *Canadian Journal of Zoology, 71*, 765–769.
Lewis, E. E., Grewal, P. R., & Gaugler, R. (1995). Hierarchical order of host cues in parasite foraging strategies. *Parasitology, 110*, 207–213.
Lewis, E. E., & Shapiro-Ilan, D. I. (2002). Host cadavers protect entomopathogenic nematodes during freezing. *Journal of Invertebrate Pathology, 81*, 25–32.
Lindegren, J. E., Hoffman, D. F., Collier, S. S., & Fries, R. D. (1979). Propagation and storage of *Neoaplectan acarpocapsae* Weiser using *Amyelois transitella* (Walker) adults. *USDA Advanced Agricola, 3*, 1–5.
Minas, R. S. (2012). *Biological characterization of one entomopathogenic nematode strain aiming to control the guava weevil (Conotracheluspsidii) in two cultivation systems.* (Doctoral Dissertation). Presented at Universidade Estadual do Norte Fluminense Darcy Ribeiro, Campos dos Goytacazes RJ, 137 pp.
Molina, J. P. A., Moino, A., Jr., & Cavalcanti, R. S. (2004). Produção *in vivo* de nematóides entomopatogênicos em diferentes insetos hospedeiros. *Arquivos do Instituto Biológico, 71*, 347–354.
Molyneux, A. S., & Bedding, R. A. (1984). Influence of soil texture and moisture on the infectivity of *Heterorhabditis* sp. D1 and *Steinernema glaseri* for larvae of the sheep blowfly, *Lucilia cuprina. Nematologica, 30*, 358–365.
Monteiro, C. M. O. (2014). *Controle de* Rhipicephalus microplus *(Acari: Ixodidae) com nematoides entomopatogênicos: Aplicação em formulação inseto–cadáver e compatibilidade com outros agentes de controle*. (Doctoral Dissertation). Presented at Universidade Federal Rural do Rio de Janeiro, RJ. 175 pp.
Monteiro, C. M. O., Furlong, J., Prata, M. C. A., Soares, A. E., Batistita, E. S. P., & Dolinski, C. (2010). Evaluation of the action of *Heterorhabditis bacteriophora* (Rhabditida: Heterorhabditidae) isolate HP88 on the biology of engorged females of *Rhipicephalus* (*Boophilus*) *microplus* (Acari: Ixodidae). *Veterinary Parasitology, 170*, 355–358.

Monteiro, C. M. O., Matos, R. S., Araújo, L. X., Campos, R., Bittencourt, V. R. E. P., Dolinski, C., et al. (2014). Entomopathogenic nematodes in insect cadaver formulations for the control of *Rhipicephalus microplus* (Acari: Ixodidae). *Veterinary Parasitology, 203*, 310–317.

Monteiro, C. M. O., Matos, R. S., Araújo, L. X., Perinotto, W. M. S., Bittencourt, V. R. E. P., Dolinski, C., et al. (2014). First report of pathogenicity of entomopathogenic nematodes of the genus *Heterorhabditis* on partially engorged females of *Dermacentor nitens* (Acari: Ixodidae). *Biologica Control, 69*, 78–81.

Monteiro, C. M. O., Prata, M. C. A., Furlong, J., Batista, E. S. P., Fazza, A. P., & Dolinski, C. (2012). The use of *Heterorhabditis bacteriophora* (Rhabditida: Heterorhabditidae) isolate HP88 for biological control of *Rhipicephalus microplus* (Acari: Ixodidae): The effect of different exposure times of engorged females to the nematodes. *Veterinary Parasitology* 185, 364–367

Monteiro, C. M. O., Prata, M. C. A., Furlong, J., Fazza, A. P., Silva, A. M. R., Andaló, V., et al. (2010). *Heterorhabditis amazonensis* (Rhabditidae: Heterorhabditidae), strain RSC–5, for biological control of the cattle tick *Rhipicephalus* (*Boophilus*) *microplus* (Acari: Ixodidae). *Parasitology Research, 106*, 821–826.

Morales-Ramos, J. A., Rojas, M. G. Shapiro-Ilan, D. I., & Tedders, W. L. (2011). *US Patent 8025027*. Automated Insect Separation System.

Morales-Ramos, J. A., Tedders, W. L., Dean, B., Shapiro-Ilan, D. I., & Rojas, M. G. (2013). US Patent 8,505,236 B1. Apparatus for packaging arthropods infected with entomopathogenic nematodes.

Perez, E. E., Lewis, E. E., & Shapiro-Ilan, D. I. (2003). Impact of host cadaver on survival and infectivity of entomopathogenic nematodes (Rhabditida: Steinernematidae, Heterorhabditidae) under desiccating conditions. *Journal Invertebrate Pathology, 82*, 111–118.

Poinar, G. O., Jr. (1990). Taxonomy and biology of Steinernematidae and Heterorhabitidae. In R. Gaugler & H. K. Kaya (Eds.), *Entomopathogenic Nematodes in Biological Control* (pp. 23–61). Boca Raton, FL: CRC Press.

Samish, M., & Glazer, I. (2001). Entomopathogenic nematodes for the biocontrol of ticks. *Trends in Parasitology, 17*, 368–371.

Shapiro, D. I., Cate, J. R., Pena, J., Hunsberger, A., & McCoy, C. W. (1999). Effects of temperature and host age on suppression of *Diaprepes abbreviatus* (Coleoptera: Curculionidae) by entomopathogenic nematodes. *Journal of Economic Entomology, 92*, 1086–1092.

Shapiro, D. I., & Glazer, I. (1996). Comparison of entomopathogenic nematode dispersal from infected hosts versus aqueous suspension. *Environmental Entomology, 25*, 1455–1461.

Shapiro, D. I., & Lewis, E. E. (1999). Comparisons of entomopathogenic nematode infectivity from infected hosts *versus* aqueous suspension. *Environmental Entomology, 28*, 907–911.

Shapiro, D. I., Lewis, E. E., Paramasivam, S., & McCoy, C. W. (2000). Nitrogen partitioning in *Heterorhabditis bacteriophora*–infected hosts and the effects of nitrogen on attraction/repulsion. 2000. *Journal of Invertebrate Pathology, 76*, 43–48.

Shapiro, D. I., & McCoy, C. W. (2000). Virulence of entomopathogenic nematodes to *Diaprepes abbreviatus* (Coleoptera: Curculionidae) in the laboratory. *Journal of Economic Entomology, 93*, 1090–1095.

Shapiro-Ilan, D. I., & Gaugler, R. (2002). Production technology for entomopathogenic nematodes and their bacterial symbionts. *Journal of Industrial Microbiology and Biotechnology, 28*, 137–146.

Shapiro-Ilan, D. I., Gaugler, R., Tedders, W. L., Brown, I., & Lewis, E. E. (2002). Optimization of inoculation for *in vivo* production of entomopathogenic nematodes. *Journal of Nematology, 34*, 343–350.

Shapiro-Ilan, D. I., Gouge, D. H., Piggott, S. J., & Fife, J. P. (2006). Application technology and environmental considerations for use of entomopathogenic nematodes in biological control. *Biological Control, 38*, 124–133.

Shapiro-Ilan, D. I., Han, R., & Dolinski, C. (2012). Entomopathogenic nematode production and application technology. *Journal of Nematology, 44*, 206–217.

Shapiro-Ilan, D. I., Han, R., & Qiu, X. (2014). Production of entomopathogenic nematodes. In J. Morales-Ramos, G. Rojas, & D. I. Shapiro-Ilan (Eds.), *Mass Production of Beneficial Organisms: Invertebrates and Entomopathogens* (pp. 321–356). Amsterdam: Academic.

Shapiro-Ilan, D. I., Lewis, E. E., Behle, R. W., & McGuire, M. R. (2001). Formulation of entomopathogenic nematode–infected–cadavers. *Journal Invertebrate Pathology, 78*, 17–23.

Shapiro-Ilan, D. I., Lewis, E. E., & Schliekelman, P. (2014). Aggregative group behavior in insect parasitic nematode dispersal. *International Journal for Parasitology, 44*, 49–54.

Shapiro-Ilan, D. I., Lewis, E. E., & Tedders, W. L. (2003). Superior efficacy observed in entomopathogenic nematodes applied in infected–host cadavers compared with application in aqueous suspension. *Journal of Invertebrate Pathology, 83*, 270–272.

Shapiro-Ilan, D. I., Morales-Ramos, J. A., Rojas, M. G., & Tedders, W. L. (2010). Effects of a novel entomopathogenic nematode–infected host formulation on cadaver integrity, nematode yield, and suppression of *Diaprepes abbreviatus* and *Aethina tumida* under controlled conditions. *Journal of Invertebrate Pathology, 103*, 103–108.

Shapiro-Ilan, D. I., Rojas, M. G., Morales-Ramos, J. A., & Tedders, W. L. (2012). Optimization of a host diet for *in vivo* production of entomopathogenic nematodes. *Journal of Nematology, 44*, 264–273.

Shapiro-Ilan, D. I., Tedders, W. L., & Lewis, E. E. (2008). *US Patent 7,374,773*. Application of entomopathogenic nematode–infected cadavers from hard–bodied arthropods for insect suppression.

Silva, E. R., Monteiro, C. M. O., Reis-Menine, C., Prata, M. C. A., Dolinski, C., & Furlong, J. (2012). Action of *Heterorhabditis indica* (Rhabditida: Heterorhabditidae) strain LPP1 on the reproductive biology of engorged females of *Rhipicephalus microplus* (Acari: Ixodidae). *Biological Control, 62*, 140–143.

Souza, J. C., Haga, A., & Souza, M. A. (2003). *Pragas da goiabeira*. Tech. Bull. no. 71, 60 pp.

Spence, K. O., Stevens, G. N., Arimoto, H., Ruiz-Vega, J., Kaya, H. K., & Lewis, E. E. (2011). Effect of insect cadaver desiccation and soil water potential during rehydratation on entomopathogenic nematode (Rhabditida: Steinernematidae and Heterorhabditidae) production and virulence. *Journal of Invertebrate Pathology, 106*, 268–273.

Stuart, R. J., & Gaugler, R. (1994). Patchiness in populations of entomopathogenic nematodes. *Journal of Invertebrate Pathology, 64*, 39–45.

Vasconcelos, V. O., Furlong, J., Freitas, G. M., Dolinski, C., Aguillera, M. M., Rodrigues, R. C. D., et al. (2004). *Steinernema glaseri* Santa Rosa strain (Rhabditida: Steinernematidae) and *Heterorhabditis bacteriophora* CCA strain (Rhabditida: Heterorhabditidae) as biological control agents of *Boophilus microplus* (Acari: Ixodidae). *Parasitology Research, 94*, 201–206.

Woodring, L., & Kaya, H. K. (1988). *Steinernematid and heterorhabditid nematodes: A handbook of techniques*. Arkansas Agricultural Experiment Station, Fayetteville, Arkansas. Series Bulletin 331, 30 pp.

Zhou, X., Kaya, H. K., Heungens, K., & Goodrich-Blair, H. (2002). Response of ants to a deterrent factor(s) produced by the symbiotic bacteria of entomopathogenic nematodes. *Applied and Environmental Microbiology, 68*, 6202–6209.

Zhu, H., Grewal, P. S., & Reding, M. E. (2011). Development of a desiccated cadaver delivery system to apply entomopathogenic nematodes for control of soil pests. *Applied Engineering in Agriculture, 27*, 317–324.

Chapter 9
Entomopathogenic Nematode Application Technology

David Shapiro-Ilan and Claudia Dolinski

9.1 Introduction

Biocontrol success when using entomopathogenic nematodes (EPNs) in the genera *Heterorhabditis* and *Steinernema* relies on a variety of factors including components of the application event itself. Successful application encompasses both abiotic and biotic influences (Grewal, 2002; Shapiro-Ilan, Gouge, & Koppenhöfer, 2002; Shapiro-Ilan, Gouge, Piggott, & Patterson Fife, 2006; Shapiro-Ilan, Han, & Dolinski, 2012). For example, a diverse array of equipment is available for EPN application including various spray technology and irrigation systems. The specific application equipment that is chosen and parameters associated with EPN distribution can have a direct impact on the level of pest suppression achieved. Additionally, the choice of nematode species, rate of application, and other concurrent management practices are critical to success. In this chapter we review recent literature on EPN application technology, discuss novel innovations, and explore opportunities for future improvement.

D. Shapiro-Ilan (✉)
Southeastern Fruit and Nut Research Laboratory, USDA-ARS, Byron, GA, USA
e-mail: david.shapiro@ars.usda.gov

C. Dolinski
Laboratório de Entomologia e Fitopatologia, Universidade Estadual do Norte Fluminense Darcy Ribeiro, Campos dos Goytacazes, RJ, Brazil
e-mail: Claudia.dolinski@censanet.com.br

© Springer International Publishing Switzerland (outside the USA) 2015
R. Campos-Herrera (ed.), *Nematode Pathogenesis of Insects and Other Pests*,
Sustainability in Plant and Crop Protection 1, DOI 10.1007/978-3-319-18266-7_9

9.2 Factors Affecting Efficacy of Application: Biotic and Abiotic

Regardless of the type of equipment used or the application target, a number of basic biotic and abiotic factors must be considered when applying EPNs.

9.2.1 *Biotic Factors*

A critical aspect to achieving biocontrol success is related to the choice of nematode species or population. Foremost, the most suitable nematode must be matched with the specific target pest. Issues to consider include virulence, host finding, and environmental tolerance and in some cases persistence (Shapiro-Ilan et al., 2002; Shapiro-Ilan, Han, & Dolinski, 2012; Shapiro-Ilan, Gouge, et al., 2006; Shapiro-Ilan, Stuart, & McCoy, 2006). Selecting the most suitable EPN can usually be addressed by simply screening a variety of candidate species and populations for possession of superior desired traits such as virulence, environmental tolerance, etc. The screening process is often accomplished by first narrowing down the candidates in laboratory comparisons; these types of assays have been successful in identifying superior EPN species or populations used in field suppression for numerous target pests, e.g., the plum curculio, *Conotrachelus nenuphar* (Herbst) (Coleoptera: Curculionidae) (Shapiro-Ilan, Leskey, & Wright, 2011; Shapiro-Ilan, Wright, Tuttle, Cooley, & Leskey, 2013) and western corn rootworm *Diabrotica virgifera virgifera* (LeConte) (Coleoptera: Chrysomelidae) (Toepfer, Peters, Ehlers, & Kuhlmann, 2008). However, the importance of confirming laboratory virulence in subsequent field trials cannot be overemphasized. An EPN that shows high virulence in the controlled environment of a laboratory could fail to control the target pest in the field due to factors that render the organism incompatible (Shapiro-Ilan, Bruck, & Lacey, 2012). A lack of understanding of the biological and ecological constraints required for pathogen persistence and proliferation in the environment is likely to lead to a discrepancy between laboratory and field efficacy (Shapiro-Ilan, Bruck, & Lacey, 2012). For example, *Steinernema feltiae* (Filipjev) Wouts, Mráček, Gerdin & Bedding (Rhabditida: Steinernematidae) was highly virulent to *C. nenuphar* in the laboratory, but failed to control the pest in peach orchards (in Georgia, USA), possibly due to unsuitable soil temperatures (Shapiro-Ilan, Mizell, Cottrell, & Horton, 2004). In conclusion, population selection based on virulence screening in the laboratory can be helpful but cannot be relied upon as the sole predictor for field efficacy. A recent emphasis on ecological considerations when selecting populations for microbial control is expected to increase pest management success rates (Shapiro-Ilan, Bruck, & Lacey, 2012).

Foraging strategy is another factor that can be considered when matching the nematode to the target host. Foraging strategies among EPN species vary along a continuum between ambushers, which generally sit and wait for a passing host, and cruisers that actively search for hosts (Lewis & Clarke, 2012). Although some species exhibit primarily ambusher-type behaviors and others are mainly cruiser types, others are considered intermediate in their foraging behavior, and thus EPNs can exhibit a combination of these behaviors to locate hosts (Lewis, 2002; Lewis & Clarke, 2012). In general, nematodes that exhibit primarily an ambush strategy, e.g., *Steinernema carpocapsae* (Weiser) Wouts, Mráček, Gerdin & Bedding (Rhabditida: Steinernematidae), may be most suitable for controlling mobile insects near the soil surface, whereas nematodes with more of a cruise strategy, e.g., *Steinernema glaseri* (Steiner) (Rhabditida: Steinernematidae) may be most suitable for suppressing less mobile insects below the soil surface (Lewis, 2002; Lewis, Gaugler, & Harrison, 1992). However, there are plenty of exceptions to this rule, e.g., in which *S. carpocapsae* causes high levels of efficacy despite the pest occurring below the soil surface (e.g., Shapiro-Ilan, Cottrell, Mizell, Horton, & Davis, 2009; Williams et al., 2013), and also one study indicates that a small proportion of *S. carpocapsae* tends to move relatively far from the application site (Bal, Taylor, & Grewal, 2014). Additionally, in some cases foraging strategy has been observed to vary depending on soil type or substrate (Kruitbos, Heritage, Hapca, & Wilson, 2010).

Also in relation to the organism selected for biocontrol purposes, the rate of application is of paramount importance. Generally, to be effective, EPNs must be applied to soil at minimum rates of 2.5×10^9 infective juvenile nematodes (IJs)/ha (or 25 IJs per treated cm^2) (Georgis, Dunlop, & Grewal, 1995; Georgis & Hague, 1991; Shapiro-Ilan et al., 2002; Shapiro-Ilan, Han, & Dolinski, 2012; Shapiro-Ilan, Gouge, et al., 2006). If the pest is especially susceptible or in a controlled environment such as in the greenhouse, lower application rates may be effective. For example, *S. carpocapsae* applied at a field rate of 12.5 IJs/cm^2, reduced black cutworm, *Agrotis ipsilon* (Hufnagel) (Lepidoptera: Noctuidae) damage in maize by more than 75 %, which was equally as effective as or more so compared with chemical insecticides that were tested (Levine & Oloumi-Sadeghi, 1992). In contrast, some insects that are less susceptible, or that are found deep below the soil surface, can require higher rates to achieve sufficient efficacy, e.g., the Diaprepes root weevil, *Diaprepes abbreviatus* (L.) (Coleoptera: Curculionidae) (McCoy, Shapiro, Duncan, & Nguyen, 2000; Shapiro-Ilan et al., 2002). Aboveground applications rates (e.g., to foliage) will vary greatly depending on the target pest and arena (Begley, 1990; Grewal, Ehlers, & Shapiro-Ilan, 2005). In one example, EPNs (*S. carpocapsae*) were applied to peach limbs infested with lesser peachtree borer, *Synanthedon pictipes* (Grote & Robinson) (Lepidoptera: Sesiidae), at a rate of 50,000 IJs per mL (resulting in 1 million IJs per infested wound) (Shapiro-Ilan, Cottrell, et al., 2010). Another successful example of foliar application was with *Heterorhabditis baujardi* (Phan, Subbotin, Nguyen & Moens) (Rhabditida: Heterorhabditidae) LPP7 and *S. carpocapsae* (All) used at a rate of 500 IJs per mL in a greenhouse against larvae of *Diatraea saccharalis* (Fabricius) (Lepidoptera: Crambidae) (Bellini & Dolinski, 2012); thus, it is clear that the range of aboveground application rates vary across target pests and arenas.

The potential for EPNs to recycle can also be considered. In general, as long as environmental conditions are amenable, nematode populations will remain high enough to provide effective pest control for approximately 2–8 weeks post-application (Duncan & McCoy, 1996; Kaya, 1990; Shapiro-Ilan et al., 2002). Consequently, seasonal re-application is often required. However, in some cases multi-season or multi-year control has been reported (Klein & Georgis, 1992; Parkman, Frank, Nguyen, & Smart, 1994; Shields, Testa, Miller, & Flanders, 1999). The potential for long-term persistence and recycling depends on various factors including soil type, ground cover, host and host density, and the particular nematode species or population (Kaya, 1990; Klein & Georgis, 1992; Shapiro, Obrycki, Lewis, & Jackson, 1999; Shapiro-Ilan et al., 2002; Shapiro-Ilan, Stuart, & McCoy, 2006).

Various biotic agents that occur within the same habitat can have positive, negative, or neutral effects on EPN applications (Kaya, 2002; Koppenhöfer & Grewal, 2005, see Chaps. 4 and 5). Antagonists may include various nematode pathogens or predators e.g., phages, bacteria, protozoans, nematophagous fungi, predacious mites and nematodes, etc. (Kaya, 2002; Ulug, Hazir, Kaya, & Lewis, 2014). Phoretic relationships, which may be considered beneficial in terms of enhancing EPN dispersal, have also been indicated with other soil organisms such as mites, earthworms, and isopods (Eng, Preisser, & Strong, 2005; Epsky, Walter, & Capinera, 1988; Shapiro, Tylka, Berry, & Lewis, 1995; Shapiro-Ilan & Brown, 2013). Interactions between EPNs and other entomopathogens may be synergistic as has been reported with *Paenibacillus popilliae* (Dutky) (Bacillales: Paenibacillaceae) (Thurston, Kaya, & Gaugler, 1994), *Bacillus thuringiensis* Berliner (Bacillales: Bacillaceae) (Koppenhöfer & Kaya, 1997), and *Metarhizium anisopliae* (Metschnikoff) Sorokin (Hypocreales: Clavicipitaceae) (Acevedo, Samuels, Machado, & Dolinski, 2007; Ansari, Shah, & Moens, 2006; Ansari, Tirry, & Moens, 2004), yet other studies indicate antagonism, e.g., with *Beauveria bassiana* (Balsamo) Vuillemin (Hypocreales: Cordycipitaceae) (Brinkman & Gardner, 2001) or *Isaria fumosorosea* (Wize) (Hypocreales: Cordycipitaceae) (Shapiro-Ilan, Jackson, Reilly, & Hotchkiss, 2004). The relationship between EPNs and other entomopathogens (antagonism, additivity, synergism) can vary depending on the timing or rate of application, nematode species, or virulence of the other entomopathogen (Acevedo et al., 2007; Barbercheck & Kaya, 1990; Koppenhöfer & Kaya, 1997; Shapiro-Ilan, Jackson, et al., 2004). Interactions among different EPN species within the soil environment may be competitive (Duncan, Dunn, Bague, & Nguyen, 2003) or the species may as coexist without apparent competition (Millar & Barbercheck, 2001). Duncan et al., (2003) reported that certain free-living bactivorous nematodes can increase insect mortality in the presence of EPNs, e.g., *Steinernema riobrave* Cabanillas, Poinar & Raulston (Rhabditida: Steinernematidae), but the interaction leads to decreased EPN reproduction. Competition for the infected-cadaver resource between EPNs and free-living bactivorous nematodes was also observed by Campos-Herrera, El-Borai, and Duncan (2012).

9.2.2 Abiotic Factors

Environmental conditions are critical to successful application of EPNs; factors of concern include protection from ultraviolet radiation, and maintaining adequate soil moisture/relative humidity and temperature (Kaya, 1990; Shapiro-Ilan, Han, & Dolinski, 2012; Shapiro-Ilan, Gouge, et al., 2006). Indeed, applications for control of aboveground pests have been limited due to environmental barriers that reduce survival and efficacy (e.g., such as UV radiation or desiccation) (Arthurs, Heinz, & Prasifka, 2004; Begley, 1990; Grewal & Georgis, 1999; Shapiro-Ilan, Han, & Dolinski, 2012; Shapiro-Ilan, Gouge, et al., 2006). Therefore, biocontrol success using EPNs has been largely achieved when EPNs are applied to soil or cryptic habitats. Additionally, given the detrimental effects of ultraviolet radiation (Gaugler & Boush, 1978), nematode applications are best implemented in the evening or early morning hours. Alternatively, exposure to ultraviolet radiation can be avoided through subsurface application (Cabanillas & Raulston, 1995); the advantages to such approaches, however, have not been observed in all studies (Schroeder et al., 1996; Wilson & Gaugler, 2004).

Soil moisture is required for EPN survival and movement. Therefore, when nematodes are applied to soil, irrigation is recommended for maintaining adequate moisture (Downing, 1994; Shapiro-Ilan, Gouge, et al., 2006; Shetlar, Suleman, & Georgis, 1988; Zimmerman & Cranshaw, 1991). In one recent report, which focused on control of *C. nenuphar*, irrigation did not increase EPN efficacy relative to non-irrigated plots, yet this situation is by far the exception and success without irrigation was attributed to adequate rainfall and a high water holding capacity in the soil (Shapiro-Ilan et al., 2013). Although irrigation is generally required it should also be noted that too much moisture in the soil may cause oxygen deprivation and restrict movement (Kaya, 1990; Koppenhöfer, Kaya, & Taormino, 1995; Wallace, 1958; Womersley, 1993). Optimum moisture levels will vary by nematode species and soil type (Koppenhöfer et al., 1995). For instance, in a sandy loam *S. carpocapsae* was infective at moisture levels as low as −5 Mpa, and had the highest infectivity at rates between −0.1 and −0.01 MPa, whereas *S. glaseri* required a minimum potential of −0.3 Mpa (Koppenhöfer et al., 1995).

Optimum temperatures for infection and reproduction will also vary among EPN species and populations (Grewal, Selvan, & Gaugler, 1994). Some nematodes such as *Heterorhabditis indica* Poinar, Karunakar & David (Rhabditida: Heterorhabditidae), *S. glaseri*, and *S. riobrave* are relatively heat tolerant while other species, such as *Heterorhabditis megidis* Poinar, Jackson & Klein (Rhabditida: Heterorhabditidae), *S. feltiae*, and *Heterorhabditis marelata* Liu & Berry (Rhabditida: Heterorhabditidae) generally more tolerant to cooler temperatures (Berry, Liu, & Groth, 1997; Grewal et al., 1994; Kung, Gaugler, & Kaya, 1991; Shapiro & McCoy, 2000; Shapiro-Ilan et al., 2011). Nematode species and populations also vary substantially in freezing tolerance, e.g., *S. feltiae* and *Heterorhabditis georgiana* Nguyen, Shapiro-Ilan & Mbata (Rhabditida: Heterorhabditidae) being relatively tolerant whereas *H. indica* and *Steinernema rarum* (de Doucet) (Rhabditida:

Steinernematidae) exhibiting poor tolerance (Shapiro-Ilan, Brown, & Lewis, 2014); freeze tolerance may be important in considering the potential for overwintering and seasonal control.

Various soil parameters contribute to the success of surface and below-ground applications. For example, a soil pH of 10 or higher is likely to be detrimental to EPN applications, whereas a range of 4–8, is not likely to have any significant effect (Kung, Gaugler, & Kaya, 1990a); thus the pH of most agricultural soils are unlikely to have an impact on EPN survival. Nonetheless, soil pH has been reported to affect native EPN distributions (Kanga, Waeyenberge, Hauser, & Moens, 2012). Soil texture can have a substantial effect on nematode movement and survival (Barbercheck, 1992; Kaya, 1990). Overall, relative to lighter soils, soils with higher clay content tend to restrict nematode movement and have potential for reduced aeration, which, can result in reduced nematode survival and efficacy (Dolinski, Pinto, Robaina, & Bellini, 2010; Georgis & Poinar, 1983; Kung, Gaugler, & Kaya, 1990b; Molyneux & Bedding, 1984). Nevertheless, exceptions to this trend have been observed and EPNs have been effective in a variety of soil types (Georgis & Gaugler, 1991; Shapiro, McCoy, Fares, Obreza, & Dou, 2000).

Success in biocontrol using EPNs can also be impacted by fertilizers and chemical pesticides. Similar to the impact of biotic agents, abiotic inputs can have positive, neutral, or negative effects on EPNs. Generally, fertilizers applied at recommended rates have little impact on EPN survival or virulence (Bednarek & Gaugler, 1997; Shapiro, Obrycki, Lewis, & Abbas, 1999; Shapiro, Tylka, & Lewis, 1996). However, high rates of chemical fertilizers (e.g., urea at 560 kg N per ha) or fresh manure can be detrimental to EPN persistence and efficacy (Bednarek & Gaugler, 1997; Shapiro, Tylka, & Lewis, 1996; Shapiro, Glazer, & Segal, 1996). Composted manure may be beneficial to EPN by reducing natural enemies, e.g., nematophagous fungi (Duncan et al., 2007). Some chemical pesticides are harmful to EPNs (e.g., abamectin, acephate, aldicarb, dodine, fenamiphos, methomyl, parathion, and teflubenuron), whereas others tend to be compatible and in some cases may be synergistic when applied with EPNs (e.g., carbaryl, chlorpyrifos, dimethoate, endosulfan, fonofos, tefluthrin, imidicloprid) (Alumai & Grewal, 2004; Koppenhöfer & Fuzy, 2008; Koppenhöfer & Grewal, 2005; Koppenhöfer & Kaya, 1998; Nishimatsu & Jackson, 1998; Shapiro-Ilan, Cottrell, & Wood, 2011). Akin to interactions with other microbial agents, the relationship between chemical pesticides and EPNs depends on the particular chemical and nematode species or population, dosages, and timing of application (Benz, 1971; Koppenhöfer & Grewal, 2005), and therefore specific combinations should be tested on a case-by-case basis.

9.3 Equipment

9.3.1 Overview

A diverse array of technology is suitable for application of EPNs including various spray equipment and irrigation systems (Shapiro-Ilan, Han, & Dolinski, 2012). Basically, EPNs can be applied with nearly all commercially available ground or aerial spray equipment, including pressurized sprayers, mist blowers, and electrostatic sprayers (Georgis, 1990). The choice of application equipment and associated parameters can have considerable impact on biocontrol efficacy. For example, sprayer type, nozzle or the variety of pumping system are some of the variables that can impact nematode efficacy following spray applications. Furthermore, the choice of application equipment used depends on the cropping system, and in each case there are a variety of handling considerations including volume, agitation, pressure and recycling time, environmental conditions, and spray distribution pattern (Grewal, 2002). Beyond the choice of application equipment, various formulations for EPNs may be used to facilitate the distribution of EPNs in aqueous suspension including activated charcoal, alginate and polyacrylamide gels, clay, diatomaceous earth, peat, sponge, vermiculite, and water dispersible granules (Georgis, 1990; Georgis et al., 1995; Shapiro-Ilan, Han, & Dolinski, 2012).

9.3.2 A Comparison of Application Equipment and Associated Parameters

Application equipment itself and associated parameters can have a substantial effect on EPN field efficacy (Brusselman et al., 2010; Bullock, Pelosi, & Killer, 1999; Curran, 1992; Hayes, Fitzpatrick, & Webster, 1999; Nilsson & Gripwall, 1999; Shields et al., 1999). For example, Curran (1992) reported that trickle irrigation was inferior to surface spray or multiple injections, and Hayes et al., (1999) reported that sprinkler irrigation was inferior to a boom sprayer. Settling of EPNs in slow moving irrigation systems (e.g., trickle) can result in poor distribution (Conner, McSorley, Stansly, & Pitts, 1998). However, unequal distribution in some irrigation systems (i.e., microjet) can be overcome through addition of extra emitters at the end of the lines (Duncan, Shapiro, McCoy, & Graham, 1999).

One concern that can affect nematode viability is the pressure differentials caused by various spray equipment. An earlier rule of thumb considered a pressure of approximately 2,000 kPa (290 psi) as a limit across EPN species (Georgis, 1990; Shapiro-Ilan, Gouge, et al., 2006). However, Fife, Derksen, Ozkan, and Grewal (2003) determined that the ability to withstand pressure varies among EPN species. The viability of *H. megidis* remained above 85 % for pressure differentials less than or equal to 1,283 kPa whereas *S. carpocapsae* and *Heterorhabditis bacteriophora* Poinar (Rhabditida: Heterorhabditidae) were able to withstand up to

2,138 kPa (310 psi). Thus, Fife, Derksen, Ozkan, and Grewal (2003) recommended a maximum operating pressure of 1,380 kPa (200 psi) for *H. megidis* and 2,000 kPa (290 psi) for *S. carpocapsae* and *H. bacteriophora*.

Specific spray components within each apparatus, such as the choice of nozzles and pumps, may also impact EPN viability or efficacy. Overall greater reductions in relative viability when using a flat fan nozzle compared with a hollow cone type, and diaphragm and roller pumps (low-capacity pumps) were found to be better compared with a centrifugal pump (due to heat buildup in the latter) (Fife, Ozkan, Derksen, & Grewal, 2006, 2007; Fife, Ozkan, Derksen, Grewal, & Krause, 2005). In another example, Beck et al., (2013) noted that an ISO 02 flat fan nozzle can clog when spraying *S. carpocapsae,* and that an ISO 04 standard flat fan nozzle is a better nozzle than the larger ISO 08 standard flat fan nozzle for spraying *S. carpocapsae* because the former provided a higher relative deposition on cauliflower leaves. *H. baujardi* LPP7, when applied though an irrigation system composed of mini-sprinklers with one-mm fan nozzles (20–35 psi), was not affected in viability, infectivity and host searching capability (Lara, Dolinski, Sousa, & Daher, 2008).

In contrast to the above, certain studies did not detect EPN efficacy differences when comparing application equipment, e.g., when targeting codling moth, *Cydia pomonella* (L.) (Lepidoptera: Tortricidae), using a lance applicator versus an airblast sprayer (Lacey, Arthurs, Unruh, Headrick, & Fritts, 2006) or targeting the peachtree borer, *Synanthedon exitiosa* (Say) (Lepidoptera: Sesiidae) using a trunk sprayer, handgun or boom sprayer (Shapiro-Ilan, Cottrell, Mizell, Horton, & Zaid, 2015). The trunk sprayer is an attractive approach because the equipment uses smart technology to only spray when passing a trunk, thereby reducing waste of nematodes (or other product) between trees (Fig. 9.1). Also, Klein and Georgis (1994) found that no adverse effects were observed for *Steinernema* spp. and *H. bacteriophora* after application through several different pumps (piston, centrifugal, roller, and diaphragm), nozzle types (Spraying Systems XR8001VS, TK-VS2, FL-5VS), and strainers (100 mesh, 50 mesh, 50 slotted), and Moreira et al. (2013) reported that several different spray nozzles with air induction (AI 11003, TTI 11003 and AD-IA 11004) were all found to be compatible for use with *S. feltiae*. Thus, it appears the relative importance of application equipment choice may vary with the cropping system and target pest.

Other factors associated with the application beyond spray equipment can sometimes be the overriding issue. For instance, Brusselman et al. (2012) compared EPN application to various parts of the plant in cabbage, cauliflower, and leek. The authors concluded that nozzle type had a minor effect on the number of nematodes delivered to hard-to-reach portions of the plant (e.g., the underside of the leaf) and that improved application techniques for directing the spray to the target site are of most importance in such cases.

Fig. 9.1 Trunk sprayer applying nematodes to the base of a peach tree in Georgia (Photo credit: Ted Cottrell, USDA-ARS)

9.4 Improved Methods of Application

A variety of mechanisms can be used to improve EPN application. One avenue to achieving superior efficacy in biocontrol is through improved formulation. Development of EPN formulations and adjuvants has made considerable progress in recent years, especially toward aboveground applications. For example, mixing EPNs with a surfactant and polymer greatly enhanced control of diamondback moth larvae, *Plutella xylostella* (L.) (Lepidoptera: Plutellidae) (Schroer & Ehlers, 2005). Working with *C. pomonella*, de Waal, Malan, and Addison (2013) tested the efficacy of a superabsorbent polymer formulation, "Zeba" and reported that the formulation combined with *Heterorhabditis zealandica* Poinar (Rhabditida: Heterorhabditidae) improved the level of control obtained at 60 and 80 % RH in the laboratory and improved survival and infection-ability of the nematodes in the field (targeting cryptic habitats on the tree). Another approach to enhancing EPN applications aboveground has been the use of a fire-gel, Barricade (Shapiro-Ilan, Cottrell, et al., 2010). Applications of *S. carpocapsae* for control of *S. pictipes,* were greatly enhanced by a follow-up application of a sprayable gel (the gel is commonly used for protecting structures from fire) (Shapiro-Ilan et al.). Recently, the gel was also shown to be useful when tank-mixed directly with the nematodes at a lower concentration (<2 % prevents nozzle clogging) and the gel can also be applied to the soil surface during ground applications to protect EPNs from UV radiation and desiccation (in lieu of irrigation) (Shapiro-Ilan et al., 2015) (Fig. 9.2).

Fig. 9.2 Barricade® gel applied to soil around a peach tree in Georgia, USA. The gel is used to protect EPNs from UV radiation and desiccation (Photo credit: Ted Cottrell, USDA-ARS)

Fig. 9.3 Application of *Heterorhabditis baujardi* LPP7 and *Steinernema carpocapsae* All, with the adjuvants Gota fix® and Joint® Oil using a hand sprayer and targeting *Diatrareae saccharalis* larvae in sugar cane in Campos dos Goytacazes, RJ, Brazil

Furthermore, EPN applications to apple tree trunks for control of codling moth, *C. pomonella*, were also enhanced when the treatments included the sprayable fire-gel as well as wood flour foam as a protecting agent (Lacey, Shapiro-Ilan, & Glenn, 2010). Improved efficacy may also be achieved by using leaf flooding and addition of surfactants to increase leaf coverage (Head, Lawrence, & Walters, 2004; Williams & Walters, 2000). On the other hand, when the adjuvants Joint® oil and Gotafix® (Fig. 9.3) were tested in combination with EPNs against *D. saccharalis*, no difference was detected compared with a no-adjuvant control (Bellini & Dolinski, 2012).

As indicated above, nematodes have been successfully applied using a wide array of equipment and biological pest control has thus been achieved in numerous cropping systems (Grewal, Ehlers, & Shapiro-Ilan, 2005; Shapiro-Ilan, Han, & Dolinski, 2012). Nonetheless, additional improvements to equipment and equipment-components can be made to improve efficacy. Examples of components that can be optimized include spray apparatus, nozzle types and pumps, spray distribution, etc. (Beck et al., 2013; Brusselman et al., 2010, 2012; Lanzoni, Ade, Martelli, Radeghieri, & Pezzi, 2014; Shapiro-Ilan, Gouge, et al., 2006). These components should be collectively adjusted for maximum pathogen survival and dispersion.

Application technology can also be advanced by developing novel EPN distribution methods. One approach that has gained a lot of attention is to apply EPNs to the target site in their infected host; thereby pest suppression is achieved by the IJs that emerge from the host cadaver (Del Valle, Dolinksi, Barreto, & Souza, 2009; Del Valle, Dolinski, Barreto, Souza, & Samuels, 2008; Shapiro-Ilan, Lewis, Behle, & McGuire, 2001; Shapiro-Ilan, Lewis, & Tedders, 2003; Shapiro-Ilan, Morales-Ramos, Rojas, & Tedders, 2010). We will not, however, discuss this approach in detail as the topic is fully covered in Chap. 8 of this volume.

Bait formulations can enhance EPN persistence and reduce the quantity of microbial agents required per unit area (Georgis, 1990; Grewal, 2002). In this vein, encapsulation of nematodes into pellets has been investigated as an efficient mechanism for application. The pellets may act as baits (for insect consumptions) or can be geared toward escape of the EPNs at the target site once moisture levels are conducive. Kaya and Nelsen (1985) first developed a calcium alginate encapsulated formulation using *H. bacteriophora* and *S. carpocapsae*. The formulation showed promise as long as adequate moisture levels were present to prevent nematode desiccation (Kaya & Nelsen, 1985). More recently, Matadamas-Ortiz, Ruiz-Vega, Vazquez-Feijoo, Cruz-Martínez, and Cortés-Martínez (2014) developed a method to mechanically formulate EPNs into pellets using different proportions of various materials including diatomaceous earth, attapulgite clay, and coatings comprised of double-distilled water and *Opuntia ficus-indica* (L.) (Caryophyllales: Cactaceae) mucilage or gelatin. The most promising combination of ingredients in terms of EPN survivability consisted of 100 % diatomaceous earth or 50:50 diatomaceous earth: attapulgite clay with *O. ficus-indica* mucilage; temperature, IJ age, and moisture levels also impacted EPN survival (Matadamas-Ortiz et al., 2014). Baits could be useful in relatively small or enclosed arenas such as urban environments. Indeed a bait was developed targeting cockroaches or social insects (e.g., ants or termites) (Chang & Gehret, 1991) yet apparently the concept has not yet advanced commercially in a substantial manner. Baits may also be useful in larger arenas particularly if an effective attractant for the target pest is incorporated. A case in point is the novel capsule plus attractant approach developed by Hiltpold, Hibbard, French, and Turlings (2012); for more information on this particular approach, reader is referred to Chap. 7 of this volume.

Biocontrol success can also be enhanced by manipulating the environment that targeted for EPN application. For example, in nursery and greenhouse applications the substrate can be optimized for EPN performance. Ansari and Butt (2011) compared six EPNs in five commercial potting media; their results indicated that potting media substantially affected dispersal and efficacy in controlling third-instar black vine weevil, *Otiorhynchus sulcatus* (Fabricus) (Coleoptera: Curculionidae). Similarly, Nielsen and Lewis (2012) demonstrated that combinations of certain soilless media and EPN species can lead to increased biocontrol efficacy. *H. bacteriophora* located hosts in a wider diversity of medium components than *S. riobrave*, although both nematode species performed well in peat moss and recycled plant material; the authors concluded that peat moss, recycled plant material and hardwood bark are conducive to EPN applications (Nielsen & Lewis, 2012).

Environmental manipulations in field applications can also significantly increase the persistence of efficacy through the addition of soil amendments such as mulch or crop residues. The persistence of *S. carpocapsae* was increased through the use of crop residue to protect the EPNs from environmental extremes (Shapiro, Obrycki, Lewis, Jackson, et al., 1999). Applications of *S. carpocapsae* and *S. feltiae* for control of overwintering *C. pomonella* were greatly enhanced through the addition of wood-chip mulch Lacey, Arthurs, Granatstein, Headrick, and Fritts (2006). Also when investigating control of diapausing *C. pomonella* using *H. zealandica*, de Waal, Malan, and Addison (2011) determined differences among various types of mulch with pine wood shavings being most favorable. In another innovative approach to habitat manipulation, citrus trees were planted in islands of soil that was more conducive to EPN control than endemic soils; 68 % more adult *D. abbreviatus* weevils were captured from native soil relative to the imported sandy soil (Duncan et al., 2013).

Entomopathogenic nematode applications can be improved by leveraging interactions with other control agents. As indicated above, certain combinations of EPNs with other agents are synergistic. Thus such combinations can be used to enhance pest control. In terms of combinations with chemical insecticides, a body of research indicates that neonicotinoids can be synergistically combined with EPNs for control of scarab pests (Koppenhöfer & Fuzy, 2008; Morales-Rodriguez & Peck, 2009). Yet recent research indicated only additive or antagonistic interactions for the control of *Helicoverpa zea* (Boddie) (Lepidoptera: Noctuidae) (Navarro, McMullen, & Stock, 2014), which again emphasizes the variation that exists among EPN and host species. A more recent example of synergy was reported by Mbata and Shapiro-Ilan (2013) between *H. bacteriophora* and chlorpyrifos in the control of the peanut burrower bug, *Pangaeus bilineatus* (Say) (Hemiptera: Cydnidae); this combination should be looked into further for suppression of *P. bilineatus* as well as other hemipterans. An example of combinations with other microbial agents that deserved further attention is the synergy between various EPNs and *Metarhizium* spp.; this combination has been reported to enhance control of *O. sulcatus* and the scarab, *Hoplia philanthus* (Füessly) (Coleoptera: Scarabaeidae) (Ansari et al., 2006; Ansari, Shah, & Butt, 2008, 2010), and thus may be of interest for control of other pests as well.

In addition to combinations with other control agents, a novel concept may be to combine EPNs with phoretic agents, e.g., earthworms (Shapiro-Ilan & Brown, 2013). Phoretic hosts enhance nematode dispersal and therefore may facilitate improved biocontrol efficacy. In greenhouse tests, the presence of earthworms caused enhanced dispersal of *S. carpocapsae* and improved control of the pecan weevil, *Curculio caryae* (Horn) (Coleoptera: Curculionidae) (Shapiro-Ilan & Brown, 2013). Conceivably, earthworms and EPNs could be sold as a unit for small scale applications (gardens, nurseries, potted plants, greenhouse, etc.). From a single package or combined application, the grower could obtain the added value of improved soil conditions from the addition of earthworms to their soil (Edwards, 1983) as well as superior biological pest control based on the earthworm–nematode relationship. Moreover, on a large scale, it may be possible to increase field populations of earthworms through cultural practices, such as addition of organic matter (Berry & Karlen, 1993). This tactic could enable increased biocontrol by native EPNs or improve EPN distribution following inundative or inoculative applications. Investigation of these possibilities is warranted.

One aspect that will improve EPN application regardless of equipment or approach used is the employment of superior nematode populations. As previously mentioned, the choice of nematodes species or population for a particular target pest is critical. However, beyond straight forward laboratory and field screening, there are a number of mechanisms that can be used to improve nematodes to levels that surpass the pool of populations "in hand". First, surveys can be implemented to discover new isolates, which can then be screened in comparison to existing populations. Such surveys have been reported extensively for EPNs (e.g., Bruck, 2004; Campos-Herrera et al., 2008; Malan, Knoetze, & Moore, 2011, Shapiro-Ilan et al., 2003). Indigenous isolates may be better adapted to the local environment (Dolinski & Moino, 2006). However, if existing or newly discovered entomopathogen populations are still not sufficient to achieve desired levels of control, another option is to improve selected candidate populations through genetic approaches.

Genetic improvement approaches can be geared toward enhancement of single or various biocontrol traits, e.g., reproductive capacity, virulence, environmental tolerance, etc. Molecular or non-molecular approaches may be used for population improvement. We refer the reader to Chap. 2 of this volume for further discussion on the topic of genetic improvement and survival mechanism. One of the non-molecular methods is directed selection for desired traits. Examples of genetic selection in EPNs encompass improvements in various traits including of host-finding (Gaugler, Campbell, & McGuire, 1989) and nematicide resistance (Glazer, Salame, & Segal, 1997). It must be noted, however, that directed selection for one trait may inadvertently select for an inferior level of another trait (Gaugler, 1987; Gaugler, Campbell, & McGuire, 1990).

Another non-molecular approach to population improvement is hybridization, i.e., the transfer of beneficial traits from one population to another accomplished through controlled breeding. Shapiro, Glazer, and Segal (1997) first demonstrated the approach in EPNs by transferring heat tolerance from one *H. bacteriophora* population to another. Given that both hermaphroditic and amphimictic forms occur in heterorhabditids, extra care must be taken to ensure that progeny nematodes in controlled crosses arise from intended mating rather than self-replication. One option to overcome this obstacle is to use marker mutations (Shapiro et al.,

1997). Only amphimictic forms exist in steinernematids and, hence, hybridization approaches are more straightforward in this genus. An example is provided in the study of Shapiro-Ilan, Stuart, and McCoy (2005), which used hybridization to develop superior environmental tolerance and virulence in *S. carpocapsae* populations. In more recent studies, the two non-molecular approaches, selection and hybridization, were used in combination for developing EPN populations with superior environmental tolerance (Mukuka, Strauch, Hoppe, & Ehlers, 2010).

Progress has also been made toward molecular approaches for improving EPNs. For example, an *H. bacteriophora* population was improved for heat tolerance through transformation using a heat shock protein that originated from *Caenorhabditis elegans* (Maupas) (Rhabditida: Rhabditidae) (Gaugler, Wilson, & Shearer, 1997). Genomic sequencing of EPNs and their bacterial symbionts (e.g., Bai et al., 2009, 2013; Bai & Grewal, 2007; Ciche, 2007; Duchaud et al., 2003; Schwartz, Antoshechkin, & Sternberg, 2011), is expected to continue expanding and will serve to enhance potential for population improvement programs using molecular or non-molecular approaches.

Once a suitable nematode population is chosen for biocontrol purposes, it is critical that the stability of that population is secured. Attenuation of beneficial traits, which may result from repeated sub-culturing, can jeopardize biocontrol efforts. Trait deterioration can be genetically based (e.g., inbreeding, drift, inadvertent selection), or stem from non-genetic factors (e.g., disease or nutrition) (Chaston et al., 2011; Hopper, Roush, & Powell, 1993; Tanada & Kaya, 1993). Deterioration in EPNs has been observed under laboratory conditions for various traits such as virulence, environmental tolerance, reproductive capacity, and host-finding (Bai, Shapiro-Ilan, Gaugler, & Hopper, 2005; Bilgrami, Gaugler, Shapiro-Ilan, & Adams, 2006; Shapiro, Glazer, & Segal, 1996; Wang & Grewal, 2002). Both the nematodes and their bacterial symbionts may experience trait loss (Bilgrami et al., 2006; Wang et al., 2007). The cause of trait deterioration in EPNs (*H. bacteriophora*) was reported to be genetically based with inbreeding depression being the prominent issue (Adhikari et al., 2009; Bai et al., 2005; Chaston et al., 2011).

To reduce trait deterioration, improved cryopreservation approaches can be helpful (Bai, Shapiro-Ilan, Gaugler, & Yi, 2004). However, cryopreservation can still be problematic and few EPN researchers or commercial producers routinely use liquid nitrogen because EPN populations vary in their adaptability to cryogenic storage, it is expensive and mechanical failure, human error, or neglect can result in complete loss of genetic material (Nugent, O'Leary, & Burnell, 1996; Shapiro-Ilan, Han, & Qiu, 2014; Wang & Grewal, 2002). Creation of homozygous inbred lines is another method for deterring EPN trait deterioration that has recently been developed (Bai et al., 2005; Shapiro-Ilan et al., 2014). Bai et al., (2005) was the first to test and validate the approach; selected *H. bacteriophora* inbred lines remained stable through serial culturing whereas their wild-type parent strains deteriorated in several traits (virulence, environmental tolerance and host-finding). Thus, Bai et al., (2005) proposed that EPN producers develop numerous inbred lines from promising candidate populations, and then select the lines that exhibit high levels of desirable traits for commercialization. In a variation of the Bai et al., (2005) approach, Anbesse, Sumaya, Dörfler, Strauch, and Ehlers (2013) reported that multiple heterorhabditid inbred lines can be created during liquid culture. Given that heterorhabditids cannot mate in liquid culture, all progeny are produced by selfing (via hermaphrodites) (Ehlers & Shapiro-Ilan, 2005), which results in automatic creation of multiple inbred lines. Both approaches have merit and may be used to produce stable EPN populations for improved biocontrol (Shapiro-Ilan et al., 2014), though only heterorhabditids can be produced using the liquid culture approach (as steinernematids mate in liquid culture).

9.5 Conclusions and Future Directions

There are many biocontrol success stories stemming from the use of EPNs (Grewal et al., 2005; Shapiro-Ilan et al., 2014). However, there have been numerous failures as well. Simply put, the key to success is that the approaches using EPNs must be cost competitive and consistently efficacious relative to alternate tactics. Toward that end, as indicated above, the use of EPNs can be improved by developing superior nematode species or populations that are more capable of suppressing the target pest. Also, EPN utility in biocontrol can be expanded by improving delivery to the target site.

For the most part, application technology for EPNs has arisen by adopting methods and equipment that is routinely used for application of chemical insecticides. Implementation using this parallel technology approach has facilitated commercialization of EPNs because growers generally do not need to learn new methodology and the equipment is already available on site. However, given that EPNs are biological organisms (as opposed to chemicals) and are thus sensitive to environmental conditions, caution must be taken in the transfer of technology. Indeed, the full impact of various application equipment and parameters on EPN survival and efficacy is continually being elucidated, and a lot of work remains. Ideally, application equipment for EPN use will eventually be optimized specifically for each nematode species and the cropping system targeted.

Conceivably, the greatest advances in improving EPN application will be based on novel formulation and delivery methods. For instance, the recent innovations in adjuvants and formulation technology have been instrumental in delving into arenas that previously may have been considered inaccessible, i.e., targeting aboveground pests (Arthurs et al., 2004; de Waal et al., 2013; Shapiro-Ilan, Morales-Ramos et al., 2010). Further investigations to enhance formulation toward aboveground application, and other novel delivery approaches for EPNs are needed, and goals should be oriented to reducing overall costs while enhancing efficacy. For maximum gains toward increasing the competitiveness of EPNs, improvements in application technology should be coordinated within an overall program that comprehensively enhances all critical aspects including population amelioration, production, and packaging technology.

References

Acevedo, J. P. M., Samuels, R. I., Machado, I. R., & Dolinski, C. (2007). Interactions between isolates of the entomopathogenic fungus *Metarhizium anisopliae* and the entomopathogenic nematode *Heterorhabditis bacteriophora*JPM4 during infection of the sugar cane borer *Diatraea saccharalis* (Lepidoptera: Pyralidae). *Journal of Invertebrate Pathology, 96*, 187–192.

Adhikari, B. N., Chin-Yo, L., Xiaodong, B., Ciche, T. A., Grewal, P. S., Dillman, A. R., et al. (2009). Transcriptional profiling of trait deterioration in the insect pathogenic nematode *Heterorhabditis bacteriophora. BMC Genomics, 10*, 609.

Alumai, A., & Grewal, P. S. (2004). Tank-mix compatibility of the entomopathogenic nematodes, *Heterorhabditis bacteriophora* and *Steinernema carpocapsae*, with selected chemical pesticides used in turfgrass. *Biocontrol Science and Technology, 14*, 725–730.

Anbesse, S., Sumaya, N. H., Dörfler, A. V., Strauch, S., & Ehlers, R.-U. (2013). Selective breeding for desiccation tolerance in liquid culture provides genetically stable inbred lines of the entomopathogenic nematode *Heterorhabditis bacteriophora. Applied Microbiology and Biotechnology, 97*, 731–739.

Ansari, M. A., & Butt, T. M. (2011). Effect of potting media on the efficacy and dispersal of entomopathogenic nematodes for the control of black vine weevil, *Otiorhynchus sulcatus* (Coleoptera: Curculionidae). *Biological Control, 58*, 310–318.

Ansari, M. A., Shah, F. A., & Butt, T. M. (2008). Combined use of entomopathogenic nematodes and *Metarhizium anisopliae* as a new approach for black vine weevil, *Otiorhynchus sulcatus* (Coleoptera: Curculionidae) control. *Entomologia Experimentalis et Applicata, 129*, 340–347.

Ansari, M. A., Shah, F. A., & Butt, T. M. (2010). The entomopathogenic nematode *Steinernema kraussei* and *Metarhizium anisopliae* work synergistically in controlling overwintering larvae of the black vine weevil, *Otiorhynchus sulcatus*, in strawberry growbags. *Biocontrol Science and Technology, 20*, 99–105.

Ansari, M. A., Shah, F. A., & Moens, T. M. (2006). Field trials against *Hoplia philanthus* (Coleoptera: Scarabaeidae) with a combination of an entomopathogenic nematode and the fungus *Metarhizium anisopliae* CLO 53. *Biological Control, 39*, 453–459.

Ansari, M. A., Tirry, L., & Moens, M. (2004). Interaction between *Metarhizium anisopliae* CLO 53 and entomopathogenic nematodes for the control of *Hoplia philanthus. Biological Control, 31*, 172–180.

Arthurs, S., Heinz, K. M., & Prasifka, J. R. (2004). An analysis of using entomopathogenic nematodes against above-ground pests. *Bulletin of Entomological Research, 94*, 297–306.

Bai, C., Shapiro-Ilan, D. I., Gaugler, R., & Hopper, K. R. (2005). Stabilization of beneficial traits in *Heterorhabditis bacteriophora* through creation of inbred lines. *Biological Control, 32*, 220–227.

Bai, C., Shapiro-Ilan, D. I., Gaugler, R., & Yi, S. (2004). Effect of entomopathogenic nematode concentration on survival during cryopreservation in liquid nitrogen. *Journal of Nematology, 36*, 281–284.

Bai, X., Adams, B. J., Ciche, T. A., Clifton, S., Gaugler, R., Hogenhout, S. A., Spieth, J., Sternberg, P. W., Wilson, R. K., & Grewal, P. S. (2009). Transcriptomic analysis of the entomopathogenic nematode *Heterorhabditis bacteriophora* TTO1. *BMC Genomics, 10,* art. No. 205.

Bai, X., Adams, B. J., Ciche, T. A., Clifton, S., Gaugler, R., Kim, K.-S., Spieth, J., Steinberg, P. W., Wilson, R. K., & Grewal, P. S. (2013). A lover and a fighter: The genome sequence of an entomopathogenic nematode *Heterorhabditis bacteriophora. PLoS ONE, 8*(7), art. No. 69618.

Bai, X., & Grewal, P. S. (2007). Identification of two down-regulated genes in entomopathogenic nematode *Heterorhabditis bacteriophora* infective juveniles upon contact with insect hemolymph. *Molecular and Biochemical Parasitology, 156*, 162–166.

Bal, H. K., Taylor, R. A. J., & Grewal, P. S. (2014). Ambush foraging entomopathogenic nematodes employ "sprinters" for long-distance dispersal in the absence of hosts. *Journal of Parasitology, 100*, 422–432.

Barbercheck, M. E. (1992). Effect of soil physical factors on biological control agents of soil insect pests. *Florida Entomologist, 75*, 539–548.

Barbercheck, M. E., & Kaya, H. K. (1990). Interactions between *Beauveria bassiana* and the entomogenous nematodes *Steinernema feltiae* and *Heterorhabditis heliothidis*. *Journal of Invertebrate Pathology, 55*, 225–234.

Beck, B., Brusselman, E., Nuyttens, D., Moens, M., Pollet, S., Temmerman, F., et al. (2013). Improving foliar applications of entomopathogenic nematodes by selecting adjuvants and spray nozzles. *Biocontrol Science and Technology, 23*, 507–520.

Bednarek, A., & Gaugler, R. (1997). Compatibility of soil amendments with entomopathogenic nematodes. *Journal of Nematology, 29*, 220–227.

Begley, J. W. (1990). Efficacy against insects in habitats other than soil. In R. Gaugler & H. K. Kaya (Eds.), *Entomopathogenic nematodes in biological control* (pp. 215–232). Boca Raton, FL: CRC Press.

Bellini, L., & Dolinski, C. (2012). Foliar application of entomopathogenic nematodes (Rhabditida: Steinernematidae and Heterorhabditidae) for the control of *Diatraea saccharalis* in greenhouse. *Semina: Ciencias Agrarias, 33*, 997–1004.

Benz, G. (1971). Synergism of micro-organisms and chemical insecticides. In H. D. Burgess & N. W. Hussey (Eds.), *Microbial control of insects and mites* (pp. 327–355). London: Academic.

Berry, E. C., & Karlen, D. L. (1993). Comparison of alternate farming systems II. Earthworm population and density and species diversity. *American Journal of Alternative Agriculture, 8*, 21–26.

Berry, R. E., Liu, J., & Groth, E. (1997). Efficacy and persistence of *Heterorhabditis marelatus* (Rhabditida: Heterorhabditidae) against root weevils (Coleoptera: Curculionidae) in strawberry. *Environmental Entomology, 26*, 465–470.

Bilgrami, A. L., Gaugler, R., Shapiro-Ilan, D. I., & Adams, B. J. (2006). Source of trait deterioration in entomopathogenic nematodes *Heterorhabditis bacteriophora* and *Steinernema carpocapsae* during *in vivo* culture. *Nematology, 8*, 397–409.

Brinkman, M. A., & Gardner, W. A. (2001). Use of diatomaceous earth and entomopathogen combinations against the red imported fire ant (Hymenoptera: Formicidae). *Florida Entomologist, 84*, 740–741.

Bruck, D. J. (2004). Natural occurrence of entomopathogens in Pacific Northwest nursery soils and their virulence to the black vine weevil, *Otiorhynchus sulcatus* (F.) (Coleoptera: Curculionidae). *Environmental Entomology, 33*, 1335–1343.

Brusselman, E., Beck, B., Pollet, S., Temmerman, F., Spanoghe, P., Moens, M., et al. (2012). Effect of the spray application technique on the deposition of entomopathogenic nematodes in vegetables. *Pest Management Science, 68*, 444–453.

Brusselman, E., Beck, B., Temmerman, F., Pollet, S., Steurbaut, W., Moens, M., & Nuyttens, D. (2010). The spray pattern of entomopathogenic nematodes. *American Society of Agricultural and Biological Engineers Annual International Meeting 2010, 6*, 5054–5067.

Bullock, R. C., Pelosi, R. R., & Killer, E. E. (1999). Management of citrus root weevils (Coleoptera: Curculionidae) on Florida citrus with soil-applied entomopathogenic nematodes (Nematoda: Rhabditida). *Florida Entomologist, 82*, 1–7.

Cabanillas, H. E., & Raulston, J. R. (1995). Impact of *Steinernema riobravis* (Rhabditida: Steinernematidae) on the control of *Helicoverpa zea* (Lepidoptera: Noctuidae) in corn. *Journal of Economic Entomology, 88*, 58–64.

Campos-Herrera, R., El-Borai, F. E., & Duncan, L. W. (2012). Wide interguild relationships among entomopathogenic and free–living nematodes in soil as measured by real time qPCR. *Journal of Invertebrate Pathology, 111*, 126–135.

Campos-Herrera, R., Gómez-Ros, J. M., Escuer, M., Cuadra, L., Barrios, L., & Gutiérrez, C. (2008). Diversity, occurrence, and life characteristics of natural entomopathogenic nematode populations from La Rioja (Northern Spain) under different agricultural management and their relationships with soil factors. *Soil Biology & Biochemistry, 40*, 1474–1484.

Chang, F. N., & Gehret, M. J. (1991). An insect bait station comprising a hydrated macrogel layer containing at least one species of entomopathogen and a hydrated water retentive compound layer which acts as a water-reservoir for the entomopathogen. Patent WO 1991017656 A1.

Chaston, J. M., Dillman, A. R., Shapiro-Ilan, D. I., Bilgrami, A. L., Gaugler, R., Hopper, K. R., et al. (2011). Outcrossing and crossbreeding recovers deteriorated traits in laboratory cultured *Steinernema carpocapsae* nematodes. *International Journal of Parasitology, 41*, 801–809.

Ciche, T. A. (2007). The biology and genome of *Heterorhabditis bacteriophora*. In *Worm Book* [the Online Review of *C. elegans* biology]. WormBook. Retrieved form http://www.wormbook.org

Conner, J. M., McSorley, R., Stansly, P. A., & Pitts, D. J. (1998). Delivery of *Steinernema riobrave* through a drip irrigation system. *Nematropica, 28*, 95–100.

Curran, J. (1992). Influence of application method and pest population size on field efficacy of entomopathogenic nematodes. *Journal of Nematology, 24*, 631–636.

de Waal, J. Y., Malan, A. P., & Addison, M. F. (2011). Evaluating mulches together with *Heterorhabditis zealandica* (Rhabditida: Heterorhabditidae) for the control of diapausing codling moth larvae, *Cydia pomonella* (L.) (Lepidoptera: Tortricidae). *Biocontrol Science and Technology, 21*, 255–270.

de Waal, J. Y., Malan, A. P., & Addison, M. F. (2013). Effect of humidity and a superabsorbent polymer formulation on the efficacy of *Heterorhabditis zealandica* (Rhabditida: Heterorhabditidae) to control codling moth, *Cydia pomonella* (L.) (Lepidoptera: Tortricidae). *Biocontrol Science and Technology, 23*, 62–78.

Del Valle, E. E., Dolinksi, C., Barreto, E. L. S., & Souza, R. M. (2009). Effect of cadaver coatings on emergence and infectivity of the entomopathogenic nematode *Heterorhabditis baujardi* LPP7 (Rhabditida: Heterorhabditidae) and the removal of cadavers by ants. *Biological Control, 50*, 21–24.

Del Valle, E. E., Dolinski, C., Barreto, E. L. S., Souza, R. M., & Samuels, R. I. (2008). Efficacy of *Heterorhabditis baujardi* LPP7 (Nematoda: Rhabditida) applied in *Galleria mellonella* (Lepidoptera: Pyralidae) insect cadavers to *Conotrachelus psidii*, (Coleoptera: Curculionidae) larvae. *Biocontrol Science and Technology, 18*, 33–41.

Dolinski, C., & Moino, A., Jr. (2006). Use of native and exotic entomopathogenic nematodes: The risk of introductions. *Nematologia Brasileira, 30*, 139–149.

Dolinski, C., Pinto, C. C. S., Robaina, R. R., & Bellini, L. L. (2010). Effect of soil texture on the mobility of *Heterorhabditis baujardi* 'LPP7' (Rhabditida: Heterorhabditidae). *Nematologia Brasileira, 34*, 123–128.

Downing, A. S. (1994). Effect of irrigation and spray volume on efficacy of entomopathogenic nematodes (Rhabditida: Heterorhabditidae) against white grubs (Coleoptera: Scarabaeidae). *Journal of Economic Entomology, 87*, 643–646.

Duchaud, E., Rusniok, C., Frangeul, L., Buchrieser, C., Givaudan, A., Taourit, S., et al. (2003). The genome sequence of the entomopathogenic bacterium *Photorhabdus luminescens*. *Nature Biotechnology, 21*, 1307–1313.

Duncan, L. W., Dunn, D. C., Bague, G., & Nguyen, K. (2003). Competition between entomopathogenic and free-living bactivorous nematodes in larvae of the weevil *Diaprepes abbreviatus*. *Journal of Nematology, 35*, 187–193.

Duncan, L. W., Graham, J. H., Zellers, J., Bright, D., Dunn, D. C., El-Borai, F. E., et al. (2007). Food web responses to augmenting the entomopathogenic nematodes in bare and animal manure-mulched soil. *Journal of Nematology, 39*(2), 176–189.

Duncan, L. W., & McCoy, C. W. (1996). Vertical distribution in soil, persistence, and efficacy against citrus root weevil (Coleoptera: Curculionidae) of two species of entomogenous nematodes (Rhabditida: Steinernematidae; Heterorhabditidae). *Environmental Entomology, 25*, 174–178.

Duncan, L. W., Shapiro, D. I., McCoy, C. W., & Graham, J. H. (1999). Entomopathogenic nematodes as a component of citrus root weevil IPM. In S. Polavarapu (Ed.), *Optimal use of insecticidal nematodes in pest management* (pp. 69–78). New Brunswick, NJ: Rutgers University.

Duncan, L. W., Stuart, R. J., El-Borai, F. E., Campos-Herrera, R., Pathak, E., Giurcanu, M., et al. (2013). Modifying orchard planting sites conserves entomopathogenic nematodes, reduces weevil herbivory and increases citrus tree growth, survival and fruit yield. *Biological Control, 64*, 26–36.

Edwards, C. A. (1983). Earthworm ecology in cultivated soils. In J. E. Satchell (Ed.), *Earthworm ecology from darwin to vermiculture* (pp. 123–138). London: Chapman and Hall.

Ehlers, R.-U., & Shapiro-Ilan, D. I. (2005). Mass production. In P. S. Grewal, R.-U. Ehlers, & D. I. Shapiro-Ilan (Eds.), *Nematodes as biological control agents* (pp. 65–79). Wallingford, UK: CABI.

Eng, M. S., Preisser, E. L., & Strong, D. R. (2005). Phoresy of the entomopathogenic nematode *Heterorhabditis marelatus* by a non-host organism, the isopod *Porcellio scaber. Journal of Invertebrate Pathology, 88*, 173–176.

Epsky, N. D., Walter, D. E., & Capinera, J. L. (1988). Potential role of nematophagous microarthropods as biotic mortality factors of entomogenous nematodes (Rhabditida: Steinernematidae, Heterorhabditidae). *Journal of Economic Entomology, 81*, 821–825.

Fife, J. P., Derksen, R. C., Ozkan, H. E., & Grewal, P. S. (2003). Effects of pressure differentials on the viability and infectivity of entomopathogenic nematodes. *Biological Control, 27*, 65–72.

Fife, J. P., Ozkan, H. E., Derksen, R. C., & Grewal, P. S. (2006). Using computational fluid dynamics to predict damage of a biological pesticide during passage through a hydraulic nozzle. *Biosystems Engineering, 94*, 387–396.

Fife, J. P., Ozkan, H. E., Derksen, R. C., & Grewal, P. S. (2007). Effects of pumping on entomopathogenic nematodes and temperature increase within a spray system. *Applied Engineering in Agriculture, 23*, 405–412.

Fife, J. P., Ozkan, H. E., Derksen, R. C., Grewal, P. S., & Krause, C. R. (2005). Viability of a biological pest control agent through hydraulic nozzles. *Transactions of the ASAE, 48*, 45–54.

Gaugler, R. (1987). Entomogenous nematodes and their prospects for genetic improvement. In K. Maramorosch (Ed.), *Biotechnology in invertebrate pathology and cell culture* (pp. 457–484). San Diego, CA: Academic.

Gaugler, R., & Boush, G. M. (1978). Effects of ultraviolet radiation and sunlight on the entomogenous nematode *Neoaplectana carpocapsae. Journal of Invertebrate Pathology, 32*, 291–296.

Gaugler, R., Campbell, J. F., & McGuire, T. R. (1989). Selection for host-finding in *Steinernema feltiae. Journal of Invertebrate Pathology, 54*, 363–372.

Gaugler, R., Campbell, J. F., & McGuire, T. R. (1990). Fitness of a genetically improved entomopathogenic nematode. *Journal of Invertebrate Pathology, 56*, 106–116.

Gaugler, R., Wilson, M., & Shearer, P. (1997). Field release and environmental fate of a transgenic entomopathogenic nematode. *Biological Control, 9*, 75–80.

Georgis, R. (1990). Formulation and application technology. In R. Gaugler & H. K. Kaya (Eds.), *Entomopathogenic nematodes in biological control* (pp. 173–194). Boca Raton, FL: CRC Press.

Georgis, R., Dunlop, D. B., & Grewal, P. S. (1995). Formulation of entomopathogenic nematodes. In F. R. Hall & J. W. Barry (Eds.), *Biorational pest control agents: Formulation and delivery* (pp. 197–205). Washington, DC: American Chemical Society.

Georgis, R., & Gaugler, R. (1991). Predictability in biological control using entomopathogenic nematodes. *Journal of Economic Entomology, 84*, 713–720.

Georgis, R., & Hague, N. G. M. (1991). Nematodes as biological insecticides. *Pesticide Outlook, 3*, 29–32.

Georgis, R., & Poinar, G. O., Jr. (1983). Effect of soil texture on the distribution and infectivity of *Neoaplectana carpocapsae* (Nematode: Steinernematidae). *Journal of Nematology, 15*, 308–311.

Glazer, I., Salame, L., & Segal, D. (1997). Genetic enhancement of nematicide resistance in entomopathogenic nematodes. *Biocontrol Science and Technology, 7*, 499–512.

Grewal, P. S. (2002). Formulation and application technology. In R. Gaugler (Ed.), *Entomopathogenic nematology* (pp. 265–287). New York: CABI.

Grewal, P. S., Ehlers, R.-U., & Shapiro-Ilan, D. I. (Eds.). (2005). *Nematodes as biological control agents*. Wallingford, UK: CABI.
Grewal, P. S., & Georgis, R. (1999). Entomopathogenic nematodes. In F. R. Hall & J. J. Menn (Eds.), *Methods in biotechnology, vol 5: Biopesticides: Use and delivery* (pp. 271–299). Totowa, NJ: Humana Press.
Grewal, P. S., Selvan, S., & Gaugler, R. (1994). Thermal adaptation of entomopathogenic nematodes – Niche breadth for infection, establishment and reproduction. *Journal of Thermal Biology, 19*, 245–253.
Hayes, A. E., Fitzpatrick, S. M., & Webster, J. M. (1999). Infectivity, distribution, and persistence of the entomopathogenic nematode *Steinernema carpocapsae* All strain (Rhabditida: Steinernematidae) applied by sprinklers or boom sprayer to dry-pick cranberries. *Journal of Economic Entomology, 92*, 539–546.
Head, J., Lawrence, A. J., & Walters, K. F. A. (2004). Efficacy of the entomopathogenic nematode, *Steinernema feltiae*, against *Bemisia tabaci* in relation to plant species. *Journal of Applied Entomology, 128*, 543–547.
Hiltpold, I., Hibbard, B. E., French, B. W., & Turlings, T. C. J. (2012). Capsules containing entomopathogenic nematodes as a Trojan horse approach to control the western corn rootworm. *Plant and Soil, 358*, 11–25.
Hopper, K. R., Roush, R. T., & Powell, W. (1993). Management of genetics of biological-control introductions. *Annual Review of Entomology, 38*, 27–51.
Kanga, F. N., Waeyenberge, L., Hauser, S., & Moens, M. (2012). Distribution of entomopathogenic nematodes in Southern Cameroon. *Journal of Invertebrate Pathology, 109*, 41–51.
Kaya, H. K. (1990). Soil ecology. In R. Gaugler & H. K. Kaya (Eds.), *Entomopathogenic nematodes in biological control* (pp. 93–116). Boca Raton, FL: CRC Press.
Kaya, H. K. (2002). Natural enemies and other antagonists. In R. Gaugler (Ed.), *Entomopathogenic nematology* (pp. 189–204). New York: CABI.
Kaya, H. K., & Nelsen, C. E. (1985). Encapsulation of steinernematid and heterorhabditid nematodes with calcium alginate: A new approach for insect control and other applications. *Environmental Entomology, 14*, 572–574.
Klein, M. G., & Georgis, R. (1992). Persistence of control of Japanese beetle (Coleoptera: Scarabaeidae) larvae with steinernematid and heterorhabditid nematodes. *Journal of Economic Entomology, 85*, 727–730.
Klein, M. G., & Georgis, R. (1994). Application techniques for entomopathogenic nematodes. In: *Proceedings of the VI international colloquium on invertebrate pathology and microbial control* (pp. 483–484). Montpellier, France: Society for Invertebrate Pathology.
Koppenhöfer, A. M., & Fuzy, E. M. (2008). Early timing and new combinations to increase the efficacy of neonicotinoid–entomopathogenic nematode (Rhabditida: Heterorhabditidae) combinations against white grubs (Coleoptera: Scarabaeidae). *Pest Management Science, 64*, 725–735.
Koppenhöfer, A. M., & Grewal, P. S. (2005). Compatibility and interactions with agrochemicals and other biocontrol agents. In P. S. Grewal, R.-U. Ehlers, & D. I. Shapiro-Ilan (Eds.), *Nematodes as biological control agents* (pp. 363–381). Wallingford, UK: CABI.
Koppenhöfer, A. M., & Kaya, H. K. (1997). Additive and synergistic interactions between entomopathogenic nematodes and *Bacillus thuringiensis* for scarab grub control. *Biological Control, 8*, 131–137.
Koppenhöfer, A. M., & Kaya, H. K. (1998). Synergism of imidacloprid and an entomopathogenic nematode: A novel approach to white grub (Coleoptera: Scarabaeidae) control in turfgrass. *Journal of Economic Entomology, 91*, 618–623.
Koppenhöfer, A. M., Kaya, H. K., & Taormino, S. P. (1995). Infectivity of entomopathogenic nematodes (Rhabditida: Steinernematidae) at different soil depths and moistures. *Journal of Invertebrate Pathology, 65*, 193–199.
Kruitbos, L. M., Heritage, S., Hapca, S., & Wilson, M. J. (2010). The influence of habitat quality on the foraging strategies of the entomopathogenic nematodes *Steinernema carpocapsae* and *Heterorhabditis megidis*. *Parasitology, 137*, 303–309.

Kung, S., Gaugler, R., & Kaya, H. K. (1990a). Influence of soil pH and oxygen on persistence of *Steinernema* spp. *Journal of Nematology, 22*, 440–445.

Kung, S., Gaugler, R., & Kaya, H. K. (1990b). Soil type and entomopathogenic nematode persistence. *Journal of Invertebrate Pathology, 55*, 401–406.

Kung, S., Gaugler, R., & Kaya, H. K. (1991). Effects of soil temperature, moisture, and relative humidity on entomopathogenic nematode persistence. *Journal of Invertebrate Pathology, 57*, 242–249.

Lacey, L. A., Arthurs, S. P., Granatstein, D., Headrick, H., & Fritts, R., Jr. (2006). Use of entomopathogenic nematodes (Steinernematidae) in conjunction with mulches for control of codling moth (Lepidoptera: Torticidae). *Journal of Entomological Science, 41*, 107–119.

Lacey, L. A., Arthurs, S. P., Unruh, T. R., Headrick, H., & Fritts, R., Jr. (2006). Entomopathogenic nematodes for control of codling moth (Lepidoptera: Tortricidae) in apple and pear orchards: effect of nematode species and seasonal temperatures, adjuvants, application equipment and post-application irrigation. *Bio logical Control, 37*, 214–223.

Lacey, L. A., Shapiro-Ilan, D. I., & Glenn, G. M. (2010). Post-application of anti-desiccant agents improves efficacy of entomopathogenic nematodes in formulated host cadavers or aqueous suspension against diapausing codling moth larvae (Lepidoptera: Tortricidae). *Biocontrol Science and Technology, 20*, 909–921.

Lanzoni, A., Ade, G., Martelli, R., Radeghieri, P., & Pezzi, F. (2014). Technological aspects of *Steinernema carpocapsae* spray application alone or mixed with *Bacillus thuringiensis aizawai* in spinach crop. *Bulletin of Insectology, 67*, 115–123.

Lara, J. C., Dolinski, C., Sousa, E. F., & Daher, R. F. (2008). Effect of mini-sprinkler irrigation system on *Heterorhabditis baujardi*LPP7 (Nematoda: Heterorhabditidae) infective juvenile. *Scientia Agricola, 65*, 433–437.

Levine, E., & Oloumi-Sadeghi, H. (1992). Field evaluation of *Steinernema carpocapsae* (Rhabditida: Steinernematidae) against the black cutworm (Lepidoptera: Noctuidae) larvae in field corn. *Journal of Entomological Science, 27*, 427–435.

Lewis, E. E. (2002). Behavioral ecology. In R. Gaugler (Ed.), *Entomopathogenic nematology* (pp. 205–224). Wallingford, UK: CABI.

Lewis, E. E., & Clarke, D. J. (2012). Nematode parasites and entomopathogens. In F. E. Vega & H. K. Kaya (Eds.), *Insect pathology* (2nd ed., pp. 395–424). Amsterdam: Elsevier.

Lewis, E. E., Gaugler, R., & Harrison, R. (1992). Entomopathogenic nematode host finding: Response to host contact cues by cruise and ambush foragers. *Parasitology, 105*, 309–319.

Malan, A. P., Knoetze, R., & Moore, S. D. (2011). Isolation and identification of entomopathogenic nematodes from citrus orchards in South Africa and their biocontrol potential against false codling moth. *Journal of Invertebrate Pathology, 108*, 115–125.

Matadamas-Ortiz, P. T., Ruiz-Vega, J., Vazquez-Feijoo, J. A., Cruz-Martínez, H., & Cortés-Martínez, C. I. (2014). Mechanical production of pellets for the application of entomopathogenic nematodes: Factors that determine survival time of *Steinernema glaseri*. *Biocontrol Science and Technology, 24*, 145–157.

Mbata, G. N., & Shapiro-Ilan, D. I. (2013). The potential for controlling *Pangaeus bilineatus* (Say) (Heteroptera: Cydnidae) using a combination of entomopathogens and an insecticide. *Journal of Economic Entomology, 106*, 2072–2076.

McCoy, C. W., Shapiro, D. I., Duncan, L. W., & Nguyen, K. (2000). Entomopathogenic nematodes and other natural enemies as mortality factors for larvae of *Diaprepes abbreviatus* (Coleoptera: Curculionidae). *Biological Control, 19*, 182–190.

Millar, L. C., & Barbercheck, M. E. (2001). Interaction between endemic and introduced entomopathogenic nematodes in conventional-till and no-till corn. *Biological Control, 22*, 235–245.

Molyneux, A. S., & Bedding, R. A. (1984). Influence of soil texture and moisture on the infectivity of *Heterorhabditis* sp. D1 and *Steinernema glaseri* for larvae of the sheep blowfly *Lucilia cuprina*. *Nematologica, 30*, 358–365.

Morales-Rodriguez, A., & Peck, D. C. (2009). Synergies between biological and neonicotinoid insecticides for the curative control of the white grubs *Amphimallon majale* and *Popillia japonica*. *Biological Control, 51*, 169–180.

Moreira, G. F., Batista, E. S. P., Campos, H. B. N., Lemos, R. E., & Ferreira, M. C. (2013). Spray nozzles, pressures, additives and stirring time on viability and pathogenicity of entomopathogenic nematodes (Nematoda: Rhabditida) for greenhouses. *PLoS One, 8*(6), e65759.

Mukuka, J., Strauch, O., Hoppe, C., & Ehlers, R.-U. (2010). Improvement of heat and desiccation tolerance in *Heterorhabditis bacteriophora* through cross-breeding of tolerant strains and successive genetic selection. *BioControl, 55*, 511–521.

Navarro, P. D., McMullen, J. G., & Stock, S. P. (2014). Effect of dinotefuran, indoxacarb, and imidacloprid on survival and fitness of two Arizona-native entomopathogenic nematodes against *Helicoverpa zea* (Lepidoptera: Noctuidae). *Nematropica, 44*, 64–73.

Nielsen, A. L., & Lewis, E. E. (2012). Designing the ideal habitat for entomopathogen use in nursery production. *Pest Management Science, 68*, 1053–1061.

Nilsson, U., & Gripwall, E. (1999). Influence of application technique on the viability of the biological control agents *Verticillium lecanii* and *Steinernema feltiae*. *Crop Protection, 18*, 53–59.

Nishimatsu, T., & Jackson, J. J. (1998). Interaction of insecticides, entomopathogenic nematodes, and larvae of the western corn rootworm (Coleoptera: Chrysomelidae). *Journal of Economic Entomology, 91*, 410–418.

Nugent, M. J., O'Leary, S. A., & Burnell, A. M. (1996). Optimised procedures for the cryopreservation of different species of *Heterorhabditis*. *Fundamental and Applied Nematology, 19*, 1–6.

Parkman, J. P., Frank, J. H., Nguyen, K. B., & Smart, G. C., Jr. (1994). Inoculative release of *Steinernema scapterisci* (Rhabditida: Steinernematidae) to suppress pest mole crickets (Orthoptera: Gryllotapidae) on golf courses. *Environmental Entomology, 23*, 1331–1337.

Schroeder, P. D., Ferguson, C. S., Shelton, A. M., Wilsey, W. T., Hoffman, M. P., & Petzoldt, C. (1996). Greenhouse and field evaluations of entomopathogenic nematodes (Nematoda: Heterorhabditidae and Steinernematidae) for control of cabbage maggot (Diptera: Anthomyiidae) on cabbage. *Journal of Economic Entomology, 89*, 1109–1115.

Schroer, S., & Ehlers, R.-U. (2005). Foliar application of the entomopathogenic nematode *Steinernema carpocapsae* for biological control of diamondback moth larvae (*Plutella xylostella*). *Biological Control, 33*, 81–86.

Schwartz, H. T., Antoshechkin, I., & Sternberg, P. W. (2011). Applications of high-throughput sequencing to symbiotic nematodes of the genus *Heterorhabditis*. *Symbiosis, 55*, 111–118.

Shapiro, D. I., Glazer, I., & Segal, D. (1996). Trait stability in and fitness of the heat tolerant entomopathogenic nematode *Heterorhabditis bacteriophora* IS5 strain. *Biological Control, 6*, 238–244.

Shapiro, D. I., Glazer, I., & Segal, D. (1997). Genetic improvement of heat tolerance in *Heterorhabditis bacteriophora* through hybridization. *Biological Control, 8*, 153–159.

Shapiro, D. I., & McCoy, C. W. (2000). Virulence of entomopathogenic nematodes to *Diaprepes abbreviatus* (Coleoptera: Curculionidae) in the laboratory. *Journal of Economic Entomology, 93*, 1090–1095.

Shapiro, D. I., McCoy, C. W., Fares, A., Obreza, T., & Dou, H. (2000). Effects of soil type on virulence and persistence of entomopathogenic nematodes in relation to control of *Diaprepes abbreviatus*. *Environmental Entomology, 29*, 1083–1087.

Shapiro, D. I., Obrycki, J. J., Lewis, L. C., & Abbas, M. (1999). The effects of fertilizers on black cutworm, *Agrotis ipsilon*, (Lepidoptera: Noctuidae) suppression by *Steinernema carpocapsae*. *Journal of Nematology, 31*, 690–693.

Shapiro, D. I., Obrycki, J. J., Lewis, L. C., & Jackson, J. J. (1999). Effects of crop residue on the persistence of *Steinernema carpocapsae*. *Journal of Nematology, 31*, 517–519.

Shapiro, D. I., Tylka, G. L., Berry, E. C., & Lewis, L. C. (1995). Effects of earthworms on the dispersal of *Steinernema* spp. *Journal of Nematology, 27*, 21–28.

Shapiro, D. I., Tylka, G. L., & Lewis, L. C. (1996). Effects of fertilizers on virulence of *Steinernema carpocapsae*. *Applied Soil Ecology, 3*, 27–34.

Shapiro-Ilan, D. I., & Brown, I. (2013). Earthworms as phoretic hosts for *Steinernema carpocapsae* and *Beauveria bassiana*: Implications for enhanced biological control. *Biological Control, 66*, 41–48.

Shapiro-Ilan, D. I., Brown, I., & Lewis, E. E. (2014). Freezing and desiccation tolerance in entomopathogenic nematodes: Diversity and correlation of traits. *Journal of Nematology, 46*, 27–34.

Shapiro-Ilan, D. I., Bruck, D. J., & Lacey, L. A. (2012). Principles of epizootiology and microbial control. In F. E. Vega & H. K. Kaya (Eds.), *Insect pathology* (2nd ed., pp. 29–72). Amsterdam: Elsevier.

Shapiro-Ilan, D. I., Cottrell, T. E., Mizell, R. F., Horton, D. L., Behle, R. W., & Dunlap, C. A. (2010). Efficacy of *Steinernema carpocapsae* for control of the lesser peachtree borer, *Synanthedon pictipes*: Improved aboveground suppression with a novel gel application. *Biological Control, 54*, 23–28.

Shapiro-Ilan, D. I., Cottrell, T. E., Mizell, R. F., Horton, D. L., & Zaid, A. (2015). Field suppression of the peachtree borer, *Synanthedon exitiosa*, using *Steinernema carpocapsae*: Effects of irrigation, a sprayable gel and application method. *Biological Control, 82*, 7–12.

Shapiro-Ilan, D. I., Cottrell, T. E., Mizell, R. F., III, Horton, D. L., & Davis, J. (2009). A novel approach to biological control with entomopathogenic nematodes: Prophylactic control of the peachtree borer, *Synanthedon exitiosa. Biological Control, 48*, 259–263.

Shapiro-Ilan, D. I., Cottrell, T. E., & Wood, B. W. (2011). Effects of combining microbial and chemical insecticides on mortality of the pecan weevil (Coleoptera: Curculionidae). *Journal of Economic Entomology, 104*, 14–20.

Shapiro-Ilan, D. I., Gardner, W. A., Fuxa, J. R., Wood, B. W., Nguyen, K. B., Adams, B. J., et al. (2003). Survey of entomopathogenic nematodes and fungi endemic to pecan orchards of the Southeastern United States and their virulence to the pecan weevil (Coleoptera: Curculionidae). *Environmental Entomology, 32*, 187–195.

Shapiro-Ilan, D. I., Gouge, D. H., & Koppenhöfer, A. M. (2002). Factors affecting commercial success: Case studies in cotton, turf and citrus. In R. Gaugler (Ed.), *Entomopathogenic nematology* (pp. 333–355). New York: CABI.

Shapiro-Ilan, D. I., Gouge, G. H., Piggott, S. J., & Patterson Fife, J. (2006). Application technology and environmental considerations for use of entomopathogenic nematodes in biological control. *Biological Control, 38*, 124–133.

Shapiro-Ilan, D. I., Han, R., & Dolinski, C. (2012). Entomopathogenic nematode production and application technology. *Journal of Nematology, 44*, 206–217.

Shapiro-Ilan, D. I., Han, R., & Qiu, X. (2014). Production of entomopathogenic nematodes. In J. Morales-Ramos, G. Rojas, & D. I. Shapiro-Ilan (Eds.), *Mass production of beneficial organisms: Invertebrates and entomopathogens* (pp. 321–356). Amsterdam: Academic.

Shapiro-Ilan, D. I., Jackson, M., Reilly, C. C., & Hotchkiss, M. W. (2004). Effects of combining an entomopathogenic fungi or bacterium with entomopathogenic nematodes on mortality of *Curculio caryae* (Coleoptera: Curculionidae). *Biological Control, 30*, 119–126.

Shapiro-Ilan, D. I., Leskey, T. C., & Wright, S. E. (2011). Virulence of entomopathogenic nematodes to plum curculio, *Conotrachelus nenuphar*: Effects of strain, temperature, and soil type. *Journal of Nematology, 43*, 187–195.

Shapiro-Ilan, D. I., Lewis, E. E., Behle, R. W., & McGuire, M. R. (2001). Formulation of entomopathogenic nematode-infected-cadavers. *Journal of Invertebrate Pathology, 78*, 17–23.

Shapiro-Ilan, D. I., Lewis, E. E., & Tedders, W. L. (2003). Superior efficacy observed in entomopathogenic nematodes applied in infected-host cadavers compared with application in aqueous suspension. *Journal of Invertebrate Pathology, 83*, 270–272.

Shapiro-Ilan, D. I., Mizell, R. F., III, Cottrell, T. E., & Horton, D. L. (2004). Measuring field efficacy of *Steinernema feltiae* and *Steinernema riobrave* for suppression of plum curculio, *Conotrachelus nenuphar*, larvae. *Biological Control, 30*, 496–503.

Shapiro-Ilan, D. I., Morales-Ramos, J. A., Rojas, M. G., & Tedders, W. L. (2010). Effects of a novel entomopathogenic nematode–infected host formulation on cadaver integrity, nematode yield, and suppression of *Diaprepes abbreviatus* and *Aethina tumida* under controlled conditions. *Journal of Invertebrate Pathology, 103*, 103–108.

Shapiro-Ilan, D. I., Stuart, R. J., & McCoy, C. W. (2005). Targeted improvement of *Steinernema carpocapsae* for control of the pecan weevil, *Curculio caryae* (Horn) (Coleoptera: Curculionidae) through hybridization and bacterial transfer. *Biological Control, 34*, 215–221.

Shapiro-Ilan, D. I., Stuart, R. J., & McCoy, C. W. (2006). A comparison of entomopathogenic nematode longevity in soil under laboratory conditions. *Journal of Nematology, 38*, 119–129.

Shapiro-Ilan, D. I., Wright, S. E., Tuttle, A. F., Cooley, D. R., & Leskey, T. C. (2013). Using entomopathogenic nematodes for biological control of plum curculio, *Conotrachelus nenuphar*: Effects of irrigation and species in apple orchards. *Biological Control, 67*, 123–129.

Shetlar, D. J., Suleman, P. E., & Georgis, R. (1988). Irrigation and use of entomogenous nematodes, *Neoaplectana* spp. and *Heterorhabditis heliothidis* (Rhabditida: Steinernematidae and Heterorhabditidae), for control of Japanese beetle (Coleoptera: Scarabaeidae) grubs in turfgrass. *Journal of Economic Entomology, 81*, 1318–1322.

Shields, E. J., Testa, A., Miller, J. M., & Flanders, K. L. (1999). Field efficacy and persistence of the entomopathogenic nematodes *Heterorhabditis bacteriophora* 'Oswego' and *H. bacteriophora* 'NC' on Alfalfa snout beetle larvae (Coleoptera: Curculionidae). *Environmental Entomology, 28*, 128–136.

Tanada, Y., & Kaya, H. K. (1993). *Insect pathology*. San Diego, CA: Academic.

Thurston, G. S., Kaya, H. K., & Gaugler, R. (1994). Characterizing the enhanced susceptibility of milky disease-infected scarabaeid grubs to entomopathogenic nematodes. *Biological Control, 4*, 67–73.

Toepfer, S., Peters, A., Ehlers, R.-U., & Kuhlmann, U. (2008). Comparative assessment of the efficacy of entomopathogenic nematode species at reducing western corn rootworm larvae and root damage in maize. *Journal of Applied Entomology, 132*, 337–348.

Ulug, D., Hazir, S., Kaya, H. K., & Lewis, E. E. (2014). Natural enemies of natural enemies: the potential top-down impact of predators on entomopathogenic nematode populations. *Ecological Entomology, 39*, 462–469.

Wallace, H. R. (1958). Movement of eelworms. *Annals of Applied Biology, 46*, 74–85.

Wang, X., & Grewal, P. S. (2002). Rapid genetic deterioration of environmental tolerance and reproductive potential of an entomopathogenic nematode during laboratory maintenance. *Biological Control, 23*, 71–78.

Wang, Y., Bilgrami, A. L., Shapiro-Ilan, D., & Gaugler, R. (2007). Stability of entomopathogenic bacteria, *Xenorhabdus nematophila* and *Photorhabdus luminescens*, during *In Vitro* culture. *Journal of Industrial Microbiology and Biotechnology, 34*, 73–81.

Williams, C. D., Dillon, A. B., Harvey, C. D., Hennessy, R., Namara, L. M., & Griffin, C. T. (2013). Control of a major pest of forestry, *Hylobius abietis*, with entomopathogenic nematodes and fungi using eradicant and prophylactic strategies. *Forest Ecology and Management, 305*, 212–222.

Williams, E. C., & Walters, K. F. A. (2000). Foliar application of the entomopathogenic nematode *Steinernema feltiae* against leaf miners on vegetables. *Biocontrol Science and Technology, 10*, 61–70.

Wilson, M., & Gaugler, R. (2004). Factors limiting short-term persistence of entomopathogenic nematodes. *Journal of Applied Entomology, 128*, 250–253.

Womersley, C. Z. (1993). Factors affecting physiological fitness and modes of survival employed by dauer juveniles and their relationship to pathogenicity. In R. Bedding, R. Akhurst, & H. Kaya (Eds.), *Nematodes for the biological control of insects* (pp. 79–88). East Melbourne, Australia: CSIRO Press.

Zimmerman, R. J., & Cranshaw, W. S. (1991). Short-term movement of *Neoaplectana* spp. (Rhabditida: Steinernematidae) and *Heterorhabditis* "HP-88" strain (Rhabditida: Heterorhabditidae) through turfgrass thatch. *Journal of Economic Entomology, 84*, 875–878.

Chapter 10
Entomopathogenic Nematode Production and Application: Regulation, Ecological Impact and Non–target Effects

Ana Piedra-Buena, Javier López-Cepero, and Raquel Campos-Herrera

10.1 Introduction

Production and commercialization of biocontrol agents is a growing market with over 225 microbial biopesticides manufactured in 30 countries (Kabaluk & Gazdik, 2007). Although the use of *Bacillus thuringiensis* Berliner (Bacillales: Bacillaceae) (BT) was the dominant product in US, Mexico and Canada, being the selected product for 75 % of the crop and forest management, in the European market decreased to 25 % in 2004, with the expansion and use of other bioagents such as entomopathogenic nematodes (EPNs) (Cuddeford, 2008). Despite the drop in sales of the conventional products during early years in 2000s, detecting a decline of 1.5 % per year for pesticides and even a 2.5 % for herbicides and fungicides (CropLife International, 2007; Research, 2006; Thakore, 2006), still, the overall market of conventional pesticides is above 90 % compared with biopesticides (Bailey, Boyetchko, & Längle, 2010). One of the critical point in the development of biopesticides, including EPN, is the connection between the research and commercialization. Firstly, the new bioproducts should overpass the characteristics of conventional pesticides products, or at least, provide successful benefits under particular scenarios. Second, the development implies producing the documents required for the permits, following the regulations that are still unclear. At this

A. Piedra-Buena
Instituto Canario de Investigaciones Agrarias, Tenerife, Spain
e-mail: apbuena@icia.es

J. López-Cepero
COPLACA, Organización de Productores de Plátanos, Tenerife, Spain
e-mail: cepero@coplaca.org

R. Campos-Herrera (✉)
University of Neuchâtel, Neuchâtel, Switzerland
e-mail: raquel.campos@unine.ch

© Springer International Publishing Switzerland 2015

R. Campos-Herrera (ed.), *Nematode Pathogenesis of Insects and Other Pests*, Sustainability in Plant and Crop Protection 1, DOI 10.1007/978-3-319-18266-7_10

moment, where IPM is the most recommended practice, and by law should be implemented in some countries, such as those belonging to EU, advancing on the clarification of those regulations and the new legal framework is urgent. Significant advances have been accomplished during the past years in the regulation and implementation of the biopesticides, which regulations and law directly affecting the EPN development for enterprises and other agents. In this chapter, issues related with the development and release of new bioproducts, such as those containing EPN, are illustrated. In particular, we cover the evolution related to the pesticides in EU, the environmental impact of their production with the example of the carbon footprint assessment and the potential non–target effects of the EPN release.

10.2 Evolution of European Community Regulations Related to Pesticides: An Example from the EU Regulation

To understand the legal frame that affects today the use of biocontrol agents, it is important to present an overview of how these advances were implemented. López-Cepero and Díaz (2013) used the EU as an example to explore how the regulation concerning pesticides evolved over time. This example perfectly illustrates how the public demands on environmental and health concerns were a dynamic variable that finally played a major role in the regulation and implementation of the applications and managements related to pesticides. Herein, we are briefly covering the most important topics related to the pest control that might affect EPN as alternative biocontrol agent.

Environment and consumers protection has not been a continuous concern for legislative European authorities. In 1967, the Directive 67/548/EEC was the first regulation for those products (mainly chemicals) that might affect human health. Its main objective was to establish the laws, regulations and administrative provisions concerning the classification, packaging and labelling of dangerous substances, including pesticide active substances. This text contained provisions for hazardous chemicals to be evaluated for their effects on health, by establishing a hazard classification system that should be included on the pesticide label to inform workers and users, and to avoid obstacles to trade. More than a decade after, Directive 78/631/EEC was the first specific rule published on plant protection products. Its aim was to approximate the laws of the different Member States related to the classification, packaging and labelling of dangerous preparations (pesticides). In those years only *acute toxicity*, which determined the effect of a single exposure, was considered, and the lethal dose 50 (LD_{50}) or lethal concentration 50 (LC_{50}) were determined, depending on if the product was solid, liquid or gaseous. According to the values obtained, products were classified as "Very Toxic", "Toxic" or "Harmful", also indicating the way of exposure: oral, dermal or inhalation.

Scientific advances and the methodologies and approaches allowed determining other toxic effects of chemicals, as those which may occur with repeated exposure at short, medium and long term, and more specific effects such as mutagenicity, carcinogenicity, and reproductive consequences, both for fertility and fetal development. To include all this, new criteria were taken into account and existing laws were modified, to classify chemicals and commercial preparations. Later, plant protection products lost their specific regulations regarding classification standards and labelling, and were eventually included in the scope of Directive 1999/45/EC of the European Parliament and of the Council for the classification and labelling of dangerous chemical preparations.

The first critical rule for the commercialization of plant protection products was Directive 91/414/EEC, which intended to harmonize procedures relating to the authorization, commercialization, use and control in the Member States. In this standard it was already recognized that plant protection products can have adverse effects and potential risks for human health and, therefore, should be regulated to ensure a high level of protection. Herein, two basic aspects were introduced: (i) *toxicological data requirements* for identifying the adverse effects of active substances on human health; (ii) these data were used to classify and label the substance and to estimate parameters that are used in *the risk assessment* of commercial preparations. In this case, both people who eat the food and the people who apply the treatment were considered, using parameters such as "Acceptable Daily Intake" (ADI), "Acceptable Operator Exposure Level" (AOEL) and "Acute Reference Dose" (ARfD).

Based on the so–called Uniform Principles, a plant protection product can only be approved for a requested use after it is demonstrated its safe use (*risk assessment*). These Principles were developed following the same scheme implemented in the EU to assess the risk of a chemical to human health, considering its potential to damage the body and produce some adverse effects (their *intrinsic toxicity*), and the known or estimated *degree or level of exposure* of the organism or population to that product. The relationship between the two aspects allows to characterize the risk or likelihood of harm occurring. Of this result depends the granting of marketing authorization and use, or the establishment of specific risk mitigation measures (risk management).

When VI Community Action Programme on Environment Council and European Parliament (2002–2012) was adopted, it was considered necessary to further reduce the impact of pesticides on human health and the environment because, despite all measures were being implemented, undesirable amounts of pesticides in the environment, water and soil and residues thereof in plant products exceeding the limits officially established could be found. Moreover, the emergence of new types of hazards, such as endocrine disruption, increased the controversy over health protection. Therefore, it was agreed to develop a "Community thematic strategy for the sustainable use of pesticides". This strategy establishes the necessary measures to achieve pesticides reduction, by means of a more sustainable use of pesticides, provided that it ensures necessary protection of agricultural crops. Hence, a significant overall reduction of health and environmental risks is expected. This general aim was developed in five objectives:

1. Minimize the risks and hazards posed by pesticides to health and the environment.
2. Improve controls on the use and distribution of pesticides.
3. Reduce the levels of harmful active substances, particularly through substituting the most dangerous for safer alternatives, including non–chemical ones.
4. Promote agricultural practices involving reduction or elimination of the use of pesticides, by raising awareness to users, promoting the use of codes of practice and the possible use of financial instruments.
5. Create a transparent system for reporting and monitoring of progress and in particular, to establish appropriate indicators.

In this context, Directive 2009/128/CE was released to achieve a reduction of the risks and negative effects of pesticides, and hence, by taking measures for improving its use and promote alternative techniques to decrease the dependence of their applications over time. The application of this community standard will involve a series of changes in the agricultural sector, as the following:

- *Need for qualification* of all professionals involved in the process (farmers, advisors, technicians, distributors and retailers), to make sure that they have acquired sufficient knowledge of the legislation, the risks and dangers associated with pesticides, means of detection and control, procedures to prepare material, emergency measures in case of accident, etc.
- *Pesticide points of sale* must have sufficient staff, available at the time of sale to provide adequate information regarding the use of pesticides, the risks to health and the environment and instructions for safe handling, also with the corresponding certificate.
- *Information and awareness of the general public* regarding the risks of pesticide use, as well as the possibility of using non–chemical alternatives and the possible cases of poisoning.
- *Mandatory periodic inspections for the equipment* for the professional use of pesticides, performed by authorized enterprises at intervals not exceeding 5 years until 2020 and 3 thereafter.
- *Aerial spraying of pesticides is prohibited*, except when there are no viable alternative or where aerial spraying has advantages from the standpoint of human health or the environment with respect to ground application of pesticides.
- In certain *sensitive areas* (those covered by Directives "Wild Birds" and "Habitats", areas accessible to general public or sensitive population –parks, public gardens, sports fields, playgrounds, etc.–), *the use of pesticides is prohibited or strictly limited.*
- *Obligation to establish as a priority Integrated Pest Management (IPM)*, mandatory *from 1 January 2014*. It will be a turning point in the field of plant pathology of the EU. All plant protection methods to eradicate pests should be considered, but prioritizing those that disturb less agricultural ecosystems and encouraging natural pest management mechanisms.

Each member state shall execute National Action Plans to measure progress achieved over time with the actions implemented to reduce risks and negative effects. These plans must contain objectives, measures and timetables to reduce risks of pesticide use on human health and the environment, *encourage the use of environmentally friendly methods or replacement techniques*, and include indicators to control the use of plant protection products containing active substances of particular concern.

The Regulation CE 1107/2009 of the European Parliament and the Council, of 21 October 2009, concerning the commercialization of plant protection products consider that the toxicological properties of active substances, particularly carcinogenicity, mutagenicity and toxicity to reproduction and endocrine disruption ability will be limiting to proceed to its authorization. This intends to achieve one of the objectives of the aforesaid thematic strategy for the sustainable use of pesticides, as was to reduce levels of active materials used, in particular by replacing the most dangerous with safer alternatives, reinforcing the high level of protection of human health and the environment. This Regulation also maintains and enhances the competitiveness of the EU chemical industry, harmonizes availability of plant protection products between farmers of the Member States and update the procedures to take into account the establishment of the European Food Safety Authority (EFSA). The Regulation contains the conditions that the active substances (as well as protectives or synergists) must accomplish to be approved, and provides a guide for the development of standards for the acceptance of co–formulants.

10.3 Impact of the Entomopathogenice Nematode Production: New Insights from the Carbon Footprint Approach

10.3.1 Carbon Footprint Definition and Main Accounting Methodologies

The carbon footprint (CF) is a useful tool for whom are concerned on measuring and reducing their environmental impact on greenhouse gases (GHG) emissions, from countries to organizations and individual citizens. The CF measures the total GHG emissions caused directly and indirectly by a person, organization, event or product. It is measured in tonnes of carbon dioxide equivalent (tCO_2e), which allows the different GHG to be compared on a like–for–like basis relative to one unit of CO_2 and measure their global warming potential. The emissions as CO_2eq are calculated by multiplying the emissions of each of the six main reported human–induced GHG considered by the Kyoto Protocol: carbon dioxide (CO_2), methane (CH_4), nitrous oxide (N_2O), hydrofluorocarbons (HFCs), perfluorocarbons (PFCs) and sulphur hexafluoride (SF_6), by its 100 year global warming potential (GWP; Carbon Trust, 2014).

Several methodologies have been developed around the world to measure, manage and report CF of products and organizations in different countries and by multiple organizations, but the most used take into account the same three scopes and follow similar steps (Table 10.1).

Usually, collecting data and looking for the corresponding emission factors are the most difficult tasks to achieve. Data collection is easier for a single industry and product, whereas becomes more complicated when several products are obtained in a same facility/by one organization –and, therefore, allocations must be performed–, or when several companies of the same sector intend to make a global emissions inventory. In addition, emission factors databases, guides and reports are still incomplete and very often it is problematic to find the emission factor of a particular item, in particular, for products resulting from biological processes (agricultural and animal products, plants from nurseries, biocontrol agents, etc.).

10.3.2 Carbon Footprint in Agriculture

According to the latest Food and Agriculture Organization (FAO) statistics, GHG emissions from agriculture (including crop and livestock production, forestry and associated land–use changes), account for a significant fraction of human–induced emissions: about 20–24 %. This means that agriculture contribution to global warming is about 5,300 million ton CO_2eq per year (FAO, 2014a, 2014b, 2014c). These emissions come from: enteric fermentation (methane from the microbial processes that take place in the digestive systems of ruminants and to a lesser extent non–ruminants), manure management (methane and nitrogen oxide from the decomposition of manure in storage and disposal systems), rice cultivation (methane from the anaerobic decomposition of organic matter in paddy fields), agriculture soils (nitrous oxide emissions linked to the application of synthetic fertilizers, animal manure applied to soils and left on pasture, crop residues returned to soils), cultivation of organic soils (nitrous oxide from cultivation of histosols under cropland and grassland), burning of crop residues and savanna (methane and nitrogen oxide from the combustion of crop residues and savanna biomass) and energy use (carbon dioxide, methane and nitrous oxide from fuel burning and electricity generation for use in agriculture and fisheries; FAO, 2014a).

Although FAO statistics do not discriminate emissions derived from pesticides use, a recent study (Audsley, Stacey, Parsons, & Williams, 2009) estimates that it would be 94 kg CO_2eq per hectare of arable crop. Arable land's definition by FAO includes land under temporary crops (double–cropped areas are counted once), temporary meadows for mowing or for pasture, land under market or kitchen gardens, and land temporarily fallow, for less than 5 years (FAO, 2013). At present, more than 1,500 millions ha (which means about 15 % of the world's land area) is used for crop production (arable land plus land under permanent crops; FAO, 2013). World's land area is 10,039 million ha, from which 38.47 % corresponds to agricultural area (FAO, 2014d; World Bank, 2014), with arable crops accounting for

Table 10.1 Summary of methodologies, covered scopes and main steps on the measure, manage and report carbon footprint (CF)

Methodology	*PAS 2050* *PAS 2050–1* *PAS 2060*	*GHG Protocol guides*	*ISO 14064* *ISO 14067* *ISO 14069*
Developed by	British Standards Institute (BSI)	World Resources Institute (WRI)- World Business Council on Sustainable Development (WCSD)	International Standards Organization (ISO)
Scopes[a]	1. All direct GHG emissions, including fuel combustion, company vehicles and fugitive emissions		
	2. Indirect GHG emissions from consumption of purchased electricity, heat or steam		
	3. Other indirect emissions, such as the extraction and production of purchased goods, services and fuels, transport-related activities in vehicles not owned or controlled by the reporting entity (such as employee commuting, transportation and distribution -up- and downstream-, and business travel), electricity-related activities not covered in Scope 2, outsourced activities, use of sold products, waste disposal, investments, leased assets and franchises, etc.		
Steps[b]	1. *Establishing the scope*		
	2. *Choose the product* (for example, EPNs)		
	(a) Choose the unit of analysis (functional unit, FU; for example, five million package)		
	(b) Identify whether a cradle-to-gate inventory is appropriate (from the raw materials needed to the finished product, ready to be sold to distributors/users)		
	3. *Boundary setting*		
	(a) Define and describe life cycle stages, depending on the production method chosen (*in vivo*, *in vitro*-solid media, *in vitro*- liquid media)		
	(b) Define specific attributable processes and relevant non-attributable processes		
	(c) Justify excluded attributable processes (including insignificance threshold)		
	(d) Define time period of inventory (usually, 1 year-time)		
	4. *Collecting data and assessing data quality*		
	5. *Allocation* (when more than a single product is derived from the process)		
	6. *Assessing uncertainty*		
	7. *Calculating inventory results:* define Global Warming Potential (GWP) values and emission factors to use		
	8. *Assurance*		
	9. *Reporting*		
	10. *Setting reduction targets and tracking*		
	11. *Verification by an independent, third party*		

[a]GHG Protocol (2014)
[b]GHG Protocol (2011)

28.43 % of agricultural area, i.e. 10.9 % of total land area, 1,098 million ha (FAO, 2013). Therefore, emissions derived from pesticides use and manufacturing in arable land would be about 103.21 million ton CO_2eq. According to estimated data of pesticides use per ha, GHG emissions from insecticide manufacturing is a 3.00 % of emissions of total pesticides (Audsley et al., 2009), i.e., 3.1 million ton CO_2eq.

The potential reduction of agricultural emissions by using EPNs instead of chemical insecticides could be calculated taking into account that GHG emissions per kg of insecticide's active ingredient (ai) varies from 18.9 kg CO_2eq per kg (Audsley et al., 2009) to 25.5 kg CO_2eq per kg (Base Carbone, 2014). Each crop, country, region and type of farming uses different amounts of pesticides: from 10.9 g ai per ha in triticale to 2.74 kg ai per ha in potato, with a mean value of 149.6 g ai per ha, i.e., 41 kg CO_2eq per ha (Audsley et al., 2009).

10.3.3 Carbon Footprint of Entomopathogenic Nematodes Production

The inventory of emissions produced along the life cycle of the EPNs production process must be developed for each specific case, taking into account that results (the CF) will vary depending on:

- Production method (*in vivo*, *in vitro*–solid media, *in vitro*– liquid media), which will affect:
  (a) EPNs yield (infective juveniles –IJs– produced by time unit)
  (b) Energy consumption by equipment (growth chambers, freezers, humidifying and aeration systems, pumps, centrifuges, vacuum filters, etc.)
  (c) Chemical products and media cultures needed
  (d) Application method (coating agents, encapsulation, packaging)
- Selected EPN (genus and species): that has an influence on:
  (a) EPNs yield (IJs produced by host/g–mL culture media)
  (b) Suitability of the production system
  (c) Type of required media
  (d) Time needed for the production process

Shapiro-Ilan, Han, and Qiuiu (2014) described extensively the different EPNs production methods (*in vivo*, *in vitro*–solid media, *in vitro*– liquid media), their advantages and disadvantages, their main problems and options to solve them, as well as future directions and perspectives, for the most produced and used EPNs species. These authors also made a comparison of the three methods in terms of cost–efficiency, investment needed, required expertise, ease of achieving quality, labor required, economy of scale and ease to adapt to new nematode species. Although for CF calculations accurate and precise data of activity are required, specific of the organization which produces the EPNs, and of the nematode

Table 10.2 Items that would take into account for production and emissions inventory of carbon footprint

	In vivo	*In vitro* solid media	*In vitro* liquid media
Petri dishes	Fabrication (F), transportation (T)[a]	F, T[a]	
Trays	F, T[a]	F, T[a]	
Pipettes	F, T[a]	F, T[a]	
Tablecloth, filter paper	F, T	F, T[a]	
Diet of insect larvae	F, T, preparation (P)		
Bacteria culture media		F, T, P	
Nutritional media for EPNs		F, T, P	F, T, P
Growth chambers	Functioning (G)	G	G
Chemical products (surface sterilization of EPNs, conservation, rinsing)		F, T	F, T, P
Humidifying and aeration systems	G	G	
Centrifuge, vacuum filters, concentration cones	G	G	
Flasks			F, T[a]
Bioreactors			G
Post–production inputs (for encapsulation, covering, packaging)	F, T	F, T	F, T

[a]Re–usable, unless disposable

species needs and characteristics, some rough estimations of CF for each production method/EPN species can be made (Table 10.2).

In general terms, and lacking from specific activity data from the sector to make an emissions inventory, as more equipment and mechanization is used for EPN production, more energy would be necessary. Also, as more consumable inputs are needed –especially if they come from abroad–, more fabrication and transport emissions would amount for the total CF. In consequence, it would seem that *in vivo* production will have the lower CF, followed by *in vitro*–solid culture method, and with the *in vitro*–liquid culture method with the highest emissions. However, it must be kept in mind that all the inventoried emissions must be referred to the FU produced in the period considered. If the production methods more based on technology produce higher amounts of FU, the individual CF of the FU will be lower, in a similar way as cost–efficiency estimations. Therefore, no assumptions can be made on CF values or comparisons without real, specific data of EPN production and, at our knowledge, at present no CF for EPN production has been published. However, provided that PAS 2050–1 (BSI, 2012) states that "it is expected that the contribution of the production of insects (for biological control of pests) has a negligible contribution (to crops CF)", whereas "the emissions resulting from (plant protection) chemical use frequently do not make a material contribution, they can be significant in some cases, especially when soil fumigation is applied", two assumptions could be made. First of all, EPN production would probably have lower CF than that of insect production, neither being significant to CF of crops; secondly, although PAS 2050–1 guidelines are general, and real, contrasted data on CF of biocontrol (EPN) agents would be required, it can be estimated that the substitution of chemical pesticides by biocontrol agents, especially for soil pests, would reduce significantly the CF of the crops and, consequently, contribute in the reduction of global GHG emissions from agriculture.

10.4 Ecological Risks of the Entomopatogenic Nematode Application

10.4.1 Indigenous Versus *Commercial Populations*

Entomopathogenic nematodes are widespread all around the world, with the exception of Antarctica (Adams et al., 2006; Griffin, Downes, & Block, 1990). More than 70 *Steinernema* and 20 *Heterorhabditis* recognized species have been described, and the list increases every year (see Chap. 1). However, our current knowledge of EPN geographic distribution is highly biased by the asymmetric sampling efforts invested in different countries and continent. For example, more than 70 % of the *Steinernema* species have been described from North America and Asia (Campos-Herrera, Barbercheck, Hoy, & Stock, 2012); only recent effort in Caribbean area/South America and Africa can allow extending the knowledge on the local species diversity and their potential in biocontrol (see Chaps. 14, 15 and 19, 20, respectively).

Learning about the natural biodiversity of this nematode group will contribute to advance in the basic knowledge of natural history, ecology and co–evolution. In addition to this, the inherent attractiveness of the applied study of the EPN resided in their contribution as biological control agents. Hence, their isolation and identification of autochthonous populations is an essential part of those studies focused in the selection of biological control agents. This screening will provide populations adapted to local conditions and target hosts at the same time that will prevent ecological imbalances produced by the application of exotic EPNs species (Millar & Barbercheck, 2001). Due to the patchy distribution of the natural populations of EPN (Stuart, Barbercheck, Grewal, Taylor, & Hoy, 2006), the metapopulation dynamics, and the spatial–temporal patterns of fluctuations (Campos-Herrera et al., 2012; see also Chap. 4), the isolation of new native populations requires an effective method of isolation and a variable but usually significant sampling effort (Hominick, 2002; Spiridonov, Moens, & Wilson, 2007; Spiridonov & Voronov, 1995; Wilson, Lewis, Yoder, & Gaugler, 2003).

In a broad sense, two types of isolation methods can be distinguished: from infected cadavers recovered in the field and by soil–bait or recovery. Historically, the location and recovery of cadaver in the field, in many cases result of an epizooty, was the first approach (Bovien, 1937). Although several isolates have been collected using this approach (summarized by Peters, 1996), is extremely rare. The most successful and well established method for EPN isolation is the "insect–trap" which employ sentinel insect as bait in a confined soil sample. Firstly described by Beding & Akhurst (1975), this method allows the isolation and the recovery of all the stages (male, female and juveniles) to establish species description, if necessary (Hominick, 2002). To increase the likelihood of new EPN species, modification of the original method were suggested to allow the detection of EPN adapted to cold temperatures or selected host (Adams & Nguyen, 2002; Hominick & Briscoe, 1990; Mráček & Bečvář, 2000; Mráček, Bečvář, Kindlmann, & Jersákova, 2005; Půža & Mráček, 2005). In addition to those, other Nematological techniques are used for the study of EPN populations, such as sucrose–centrifugation or sieving and decanting (Hominick, 2002). Despite their high recovery efficiency, their use has been strongly limited due to the complexity nature of a mixed Nematological sample, making highly difficult to distinguish EPN's IJ from other Rhabditids. Only recent molecular biology advances are allowing identifying and quantifying those organisms from the pool of nematodes in a species–specific manner (Campos-Herrera, El-Borai, Stuart, Graham, & Duncan, 2011; Campos-Herrera et al., 2011, 2013, 2015; Torr, Spiridonov, Heritage, & Wilson, 2007). However, although these studies allow the advance on our understanding of the ecology and biology of this organism, these don't allow the establishment of new cultures and hence, insect bait in parallel is desirable (Campos-Herrera et al., 2015).

Despite the efforts invested in the isolation and ecological characterization of native isolates, numerous commercial products are available in the market all around the word (see revision by Kaya et al., 2006), although the demand can fluctuate over years and might produce discontinuing of some of these products (Dolinski, Choo, & Duncan, 2012). Most of these products are used following an augmentation strategy, and most of the cases, the survival of the selected EPN species is very low, moving to extinction in few weeks, due to environmental limitations (Duncan et al., 2003, 2013; Gaugler, Campbell, Selvan, & Lewis, 1992). However, there are some examples of establishment of these foreigner populations. For example, *S. scapterisci*, originally from Uruguay, was introduced in Florida as part of the program to control crickets, and it has well established in the target area (grassland) (Parkman, Frank, Nguyen, & Smart, 1993; Parkman & Smart, 1996) and even extended to other non–selected crops (citrus) (Campos-Herrera, El-Borai et al., 2011). The impact of these introductions in the natural fauna is unknown and is particularly critical in areas where no information is available about the natural populations.

International law plays an important role in establishing mechanisms to regulate both the introduction and implementation of alien species in a country, within their biological control programs, but also the genetic resource of new populations (Akhurst & Smith, 2002; Hokkanen & Menzler-Hokkanen, 2007; Hominick, Reid, Bohan, & Briscoe, 1996; Smith, 2000). The Convection of Biological Diversity (June 13th, 1992), ratified in December 21st, 1993 was supported by 188 countries and signed by 168 of those. The objectives set for this meeting were the conservation of biodiversity, compatible use of its components and equitable sharing of benefits arising from the utilization of genetic resources. Within this convention was ratified that the country is sovereign on the genetic resources of their state. Today, almost all the countries are CBD party, with the significant exception of US (CBD, 2014). Into this context, The Cartagena Protocol on Biosafety was a subsequent international agreement which objective was "to ensure the safe handling, transport and use of living modified organisms (LMOs) resulting from modern biotechnology that may have adverse effects on biological diversity, taking also into account risks to human health" (CBD, 2014). Firstly signed on January, 29th, 2000, it entered into force on September 11th, 2003, and today, a total of 167 countries have ratified it (CBD, 2014). In addition to those, Smith (2000) proposed a code of conduct for studies on biodiversity and the use of genetic resources within the law and with active participation from both sides, which means transparency in all processes that develop between the countries which have the resources and the country investigating these resources. The Nagoya Protocol was the international development of this proposal, produced under the framework of the CBD. The general objective of the Nagoya Protocol was "sharing the benefits arising from the utilization of genetic resources in a fair and equitable way, including by appropriate access to genetic resources and by appropriate transfer of relevant technologies". Recently, the Nagoya Protocol entered into force (October 12th, 2014), and at the moment, a total of 53 countries plus the European Union have ratified the protocol (CBD, 2014). These regulations and protocols, together with the production and applications provide a legal framework for the use of EPN as biological control in safe and rational approach, although still many efforts have to be invested to ensure the protection at different scales.

10.4.2 International Regulations for Production and Environmental Risks: Legal Development and Application

International regulation affects differently the EPN commercialization, depending on the nation or group of nations. For example, in Europe, the potential problems derived from the use of EPN as biological control was early revised in the 90s (Boemare, Laumond, & Mauleon, 1996; Ehlers, 1996; Ehlers & Hokkanen, 1996; Richardson, 1996). This first analysis concluded that the implementation of EPN was safe and low risk was identified, so that a search for marketing was required. However, the introduction of new species should be regulated (see revision by Akhurst & Smith, 2002). Lately, the European Commission aimed to unify the procedures for the distribution of these products. The European project REBECA (Regulation of Biological Control Agents, Specific Support Action, SSPE–CT–2005–022709, http://www.rebeca-net.de/, project coordinator

R.-U. Ehlers) provided a comprehensive revision of the current situation not only concerning Europe and associated countries but also important producers of biological control agents and commercializers such as US, Canada and Australia. This revision provided insights in the risk assessment, production, commercialization and application of these beneficial organisms, and presented a possible new frame for legislation and implementation, including EPN (Loomans, 2007). In Europe, microbial (which also include EPN) are now regulated under the directive CE1107/2009 (see details in Sect. 10.1 in this chapter). Still, the requirements of each of the European countries are not yet homogenized and regulation is not the same in all these countries (Ehlers, 2007; Loomans, 2007). Following the REBECA evaluation of the regulations of beneficial organisms' production and release, Loomans (2007) established three different categories for the status of implementation:

- Regulation implemented in some degree (Austria, Czech Republic, Denmark, Hungary, Norway, Slovenia, Sweden, Switzerland and UK)
- Regulation under development (Finland, Germany, Ireland, Netherlands and Spain).
- No regulation developed neither implemented (Belgium, France, Greece, Italy, Poland, Portugal) nor contact established to develop the report (Cyprus, Estonia, Ireland, Latvia, Lithuania, Luxembourg, Malta and Slovakia).

For the risk assessment, REBECA proposed a "Hierarchical System of Risk Screening" with several case studies, including some examples for EPN. For each organism, data available related to the taxonomy and biology, human health risk and environmental risk is described, drawing a conclusion about its potential as alien invasive species. In addition to this, REBECA Action developed several recommendations for the safe release of new biocontrol agents, providing documents that could "guide" the Expert Group selected for the implementation. These documents included "Dossier Application Form", "Guidance Document" and "Environmental Risk Assessment –ERA– Methods" (Deliverable 22; Ehlers, 2007).

In other countries outside the EU, such as US, Canada, New Zealand and Australia, the situation is different. For implementing regulatory procedures for the import and release of invertebrate biological control agents, EU is still far behind the other countries (Hunt et al., 2007). Until the late 90s, import of EPNs to US had virtually no legal regulation and enforcement occurred with few control measures. Nickle, Drea, and Coulson (1988) proposed a series of guidelines to understand the properties and characteristics of imported products, but were not very successful. In the mid-90s, the policy on legal procedures for import–export of exotic species developed complex procedures for their regulation (Rizvi, Hennessey, & Knott, 1996), resulting in a more or less lengthy processing of the necessary permits for acceptance. Here were involved different government institutions that overseen the effective enforcement of quarantine laws, export–import and application of products not tested. Today, the Animal and Plant Health Inspection Service and Plant Protection and Quarantine (APHIS–PPQ) of the United States Department of Agriculture's (USDA) has a well defined system to establish the permits to importation and release of beneficial organisms, including EPN, to the environment (Hunt et al., 2007). In addition to the several forms demanded by the APHIS–PPQ, if the beneficial organisms is non indigenous and has not been released in the past in US, it will be mandatory to specify the biological data following the North American Plant Protection Organization (NAPPO) guidelines (NAPPO, 2000). Similarly, in Canada, the Plant Health Division of the Canadian Food Inspection Agency (CFIA–PHD) is in charge of the implementation and regulations, overseen by the Ministry of Agriculture and Agri–Food Canada (AAFC). In Canada, the application procedure also requires the permit for importation (even for scientific reasons) and release approval (Hunt et al., 2007).

The requirements for the introduction of exotic species in Australia and New Zealand were early reviewed by Bedding, Tyler, and Rocherster (1996). Australia regulates the beneficial biological controls as they regulate entry of exotic species as possible impact in the native flora and fauna (Delfosse, 2005; Harrison, Moeed, & Sheppard, 2005). Then, applications submitted to the Australian Quarantine Inspection Service (AQIS) are derived to Biosecurity Australia, which evaluate the characteristics of the import, the host–specificity test and the impact of the release (Hunt et al., 2007). Nowadays, the Australian Government, Department of Agriculture (2014) has released the "Biosecurity guidelines for the introduction of exotic biological control agents for the control of weeds and plant pests", where is described in detail the appropriate protocols required for approvals of new introductions. Similarly, in New Zealand, the Hazardous Substances and New Organisms (HSNO) Act (approved in 1998) controls the introductions of all organisms not indigenous for the country, including the biological control agents, such as EPN. The Environmental Risk Management Authority (ERMA New Zealand), developed under the HSNO Act, is in charge of implementing all processes and hence, the import and release applications of these organisms (Hunt et al., 2007).

Applying the symbiotic bacteria is also possible, once tested its viability. The maximum survival of the bacteria in the environment has been successfully detected during 1 week, both in soil and freshwater (Morgan, Kuntzelmann, Tavernor, Ousley, & Winstanley, 1997), which provides a framework for control a target pest. As a result, attempt to use formulations based in just the bacteria has been recorded. For example, the patent registered for the use of *X. nematophila* against *Solenopsis invicta* Buren (Hymenoptera: Formicidae) (Dudney, 1997).

In addition to the formulation of naturally enhanced EPN, it is also possible to engineer either the nematode or the bacteria to produce a resistant line for certain purposes (see Chap. 2 for details), or combine certain genes of these bacteria for building transgenic plants. For example, some genes from *Xenorhabdus* and *Photorhabdus* have been recombined into plants to confer them resistance to insects (Bowen & Ensing, 1998; East, Cao, & Akhurst, 1998). In this case, the resulting plants are regulated by legislation concerning transgenic plants, since the expression of certain proteins in food products is being widely questioned for possible allergenic activity. Therefore, caution should be taken about asserting that transgenic plants with genes of bacterial symbionts of EPN can be produced without health or environmental risks. In general, the production of these organisms is highly questioned by the general public and government agencies exert a decisive control pressure, so that their future is in debate.

10.4.3 Impact on Non–target Organisms: Nematodes

Impact on non–target organisms can have different nature, such as competition and displacement but also the direct mortality. For the selection of a new biological control agent, a balance between the host–specificity and the broad–host range is considered a challenge. For one side, if a selected nematode is highly specific, i.e. *Steinernema scapterisci* Nguyen & Smart (Rhabditida: Steinernematidae) and *Steinernema neocurtillae* Nguyen & Smart (Rhabditida: Steinernematidae), the possible risk associated with the non–target effects will be reduced to the minimum; however, from a market and sales perspective, a broad range of effect is also desirable, so the final product will be able to be sold for targeting many different pests. Hence, the most ubiquitous EPN, i.e. *Steinernema feltiae* (Filipjev) Wouts, Mráček, Gerdin & Bedding (Rhabditida: Steinernematidae), *Steinernema carpocapsae* (Weiser) Wouts, Mráček, Gerdin & Bedding (Rhabditida: Steinernematidae) and *Heterorhabditis bacteriophora* Poinar (Rhabditida: Heterorhabditidae) are the most broadly developed for bioagent companies (see Kaya et al., 2006). In general, indirect effect such as displacement might affect non–target organisms even those that have been tested as extremely species–specific (Pearson & Callaway, 2005). The final negative impact will depend on the ecosystem in which the nematode will be introduced, it will be driven by the biotic and abiotic forces that shape the community and limit its persistence (Campos-Herrera et al., 2012).

Native species of EPNs are considered more suitable for introduction in biological control programs because they are adapted to the environmental conditions of the area. However, even if the native species are reared and prepared for an augmentative application, it is necessary to take into account two main points: (1) the high pathogenicity described for EPN with at least 17 orders of insects (Smith, Miller, & Simser, 1992) and (2) the impact of the nematode–bacteria complex in other beneficial organisms that may be eventually affected. Numerous are the studies focused in the possible activity of selected EPN species (see revision by Georgis et al., 2006 and Kaya et al., 2006). However, most of those studies focused their effort in testing EPN against insect pests. In another hand, studies on post–application biology and the impact on other organisms are rare, despite its relevance, although new molecular techniques and multidisciplinary studies are helping to advance in this direction (see Chaps. 3, 4 and 13). Naturally, there are pathogens that exert control over the populations of other groups, such as other EPNs species (displacement by habitat or host preference) and insect predators and parasites (Stuart et al., 2006; also see Chap. 4). Therefore, it is critical to evaluate the impact of EPNs and their mutualistic bacteria in non–target organisms, both vertebrate and invertebrate, to assess the potential ecological risk of their implementation at large scale.

In natural conditions, the EPN natural host range is more restricted than that detected in laboratory conditions (Peters, 1996). The studies concerning the impact of EPN in other non–target arthropods, non–arthropod invertebrates and vertebrates was revised in depth by Akhurst & Smith (2002). Herein, we will enumerate some of these examples in relation to these three groups:

Non–target Arthropods Insects acting as predators and parasitoids may be affected by the EPNs, in particular, in aerial applications that are currently advancing in the implementation. Some of these species can be directly affected if the nematodes cause the death of the parasitized host or even kill the adults. This negative effect was observed in laboratory studies (Kaya, 1978, 1984; Triggiani, 1985). Predators can also die by EPN direct infection, although the effect on these organisms depends on the development of the insect inside the host (Georgis, Kaya, & Gaugler, 1991; Lemiere, Coderre, Vincent, & Bélair, 1996; Ropek & Jaworska, 1994). In addition to this, EPN can affect these insects if they significantly reduce the available hosts. Indirectly observation of this phenomenon was reported by Battisti (1994), who observed a reduction in the emergence of parasitoids after the application of *S. feltiae*. Interestingly, some insect might serve as phoretic host, helping EPN to spread to other areas. For example, infected mole crickets (*Scapteriscus* spp.) can serve as spreaders of the nematode *S. scapterisci* (Parkman et al., 1993), which can serve as biocontrol for *Popillia japonica* Newman (Coleoptera: Scarabaeidae) (Lacey, Bettencourt, Garrett, Simões, & Gaugler, 1993). Also, the adult of the large pine weevils, *Hylobius abietis* L. (Coleoptera: Curculionidae) were described as potential phoretic host to EPN (Kruitbos, Heritage, & Wilson, 2009), even if they can serve for infection (Girling, Ennis, Dillon, & Griffin, 2010).

In addition to this, certain arthropods such as Collembola, Symphyla, Diplopoda, arachnids, and crustaceans can serve as unusual host for EPN development, at least, potentially, as showed in some laboratory studies (Poinar, 1989). Other researchers have described the negative impact of EPNs in terrestrial isopods, diplopods, and members of the order Acarina (Hill, 1998; Ishibashi, Young, Nakashima, Abiru, & Haraguchi, 1987; Jaworska, 1991; Mauleon, Barré, & Panoma, 1993; Samish & Glazer, 1991). However, those arthropods might serve as phoretic host of EPN, as described for isopods by Eng, Preisser, and Strong (2005).

Non–arthropods Invertebrates Earthworms have been the focus of several studies in relationship with EPN. Although early studies showed that under certain circumstance, EPN can develop in previously damaged earthworms (Capinera, Blue, & Wheeler, 1982; Nüutinen, Tyni-Juslin, Vänninen, & Vainio, 1991; Potter, Spicer, Redmond, & Powell, 1994), the interaction between EPN and earthworm results in most of the cases in a phoretic association that help nematodes to spread. Shapiro, Berry, and Lewis (1993) and Shapiro, Tylka, Berry, and Lewis (1995) studied the ability of earthworms to disperse the EPNs in soil. They observed that IJs could be "transported" attached to the setae of his body, and even looked inside his gut. In regard to the internal passage, Campos-Herrera, Trigo, and Gutiérrez (2006) demonstrated that EPN can be ingested by earthworms and the casts can have some viable EPNs, but only in rare occasions. Therefore, this association is more likely due to the external attachment than as a mix of soil digested by the earthworms. More recently, Shapiro-Ilan and Brown (2013) provided a complete study that demonstrated how earthworms can improve nematode dispersal throughout the soil, suggesting this phoretic association as a natural mechanism to enhance biocontrol.

The impact in gastropods has been also evaluated, with variable results. For example, Li, Deng, Zhang, and Yang (1986) found that 97.5 % of individuals of the snail *Oncomelania hupensis* Gredler (Sorbeoconcha: Pomatiopsidae) were infected when exposed to a high dose of 300 IJs/cm^2 of *S. feltiae, Steinernema glaseri* (Steiner) Wouts, Mráček, Gerdin & Bedding (Rhabditida: Steinernematidae) and *H. bacteriophora*. Later, Jaworska (1993) noted that EPN could develop in the wild slugs *Deroceras agreste* L. (Pulmonata: Agriolimacidae) and *Deroceras reticulatum* OF Müller (Pulmonata: Agriolimacidae) in laboratory tests. They observed a high sensitivity to the attack by *S. carpocapsae, S. feltiae* and *H. bacteriophora* with 100 % mortality after 4–10 days of exposure, depending of the slug species. However, Wilson, Glen, Hughes, Pearce, and Rodgers (1994) found that EPNs could not infect *D. reticulatum* under laboratory conditions. It is possible that the mortality caused by EPN were observed in un–naturally high concentration, and hence, the likelihood of this negative effect is only limited to these gastropods close to a new IJs emerging cadaver.

As other members of the soil biota, nematodes can be also non–target effect of EPN applications. This possible effect has been evaluated in the context of biocontrol potential of plant–parasitic nematodes (PPN). For example, Grewal, Lewis, and Venkatachari (1999) found that the applications of *Xenorhabdus* sp. reduced egg hatch of *Meloidogyne incognita* (Kofoid & White) Chitwood, (Tylenchida: Heteroderidae) in the roots of plants. Also, Lewis, Grewal, and Sardanelli (2001) found that the application of *S. feltiae* diminished populations of *M. incognita*, reducing nodule formation, production and hatching of eggs per plant, but did not affect egg production per female. Jagdale, Somasekhan, Grewal, and Klein (2002) also observed that the activity of the nematode *S. carpocapsae* and *Xenorhabdus nematophila* Thomas & Poinar, (Enterobacteriales: Enterobacteriaceae) associated bacteria led to reductions in plant–parasitic nematodes, while producing no reduction in the populations of other free–living nematodes, results that were in agreement with those observed in similar host–PPN–EPN systems by Somasekhar, Grewal, de Nardo, and Stinner (2002), Perez & Lewis (2004) and Vyas, Maghodia, Patel, and Patel (2004). It has been speculated that EPN can serve as signal to systemic induced resistance (Jagdale & Grewal, 2008; Jagdale, Kamoun, & Grewal, 2009), but still mechanisms and detailed phenotypic is needed to ensure this possible plant–EPN interaction.

Vertebartes Reported only rare negative effects caused by EPN. For example, the adult frog *Bufo marinus* (Anura: Bufonidae) were unaffected, although some tadpoles were infected, detecting the presence of EPNs in the digestive tract and were able to release the symbiotic bacteria and developing females (Kermarrec & Mauleón, 1985; Kermarrec, Mauleón, Sirjusingh, & Baud, 1991; Poinar &

Thomas, 1988). Studies injecting EPNs orally, intradermally, subcutaneously and intraperitoneally to mice, chickens, rabbits and Guinea pigs have shown no lethal effects. The body temperature of 37 °C prevents nematode reproduction, and a few days of injection are eliminated by the immune system (Georgis, 1992). Furthermore, the thickness of the walls of the digestive tract of vertebrates prevents IJs penetration into the bloodstream (Poinar, 1989). Only studies with mice by Kobayashi, Okano, and Kirihara (1987) showed that after subcutaneous injection of 2×10^4 IJs produced ulcers in the skin, but at lower concentrations such as 10^3 IJ, this condition was not developed. Also, Park, Kim, Kim, Yang, and Kim (2001) evaluated the effect of oral administration of *S. carpocapsae* in mice and found no lethal effects although nematodes were found in the internal organs. In humans the development of allergies has only been detected in individuals working in companies producing EPN due to the continuous and prolonged contact with these nematodes (Akhurst & Smith, 2002).

In general, the impact of EPNs on organisms considered "non–target" is limited, with infections only occurring when these organisms are exposed to very high concentrations and under laboratory conditions. However, the broad host range detected in these studies points to the need for field experiments to ensure the safety of EPNs on other species and, therefore, avoid imbalances in the soil ecosystem food webs. Barbercheck & Miller (2000) conducted a review of the potential environmental impact of EPNs, indicating that the beneficial species most affected by the implementation of EPNs are those in which some stage developed in the soil. However, they have only been associated with small reductions in the populations of beneficial organisms when applying EPNs (Ishibashi et al., 1987; Ropek & Jaworska, 1994).

10.4.4 Impact on Non–target Organisms: Bacteria

The bacteria *Xenorhabdus* and *Photorhabdus* are intimately and mutualistically–symbiotically associated with EPN (Boemare, 2002). Production of a large number of toxic metabolites for other bacteria, fungi and even nematodes has been reported (Webster, Chen, Hu, & Li, 2002). These chemical compounds might produce changes in the composition of the rhizosphere (Webster, 2000), that negatively affects the soil community (Akhurst, 1982; De Nardo, Grewal, McCartney, & Stinner, 2006; Han & Ehlers, 1999; Li, Chen, & Webster, 1997; Maxwell, Chen, Webster, & Dunphy, 1994; Ng & Webster, 1997). However, a recent study has explored this negative impact as a possible mechanism of control of other plant pathogens (Fang, Zhang, Tang, Wang, & Zhang, 2014). From a "vertebrate" perspective, this impact can be considerdd very limited. For example, when these bacteria were applied orally or injected intradermally, subcutaneously and intraperitoneally to Guinea pigs, mice and rats, no signs of pathogenicity or toxicity were observed (Obendorf, Peel, Akhurst, & Miller, 1983; Poinar, Thomas, Presser, & Hardy, 1982). Only one person developed an allergy to the bacteria *X. nematophila* although, as in the case of nematodes, was a worker who had been in continuous contact with the bacteria (Akhurst & Smith, 2002). Therefore, although there have been occasional episodes of allergy processes, provided in people who have had a very close contact with these organisms may be considered that bacteria, like nematodes, have a very low risk for human health.

Interestingly, the species *Photorhabdus asymbiontica* Fisher le Saux, Villard, Brunel, Normand & Boemare (Enterobacteriales: Enterobacteriaceae) is known to be a human pathogen (Farmer et al., 1989; Peel et al., 1999). There are two subspecies, *P. asymbiontica* subsp. *asymbiontica* and *P. asymbiontica* subsp. *australis*, corresponding to two clinic strains from US and Australia, respectively (Akhurst & Smith, 2002). For several years, the nematode associated with this bacterial species was unknown. However, in 2006 it was reported the isolation of a new *Heterorhabditis* species isolated from a sandy soil from a beach in New South Wales (Australia), area where an infection was previously reported (Gerrard et al., 2006). Subsequent studies on the nematode–bacteria complex resulted in the description of a new species, *Heterorhabditis gerrardi* Plichta, Joyce, Clarke, Waterfield & Stock (Rhabditida: Heterorhabditidae). This bacteria has been reported to invades soft tissues and produce bacteraemic infections in human (Akhurst & Smith, 2002; Gerrard et al., 2006), although also are infectious to insects (Plichta, Joyce, Clarke, Waterfield, & Stock, 2009). An interesting difference with the other bacteria of the same genus is the typical coloration acquired by the cadaver. In this case, the cadavers turn grey with pick spots, whereas the colour for other species of the same genus is in most cases reddish.

In addition to the mutualistic bacteria associated with EPN, Poinar (1979) described other bacteria on the surface of the EPNs that sometimes, due to incomplete disinfection of the surface, might reproduce concomitantly inside the cadaver. For example, Bedding & Molyneux (1982) noted that the bacteria attached to the J2 cuticle of the *Heterorhabditis* can be introduced in the insect while the process of entering in the host, and can became even established inside the host (Molyneux, Bedding, & Akhurst, 1983). This phenomenon has been also reported for steinernematids, in particular with the cuticle of the J2 of *S. scapterisci* (Bonifassi et al., 1999). Despite this possible contamination coming from IJs, there have been reported bacteria from the genus *Paenibacillus* that are associated with the EPN, both *Heterorhabditis* and *Steinernema.* In this case, EPNs and those non–entomopathogenic bacteria are associated thanks to the endospores linked to the cuticle (El-Borai, Duncan, & Preston, 2005; Enright, McInerney, & Griffin, 2003). These bacteria are able to reproduce concomitantly with the nematode and associated symbiotic–mutualistic bacteria inside the host, but the specificity varies with the bacterial species. Then, whereas *Paenibacillus nematophilus* Enright, McInerney, & Griffin (Bacillales: Paenibacillaceae) can attach a variety of heterorhabditids species and also some strongylid (animal parasites) species (Enright & Griffin, 2004), the bacteria *Paenibacillus* sp. associated with *Steinernema diaprepesi* Nguyen & Duncan (Rhabditida: Steinernematidae) is more species–specific (El-Borai et al., 2005). These bacteria have a detrimental effect in the final fitness of the EPN, since highly encumbered nematodes showed a limited ability to migrate, and hence, to locate the host (El-Borai et al., 2005; Enright & Griffin,

2005). To date, the role that *Paenibacillus* spp. might play in the EPN population dynamics in nature is still poorly understood. Duncan et al. (2007) observed that this interaction *S. diaprepesi–Paenibacillus* sp. was detected in the infected insect with this nematode species. Advances in the development of molecular tools allow now evaluating these relationships in nature (Campos-Herrera, El-Borai et al., 2011). Recently, first insights on the natural relationships between EPN and those bacteria were reported in a regional survey in Florida citrus groves (Campos-Herrera et al., 2013). The abundance of *Paenibacillus* sp. was related with its species–specific nematode *S. diaprepesi*. In addition to this, regional preference, linked to EPN distribution was described for the first time. Advances in molecular tools and integration of the data with regional and temporal dynamics might provide the comprehensive knowledge required to establish the real risk of EPN and their associated bacteria in nature in a long–term.

10.5 Conclusions and Future Directions

Regulation and implementation is taking the incredible task to organize and place in a legal context the new necessities of our society. Production and commercialization of new bioagents, such as EPNs, requires a complex and international frame to be established. It is expected that product based on microorganisms and other biological agents will experience a significant increase in the coming years. This is thanks to the changes in the public demand during the last decades. Now, costumers are aware about the right for asking better products in terms of quality but also in terms of environmentally friendly products. Both researchers and companies should be ready to accomplish the new challenges in the new global market framed by emerging national and international regulations.

The new measures taken to reduce the pollutants are in line with accounting for the impact that our activities have in the environment and also the economy. Also, estimating the ecological impact of the companies producing EPN might help to support those approaches. The estimation of the CF is one of the main options for a company to show its compromise with the environment. For a long–term perspective, the trend is to reduce the ecological impact of the EPN production and commercialization, and hence, the local–regional scale of naturally adapted, customized product will increase the demand. At the same time, advance in the knowledge of the non–target effects of the EPN application can benefit of the new molecular and geostatistical approaches (Campos-Herrera et al., 2013), so we can assess the impact in the whole community, not just in terms of affecting the biodiversity of multiples groups but also in impact on functional drivers.

Acknowledgements The authors would like to thank their colleagues from the Instituto Canario de Investigaciones Agrarias (Spain), Organización de Productores de Plátanos COPLACA (Spain), and University of Neuchàtel (Switzerland) for their support. The authors also acknowledge A.L.C.P.B. and R.B.P. for their time while preparing this chapter.

References

Adams, B. J., Fodor, A., Koppenhöfer, H. S., Stackenbrandt, E., Stock, S. P., & Klein, M. G. (2006). Biodiversity and systematic of nematode–bacterium entomopathogens. *Biological Control, 38*, 4–21.

Adams, B. J., & Nguyen, K. B. (2002). Taxonomy and systematics. In R. Gaugler (Ed.), *Entomopathogenic nematology* (pp. 1–33). Wallingford, UK: CABI Publishing.

Akhurst, R. J. (1982). Antibiotic activity of *Xenorhabdus* spp., bacteria symbiotically associated with insect pathogenic nematodes of the families Heterorhabditidae and Steinernematidae. *Journal of General Microbiology, 128*, 3061–3065.

Akhurst, R. J., & Smith, K. (2002). Regulation and safety. In R. Gaugler (Ed.), *Entomopathogenic nematology* (pp. 311–332). Wallingford, UK: CABI Publishing.

Audsley, E., Stacey, K., Parsons, D. J., & Williams, A. G. (2009). *Estimation of the greenhouse gas emissions from agricultural pesticide manufacture and use*. Cranfield, UK: Cranfield University Report.

Australian Government, Department of Agriculture. (2014). Agriculture. http://www.agriculture.gov.au/Style%20Library/Images/DAFF/__data/assets/pdffile/0008/2395160/Revised-guidelines-introduction-exotic-BCAs-update06-2014.pdf. Revised November 23, 2014.

Bailey, K. L., Boyetchko, S. M., & Längle, T. (2010). Social and economic drivers shaping the future of biological control: A Canadian perspective on the factors affecting the development and use of microbial biopesticides. *Biological Control, 52*, 221–229.

Barbercheck, M. E., & Millar, L. C. (2000). Environmental impacts of entomopathogenic nematodes used for biological control in soil. In P. A. Follett & J. J. Duan (Eds.), *Nontarget effects of biological control* (pp. 287–308). Boston: Kluwer Academic Publishers.

Base Carbone. (2014). Bilans GES. Centre de resources sur les bilans de gaz à effet de serre. Database consultation. http://www.bilans-ges.ademe.fr/fr/basecarbone/donnees-consulter/liste-element?recherche=insecticide. Revised November 23, 2014.

Battisti, A. (1994). Effects of entomopathogenic nematodes on the spruce web–spinning sawfly *Cephalcia arvensis* Panzer and its parasitoids in the field. *Biocontrol Science and Technology, 4*, 95–102.

BCC Research. (2006). The new biopesticide market. http://www.bccresearch.com/chemicals

Bedding, R. A., & Akhurst, R. J. (1975). A simple technique for the detection of insect parasitic rhabditids nematodes in soil. *Nematologica, 21*, 109–110.

Bedding, R. A., & Molyneux, A. S. (1982). Penetration of insect cuticle by infective larvae of *Heterorhabditis* spp. (Heterorhabditidae: Nematoda). *Nematologica, 28*, 354–359.

Bedding, R. A., Tyler, S., & Rocherster, N. (1996). Legislation on the introduction of exotic entomopathogenic nematodes into Australia and New Zealand. *Biocontrol Science and Technology, 6*, 465–475.

Boemare, N. E. (2002). Biology, taxonomy and systematics of *Photorhabdus* and *Xenorhabdus*. In R. Guagler (Ed.), *Entomopathogenic nematology* (pp. 35–56). Wallingford, UK: CABI Publishing.

Boemare, N. E., Laumond, C., & Mauleon, H. (1996). The nematode–bacterium complexes: Biology, life cycle and vertebrate safety. *Biocontrol Science and Technology, 6*, 333–345.

Bonifassi, E., Fischer-Le Saux, M., Boemare, N., Lanois, A., Laumond, C., & Smart, G. (1999). Gnotobiological study of infective juveniles and symbionts of *Steinernema scapterisci*: A model to clarify the concept of the natural occurrence of monoxenic associations in entomopathogenic nematodes. *Journal of Invertebrate Pathology, 74*, 164–172.

Bovien, P. (1937). Some types of association between nematodes and insects. *Videnskabelige Meddelelser Fra Dansk Naturahistorisk Forening, Khobenhavn, 101*, 1–114.

Bowen, D. J., & Ensing, J. C. (1998). Purification and characterization of a high–molecular–weight insecticidal protein complex produced by the entomopathogenic bacterium *Photorhabdus luminescens*. *Applied and Environmental Microbiology, 64*, 3029–3035.

BSI. (2012). PAS 2050–1:2012. Publicly Available Specification. Assessment of life cycle greenhouse gas emissions from horticultural products – Supplementary requirements for the cradle to gate stages of GHG assessments of horticultural products undertaken in accordance with PAS 2050. BSI Standards Limited. London, UK: British Standards Institute.

Campos-Herrera, R., Barbercheck, M., Hoy, W. H., & Stock, S. P. (2012). Entomopathogenic nematodes as a model system for advancing the frontiers of ecology. *Journal of Nematology, 44*, 162–176.

Campos-Herrera, R., El-Borai, F. E., Stuart, R. J., Graham, J. H., & Duncan, L. W. (2011). Entomopathogenic nematodes, phoretic *Paenibacillus* spp., and the use of real time quantitative PCR to explore soil food webs in Florida citrus groves. *Journal of Invertebrate Pathology, 108*, 30–39.

Campos-Herrera, R., Jaffuel, G., Chiriboga, X., Blanco-Pérez, R., Fesselet, M., Půža, V., Mascher, F., Turlings, T.C.J. (2015). Traditional and molecular detection methods reveal intense interguild competition and other multitrophic interactions associated with native entomopathogenic nematodes in Swiss tillage soils. *Plant and Soil, 389,* 237–255.

Campos-Herrera, R., Johnson, E. G., El-Borai, F. E., Stuart, R. J., Graham, J. H., & Duncan, L. W. (2011). Long–term stability of entomopathogenic nematode spatial patterns measured by sentinel insects and real–time PCR assays. *Annals of Applied Biology, 158*, 55–68.

Campos-Herrera, R., Pathak, E., El-Borai, F. E., Stuart, R. J., Gutiérrez, C., Rodríguez-Martín, J.A., Graham, J.H., Duncan, L.W. (2013). Geospatial patterns of soil properties and the biological control potential of entomopathogenic nematodes in Florida citrus groves. *Soil Biology and Biochemistry, 66*, 163–174.

Campos-Herrera, R., Trigo, D., & Gutiérrez, C. (2006). Phoresy of the entomopathogenic nematode *Steinernema feltiae* by the earthworm *Eisoenia fetida. Journal of Invertebrate Pathology, 92*, 50–54.

Capinera, J. L., Blue, S. L., & Wheeler, G. S. (1982). Survival of earthworms exposed to *Neoaplectana carpocapsae* nematodes. *Journal of Invertebrate Pathology, 39*, 419–421.

Carbon Trust. (2014). *Product carbon footprinting for beginners*. London, UK: Carbon Trust.

CBD (2014). The convention on biological diversity. http://www.cbd.int/. Revised November 23, 2014.

CropLife International. (2007). Annual report 2006/2007. http://www.croplife.org/library/attachments/5650a1df50df403daecdd0e607e878ce/7/Annual_Report_2006\T1\textendash2007.pdf

Cuddeford, V. (2008). Biocontrol files, Issue 13, March 2008, Co-published by World Wildlife Fund, the Biocontrol Network, and Agriculture and Agri-Food Canada, pp. 1–8. www.biocontrol.ca

Delfosse, E. S. (2005). Risk and ethics in biological control. *Biological Control, 35*, 319–329.

De Nardo, E. A. B., Grewal, P. S., McCartney, D., & Stinner, B. R. (2006). Non-target effects of entomopathogenic nematodes on soil microbial community and nutrient cycling processes: A microcosm study. *Applied Soil Ecology, 34*, 250–257.

Dolinski, C., Choo, H. Y., & Duncan, L. W. (2012). Grower acceptance of entomopathogenic nematodes: Case studies on three continents. *Journal of Nematology, 44*, 226–235.

Dudney, R. A. (1997). Use of *Xenorhabdus nematophilus* Im/1 and 19061/1 for the fire ant control. US Patent US 5,616,318 970,401 (Aplicación informática, US 95–488,820 950,609).

Duncan, L. W., Graham, J. H., Dunn, D. C., Zellers, J., McCoy, C. W., & Nguyen, K. (2003). Incidence of endemic entomopathogenic nematodes following application of *Steinernema riobrave* for control of *Diaprepes abbreviatus. Journal of Nematology, 35*, 178–186.

Duncan, L. W., Graham, J. H., Zellers, J., Bright, D., Dunn, D. C., El-Borai, F. E., et al. (2007). Food web responses to augmenting the entomopathogenic nematodes in bare and animal manure–mulched soil. *Journal of Nematology, 39*, 176–189.

Duncan, L. W., Stuart, R. J., El-Borai, F. E., Campos-Herrera, R., Pathak, E., & Graham, J. H. (2013). Modifying orchard planting sites conserves entomopathogenic nematodes, reduces weevil herbivory and increases citrus tree growth, survival and fruit yield. *Biological Control, 64*, 26–36.

East, P. D., Cao, A., & Akhurst, R. J. (1998). Toxin genes from the bacterium *Xenorhabdus nematophilus* and *Photorhabdus luminescens*. International Patent Application PCT/AU98/00562.

Ehlers, R.-U. (1996). Current and future use of nematodes in biocontrol: Practice and commercial aspects with regard to regulatory policy issues. *Biocontrol Science and Technology, 6*, 303–316.

Ehlers, R.-U. (2007). Sustainable management of Europe's natural resources. Final report (pp. 1–50). REBECA, Specific Support Action, project no. SSPE–CT–2005–022709.

Ehlers, R.-U., & Hokkanen, H. M. T. (1996). Insect control with non-endemic entomopathogenic nematodes (*Steinernema* and *Heterorhabditis* spp.): Conclusions and recommendations for a combined OECD and COST workshop on scientific and regulatory policy issues. *Biocontrol Science Technology, 6*, 295–302.

El-Borai, F. E., Duncan, L. W., & Preston, J. F. (2005). Bionomics of a phoretic association between *Paenibacillus* sp. and the entomopathogenic nematode *Steinernema diaprepesi. Journal of Nematology, 37*, 18–25.

Eng, M. S., Preisser, E. L., & Strong, D. R. (2005). Phoresy of the entomopathogenic nematode *Heterorhabditis marelatus* by a non-host organism, the isopod *Porcellio scaber. Journal of Invertebrate Pathology, 88*, 173–176.

Enright, M. R., & Griffin, C. T. (2004). Specificity of association between *Paenibacillus* spp. and the entomopathogenic nematodes, *Heterorhabditis* spp. *Microbial Ecology, 48*, 412–421.

Enright, M. R., & Griffin, C. T. (2005). Effects of *Paenibacillus nematophilus* on the entomopathogenic nematode *Heterorhabditis megidis. Journal of Invertebrate Pathology, 88*, 40–48.

Enright, M. R., McInerney, J. O., & Griffin, C. T. (2003). Characterization of endospore-forming bacteria associated with entomopathogenic nematodes, *Heterorhabditis* spp., and description of *Paenibacillus nematophilus* sp. nov. *International Journal of Systematic and Evolutionary Microbiology, 53*, 435–441.

Fang, X. L., Zhang, M. R., Tang, Q., Wang, Y. H., & Zhang, X. (2014). Inhibitory effect of *Xenorhabdus nematophila* TB on plant pathogens *Phytophthora capsici* and *Botrytis cinerea* in vitro and in planta. *Scientific Reports, 4*, article no. 4300.

FAO. (2013). *FAO statistical yearbook 2013. World, food and agriculture*. Rome: Food and Agriculture Organization of the United Nations.

FAO. (2014a). *Food and nutrition in numbers 2014*. Rome: Food and Agriculture Organization of the United Nations.

FAO. (2014b). Greenhouse gas emissions from agriculture, forestry and other land use. FAO Infographics, 1 p. Downloadable from: http://www.fao.org/resources/infographics/infographics\T1\textendashdetails/en/c/218650/

FAO. (2014c). Emissions from agriculture. Food and Agriculture Organization of the United Nations. Statistics Division web page. http://faostat3.fao.org/download/G1/GT/E. Revised October 15, 2014.

FAO. (2014d). Agricultural and arable area 2011. Food and Agriculture Organization of the United Nations. Statistics Division web page. http://faostat3.fao.org/download/E/EL/E. Revised October 15, 2014.

Farmer, J. J., Jorgernsen, P. A., Grimount, P. A. D., Akhurst, R. J., Poinar, G. O., Ageron, E., et al. (1989). *Xenorhabdus luminescens* (DNA hybridization group 5) from human clinical specimens. *Journal of Clinical Microbiology, 27*, 1594–1600.

Gaugler, R., Campbell, J. F., Selvan, S., & Lewis, E. E. (1992). Large-scale inoculative releases of the entomopathogenic nematode *Steinernema glaeri*: Assessment 50 years later. *Biological Control, 2*, 181–187.

Georgis, R. (1992). Present and future for entomopathogenic nematodes products. *Biocontrol Science and Technology, 2*, 83–99.

Georgis, R., Kaya, H. K., & Gaugler, R. (1991). Effect of steinernematid and heterorhabditid nematodes (Rhabditida: Steinernematidae and Heterorhabditidae) on nontarget arthropods. *Environmental Entomology, 20*, 815–822.

Georgis, R., Koppenhöfer, A. M., Lacey, L. A., Bélair, G., Duncan, L. W., Grewal, P. S., et al. (2006). Successes and failures in the use of parasitic nematodes for pest control. *Biological Control, 38*, 103–123.

Gerrard, J. G., Joyce, S. A., Clarke, D. J., French-Constant, R. H., Nimmo, G. R., Looke, D. F. M., et al. (2006). Nematode symbiont for *Photorhabdus asymbiotica. Emerging Infectious Diseases, 12*, 1562–1564.

GHG Protocol. (2011). Product life cycle accounting and reporting standard. World Resources Institute (WRI)–World Business Council for Sustainable Development (WBCSD). Downloadable at: http://www.ghgprotocol.org/standards/product\T1\textendashstandard

GHG Protocol. (2014). FAQ. GHG Protocol. http://www.ghgprotocol.org/calculation\T1\textendashtools/faq. Revised February 13, 2014.

Girling, R. D., Ennis, D., Dillon, A. B., & Griffin, C. T. (2010). The lethal and sub-lethal consequences of entomopathogenic nematode infestation and exposure for adult pine weevils, *Hylobius abietis* (Coleoptera: Curculionidae). *Journal of Invertebrate Pathology, 104*, 195–202.

Grewal, P. S., Lewis, E. E., & Venkatachari, S. (1999). Allelopathy: A possible mechanism of suppression of plant–parasitic nematodes by entomopathogenic nematodes. *Nematology, 1*, 735–743.

Griffin, C. T., Downes, M. J., & Block, W. (1990). Test of Antarctic soils for insect parasitic nematodes. *Antarctic Science, 2*, 221–222.

Han, R., & Ehlers, R.-U. (1999). Trans-specific nematicidal activity of *Photorhabdus luminescens. Nematology, 1*, 735–743.

Harrison, L., Moeed, A., & Sheppard, A. (2005). Regulation of the release of biological control agents of arthropods in New Zealand and Australia. In M. S. Hoddle (Ed.), *Proceedings of the second international symposium on biological control of arthropods, Davos, Switzerland, 12–16 September 2005* (FHTET–2005–08, pp. 715–725). Morgantown, WV: United States Department of Agriculture, Forest Service.

Hill, D. E. (1998). Entomopathogenic nematodes as control agents of developmental stages of the black-legged tick, *Ixodes scapularis. Journal of Parasitology, 84*, 1124–1127.

Hokkanen, H., & Menzler-Hokkanen, I. (2007). Comparison of benefits/risks between different groups of plant protection products. Deliverable 26 (pp. 1–15), REBECA, Specific Support Action, project no. SSPE–CT–2005–022709.

Hominick, W. M. (2002). Biogeography. In R. Gaugler (Ed.), *Entomopathogenic nematology* (pp. 115–143). Wallingford, UK: CABI Publishing.

Hominick, W. M., & Briscoe, B. R. (1990). Survey of 15 sites over 28 months for entomopathogenic nematodes (Rhabditida: Steinernematidae). *Parasitology, 100*, 289–294.

Hominick, W. M., Reid, A. P., Bohan, D. A., & Briscoe, B. R. (1996). Entomopathogenic nematodes: Biodiversity, geographical distribution and the Convection on Biological Diversity. *Biocontrol Science and Technology, 6*, 317–331.

Hunt, E. J., Kuhlmann, U., Sheppard, A., Qin, T. K., Barratt, B. I. P., Harrison, L., et al. (2007). Review of invertebrate biological control agent regulation in Australia, New Zealand, Canada and the USA: Recommendations for a harmonized European system. *Journal of Applied Entomology, 132*(2), 89–123.

Ishibashi, N., Young, F. Z., Nakashima, M., Abiru, C., & Haraguchi, N. (1987). Effects of application of DD–136 on silkworm, *Bombyx mori* predatory insect, *Agriosphodorus dohrni*, parasitoid, *Trichomalus apanteloctenus*, soil mites, and other non–target soil arthropods, with brief notes on feeding behaviour and predatory pressure of soil mites, tardigrades and predatory nematode on DD–136 nematodes. In N. Ishibashi (Ed.) *Recent advances in biological control of insect pests by entomogenous nematodes in Japan* (pp. 158–164). Ministry of Education, Japan, Grant n° 59860005 (in Japanese, abstract in English).

Jagdale, G. B., & Grewal, P. S. (2008). Influence of the entomopathogenic nematode *Steinernema carpocapsae* infected host cadavers or their extracts on the foliar nematode *Aphelenchoides fragariae* on Hosta in the greenhouse and laboratory. *Biological Control, 44*, 13–23.

Jagdale, G. B., Kamoun, S., & Grewal, P. S. (2009). Entomopathogenic nematodes induce components of systemic resistance in plants: Biochemical and molecular evidence. *Biological Control, 51*, 102–109.

Jagdale, G. B., Somasekhan, N., Grewal, P. S., & Klein, M. G. (2002). Suppression of plant–parasitic nematodes by application of life and dead infective juveniles of an entomopathogenic nematode, *Steinernema carpocapsae*, on boxwood (*Buxus* spp.). *Biological Control, 24*, 42–49.

Jaworska, M. (1991). Infection of the terrestrial isopod *Porcellio scaber* Larr. and millipede *Blaniulus guttulatus* Bosc. with entomopathogenic nematodes (Nematoda: Rhabditidae) in laboratory conditions. *Folia Horticulturae, 3*, 115–120.

Jaworska, M. (1993). Laboratory infection of slugs (Gastropoda: Pulmonata) with entomopathogenic nematodes (Rhabditida: Nematoda). *Journal of Invertebrate Pathology, 61*, 223–244.

Kabaluk, T., & Gazdik, K. (2007). Directory of microbial pesticides for agricultural crops in OECD countries. Agriculture & Agri-Food Canada. http://www4.agr.gc.ca/resources/prod/doc/pmc/pdf/micro_e.pdf> .

Kaya, H. K. (1978). Infectivity of *Neoaplectana carpocapsae* and *Heterorhabditis heliothidis* to pupae of the parasite *Apanteles militaris. Journal of Nematology, 10*, 241–244.

Kaya, H. K. (1984). Effect of the entomogenous nematode *Neoaplectana carpocapsae* on the tachinid parasite *Compsilura concinnata* (Diptera: Tachinidae). *Journal of Nematology, 16*, 9–13.

Kaya, H. K., Aguillera, M. M., Alumai, A., Choo, H. Y., de la Torre, M., Fodor, A., et al. (2006). Status of entomopathogenic nematodes and their symbiotic bacteria from selected countries or regions of the world. *Biological Control, 38*, 134–155.

Kermarrec, A., & Mauleón, H. (1985). Potential noxiousness of the entomogenous nematode *Neoaplectana carpocapsae* Weiser to the Antillan toad *Bufo marinus* L. *Mededelingen Faculteit Landbouwkundige University of Gent, 50*, 831–838.

Kermarrec, A., Mauleón, H., Sirjusingh, C., & Baud, L. (1991). Experimental studies on the sensitivity of tropical vertebrates (toad, frogs and lizards) to different species of entomopathogenic nematodes of the genera *Heterorhabditis* and *Steinernema*. In C. Pavis & A. Kermarrec (Eds.), *Rencontres Caraïbes en Lutte Biologique* (pp. 193–204). Paris: Institute National de la Recherche Agronomique.

Kobayashi, M., Okano, H., & Kirihara, S. (1987). The toxicity of steinernematid and heterorhabditid nematodes to male mice. In N. Ishibashi (Ed.), *Recent advances in biological control of insect pests by entomogenous nematodes in Japan* (pp. 153–157). Japan: Saga University.

Kruitbos, L. M., Heritage, S., & Wilson, M. J. (2009). Phoretic dispersal of entomopathogenic nematodes by *Hylobius abietis. Nematology, 11*, 419–427.

Lacey, L. A., Bettencourt, R., Garrett, F. J., Simões, N. J., & Gaugler, R. H. (1993). Factors influencing parasitism of adult Japanese beetles, *Popillia japonica* (Col.: Scarabaeidae) by entomopathogenic nematodes. *Entomophaga, 38*, 501–509.

Lemiere, S., Coderre, D., Vincent, C., & Bélair, G. (1996). Lethal and sublethal effects of the entomogeus nematodes, *Steinernema carpocapsae*, on the coccinellid *Harmonia axyridis. Nematropica, 26*, 284–285.

Lewis, E. E., Grewal, P. S., & Sardanelli, S. (2001). Interactions between the *Steinernema feltiae–Xenorhabdus bovienii* insect pathogen complex and the root–knot nematode *Meloidogyne incognita. Biological Control, 21*, 55–62.

Li, J. X., Chen, G. H., & Webster, J. M. (1997). Nematophin, a novel antimicrobial substance produced by *Xenorhabdus nemathophilus* (Enterobacteriaceae). *Canadian Journal of Microbiology, 43*, 770–773.

Li, P. S., Deng, C. S., Zhang, S. G., & Yang, H. W. (1986). Laboratory studies on the infectivity of the nematode *Steinernema glaseri* to *Oncomelania hupensis*, a snail intermediate host of blood fluke, *Schistosoma japonicum. Chinese Journal of Biological Control, 2*, 50–53.

Loomans, A. J. M. (2007). Regulation of invertebrate Biological Control Agents in Europe: Review and recommendations in its pursuit of a harmonised regulatory system. Report EU project REBECA (pp. 1–26), Specific Support Action, project no. SSPE–CT–2005–022709.

López-Cepero, J., & Díaz, M. (2013). *La producción integrada como herramienta para una producción sostenible: análisis crítico y aplicación a Canarias*. Graduate thesis, Universidad de La Laguna, Spain.

Mauleon, H., Barré, N., & Panoma, S. (1993). Pathogenicity of 17 isolates of entomophagous nematodes (Steinernematidae and Heterorhabditidae) for the ticks *Amblyomma variegatum* (Fabricius), *Boophilus microplus* (Canestrini) and *Boophilus annulatus* (Say). *Experimental and Applied Acarology, 17*, 831–838.

Maxwell, P. W., Chen, G., Webster, J. W., & Dunphy, G. B. (1994). Stability and activities of antibiotics produced during infection of the insect *Galleria mellonella* by two isolates of *Xenorhabdus nematophilis*. *Applied and Environmental Microbiology, 60*, 715–721.

Millar, L. C., & Barbercheck, M. E. (2001). Interaction between endemic and introduced entomopathogenic nematodes in conventional–till and no till corn. *Biological Control, 22*, 235–245.

Molyneux, A. S., Bedding, R. A., & Akhurst, R. J. (1983). Susceptibility of larvae of the sheep blowfly *Lucilia cuprina* to various *Heterorhabditis* spp., *Neoaplectana* spp., and an undescribed steinernematid (Nematoda). *Journal of Invertebrate Pathology, 42*, 1–7.

Morgan, J. A. W., Kuntzelmann, V., Tavernor, S., Ousley, M. A., & Winstanley, C. (1997). Survival of *Xenorhabdus nematophilus* and *Photorhabdus luminescens* in water and soil. *Journal of Applied Microbiology, 83*, 665–670.

Mráček, Z., & Bečvář, S. (2000). Insect aggregations and entomopathogenic nematode occurrence. *Nematology, 2*, 297–301.

Mráček, Z., Bečvář, S., Kindlmann, P., & Jersákova, J. (2005). Habitat preference for entomopathogenic nematodes, their insect hosts and new faunistic records for the Czech Republic. *Biological Control, 34*, 27–37.

NAPPO. (2000). Guidelines for petition for release of exotic entomophagous agents for the biological control. Regional standards for phytosanitary measures #12. http://www.nappo.org/Standards/OLDSTDS/RSPM12\T1\textendashe.pdf

Ng, K. K., & Webster, J. M. (1997). Antimycotic activity of *Xenorhabdus bovienii* (Enterobacteriaceae) metabolites against *Phytophthora infestans* on potato plants. *Canadian Journal of Plant Pathology, 19*, 125–132.

Nickle, W. R., Drea, J. J., & Coulson, J. R. (1988). Guidelines for introducing beneficial insect–parasitic nematodes intro the United States. *Annals of Applied Nematology, 2*, 50–56.

Nüutinen, V., Tyni-Juslin, J., Vänninen, I., & Vainio, A. (1991). The effects of four entomopathogenic fungi and an entomopathogenic nematode on the hatching of earthworm (*Aporrectodea caliginosa*) cocoons in laboratory. *Journal of Invertebrate Pathology, 58*, 147–149.

Obendorf, D. L., Peel, B., Akhurst, R. J., & Miller, L. A. (1983). Non-susceptibility of mammals to the entomopathogenic bacterium *X. nemathophilus*. *Environmental Entomology, 12*, 368–370.

Park, Y., Kim, M., Kim, J., Yang, K.-U., & Kim, Y. (2001). Toxicological analysis of the entomopathogenic nematode, *Steinernema carpocapsae* and the symbiotic bacteria, *Xenorhabdus nematophilus* on beneficial insects and mammals. *Korean Journal of Applied Entomology, 40*, 259–264.

Parkman, J. P., Frank, J. H., Nguyen, K. B., & Smart, G. C., Jr. (1993). Dispersal of *Steinernema scapterisci* (Rhabditida: Steinernematidae) after inoculative applications for mole cricket (Orthoptera: Gryllotalpidae) control in pastures. *Biological Control, 3*, 226–232.

Parkman, J. P., & Smart, G. C. (1996). Entomopathogenic nematodes, a case of study: Introduction of *Steinernema scapterisci* in Florida. *Biocontrol Science and Technology, 6*, 413–419.

Pearson, D. E., & Callaway, R. M. (2005). Indirect nontarget effects of host–specific biological control agents: Implications for biological control. *Biological Control, 35*, 288–298.

Peel, M. M., Alferson, D. A., Gerrad, J. G., Davis, J. M., Robson, H. M., McDougall, R. J., et al. (1999). Isolation, identification and molecular characterization of strains of *Photorhabdus luminescens* from infected humans in Australia. *Journal of Clinical Investigation, 37*, 3647–3653.

Pérez, E. E., & Lewis, E. E. (2004). Suppression of *Meloidogyne incognita* and *Meloidogyne hapla* with entomopathogenic nematodes on greenhouse peanuts and tomatoes. *Biological Control, 30*, 336–341.

Peters, A. (1996). The natural host range of *Steinernema* and *Heterorhabditis* spp. and their impact on insect populations. *Biocontrol Science and Technology, 6*, 389–402.

Plichta, K. L., Joyce, S. A., Clarke, D., Waterfield, N., & Stock, S. P. (2009). *Heterorhabditis gerrardi* n. sp. (Nematoda: Heterorhabditidae): The hidden host of *Photorhabdus asymbiotica* (Enterobacteriaceae: γ-Proteobacteria). *Journal of Helminthology, 83*, 309–320.

Poinar, G. O., Jr. (1979). *Nematodes for biological control of insect*. Boca Ratón: CRC Press.

Poinar, G. O., Jr. (1989). Non-insect hosts for the entomogenous rhabditoid nematodes *Neoaplectana* (Steinernematidae) and *Heterorhabditis* (Heterorhabditidae). *Revue de Nématologie, 12*, 423–428.

Poinar, G. O., Jr., & Thomas, G. M. (1988). Infection of frog tadpoles (Amphibia) by insect parasitic nematodes (Rhabditida). *Experientia, 44*, 528–531.

Poinar, G. O., Jr., Thomas, G. M., Presser, S. B., & Hardy, J. L. (1982). Inoculation of entomogenous nematode, *Neoaplectana* and *Heterorhabditis* and their associated bacteria, *Xenorhabdus* spp. into chicks and mice. *Environmental Entomology, 11*, 137–138.

Potter, D. A., Spicer, P. G., Redmond, C. T., & Powell, A. J. (1994). Toxicity of pesticides to earthworms in Kentucky bluegrass turf. *Bulletin of the Environmental Contamination and Toxicology, 52*, 176–181.

Půža, V., & Mráček, Z. (2005). Seasonal dynamics of entomopathogenic nematodes of the genera *Steinernema* and *Heterorhabditis* as a response to abiotic factors and abundance of insect hosts. *Journal of Invertebrate Pathology, 89*, 116–122.

Richardson, P. N. (1996). British and European legislation regulating rhabditid nematodes. *Biocontrol Science and Technology, 6*, 449–463.

Rizvi, S. A., Hennessey, R., & Knott, D. (1996). Legislation on the introduction of exotic nematodes. *Biocontrol Science and Technology, 6*, 477–480.

Ropek, D., & Jaworska, M. (1994). Effect of an entomopathogenic nematode, *Steinernema carpocapsae* Weiser (Nematoda: Steinernematidae), on a carabid beetles in field trials with annual legumes. *Anzeiger für Schädlingskunde Pflanzenschutz Umweltschutz, 67*, 97–100.

Samish, M., & Glazer, I. (1991). Killing ticks with parasitic nematodes of insects. *Journal of Invertebrate Pathology, 58*, 281–282.

Shapiro, D. I., Berry, E. C., & Lewis, L. C. (1993). Interactions between nematodes and earthworms: Enhanced dispersal of *Steinernema carpocapsae. Journal of Nematology, 25*, 189–192.

Shapiro, D. I., Tylka, G. L., Berry, E. C., & Lewis, L. C. (1995). Effects of earthworm on the dispersal of *Steinernema* spp. *Journal of Nematology, 27*, 21–28.

Shapiro-Ilan, D., & Brown, I. (2013). Earthworms as phoretic hosts for *Steinernema carpocapsae* and *Beauveria bassiana*: Implications for enhanced biological control. *Biological Control, 66*, 41–48.

Shapiro-Ilan, D. I., Han, R., & Qiu, X. (2014). Production of entomopathogenic nematodes. In J. A. Morales-Ramos, M. Guadalupe Rojas, & D. Shapiro-Ilan (Eds.), *Mass production of beneficial organisms* (pp. 321–355). London: Elsevier Inc.

Smith, D. (2000). Legislation affecting the collection, use and safe handling of entomopathogenic microbes and nematodes. In A. Navon & K. R. S. Ascher (Eds.), *Bioassays of entomopathogenic microbes and nematodes* (pp. 295–314). Wallingford, UK: CAB International.

Smith, K. A., Miller, R. W., & Simser, D. H. (1992). Entomopathogenic nematode bibliography: Heterorhabditid and steinernematid nematodes. In N. G. Wyatt (Ed.), *A publication of the Nematode Subcommittee of the Southern Project S–240 development of entomopathogens as control agents for insect pests* (Southern Cooperative Series Bulletin). Fayetteville, AR: University of Arkansas, Arkansas Agriculture Experiment Station.

Somasekhar, N., Grewal, P. S., de Nardo, E. A. B., & Stinner, B. R. (2002). Non-target effects of entomopathogenic nematodes on the soil nematode community. *Journal of Applied Ecology, 39*, 735–744.

Spiridonov, S. E., Moens, M., & Wilson, M. J. (2007). Fine scale spatial distributions of two entomopathogenic nematodes in a grassland soil. *Applied Soil Ecology, 37*, 192–201.

Spiridonov, S. E., & Voronov, D. A. (1995). Small-scale distribution of *Steinernema feltiae* juveniles in cultivated soil. In C. T. Griffin, R. L. Gwynn, & J. P. Masson (Eds.), *COST 819. Biotechnology: Ecology and transmission strategies of entomopathogenic nematodes*(pp. 36–41). Brussels, Belgium: European Communities.

Stuart, R. J., Barbercheck, M. E., Grewal, P. S., Taylor, R. A. J., & Hoy, C. W. (2006). Population biology of entomopathogenic nematodes: Concepts, issues, and models. *Biological Control, 38*, 80–102.

Thakore, Y. (2006). The biopesticide market for global agricultural use. *Industrial Biotechnology, 2*, 194–208.

Torr, P., Spiridonov, S. E., Heritage, S., & Wilson, M. J. (2007). Habitat associations of two entomopathogenic nematodes: A quantitative study using real–time quantitative polymerase chain reactions. *Journal of Animal Ecology, 76*, 238–245.

Triggiani, O. (1985). Influence of *Steinernema* and *Heterorhabditis* (Rhabditida) on the parasitoid *Apanteles ultor* Rhd. (Hymenoptera: Braconidae). *La difesa delle Piante, 8*, 293–300.

Vyas, R. V., Maghodia, A. B., Patel, B. A., & Patel, D. J. (2004). Interaction between root–knot nematodes and *Steinernema riobrave* on okra (*Abelmoschus esculentum*). *International Journal of Nematology, 14*, 186–190.

Webster, J. M. (2000). The role and potential of secondary metabolites of the bacterial symbionts: Insect pathogens and insect parasitic nematodes. *IOBC/WPRS Bulletin, 23*, 93–95.

Webster, J. M., Chen, G., Hu, K., & Li, J. (2002). Bacterial metabolites. In R. Gaugler (Ed.), *Entomopathogenic nematology* (pp. 99–114). Wallingford, UK: CABI Publishing.

Wilson, M. J., Glen, D. M., Hughes, L. A., Pearce, J. D., & Rodgers, P. B. (1994). Laboratory tests of the potential of entomopathogenic nematodes for the control of field slugs (*Deroceras reticulatum*). *Journal of Invertebrate Pathology, 64*, 182–187.

Wilson, M. J., Lewis, E. E., Yoder, F., & Gaugler, R. (2003). Application pattern and persistence of the entomopathogenic nematode *Heterorhabditis bacteriophora. Biological Control, 26*, 180–188.

World Bank. (2014). World development indicators: Agricultural inputs. http://wdi.worldbank.org/table/3.2. Revised November 14, 2014.

Part III
Entomopathogenic Nematode Exploitation: Case-Studies in Crop Protection in Different Crops and Countries

Chapter 11
New York Case Study: Biological Control of *Otiorhynchus ligustici* with Native Persistent Entomopathogenic Nematodes Using a More Classical Approach

Elson J. Shields and Antonio M. Testa

11.1 History and Economic Importance of Alfalfa Snout Beetle in New York

Alfalfa snout beetle (ASB), *Otiorhynchus ligustici* (L.) (Coleoptera: Curculionidae) is a serious pest of alfalfa and clover production in Northern New York (NY), USA. Currently the infestation area covers >200,000 ha across nine NY counties and a small portion of Canada across the St Lawrence River from the NY infestation (Fig. 11.1). Within the infested area, entire fields of alfalfa and clover can be destroyed in a single year from root feeding by the larvae. This flightless parthenogenetic (clonal reproduction) European insect was introduced into the Port of Oswego, NY on shipping ballast between 1848 and 1896, when it was collected by Wickham (Claassen & Palm, 1935; Shields, Testa, Neumann, Flanders, & Schroeder, 2009). *O. ligustici* became an economic pest after alfalfa and clover were introduced into the area in the 1920s. In 1935, the infestation was reported to cover 235 ha, growing to 1,200 ha by 1941. By the 1970s, the infestation area had grown to 6,400 ha and by 2006, there were over 200,000 ha infested.

Alfalfa and clover production are critical to efficient and profitable dairy production, contributing an economical source of protein and nutrients to the dairy ration. Within the nine county infested area, dairy production is the major agricultural enterprise valued at $435 million USD. In NYS, dairy production is valued at 2560 millon USD and should the beetle spread into surrounding areas and states, those areas will also be at risk (Ag Statistics, 2013).

The beetle, *O. ligustici,* has a 2-year lifecycle on alfalfa and clover with 98 % of its lifecycle in contact with soil. Adults emerge in early spring, feed for

E.J. Shields (✉) • A.M. Testa
Department of Entomology, Cornell University, Ithaca, NY, USA
e-mail: es28@cornell.edu; at28@cornell.edu

© Springer International Publishing Switzerland 2015

R. Campos-Herrera (ed.), *Nematode Pathogenesis of Insects and Other Pests*,
Sustainability in Plant and Crop Protection 1, DOI 10.1007/978-3-319-18266-7_11

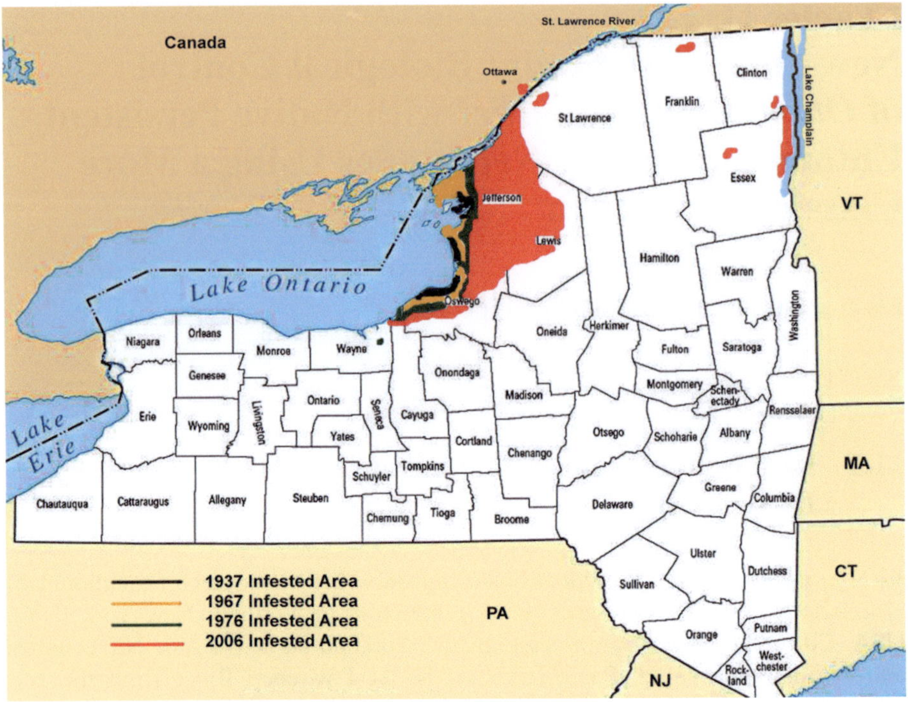

Fig. 11.1 Area of New York State and Canada infested with alfalfa snout beetle, *Otiorhynchus ligustici*. This map represents the entire known North American infestation of this European insect

approximately 3 weeks and then oviposit for 4–6 weeks during a walking dispersal. Adults are in soil contact during the cool periods after emergence at the soil–duff interface and during oviposition when they lay eggs in the top layer of soil around the base of the plant. Eggs hatch in *ca.* 30 days and the larvae feed on roots until mature in the late fall. Mature larvae move down through the soil profile to a depth of 45–60 cm where they spend the next 18 months, pupating and molting to adults after 9–10 months. Adults emerge the following spring when soil temperatures increase to *ca.* 5 °C. Adults and larvae are in the soil profile for 21 months and within the upper 30 cm of the soil profile for 4–5 months during the growing season (Lincoln & Palm, 1941; Palm, 1935).

Multiple control strategies have been implemented over the past 80+ years including area wide poison baiting programs and none of the programs have been effective (Shields et al., 2009). More recently, the use of foliar insecticides has been shown to be largely ineffective in controlling this insect because of a non–feeding migratory behavior once oviposition begins. A farmer can effectively reduce the adult beetle population in his fields with foliar insecticides and then still lose his stand as non–feeding ovipositing beetles migrate in from neighboring fields.

11.2 Biological Control with Entomopathogenic Nematodes

Entomopathogenic nematodes (EPNs) have been shown to be effective in killing closely related *Otiorhynchus* sp. such as Black vine weevil, *Otiorhynchus sulcatus* (Fabricius) (Coleoptera: Curculionidae) (e.g. Shanks & Agudelo-Silva, 1990) and Strawberry root weevil, *Otiorhynchus ovatus* L. (Coleoptera: Curculionidae) (e.g. Georgis, Poinar, & Wilson, 1982). Since the *O. ligustici* lifecycle is so tightly connected with the soil interface and other *Otiorhynchus* larvae are EPN sensitive, a biological control program utilizing EPNs indicated promise. The relatively low value of alfalfa and clover per acre and the high cost of commercial nematodes that would require an annual application, strongly suggested that the use of commercial EPNs as a biopesticide was not an economically viable option.

The concept of a more classical approach for the utilization of EPNs was intriguing given the low–value of alfalfa. Could an adapted, persistent efficacious EPN be applied a single time in an inoculative approach and result in a multiple year suppression of alfalfa snout beetle? For this approach to be viable, a number of questions needed to be answered.

1. Were adapted, persistent, effective populations of EPNs available for the Northern NY region?
2. Would these populations persist in the field for multiple years from a single inoculation at levels to provide *O. ligustici* population suppression?
3. Would these populations persist across the common alfalfa/clover/grass and corn rotation at a high enough level to provide *O. ligustici* population suppression once the field is rotated back into alfalfa/clover and is reinvaded by *O. ligustici*?
4. Can the EPNs be easily applied utilizing common pesticide sprayers available in most agricultural operations?
5. How much of the field needs to be inoculated and at what dose?
6. Can an economic and farmer friendly rearing method be developed which will allow farmers to rear and apply their own EPNs?

11.3 Availability of Adapted, Persistent Effective Populations of Entomopathogenic Nematodes for Northern New York

After an intensive EPN survey of the *O. ligustici* infested area in New York in 1991–1992, populations of *Steinernema carpocapsae* (Weiser) Wouts, Mráček, Gerdin & Bedding (Rhabditida: Steinernematidae) (population NY 001), *Steinernema feltiae* (Filipjev) Wouts, Mráček, Gerdin & Bedding (Rhabditida: Steinernematidae) (population NY 004–2E), and *Heterorhabditis bacteriophora* Poinar (Rhabditida: Heterorhabditidae) (population Oswego) were isolated. *S. feltiae* was re–isolated from the field in 2004 (population NY 04).

Laboratory tests at Cornell University showed all three native nematode species were effective in killing *O. ligustici* larvae and the two *Steinernema* sp. were also effective at killing adults (Neumann, 2003; Neumann & Shields, 2004; Schroeder, Ferguson, Shields, & Villani, 1994). Laboratory bioassays also indicated that adult beetles were infected and killed by both of the native *Steinernema* sp. at a very low temperature of 6 °C with the insect mortality rate over 80 %. Infection by the native *H. bacteriophora* was initiated at 8 °C with a 25 % infection rate to 100 % at 10 °C (Neumann, 2003). These results along with insect field emergence data (Schroeder, Ferguson, & Shields, 1995) suggested that the adult beetles during and after emergence were vulnerable to attack during times when they were in soil contact (during periods of cool temperatures, resting and oviposition). These results also suggested that both the adult and larval stages were vulnerable to nematode infection.

An early EPN field persistence study provided interesting data that indicated a soil partitioning behavior among nematode species when more than one species was present (Ferguson, Schroeder, & Shields, 1995). This suggested that a multi–species EPN inoculation approach might be more effective than a single EPN species application approach. Since *O. ligustici* larvae and adults can be found throughout the upper 45–60 cm of the soil profile at different times during their lifecycle, a multi–species approach might expose the larvae and adults to EPN infection across a wider range of soil depths than simply utilizing a single EPN species. Subsequent field studies verified that all three native NY nematode species were also effective in killing *O. ligustici* larvae in the field (Neumann & Shields, 2008; Shields, Testa, Miller, & Flanders, 1999). Both of these studies reported field persistence of 24 months for all three EPN species from a single inoculation, the total length of the study. In contrast, population levels of the commercial population of *H. bacteriophora* (NC) declined rapidly and persisted in the field plot less than a year (Shields et al., 1999).

The species *S. carpocapsae* dominates the top most soil layer, when present, but its soil profile limitation is 5–6 cm from the surface, allowing insects to escape EPN infection when they move below 5 cm (Georgis & Poinar, 1983; Moyle & Kaya, 1981; Schroeder & Beavers, 1987). The thinking was then "If *S. carpocapsae* was present in the EPN mix with its low temperature activity threshold, both adult beetles and small larvae would be under threat of infection from this aggressive EPN while on the soil surface or within 5 cm of the surface. Since both *S. feltiae* and *H. bacteriophora* use a more active searching strategy (Grewal, Selvan, & Gaugler, 1994) and range down through deeper layers of the soil profile in search for host (Ferguson et al., 1995), matching either of these species with *S. carpocapsae* would target larvae moving below the top 5 cm in the soil." Taking the thought a bit further, "would all three EPN species coexist in the profile across multiple years and provide a higher level of biological control?"

A laboratory study using sand columns showed that the three species partition the soil profile when applied in combination. When applied together, *S. carpocapsae* dominating the top 6.5 cm and *H. bacteriophora* dominating the lower level of the sand column (>19.5 cm) (Neumann & Shields, 2006). *S. feltiae* was sandwiched in between the two other species suggesting that it may be challenged to coexist in the field in a three species mix. A subsequent field study confirmed that any combination of two species tested coexisted in the field for three growing seasons from a single inoculation, but in field plots with a three species combination, *S. feltiae* disappeared within the first growing season (Neumann & Shields, 2008). Also, the timing of EPN infection on *O. ligustici* larvae in the field appeared to be different according to the EPN species mix, resulting in different levels of feeding damage to the alfalfa roots (Neumann & Shields, 2008).

Alfalfa plants in all nematode treatments (irrespective of nematode mixtures/combinations) had less root feeding damage than the untreated controls and all treatments except *S. carpocapsae* alone, had significantly lower number of surviving larvae. However, even the *S. carpocapsae* alone treatment had a significantly lower number of larvae surviving than the untreated control plot. Between the single nematode applications made, *H. bacteriophora* was the least effective in preventing root feeding damage by *O. ligustici* larvae. All nematode combinations allowed comparative root feeding damage by *O. ligustici* larvae except *S. carpocapsae* × *H. bacteriophora* which allowed a higher level of damage before killing larvae. These data suggests that nematode combinations *S. carpocapsae* × *S. feltiae*, and *S. feltiae* × *H. bacteriophora* were superior because *O. ligustici* larvae were infected when they are smaller, preventing the higher level of root feeding damage (Neumann & Shields, 2008).

11.4 Natural Spread of Persistent Entomopathogenic Nematodes

When designing an *O. ligustici* biological control program around the concept of inoculating persistent adapted EPNs a single time for multiple year pest suppression, the ability of each EPN species to move and spread directly impacts the inoculation strategy. For example, if EPN movement is extremely limited, then EPNs would have to be inoculated into the system with a "total coverage" focus similar to a pesticide. However, if EPNs have the ability to disperse or be moved passively significant distances within a growing season, areas of inoculation could be reduced, resulting in a less expensive application.

Under greenhouse conditions, *H. bacteriophora* 'Oswego', moved 26 cm in 7 days after application (Schroeder, Villani, Ferguson, Nyrop, & Shields, 1993). Under field conditions, *Heterorhabditis heliothidis* (Khan, Brooks & Hirschmann) (Rhabditida: Heterorhabditidae) was detected in neighboring untreated plots within three weeks of application (Shanks & Agudelo-Silva, 1990). All three NY native species were detected in neighboring plots where they were not applied in 12 months suggesting a minimum of 3 m movement (Neumann & Shields, 2011). Pre–sampling the plot area before EPN application detected no natural EPN populations therefore the presence of EPN species which were not applied was concluded as a function of movement. This rapid movement was suggested to be a result of the movement of nematode–infected adults before death as well as the natural movement of EPN infective juveniles (IJs) in the soil.

Considerations about the possible large scale movement of EPNs, were triggered by a farm wide collapse of the *O. ligustici* population in 2002, a location where field research had been conducted for 10 years. Prior to the *O. ligustici* population collapse, adult beetle numbers often exceeded a million beetles per hectare. After the population collapse, adult beetles could only be found on the neighboring farm and larvae were rare in the farm's alfalfa fields between 2002 and 2006. In 2006, a large farm wide sampling program for EPNs indicated that *H. bacteriophora* 'Owego' population, introduced into seven different research plots around the farm over a 10 year period had been moved into every field, even fields where no research plot had ever been established. Prior to its introduction into research plots, *H. bacteriophora* had never before been detected on the farm. These results suggest large scale and long distance movement of EPNs through soil movement by farm machinery around the farm during normal farming practices (Shields et al., 2009). Two subsequent experiments have documented unassisted EPN movement (all three species) of 1–1.5 m in a single growing season in the field and a longer distance movement of 25–45 m within soil when the field is plowed (Shields, unpublished) (Fig. 11.2 also the Fig. 6.1 in Chap. 6 in this volume).

11.5 Application Techniques, Entomopathogenic Nematode Rates and Application Patterns

EPNs are frequently applied in large amounts of water or the application is followed by irrigation to assist the IJs with soil penetration before dying from UV exposure or desiccation (Gaugler, Bednarek, & Campbell, 1992; Gaugler & Boush, 1978). Upscaling these practices to the typical agricultural field without irrigation presents a challenge, so a series of studies were conducted to adapt application to the large scale field crop environment.

11.5.1 Water rates

Water rates and methods for EPN field application have evolved in our research over the past 20 years. The first study utilized a watering can and the equivalent of 53,000 L per Ha (8 L/1.5 m^2) to apply EPNs at rates ranging from 2 to 16 $\times$ 10^9 IJs per ha (Ferguson et al., 1995). The 60 day establishment rate ranged from 90 % of the soil cores testing positive with *H. bacteriophora* 'Oswego' population to 45 % of the soil cores positive with *S. feltiae* 'NY 004–2E' population and 50 % of the soil cores testing positive with *S. carpocapsae* 'NY 001' population.

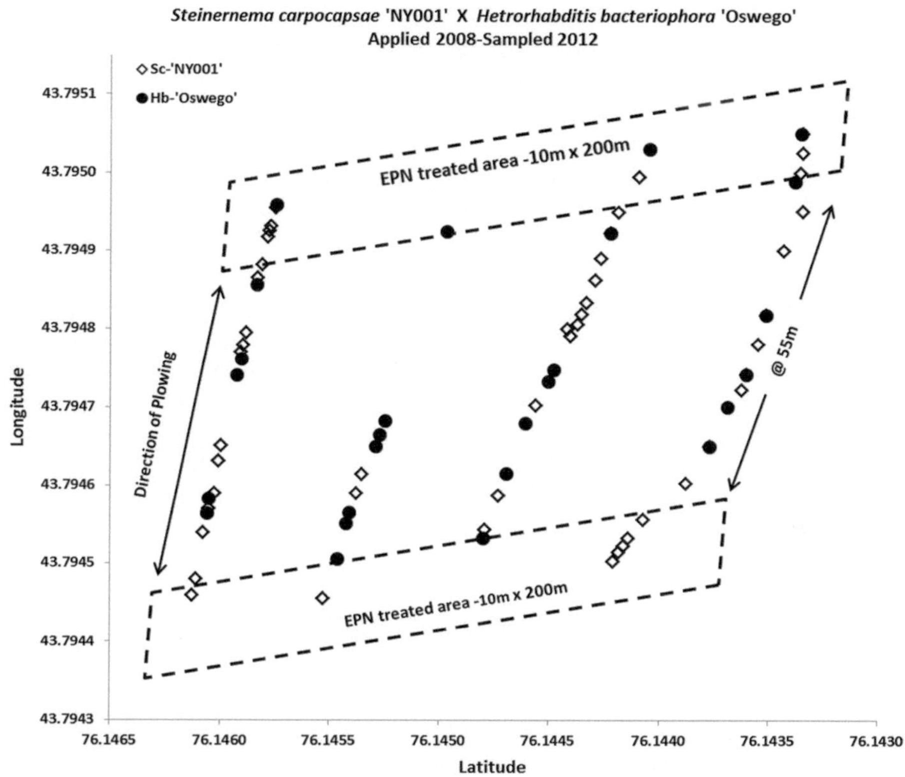

Fig. 11.2 Areas within the treated areas were inoculated with *Steinernema carpocapsae* 'NY 001' and *Heterorhabditis bacteriophora* 'Oswego' (2.5×10^8 IJs/ha) during July 2008 when the field was alfalfa/grass mixture. The field was plowed in both 2009 and 2010 and planted to Triticale both years. Then plowed and replanted to alfalfa in 2011. The field was alfalfa when sampled for EPN movement in June 2012. Both EPN species had been moved a minimum of 25 m between application and subsequent sampling, presumably by soil movement during plowing

A second study utilized a two–person hand–carried, CO_2 powered spray boom, 3.5 m in length, to evaluate the application efficacy of two populations of *H. bacteriophora* (Oswego, NC) using two different spray nozzles (nozzle types: flat fan 8006 vs fertilizer stream 0006) and two different rates of EPNs (2.5 and 15×10^9 IJs/ha). Spray nozzles were separated by 30 cm on the spray boom. EPNs were suspended in 11.5 L and applied to a 3 m × 3 m plot. Application was followed with 57 L of water applied with flat fan nozzles to assist the EPNs with soil penetration. Total water applied to the plot was the equivalent of 36,600 L/ha. The 60 day establishment rate using 2.5×10^9 IJs per ha, ranged from 70 % to 90 % of the soil cores testing positive with *H. bacteriophora* 'Oswego' with the lower rate from the stream nozzle application. EPN population distribution was more uniform in plots where flat fan nozzles were used in the application compared to the stream nozzles which applied the EPNs in concentrated streams separated by 30 cm. Over time, the plots in which the EPNs were applied with stream nozzles, the distribution of the EPNs became more uniform as the nematodes redistributed in search for hosts. At the 15 × 10^9 IJs per ha rate, establishment was 100 % of the soil cores positive and no difference was observed between the two application methods (Shields et al., 1999).

A third study focused on water carrier application rates. *H. bacteriophora* 'Oswego' at two rates (2.5 and 15 × 10^9 IJs/ha) was suspended in 1,000, 2,000 and 7,000 L/ha and applied through fertilizer stream nozzles (type: 0006) or flat fan nozzles (type: 8006) onto the soil surface of a recently harvested alfalfa field. Water carrier volume did not have a significant impact on EPN establishment at either EPN concentration tested. In addition, there was no significant difference in establishment between the stream nozzle and the flat fan nozzles used in the application. In all cases, EPN establishment ranged from 35 % to 40 % of the soil cores which tested positive. However, when applying EPNs through a plant canopy, the fertilizer stream nozzles penetrate the canopy better with very little spray residue remaining on the plant canopy, unlike applications with flat fan nozzles (Shields, unpublished).

11.5.2 Entomopathogenic Nematode Application Rate

With a focus on inoculating the soil with native EPN populations adapted to persist across winter and expecting recycling after establishment, the typical biopesticide rate of 2.5 × 10^9 IJs/ha was too expensive if a lower rate of EPNs would become established. The rate of 2.5 × 10^9 IJs/ha is frequently recommended when using EPNs in an inundative or biopesticide application (Kaya & Gaugler, 1993). However, with a focus on inoculating the soil with native EPN populations adapted to persist across winter and expecting recycling after establishment, a lower application rate would be more cost effective if the EPNs became established at these lower rates.

A small field trial was conducted to investigate the possibility of reducing the EPN application rate while retaining an effective establishment rate. Three rates of *H. bacteriophora* 'Oswego' were used (2.5 × 10^9/ha, 1.25 × 10^9/ha and 0.63 × 10^9/ha) applied in 1,000 L/ha water through fertilizer stream nozzles (type 0006) spaced 30 cm apart on the spray boom. The EPNs were applied to an *O. ligustici* infested alfalfa field, harvested 10 days prior to application with 15 cm regrowth to shade the soil surface. Application was initiated at sunset to allow time for the IJs to enter the soil without exposure to UV. Sixty days after application, 35–40 % of the soil cores tested positive for *H. bacteriophora* 'Oswego' independent of the rate of EPN application and all treatments were not statistically different ($P < 0.01$). Twelve months later, the incidence of *H. bacteriophora* 'Oswego' positive soil cores for had increased to 50–60 %. This increased incidence is thought to be a function of nematode recycling in hosts and a more uniform distribution from EPN movement into the areas between the strips of application through the stream nozzles (Shields, unpublished).

11.6 Moving Toward an Area Wide Biological Control Program

Research to this stage, supported the idea of utilizing persistent adapted EPNs in an inoculative approach in an area–wide biological control program focused on alfalfa snout beetle in alfalfa fields. Research supported the concept of applying a two–species mixture of native NY EPN species, retaining their ability to persist for multiple years under NY conditions at a reduced rate (0.63×10^9 IJs/ha) using commercial pesticide sprayers (filters and screens removed) fitted with fertilizer stream nozzles spaced at the standard spacing of 55 cm. Water carrier volumes could be reduced to 500–1,000 L/ha when fertilizer stream nozzles were used without significant impact on EPN establishment.

The two remaining questions were: (1) Can an inexpensive and farmer–friendly rearing method be developed to decrease the expense of the biocontrol nematodes and (2) can the inoculated EPNs persist multiple years in a continuous alfalfa/grass field system, continue to persist for a 4 year corn rotation at high enough numbers to attack invading alfalfa snout beetle when the field was returned to alfalfa?

11.6.1 Developing a Farmer–Friendly and Inexpensive Rearing Method

With a goal of having farmers rear and apply EPNs on their own farm, research was initiated to develop a cost effective, low labor rearing technique. Due to the complexities of rearing EPNs using artificial diet (*in vitro*), this method was viewed as unsuitable. EPN rearing using insects as host is often utilized in the laboratory for small quantities of IJs, but the methodology is labor intensive and temperature sensitive (Flanders, Miller, & Shields, 1996; Gaugler & Georgis, 1991; Grewal, Lewis, Gaugler, & Campbell, 1994). In addition, when IJs enter free water in any large quantity, death accelerates from lack of oxygen due to the poor exchange rate of the water. After many false starts, a cost effective, low labor rearing method was discovered from a last–ditch idea. In the US, wax moth larvae, *Galleria mellonella* L. (Lepidoptera: Pyralidae) are sold as fish bait and delivered in 500 mL plastic containers filled with either wood shavings or saw dust from multiple companies. Approximately 250 larvae are in each container with small holes are punched in the lid for ventilation. The lid is removed and 20 ml of water containing 15,000 IJs are injected onto the wood shaving or sawdust in a circular pattern. The lid is then replaced and the container is incubated at room temperature (22–25 °C). Death of the *Galleria* larvae is noted within 24–36 h and all of the dead larvae accumulate on top of the wood shavings or saw dust. After 12–15 days, IJs emerge from the cadavers and enter the surrounding wood shavings or saw dust where they survive in high numbers for several days due to the improved oxygen exchange (Fig. 11.3).

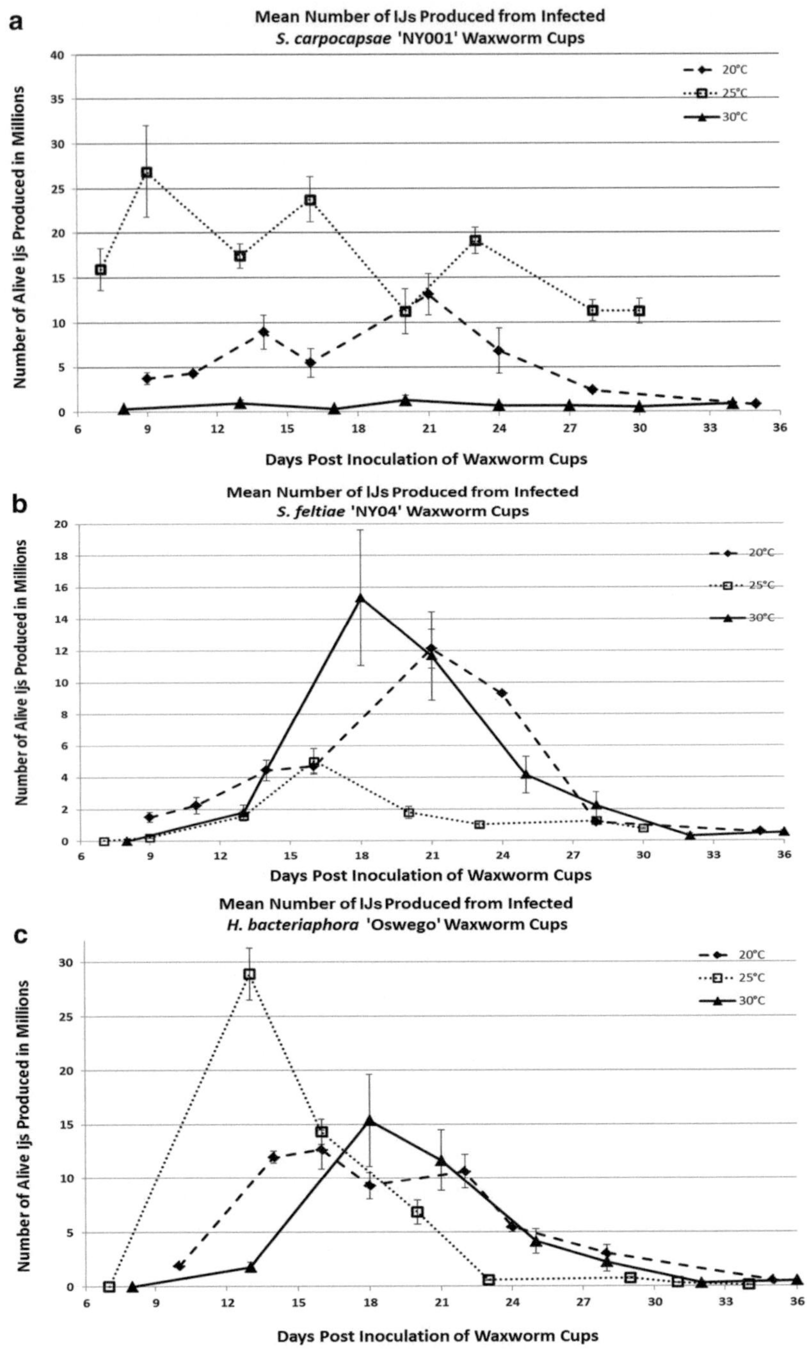

Fig. 11.3 Number of alive IJs produced in the low–labor farmer–friendly rearing method

To remove the IJs from the wood shavings or saw dust and other biological material, the contents of the cup is inverted on a wire screen (20 mesh, 841 μm openings) and the IJs are washed into a lower container with non–chlorinated water. This solution of nematodes is then poured through a second finer screen (40 mesh, 400 μm opening) to remove any remaining debris. The number of EPNs present and alive in the nematode wash solution can be estimated using a dissecting microscope and the standard serial dilution methodology. This solution of water and nematodes can then be dumped into a spray tank filled with water for field application. However, all screens and filters have to be removed in the sprayer to allow nematode IJs to flow through the sprayer unimpeded.

Despite the fact that EPN IJ production yields are influenced by incubation temperatures, this method has been utilized by numerous Northern NY dairy farmers to rear their own EPNs for release on their own farm. At 25 °C incubation temperature, IJ yields of *S. carpocapsae* 'NY 001' and *H. bacteriophora* 'Oswego' are about 2.5–3.0 × 10^{10} IJs per container and *S. feltiae* 'NY 04' produces 1.5–2.0 × 10^{10} IJs. This *Galleria*–based rearing method has been used by the Shields' lab to rear more than 2.0 × 10^{11} IJs over the past 5 years for field release in an area–wide biological control program focused on alfalfa snout beetle. The cost of this rearing method is between $250 and $300 USD 1.0 × 10^9 IJs (excluding labor) and it is simple enough for on–farm rearing (Shields, unpublished).

Since the focus of the area wide EPN biological control program against *O. ligustici* is the inoculation of EPN populations with the ability to persist under field conditions for multiple years, rearing strategies to retain these important genes in the EPN cultures was very important. The strategies utilized to retain these traits are discussed in Chap. 6.

11.6.2 Instituting an Area Wide Biological Control Program

Growers losing their alfalfa from *O. ligustici* feeding were interested in initiating a biological control program as soon as possible and so a pilot program was initiated. During the winter 2006–2007, we were able to make the following conclusions from the research: (1) the native NY EPN populations that were isolated and in culture were efficacious on *O. ligustici*, (2) they persisted in the field from a single application for multiple years, (3) they could be applied to the soil surface through a slightly modified commercial pesticide sprayer, and (4) the carrier volume could be reduced to between 500 and 1,000 L per ha without significant impact on establishment. The EPN species mix of *S. carpocapsae* × *S. feltiae* prevented the most root feeding damage by attacking the larvae when they were a smaller size (Neumann & Shields, 2008). Further more, rearing of both species was very similar regarding the time to IJ emergence and number of IJs produced, reducing the complexity of rearing two different species for a combined application. Additionally, a cost effective, low–labor rearing method to rear large quantities of IJs, to initiate either a larger pilot study or have farmers rear/apply their own EPNs had been developed. Moving the research to a larger scale with multiple geographic locations would help us to answer a frequently grower asked question "Will the EPNs applied to alfalfa persists across crop rotation with alternative crops (corn, soybeans)?"

11.7 The Pilot Program

In 2007–2009, a pilot program was launched to examine the pitfalls and problem areas, inherent to the scaling up of a small scale research program. The foci of the program was to implement EPN application on a larger scale using commercial sprayers, testing the large scale EPN rearing protocol and begin the documentation of multi–year EPN persistence, across a wide array of soil types used to grow alfalfa and across crop rotations.

In 2007, larger sized plots (4 m × 16 m, four replicates) were established on six different farms, located in three different counties (Lewis, St. Lawrence, Franklin). In Lewis Co. a mixture of *S. carpocapsae* 'NY001' × *H. bacteriophora* 'Oswego' was applied, in St. Lawrence Co. a mixture of *S. feltiae* 'NY04' × *H. bacteriophora* 'Oswego' was applied and in Franklin Co. a mixture of *S. carpocapsae* 'NY001' × *S. feltiae* 'NY04' was applied. At all locations, EPNs in a two–species mix were applied at 1.25×10^8 IJs per species (total IJs $= 2.5 \times 10^8$ IJs) through a 3.5 m 2-person CO_2 powered hand boom fitted with 0006 stream nozzles separated by 30 cm utilizing 1,000 L water carrier per ha. The length of all plots (16 m) was aligned perpendicular to the direction the field was tilled/plowed so subsequent tillage operations would assist with EPN movement throughout the field.

Five additional sites (Jefferson –1, Lewis – 1, St. Lawrence – 2, Franklin – 1) were established using a larger scale sprayer mounted on a pickup truck fitted with fertilizer stream nozzles (0006) separated by 60 cm to inoculate four locations within each field (2,000 m^2 comprising areas of 10 m × 200 m, total $= 8{,}000$ m^2 per field). The length of all plots (200 m) was aligned perpendicular to the direction the field was tilled/plowed so subsequent tillage operations could assist EPN movement throughout the field. Water carrier quantity was 500 L/ha. EPN species mixtures in the truck applications were the same in each county as applied with the hand boom. The Jefferson county site also used the same mixture as the Lewis county hand boom site. In all 11 locations, EPN establishment was verified through soil cores, but establishment rates were lower than expected (5–15 % soil cores positive). The extremely dry year in 2007 was thought to have a significant impact on the establishment level. The following year (2008), all sites had increased to 25–70 % of the soil cores positive for EPNs, indicating establishment and recycling on the host had occurred in the fields. Rearing problems and low establishment rates across all sites with *H. bacteriophora* 'Oswego' was consistent and EPN combination with *H. bacteriophora* 'Oswego' were eliminated for 2008 and beyond (Fig. 11.4).

In 2008 and 2009, a total of 80 fields were inoculated, distributed across six of the nine *O. ligustici* infested counties. At each location, the EPN species mix

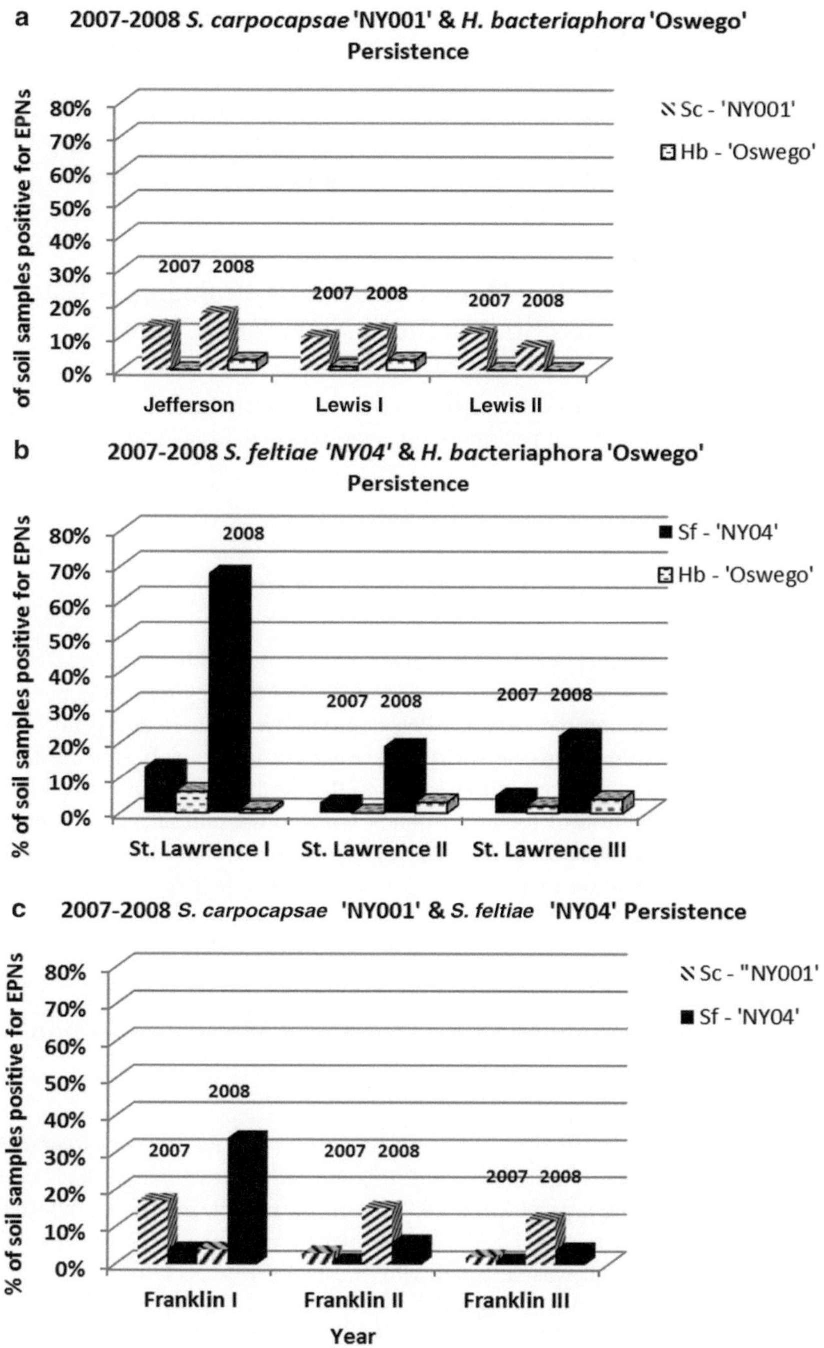

Fig. 11.4 Early establishment levels for native NY EPNs inoculated into alfalfa fields under dry conditions

used was *S. carpocapsae* 'NY001' × *S. feltiae* 'NY04', applied at a rate of 1.25×10^8 IJs/species/ha in 500 L water/ha (2.5×10^8 IJs total IJs). These 80 fields were inoculated identically to the truck applications in 2007 using similar treated area dimensions and orientations. Additionally, four farmers reared their own EPNs and inoculated one of their own fields. Establishment in all fields 60 d after application ranged between 23 % and 35 % of the soil cores testing positive for EPNs.

11.8 Initiating an Area–Wide Farm–Scale Biological Control Program Across Six Northern New York Counties

Farmers within the *O. ligustici* infested area were interested in expanding the pilot program so they could begin treating their own fields. A farm–scale program was initiated in 2010.

11.8.1 Entomopathogenic Nematodes Choice

A review of the research data and rearing experience indicated that the best nematode combination was *S. carpocapsae* × *S. feltiae* for the area–wide biological control program. This combination of EPNs could infect adults in the spring and the larvae before a significant amount of root feeding could occur. If root feeding by *O. ligustici* larvae was reduced, direct stress to the alfalfa plant from root loss was reduced and the reduction of feeding wounds reduced the entrance zones for plant pathogens. In addition, both EPN species responded in a similar manner to the same rearing conditions, making them compatible for on–farm rearing by inexperienced growers.

11.8.2 Entomopathogenic Nematodes and Water Application Rates

The focus of this program was to inoculate fields with native persistent EPNs, adapted to the northern NY climate, and this meant that the rate of EPNs could be reduced to below the typical 2.5×10^9 IJs per ha. In fact, research supported an even lower application rate of 1.25×10^8 IJs per ha per species (2 species = 2.50 $\times 10^8$ total IJs/ha) which was selected. EPN costs were $150 USD for 5.0×10^8 IJs (farmer reared). However, if the EPNs were purchased from the Shields' lab at Cornell, the cost was $300 USD for 5.0×10^8 IJs. These costs equate to nematode costs of $75/ha (farmer reared IJs) or $150/ha (purchased from Cornell) with actual application costs not included. EPNs were applied to fields in water carrier rates of between 500 and 1,000 L/ha through a wide range of on–farm application equipment. Application equipment varied from large commercial pesticide sprayers with all screens and filters removed and nozzles changed to fertilizer stream nozzles (0006–0015), smaller sprayers with open nozzle bodies dribbling a stream of water and home–made gravity dispersed applicators composed of a water tank and a pipe with holes drilled in it.

11.8.3 Application Strategies and Cost Saving

Some farmers were interested in treating all of their acreage directly for faster results at $75–150/ha. They applied EPNs to entire fields in streams on a 0.55 m spacing, similar to a pesticide application. However, many farmers were interested in utilizing a reduced cost strategy by relying on the natural movement of EPNs to fill in the areas where EPNs were not applied. Natural movement includes the physical dispersal of the EPN, movement of infected insects before they die and the movement of soil with EPNs to other parts of the field during farming practices. Previous research showed that all three species of native EPNs move 1–1.5 m in a single growing season and are moved 25–45 m by farm equipment when the field is plowed (Fig. 11.2, Chap. 6, Fig. 6.1). Some farmers treated strips of the field oriented perpendicular to the direction of field tillage, allowing the soil movement during soil tillage to move the EPNs into areas where they were not applied. Other farmers utilized a commercial sprayer in a unique way to capitalize on both the natural movement of EPNs and the reduced cost from treating only a portion of the field.

In the US, nozzles on a commercial sprayer are mounted ca.0.55 m apart. If all nozzle bodies are equipped with fertilizer stream nozzles, a concentrated stream of nematodes are applied to the soil surface at 0.55 m spacing at a cost of $75–150/ha. Within 30 days, the zones between the applied streams are occupied with applied EPNs. If the concentrated streams of EPNs was applied at 1.65 m apart by blocking two nozzles, allowing only every third nozzle to apply EPNs, the sprayer could travel across the entire field with only 33 % of the field actually treated, thus reducing the cost of nematodes to $25–50/ha. In order for the EPNs to colonize the entire area between the concentrated streams separated by 1.65 m, the nematodes only needed to disperse 0.80 m from each stream to be present in the entire field. EPNs occupied the zones between the applied streams within 60 days (Bal, Michael, & Grewal, 2014; Bal, Taylor, & Grewal, 2014). Some farmers, after listening to the research on nematode movement, chose to apply streams of nematodes that were spaced by 3.30 m and travel over the whole field resulting in a treatment of only 16 % of the field. EPNs occupied the zones between the applied concentrated streams within 1–2 growing seasons and the EPN cost was reduced to $12–24/ha. The full field application and the 33 % application rate was recommended for fields with large to moderate populations of *O. ligustici* whereas the 16 % application rate was recommended for fields with moderate to low *O. ligustici* populations.

Regardless of the equipment utilized, 100 % of the fields tested positive for EPN establishment ranging from 15 % to 30 % of the soil cores yielding a positive EPN result tested through the use of bioassays with *Galleria*, 30–60 days after application. To date, native and persistent EPNs have been successfully applied to ca. 200 fields and ca. 4,000 ha across 6 northern NY counties within the ASB infested area and have persisted from that single application for 7 years.

11.9 Entomopathogenic Nematodes Persistence Across Years and Crop Rotation

Fifty one fields were selected from the 87 total fields inoculated with EPNs in 2007–2009 to track EPN persistence across years and crop rotations. The selected fields represented a wide array of soil types ranging from the clay loams to sandy soils. These fields also represented different crop rotation ranging from continuous alfalfa/grass mixture to an alfalfa/row crop rotation (4 years – corn or soybeans). GPS locations were recorded for the inoculation zone at the time of the EPN application, allowing those areas to be relocated for subsequent sampling in the following years. Fields were not sampled prior to EPN inoculation because previous research has indicated that if EPNs were naturally present in these soils, they are present below 5 % of positive soil cores. Each field was sampled annually between June 1 and October 15th by taking fifty – 2 cm diameter soil cores to a depth of 20 cm, on two different transects within the EPN treated zones (total/field = 100 individual samples). Each sample was divided and placed in two different containers. The top 5 cm of the soil core was placed into a 130 mL container with a lid and the lower 15 cm of the soil core was placed in a 260 mL container with a lid. The samples were then returned to the laboratory and tested for the presence of EPNs using the *Galleria* bait method (Bedding & Akhurst, 1975; Fan & Hominick, 1991), which involves placing five larvae in the smaller container and 10 larvae in the larger container and incubated for 7 days at 22 °C. EPNs will migrate from the soil into the insect host and infect it, thus confirming the presence of EPNs in the soil core.

11.9.1 Entomopathogenic Nematodes Persistence in Continuous Alfalfa–Grass Mixture

The four fields illustrated in Fig. 11.5 demonstrate the range of EPN response typical of all fields sampled which remain in a continuous cropping of alfalfa–grass. Two important points to draw from the graphs are that (1) EPNs from this single introduction persisted across multiple years in a continuous cropping of alfalfa–grass, recycling in the multitude of host invading the alfalfa–grass ecosystem, and

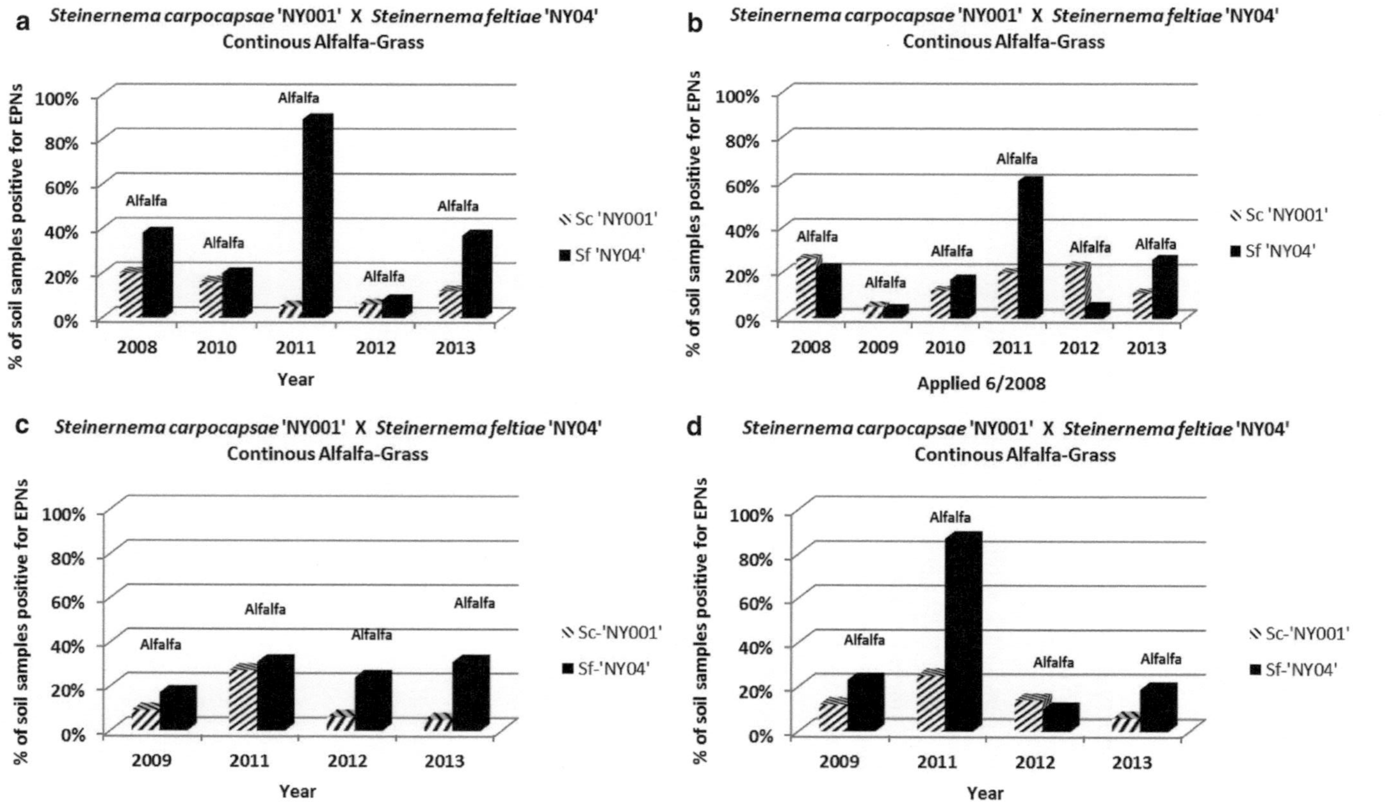

Fig. 11.5 Four different fields in Northern NY. Fields A & B were inoculated in 2008 with 50 million IJs of *Steinernema carpocapsae* 'NY 001' and 50 million IJs of *Steinernema feltiae* 'NY 04' per acre. Fields C & D were inoculated with the same EPN rate in 2009. EPN population frequency was measured once per year during the growing season. EPN population appears stable and responds to invasion of host insects

(2) EPN populations rise and fall in response to the various levels of insect host availability. Data was not collected to suggest the array of host being utilized by the two species of EPNs to recycle or persist in these fields. These graphs suggest that the residual population of EPNs maintain a population in the range of 10–20 % of the soil cores testing positive for EPNs. At that level, the EPN population appears to be stable, persist long–term and is capable of responding to host invasion.

11.9.2 Entomopathogenic Nematodes Persistence Across a Crop Rotation

Multi–year EPN persistence within a continuous alfalfa–grass cropping system was expected due to the wide array of susceptible hosts feeding within that cropping system. However, high EPN persistence was not expected across a corn rotation due to the more limited number of hosts supported within the corn ecosystem. However, when rotated to corn, EPN population responded to insect invasion, within the corn cropping years. During the second year of corn production, a large increase in EPN numbers was observed, which was thought to be a response to the higher level of corn rootworm larvae, *Diabrotica virgifera virgifera* LeConte (Coleoptera: Chrysomelidae), found in second year corn fields (Fig. 11.6). These results, a sub sample of numerous fields, indicate our preconceived idea of EPN loss during corn culture was in error. The three conclusions that can be drawn from these graphs is (1) native EPN populations inoculated in the alfalfa–grass ecosystem easily persisted during multiple years of a corn crop growing in rotation with alfalfa–grass, (2) EPN populations responded to the corn–specific herbivore invasion, and (3) EPN populations were equal to or higher after 5 years of corn than the final year of the alfalfa–grass mixture.

In these fields (Fig. 11.7), the EPN population was high enough after the first year of alfalfa, that when alfalfa snout beetle invades during the second year, the EPN population will be high enough to respond quickly to the insect invasion and minimize the root feeding damage on the alfalfa stand.

11.10 Conclusions and Future Directions

In the small plot format, it was easy to demonstrate the effectiveness of native EPNs in reducing *O. ligustici* populations and alfalfa stand retention from reduced feeding. In moving to a larger scale, these direct measures of effectiveness have been difficult to document. We have demonstrated native EPN persistence at a significant level for multiple years from a single inoculation. In addition, EPN soil populations increase significantly in some years, suggesting they are responding to an insect invasion. We were very surprised that EPN populations actually increased when

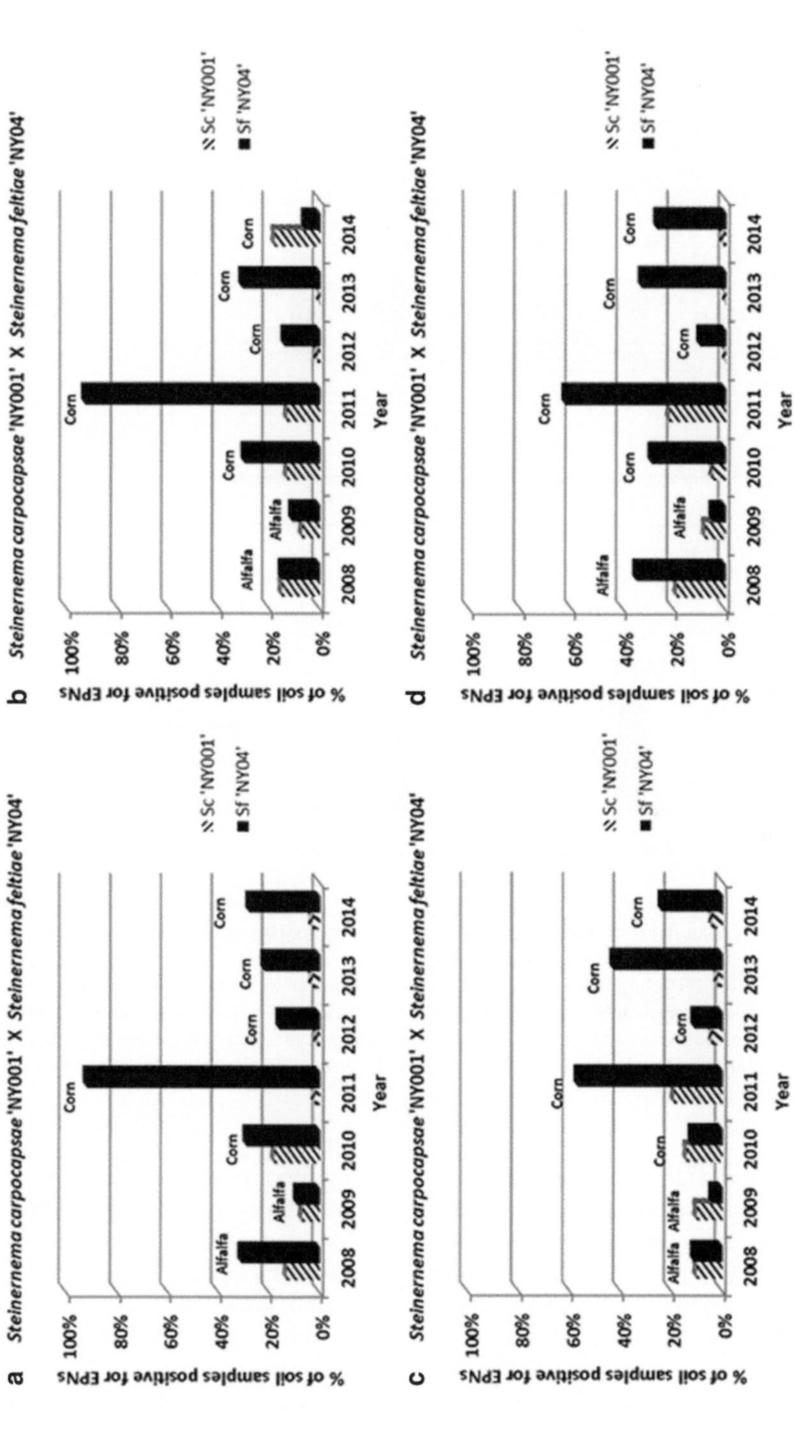

Fig. 11.6 Four different fields in Northern NY where native EPNs were applied once in 2008. Entomopathogenic nematodes (EPNs) population frequency was measured once per year during the growing season. EPN recycling on *Diabrotica virgifera virgifera*, the corn root worm (CRW) larvae during the corn years is inferred based on insect biology but not directly measured

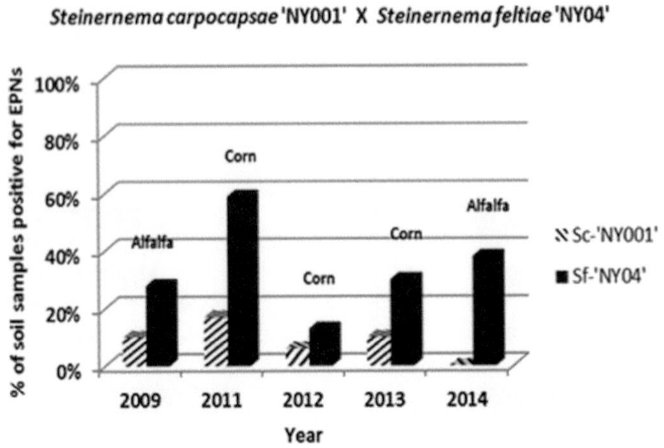

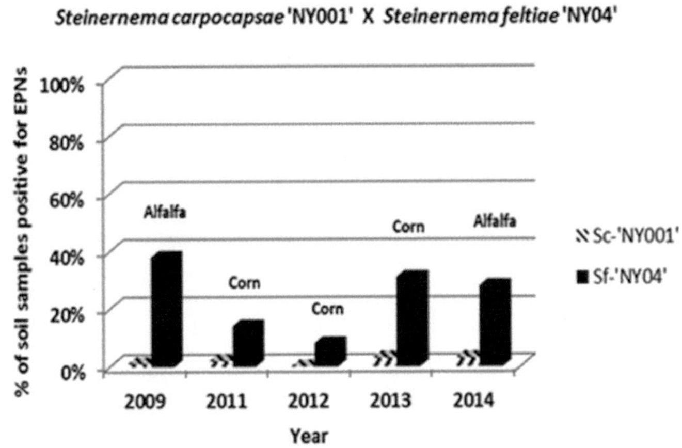

Fig. 11.7 Two different fields in Northern NY where native entomopathogenic nematodes (EPNs) were applied once in 2009. EPN population frequency was measured once per year during the growing season. EPN recycling on CRW larvae during the corn years is inferred based on insect biology but not directly measured. EPN populations during the corn portion of the crop rotation maintain their levels, appear to be stable and are present at a significant level when the field is rotated back into alfalfa. An alfalfa snout beetle (ASB), *Otiorhynchus ligustici* invasion would be met by a significant level of persisting EPNs

the fields were rotated to corn or soybeans indicating that insects associated with these crops were invading and thus providing recycling hosts for the native EPNs. In contrast, we expected the EPN population to slowly decline during the years of corn and soybeans rotations and increase when alfalfa was replanted. While we have been less than successful in demonstrating increased alfalfa stand retention on these larger scale efforts, farmers report a sharp reduction in *O. ligustici* adult beetles during spring migration after most of the fields have been inoculated in an area with native EPNs. Efforts are continuing to demonstrate the direct impact of native EPNs as biological control with improved alfalfa stand retention with the establishment of 0.2 ha plots in a replicated demonstration format with untreated checks. These demonstration sites are located across five of the NNY *O. ligustici* infested counties, inoculated with EPNs during the first production year (second year of stand life) and will be maintained for the life of the alfalfa stand (usually 4 years). Data collected includes alfalfa stand counts, and nematode population level and persistence data.

With the successes in controlling *O. ligustici* with native EPNs in the more classical approach of a single inoculation with multi–year persistence, our research effort has broadened to other cropping systems attacked by other *Otiorhynchus* species. A widespread black vine weevil, *O. sulcatus*, infestation in 80 ha of upland cranberry production causing widespread economic losses was reduced to sub–economic levels with a single inoculation of *S. carpocapsae* × *S. feltiae* (see Chap. 6). A second research effort is focused on black vine, *O. sulcatus* and strawberry root weevil, *O. ovatus,* control in strawberries. In late summer 2013, a heavily infested series of NY strawberry fields totaling 4 ha were treated with a single inoculation of *S. carpocapsae* × *S. feltiae* using a concentration of 5.0×10^8 IJs per species per ha (total IJs = 1.0×10^8/ha). EPN populations were monitored 30 days after application to verify establishment and during the spring and summer of 2014. The EPN population established, persisted and actively reduced the economically damaging black vine weevil population attacking the strawberries. The EPN and black vine weevil populations will be continued to be monitored for the next several years. We are continuing to explore other agriculture cropping systems where the inoculation of a mix of persistent EPNs would be beneficial.

References

Bal, H. K., Michael, A., & Grewal, P. S. (2014). Genetic selection of the ambush foraging entomopathogenic nematode *Steinernema carpocapsae* for enhanced dispersal and its associated trade-offs. *Evolutionary Ecology, 28*, 923–939.

Bal, H. K., Taylor, R. A. J., & Grewal, P. S. (2014). Ambush foraging entomopathogenic nematodes employ sprinting emigrants for long distance dispersal in the absence of hosts. *Parasitology, 100*, 422–432.

Bedding, R. A., & Akhurst, R. J. (1975). A simple technique for the detection of insect parasitic rhabditid nematodes in soil. *Nematologica, 21*, 109–110.

Claassen, P. W., & Palm, C. E. (1935). The alfalfa snout beetle, *Brachyrhinus ligustici* L., a new insect pest in New York State. *Journal of Economic Entomology, 28*, 417–420.

Fan, X., & Hominick, W. M. (1991). Efficiency of the *Galleria* (wax moth) baiting technique for recovering infective stages of entomopathogenic rhabditids (Heterorhabditidae and Steinernematidae) from sand and soil. *Revue de Nématologie, 14*, 381–387.

Ferguson, C. S., Schroeder, P. C., & Shields, E. J. (1995). Vertical distribution, persistence and activity of entomopathogenic nematodes (Nematoda: Heterorhabditidae and Steinernematidae) in alfalfa snout beetle (Coleoptera: Curculionidae) infested fields. *Environmental Entomology, 24*, 149–158.

Flanders, K. L., Miller, J. M., & Shields, E. J. (1996). *In vivo* production of *Heterorhabditis bacteriophora* 'Oswego' (Rhabditida: Heterorhabditidae), a potential biological control agent for soil-inhabiting insects in temperate regions. *Journal of Economic Entomology, 89*, 373–380.

Gaugler, R., Bednarek, A., & Campbell, J. F. (1992). Ultraviolet inactivation of heterorhabditid and steinernematid nematodes. *Journal of Invertebrate Pathology, 59*, 155–160.

Gaugler, R., & Boush, G. M. (1978). Effects of ultraviolet radiation and sunlight on the entomogenous nematode, *Neoaplectana carpocapsae*. *Journal of Invertebrate Pathology, 32*, 291–296.

Gaugler, R., & Georgis, R. (1991). Culture method and efficacy of entomopathogenic nematodes (Rhabditida: Steinernematidae and Heterorhabditidae). *Biological Control, 1*, 269–274.

Georgis, R., & Poinar, G. O., Jr. (1983). Effects of soil texture on the distribution and infectivity of *Neoaplectana carpocapsae* (Nematoda: Steinernematidae). *Journal of Nematology, 15*, 308–311.

Georgis, R., Poinar, G. O., Jr., & Wilson, A. B. (1982). Susceptibility of strawberry root weevil, *Otiorhynchus ovatus,* to neoaplectanid and heterorhabditid nematodes. *IRCS Medice Science, 10*, 42.

Grewal, P. S., Lewis, E. E., Gaugler, R., & Campbell, J. F. (1994). Host finding behavior as a predictor of foraging strategy in entomopathogenic nematodes. *Parasitology, 108*, 207–215.

Grewal, P. S., Selvan, S., & Gaugler, R. (1994). Thermal adaptation of entomopathogenic nematodes: Niche breadth for infection, establishment and reproduction. *Journal of Thermal Biology, 19*, 245–253.

Kaya, H. K., & Gaugler, R. (1993). Entomopathogenic nematodes. *Annual Review of Entomology, 38*, 181–206.

Lincoln, C., & Palm, C. E. (1941). *Biology and ecology of the alfalfa snout beetle*. Ithaca, NY: Cornell University Agricultural Experiment Station Memoir 236.

Moyle, P. L., & Kaya, H. K. (1981). Dispersal and infectivity of the entomogenous nematode, *Neoaplectana carpocapsae* Weiser (Rhabditida: Steinernematidae) in sand. *Journal of Nematology, 13*, 295–300.

Neumann, G. (2003). *Naturally occurring biological control of the alfalfa snout beetle*, Otiorhynchus ligustici *(L.) in Hungary and in New York State*. MSc. thesis, Cornell University, 84 pp.

Neumann, G., & Shields, E. J. (2004). Survey for entomopathogenic nematodes and entomopathogenic fungi in alfalfa snout beetle, *Otiorhynchus ligustici* (L.) (Coleoptera: Curculionidae) in Hungary and New York. *The Great Lakes Entomologist, 37*, 152–158.

Neumann, G., & Shields, E. J. (2006). Interspecific interactions among three entomopathogenic nematodes, *Steinernema carpocapsae*, *S. feltiae*, and *Heterorhabditis bacteriophora*, with different foraging strategies for hosts in multi-piece sand columns. *Environmental Entomology, 35*, 1578–1583.

Neumann, G., & Shields, E. J. (2008). Multiple-species natural enemy approach for the biological control of the alfalfa snout beetle, *Otiorhynchus ligustici* L. (Coleoptera: Curculionidae), using entomopathogenic nematodes. *Journal of Economic Entomology, 101*, 1533–1539.

Neumann, G., & Shields, E. J. (2011). Long term field persistence and coexistence in three entomopathogenic nematodes, *Steinernema carpocapsae* Weiser, *S. feltiae* Filipjev, and *Heterorhabditis bacteriophora* Poinar. *The Great Lakes Entomologist, 44*, 42–52.

NYS Ag Statistics. (2013). http://www.nass.usda.gov/Statistics_by_State/New_York/

Palm, C. E. (1935). *Alfalfa snout beetle*, Brachyrhinus ligustici *L.* Ithaca, NY: Cornell University Agricultural Experiment Station Bulletin 629.

Schroeder, P. C., Ferguson, C. S., Shields, E. J., & Villani, M. G. (1994). Pathogenicity of rhabditid nematodes (Nematode: Heterorhabditidae and Steinernematidae) to alfalfa snout beetle (Coleoptera: Curculionidae) larvae. *Journal of Economic Entomology, 87*, 917–922.

Schroeder, P. C., Ferguson, C. S., & Shields, E. J. (1995). Spring activity of adult alfafa snout beetle (Coleoptera: Curculionidae): Upward movement from overwintering sites, dispersal and oviposition. *Journal of Economic Entomology, 88*, 859–863.

Schroeder, P. C., Villani, M. G., Ferguson, C. S., Nyrop, J. P., & Shields, E. J. (1993). Behavioral interactions between Japanese beetle (Coleoptera: Scarabaeidae) grubs and an entomopathogenic nematode (Nematode: Heterorhabditidae) within turf microcosms. *Environmental Entomology, 22*, 595–600.

Schroeder, W. J., & Beavers, J. B. (1987). Movement of the entomogenous nematodes of the families Heterorhabditidae and Steinernematidae in soil. *Journal of Nematology, 19*, 257–259.

Shanks, C. H., & Agudelo-Silva, F. (1990). Field pathogenicity and persistence of heterorhabditid and steinernematid nematodes infecting black vine weevil larvae (Coleoptera: Curculionidae) in cranberry bogs. *Journal of Economic Entomology, 83*, 107–110.

Shields, E. J., Testa, A., Miller, J. M., & Flanders, K. L. (1999). Field efficacy and persistence of the entomopathogenic nematodes *Heterorhabditis bacteriophora* 'Oswego' and *H. bacteriophora* 'NC' on alfalfa snout beetle larvae (Coleoptera: Curculionidae). *Environmental Entomology, 28*, 128–136.

Shields, E. J., Testa, A., Neumann, G., Flanders, K. L., & Schroeder, P. C. (2009). Biological control of alfalfa snout beetle with a multi-species application of locally-adapted persistent entomopathogenic nematodes: The first success. *American Entomologist, 55*, 250–257.

Chapter 12
Entomopathogenic Nematodes in Turfgrass: Ecology and Management of Important Insect Pests in North America

Albrecht M. Koppenhöfer, Olga S. Kostromytska, Benjamin A. McGraw, and Lemma Ebssa

12.1 Introduction: The Turfgrass System

Several genera of grasses are capable of forming a mat of intertwined plants to form a solid ground cover with an extensive root mass. They can also regenerate from the crown after defoliation. The about 50 grass species amenable to use in turfgrass systems are further on able to form a high shoot density under the continuous mowing regimes characteristic for turfgrass systems (Christians, 1998). These properties allow turfgrasses to provide a hard–wearing permanent or semi–permanent ground cover that can be used for various recreational spaces in urban and suburban environments including lawns, parks, golf courses, and athletic fields. Other areas in which turfgrasses are grown include cemeteries, roadsides and sod farms. In the USA, turfgrass areas cover about 20 million ha and the size of the turfgrass industry is estimated at $40,000 million per year (National Turfgrass Federation, 2009). Besides their recreational uses, turfgrasses control soil erosion, capture and clean run–off water from urban areas, provide soil improvement and restoration, moderate temperature, reduce glare and noise, reduce pests, pollen and human disease exposure, create good wildlife habitats, and improve physical and mental health of urban populations (Beard & Green, 1994).

A.M. Koppenhöfer (✉) • O.S. Kostromytska
Department of Entomology, Rutgers University, New Brunswick, NJ, USA
e-mail: koppenhofer@aesop.rutgers.edu; kolgaent@rci.rutgers.edu

B.A. McGraw
Department of Plant Science, Pennsylvania State University, University Park, PA, USA
e-mail: mcgrawba@delhi.edu

L. Ebssa
Department of Entomology, Institute for Health, Health Care Policy, and Aging Research, Rutgers School of Environmental and Biological Sciences, Rutgers University, New Brunswick, NJ, USA
e-mail: lemma@rci.rutgers.edu

© Springer International Publishing Switzerland 2015
R. Campos-Herrera (ed.), *Nematode Pathogenesis of Insects and Other Pests*,
Sustainability in Plant and Crop Protection 1, DOI 10.1007/978-3-319-18266-7_12

Between the different types of turf maintenance systems, great variations exist in value, input, demands, damage thresholds, and, consequently, tolerances for pests. Permanent turf provides habitat for many invertebrate species, most of which feed on vegetation and detritus without causing obvious damage or loss of productivity. However, amenity turf is under constant critical scrutiny from the public, and playing and safety standards on athletic field and golf courses are very high. Consequently, damage thresholds are generally low and therefore a large number of insect species are regarded as pests, particularly in higher value areas where tolerance for pests is the lowest. As a result of these high standards and expectations, application of synthetic insecticides has been the primary method used to control insect pests in turfgrass (Held & Potter, 2012; Racke, 2000).

Concerns about health risks and environmental hazards of pesticides have led to pesticide legislation in the USA and Canada that have resulted in the loss of many insecticides, especially organophosphates, for turfgrass uses (Bélair, Koppenhöfer, Dionne, & Simard, 2010). The chemical industry has been able to respond to these losses with the development of new active ingredients from several new insecticide classes that are considered low–risk insecticides (e.g., neonicotinoids, oxadiazines, anthranilic diamides). Nonetheless, in Canada, the use of pesticides for cosmetic purposes is banned in two provinces (Québec and Ontario) and more than 152 municipalities have adopted by–laws restricting or banning the use of landscape pesticides. Canadian golf courses may still use pesticides but have to comply with new stricter regulations.

Compared to Canada and the European Union, in the USA the regulatory process for pesticides is less cumbersome and faster, the turf market larger and legislatively more uniform, and hence more insecticides, particularly the newer types, are available (Bélair et al., 2010). As a result, synthetic insecticides remain the mainstay for insect control in turfgrass in the USA. Nonetheless, public concerns about insecticide use persist, recently driven by findings that the widely used group of neonicotinoid insecticides may contribute to the decline of or have detrimental effects on endemic pollinator populations and commercial honeybees (Charles et al., 2014).

Individual states in the USA can impose additional requirements on pesticide registration and the states of California and New York generally have the highest standards for registration. Additional restrictions can be imposed by individual municipalities, e.g., if a greater risk of contamination of the water supply exists. There has also been a general trend to restrict pesticide use in schools including on turfgrass area on the school grounds which is guided and supported by the US Environmental Protection Agency (2014). In 2014, 39 states and the District of Columbia had some type of school pest management law or regulation in place, and 23 states have a school IPM law or regulation (National Pest Management

Association, 2014). Regulations range from promoting non–chemical methods, allowing chemical use only in the case of pest emergencies, to complete bans of pesticide use.

Likely future restrictions on insecticide use and a growing interest in organic lawn care could increase opportunities for greater use of biorational and biological insect control products including entomopathogenic nematodes (EPNs). Several important insect turf pests are amenable to control by EPNs. Pest that have received the most attention as targets for EPNs include the white grub complex (Coleoptera: Scarabaeidae), mole crickets, *Scapteriscus* spp. (Orthoptera: Gryllotalpidae), billbugs, *Sphenophorus* spp. (Coleoptera: Curculionidae), and the black cutworm, *Agrotis ipsilon* (Hufnagel) (Lepidoptera: Noctuidae). Other pests that have been controlled experimentally with nematodes include annual bluegrass weevil, *Listronotus maculicollis* (Kirby) (Coleoptera: Curculionidae), cutworms and armyworms (Lepidoptera: Noctuidae), sod webworms (Lepidoptera: Pyralidae), and crane flies, *Tipula* spp. (Diptera: Tipulidae).

12.2 Entomopathogenic Nematode Ecology in Turfgrass

Studies involving EPN ecology in turfgrass have been conducted almost exclusively in the temperate moist climate of the northeastern USA, primarily in New Jersey and Ohio. In these studies, natural EPN populations usually consisted primarily of *Heterorhabditis bacteriophora* Poinar (Rhabditida: Heterorhabditidae) and *Steinernema carpocapsae* (Weiser) Wouts, Mráček, Gerdin & Bedding (Rhabditida : Steinernematidae); occasionally *Steinernema feltiae* (Filipjev) Wouts, Mráček, Gerdin & Bedding (Rhabditida: Steinernematidae) and *Steinernema glaseri* (Steiner) Wouts, Mráček, Gerdin & Bedding (Rhabditida: Steinernematidae) were also found (Alumai, Grewal, Hoy, & Willoughby, 2006; Campbell, Lewis, Yoder, & Gaugler, 1995, 1996; Campbell, Orza, Yoder, Lewis, & Gaugler, 1998; Koppenhöfer & Fuzy, 2009; McGraw & Koppenhöfer, 2009; Stuart & Gaugler, 1994). *Steinernema scarabaei* Stock & Koppenhöfer (Rhabditida: Steinernematidae) was not found in these studies that used wax moth larvae to bait infective juveniles (IJs) from soil samples, but this species has been found regularly infecting white grub larvae when larger numbers of larvae were collected (A. Koppenhöfer, personal observation; R. Cowles, personal communication).

Most of the above studies were conducted in turf areas maintained under lawn–like conditions (i.e., weekly mowing at 3.8–10 cm, natural soil, limited input of pesticides). However, it is likely that the type of turf system (e.g., golf course, home lawn) and maintenance intensity affect EPN populations. Thus, when baiting soil samples from numerous golf courses in Ohio, Alumai et al. (2006) found natural EPN in 0 % of the putting greens (daily mowing at around 3 mm, sand–based or topdressed with sand, most intense pesticide input), 43 % of the fairways (mowed 3–4 times weekly at around 12 mm height, natural soil, intermediate level of pesticide input), and 57 % of the roughs (lawn–like conditions).

In New Jersey turfgrass sites, natural populations of *H. bacteriophora* and *S. carpocapsae* can be recovered throughout the year (Campbell et al., 1995, 1996, 1998; Koppenhöfer & Fuzy, 2009; McGraw & Koppenhöfer, 2009) including during winter (Elmowitz, Ebssa, & Koppenhöfer, 2013). In lower maintenance sites (mowed 1–2 times weekly at 38 mm height, natural soil, limited level of pesticide input) EPN populations tended to be relatively stable with no clear relationship between potential host species and *S. carpocapsae* population distribution but an inverse relationship between populations of *H. bacteriophora* and a potential host, Japanese beetle, *Popillia japonica* Newman (Coleoptera: Scarabaeidae), larvae (Campbell et al., 1995, 1998). However, on golf course fairways with limited insecticide use *S. carpocapsae* and *H. bacteriophora* exhibited a distinct seasonality with high densities in the weeks immediately following high densities of a host, *L. maculicollis* larvae and pupae; dispersion of *L. maculicollis* larvae and pupae had little influence on EPN dispersion; and distribution of both EPN species dynamically cycled between aggregated and uniform throughout the season (McGraw & Koppenhöfer, 2009).

Even though turfgrass can be a fairly uniform habitat, native EPN populations in turfgrass are spatially very patchy with *H. bacteriophora* populations tending to be patchier than *S. carpocapsae* populations, at least at a larger scale (Campbell et al., 1995, 1996, 1998). *S. carpocapsae* was recovered primarily near the soil surface and *H. bacteriophora* more uniformly throughout the soil profile (Campbell et al., 1996). *S. carpocapsae* populations are likely more contiguous due to the species greater activity near the soil surface where it tends to infect more mobile hosts that can disperse before succumbing to an infection, thereby creating new infection foci over greater distances.

When *H. bacteriophora* was applied to lower maintenance type turf, its densities declined quickly and returned to the aggregated pattern typical of natural populations within weeks, whether they were applied uniformly or in patches (Campbell et al., 1998; Wilson, Lewis, Yoder, & Gaugler, 2003). Ebssa and Koppenhöfer (2011) applied *S. carpocapsae*, *S. feltiae*, *Steinernema riobrave* Cabanillas, Poinar & Raulston (Rhabditida: Steinernematidae) or *H. bacteriophora* (commercial products) to greens (soil and sand based), fairways, and roughs and baited for 14 days; recovery of all species declined rapidly but there was no consistent effect of turf type with persistence often higher on the greens. Wilson and Gaugler (2004) found short persistence of *H. bacteriophora* whether applied to the surface or subsurface (i.e., at 5 cm depth by removing 5–cm thick sod/soil strips before application and then replacing the strips); poor persistence was positively correlated with mite and Collembola densities in plots with surface application but not in plots with subsurface application. Thus, it could be that IJs applied in large densities could be quickly decimated by natural enemies capable of a functional response. And diversity and density of these natural enemies are likely higher in lower maintenance turf areas. However, once EPN are established in the soil, lower maintenance turf may provide better condition for long–term persistence of EPN populations through greater host diversity and density and better buffering of environmental extremes by the higher turf canopy.

12.3 Case Studies

12.3.1 White Grubs (Coleoptera: Scarabaeidae)

A complex of white grub species is the most widespread and destructive insect pest of turf in the transition and cool–season turf grass zones of the USA and in southern portions of Canada (Potter, 1998; Vittum, Villani, & Tashiro, 1999). The extensive feeding activity of the larger larvae can kill large areas of grass especially under warm, dry conditions. In addition, vertebrate predators (e.g., skunks, racoons, crows, etc.) can tear up the turf to feed on the grubs even at larval densities that by themselves would not cause damage (Potter, 1998; Vittum et al., 1999). The majority of white grub species have a very similar 1–year life–cycle [(e.g., *P. japonica*), oriental beetle (*Anomala orientalis* [Waterhouse]), European chafer (*Rhizotrogus majalis* [Razoumowsky), masked chafer (*Cyclocephala* spp.)]. Around the latitude of New Jersey, adult flight occurs between June and August. Eggs are laid among turfgrass roots and hatch in 2–3 weeks. First and second larval stages each lasts about 3 weeks and third stage larvae start to appear in late August. After overwintering deeper in the soil between mid–/late October and mid–April, the larvae feed for 4–6 weeks in spring before pupating in the soil.

As soil insects, white grubs share their natural soil and rhizosphere habitats with EPNs. At least five species of entomopathogenic nematodes, *Steinernema arenarium* (Artyukhovsky) (Rhabditida: Steinernematidae), *S. glaseri, Steinernema kushidai* Mamiya (Rhabditida: Steinernematidae), *S. scarabaei*, and *Heterorhabditis megidis* Poinar, Jackson & Klein (Rhabditida: Heterorhabditidae), were originally collected and described from naturally infected white grubs and many more species have been documented to use white grubs as natural hosts (Peters, 1996). However, as a result of their coevolution with soil pathogens, white grubs have developed defense mechanisms including infrequent carbon dioxide output, sieve–plates over their spiracles, frequent defecation, defensive and evasive behaviors, a dense peritrophic membrane, and a strong immune response that make them relatively resistant to infection by EPN (Grewal, Koppenhöfer, & Choo, 2005 and references therein).

In early studies, *Heterorhabditis* spp. and *S. glaseri* were generally more effective than *S. feltiae* and *S. carpocapsae* (Klein, 1993). Georgis and Gaugler (1991) analysed 82 field trials conducted against *P. japonica* and concluded that *H. bacteriophora* populations (at 2.5×10^9 infective juvenile nematodes [IJs]/ha) used under the right conditions were as effective as the standard chemical insecticides used at the time (organophosphates and carbamates), whereas *S. carpocapsae* was ill–adapted for white grub control. More recent studies have shown that white grub species differ in their susceptibility to EPN and that the relative virulence of different EPN species also varies among white grub species (Grewal, Grewal, Malik, & Klein, 2002; Koppenhöfer & Fuzy, 2003a; Koppenhöfer, Fuzy, Crocker, Gelernter, & Polavarapu, 2004; Koppenhöfer, Grewal, & Fuzy, 2006). Among white grub species that are important pests of turfgrass in the USA, *P. japonica* appears to be the most EPN–susceptible species, whereas larvae of other white grub species including *Cyclocephala* spp., *A. orientalis*, *R. majalis*, or Asiatic garden beetle, *Maladera castanea* (Arrow), appear to be less susceptible to the commonly used EPN species (Cappaert & Koppenhöfer, 2003; Grewal et al., 2002; Koppenhöfer et al., 2004, 2006; Koppenhöfer, Brown et al., 2000; Koppenhöfer, Cowles, Cowles, Fuzy, & Baumgartner, 2002; Koppenhöfer & Fuzy, 2003a, 2003b; Koppenhöfer, Wilson, Brown, Kaya, & Gaugler, 2000; Simard, Belair, & Brodeur, 2001).

Grewal et al. (2005) give an extensive summary of studies on EPN efficacy against white grubs using as a qualifier for 'good' control at least 70 % control at a rate of 2.5×10^9 IJs/ha. Good control of *P. japonica* has been achieved with *S. scarabaei* AMK001 (100 %) (Koppenhöfer & Fuzy, 2003a), *H. bacteriophora* GPS11 (34–97 %) (Grewal, Power, Grewal, Suggars, & Haupricht, 2004), *H. bacteriophora* TF (65–92 %) (Koppenhöfer & Fuzy, 2003a; Koppenhöfer et al., 2000, 2002; Koppenhöfer, Wilson et al., 2000), and *Heterorhabditis zealandica* Poinar (Rhabditida: Heterorhabditidae) X1 (73–98 %). *S. scarabaei* is the only nematode species that has provided high field control of *A. orientalis* (87–100 %) (Koppenhöfer & Fuzy, 2003a, 2003b, 2009), *M. castanea* (71–86 %), and *R. majalis* (89 %) (Cappaert & Koppenhöfer, 2003). And good control of northern masked chafer, *Cyclocephala borealis* Arrow, has been observed with *H. zealandica* X1 (72–96 %), *S. scarabaei* (84 %), and *H. bacteriophora* GPS11 (47–83 %) (Grewal et al., 2004; Koppenhöfer & Fuzy, 2003a).

To address these differences in EPN–susceptibility among white grubs species and differences in virulence to white grubs among EPN species, several parallel studies examined EPN virulence (Koppenhöfer et al., 2006), EPN infectivity and infection routes (Koppenhöfer, Grewal, & Fuzy, 2007), EPN attraction to hosts (Koppenhöfer & Fuzy, 2008a; Koppenhöfer, Ebssa, & Fuzy, 2013), and white grub aggressive and evasive behaviours in response to EPN attack (Grewal, unpublished data) in combinations of the same white grub species (*P. japonica*, *A. orientalis*, *C. borealis*, *R. majalis*) and EPN species and populations thereof (*S. glaseri* NC, *S. scarabaei* AMK001, *H. zealandica* X1, *H. bacteriophora* GPS11). The high efficacy of *S. scarabaei* in field studies can be ascribed to its outstanding virulence, high infectivity, and limited elicitation of host defensive behaviors; all these factors were somewhat correlated with the species' degree of efficacy against different white grub species (*P. japonica* > *A. orientalis* = *R. majalis* > *C. borealis*). The low dispersal rate of *S. scarabaei* clearly is not a limiting factor in the turf system where white grub density can be very high (>100 per m^2 concentrated in top 5 cm of soil). The lower efficacy of *S. scarabaei* against *C. borealis* is due to its lower virulence and infectivity in this host. The low efficacy of the other studied EPN species against *A. orientalis* and especially against *R. majalis* is due to a combination of low virulence (especially to *R. majalis*) or low infectivity (*H. bacteriophora*, *S. glaseri*). Good efficacy of *H. bacteriophora* and *H. zealandica* against *P. japonica* can be ascribed to high virulence (*H. bacteriophora*), high infectivity (*H. zealandica*), and high dispersal rates (*H. bacteriophora*, *H. zealandica*). Since populations of the same EPN species can differ to some extent in factors affecting virulence and efficacy (e.g., Grewal et al., 2002; Koppenhöfer & Fuzy, 2003a) it is quite likely that using different EPN population would have affected the outcome of the above studies accordingly.

Persistence in the first few weeks after application should also play an important role in EPN efficacy. In laboratory studies in a sandy loam at moderate soil moisture levels (−10 to −100 kPa soil water potential), *S. scarabaei* and *S. glaseri* showed excellent persistence (no significant reduction in recovery by baiting over 28 days), whereas *H. bacteriophora* and *H. zealandica* recovery declined quickly (losses of 65–75 % after 7 days and 85–92 % after 14 days) (Koppenhöfer & Fuzy, 2006, 2007).

Larval stages of white grubs also may differ in their susceptibility but the effect varies with white grub and EPN species. Against *P. japonica*, efficacy is higher against second instars than third instars for *H. bacteriophora* (Grewal et al., 2004; Koppenhöfer & Fuzy, 2004; Power, An, & Grewal, 2009), but there is no significant effect for *S. scarabaei* or *H. zealandica*. For *A. orientalis*, susceptibility decreases from second to third instars for *H. bacteriophora* (Koppenhöfer & Fuzy, 2004, 2008c) and *Heterorhabditis* sp. Gyeongsan, *S. carpocapsae*, *S. glaseri*, and *Steinernema longicaudum* Shen & Wang (Rhabditida: Steinernematidae) (Lee et al., 2002) and from young third instars to older third instars for *H. bacteriophora*; but there is no difference in susceptibility between second and third instars for *S. scarabaei* or *S. glaseri*. *M. castanea* susceptibility to *S. scarabaei* increases from second to third instar.

Host density did not play a significant role in EPN efficacy in greenhouse pot experiments using combinations of different white grub species with different EPN species representing systems with somewhat resistant host to highly susceptible hosts (Ebssa, Fuzy, Bickerton, & Koppenhöfer, 2012). Any exhaustion of available EPN IJ populations to less lethal levels by high host numbers may have been counteracted by other factors such as increased chances for IJ–host contact and increased host susceptibility due to stress via reduced food resources and increased aggression between larvae.

EPN, especially *H. bacteriophora*, become increasingly ineffective for white grub control as soil temperature drops below 20 °C (Georgis & Gaugler, 1991). Thatch, an accumulation of organic matter between the soil and turfgrass foliage, restricts IJ downward movement and its thickness is negatively related to EPN efficacy. Irrigation volume and frequency and soil moisture are positively related to efficacy (Georgis & Gaugler 1991; Grewal et al., 2004) with a minimum of 7.4 mm of post–application irrigation required for establishment of the IJs in turfgrass (Shetlar, Suleman, & Georgis, 1988). Georgis and Gaugler concluded that *H. bacteriophora* was more effective against *P. japonica* in fine–textured soils, probably because finer soils retain moisture better and restrict IJ movement to the upper soil layers where most of the white grubs can be found.

EPN application timing for white grub control should consider presence of the most susceptible white grub stages but also other environmental conditions, particularly soil temperatures. For example, in the northeastern USA, *H. bacteriophora* applications against *P. japonica* or *A. orientalis* will tend to be more effective if applied between mid–August and early September due to the presence of the more susceptible younger larvae and a longer period after application before soil temperatures become too cool for good IJ activity. The efficacy of application before mid–August could be limited by hot and dry conditions especially where irrigation is limited and due to limited recycling of the EPN in the smallest larval stages (Power et al., 2009).

EPN persistence beyond a season following application against third instar white grubs has been reported for *H. bacteriophora* and *S. scarabaei*, suggesting the potential impact of EPNs on multiple generations of white grubs (e.g., Klein & Georgis, 1992; Koppenhöfer & Fuzy, 2009). Due to its excellent adaptation to white grubs as hosts (high virulence and recycling in hosts) and outstanding persistence of individual IJs, *S. scarabaei* provided not only high control of *A. orientalis* within 1 month of application at low application rates (77–100 % control at $0.25–2.5 \times 10^9$ IJs/ha), but provided additional control of overwintered larvae in the following spring (86–100 % control at $0.1–2.5 \times 10^9$ IJs/ha) and persisted and suppressed *A. orientalis* for up to 4 years after application. Thus, this species could be an excellent candidate for long–term suppression of white grubs with periodic reapplication of low rates.

12.3.2 Mole Crickets (Orthoptera: Gryllotalpidae)

Mole crickets, *Scapteriscus* spp., were accidentally introduced into Florida from South America around 1900 and have become one of the most important turfgrass insect pests throughout the coastal plain region of the southeastern USA from eastern Texas through North Carolina (Potter, 1998; Vittum et al., 1999). Adults and nymphs cause damage by feeding on grass roots and shoots and through their extensive tunneling activity. There is one generation per year. After egg laying in spring, the adults die off, and the crickets develop through the nymphal stages during summer. Overwintering occurs primarily in the nymphal (southern mole cricket, *Scapteriscus borellii* GiglioTos) or adult (tawny mole cricket, *Scapteriscus vicinus* Scudder) stage.

First attempts at controlling mole crickets with EPN using *S. carpocapsae* provided on average 58 % control at 2.5×10^9 IJs/ha (Georgis & Poinar, 1994). Superior activity was later found with *S. scapterisci* and *S. riobrave* (average 75 % control at 2.5×10^9 IJs/ha). While *S. riobrave* only provides curative control of mole crickets because it does not reproduce in them, *S. scapterisci* proved to be an excellent agent for inoculative releases due to its ability to recycle and spread in mole crickets (Parkman & Smart, 1996).

Introduction of *Steinernema scapterisci* Nguyen & Smart (Rhabditida: Steinernematidae) into the USA for mole cricket control is the first successful use of an EPN in classical biological control (Parkman & Smart, 1996; Frank & Walker, 2006). The species was isolated from mole crickets in Uruguay and Argentina. After laboratory studies indicated excellent control potential and non–target safety of *S.*

scapterisci, it was successfully established after inundative applications, application of *S. scapterisci*–infested cricket cadavers, and using electronic mating callers to attract mole crickets to the site of application (Parkman, Hudson, Frank, Nguyen, & Smart, 1993). These studies also showed that *S. scapterisci* is more effective against *S. borellii* than *S. vicinus*, least effective against the short–winged mole cricket (*S. abbreviatus* Scudder), and most effective against adult mole crickets and ineffective against small nymphs (Parkman & Frank, 1992).

Steinernema scapterisci is an ideal control agent for pastures and turfgrass areas that can tolerate some mole cricket damage (Frank & Walker, 2006). In the turf market, *S. scapterisci* has been applied to low profile and environmentally sensitive areas on golf courses, sod farms, and recreational areas at a rate of 2.5×10^9 IJs/ha. In more damage sensitive areas, *S. scapterisci* use has been limited due to the competition from the more effective but similarly expensive insecticide fipronil and often by the need for application of nematicides for the control of plant–parasitic nematodes (a widespread, major problem in southern USA turf). Success of *S. scapterisci* in this market segment has also been hampered by the fact that it has to be applied in spring or fall when adults are present, whereas control measures are typically necessary in summer against nymphs. In pastures, the potentially biggest market for *S. scapterisci*, it has been applied using slit injectors in strips covering 12.5 % of the area from which it then spread over a period of several years. This approach reduced the cost to well below that of chemical insecticides that provide only short–term suppression.

In areas in Florida where both *S. scapterisci* and a parasitoid wasp, *Larra bicolor* F. (Hymenoptera: Crabronidae), also introduced from South America, have become established, mole crickets populations are significantly suppressed and cause much less problems (Frank & Walker, 2006). While *S. scapterisci* could probably be used for mole cricket management throughout the area of distribution of mole crickets in the USA, it is not clear how far north it will be able to overwinter and with that persist effectively. It should have the potential to become established throughout much of the mole crickets' present area of distribution in the USA. However, because of the nematode's slow spread from inoculation sites, widespread use of a commercial product or a large–scale inoculation program would be necessary to accelerate its spread.

After going through several periods of not being commercially available, a product was available since 2003 and appeared to become a success story especially for uses in low maintenance turf and in pastures. Nonetheless, demand for *S. scapterisci* has not been high enough to sustain the commercial product and it was therefore discontinued in 2013. The future will have to tell whether this potential success story can be picked up again.

12.3.3 Black Cutworm (Lepidoptera: Noctuidae)

Agrotis ipsilon is a perennial problem on the close cut bentgrass of golf course greens and tees throughout the world (Potter, 1998; Vittum et al., 1999). The larvae dig burrows in the thatch or soil and emerge at night to eat the grass blades and stems around the burrow. Feeding by the older larvae creates dead patches, sunken areas, or pockmarks making the turf unattractive and disrupting the uniformity and smoothness of the putting surface. The black cutworm has multiple generations per year.

In field efficacy testing of EPN products in the 1980s and early 1990s, *H. bacteriophora* did not provide satisfactory control of *A. ipsilon* larvae (average 62 %), whereas *S. carpocapsae* was highly effective (average 95 %) (Georgis & Poinar, 1994). In detailed laboratory studies that included seven EPN species: *H. bacteriophora*, *H. megidis*, *H. indica*, *S. carpocapsae*, *S. riobrave*, *S. feltiae*, and *Steinernema kraussei* (Steiner) Travassos (Rhabditida: Steinernematidae), Ebssa and Koppenhöfer (2012) found that (1) *H. megidis* was the most virulent species in small containers with soil and with diet as food, (2) *S. carpocapsae* tended to cause the highest mortality followed by *H. bacteriophora*, *H. megidis*, and *S. riobrave* in pots with grass, and (3) fourth and/or fifth instar larvae were the most susceptible stage to most EPN species and pupae the least susceptible. In recent field studies (Ebssa & Koppenhöfer, 2011), *S. carpocapsae* was the best performing species due to a combination of high (average 83 %) and most consistent (70–90 %) control rates (at 7 days after treatment) and highest speed of kill (average 68 % at 4 days after treatment); *S. feltiae* and *H. bacteriophora* often provided similar control but were less consistent. In additional studies (Ebssa & Kostromytska, unpublished data), (1) combinations of two EPN species at half rate each did not provide significantly better control than the better of the two species alone at full rate, (2) syringing (i.e., twice daily small amount of irrigation) provided some limited improvement of *S. carpocapsae* efficacy under warm, sunny conditions, and (3) split application (two applications at half rate 3 days apart) significantly improved *S. carpocapsae* efficacy (by about 20 %).

Despite this high efficacy, EPN are not widely used for *A. ipsilon* control because damage thresholds on golf course tees and especially greens are so low that golf course superintendents will prefer to use chemical insecticides that provide even better and more consistent control at a lower cost than *S. carpocapsae*. This will likely continue until expectations and attitude of their clientele changes or synthetic insecticides availability declines.

12.3.4 Weevils (Coleoptera: Curculionidae)

Billbugs, *Sphenophorus* spp., are important turfgrass pests throughout much of the USA (Potter, 1998; Vittum et al., 1999). Damage is caused by the young larvae feeding inside the stem and crown and the older larvae feeding externally on the crown and belowground parts of the plant. The bluegrass billbug, *Sphenophorus parvulus* Gyllenhal, is an important pest of cool–season grasses in the northern half of the USA. It overwinters in the adult stage, becomes active around late April, and most egg laying occurs between early May and early July. The older larvae are most abundant in the soil from around early July to early August, and damage usually becomes apparent from late June into August. Studies on other billbug species that may damage cool–season grasses are more limited. The hunting billbug, *Sphenophorus venatus vestitus* Chittenden, is a pest of warm season grasses in the southern USA, but in Japan, it is the most important insect pest on golf courses. In the northern parts of its range, *S. venatus vestitus* has one generation per year with a life cycle similar to that of *S. parvulus*. In the southern parts of its range, it primarily overwinters in the adult stage with some larvae overwintering, and it can have several overlapping generations per year.

No detailed studies on billbug–EPN interaction have been published. In field tests in Ohio, USA, targeting the larvae in the soil, control of *S. parvulus* by *S. carpocapsae* (average 78 %) and *H. bacteriophora* (average 74 %) was similar to that by standard insecticides at the time (organophosphates, carbamate) (Georgis & Poinar, 1994) but use of EPN products containing *S. carpocapsae* and *H. bacteriophora* against billbugs is limited in the USA due to the availability of several newer effective insecticides for this use that are easier to use, more effective, and cheaper. In golf courses in Japan, *S. carpocapsae* was very effective for control of *S. venatus vestitus* (average 84 %) (Smith, 1994; Kinoshita & Yamanaka, 1998), in part due to favorable environmental conditions (temperature and rainfall) and the adoption of EPN–friendly application protocols, i.e., immediate watering after spraying and generally very careful following of label instructions. However, *S. carpocapsae* sales for billbug control in Japan significantly declined after the registration of imidacloprid for turfgrass uses.

The annual bluegrass weevil, *L. maculicollis*, is a severe pest of short–mown golf course turf (greens, approaches, fairways, tee boxes) in the Northeastern USA and Eastern Canada (Potter, 1998; Vittum et al., 1999). The adults overwinter in protected habitats adjacent to short–mown playing surfaces. In spring, the adults walk on to short–mown areas and place eggs between the sheaths of turfgrass stems. Young larvae feed inside the stems, older larvae feed externally on the crown. Damage is often most apparent in late spring in areas with high percentages of annual bluegrass, *Poa annua* (Cyperales: Poaceae). *L. maculicollis* populations complete 1–3 generations per year. Typically the first generation is the densest and most destructive, but insecticide applications often need to be repeated throughout the summer. Over–reliance on synthetic insecticides, particularly pyrethroids, has led to the development of resistant populations (Ramoutar, Alm, & Cowles, 2009) on an increasing number of golf courses throughout the Northeastern USA (McGraw & Kostromytska, unpublished data). Highly resistant populations are less susceptible to most presently available insecticides (Koppenhöfer et al., 2012).

Studies on golf courses in New Jersey, USA showed that native EPN populations (*S. carpocapsae, H. bacteriophora*) were common on insecticide–free fairways and caused up to 50 % generational mortality of weevil larvae. However, they failed to generate a functional response, i.e., infection rates did not increase proportionally with increasing weevil larval densities, suggesting unreliability in reducing densities in a conservation biological control approach (McGraw & Koppenhöfer, 2009). The period during which weevil stages are susceptible to EPN infection (fourth– and fifth–instar larvae feeding externally on plant and pupae in soil) is likely too short to allow EPN progeny emerging from the first infected larvae to have a significant impact within the same generation, at least for the more synchronized spring generation which causes most of the damage. During summer larval population may typically be too low and dispersed to allow for a significant impact of EPNs.

Under laboratory conditions *S. carpocapsae, S. feltiae* and *H. bacteriophora* showed promising control of fourth– and fifth–instar *L. maculicollis* larvae (McGraw & Koppenhöfer, 2008). But in numerous field trials with multiple EPN species, rates, and species combinations, EPN performance was variable and EPN persistence limited (McGraw, Cowles, Vittum, & Koppenhöfer, 2010). Control rates for the overall most promising and consistent species, *S. carpocapsae*, averaged 51 % but reached up to 98 % in some field trials.

Ongoing research on EPN use against *L. maculicollis* indicates that split applications of *S. carpocapsae* (about 1 week apart) can improve control rates and that simultaneous application of *S. carpocapsae* and *H. bacteriophora* and imidacloprid (applied for white grub control) have an additive effect on larval mortality (O. Kostromytska, unpublished data). Because the destructive nature of *L. maculicollis* necessitates high levels of control, the willingness by turf managers to adopt inherently variable biologically–based controls is low, regardless of the perceived environmental benefits. Yet, the dearth of effective insecticides for insecticide resistant *L. maculicollis* populations may increase opportunities for EPN use.

12.3.5 Other Insect Pests

The larvae of two crane fly species, *Tipula paludosa* Meigen (Diptera: Tipulidae) (European crane fly) and *Tipula oleracea* L. (Diptera: Tipulidae) (common crane fly), are important turfgrass pests in parts of North America and are susceptible to heterorhabditid nematodes and particularly to *S. feltiae* (Ehlers & Gerwien, 1993; Simard, Bélair, Gosselin, & Dionne, 2006). Both species can be controlled with *S. carpocapsae* and *S. feltiae* if they are applied against the young larval stages in October; larval susceptibility, at least to *S. feltiae*, decreases with larval development (Peters & Ehlers, 1994).

Various species of sod webworms, cutworms, and armyworms are susceptible to *S. carpocapsae* and *H. bacteriophora* (Georgis & Poinar, 1994) but no detailed studies have been conducted to date for these insect pests.

12.4 Combination with Other Control Agents

The goal of combining EPN with other control agents is to provide effective curative pest control while reducing the use of more hazardous synthetic insecticides, increase consistency and level of pest control, and lower costs by being able to use reduced rates of chemicals and EPN. Much of the work on combining EPN with other control agents has been conducted with white grubs and in turfgrass.

Combination of two EPN species against third–instar white grubs has generally resulted in additive effects on mortality (*H. bacteriophora* with *S. kushidai* or *S. glaseri* against *Cyclocephala hirta* LeConte (Coleoptera: Scarabaeidae); *S. glaseri* and *S. kushidai* against *A. orientalis*) (Koppenhöfer, Wilson et al., 2000). Combination of *S. carpocapsae*, *S. feltiae*, and *H. bacteriophora* against *A. ipsilon* and *L. maculicollis* generally resulted in additive effects (McGraw, Ebssa, & Kostromytska, unpublished data). Combination of two EPN species against one target is likely to lead to competitive exclusion of one of the EPN species in the long term unless alternative hosts and differences in susceptibility of the hosts to the two EPN species facilitate coexistence (Koppenhöfer & Kaya, 1996).

Combinations of *Bacillus thuringiensis* ssp. *japonensis* (Btj) with EPN have resulted in additive and synergistic effects on mortality of third instars of different white grub species in laboratory, greenhouse, and field experiments (Koppenhöfer, Choo, Kaya, Lee, & Gelernter, 1999; Koppenhöfer, Wilson et al., 2000); these combinations may be more feasible against scarab species with intermediate to high Btj–sensitivity (e.g., *C. borealis*, *P. japonica*, *A. orientalis*) and when applied against younger more Btj– and/or EPN–susceptible larvae. Infection of third–instar *C. hirta* with the milky disease bacterium, *Paenibacillus popilliae* Dutky (Bacillales: Paenibacillaceae), facilitated IJ penetration into the host's midgut and resulted in additive (*H. bacteriophora*) or synergistic (*S. glaseri*) effects on larval mortality in a greenhouse experiment (Thurston, Kaya, Burlando, & Harrison, 1993; Thurston, Kaya, & Gaugler, 1994). However, the lack of in–vitro production methods for *P. popilliae* and the slow establishment of milky disease in white grub populations limit the feasibility of this combination.

The EPN species *H. megidis* and *S. glaseri* and the fungus *Metarhizium anisopliae* (Metschnikoff) (Hypocreales: Clavicipitaceae) showed a strong synergistic interaction in second– and third–instar *Hoplia philanthus* (Füessly) (Coleoptera: Scarabaeidae) in laboratory, greenhouse, and field studies when the fungus was applied 3–4 weeks before the EPN; combination with *M. anisopliae* had no effect on *S. glaseri* reproduction but negatively affected *H. megidis* reproduction at higher *M. anisopliae* application rates (Ansari, Tirry, & Moens, 2004; Ansari, Shah, Tirry, & Moens, 2006). Combinations of *M. anisopliae* or *Beauveria bassiana* (Balsamo) (Hypocreales: Ophiocordycipitaceae) against third–instar southern masked chafer, *C. lurida* Bland, in laboratory and greenhouse experiments generally resulted in additive mortality irrespective of application interval but had not negative effects on EPN reproduction (Wu, Youngman, Kok, Laub, & Pfeiffer, 2014).

Combinations of the organophosphate diazinon with *S. kushidai* resulted in additive mortality of third–instar *C. hirta* and *A. orientalis* in greenhouse experiments (Koppenhöfer, Wilson et al., 2000), combinations of the insect growth regulator halofenozide and *Heterorhabditis marelatus* Liu & Berry (Rhabditida: Heterorhabditidae) caused additive *P. japonica* mortality in the laboratory (Mannion, Winkler, Shapiro, & Gibb, 2000), and the organophosphate chlorpyrifos and *Heterorhabditis* sp. Gyeongsan or *S. longicaudum* interacted synergistically on

A. orientalis mortality in field experiments (Lee et al., 2002). Combinations of *H. bacteriophora* and the anthranilic diamide chlorantraniliprole resulted in mostly synergistic but also additive mortality of third–instar *A. orientalis*, *P. japonica*, and *C. borealis* in greenhouse and field experiments (Koppenhöfer & Fuzy, 2008b).

Best documented among EPN–synthetic insecticide combinations are the interactions between EPN and neonicotinoid insecticides, especially imidacloprid, in laboratory, greenhouse, and field experiments. Imidacloprid mostly interacted synergistically in combinations with several EPN species (*S. glaseri*, *H. bacteriophora*, *H. marelatus*, *H. megidis*) in third instars of *C. borealis*, *C. hirta*, *Cyclocephala pasadenae* Casey, *P. japonica*, and *A. orientalis* (Koppenhöfer et al., 2000, 2002; Koppenhöfer, Wilson et al., 2000b; Koppenhöfer & Fuzy, 2008c). However, two rather scarab–specific species, *S. kushidai* and *S. scarabaei*, generally did not interact with imidacloprid against several white grub species (Koppenhöfer, Wilson et al., 2000; Cappaert & Koppenhöfer, 2003; Koppenhöfer & Fuzy, 2003a, 2003b). Imidacloprid–*H. bacteriophora* combinations provide more consistent synergism against third instars but higher control rates against second instars and early third instars in laboratory and field trials using rates as low as 25 % and 50 % of the full rates for *H. bacteriophora* and imidacloprid, respectively (Koppenhöfer & Fuzy, 2008c). The EPN–imidacloprid interaction is primarily based on reduced defensive and evasive larval behaviors resulting in increased host attachment and penetration by IJs (Koppenhöfer, Grewal, & Kaya, 2000). Neonicotinoid–EPN combination generally had no negative effect on EPN–reproduction in hosts and resulted in higher IJ densities in the soil following application due to the greater number of infected white grubs (Koppenhöfer, Cowles, Cowles, Fuzy, & Kaya, 2003).

Only the interactions of nematodes with Btj, imidacloprid, and, to a more limited extent, chlorantraniliprole have been sufficiently studied in greenhouse and field trials to allow for recommendations to be made at this time. Against medium to large third instars all three combinations required relatively high rates of both agents, limiting their economic feasibility. All three combinations should be more effective or equally effective at lower rates when applied against younger larvae, i.e., young third instars or second instars (Koppenhöfer et al., 1999; Koppenhöfer & Fuzy, 2008c).

12.5 Conclusion and Future Research

Use of EPN for turfgrass insect management in the USA remains very limited despite considerable efforts in research and development over the last decades. The major reason for this has been competition from synthetic insecticides that are easier to use and cheaper. Greater sensitivity to extremes of temperature, moisture, and pressure during shipping, storage, application, and/or post–application; short product shelf life; and up to ten times higher price compared to synthetic insecticides that have gone off patent will continue to limit EPN use for commercial turfgrass markets (golf courses, athletic fields, professional landscapers, sod growers) as long as effective synthetic insecticides will continue to be available. Somewhat better is the situation in the consumer market because many homeowners value reduced exposure to synthetic insecticides higher than product efficacy, ease of use, and price.

EPN efficacy could be improved by isolating and developing more virulent populations from field populations, increasing virulence through biotechnology, increasing our understanding on how to use them most effectively, and improving application technologies. Cost could be reduced by improving production technologies. Ease–of–use could be improved by increasing shelf–life and tolerance to high temperatures of formulations. Most such improvements would likely be only gradual, but they may suffice to significantly increase EPN use at least where pesticide regulations, local ordinances, public opinion, or even widespread insecticide resistance already impinge on the use of synthetic insecticides.

EPN use could also increase through better exploitation of their ability to recycle in hosts. Species that have a potential for long–term suppression generally appear to be rather host–specific, but if their hosts are key pests they can still be successful. Two examples for such systems are mole cricket management with *S. scapterisci* and white grub management with *S. scarabaei*. But even though *S. scapterisci* appeared be on its way to become a success story, sales were stopped in 2013 due to limited demand. And attempts at *in vitro* production of *S. scarabaei* have thus far been unsuccessful. Because host–specific pathogens depend on the presence of hosts for recycling and may become patchy in distribution over time, their use may only be feasible in areas with some tolerance for occasional pest damage.

Ultimately, significant increase in EPN use, whether as biopesticides or inoculative agents, will likely happen only through education and legislations. Major changes in insecticides use pattern will have to be encouraged through legislative incentives, regulations, and restrictions in a similar fashion as they have been happening in Canada since the late 1990s (Bélair et al., 2010).

References

Alumai, A., Grewal, P. S., Hoy, C. W., & Willoughby, D. A. (2006). Factors affecting the natural occurrence of entomopathogenic nematodes in turfgrass. *Biological Control, 36*(3), 368–374.

Ansari, M. A., Shah, F. A., Tirry, L., & Moens, M. (2006). Field trials against *Hoplia philanthus* (Coleoptera: Scarabaeidae) with a combination of an entomopathogenic nematode and the fungus *Metarhizium anisopliae* CLO 53. *Biological Control, 39*(3), 453–459.

Ansari, M. A., Tirry, L., & Moens, M. (2004). Interaction between *Metarhizium anisopliae* CLO 53 and entomopathogenic nematodes for the control of *Hoplia philanthus. Biological Control, 31*(2), 172–180.

Beard, J. B., & Green, R. L. (1994). The role of turfgrass in environmental protection and their benefits to humans. *Journal of Environmental Quality, 23*(3), 452–460.

Bélair, G., Koppenhöfer, A. M., Dionne, J., & Simard, L. (2010). Current and potential use of pathogens in the management of turfgrass insects as affected by new pesticide regulations in North America. *International Journal of Pest Management, 56*(1), 51–60.

Campbell, J. F., Lewis, E., Yoder, F., & Gaugler, R. (1995). Entomopathogenic nematode (Heterorhabditidae and Steinernematidae) seasonal population dynamics and impact on insect populations in turfgrass. *Biological Control, 5*(4), 598–606.

Campbell, J. F., Lewis, E., Yoder, F., & Gaugler, R. (1996). Entomopathogenic nematode (Heterorhabditidae and Steinernematidae) spatial distribution in turfgrass. *Parasitology, 113*(5), 473–482.

Campbell, J. F., Orza, G., Yoder, F., Lewis, E., & Gaugler, R. (1998). Spatial and temporal distribution of endemic and released entomopathogenic nematode populations in turfgrass. *Entomologia Experimentalis et Applicata, 86*(1), 1–11.

Cappaert, D. C., & Koppenhöfer, A. M. (2003). *Steinernema* scarabaei, an entomopathogenic nematode for control of the European chafer. *Biological Control, 28*(3), 379–386.

Charles, H., Godfray, J., Blacquière, T., Field, L. M., Hails, R. S., Petrokofsky, G., et al. (2014). A restatement of the natural science evidence base concerning neonicotinoid insecticides and insect pollinators. *Proceedings of the Royal Society B, 281*, 20140558.

Christians, N. E. (1998). *Fundamentals of turfgrass management*. Chelsea, MI: Ann Arbor Press.

Ebssa, L., Fuzy, E. M., Bickerton, M. W., & Koppenhöfer, A. M. (2012). Host density effects on efficacy of entomopathogenic nematodes against white grub (Coleoptera: Scarabaeidae) species. *Biocontrol Science and Technology, 22*(1), 117–123.

Ebssa, L., & Koppenhöfer, A. M. (2011). Efficacy and persistence of entomopathogenic nematodes for black cutworm control in turfgrass. *Biocontrol Science and Technology, 21*(7), 779–796.

Ebssa, L., & Koppenhöfer, A. M. (2012). Entomopathogenic nematodes for black cutworm management: Effect of instar, nematode species, and nematode production method. *Pest Management Science, 68*(6), 947–957.

Ehlers, R.–U., & Gerwien, A. (1993). Selection of entomopathogenic nematodes (Steinernematidae and Heterorhabditidae, Nematoda) for the biological control of cranefly larvae *Tipula paludosa* (Tipulidae, Diptera). *Zeitschrift fur Pflanzenkrankheiten und Pflanzenschutz, 100*(4), 343–353.

Elmowitz, D. E., Ebssa, L., & Koppenhöfer, A. M. (2013). Overwintering behavior of the entomopathogenic nematodes *Steinernema scarabaei* and *Heterorhabditis bacteriophora* (Rhabditida: Steinernematidae and Heterorhabditidae) and their white grub hosts (Coleoptera: Scarabaeidae). *Entomologia Experimentalis et Applicata, 148*(3), 246–258.

Frank, J. H., & Walker, T. J. (2006). Permanent control of pest mole crickets (Orthoptera: Gryllotalpidae) in Florida. *American Entomologist, 52*(3), 139–144.

Georgis, R., & Gaugler, R. (1991). Predictability in biological control using entomopathogenic nematodes. *Journal of Economic Entomology, 84*(3), 713–720.

Georgis, R., & Poinar, G. O., Jr. (1994). Nematodes as bioinsecticides in turf and ornamentals. In A. Leslie (Ed.), *Integrated pest management for turf and ornamentals* (pp. 477–489). Boca Raton, FL: CRC Press.

Grewal, P. S., Grewal, S. K., Malik, V. S., & Klein, M. G. (2002). Differences in the susceptibility of introduced and native white grub species to entomopathogenic nematodes from various geographic localities. *Biological Control, 24*(3), 230–237.

Grewal, P. S., Koppenhöfer, A. M., & Choo, H. Y. (2005). Lawn, turfgrass, and pasture applications. In P. S. Grewal, R.–U. Ehlers, & D. I. Shapiro–Ilan (Eds.), *Nematodes as biocontrol agents* (pp. 115–146). Wallingford, UK: CABI Publishing.

Grewal, P. S., Power, K. T., Grewal, S. K., Suggars, A., & Haupricht, S. (2004). Enhanced consistency in biological control of white grubs (Coleoptera: Scarabaeidae) with new strains of entomopathogenic nematodes. *Biological Control, 30*(1), 73–82.

Held, D. W., & Potter, D. A. (2012). Prospects for managing turfgrass pests with reduced chemical inputs. *Annual Review of Entomology, 57*, 329–354.

Kinoshita, M., & Yamanaka, S. (1998). Development and prevalence of entomopathogenic nematodes in Japan. *Japanese Journal of Nematology, 28*, 42–45.

Klein, M. G. (1993). Biological control of scarabs with entomopathogenic nematodes. In R. Bedding, R. Akhurst, & H. Kaya (Eds.), *Nematodes and the biological control of insect pests* (pp. 49–58). East Melbourne, Australia: CSIRO Press.

Klein, M. G., & Georgis, R. (1992). Persistence of control of Japanese beetle (Coleoptera: Scarabaeidae) larvae with steinernematid and heterorhabditid nematodes. *Journal of Economic Entomology, 85*(3), 727–730.

Koppenhöfer, A. M., Alm, S. R., Cowles, R. A., McGraw, B. A., Swier, S., & Vittum, P. J. (2012, March). Controlling annual bluegrass weevil: Optimal timing and rates. *Golf Course Management, 84*, 98–104

Koppenhöfer, A. M., Brown, I. M., Gaugler, R., Grewal, P. S., Kaya, H. K., & Klein, M. G. (2000). Synergism of entomopathogenic nematodes and imidacloprid against white grubs: Greenhouse and field evaluation. *Biological Control, 19*(3), 245–251.

Koppenhöfer, A. M., Choo, H. Y., Kaya, H. K., Lee, D. W., & Gelernter, W. D. (1999). Increased field and greenhouse efficacy with a combination of entomopathogenic nematode and *Bacillus thuringiensis* against scarab grubs. *Biological Control, 14*(1), 37–44.

Koppenhöfer, A. M., Cowles, R. S., Cowles, E. A., Fuzy, E. M., & Baumgartner, L. (2002). Comparison of neonicotinoid insecticides as synergists for entomopathogenic nematodes. *Biological Control, 24*(1), 90–97.

Koppenhöfer, A. M., Cowles, R. S., Cowles, E. A., Fuzy, E. M., & Kaya, H. K. (2003). Effect of neonicotinoid insecticide synergists on entomopathogenic nematode fitness. *Entomologia Experimentalis et Applicata, 106*(1), 7–18.

Koppenhöfer, A. M., Ebssa, L., & Fuzy, E. M. (2013). Storage temperature and duration affect *Steinernema scarabaei* dispersal and attraction, virulence, and infectivity to a white grub host. *Journal of Invertebrate Pathology, 112*(2), 129–137.

Koppenhöfer, A. M., & Fuzy, E. M. (2003a). *Steinernema scarabaei* for the control of white grubs. *Biological Control, 28*(1), 47–59.

Koppenhöfer, A. M., & Fuzy, E. M. (2003b). Biological and chemical control of the Asiatic garden beetle, *Maladera castanea* (Coleoptera: Scarabaeidae). *Journal of Economic Entomology, 96*(4), 1076–1082.

Koppenhöfer, A. M., & Fuzy, E. M. (2004). Effect of white grub developmental stage on susceptibility to entomopathogenic nematodes. *Journal of Economic Entomology, 97*(6), 1842–1849.

Koppenhöfer, A. M., & Fuzy, E. M. (2006). Effect of soil type on infectivity and persistence of the entomopathogenic nematodes *Steinernema scarabaei*, *Steinernema glaseri*, *Heterorhabditis zealandica*, and *Heterorhabditis bacteriophora*. *Journal of Invertebrate Pathology, 92*(1), 11–22.

Koppenhöfer, A. M., & Fuzy, E. M. (2007). Soil moisture effects on infectivity and persistence of the entomopathogenic nematodes *Steinernema scarabaei*, *Steinernema glaseri*, *Heterorhabditis zealandica*, and *Heterorhabditis bacteriophora*. *Applied Soil Ecology, 35*(1), 128–139.

Koppenhöfer, A. M., & Fuzy, E. M. (2008a). Attraction of four entomopathogenic nematode species to four white grub species. *Journal of Invertebrate Pathology, 99*(2), 227–234.

Koppenhöfer, A. M., & Fuzy, E. M. (2008b). Early timing and new combinations to increase the efficacy of neonicotinoid–entomopathogenic nematodes (Rhabditida: Heterorhabditidae) combinations against white grubs (Coleoptera: Scarabaeidae). *Pest Management Science, 64*(7), 725–735.

Koppenhöfer, A. M., & Fuzy, E. M. (2008c). Effects of the anthranilic diamide insecticide, chlorantraniliprole, on *Heterorhabditis bacteriophora* (Rhabditida: Heterorhabditidae) efficacy against and reproduction in white grubs (Coleoptera: Scarabaeidae). *Biological Control, 45*(1), 93–102.

Koppenhöfer, A. M., & Fuzy, E. M. (2009). Long–term effects and persistence of *Steinernema scarabaei* applied for suppression of *Anomala orientalis* (Coleoptera: Scarabaeidae). *Biological Control, 48*(1), 63–72.

Koppenhöfer, A. M., Fuzy, E. M., Crocker, R., Gelernter, W., & Polavarapu, S. (2004). Pathogenicity of *Steinernema scarabaei*, *Heterorhabditis bacteriophora* and *S. glaseri* to twelve white grub species. *Biocontrol Science and Technology, 14*(1), 87–92.

Koppenhöfer, A. M., Grewal, P. S., & Fuzy, E. M. (2006). Virulence of the entomopathogenic nematodes *Heterorhabditis bacteriophora*, *H. zealandica*, and *Steinernema scarabaei* against five white grub species (Coleoptera: Scarabaeidae) of economic importance in turfgrass in North America. *Biological Control, 38*(3), 397–404.

Koppenhöfer, A. M., Grewal, P. S., & Fuzy, E. M. (2007). Differences in penetration routes and establishment rates of four entomopathogenic nematode species into four white grub species. *Journal of Invertebrate Pathology, 94*(2), 184–195.
Koppenhöfer, A. M., Grewal, P. S., & Kaya, H. K. (2000). Synergism of entomopathogenic nematodes and imidacloprid against white grubs: The Mechanism. *Entomologia Experimentalis et Applicata, 94*(3), 283–293.
Koppenhöfer, A. M., & Kaya, H. K. (1996). Coexistence of entomopathogenic nematodes species (Rhabditida: Steinernematidae, Heterorhabditidae) with different foraging behavior. *Fundamental and Applied Nematology, 19*(2), 175–183.
Koppenhöfer, A. M., Wilson, M. G., Brown, I., Kaya, H. K., & Gaugler, R. (2000). Biological control agents for white grubs (Coleoptera: Scarabaeidae) in anticipation of the establishment of the Japanese beetle in California. *Journal of Economic Entomology, 93*(1), 71–80.
Lee, D. W., Choo, H. Y., Kaya, H. K., Lee, S. M., Smitley, D. R., Shin, H. K., et al. (2002). Laboratory and field evaluation of Korean entomopathogenic nematode isolates against the oriental beetle *Exomala orientalis* (Coleoptera: Scarabaeidae). *Journal of Economic Entomology, 95*(5), 918–926.
Mannion, C. M., Winkler, H. E., Shapiro, D. I., & Gibb, T. (2000). Interaction between halofenozide and entomopathogenic nematode *Heterorhabditis marelatus* for control of Japanese beetle (Coleoptera: Scarabaeidae) larvae. *Journal of Economic Entomology, 93*(1), 48–53.
McGraw, B. A., Cowles, R., Vittum, P. J., & Koppenhöfer, A. M. (2010). Field evaluation of entomopathogenic nematodes for the biological control of the annual bluegrass weevil, *Listronotus maculicollis* (Coleoptera: Curculionidae) in golf course turfgrass. *Biocontrol Science Technology, 20*(2), 149–163.
McGraw, B. A., & Koppenhöfer, A. M. (2008). Evaluation of two endemic and five commercial entomopathogenic nematode species (Rhabditida: Heterorhabditidae and Steinernematidae) against annual bluegrass weevil (Coleoptera: Curculionidae) larvae and adults. *Biological Control, 46*(3), 467–475.
McGraw, B. A., & Koppenhöfer, A. M. (2009). Population dynamics and interactions between endemic entomopathogenic nematodes and annual bluegrass weevil populations in golf course turfgrass. *Applied Soil Ecology, 41*(1), 77–89.
National Pest Management Association. (2014). *Overview of state school pest management requirements*. Fairfax, VA: National Pest Management Association. https://npmapestworld.org/publicpolicy/StateSchoolRegs.cfm [revised 4/11/2014].
National Turfgrass Federation. (2009). *The national turfgrass research initiative*. Beltsville, MD: National Turfgrass Federation. http://turfresearch.org/pdf/turfinitiative.pdf [revised 21/08/2014].
Parkman, J. P., & Frank, J. H. (1992). Infection of sound–trapped mole crickets, *Scapteriscus* spp., by *Steinernema scapterisci. Florida Entomologist, 75*(1), 163–165.
Parkman, J. P., Hudson, W. G., Frank, J. H., Nguyen, K. B., & Smart, G. C., Jr. (1993). Establishment and persistence of *Steinernema scapterisci* (Rhabditida: Steinernematidae) in field populations of *Scapteriscus* spp. mole crickets (Orthoptera: Gryllotalpidae). *Journal of Entomological Science, 28*(2), 182–190.
Parkman, J. P., & Smart, G. C., Jr. (1996). Entomopathogenic nematodes, a case study: Introduction of *Steinernema scapterisci* in Florida. *Biocontrol Science and Technology, 6*(3), 413–419.
Peters, A. (1996). The natural host range of *Steinernema* and *Heterorhabditis* spp. and their impact on insect populations. *Biocontrol Science and Technology, 6*(3), 389–402.
Peters, A., & Ehlers, R.–U. (1994). Susceptibility of leatherjackets (*Tipula paludosa* and *T. oleracea*; Tipulidae: Nematocera) to the entomopathogenic nematode *Steinernema feltiae. Journal of Invertebrate Pathology, 63*(2), 163–171.
Potter, D. A. (1998). *Destructive turfgrass insects: Biology, diagnosis, and control*. Chelsea, MI: Ann Arbor Press.

Power, K. T., An, R., & Grewal, P. S. (2009). Effectiveness of *Heterorhabditis bacteriophora* strain GPS11 applications targeted against different instars of the Japanese beetle *Popillia japonica. Biological Control, 48*(3), 232–236.

Racke, K. D. (2000). Pesticides for turfgrass pest management: Uses and environmental issues. In J. M. Clark & M. P. Kenna (Eds.), *Fate and management of turfgrass chemicals* (pp. 45–64). Washington, DC: American Chemical Society.

Ramoutar, D., Alm, S. R., & Cowles, R. S. (2009). Pyrethroid resistance in populations of *Listronotus maculicollis* (Coleoptera: Curculionidae) from southern New England golf courses. *Journal of Economic Entomology, 102*(1), 388–392.

Shetlar, D. J., Suleman, P. E., & Georgis, R. (1988). Irrigation and use of entomogenous nematodes, *Neoaplectana* spp. and *Heterorhabditis heliothidis* (Rhabditida: Steinernematidae and Heterorhabditidae), for control of Japanese beetle (Coleoptera: Scarabaeidae) grubs in turfgrass. *Journal of Economic Entomology, 81*(5), 1318–1322.

Simard, L., Belair, G., & Brodeur, J. (2001). Susceptibility of the European chafer (Coleoptera: Scarabaeidae) to entomopathogenic nematodes (Rhabditida: Steinernematidae, Heterorhabditidae). *Supplement to the Journal of Nematology, 33*(4S), 297–301.

Simard, L., Bélair, G., Gosselin, M. E., & Dionne, J. (2006). Virulence of entomopathogenic nematodes (Rhabditida: Steinernematidae, Heterorhabditidae) against *Tipula paludosa* (Diptera: Tipulidae), a turfgrass pest on golf courses. *Biocontrol Science and Technology, 16*(8), 789–801.

Smith, K. A. (1994). Control of weevils with entomopathogenic nematodes. In K. A. Smith & M. Hatsukade (Eds.), *Control of insect pests with entomopathogenic nematodes* (pp. 1–13). Republic of China in Taiwan: Food and Fertilizer Technology Center.

Stuart, R. J., & Gaugler, R. (1994). Patchiness in populations of entomopathogenic nematodes. *Journal of Invertebrate Pathology, 64*(1), 39–45.

Thurston, G., Kaya, H. K., & Gaugler, R. (1994). Characterizing the enhanced susceptibility of milky disease–infected scarabaeid grubs to entomopathogenic nematodes. *Biological Control, 4*(1), 67–73.

Thurston, G. S., Kaya, H. K., Burlando, T. M., & Harrison, R. E. (1993). Milky disease bacterium as a stressor to increase susceptibility of scarabaeid larvae to an entomopathogenic nematode. *Journal of Invertebrate Pathology, 61*(2), 167–172.

US Environmental Protection Agency. (2014). *Integrated pest management (IPM) in schools.* http://www.epa.gov/opp00001/ipm/ [revised 4/11/2014].

Vittum, P. J., Villani, M. G., & Tashiro, H. (1999). *Turfgrass insects of the United States and Canada*. Ithaca, NY: Cornell University Press.

Wilson, M., & Gaugler, R. (2004). Factors limiting short–term persistence of entomopathogenic nematodes. *Journal of Applied Entomology, 128*(4), 250–253.

Wilson, M. J., Lewis, E. E., Yoder, F., & Gaugler, R. (2003). Application pattern and persistence of the entomopathogenic nematode *Heterorhabditis bacteriophora. Biological Control, 26*(2), 180–188.

Wu, S., Youngman, R. R., Kok, L. T., Laub, C. A., & Pfeiffer, D. G. (2014). Interaction between entomopathogenic nematodes and entomopathogenic fungi applied to third instar southern masked chafer white grubs, *Cyclocephala lurida* (Coleoptera: Scarabaeidae), under laboratory and greenhouse conditions. *Biological Control, 76*(1), 65–73.

Chapter 13
It Takes a Village: Entomopathogenic Nematode Community Structure and Conservation Biological Control in Florida (U.S.) Orchards

Raquel Campos-Herrera, Fahiem E. El-Borai, and Larry W. Duncan

13.1 Root Weevils in Florida Citrus, Blueberries and Peaches

Root weevils in Florida comprise several species, *Diaprepes abbreviatus* L. (Coleoptera: Curculionidae), *Pachnaeus litus* Schoenherr (Coleoptera: Curculionidae), *Pachnaeus opalus* Schoenherr (Coleoptera: Curculionidae), *Artipus floridanus* Horn (Coleoptera: Curculionidae), *Myllocerus undecimpustulatus* Faust (Coleoptera: Curculionidae) and *Naupactus godmani* (Crotch) (Coleoptera: Curculionidae) being those which are commonly encountered in citrus orchards. The first two of these species are the most economically important arthropod pests of citrus other than *Diaphorina citri* Kuwayama (Hemiptera: Psyllidae), which vectors the devastating disease huanglongbing, *Candidus liberobacter asiaticus* Jagoueix (Rhizobiales: Rhizobiaceae) (Duncan, McCoy, Stansly, Graham, & Mizell, 2001; McCoy, 1999). We focus here on the most widely studied of these weevils, *D. abbreviatus* (commonly called *Diaprepes* root weevil); however, all of the species have similar life cycles, host plant interactions, and susceptibility to entomopathogenic nematodes (EPNs).

R. Campos-Herrera
University of Neuchâtel, Neuchâtel, Switzerland
e-mail: raquel.campos@unine.ch

F.E. El-Borai
Citrus Research and Education Center, University of Florida, (US), Lake Alfred, FL, USA

Plant Protection Department, Faculty of Agriculture, Zagazig University (Egypt), Ash Sharqia Governorate, Egypt
e-mail: fahiem@ufl.edu

L.W. Duncan (✉)
Citrus Research and Education Center, University of Florida, (US), Lake Alfred, FL, USA
e-mail: lwduncan@ufl.edu

© Springer International Publishing Switzerland 2015
R. Campos-Herrera (ed.), *Nematode Pathogenesis of Insects and Other Pests*, Sustainability in Plant and Crop Protection 1, DOI 10.1007/978-3-319-18266-7_13

Fig. 13.1 Damage to structural roots of young citrus tree (**a**) and lower trunk of blueberry (**b**) caused by feeding of *Diaprepes abbreviatus* larvae. The channels caused by feeding on the cortical tissue provide infection sites for *Phytophthora* spp.

The *Diaprepes* root weevil is a highly polyphagous native of the Caribbean that was first detected in Florida in the mid–1960s. Adult weevils feed on young leaves where the females lay their eggs in masses between two leaves secured by an adhesive substance. Upon hatching, the neonate larvae fall to the soil where they will develop through pupation during several months. Adults emerge from the pupal chambers and from soil throughout the warm months, but peak emergence occurs in spring and sometimes again in autumn. In citrus, the larvae feed on progressively larger roots as they develop and grow. Wounds created by root feeding provide access for plant parasitic *Phytophthora* spp., creating a pest–disease complex (Graham, Bright, & McCoy, 2003). On larger roots, the larvae feed on the outer bark, including the cambium. Feeding causes deep grooves in the root that do not heal but merely form callus along the margins of the groves (Fig. 13.1a). In habitats that support large weevil populations, the damage can cause widespread tree decline and death in just 1 or 2 years. Even where weevil populations are small enough to go unnoticed by growers, the damage accumulates and older orchards can suddenly exhibit widespread tree decline when root damage reaches a threshold.

Soil applied, halogenated hydrocarbon insecticides provided adequate management of *Diaprepes* root weevil until they were deregistered. Subsequently, the weevil became a major pest because the continuous cycling of weevil stages between canopy and soil is difficult to disrupt with pesticides that have short residual activity in the tree canopy. There are no effective pesticides registered for managing larvae in soil. As a consequence, biological control of *Diaprepes* root weevil larvae in citrus soil has been actively studied for more than 30 years (Beavers, McCoy, & Kaplan, 1983; McCoy & Boucias, 1989; see Sect. 13.4 in this chapter) and the importance of native entomopathogenic nematodes (EPNs) in modulating the occurrence of root weevils in Florida became evident during the course of that research (Sect. 13.3). The need to manage root weevils acquired new urgency with the introduction of huanglongbing in Florida. The bacterial disease was detected in the state in 2005 (Halbert, 2005) and has reduced the citrus acreage and fruit production by one–third in less than a decade. Infected trees lose as much as 40 % of their fibrous roots prior to development of aboveground symptoms (Graham, Johnson, Gottwald, & Irey, 2013). As a result of this root loss, the tolerance of trees to subterranean pests and root diseases has decreased substantially, and better methods to manage concomitant root problems are urgently needed.

Although citrus continues to dominate Florida agriculture, newly introduced perennial fruit crops are increasing in importance. High yielding peach and blueberry cultivars adapted to low chill conditions of subtropical winters are now grown on more than 2,000 and 800 ha, respectively. Both crops are reported to be suitable hosts for *Diaprepes* root weevil (Olmstead et al., 2013; Williamson, Lyrene, & Olmstead, 2012) and sporadic, but serious damage by weevils is known to occur (Fig. 13.1b). As these new crops are increasingly planted into former citrus orchards or adjacent to weevil infested, declining citrus orchards, they may become preferred hosts of the pests. Moreover, peach root and stem borer, *Synanthedon exitosa* Say (Lepidoptera: Sesiidae), and several root–knot nematodes (*Meloidogyne* spp.) are prevalent pests of peach that may reduce the tolerance of trees to weevil injury. Although blueberry has fewer known serious root pests and diseases in Florida, both it and peach are susceptible to root and crown rot (*Phytophthora* spp.) that is greatly facilitated in citrus by root weevil feeding and various EPN species have been studied for control of beetle pests of blueberries in New Jersey and California (Haviland & Hernandez, 2012; Polavarapu, Koppenhoefer, Barry, Holdcraft, & Fuzy, 2007). Both crops may benefit from the diverse, abundant EPN communities found throughout much of the peninsula (Campos-Herrera, Pathak, El–Borai, Stuart et al., 2013), because perennial crops tend to favor EPN prevalence (Tarasco et al., 2014). Nevertheless, the effects of different cultural practices and crop germplasm on subterranean food webs are mostly unknown (Campos-Herrera, Pathak, El–Borai, Schumann et al., 2013). For example, blueberry soil pH is adjusted to levels (4.0–5.5) well below those common in Florida orchard soils, and the effects of practices such as these on native EPNs and natural biological control of weevils remain to be determined (Williamson et al., 2012).

13.2 Native Entomopathogenic Nematode Diversity and Spatial Structure

The biogeography of EPN species has been the focus of several reviews (Adams et al., 2006; Hominick, 2002; Hominick, Reid, Bohan, & Briscoe, 1996). EPN occurrence is global, both in natural and agricultural areas, except in Antarctica (Griffin, Downes, & Block, 1990). The known distribution of many species is restricted to the type locality (Adams & Nguyen, 2002; Hominick, 2002), whereas others, such as *Steinernema feltiae* (Filipjev) Wount, Mráček, Gerdin & Bedding (Rhabditidae: Steinernematidae), and *Steinernema carpocapsae* (Weiser) Wount, Mráček, Gerdin & Bedding (Rhabditidae: Steinernematidae), are widely dispersed. In addition to the various adaptive abilities of EPN species, our knowledge of EPN biogeography is heavily biased by historically strong sampling efforts in Europe and North America compared to virtually none in vast areas of Africa, South America, and Asia (Campos-Herrera, Barbercheck, Hoy, & Stock, 2012). The global market for commercial EPN products could confound efforts to understand the natural occurrence of EPN if exotic species establish sympatry with natives, although there is no evidence that introduced EPNs have displaced native species (see Chap. 10). Sampling methodology also affects EPN detection. Whereas all life stages are easily retrieved by the widely utilized insect–bait technique (Bedding & Akhurst, 1975), methods that isolate nematodes from soil (sieving and decanting or sucrose–centrifugation) recover just the infective juvenile stages (IJs) which are difficult to identify accurately based on morphology. Alternatively, real time qPCR can be used to identify and quantify IJs and primer/probe sets are now available for more than 20 tropical and temperate species (Campos-Herrera et al., 2015; Campos-Herrera, Ali, Díaz, & Duncan, 2013; Torr, Spiridonov, Heritage, & Wilson, 2007). This technique is especially powerful for measuring nematode community diversity. Indeed, qPCR was used recently to show that the progeny from single cadavers are frequently mixtures of several EPN and free living nematode species (Campos-Herrera et al., 2015).

Hominick (2002) disputed early studies of EPN biogeography that portrayed steinernematids as generally associated with temperate regions as opposed to heterorhabditids purported to occur mostly in warmer areas. Rather, he emphasized later data suggesting that EPN distribution depends more on species than genus. Reproductive strategy (amphimixis *versus* hermaphroditism) is the salient behavioral difference between the two genera that have otherwise converged so remarkably. However, variation within each genus of traits such as morphology (size, sheath retention, etc.), survival strategy, foraging behavior, and host specialization are more likely to drive their speciation and global distribution (Hominick, 2002; Tarasco et al., 2014). These traits represent adaptation not only to specific habitats but also to the stability of those habitats. Stability is thought to favor the diversity of biological control organisms in nature (Fuxa, 1995). At present, the trend for EPN in terms of climatic stability is the opposite, with countries such as Czech Republic, Italy, Germany, or USA accounting for the highest EPN species richness (Adams et al., 2006; Campos-Herrera, Barbercheck et al., 2012; Hominick, 2002). However, this pattern could change as EPN are studied more intensively in long neglected tropical areas, as was recently shown for the Phylum Nematoda (Porazinska, Giblin-Davis, Powers, & Thomas, 2012). In addition to climate, cropping systems affect habitat stability with respect to the crop longevity and its cultural practices. Forests, timber lots, pastures, orchards, and other perennial systems provide edaphic stability that does not exist in annual cropping systems, especially those in which several crops are produced annually (Tarasco et al., 2014; Fig. 13.2). Pesticide application intensity may also affect soil communities. In La Rioja, Spain, the natural occurrence of EPN was higher in natural areas and organic orchards and vineyards than in conventional orchards and vineyards. EPN were not detected in fields of organic or conventional annual crops (Campos-Herrera et al., 2008; Campos-Herrera, Piedra-Buena, Escuer, Montalban, & Gutiérrez,

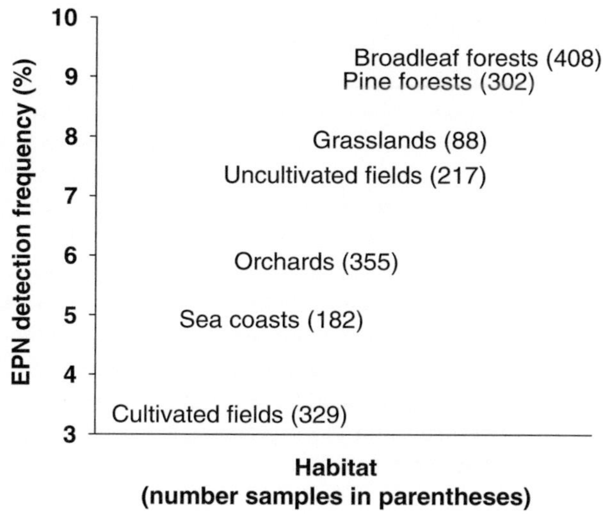

Fig. 13.2 The relationship between habitat and the detection frequency of entomopathogenic nematodes surveyed in Italy during 1991–2010 (Redrawn with permission from Tarasco et al. (2014))

2010). Tillage also adversely affected some EPN species in the southeastern and central USA (Lawrence, Hoy, & Grewal, 2006; Millar & Barbercheck, 2001; 2002) and EPN were rarely encountered in long–term tillage trials in Switzerland (Campos-Herrera et al., 2015). Increased EPN abundance is frequently measured on the undisturbed borders of tilled fields, further demonstrating the potential of agricultural practices to conserve biocontrol services provided by EPN (Campos-Herrera et al., 2007; Lawrence et al., 2006).

For any cropping system in a given climatic zone, differences in agronomic/horticultural practices and differences in soil physical properties also affect the well–being of EPN in species–specific ways. Indeed, with the exception of tillage, edaphic properties are generally found to surpass crop management practices in their effects on nematode distribution (Neher, 1999). Fine textured soils, especially those with abundant clay content, generally support less EPN activity than do sandy or loamy soils (Barbercheck, 1992; Campos-Herrera & Gutiérrez, 2009; Duncan et al., 2013; El-Borai, Stuart, Campos-Herrera, Pathak, & Duncan, 2012; Hara, Gaugler, Kaya, & Lebeck, 1991; Hazir, Keskin, Stock, Kaya, & Özcan, 2003; Kaspi et al., 2010). Soil water potential (Duncan & McCoy, 2001), salinity (Nielsen, Spence, Nakatani, & Lewis, 2011), pH (Kung, Gaugler, & Kaya, 1990), the concentration of fertilizer elements and pesticide residues (Moore et al., 2013) are just a few of the edaphic variables shown to influence EPN (Barbercheck & Duncan, 2004). Multivariate and/or spatial analytical methods are used increasingly to identify those soil properties that may be most important in modulating the natural distribution of EPN. Canonical correspondence analysis (CCA) and redundancy analysis (RA) are two of several ordination techniques long used by ecologists to reduce large numbers of environmental variables to just the few that best explain community variation across space or time. Canonical correspondence analysis revealed P and K content and C:N ratio from among 16 measured edaphic factors as strong predictors of EPN abundance in 600 samples from six Ohio USA habitats (Hoy, Grewal, Lawrence, Jagdale, & Acosta, 2008). Elevation and soil pH were related to EPN distribution in a CCA of 251 sites across Cameroon (Kanga, Waeyenberge, Hauser, & Moens, 2012). Whereas ordination methods measure relationships of variables between samples, Spatial Analysis by Distance Indices (SADIE) is a method requiring georeferenced data that provides stochastic estimates of both the spatial aggregation of variables (e.g., Stuart, Barbercheck, Grewal, Taylor, & Hoy, 2006) and whether the spatial patterns of two variables are associated or dissociated at one or more points in time (Perry & Dixon, 2002; Perry & Hewiit, 1991). Wilson, Lewis, Yoder, and Gaugler (2003) used SADIE to characterize the spatial patterns of EPNs that subsequently developed when they were applied either in patches or uniformly to soil. *Steinernema feltiae* and *Steinernema affine* (Artykhovsky) Wount, Mráček, Gerdin & Bedding (Rhabditidae, Steinernematidae) were shown by SADIE to be highly aggregated and closely associated in soil (Spiridonov, Moens, & Wilson, 2007). The highest aggregation was exhibited by younger IJs, whereas older IJs (determined by lipid content and retention of second stage cuticle) were randomly distributed in soil. SADIE revealed significant spatial associations between EPN and nematophagous fungi that prey on EPN (Campos-Herrera, El-Borai, & Duncan, 2012). EPN have also been shown by SADIE and other analyses to be naturally associated at both local and landscape scales with certain free living nematodes that share similar life history traits and which may even compete with EPN for resources in the insect cadaver (Campos-Herrera, El-Borai et al. 2012; Campos-Herrera, Pathak, El-Borai, Stuart et al., 2013; Park, Jagdale, Cho, Grewal, & Hoy, 2014).

Identifying habitat characteristics that are conducive to the occurrence and/or biological control efficacy of particular EPN species offers the possibility of developing cultural practices that conserve and enhance pest management provided by naturally occurring EPN (see Chap. 6). However, the potential importance of EPN conservation is difficult to assess because little is known about the relative contribution of EPN to the natural regulation of economically important arthropod pests in most agricultural systems. The frequency with which EPN are detected in soil samples is often very low, suggesting that naturally occurring EPN contribute relatively little to pest regulation in many systems (Fig. 13.2; Campos-Herrera et al., 2015; Kanga et al., 2012; Kaspi et al., 2010; Lawrence et al., 2006; Malan, Knoetze, & Moore, 2011; Millar & Barbercheck, 2001, 2002). Still, nothing is known about the relationship between detection frequency or IJ abundance and the rate of arthropod control for any sampling method or soil habitat. Some understanding of these relationships is necessary because implementing cultural practices to conserve or enhance EPN services will likely require an economic investment by growers (Duncan et al., 2013). Cropping systems in which native EPN are known to be key mortality factors for important pests would likely provide the greatest opportunity to identify and implement profitable conservation tactics. A method proposed to identify such systems is to study those in which the importance of a pest species varies predictably by location, perhaps as a result of natural enemies (Stirling, McKenry, & Mankau, 1979). The classic example of such an approach with EPN was the search for biological control agents of *Scapteriscus* mole crickets in Uruguay where *Steinernema scapterisci* Nguyen & Smart (Rhabditida. Steinernematidae) was discovered and shown to be one of several antagonists that severely regulated mole crickets in that country compared to countries where the pest was introduced without its antagonists (Dolinski, Choo, & Duncan, 2012). A somewhat analogous situation involves citrus root weevils in orchards across the Florida peninsula.

13.3 Native Entomopathogenic Nematode Efficacy Against Root Weevils

In Florida, citrus is grown in two ecoregions, the *central ridge* and the *flatwoods*. Both regions have very sandy soil (often 85–98 % sand) although flatwoods soils are somewhat finer textured. Trees in flatwoods orchards are frequently grown on bedded soil to provide adequate rooting volume because soils are poorly drained due to shallow (<1 m) groundwater. By contrast, the root system depths of trees on the deep, coarse sands of the central ridge frequently exceed 5 m. The size of root weevil populations and the damage they cause to citrus trees differs in the two regions (Futch, Duncan, & Zekri, 2005). Root weevil population density can be high enough is some flatwoods orchards to kill large numbers of young trees, especially in poorly drained areas (McCoy, 1999). Root weevils are less abundant and often go undetected for many years in orchards on the central ridge. Nevertheless, the continuous damage to roots caused by few larvae on the central ridge accumulates and can eventually debilitate trees many years old.

The survival rate of *Diaprepes* weevils in soil is very low. McCoy, Stuart, and Nigg (2003) estimated survival to range between 0.7 and 1.6 % from the time neonate larvae fall to the ground until adult weevils emerges from soil. Indigenous EPN were first reported to be major subterranean enemies of *Diaprepes* root weevils in citrus orchards by Beavers et al. (1983) who found as high as 70 % infection by nematodes of caged larvae that were buried for three weeks beneath citrus trees in central Florida. Beavers et al. also reported that *Diaprepes* larvae were killed by EPNs in 45 % of soil samples from 55 Florida orchards and noted the unusually high rate of EPN incidence in Florida compared to that from his concurrent survey in Puerto Rico that detected almost no EPN (Roman & Beavers, 1983). McCoy, Shapiro, Duncan, and Nguyen (2000) reported average weekly weevil parasitism rates by EPNs to range from 11 to 33 % at three orchard sites in the flatwoods, concluding that EPN and ants were the key weevil mortality factors in these soils. Duncan et al. (2003) monitored the weekly mortality rates of caged, buried *Diaprepes* root weevil larvae in an orchard on the central ridge and another in the flatwoods. During 2 years, native *Steinernema diaprepesi* Nguyen & Duncan (Rhabditida: Steinernematidae), *Heterorhabditis indica* Poinar, Kuranakar & David (Rhabditida: Heterorhabditidae), and *Heterorhabditis zealandica* Poinar (Rhabditida: Heterorhabditidae), killed as many as 50 % of the buried larvae each week at the central ridge site, whereas only *H. indica* was detected in the flatwoods orchard and mortality caused by the nematode never exceeded 8 %. Because the central ridge site had few weevils and the site in the flatwoods had abundant weevils, Duncan et al. speculated that the regional patterns of *Diaprepes* root weevil abundance might be caused in part by natural control exerted by native EPN which appeared to be more abundant and species diverse on the central ridge.

The possibility that *Diaprepes* abundance in Florida is modulated by EPN suggests an opportunity to increase rates of natural weevil control by identifying habitat features that affect the insecticidal efficacy of key EPN species, either directly or by their influence on natural enemies of EPN. Campos-Herrera, Pathak, El-Borai, Stuart et al. (2013) designed or utilized more than 20 sets of molecular primers and probes to explore this possibility by using real time quantitative PCR (qPCR) to characterize food webs in orchards comprising diverse habitats in both ecoregions. A major advantage of characterizing soil communities using qPCR is that the method permits the measurement of DNA from microorganisms recovered in nematode samples. Some of these microorgnisms are only facultatively nematophagous, existing primarily as saprophytes. Traditional methods of culturing and measuring these organisms in soil cannot distinguish between saprophagy and nematophagy (Jaffee, Strong, & Muldoon, 1996). Even when cultured from nematode samples, species recovery is biased by the culturing conditions (Duncan et al., 2007). Direct measurement of DNA recovered from samples of nematodes increases the likelihood that the detected organisms are from DNA on or inside of nematodes and that quantification is not influenced by culturing conditions (Pathak et al., 2012). Molecular sets used by Campos-Herrera, Pathak, El-Borai, Stuart et al. (2013) included those to measure 13 species of EPN (Campos-Herrera, Johnson et al., 2011; Campos-Herrera, Pathak, El-Borai, Stuart et al. 2013; Campos-Herrera, El-Borai, Stuart, Graham, & Duncan, 2011), two ectophoretic bacteria, *Paenibacillus* sp. and *Paenibacillus nematophilus* Enright, McInerney & Griffin (Bacillales: Paenibacillaceae), which limit the mobility of the EPN infective juveniles (Campos-Herrera, El-Borai et al. 2011; El-Borai, Duncan, & Preston, 2005; Enright & Griffin, 2005), the free–living nematodes (FLN) *Acrobeloides*–group, reported to compete with the EPNs for the cadaver (Campos-Herrera, El-Borai & Duncan, 2012), and *Phytophthora nicotianae* Breda de Haan (Pythiales: Phythiaceae) the citrus pathogen associated with *D. abbreviatus* root damage (Graham et al., 2003; Huang, Li, Xiao, & Wang, 2010). Pathak (2014) also evaluated the role of seven nematophagous fungi (NF) (Atkins, Clark, Pande, Hirsch, & Kerry, 2005; Pathak et al., 2012; Zhang, Liu, Zhu, & Chen, 2006) in the same samples. More than 50 abiotic properties were also measured at each site. At least one of seven EPN species were detected at each of the 53 citrus orchards sampled. Four EPN species were encountered frequently in either the central ridge or the flatwoods or in both ecoregions. The survey supported the speculation (Duncan et al., 2003) that EPN diversity and species richness are greater on the central ridge than in the flatwoods. Moreover, SADIE analysis confirmed that the majority of commonly encountered EPN species had strong affinities for either ecoregion – *S. diaprepesi* and *H. zealandica* for the central ridge and an undescribed *Steinernema* sp. for the flatwoods. Only *H. indica,* the most frequently encountered species, occurred independently of these regions. However, in contrast to speculation that EPNs are more abundant on the central ridge, Campos-Herrera, Pathak, El-Borai, Stuart et al. (2013) found no difference in the average EPN abundance in either region.

Compared to other reports (e.g., Fig. 13.2), the ubiquitous occurrence of EPN encountered by Campos-Herrera, Pathak, El-Borai, Stuart et al. (2013) confirmed that Florida citrus orchards are especially favorable for EPN communities. This view is reinforced by results of a second survey of EPN in 91 natural areas adjacent to Florida orchards in which, although seven species including those that are dominant in orchards were detected, just 62 % of the sites contained detectable EPN (Fig. 13.3a; Campos-Herrera et al., unpublished). The differences in detection frequency suggest that the greater diversity of vegetation in the natural areas produced habitats that varied in suitability for EPN in contrast to the more uniformly favorable orchard conditions. Indeed, the abundance of EPN when detected in natural areas was similar to that in orchards (not shown). But do EPN modulate root weevil abundance differently on the central ridge than in the flatwoods and what is the mechanism, if EPN are not more abundant on the central ridge? Results of laboratory studies indicate that EPN efficacy varies regionally because of differences in both the EPN communities (species composition and species diversity) and the soils. The two commonly encountered Steinernematids, *S. diaprepesi* and *Steinernema* sp., killed weevil larvae and protected citrus seedlings much more

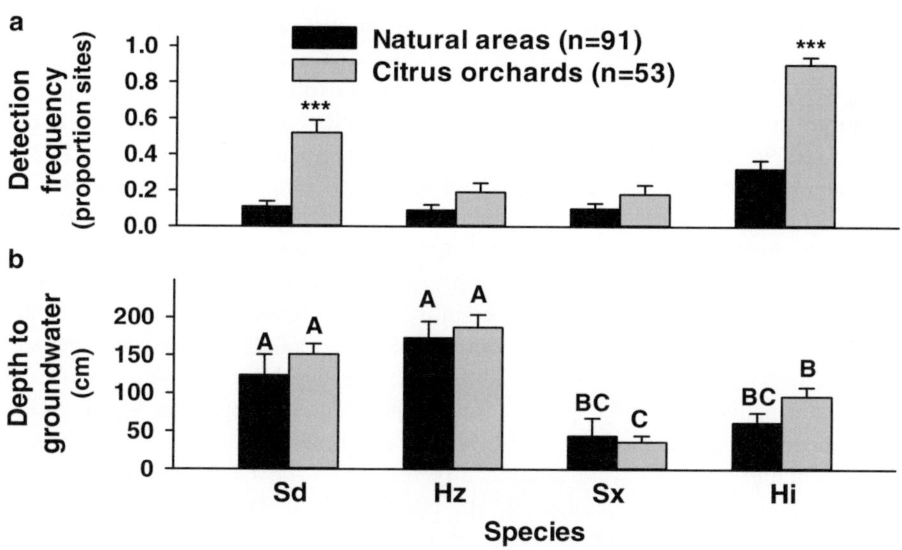

Fig. 13.3 Florida citrus orchards are generally more conducive to the occurrence of EPNs than the more heterogeneous natural habitats (**a**), whereas groundwater depth (soil moisture) is intimately related to the occurrence of the most commonly encountered native EPNs in both natural areas and citrus orchards (**b**). Sd *Steinernema diaprepesi*, Hz *Heterorhabditis zealandica*, Sx *Steinernema* sp., Hi *H. indica*

effectively in pot studies than did *H. indica* or *H. zealandica* (El-Borai et al., 2012; El-Borai & Duncan, 2007). Whereas these effective steinernematids were nearly always detected in orchards on the central ridge, they occurred in fewer than half of the orchards in flatwoods. Orchards on the central ridge are also more likely than those in flatwoods to have two or three evenly abundant species of EPN, whereas *H. indica* dominates EPN communities in flatwoods and is the only species in nearly half of the orchards there (Campos-Herrera, Pathak, El-Borai, Stuart et al., 2013; Duncan et al., 2003, 2013). Jabbour, Crowder, Aultman, and Snyder (2011) found that EPN species richness increases their biocontrol efficacy, much as has been shown for arthropod predators if they are not mutually antagonistic (Snyder, Snyder, Finke, & Straub, 2006). Finally, the coarser texture of soil on the central ridge was shown by El-Borai et al. (2012) to be more conducive than finer textured flatwoods soils to EPN efficacy against root weevil larvae and this relationship between texture and EPN efficacy is supported by a great deal of research (see Chap. 4).

Controlled experiments are needed to understand why EPN communities form in characteristic ways in these different habitats. Duncan et al. (2007) showed that the application of composted animal manure reduced the incidence of some nematophagous fungal species (NF) at the same time it increased the incidence of native EPNs recovered from sentinel weevil larvae. During 2 years in that study, the NF spatial patterns were inversely related to some EPN species and positively related to others. In the laboratory, aquatic endoparasitic NF were less virulent to *H. indica* than to other EPN species and trapping fungi were less virulent to *S. diaprepesi* (El-Borai, Campos-Herrera, Stuart, & Duncan, 2011). Results of these two studies support the hypothesis that abiotic properties from a selected habitat may initiate trophic cascades extending to the health of the trees if the soil properties affect particular NF species that in turn affect EPN species that are more or less virulent to *Diaprepes* root weevil. For example, the generally wetter soil in flatwoods could be more favorable to aquatic endoparasitic NF that attack most EPN species, thereby providing a competitive advantage to the relatively un–preyed upon *H. indica*. Indeed, a redundancy analysis of relationships between abiotic and biotic variables in a survey of 53 citrus orchards revealed three physical properties (electrical conductivity, available water–holding content and K content) that explained significant variability in the abundance of five of the six NF species measured and 65 % of the variation in total NF abundance (Pathak, 2014). However, the analysis revealed no evidence that NF communities affected the spatial patterns of EPN species. Rather, just four of more than 50 measured edaphic properties (depth to groundwater, water holding capacity and clay and organic matter content) were significantly related to the EPN pattern (Campos-Herrera, Pathak, El-Borai, Stuart et al., 2013) (Fig. 13.3b). Because soil water potential is affected by each of these soil properties, ongoing laboratory research measures the responses by each native EPN species to moisture–related properties (soil water potential, desiccation tolerance, hypertonic stress) that characterize the two ecoregions. The different responses of the EPN species to these various physical conditions (El-Borai et al., unpublished) strongly support the causality between water potential and EPN spatial patterns derived by Campos-Herrera, Pathak, El-Borai, Stuart et al. By encompassing additional approaches such as comparative proteomic analysis for each species–habitat combination (Fig. 13.4), this research seeks not only to identify cultural practices that can maintain or increase the services of specific EPN (Sect. 13.5), but also to reveal some of the physiological processes by which closely related species such as *S. diaprepesi* and *Steinernema* sp. have adapted differently to different habitats.

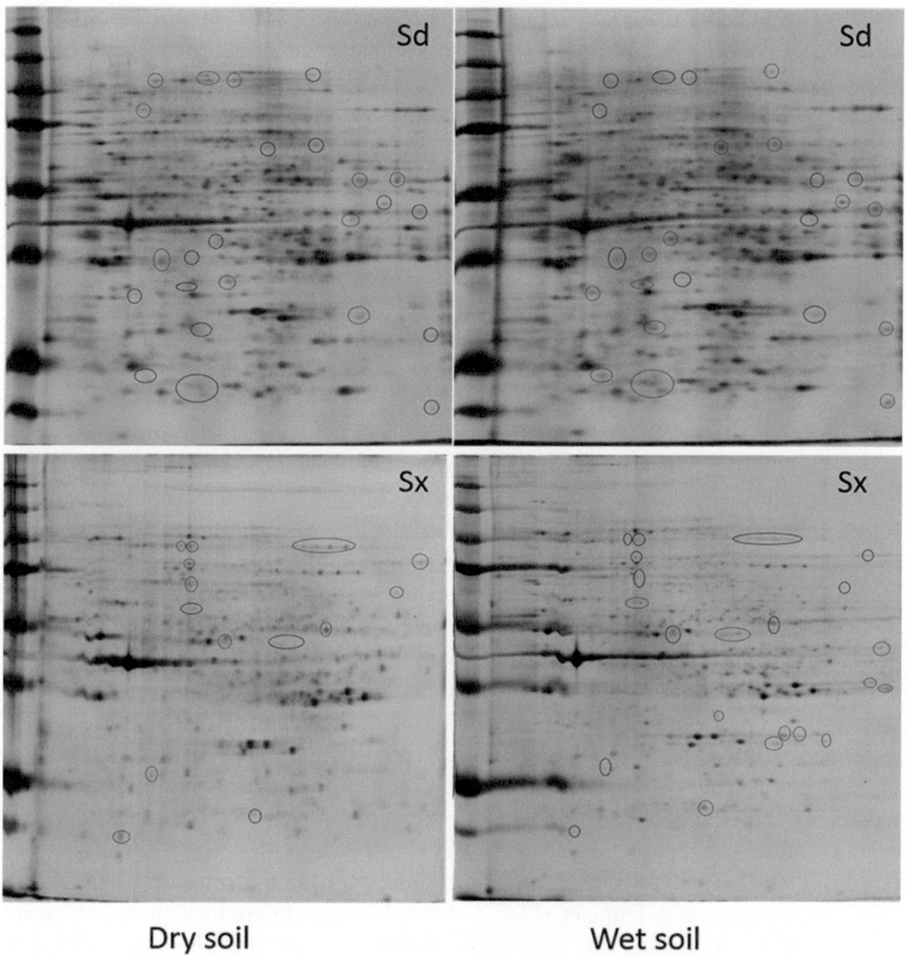

Fig. 13.4 Comparative proteomic analysis revealing differences in protein expression by two closely related species (Sd *Steinernema diaprepesi*, Sx *Steinernema* sp.) after 48 h in well–drained (6 % moisture) and saturated (18 % moisture) sandy soil in the laboratory. Detected proteins are being identified by LC–MS–MS to understand mechanisms of habitat adaptation. These traits may be useful in guiding the selection of EPN species for use in different ecoregions. *Circled spots* are proteins that are over– or under–expressed in one moisture condition compared to the other

13.4 The Efficacy and Post-application Biology of Commercial Entomopathogenic Nematodes: Non-target Effects

Following the loss of persistent organochlorine pesticides, one of the few methods recommended to control weevil larvae in Florida citrus is periodic application of EPNs (Bullock, Pelosi, & Killer, 1999; Duncan et al., 2007; Duncan & McCoy, 1996; Duncan, McCoy, & Terranova, 1996; McCoy, Stuart, Duncan, & Nguyen, 2002). The first commercial products used *S. carpocapsae, Heterorhabditis bacteriophora* Poinar (Rhabditida: Heterorhabditidae) and *H. indica* (Adair, 1994; Figueroa & Roman, 1990; Schroeder, 1987), but *Steinernema riobrave* Cabanillas, Poinar & Raulston (Rhabditida: Steinernematidae), was later shown to provide greater control of *D. abbreviatus* than these other EPNs (Bullock et al., 1999; Duncan et al., 1996; Schroeder, 1994). Larval suppression in the field ranged from 46 % at recommended rates (25 IJs/cm^2 soil surface) (Duncan et al., 2003) to 90 % at higher rates (Bullock et al., 1999; Duncan et al., 1996; Duncan & McCoy, 1996). Use of *S. riobrave* at recommended rates also reduced adult emergence from soil during 12 months by half (Duncan et al., 2007). Nevertheless, the efficacy of *S. riobrave* for weevil control has varied widely due to factors such as product quality (producers and products have changed over time), application rates and methods, and edaphic conditions (Jenkins et al., 2008; Jenkins, Shapiro-Ilan, & Goenaga, 2007, McCoy et al., 2000, 2002). EPN use in Florida citrus peaked in 1999 when ca. 19,000 ha of citrus were treated with *S. riobrave* (Shapiro-Ilan, Gouge, & Koppenhöfer, 2002). Thereafter, reduced product quality undermined confidence in and use of commercial EPNs. The eventual reintroduction of high quality EPN products in Florida closely preceded the arrival of huanglongbing, a disease which drastically increased the cost of pest management, reduced fruit production and thereby further reduced the market for EPN products, which were withdrawn in 2011 (Dolinski et al., 2012). The recent discovery that huanglongbing kills much of the citrus root system before symptoms occur aboveground has rekindled the concern about root weevils and created new demand for EPN products (Duncan, 2014).

Commercial EPN generally function in much the same way as do chemical insecticides – they kill insects for a short time following application, after which reapplication is necessary. Whereas modern chemical pesticides degrade relatively quickly after application, augmented EPN are soon killed by natural enemies (Ishibashi & Kondo, 1986; Koppenhöfer, Jaffee, Muldoon, Strong, & Kaya, 1996). El-Borai, Brentu, and Duncan (2007) showed in the laboratory that EPN were killed at higher rates and NF were more abundant when the EPN were added to orchard soil previously augmented with EPN than when added to virgin soil. This observation suggests that the abundance of native EPN (and insect control) can be reduced by natural enemies that increase when commercial nematodes are applied to soil. Some evidence for this non–target effect was reported by Duncan et al. (2003, 2007) who found that sentinel weevils were killed at higher rates by EPNs in field plots immediately following EPN amendments, but afterwards for several weeks

NF increased and sometimes sentinel weevil mortality declined compared to that in non–amended plots (Duncan et al., 2007). Such a non–target effect is unlikely to be of practical concern except in habitats where native EPN are especially active, such as on the central ridge where reduced activity by native EPN could obviate arthropod control achieved by commercial EPN (Duncan et al., 2003). Regardless, the non–target effect of commercial EPN on native EPN is brief and should not be especially detrimental unless EPN amendments occur immediately before a major period of arthropod recruitment into the soil (Duncan et al., 2007).

13.5 Creating Habitats to Exploit the Services of Native Entomopathogenic Nematodes

The possibility of exploiting EPN services by selecting or developing cultural practices that favor biological control by these nematodes has been the subject of some speculation (Barbercheck & Hoy, 2005; Lewis, Campbell, & Gaugler, 1998; Stuart et al., 2006; Stuart, El-Borai, & Duncan, 2008; Toepfer et al., 2009) and limited research. A recent study of biological control in urban habitats (vacant lots and gardens) revealed resilient food webs that provided considerable biological control activity which varied significantly by organism (ants, microbial, EPN) and land use (Yadav, Duckworth, & Grewal, 2012). Because habitat heterogeneity helps conserve biological control organisms aboveground, Jabbour and Barbercheck (2008) studied persistence and movement of EPN in plots with one or several plant species and concluded that soil bulk density and water potential, and plant (root) density had greater influence than the complexity of plant species. A similar observation involved increased biological control by EPN when 'root–routeways' connected insects in soil compared to soil in which simulated roots (twigs) were absent (Ennis, Dillon, & Griffin, 2010). Moore et al. (2013) adjusted an IPM program to avoid nematicide use and thereby dramatically increased natural control of false codling moth, *Thaumatotibia leucotreta* Meyrick (Lepidoptera, Tortricidae) by *H. zealandica*. Simply selecting the species best able to persist in a given habitat can sometimes conserve the beneficial effects of augmenting EPN (Williams, Fickle, Grewal, & Dutcher, 2010).

Animal manure mulch affected EPN to some extent in several reports. Bednarek and Gaugler (1997) detected more abundant EPN in plots of rotated field crops amended with animal manure compared to plots with chemical fertilizer. In Florida citrus, higher than recommended rates of composted manure had a similar positive affect on EPN abundance (Duncan et al., 2007). Although the manure decreased EPN antagonists (i.e., NF) in citrus soils, an equally plausible mechanism for increased EPN is the likelihood that manure increased the numbers of arthropod hosts of EPN as speculated by Bednarek and Gaugler (1997). When a much lower, recommended rate of pelleted chicken manure was used to amend orchard soil under organic management, EPN were not greatly affected although *H. zealandica* and free living nematodes increased occasionally (Campos-Herrera, El-Borai, & Duncan, 2015).

An advanced citrus production system (APS) that employs daily fertigation has been shown to be highly resource efficient and effective at bringing young trees into early production compared to conventional citriculture methods (CC). However, the increased frequency with which APS supplies water and nutrients to trees produces major changes in the soil which must alter the dynamics of soil food webs. In an orchard on the central ridge, Campos-Herrera, Pathak, El-Borai, Schumann et al. (2013) and Campos-Herrera, El-Borai, Ebert, Schumann, and Duncan (2014) showed that APS reduced native *S. diaprepesi* while increasing the prevalence of the disease complex comprising citrus root weevils (*D. abbreviatus* and *A. floridanus*) and the root pathogen *P. nicotianae* (Fig. 13.5). Wetter soil under APS may have reduced *S. diaprepesi*, which appears to be better adapted to drier soil conditions and less well adapted to wetter conditions than some species (Campos-Herrera, Pathak, El-Borai, Stuart et al., 2013; El-Borai, unpublished). Equally intriguing

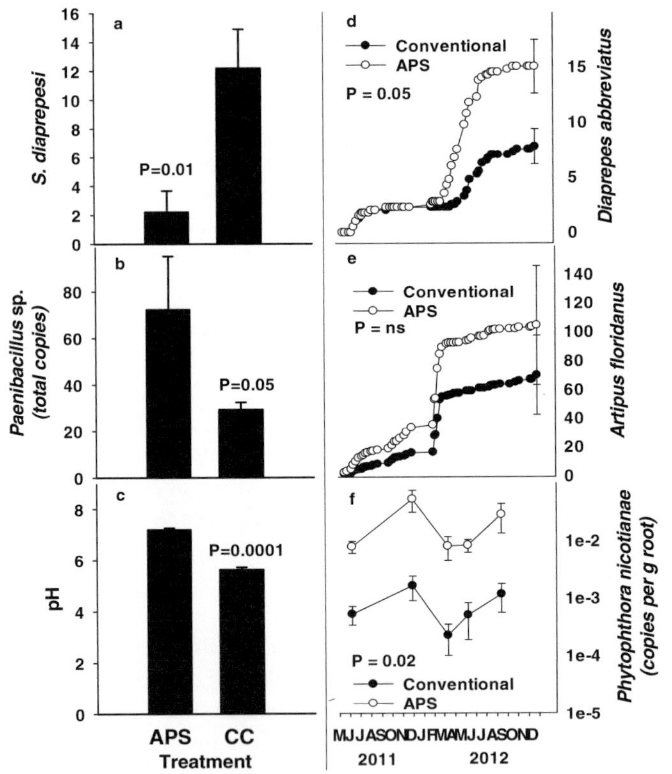

Fig. 13.5 The effects of an advanced citrus production system (*APS*) compared to a conventional citriculture system (*CC*) on the abundance of *Steinernema diaprepesi* (**a**), the bacterium *Paenibacillus* sp. that is a phoretic ectoparasitic bacteria of *S. diaprepesi* (**b**), soil pH (**c**), the root weevils *Diaprepes abbreviatus* (**d**) and *Artipus floridanus* (**e**), and the plant parasitic oomycete *Phytophthora nicotianae* (**f**) (Reprinted from Campos-Herrera, Pathak, El-Borai, Schumann et al. (2013) and Campos-Herrera et al. (2014) with permission)

is that APS increased soil pH from 5.5 to 7.0 and increased *Paenibacillus* sp. (Fig. 13.5b–c). Laboratory studies have shown that the bacterium readily adheres to the *S. diaprepesi* cuticle at pH 7.0, but not at pH 5.5 (El-Borai, unpublished). Thus, it is possible that, compared to conditions in CC, *S. diaprepesi* mobility is impaired in APS to the extent that fewer weevils are killed and fewer EPN produced. The effects of APS on weevil biological control were great enough that additional studies are warranted in which properties such as soil pH and soil moisture are varied by choices of fertilizers and irrigation methods.

Soil texture has major effects on nematodes because it modulates many other physicochemical attributes such as bulk density (pore size), water holding capacity, and cation exchange. In addition to greater groundwater depth, the coarse texture of soils on the central ridge is an important difference from the finer textured soils of the flatwoods. Some growers use soil from the central ridge to fill tree planting holes in flatwoods orchards to improve drainage. This practice has a high likelihood of altering the natural EPN communities because it alters the porosity and water potential in those orchards. Duncan et al. (2013) monitored soil food webs beneath 100 trees in a flatwoods orchard with very shallow loamy soil, heavily infested with *Diaprepes* root weevil, in which half the trees were planted in holes filled with coarse sand from the central ridge. The orchard was unusually depauperate, containing only *H. indica* that were rarely detected (Duncan et al., 2003). Different species of EPN were introduced beneath some trees in both soils and communities were permitted to equilibrate for 2 years before they were monitored for two additional years. When nematodes were extracted from soil samples and characterized by qPCR, EPN species richness and diversity were always highest in sand (Fig. 13.6). Although soil type had no effect on EPN abundance (Fig. 13.6a), nevertheless, EPN efficacy against caged and buried weevil larvae was higher in sand than in loam (Fig. 13.6b). The numbers of adult weevils that emerged from soil were inversely related to the infection rate of buried weevils and were 68 % greater in loamy than in sandy soil. Tree mortality in sandy soil was 14 % that of trees in loamy soil. The two soil habitats used in this experiment supported EPN communities with virtually the same structure and biocontrol efficacy as those characterized in the different ecoregions surveyed by Campos-Herrera, Pathak, El-Borai, Stuart et al. (2013). More remarkable, is that 2 years after the experiment concluded and 6 years after the trees were planted, nematode samples (El-Borai, unpublished) taken from the shallow soil of this orchard revealed that the nematode communities in both sandy and loamy soils had evolved to become heavily dominated by *H. indica* and *Steinernema* sp. (Fig. 13.7), the two species best adapted to soils with shallow depth to groundwater both in Florida orchards and natural areas (Fig. 13.3b). Despite increasing the species richness (and biological control) in the orchard, the introduction of coarse sand was insufficient to prevent the displacement of species such as *S. diaprepesi* and *H. zealandica* in a habitat with shallow groundwater. Nevertheless, the presence of sand dramatically increased the abundance and efficacy of the two locally adapted species.

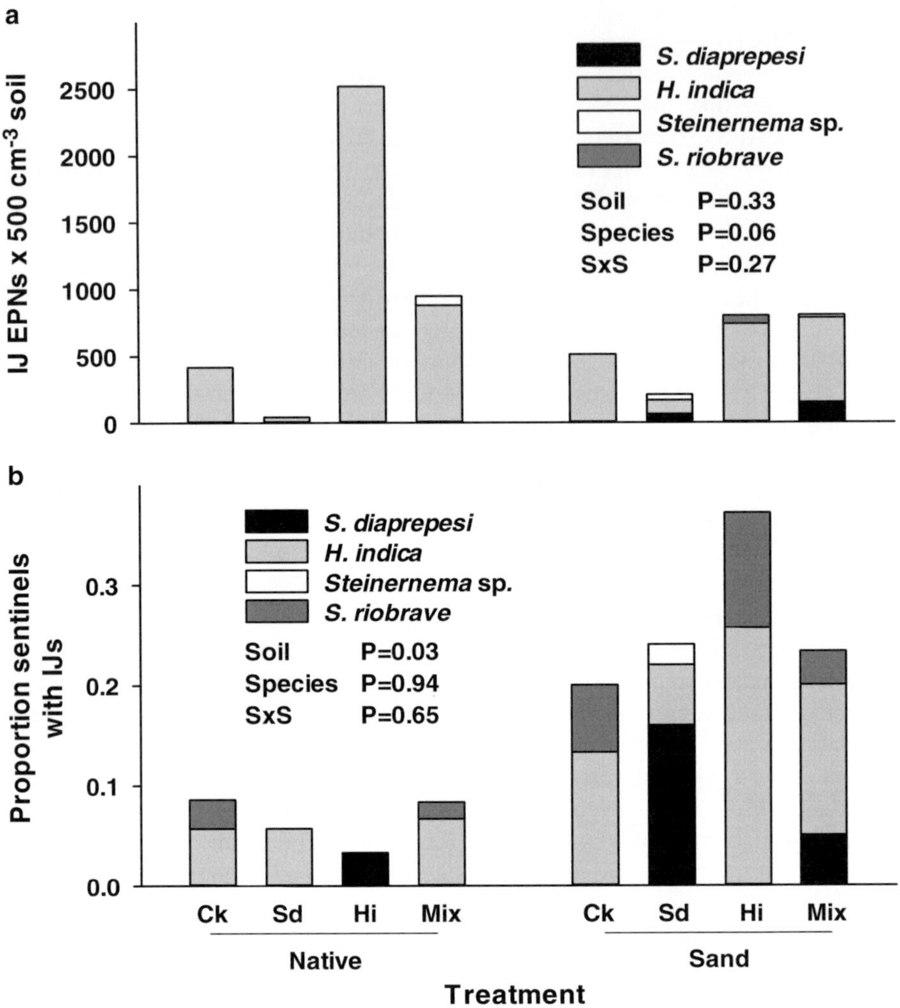

Fig. 13.6 Flatwoods and central ridge soils have no effect on the number of EPN (**a**), but species richness and diversity (**a**, **b**) and efficacy against sentinel weevil larvae (**b**) are favored by soil from the central ridge that was introduced into planting holes of trees in a flatwoods orchard. Treatments: CK, no augmented EPN; Sd *Steinernema diaprepesi*, Hi *Heterorhabditis indica*, *Mix* augmented with Sd, Hi *H. zealandica*, *Steinernema* sp., *S. riobrave* (Reprinted from Duncan et al. (2013) with permission)

13.6 Conclusions and Future Directions

The citrus orchards in Florida are ideal habitats for studying methods to optimize cultural practices that either conserve or enhance biological control because the EPN vary regionally in diversity, abundance and activity against important pests. What is it about APS that impedes EPN control of *Diaprepes* root weevil on the central

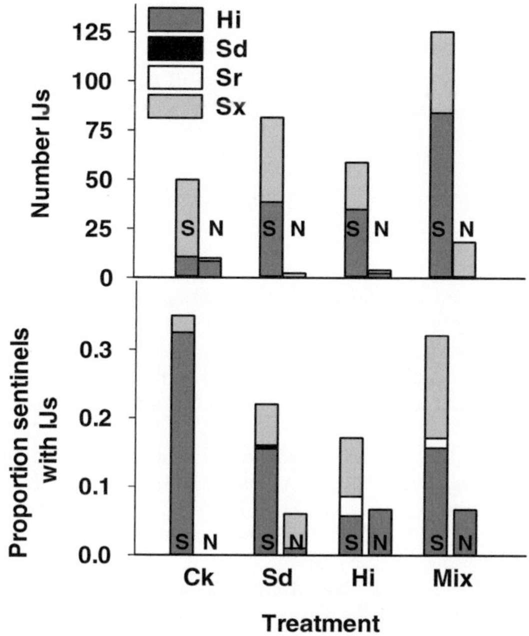

Fig. 13.7 Six years after sand (*S*) from the central ridge was used to fill planting holes of half the young trees in a flatwoods orchard, while the remaining trees were planted into the native soil (*N*), the number of IJ EPNs (panel **a**) and the EPNs that emerged from sentinel weevil larvae (**b**) were primarily species that are normally associated with wetter conditions in flatwoods soils, rather than species from the central ridge that were initially introduced into planting holes of both soils. Notice, however, that the EPN abundance and efficacy was favored by the sandy soil introduced into this orchard. Treatments: CK, no augmented EPN; Sd *Steinernema diaprepesi*, Hi *Heterorhabditis indica*, *Mix* augmented with Sd, Hi *H. zealandica*, *Steinernema* sp., *S. riobrave*

ridge? How can soil be changed to make EPN in flatwoods orchards as effective as those on the central ridge? What combination of species is likely to provide the greatest efficacy in a particular habitat and why?

New molecular methods of measuring EPN and more powerful methods to identify key environmental variables that might modulate EPN effectiveness have increased the likelihood of discovering soil properties that can be exploited to develop effective conservation tactics. The ease with which EPN are isolated, maintained and increased in the laboratory, facilitates controlled experiments in both the laboratory and field to demonstrate direct causality or other mechanisms that underlay the relationships found in natural surveys. Alternatively, identifying the physiological mechanisms by which different species or population adapt to different habitats will open the possibility of selecting or modifying genomes in ways that change where an EPN might be employed to bolster the resistance in a habitat to a crop pest. Finally, if conservation of EPN is to become a recognized IPM option, sustained efforts to document the economic value and verify the mechanism of conservation tactics may be as important as the research needed to develop effective tactics. As difficult as it may be to identify edaphic properties responsible for increased EPN efficacy, it is more challenging to prove that crop responses to new cultural practices are due to EPN and no other component of the soil food web.

References

Adair, R. C. (1994). A four–year field trial of entomopathogenic nematodes for control of *Diaprepes abbreviatus* in a Flatwoods citrus grove. *Proceedings of the Florida State Horticultural Society, 107*, 63–68.

Adams, B. J., Fodor, A., Koppenhöfer, H. S., Stackenbrandt, E., Stock, S. P., & Klein, M. G. (2006). Biodiversity and systematic of nematode–bacterium entomopathogens. *Biological Control, 38*, 4–21.

Adams, B. J., & Nguyen, K. B. (2002). Taxonomy and systematics. In R. Gaugler (Ed.), *Entomopathhogenic nematology* (pp. 1–33). Wallingford, UK: CABI Publishing.

Atkins, S. D., Clark, I., Pande, S., Hirsch, P. R., & Kerry, B. R. (2005). The use of real–time PCR and species–specific primers for the identification and monitoring of *Paecilomyces lilacinus. FEMS Microbiology Ecology, 51*, 257–264.

Barbercheck, M., & Duncan, L. W. (2004). Abiotic factors. In R. Gaugler & B. Bilgrami (Eds.), *Nematode biology* (pp. 309–343). St. Albans, UK: CAB International.

Barbercheck, M. E. (1992). Effect of soil physical factors on biological–control agents of soil insect pests. *Florida Entomologist, 75*(4), 539–548.

Barbercheck, M. E., & Hoy, C. W. (2005). A systems approach to conservation of nematodes. In P. S. Grewal, R.-U. Ehlers, & D. I. Shapiro-Ilan (Eds.), *Nematodes as biocontrol agents* (pp. 331–347). Wallingford, UK: CABI Publishing.

Beavers, J. B., McCoy, C. W., & Kaplan, D. T. (1983). Natural enemies of subterranean *Diaprepes–abbreviatus* (Coleoptera, Curculionidae) larvae in Florida. *Environmental Entomology, 12*(3), 840–843.

Bedding, R. A., & Akhurst, R. J. (1975). A simple technique for the detection of insect parasitic rhabditids nematodes in soil. *Nematologica, 21*, 109–110.

Bednarek, A., & Gaugler, R. (1997). Compatibility of soil amendments with entomopathogenic nematodes. *Journal of Nematology, 29*(2), 220–227.

Bullock, R. C., Pelosi, R. R., & Killer, E. E. (1999). Management of citrus root weevils (Coleoptera: Curculionidae) on Florida citrus with soil–applied entomopathogenic nematodes (Nematoda: Rhabditida). *Florida Entomologist, 82*, 1–7.

Campos-Herrera, R., Ali, J. G., Díaz, B. M., & Duncan, L. W. (2013). Analyzing spatial patterns linked to the ecology of herbivores and their natural enemies in the soil. *Frontiers in Plant Science, 4*(378), 1–18.

Campos-Herrera, R., Barbercheck, M., Hoy, W. H., & Stock, S. P. (2012). Entomopathogenic nematodes as a model system for advancing the frontiers of ecology. *Journal of Nematology, 44*, 162–176.

Campos-Herrera, R., El-Borai, F. E., & Duncan, L. W. (2012). Wide interguild relationships among entomopathogenic and free–living nematodes in soil as measured by real time qPCR. *Journal of Invertebrate Pathology, 111*, 126–135.

Campos-Herrera, R., El-Borai, F. E., & Duncan, L. W. (2015). Modifying soil to enhance biological control of belowground dwelling insects in citrus groves under organic agriculture in Florida. *Biological Control, 84*, 53–63.

Campos-Herrera, R., El-Borai, F. E., Ebert, T. A., Schumann, A., & Duncan, L. W. (2014). Management to control citrus greening alters the soil food web and severity of a pest–disease complex. *Biological Control, 76*, 41–51.

Campos-Herrera, R., El-Borai, F. E., Stuart, R. J., Graham, J. H., & Duncan, L. W. (2011). Entomopathogenic nematodes, phoretic *Paenibacillus* spp., and the use of real time quantitative PCR to explore soil food webs in Florida citrus groves. *Journal of Invertebrate Pathology, 108*, 30–39.

Campos-Herrera, R., Escuer, M., Labrador, S., Robertson, L., Barrios, L., & Gutiérrez, C. (2007). Distribution of the entomopathogenic nematodes from La Rioja (Northern Spain). *Journal of Invertebrate Pathology, 95*, 125–139.

Campos-Herrera, R., Gómez-Ros, J. M., Escuer, M., Cuadra, L., Barrios, L., & Gutiérrez, C. (2008). Diversity, occurrence, and life characteristics of natural entomopathogenic nematode populations from La Rioja (Northern Spain) under different agricultural management and their relationships with soil factors. *Soil Biology and Biochemistry, 40*, 1474–1484.

Campos-Herrera, R., & Gutiérrez, C. (2009). Screening for Spanish isolates of Steinernematid nematodes as biological control agents through laboratory and greenhouse microcosm study. *Journal of Invertebrate Pathology, 100*, 100–105.

Campos-Herrera, R., Johnson, E. G., El-Borai, F. E., Stuart, R. J., Graham, J. H., & Duncan, L. W. (2011). Long–term stability of entomopathogenic nematode spatial patterns measured by sentinel insects and real–time PCR assays. *Annals of Applied Biology, 158*, 55–68.

Campos-Herrera, R., Pathak, E., El-Borai, F. E., Schumann, A., Abd-Elgawad, M. M. M., & Duncan, L. W. (2013). New citriculture system suppresses native and augmented entomopathogenic nematodes. *Biological Control, 66*, 183–194.

Campos-Herrera, R., Pathak, E., El-Borai, F. E., Stuart, R. J., Gutiérrez, C., Rodríguez-Martín, J. A., et al. (2013). Geospatial patterns of soil properties and the biological control potential of entomopathogenic nematodes in Florida citrus groves. *Soil Biology and Biochemistry, 66*, 163–174.

Campos-Herrera, R., Piedra-Buena, A., Escuer, M., Montalban, B., & Gutiérrez, C. (2010). Effect of seasonality and agricultural practices on occurrence of entomopathogenic nematodes and soil characteristics in La Rioja (Northern Spain). *Pedobiologia, 53*, 253–258.

Campos-Herrera, R., Jaffuel, G., Chiriboga, X., Blanco-Pérez, R., Fesselet, M., Půža, V., et al. (2015). Traditional and molecular methods reveal intense interguild competition and other multitrophic interactions associated with native entomopathogenic nematodes in Swiss tillage soils. *Plant and Soil, 389*, 237–255.

Dolinski, C., Choo, H. Y., & Duncan, L. W. (2012). Grower acceptance of entomopathogenic nematodes: Case studies on three continents. *Journal of Nematology, 44*, 226–235.

Duncan, L. W. (2014). Can *Diaprepes* root weevil be managed with physical barriers? *Citrus Industry, 95*(6), 14–16.

Duncan, L. W., Graham, J. H., Dunn, D. C., Zellers, J., McCoy, C. W., & Nguyen, K. (2003). Incidence of endemic entomopathogenic nematodes following application of *Steinernema riobrave* for control of *Diaprepes abbreviatus*. *Journal of Nematology, 35*, 178–186.

Duncan, L. W., Graham, J. H., Zellers, J., Bright, D., Dunn, D. C., El-Borai, F. E., et al. (2007). Food web responses to augmenting the entomopathogenic nematodes in bare and animal manure–mulched soil. *Journal of Nematology, 39*, 176–189.

Duncan, L. W., & McCoy, C. W. (1996). Vertical distribution in soil, persistence, and efficacy against citrus root weevil (Coleoptera: Curculionidae) of two species of entomogenous nematodes (Rhabditida, Steinernematidae, Heterorhabditidae). *Environmental Entomology, 25*, 174–178.

Duncan, L. W., & McCoy, C. W. (2001). Hydraulic lift increases herbivory by *Diaprepes abbreviatus* larvae and persistence of *Steinernema riobrave* in dry soil. *Journal of Nematology, 33*, 142–146.

Duncan, L. W., McCoy, C. W., Stansly, P. A., Graham, J. H., & Mizell, R. F. (2001). Estimating the relative abundance of citrus root weevils with modified Tedders traps. *Environmental Entomology, 30*, 939–946.

Duncan, L. W., McCoy, C. W., & Terranova, A. C. (1996). Estimating sample size and persistence of entomogenous nematodes in sandy soils and their efficacy against the larvae of *Diaprepes abbreviatus* in Florida. *Journal of Nematology, 28*, 56–67.

Duncan, L. W., Stuart, R. J., El-Borai, F. E., Campos-Herrera, R., Pathak, E., Giurcanu, M., et al. (2013). Modifying orchard planting sites conserves entomopathogenic nematodes, reduces weevil herbivory and increases citrus tree growth, survival and fruit yield. *Biological Control, 64*, 26–36.

El-Borai, F. E., Brentu, C. F., & Duncan, L. W. (2007). Augmenting entomopathogenic nematodes in soil from a Florida citrus orchard: Non–target effects of a trophic cascade. *Journal of Nematology, 39*(2), 203–210.

El-Borai, F. E., Campos-Herrera, R., Stuart, R. J., & Duncan, L. W. (2011). Substrate modulation, group effects and the behavioural responses of entomopathogenic nematodes to nematophagous fungi. *Journal of Invertebrate Pathology, 106*, 347–356.

El-Borai, F. E., & Duncan, L. W. (2007). Suppression of *Diaprepes abbreviatus* in potted citrus by combinations of entomopathogenic nematodes with different lifespans. *Nematropica, 37*, 33–41.

El-Borai, F. E., Duncan, L. W., & Preston, J. F. (2005). Bionomics of a phoretic association between *Paenibacillus* sp. and the entomopathogenic nematode *Steinernema diaprepesi. Journal of Nematology, 37*, 18–25.

El-Borai, F. E., Stuart, R. J., Campos-Herrera, R., Pathak, E., & Duncan, L. W. (2012). Entomopathogenic nematodes, root weevil larvae, and dynamic interactions among soil texture, plant growth, herbivory, and predation. *Journal of Invertebrate Pathology, 109*, 134–142.

Ennis, D. E., Dillon, A. B., & Griffin, C. T. (2010). Simulated roots and host feeding enhance infection of subterranean insects by the entomopathogenic nematode *Steinernema carpocapsae. Journal of Invertebrate Pathology, 103*(2), 140–143.

Enright, M. R., & Griffin, C. T. (2005). Effects of *Paenibacillus nematophilus* on the entomopathogenic nematode *Heterorhabditis megidis. Journal of Invertebrate Pathology, 88*, 40–48.

Figueroa, W., & Roman, J. (1990). Biocontrol of the sugarcane rootstalk borer, *Diaprepes abbreviatus* (L.), with entomophilic nematodes. *Journal of Agriculture of the University of Puerto Rico, 74*, 395–404.

Futch, S. H., Duncan, L. W., & Zekri, M. (2005). Validation of an area–wide extension program to estimate the seasonal abundance of adult citrus root weevils with un–baited pyramidal traps. *Proceedings of the Florida State Horticultural Society, 117*, 143–147.

Fuxa, J. R. (1995). Biorational pest control agents: Formulation and delivery. In F. R. Hall & J. W. Barry (Eds.), *Developed from symposia sponsored by the division of agrochemicals at the 207th national meeting of the American Chemical Society*. San Diego, CA: American Chemical Society.

Graham, J. H., Bright, D. B., & McCoy, C. W. (2003). *Phytophthora–Diaprepes* weevil complex: *Phytophthora* spp. relationship with citrus rootstocks. *Plant Disease, 87*, 85–90.

Graham, J. H., Johnson, E. G., Gottwald, T., & Irey, M. (2013). Pre–symptomatic fibrous root decline in citrus trees caused by Huanglongbing and potential interaction with *Phytophthora* spp. *Plant Disease, 97*, 1195–1199.

Griffin, C. T., Downes, M. J., & Block, W. (1990). Test of antarctic soils for insect parasitic nematodes. *Antarctic Science, 2*, 221–222.

Halbert, S. E. (2005). *Pest alert: Citrus greening/huanglongbing*. Available from: http://www.freshfromflorida.com/Divisions-Offices/Plant-Industry/Plant-Industry-Publications/Pest-Alerts/Pest-Alerts-Citrus-Greening

Hara, A. H., Gaugler, R., Kaya, H. K., & Lebeck, L. M. (1991). Natural–populations of entomopathogenic nematodes (Rhabditida, Heterorhabditidae, Steinernematidae) from the Hawaiian–islands. *Environmental Entomology, 20*(1), 211–216.

Haviland, D. R., & Hernandez, N. M. (2012). Development of management programs for white grubs in California blueberries. Conference: Proceedings of the 11th North American Research and Extension Workers Conference, Kalamazoo, Michigan, USA, 25–28 July 2010. *International Journal of Fruit Science, 12*(1/3), 114–123.

Hazir, S., Keskin, N., Stock, S. P., Kaya, H. K., & Özcan, S. (2003). Diversity and distribution of entomopathogenic nematodes (Rhabditida: Steinernematidae and Heterorhabditidae) in Turkey. *Biodiversity and Conservation, 12*(2), 375–386.

Hominick, W. M. (2002). Biogeography. In R. Gaugler (Ed.), *Entomopathhogenic nematology*(pp. 115–143). Wallingford, UK: CABI Publishing.

Hominick, W. M., Reid, A. P., Bohan, D. A., & Briscoe, B. R. (1996). Entomopathogenic nematodes: Biodiversity, geographical distribution and the convection on biological diversity. *Biocontrol Science and Technology, 6*, 317–331.

Hoy, C. W., Grewal, P. S., Lawrence, J. L., Jagdale, G., & Acosta, N. (2008). Canonical correspondence analysis demonstrates unique soil conditions for entomopathogenic nematode species compared with other free–living nematode species. *Biological Control, 46*(3), 371–379.

Huang, J., Li, C., Xiao, C., & Wang, G. (2010). Detection of *Phytophthora nicotianae* in soil with real–time quantitative PCR. *Journal of Phytopathology, 158*, 15–21.

Ishibashi, N., & Kondo, E. (1986). *Steinernema feltiae* (DD–136) and *S. glaseri*: Persistence in soil and bark compost and their influence on native nematodes. *Journal of Nematology, 18*, 310–316.

Jabbour, R., & Barbercheck, M. E. (2008). Soil and habitat complexity effects on movement of the entomopathogenic nematode *Steinernema carpocapsae* in maize. *Biological Control, 47*(2), 235–243.

Jabbour, R., Crowder, D. W., Aultman, E. A., & Snyder, W. E. (2011). Entomopathogen biodiversity increases host mortality. *Biological Control, 59*, 277–283.

Jaffee, B. A., Strong, D. R., & Muldoon, A. E. (1996). Nematode–trapping fungi of a natural shrubland: Tests for food chain involvement. *Mycologia, 88*(4), 554–564.

Jenkins, D. A., Shapiro-Ilan, D., & Goenaga, R. (2007). Virulence of entomopathogenic nematodes against *Diaprepes abbreviatus* in an Oxisol. *Florida Entomologist, 90*, 401–403.

Jenkins, D. A., Shapiro-Ilan, D., & Goenaga, R. (2008). Efficacy of entomopathogenic nematodes *versus Diaprepes abbreviatus* (Coleoptera: Curculionidae) larvae in a high clay–content Oxisol soil: Greenhouse trials with potted *Litchi chinensis. Florida Entomologist, 91*, 75–78.

Kanga, F. N., Waeyenberge, L., Hauser, S., & Moens, M. (2012). Distribution of entomopathogenic nematodes in Southern Cameroon. *Journal of Invertebrate Pathology, 109*(1), 41–51.

Kaspi, R., Ross, A., Hodson, A., Stevens, G. N., Kaya, H. K., & Lewis, E. E. (2010). Foraging efficacy of the entomopathogenic nematode *Steinernema riobrave* in different soil types from California citrus groves. *Applied Soil Ecology, 45*, 243–253.

Koppenhöfer, A. M., Jaffee, B. A., Muldoon, A. E., Strong, D. R., & Kaya, H. K. (1996). Effect of nematode–trapping fungi on an entomopathogenic nematode originating from the same field site in California. *Journal of Invertebrate Pathology, 68*, 246–252.

Kung, S. P., Gaugler, R., & Kaya, H. K. (1990). Influence of soil–pH and oxygen on persistence of *Steinernema* spp. *Journal of Nematology, 22*(4), 440–445.

Lawrence, J. L., Hoy, C. W., & Grewal, P. S. (2006). Spatial and temporal distribution of endemic entomopathogenic nematodes in a heterogeneous vegetable production landscape. *Biological Control, 37*, 247–255.

Lewis, E. E., Campbell, J. F., & Gaugler, R. (1998). Chapter 13 – A conservation approach to using entomopathogenic nematodes in turf and landscapes. In: P. Barbosa (Ed.), *Conservation biological control* (pp. 235–254). Academic.

Malan, A. P., Knoetze, R., & Moore, S. D. (2011). Isolation and identification of entomopathogenic nematodes from citrus orchards in South Africa and their biocontrol potential against false codling moth. *Journal of Invertebrate Pathology, 108*, 115–125.

McCoy, C. W. (1999). Arthropod pests of citrus roots. In L. W. Timmer & L. W. Duncan (Eds.), *Citrus health management* (pp. 149–156). St. Paul, MN: APS Press.

McCoy, C. W., & Boucias, D. G. (1989). Selection of *Beauveria bassiana* pathotypes as potential microbial control agents of soil–inhabiting citrus weevils (*Artipus floridanus, Pachnaeus opalus, Pantomorus cervinus, Diaprepes abbreviatus*). *Memórias do Instituto Oswaldo Cruz, 84*, 75–80.

McCoy, C. W., Shapiro, D. I., Duncan, L. W., & Nguyen, K. (2000). Entomopathogenic nematodes and other natural enemies as mortality factors for larvae of *Diaprepes abbreviatus* (Coleoptera: Curculionidae). *Biological Control, 19*, 182–190.

McCoy, C. W., Stuart, R. J., Duncan, L. W., & Nguyen, K. (2002). Field efficacy of two commercial preparations of entomopathogenic nematodes against larvae of *Diaprepes abbreviatus* (Coleoptera: Curculionidae) in alfisol type soil. *Florida Entomologist, 85*, 537–544.

McCoy, C. W., Stuart, R. J., & Nigg, H. N. (2003). Seasonal life stage abundance of *Diaprepes abbreviatus* in irrigated and non–irrigated citrus plantings in central Florida. *Florida Entomologist, 86*(1), 34–42.

Millar, L. C., & Barbercheck, M. E. (2001). Interaction between endemic and introduced entomopathogenic nematodes in conventional–till and no till corn. *Biological Control, 22*, 235–245.

Millar, L. C., & Barbercheck, M. E. (2002). Effects of tillage practices on entomopathogenic nematodes in a corn agroecosystem. *Biological Control, 25*, 1–11.

Moore, S., Coombes, C., Manrakhan, A., Kirkman, W., Hill, M., Ehlers, R.-U., et al.. (2013). Subterranean control of an arboreal pest: EPNs and EPFs for FCM. In: J. A. Jehle, R. Bazok, N. Crickmore, M. Lopez-Ferber, I. Glazer, E. Quesada-Moraga, & M. Traugott (Eds.), *IOBC/WPRS Bulletin 90*, 247–250.

Neher, D. A. (1999). Nematode communities in organically and conventionally managed agricultural soils. *Journal of Nematology, 31*(2), 142–154.

Nielsen, A. L., Spence, K. O., Nakatani, J., & Lewis, E. E. (2011). Effect of soil salinity on entomopathogenic nematode survival and behavior. *Nematology, 13*, 859–867.

Olmstead, M., Chaparro, J., Williamson, J. G., Rouse, R., Mizell, R., Harmon, P., et al. (2013). *Florida subtropical peaches: Production practices*. Horticultural Sciences Department, Florida Cooperative Extension Service, Institute of Food and Agricultural Sciences, University of Florida. (HS1109). Original publication date July 2007. Revised August 2013. http://edis.ifas.ufl.edu

Park, J.-J., Jagdale, G. B., Cho, K., Grewal, P. S., & Hoy, C. W. (2014). Spatial association between entomopathogenic and other free–living nematodes and the influence of habitat. *Applied Soil Ecology, 76*, 1–6.

Pathak, E. (2014). *Diversity of nematophagous fungi populations and their relevance to the regulation of entomopathogenic nematodes in Florida citrus orchards*. Ph.D. thesis, University of Florida, Gainesville.

Pathak, E., El-Borai, F. E., Campos-Herrera, R., Johnson, E. G., Stuart, R. J., Graham, J. H., et al. (2012). Use of real–time PCR to discriminate predatory and saprophagous behavior by nematophagous fungi. *Fungal Biology, 116*, 563–573.

Perry, J. N., & Dixon, P. M. (2002). A new method to measure spatial association for ecological count data. *Ecoscience, 9*, 133–141.

Perry, J. N., & Hewiit, M. (1991). A new index of aggregation for animal counts. *Biometrics, 47*, 1505–1518.

Polavarapu, S., Koppenhoefer, A. M., Barry, J. D., Holdcraft, R. J., & Fuzy, E. M. (2007). Entomopathogenic nematodes and neonicotinoids for remedial control of oriental beetle, *Anomala orientalis* (Coleoptera: Scarabaeidae), in highbush blueberry. *Crop Protection, 26*(8), 1266–1271.

Porazinska, D. L., Giblin-Davis, R. M., Powers, T. O., & Thomas, W. K. (2012). Nematode spatial and ecological patterns from tropical and temperate rainforests. *PLoS ONE, 7*(9), e44641.

Roman, J., & Beavers, J. B. (1983). A survey of Puerto–Rican soils for entomogenous nematodes which attack *Diaprepes–abbreviatus* (L). (Coleoptera: Curculionidae). *Journal of Agriculture of the University of Puerto Rico, 67*(3), 311–316.

Schroeder, W. J. (1987). Laboratory bioassays and field trials of entomogenous nematodes for control of *Diaprepes abbreviatus. Environmental Entomology, 16*, 987–989.

Schroeder, W. J. (1994). Comparison of two steinernematid species for control of the root weevil *Diaprepes abbreviatus. Journal of Nematology, 26*, 360–362.

Shapiro-Ilan, D. I., Gouge, D. H., & Koppenhöfer, A. M. (2002). Factors affecting commercial success: Case studies in cotton, turf and citrus. In R. Gaugler (Ed.), *Entomopathogenic nematology* (pp. 333–356). New York: CAB International.

Snyder, W. E., Snyder, G. B., Finke, D. L., & Straub, C. S. (2006). Predator biodiversity strengthens herbivore suppression. *Ecology Letters, 9*, 789–796.

Spiridonov, S. E., Moens, M., & Wilson, M. J. (2007). Fine scale spatial distributions of two entomopathogenic nematodes in a grassland soil. *Applied Soil Ecology, 37*, 192–201.

Stirling, G. R., McKenry, M. V., & Mankau, R. (1979). Biological control of root–knot nematodes (*Meloidogyne* spp.) on peach. *Phytopathology, 69*, 806–809.

Stuart, R. J., Barbercheck, M. E., Grewal, P. S., Taylor, R. A. J., & Hoy, C. W. (2006). Population biology of entomopathogenic nematodes: Concepts, issues, and models. *Biological Control, 38*, 80–102.

Stuart, R. J., El-Borai, F. E., & Duncan, L. W. (2008). From augmentation to conservation of entomopathogenic nematodes: Trophic cascades, habitat manipulation and enhanced biological control of *Diaprepes abbreviatus* root weevils in Florida citrus groves. *Journal of Nematology, 40*(2), 73–84.

Tarasco, E., Clausi, M., Rappazzo, G., Panzavolta, T., Curto, G., Sorino, R., et al. (2014). Biodiversity of entomopathogenic nematodes in Italy. *Journal of Helminthology, 89*, 359–366.

Toepfer, S., Haye, T., Erlandson, M., Goettel, M., Lundgren, J. G., Kleespies, R. G., et al. (2009). A review of the natural enemies of beetles in the subtribe *Diabroticina* (Coleoptera: Chrysomelidae): Implications for sustainable pest management. *Biocontrol Science and Technology, 19*(1), 1–65.

Torr, P., Spiridonov, S. E., Heritage, S., & Wilson, M. J. (2007). Habitat associations of two entomopathogenic nematodes: A quantitative study using real–time quantitative polymerase chain reactions. *Journal of Animal Ecology, 76*(2), 238–245.

Williams, R. N., Fickle, D. S., Grewal, P. S., & Dutcher, J. (2010). Field efficacy against the grape root borer *Vitacea polistiformis* (Lepidoptera: Sesiidae) and persistence of *Heterorhabditis zealandica* and *H. bacteriophora* (Nematoda: Heterorhabditidae) in vineyards. *Biological Control, 53*(1), 86–91.

Williamson, J. G., Lyrene, P. M., & Olmstead, J. W. (2012). *Blueberry gardener's guide.* Horticultural Sciences Department, Florida Cooperative Extension Service, Institute of Food and Agricultural Sciences, University of Florida. (Circular 1192). Original publication date April 2004. Revised January 2012. http://edis.ifas.ufl.edu

Wilson, M. J., Lewis, E. E., Yoder, F., & Gaugler, R. (2003). Application pattern and persistence of the entomopathogenic nematode *Heterorhabditis bacteriophora. Biological Control, 26*, 180–188.

Yadav, P., Duckworth, K., & Grewal, P. S. (2012). Habitat structure influences below ground biocontrol services: A comparison between urban gardens and vacant lots. *Landscape and Urban Planning, 104*(2), 238–244.

Zhang, L. M., Liu, X. Z., Zhu, S. F., & Chen, S. Y. (2006). Detection of the nematophagous fungus *Hirsutella rhossiliensis* in soil by real–time PCR and parasitism bioassay. *Biological Control, 36*, 316–332.

Chapter 14
Entompathogenic Nematodes in Cuba: From Laboratories to Popular Biological Control Agents for Pest Management in a Developing Country

Mayra G. Rodríguez Hernández

14.1 Introduction

The use of biological control of agrarian pest in Cuba is a practice that has accompanied the farmers for more than seven decades, since the first works developed late in 1920s decade and early in 1930. The Cuban agriculture currently experiences two extreme food–production models: an intensive model with high inputs: (1) used in state farms (enterprises), and (2) began at the onset of the so–called special period (during the economically critical years of the 1990s), oriented towards an agro–ecology and based on low inputs (Altieri & Funes-Monzote, 2012); the latter are mainly used by small farmers in countryside or in urban and peri–urban agriculture systems. The small farmers produce more than 80 % of the food in Cuba, and most of them use biological control agents for pest management, bringing to the consumers a healthy offer of vegetables, fruits, grains, and tubers. In this frame, the entomopathogenic nematodes (EPN) maybe are one of biological control agents more recently generalized in Cuba with nearly 20 years of intensive use. They were initially applied in the *Citrus* crop and more recently have been extended to other crops. This chapter attempts to offer a brief view of research results and use of these organisms in Cuba, and the challenges for research workers, farmers and stakeholders related to the agriculture sector.

M.G. Rodríguez Hernández (✉)
Departamento de Proteccion Vegetal, Centro Nacional de Sanidad Agropecuaria (CENSA), San José de las Lajas, Provincia Mayabeque, Cuba
e-mail: mrguez@censa.edu.cu

© Springer International Publishing Switzerland 2015

R. Campos-Herrera (ed.), *Nematode Pathogenesis of Insects and Other Pests*, Sustainability in Plant and Crop Protection 1, DOI 10.1007/978-3-319-18266-7_14

14.2 A Brief History of Entomopathogenic Nematodes in Cuba

The EPNs have been studied in Cuba since 1977, when Dr. Magda Montes (Fig. 14.1a) found nematodes emerging from dead larvae and pupae of the citrus green–blue weevil, *Pachnaeus litus* Germar (Coleoptera: Curculionidae). These nematodes were assigned to the *Neoaplectana* genus by the author, suggesting to be possible their reproduction in *Galleria mellonella* L. (Lepidoptera: Pyralidae) for being used in *P. litus* control in citrus nursery (Montes, 1978; Montes et al., 2014). To study this organism, a research team was created at the former National Station for Citrus Sanity, institution that was later integrated to the current Research Institute of Tropical Fruits (www.fruticulturacubana.co.cu), with Dr. Eva Arteaga (Fig. 14.1b) joining this team. This specialist and Dr. Montes were the pioneers of the Entomopathogenic Nematology in Cuba, and together with Dr. Lourdes Sánchez Portales (Fig. 14.1c), their contributions were essential for the advances in the development and use of EPN being achieved in the country. In a country with a tradition in the pest management with biological control agents, the EPN has been gaining space in the agriculture context during the last 20 years. Cuban scientific institutions and universities have approached different aspects in the study of these biological control organisms, i.e. mass productions and application, and the contributions of each group have allowed an efficient use of selected populations of nematodes for pest management in the context of a diverse agro–ecosystems (Rodríguez, Hernández-Ochandía, & Gómez, 2012).

Since 70s and 80s, strong international cooperation with other EPN researchers was established. For example, Dr. Zdenek Mrácek (Institute of Entomology, Czechoslovak Academy of Science, former Czechoslovskia, actual Czech Academy of Sciences, Czech Republic) and Noël Boemare (UMII/INRA, Montpellier, France). They worked in taxonomical studies of Cuban population of *Heterorhabditis indica* Poinar, Karunakar & David (Rhabditida: Heterorhabditidae

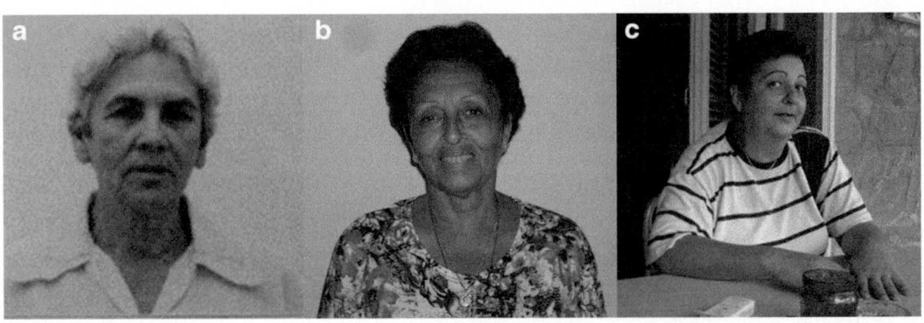

Fig. 14.1 Professors Dra. Magda Montes (**a**) and Dra. Eva Arteaga (**b**), pioneers of the Entomopathogenic Nematology in Cuba. (**c**) Dra. Lourdes Sánchez Portales (1956–2004), leader of the research work conducting to EPN generalization in Cuba

and *Steinernema cubana* Mráček, Hernández & Boemare (Rhabditida: Steinernematidae) (Mráček, Arteaga, & Boemare, 1994). Other international cooperation was established between Cuban specialists from the National Center for Plant and Animal Health (CENSA) and the Bolivian Non–Governmental Organization PROBIOMA (www.probioma.org.bo) in the production of biological control agents at the end of the 90s and the beginning of this century. EPNs were among the organisms produced and used in Santa Cruz de la Sierra Province, as a modest contribution from Cuba to the development of biological control in the Latin America area. Recently, Cuban specialists from the National Center for Plant and Animal Health (CENSA) started a fruitful cooperative work with researchers from Venezuela. One of the collaborative grant granted with an international collaboration grant (Research Project "Producción de nematodes entomopatógenos para el manejo de plagas agricolas") has as main objective to obtain and develop native populations of entomopathogenic nematodes for pest management in some areas in Venezuela and is linked to National Institute for Agricultural Research (INIA) (www.inia.gov.ve) at Aragua State. In addition to this, another research project is being developed between CENSA and the Venezuelan Institute for Scientific Research (IVIC) (Zulia State) (www.ivic.gob.ve) in order to obtain commercial products based on EPN by fermentation technologies for Cuba and Venezuela.

Herein, we briefly describe the main achievement by the Cuban research on EPN regional distribution, taxonomy and basic knowledge on the relationship between the nematode and the bacteria.

14.2.1 Prospection and Identification

In the first prospection, between 1992 and 1994, 251 samples from 27 different crops in 9 provinces (64 % of the provinces of Cuba) were collected and examined. The isolates found were six *Heterorhabditis* and one of *Steinernema* from citrus (*Citrus* spp.), coffee (*Coffea* spp.) and guava (*Psidium guajava* L.) crops, representing the 2.79 % of positive samples (Sánchez, 2002). A later prospection study conducted by Pozo, López, and Martínez (2003) in Villa Clara Province (central region) focused their efforts in other crops such as sweet potato (*Ipomoea batatas* (L.) Lam), rice (*Oriza sativa* L.), banana – plantain (*Musa* sp.) and sorghum (*Sorghum bicolour* (L.) Moench). They found two isolates of *Heterorhabditis* in sorghum, which were denominated as CIAP–DEY–6 and CIAP–DEY–7. Prospection results published in Cuba evidenced that *Heterorhabditis* is the genus appearing more frequently (Table 14.1).

In the studies conducted by Sánchez (2002), she verified that the population HC1 showed morphometric variables compatible with *Heterorhabditis bacteriophora* Poinar (Rhabditida: Heterorhabditidae) and *H. indica*; however, in the diversity analysis by RAPD markers, using *H. bacteriophora* (population HP88) and *H. indica* (Cuban population) as a references and other Cuban population of *Heterorhabditis* spp. the similarity between HC1 (Cuba) and HP88 (*H. bacteriophora*)

Table 14.1 Species and populations of entomopathogenic nematodes found in Cuba

Specie	Population code	Locality	Province	Reference
H. indica	P_2M	Alquizar	Habana	Montes (1978); Stack et al. (2000)
Heterorhabditis bacteriophora	HC1	San José de las Lajas	Mayabeque	Sánchez (2002)
Heterorhabditis sp.	CIAP DEY–7; CIAP–DEY–6	Santa Clara	Villa Clara	Pozo et al. (2003)
Heterorhabditis sp.	HI–24	nd	Villa Clara	Castellanos, González, and Jacomino (2000)
Steinernema cubanum	nd	nd	Pinar del Río	Mráček et al. (1994)
Steinernema sp.	nd	Jibacoa	Villa Clara	Castellanos (2000)
Steinernema sp.	SC1	Buey Arriba	Granma	Sánchez (2002)

Nd no declared

obtained by Sánchez in molecular study made her to assert that the population HC1 belonged to *H. bacteriophora*; however, it is necessary to conclude the sequencing of the nematode and the symbiotic bacterium for the definitive identification of this isolate and to be able to make a safe deposit in international collection and to include the sequence in DNA databases.

Cuban research has also advanced in the knowledge related to the mutualistic bacteria of the EPN. For example, in the National Research Institute of Plant Health (INISAV), the associated bacteria of *H. indica* P_2M and *S. cubanum* were identified as *Photorhabdus luminescens* Fisher le Saux, Villard, Brunel, Normand & Boemare (Enterobacteriales: Enterobacteriaceae) P–01 and *Xenorhabdus* sp. X–01, respectively (Pérez, Márquez, & Gómez, 2006) thanks to the evaluation of protease, lipase, and antibiotic activities. According to Fischer-Le Saux, Arteaga-Hernández, Mrácek, and Boemare (1999), the symbiotic bacteria of *S. cubanum* is *Xenorhabdus poinarii* Akhurst & Boemare (Akhurst) (Enterobacteriales: Enterobacteriaceae). At CENSA, the morphological, biological, serological and biochemical studies of the symbiotic bacterium of the population HC1 were made in the first decade of the present century by Sánchez (2002), and the population was shown to have a typical characteristic of *P. luminescens* with 37 °C as a maximum temperature for growth. This bacterium shows a great bioluminescence after 4 min, with high lipolytic, proteolytic and antibiotic activities. It was as pathogenic as the bacterium strain HP88, causing death in 24 h to 69 % of larvae until a concentration of 10^4 CFU/mL–1 (Martín, 2007). However, more specific studies must be developed in the future to define the subspecies of *P. luminiscens*.

Other studies about symbiotic bacterium were developing in Cuba. The quality indicators for phase I of *P. luminescens* strain P_2M were determined by Márquez, Fernández-Larrea, and Arteaga (1997). In addition to this, some advanced tech-

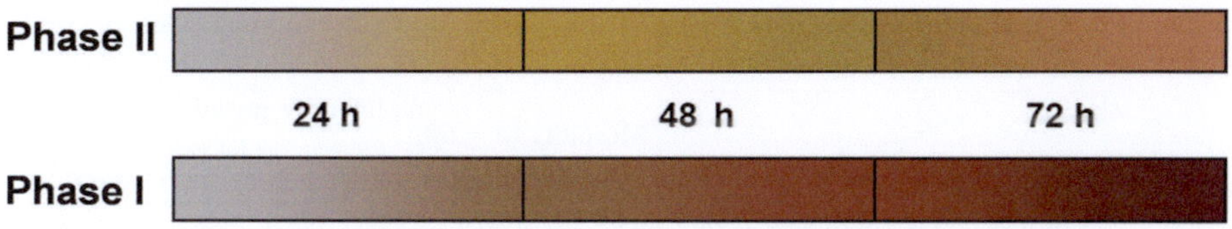

Fig. 14.2 Diagrammatic scale for color changes caused by *Heterorhabditis bacteriophora* population HC1 in dead *Galleria mellonella* larvae at the first 72 h. (*Upper bar*: color changes in larva inoculated with nematodes where *Photorhabdus luminescens* is in phase II; *lower bar*: color changes in larva inoculated with nematodes where *P. luminescens* is in phase I)

niques were used for phases bacteria studies (Gómez, 2002; Martín, 2007) using scanning microscopy and a total protein (PAGE), respectively, found differences between phase I and II.

Some changes were associated by Sánchez, Christopher, Rodríguez, Gómez, and Soler (2003) to phase bacteria change. They observed that the continued mass production of the strain HC1 of *H. bacteriophora* in the laboratory for 3 years, caused changes in its characteristics related to the color in the dead *G. mellonella* larvae and decreased the number of infective juveniles (IJs) produced in each *G. mellonella* larvae and bioluminescence. The color changes in the dead *G. mellonella* larvae infected by strain HC1 in both phases (I and II) were used by Sánchez (2002) to structure a diagrammatic scale (Fig. 14.2) being used in the mass production laboratories in the island for monitoring the production by giving the workers indication of possible changes in bacteria phases. The diagrammatic scalier present a practical interrelationship between laboratories results and production of nematodes for pest control in cottage laboratories in the country.

14.2.2 Selection of Populations

Information on EPN selection studies in the Cuban literature is scarce. Only Sánchez (2002) carried out some bioassays recommended by Glazer and Lewis (2000) and using an international reference population. There are other studies comparing isolates against specific targets, but it has not been possible for the researchers to do the bioassays on penetration, exposure time and one–on–one. Sánchez compared six Cuban isolates, and included the *H. bacteriophora* population HP88 (donated by MicroBioGropu Ltd, UK) as a reference population. The population HC1 showed the best results compared with other populations and the reference one (Table 14.2). Sánchez selected this isolate for further efficacy studies against several target insects under semi–controlled conditions in the laboratory and in the field. The yields obtained made *in vivo* cultures very attractive for their use in the CREE (standing for Centers for the Reproduction of Entomophagous and Entomopathogenic Organisms) using *G. mellonella* as host. This population was compared with the Cuban populations MC1, MC2 and MC3 by using some of the

Table 14.2 Behavior of Cuban *Heterorhabditis* population in the preliminary selection study (Sánchez, 2002). (Experiment replicate three times)

Isolates	Locality/Province	Mortality (n = 60) 48 h	72 h	Infective juveniles/*G. mellonella* larvae (n = 30) (230 ± 2 mg fresh weigh)
HC1	San José de las Lajas/Mayabeque	93.3 a	100 a	281,588 a
HC2	Alquízar/Artemisa	27.3 d	53.7 d	76,624 d
HC3	Güira de Melena/Artemisa	63.5 c	70.2 c	42,850 d
MC1	Jagüey Grande/Matanzas	83.4 b	86.7 b	183,181 ab
MC2	Jagüey Grande/Matanzas	83.4 b	90 ab	171,785 bc
MC3	Jovellanos/Matanzas	80.1 b	83.4 b	88,401 c
HP–88	International reference	86.7 ab	96.6 a	243,200 a

Media in the same column without letter in common differ at $P < 0.05$

standard assays referred by Glazer and Lewis. The result about population selection, using the trials of Exposure time assay, One–on–one assay and Sand column assay made by Sánchez, indicated that the HC1 population had superior acting that other Cuban populations, with inferior ET_{50} (min) and high percent of mortality on *G. mellonella.*

The bioassays emphasize the potential of measuring quantitative behavioral responses as specific criteria for nematode virulence (Glazer & Lewis, 2000). These types of bioassays were simple, and allowed Sánchez (2002) to select HC1 as an adequate population for biological control purposes because of the best behavior showed by it in the assays.

14.3 Production Technologies and Formulation of Entomopathogenic Nematodes

EPN mass production in Cuba has strong links with the sugarcane industry. Early in the twentieth century, the extensive Cuban plains were planted with sugar cane. For the management of the key pest of this crop, the sugarcane borer (*Diatraea saccharalis* Fabr. (Lepidoptera: Pyralldae)), a program for the mass rearing and release of *Lixophaga diatraea* Towns (Diptera: Tachinidae) (Fuentes, Llanes, Méndez, & González, 1998) was implemented, which was reproduced exclusively by using *G. mellonella*. The experience gained in the *G. mellonella* mass rearing together with the availability of local substrates offered the good starting point for EPN development *in vivo*. During the 90s, Dr. Sanchez launches the collaboration, transferring not only the knowledge but the native population with the most promising results. Besides, these cottage laboratories (CREE) some of them with more than 10 years of experience had a prestige and identified clients that could be potential users of EPN.

14.3.1 *In Vivo* Reproduction: The Cuban Experience in Cottage Laboratories (CREE)

Due to the expensive production of *G. mellonella* as host, the evaluations of alternative host were under study in the early in 90s. The rice moth (*Corcyra cephalonica* Stainton) (Lepidoptera: Pyralidae) was evaluated for mass production of *H. indica* (ex *H. helothidis*) population P_2M (Pérez, Vázquez, & Mollineda, 1991). Although the results showed that 80IJ per rice moth larvae caused rates of mortality of 95 %, gathering 7,000–10,800 IJ per larva, and showed activity against *G. mellonella* but low yields compared with those obtained with *G. mellonella* resulted in an unprofitable for commercial. Another attempt in search of new alternatives host for EPN reproduction, Pozo-Velázquez, Sandi-St. Luis, Valdés-Herrera, and Alizar-Saavedra (unpublished) studied the possible use of *Spodoptera frugiperda* (Smith) (Lepidoptera: Noctuidae) with *Sorghum halepense* Pers. (Poales: Poaceae), as natural diet. Alizar, St Louis, Cárdenas, Valdés, and Pozo (2010) determined the V instar of *Spodoptera frugiperda* Walker (Lepidoptera: Noctuidae) was the most productive instar when was inoculated with 250–500 IJs per larva. This option could be used in those CREE with no access to substrates for *G. mellonella* production but the quality of nematodes must be quantified, mainly the lipid content.

Massive reproduction did not begin until 1994, when the HC1 population of *H. bacteriophora* and the basic methodology for mass rearing in *G. mellonella* were transferred from CENSA to the CREE "Pablo Noriega" (Quivicán Municipality) of the former Ministry of Sugarcane (Sánchez, 1997). This methodology for EPN mass production was based on the method used by Dutky, Thompson, and Canwell (1964) modified by the CENSA's research team and including some quality control elements. The efficiency of *in vivo* culture production relies on the quality of the host insects, and therefore, an adequate host diet is important in the process. The standard diet for *G. mellonella* contains, among other components, milk powder and glycerin (Kulkarni, Kumar, Kumar, & Paunikar, 2012), but, as these are foods or components of the human diet in Cuba, they cannot be used for this purpose. Thus, several economic alternative substrates were assessed, and today different mixes of cereals and byproducts from the sugarcane industry are used (Alemán & Goicoechea, 1993; Enrique, Sánchez, Rodríguez, Gómez, & Valle, 2006).

A critical step of the *in vivo* production of EPN is the rate of insect infection or inoculation. Very low dosage results in low mortality of the host insect, but extreme high dosage may result in failed infections (Woodring & Kaya, 1988). Sánchez (2002) determined that the optimum dosage of the Cuban HC1 population for *in vivo* productions was between 20 and 40 IJs per larva of *G. mellonella* (weigh ≥ 0.2 g) to obtain yields of about 250,000 IJs/larva. According to the *in vivo* methodology established in the CREE, the producers must evaluate the stock culture conditions periodically. Color changes in *G. mellonella* (Fig. 14.2) larvae or yields are signs of its deterioration. In addition to this, the methodology for harvesting was based on the nematode migration to the water, so–called "White trap" (White, 1927), but using trays and glass pieces (Fig. 14.3). In some laboratories, additional operations

Fig. 14.3 "White Trap" using trays and glass pieces in a wood cabinet to avoid contamination of *Galleria mellonella* larvae and nematode suspensions

such as separation of EPN, using vacuum pump and sieves, and cleanup the solution are followed for formulation, but the expensive equipment limit the generalization of them.

At the beginning of 2000, the EPNs were reproduced in more than 60 CREEs belonging to equal number of sugarcane factories in Cuba. In 2002, the sugarcane industry, with 156 factories, was reorganized and 71 factories were disassembled (Pacheco, 2010). Some of the CREE working at these factories were closed, and today the EPNs are reproduced in 33 CREEs belonging to the sugarcane industry, with yields of some 700 million of IJs per month (INICA, 2011) (www.azcuba.cu). The methodology with alternative substrates and the population HC1 were successfully transferred to laboratories also producing biological control agents for coffee berry borer management in Granma Province (García, Rodríguez, Cabrera, Gómez, & Rodríguez, 2007).

During the last 20 years, the EPN *in vivo* mass production technology has been sufficient to obtain some quantities of IJs for local applications in more than 60 municipalities of the island, but nowadays, the current demand surpasses the offer. The EPNs have become one of most popular biological control agents in some agricultural areas in Cuba, for the high number of target pests that farmers can manage with these organisms. In consequence with the above situation and the fact that there are some facilities in Cuba for the solid (in some cottages laboratories) and liquid fermentation (in industrial factories), it is in progress a joint project conducted by a multidisciplinary team from CENSA and IVIC (*Desarrollo de bioplaguicidas a base de nematodos entomopatógenos para el manejo de plagas del sector agrario en Venezuela y Cuba*) to obtain technologies for the solid and liquid fermentation of selected EPN populations.

14.3.2 *In Vitro Reproduction: Modest Results in Solid Culture*

The monoaxenic culture of EPN using solid and liquid fermentation was developed in the 60s (Friedman, 1990). Since the initial works, a wide variety of media has been used in order to provide adequate conditions for the nematodes–bacteria complex, but many of them, such as animal liver or kidneys, corn oil, yeast, cholesterol, soybean, animal fat and egg yolk, among others, are highly demanded human foods with high prices in international markets. Therefore, their use for producing biological control agents is not allowed by the Cuban authorities. Consequently, several alternative components were evaluated in Cuba in a solid media in plates and three–dimensional system (using sponges) in flasks (Sánchez, 2002; Valdés, Lobaina, Márquez, Gómez, & Escobar, 2006).

The studies made at CENSA were successful. As a result, the Cuban Patent Office (www.ocpi.cu) accepted the document protecting the results obtained with byproducts from the fish industry (fish fat), potato, sweet potato, corn and soybean in different combinations (Sánchez, Soler, Gómez, & Martín, 2006). In another institute, INISAV, Valdés et al. (2006) obtained good results in combining chicken or pork liver with molasses, starch and rice dust (byproduct from the rice industry) in a bi–dimensional system for *H. indica*–*P. luminescens* complex reproduction. So far, the Cuban results in EPN solid fermentation are modest, but there are evidences of the potential of different raw materials in the island, and of the possibility of these materials to be used in cottage laboratories with the know–how in solid fermentation.

14.3.3 *Formulation*

Sánchez et al. (2001) determined the possible formulation of 5 million of IJs of HC1 population of *H. bacteriophora* in sponge (12 cm $\times$ 7 cm), in good conditions for two and a half months (24 $\pm$ 1 °C). Nevertheless, the viability of IJs depends of their lipid reserves and water quality. In spite of the advances that this type of formulation represent for nematode quality during preservation and transportation, it can be used only by a limited number of laboratories because of the lack of sponges and polyethylene bags in many local markets in Cuba. To avoid these difficulties, some laboratories transfer the juveniles in different types of recipients with clean water to the farmers, but handling of these "formulations" is not easy, and in sometimes the juveniles die during transportation due to the shaking effects in big recipients and to the lack of oxygen.

14.4 Use of Nematodes in the Field: Experimental Trials, Use and Perception of the Farmers

In Cuba has dedicated 6,619,500 ha for agricultural purposes, comprising in most part sugarcane, coffee, citrus–fruits and banana–plantains as permanents crops, and rice, tobacco, vegetables and tuber as temporal crops (ONEI, 2012). Depending on the national relevance of some of these crops field trials using EPNs have been developed with more or less intensity. Herein, we present a brief overview on the research performed from the laboratory to the field with selected key insect per crop.

*14.4.1 Citrus (*Citrus *spp.) (Sapindales: Rutaceae)*

Citrus is an important export crop for Cuba (Martínez, Barrios, Rovesti, & Santos, 2006). In the country, citrus and fruits occupied 169,600 ha (ONEI, 2012), and historically they received a special attention of research concerning pest control. The blue–green weevil (*P. litus*) (adults and larvae) is a significant pest in the country. The first EPN evaluations were conducted more than 30 years ago, when Artega, Montes, Broche, and Chang (1984) determined the impact of applying *H. indica* P_2M, using a back–pack sprayer or by applying infested *G. mellonella* larvae, against *P. litus* occurring in nurseries of *Citrus* spp. The mortality rate was 74–82 % with the application of the IJs suspension, while it was 61–80 % when the nematodes were applied inside *G. mellonella* cadavers. In some trials carried out at the Citrus Enterprise "Victoria de Girón", where the population P_2M was applied in citrus nurseries, there was 70–95 % of initial control, and 5 years later of its use, the general average control was 78 % (Montes & Montejo, 1990).

*14.4.2 Coffee (*Coffea *spp.) (Rubiales: Rubiaceae)*

Coffea species represent exportation crop for Cuba and are grown as a principal crop, mainly in the mountain areas (Martínez et al., 2006), with 135,500 ha (ONEI, 2012). Until the middle of the last decade of the twentieth century, some pests like mealy bugs constituted important coffee pests (Martínez & Suris, 2000). In 1995, the coffee berry borer (CBB) (*Hypothenemus hampei* (F) (Coleoptera: Curculionidae: Scolytinae) arrived in Cuba and nowadays are present in all coffee regions (Cintrón & Grillo, 2006). The pest could cause up to 50 % losses in yield becoming now the most important coffee pest in the country. Applications of *H. bacteriophora* HC1 were evaluated for the management of the mealy bugs in coffee (*Coffea* spp.) in the mountain of Santiago de Cuba (Eastern region). In this study, Rodríguez, Martínez, Sánchez, and Rodríguez (1998) used a suspension of 8×10^4 IJs/L

(*H. bacteriophora* population HC1) for spraying the soil under tree canopies against mealy bugs. Four month later, the mealy bugs' populations were not significant.

Some field trials were developed by Rodríguez, García et al. (2008) in *Coffea arabica* L. cv caturra grown in "Victoria" Farm in Buey Arriba Municipality (Granma Province). They used four concentrations and a control (no nematodes) for treating the berries over the soil surface under the trees. Fifteen days later, more than 30 % of mortality (adults, pupae and larvae stages) was recorded inside the evaluated berries, comparing with control. These berries are a source for new infestations in the field (Baker, Ley, Balbuena, & Barrera, 1992), and treating them can result in a reduction of the coffee berry borer infestation index. In "El Cedro" Farm in Buey Arriba Municipality (Granma Province), another field experiment was conducted for the coffee berry borer management. When *H. bacteriohora* HC1 (2×10^5 IJs/plant) and *Cephalonomia stephanoderis* Betrem (Hymenoptera: Bethylidae) were applied in the same plots, the coffee berry borer infestation index was reduced from 4.5 % to 2 %, an index accepted by the sanitary authorities in Cuba for adequate coffee yields.

The satisfactory results obtained with the use of EPN in these experiments made the Direction of Plant Health in Cuba suggest to the coffee regions to prepare cottage laboratories (CREE) for EPN production using *G. mellonella* and include EPN in CBB management program. In coffee agro–ecosystems (Buey Arriba Municipality, Granma Province) the EPN were observed three months after being applied (using *G. mellonella* bait techniques) indicating that they founded proper ecological conditions for recycling in soils, main hosts (Rodríguez, Hernández, Borrero, Gómez, & Enrique, 2011)

14.4.3 Grains and Corn

In Cuba, about 123,914 ha are dedicated to grain crops and 6,819 ha to corn (ONEI, 2012), in correspondence with the importance of grains (mainly common bean, soybean and chickpea) and corn in the Cuban diet.

Chickpea (*Cicer arietinum* L.) (Leguminosae: Fabaceae): In some areas in Cuba, the insects founded with the highest frequency and distribution were *Heliothis virescens* (F.) (Lepidoptera: Noctuidae) and *S. frugiperda* (Pérez, 2013). In Pinar del Río Province (western region), a preliminary field trial was carried out in 0.2 ha with chickpea with high infestation of *H. virescens* larvae. The farmers applied 5×10^4 IJs/m^2 of *H. bacteriophora* HC1. After 48 h, the application showed an efficacy of 90 % (Anaiza Echevarría, personal communication,). This trial, but in a bigger area, is being repeated by the personnel originally involved to be used as a reference trial by the private farmers developing the chickpea crop in the region.

Corn (*Zea mays* L.) (Poales: Poaceae): corn or maize is a popular plant produced for human and animal food in Cuba. The most important pest for corn is the fall army worm, *S. frugiperda,* because of its severe damages caused by larvae stage (Martínez et al., 2006). In the central zone of Cuba, in Villa Clara Province, applications of 500 IJs/plant of *H. indica* CIAP DEY–7 were evaluated for *S. frugiperda* control in a field trial and reached 72 % of efficacy (Rojas, Gómez-Sousa, & Barreada, 2001). Later, another successful control of this pest was achieved by Marrero (2006) in a field trial by using *Heterorhabditis* sp. CIAP–DEY–6 and CIAP–DEY–7 and *H. indica* P_2M. Yields in the treated plots were higher than in the untreated plots in 1.2 t/ha (dry corn) and 3 t/ha (wet corn). Also, other field experiment using *H. bacteriophora* HC1 (10^4 IJs/m^2) was conducted in the farm "La Asunción" (Property of Mr. Rolando Muñóz), San José de las Lajas Municipality (Mayabeque Province). A significant reduction of *S. frugiperda* was observed (Rodríguez, Enrique et al., 2008). Due to the positive results obtained in this reference farm trial, several farmers in José de las Lajas Municipality are systematically using EPNs for fall army worm management.

14.4.4 Vegetables (Tomato, Cucumber and Cabbage)

In Cuba, tomato, cucumber and cabbage occupied near 70,000 ha each year and represent important crops due high demand for population consume.

Tomato (*Solanum lycopersicum* L.) (Solanales: Solanaceae): Tomato is the most important horticultural crop in Cuba. The crop is grown in 54,955 ha in open fields, with a total production of 601,000 t (ONEI, 2012). Additionally, it is also important the production obtained in 167 ha of covered crops with yields between 75 and 83 t/ha per year (GEF, 2013). Tomato pinworm (*Keiferia lycopersicella* Walsingham) (Lepidoptera: Gelechiidae) is distributed and generalized in all the covered crops throughout the country; its larvae caused damages in leaves, stems and green fruits (Martínez et al., 2006). A field experiment was carried out in tomato protected crops by Sierra, Pozo, González, and Pérez (2014) to evaluate the effect of applications of *Bacillus thuringiensis, Beauveria bassiana* and *H. bacteriophora* on *K. lycopersicella*. They determined that the biological control agents must be applied 5 and 15 days after planting. The best result was reached by *H. bacteriophora* with only 9 % of plant affected, compared with 43 % in control plots.

Cucumber (*Cucumis sativus* L.) (Cucurbitales: Cucurbitaceae): In Cuba, cucumber consumption has been a tradition to consume in fresh salad, and it is mainly grown in urban and peri–urban agricultural systems. Due to its short vegetative cycle, the crop is producing during the whole year. The melon worm (*Diaphania hyalinata* L. [Lepidoptera: Crambidae]) is an important pest in melon, cucumber and squash and it is distributed in all the country and the larvae causing damages in leaves and fruits, mainly in summer (Martínez et al., 2006). In a field trial in cucumber organic production area, Pozo, Valdés, Mora, and Cárdenas (2004) evaluated the effect of *H. indica* populations P_2M and *Heterorhabditis* sp. CIAP–DEY–6 and CIAP–DEY–7 on *D. hyalinata.* The technical efficacy was shown to be different with each population, which was 68 % (P_2M) and 62 % (CIAP–DEY–6), respectively.

Cabbage (*Brassica oleracea capitata* B.) (Brassicales: Brassicaceae): this crucifer crop is the most extended in Cuba, growing in gardens and farms in urban and peri–urban agricultural areas, with an optimum season between September to December (Benítez et al., 2010; Martínez et al., 2006). Cabbage and other crucifers are grown in the country in some 6,322 ha (FAOSTAT, 2013) and the diamond back moth (DBM) (*Plutella xylostella* L. [Lepidoptera: Plutellidae]), is distributed in all the country and their larvae causing serious damages and decreasing the quality of fruits and yields. Another important pest in the cabbage crop in Cuba is the Cabbage aphid (*Brevicoryne brassicae* L., [Hemiptera: Aphididae)). It may cause plant growth reduction and even plant death. It is a virus vector.

In a field trial, Marrero (2006) obtained 72 % of efficacy with *Heterorhabditis* sp. CIAP–DEY–6 and CIAP–DEY–7 and *H. indica* P_2M on *P. xylostella.* The infestation index was reduced in the treated plot to 0.4, compared with 1.46 in the untreated control plot. Another experiment was carried out by Rodríguez, Enrique et al. (2013) in "Doña Amalia" Farm, in the western Province of Mayabeque, shown the reduction of DBM population with EPN applications. In different localities the DBM and cabbage aphid appear as concomitant pests in cabbage crops. In a field experiment, *Bacillus thuringiensis* Berliner (Bacillales: Bacillaceae), *H. bacteriophora* HC1 and *L. lecanii* were used for *P. xylostella* and *B. brassicae* management with good results. The nematodes and fungi or bacteria were applied together without detrimental effects over nematodes (Richards, 2002). Casanova et al. (2010) evaluated the combined use of *Tetrastichus howardi* Olliff (Hymenoptera: Eulophidae) and *H. bacteriophora* population HC1 in cabbage for *P. xylostella* control in Matanzas Province. In plots of 4.5 m × 10 m, they applied 4.5×10^4 IJs per plot of 45 specimens of *T. howardi.* They observed in the plots with both biocontrol agents that the parasitism in *P. xylostella* was 96 % with a significant reduction in the percentage of damaged leaves, suggesting the possible use of both agents combined in cabbage fields.

14.4.5 Tobacco (Nicotiana tabacum *L.*) *(Solanales: Solanaceae)*

Tobacco is a major crop, leader of agricultural export products, in the country. Each year, more than 66,000 ha of land are used for tobacco production (Comisión Nacional de Recursos Genéticos, 2007), and the Lepidoptera complex constitutes one of the most important tobacco pests (Martínez et al., 2006). Within the Lepidoptera, the *Heliothis* complex is an important pest of tobacco, mainly in the eastern region (Rivas, 2012). The technologies implemented in Cuba for tobacco production includes the production of plantlets in nurseries for being afterwards replanted in the field, being necessary the monitoring and pest management in both areas. In a tobacco nursery, Marrero (2006) evaluated the effect of *H. indica* P_2M and *Heterorhabditis* sp. CIAP–DEY–6 and CIAP–DEY–7 against *H. virescens* with good results, and suggested the use of EPN during this phase of the tobacco technology. Gutierréz (2008) studied the effect of *H. indica* P_2M against *H. virescens* using three concentrations (450, 1,200 and 2,000 IJs per plant). All the treatments were successful and the efficacy increased with the increment of the concentration, with high impact on the III and IV instars of the insect. The economic analysis showed that the use of EPN on tobacco were favorable to the farmers. *H. virescens*, *Heliothis tergeminus* (Felder & Rogenhofer) (Lepidoptera: Noctuidae), and *Manduca sexta* (L.) (Lepidoptera: Sphingidae) constitute a lepidopteron complex affecting the Cuban tobacco cultivar IT–2004 in the eastern region and the highest control efficacy was obtained when a mix of *B. thuringiensis* (Cuba population 24) and *H. bacteriophora* HC1 (Rivas, 2012) was used in open field trial.

14.4.6 Sweet Potato (Ipomoea batatas *Lam.*) *(Solanales: Convolvulaceae)*

The sweet potato is a strategic crop for Cuba because of its high level of consumption by the population and its short vegetative cycle that allows having several harvests in the year. It represents one of the most important food options in periods after extreme meteorological events. Its production and quality are affected by pests like the sweet potato weevil (*Cylas formicarius* Fab. [Coleoptera: Curculionidae]) and (*Typophorus nigritus* Fab. [Coleoptera: Chrysomelidae]). The viability and presence of *Heterorhabditis* sp. was evaluated in *I. batatas*. With a rate of application of 800 IJs/m^2, Liens, Andino, Expósito, and Jiménez (1998) found that nematode survival (at 10 cm) between 30 and 35 days after applications.

The impact of two Cuban bio–products and the EPNs on the quality and production of sweet potato were determined in a field experiment in San José de las Lajas (Mayabeque Province), with the aim of offering elements to the farmers for their use in production. In field conditions, three concentrations (50,000, 100,000 and 150,000 IJs/m^2) of HC1 population of *H. bacteriophora* of two national bio–nutrients (EcomiC® and FitomaS–E®) were evaluated. The results of the vegetative parameters and yields in the treated plots were statistically superior (11.2 t per ha in plots without treatments and 12.7 t per ha in treated plots). All the doses of EPNs caused significant reductions in the percentage of sweet potato weevil affectation in the treated plots (3–6 %) in relation with the untreated plots (>30 %) (Rodríguez, García et al., 2008).

Sweet potato leaf beetle *Typophorus nigritus* Fab. (Coleoptera: Chrysomelidae) is a re–emergent pest in Cuba. It was found more than 20 years ago, in the 90s, and since 2002, the populations have increased, and nowadays is an important pest in sweet potato (*I. batatas*) (Castellón, 2011). The applications of EPNs (*H. indica*) (1–3 millions of IJs/ha) on the sweet potato leaf beetle was evaluated in a field trial developed in the central part of the island (Villa Clara Province), in a place where this pest is more important than *C. formicarius*. The results indicated that the application of EPNs at planting and 40 days later, reduced significantly the attack index of the pest from 22.5 % (control without EPNs) to 10.5 (million/ha) and 2.5 % (3 million/ha).

14.4.7 Flowers

The ornamental and flower plants represent an emergent sector in the Cuban agriculture. Their productions are concentrated in areas near the cities. The white ginger lily *Hedychium coronarium* Koenig (Zingiberales: Zingiberaceae) is referred to as the Cuban National Flower (Roig, 1965), and has a preferential place in the Cuban culture and traditions, as ornamental and marketable plant or with a religious meaning. Its production is assumed by farmers in the urban and peri–urban agriculture systems, and the guidelines of this technological system establish that the chemical pesticides are forbidden, and the farmers can use biological control agents and agronomic practices for pest management (Companioni, Ojeda, Páez, & Murphy, 2001).

In 2008, colonies of *Dysmicoccus brevipes* (Cokerell) (Hemiptera: Pseudococcidae) seriously affecting the white ginger lily plantations were found in Havana City (Hernández & Martínez, 2012), and farmers demanded the biological control for this pest. In "Barroso" Farm, in Havana City a field trial was developed to study the effectiveness of EPNs in the management of *D. brevipes* in *H. coronarium.* The EPN *H. bacteriophora* population HC1 (2×10^5 IJs/plant) was used and the mealy bug population decreased from 100 to 150 specimens per plant to 9.6, with significant differences with the untreated control (Rodríguez, Hernández-Sabourin et al., 2013). Another popular plant in Cuba is the sunflower *Helianthus annuus* L. (Asterales: Asteraceae), mainly by its religious and ornamental uses, and is also produced in urban and peri–urban areas. The American sunflower moth *Homeosoma electellum* (Hulst) (Lepidoptera; Pyralidae) is the main pest of sunflower in the country causing low yields and limiting the sowing to the spring season (Limonte et al., 2010). The EPN *H. indica* P_2M was evaluated against *H. electellum* using the sunflower genotype CIAP JE–94 under field conditions. A suspension of 6 mL of media containing 312 IJs/mL of *H. indica* was applied to the flowering head with a manual backpack sprayer. The American sunflower moth was susceptible to EPN in the experiment, the yield losses were low in the treated plants compared with the untreated control, but the percentage of control was higher in winter than in spring.

14.4.8 Social Insects

Sánchez (2002) conducted a trial in San José de las Lajas Municipality (Mayabeque Province) to evaluate the effect of three applications (once per week) of 10^6 IJs/m^2 at the entrance of leaf cutting ant (*Atta* spp.) (Hymenoptera: Formicidae) nest. The leaf cutting ant abandoned the habitual routes, and three weeks after the applications no insect could be observed in the field for a short period but later the insect come again. The author pointed out that the results with EPN against social insects like leaf cutter ants in Cuba were erratic and need more research.

14.5 Entopathogenic Nematodes in Cuba: Farmer's Training and Perception

Since the 70s, the EPNs have been becoming, step by step, popular biological control agents in our country, even when only few pest–crop systems have been evaluated in the field in Cuba. However, farmers and technicians have made unsophisticated field trials and observations, and the positive results on EPN use against different pest have been communicated orally or by using other media as farmer's magazines, workshops, practical demonstrations and farmer's interviews on local television and newspapers. In Cuba, the farmers received, systematically, technical assistance from researchers, extension services and technician from research centers or agricultural authorities. Several printer materials with information about biological control and good agricultural practices are edited each year and distributed, with support of local or national institutions. The Ministry of Agriculture, the Non–Governmental Organization ANAP (Small Farmers Association), Universities and Research Centers have an important function in farmers training.

Vidal and Llanes (1999) reported in a farmer´s magazine that EPN was being used with good efficacy on several pests in La Habana, Matanzas, Cienfuegos, Las Tunas, Holguín, Guantánamo, and Pinar del Río Provinces. Farmers from different municipalities have informed in several workshops or interviews, that the EPN have been used in yam or cocoyam *Xanthosomas sagittifolium* (L.) Schott (Alismatales: Araceae), sweet potato, rice, cabbage, watercress *Nasturtium officinale* Aiton (*Brassicales: Brassicaceae*), banana – plantain (*Musa* spp.), sugar cane, common bean *Phaseolus vulgaris* L. (Fabales: Fabaceae), cucumber, guava *Psidium guajava* L. (Myrtales: Myrtaceae), grape vine *Vytis vinifera* L. (Vitales: Vitaceae), pineapple *Ananas comosus* L. (Poales: Bromeliaceae), and grasses. Beside, Vázquez et al. (2010) studied the use of biological control agents throughout Cuba, and reported the use of EPN in several crop–pest combinations with good results.

Regarding EPN and its use by farmers, the Science and Technology Ministry (Havana Province) granted an award to research team work in 2007 because of the usefulness this biological control agent had for the farmers in this province (Vázquez, 2008 in http://www.elhabanero.cubaweb.cu). A survey on EPN use by CENSA was recently performed and farmers from 11 provinces answered it. The two most recurrent pests in the use of EPN were the diamondback moth and the sweet potato weevil. In spite of the several field trials with EPN performed in Cuba, details about concentration and application frequency are scarce. Nevertheless, the farmers, in their innovative experimentation, a common activity in our country, have determined to be necessary two or three applications each cycle in temporary crops like sweet potato, cabbage, squash, and tomato, whereas in permanent crops (coffee, pineapple), the frequency must be according to the key pest life cycle, among other factors. Aspects related to concentration, frequency, and effect of soil type and other ecological aspects on efficacy of EPN require more basic research in Cuba. For it, international protocols must be used in different crops and agro–ecological conditions to obtain valuable information for researchers, farmers and the directive personnel of agriculture. Multidisciplinary and inter–institutional teams must be conformed to improve the use EPN, alone or in combination with other biological control agents and natural products in Cuba.

14.6 Conclusions and Future Directions

The Entomopathogenic Nematology have a modest but solid development level in Cuba, and the trafficked road since the pioneers Professors Magda Montes and Eva Arteaga until today, offers valuable scientific data and practical evidences, which make EPN become one of the most popular biological control agents among farmers and extension workers. However, several challenges for scientists, farmers and some agriculture stakeholders for the coming years still remind. Among them are:

- To develop molecular biology studies to obtain selected populations (nematodes and bacteria) DNA sequences and their deposit in international database.
- To develop studies on toxins and metabolites from symbiotic bacteria of selected nematodes, looking for new applications in agriculture or other public sectors.
- To implement the infrastructure and material conditions for EPN collections for scientific, productive and education purposes.
- To obtain the technologies for in vitro culture in two productive scales: solid culture for improving the installed capacities in CREE for EPN productions and liquid culture technology for being used in industrial factories established in Cuba.
- To obtain formulations to improve the shelf life and applications conditions.
- To continue studies related to the field use of EPN: concentrations, frequency and application time, equipment, compatibility with other biological control agents, natural oils, and bio–fertilizers of common use in Cuba.
- To increase the theoretical and practical education of professionals, biological control producers, and biology and agronomy students to achieve a better understand and use of EPN, in order to warranty that the new generations in a country like Cuba, where EPN is accepted by farmers as one of the most useful biological control agents, the future professionals and technicians used EPN.
- To continue increasing the knowledge about EPN of farmers and stakeholders in the agricultural sector, using the massive media of communications more frequently. To prepare printed material for farmers and agricultural extension workers as a support of farmer's school and workshops.

- To create a national network between scientific and academic institutions working on EPN, to take advantage of the strengths of each partner.
- To increase international collaborative works between laboratories and universities from different parts of the world, for training scientific workers and students on EPN.
- To create a webpage with relevant information about EPN in Cuba and use it for communicating the Cuban results on the basic and applied research. To increase significantly the number of Cuban papers to be published in international peer review journals.

The future of EPN in Cuba is bright, but the proper use of these biological control organisms depends on the research and technological advancements in the next years. The Environmental Law in the country and the social concern about sustainability of our agriculture represents opportunities for increasing the use of EPN and other biological control agents in a frame of Integrated and/or Agro–ecological Pest Management. In the next years, the research teams in Cuba must increase the efforts to reach the in vitro culture technology and the levels of IJs produced to be able to satisfy the high demand of EPN by farmers. Crops like corn, grains, vegetables, coffee and sweet potato must be priorities for their functions in food security and exportation. The Cuban scientific workers are open to learn more from other researchers in the world and to collaborate and interchange in Entomopathogenic Nematology.

Acknowledgments The author is really grateful to Dr. Eduardo Sistachs Rodríguez for his contribution with the English text revision and suggestions, and Dr. Raquel Herrera–Campos for her suggestions. Thanks are due to Mrs. Nery Hernández for supplying original papers from different Cuban libraries. This contribution was made under the bilateral Project "*Desarrollo de bioplaguicidas a base de nematodos entomopatógenos para el manejo de plagas del sector agrario en Venezuela y Cuba*" CENSA–IVIC (financially supported by Convenio de Cooperación Integral Cuba–Venezuela) and the Cuban Project: "*Diagnóstico y manejo de plagas en granos con énfasis en el desarrollo y uso de productos bioactivos*" (supported by the National Scientific Program for Animal and Plant Health, Cuba)

References

Alemán, J., & Goicoechea, J. L. (1993). Nueva dieta para *Galleria mellonella* Linneo. *Revista de Protección Vegetal, 9*, 309–310.

Alizar, T., St Louis, L., Cárdenas, M., Valdés, R., & Pozo, E. (2010). Determinación del instar óptimo de *Spodoptera frugiperda* en la reproducción de *Heterorhabditis indica*. *Centro Agrícola, 37*(2), 19–25.

Altieri, M. A., & Funes-Monzote, F. R. (2012, January). The paradox of Cuban agriculture. *Monthly Review, 63*(8). Available in http://monthlyreview.org/2012/01/01/the-paradox-of-cuban-agriculture/ [revised 20/10/2014]

Artega, E., Montes, M., Broche, R., & Chang, B. (1984). Utilización de *Heterorhabditis heliothidis* (Heterorhabditidae) contra *Pachnaeus litus* (Coleoptera, Curculionidae) en Cuba. *Ciencia y Técnica en Agricultura de Cítricos y otros frutales, 7*(2), 79–85.

Baker, P. S., Ley, S., Balbuena, R., & Barrera, G. J. F. (1992). Factors affecting the emergence of *Hypothenemus hampei* (Coleoptera: Scolytidae) from coffee berries. *Bulletin of Entomological Research, 82*(2), 145–150.

Benítez, M. E., Rivero, P. P., Gil, J. F., Marrero, C. Martínez, Y., & Soto, J. A. (2010). *Manual Técnico para el cultivo y la producción de semillas de la col en Cuba. Instituto de Investigaciones Fundamentales en Agricultura Tropical «Alejandro de Humboldt»*. 20 pp. La Habana, Cuba: Editora Agroecológica. Asociación Cubana de Técnicos Agrícolas y Forestales (ACTAF).

Casanova, Y., Díaz, M., Naranjo, F., Álvarez, J. F., Barroso, G., Albuernes, F., et al. (2010). Evaluación de las potencialidades parasíticas de *Tetrastichus howardi* (Olliff) y efectividad combinada con *Heterhorabditis bacteriophora* contra *Plutella xylostella* Lin. en col. Congreso Científico del Instituto Nacional de Ciencias Agrícolas (INCA) (www.inca.edu.cu). CD ISBN: 978-959-7023-48-7.

Castellanos, L. (2000). Efectividad de los nematodos entomopatógenos *Heterorhabditis bacteriophora* (HC1) y *Steinernema* sp. (Sc1) en el control de insectos del orden Homoptera (pulgones, cóccidos y moscas blancas), en condiciones de laboratorio. *Centro Agrícola, 1*, año 27, 25–30.

Castellanos, L. L., González, J. B., & Jacomino, M. (2000). Efectividad del aislado HI–24 de *Heterorhabditis* sp., en el control biológico del pulgón negro del plátano (*Pentalonia nigronervosa* Cql. (Homoptera: Aphididae), en condiciones de laboratorio. *Centro Agrícola, 27*(3), 19–22.

Castellón, M. C. (2011). *Estudios biológicos y elementos para el manejo de* Typophorus nigritus *Fabricius (Coleoptera: Chrysomelidae) en plantaciones de boniato (Ipomoea batatas (L.) Lam* (100 p.) [PhD thesis dissertation]. Universidad Central "Marta Abreu" de Las Villas (UCLV), Villa Clara, Cuba.

Cintrón, B., & Grillo, H. (2006). Caracterización de la dinámica poblacional de la broca del café *(Hypothenemus hampei* (Ferrari) (Curculionidae: Scolytinae) durante el desarrollo de los frutos. *Centro Agrícola, 33*(3), 55–60.

Comisión Nacional de Recursos Genéticos. (2007). *Cuba: Segundo Informe Nacional sobre Recursos Fitogenéticos para la Agricultura y la Alimentación (RFAA)* (189 p.). La Habana, Cuba: Unidad de Producciones Gráficas del MINREX.

Companioni, N., Ojeda, Y., Páez, E., & Murphy, C. (2001). La agricultura urbana en Cuba. In F. Funes, L. García, M. Bourque, N. Pérez, & P. Rosset (Eds.), *Transformando el campo cubano. Avances de la agricultura sostenible* (pp. 94–109). La Habana, Cuba: Asociación Cubana de Técnicos Agrícolas y Forestales (ACTAF).

Dutky, S. R., Thompson, J. V., & Canwell, G. E. (1964). A technique for mass propagation of the DD–136 nematodes. *Journal of Insect Pathology, 6*, 417.

Enrique, R., Sánchez, L., Rodríguez, M. G., Gómez, L., & Valle, Z. (2006). *Dietas alternativas para la cría de* G. mellonella. *Influencia sobre el rendimiento – peso de larvas de Galleria mellonella y recobrado de juveniles infectivos* (20 p.). Ciudad de la Habana, Cuba: Centro Nacional de Derecho de Autor (CENDA). Número de depósito CENDA2874–2006.

FAOSTAT. (2013). *Food and Agriculture Organization of United Nations Home page*. Available in: http://faostat.fao.org/site/567/default.aspx#ancor [revised 28/05/2014]

Fischer-Le Saux, M., Arteaga-Hernández, E., Mrácek, Z., & Boemare, N. E. (1999). The bacterial symbiont *Xenorhabdus poinarii* (Enterobacteriaceae) is harbored by two phylogenetic related host nematodes: the entomopathogenic species *Steinernema cubanum* and *Steinernema glaseri* (Nematoda: Steinernematidae). *FEMS Microbiology Ecology, 29*, 149–157.

Friedman, M. J. (1990). Commercial production and development. In R. Gaugler & H. K. Kaya (Eds.), *Entomopathogenic nematodes in biological control* (pp. 153–172). Boca Raton, FL/Ann Arbor, MI/Boston: CRC Press.

Fuentes, A., Llanes, V., Méndez, F., & González, R. (1998). El control biológico en la agricultura sostenible y su importancia en la protección de la caña de azúcar en Cuba. *Phytoma, 95*, 24–26.

García, M., Rodríguez, Y., Cabrera, D., Gómez, L., & Rodríguez, M. G. (2007). Producción de nematodos entomopatógenos en el Centro Nacional de Referencia Fitosanitaria para la Montaña en Cuba. *Revista de Protección Vegetal, 22*(2), 131–133.

Glazer, I., & Lewis, E. E. (2000). Bioassays for entomopathogenic nematodes. In A. Navon & K. R. S. Ascher (Eds.), *Bioassays of entomopathogenic microbes and nematodes* (pp. 229–247). Wallingford, UK: CAB International.

Gómez, M. (2002). *Biología y Patogenicidad de* Heterorhabditis indica P_2M (80 p.) [MSc Tesis]. La Habana, Cuba: Instituto de Fruticultura Tropical.

Grupo Empresarial Frutícola (GEF). (2013). *Informe de cultivos protegidos* (10 p.). La Habana, Cuba: Ministerio de la Agricultura.

Gutierréz, J. A. (2008). *Dosis letal y tiempo letal en tratamiento con* Heterorhabditis indica P_2M *para el control de Heliothis virescens en tabaco* [MSc Tesis]. Las Villas, Cuba: Universidad Central "Marta Abreu".

Hernández, I., & Martínez, M. A. (2012). *Dysmicoccus brevipes* Cokerell (Hemiptera: Pseudococcidae) nuevo informe para *Hedychium coronarium* Koenig, Flor de la Mariposa, en Cuba. *Revista de Protección Vegetal, 27*(1), 54–55.

Instituto Nacional de Investigaciones de la Caña de Azúcar (INICA). (2011). *Informe de la IX Reunión Nacional de variedades, semilla y sanidad vegetal. 23–25 febrero 2011* (152 p.). Sancti Spiritus, Cuba: Instituto Nacional de Investigaciones de la Caña de Azúcar.

Kulkarni, N., Kumar, D., Kumar, V., & Paunikar, S. (2012). Effect of economical modification in artificial diet of greater wax moth *Galleria mellonella* (Lepidoptera: Pyralidae). *Indian Journal of Entomology, 74*(4), 369–374.

Liens, B. R., Andino, M., Expósito, I., & Jiménez, C. (1998). Permanencia del *Heterorhabditis* sp. en el suelo a diferentes profundidades. *Centro Agrícola, 25*(1), 43–44.

Limonte, A., Grillo, H., Hernández, U., Pozo, E., Valdés, R., & Cárdenas, M. (2010). Nematodos entomopatógenos en el control de *Homeosoma electellum* (HULST) en campos de girasol. *Revista Centro Agrícola, 37*(2), 13–17.

Márquez, M. E., Fernández-Larrea, O., & Arteaga, E. (1997). Indicadores para determinar la fase primaria de *Photorhabdus luminescens. Revista de Protección Vegetal, 12*(2), 85–88.

Marrero, M. A. (2006). *Nematodos entomopatógenos (*Heterorhabditis *spp.) para el control de* Spodoptera frugiperda *(J. E. Smith),* Plutella xylostella *(Linnaeus.) y* Heliothis virescens *(Fabricius)* [MSc Tesis]. *Centro Agrícola* año *33*(2), 90.

Martín, D. (2007). *Identificación, caracterización y reproducción masiva del simbionte bacteriano de* Heterorhabditis bacteriophora *cepa HC1* (80 p.) [MSc Tesis]. La Habana, Cuba: Universidad de La Habana.

Martínez, E., Barrios, B., Rovesti, R., & Santos, R. (Eds). (2006). *Manejo Integrado de Plagas. Manual Práctico. Centro Nacional de Sanidad Vegetal (CNSV), Cuba* (482 p.). Tarragona, Cataluña, Spain: Entrepueblos, España & Gruppo di Volontariato Civile (GVC), Italia.

Martínez, M. A., & Suris, M. (2000). Bases bioecológicas para el manejo de cochinillas harinosas en el cultivo del café en Cuba. *Revista Manejo Integrado de Plagas (Costa Rica), 57*, 58–64.

Montes, M. (1978). Informe sobre un nematodo del género *Neoaplectana* como enemigo natural de las larvas del picudo verde azul *Pachnaeus litus* (Coleoptera, Curculionidae). *Ciencia y Técnicas Agricolas en Cítricos y otros frutales, 3*(1), 43–45.

Montes, M., Broche, R., Hernández, M. R., Gómez, M., Hernández, D., & Rodríguez, J.-L. (2014). *Curculiónidos que atacan los cítricos en Cuba y sus enemigos naturales*. http://www.fao.org/docs/eims/upload/cuba/5380/UltimoCurculiónidosTrabajoFAO.pdf [revised 26/03/2014]

Montes, M., & Montejo, R. (1990). Lucha biológica contra *P. litus* (Coleoptera, Curculionidae) con la utilización de *Heterorhabditis heliothidis*. In C. Pavis & A. Kermarrec (Eds.), *Rencontres caraibes en lute biologique. Proceedings of the Caribbean meetings on biological control, Guadeloupe, 5–7 November 1990* (pp. 157–160). Paris: Institut National de la Recherche Agronomique.

Mráček, Z., Arteaga, E., & Boemare, N. E. (1994). *Steinernema cubana* sp. n (Nematoda: Rhabditidae: Steinernematidae) and preliminary characterisation of its associated bacterium. *Journal of Invertebrate Pathology, 64*, 123–129.

Oficina Nacional de Estadísticas e Información (ONEI). (2012). *República de Cuba. Anuario Estadística de Cuba 2011*. Agricultura, Ganadería, Silvicultura y Pesca. 29 p.

Pacheco, U. (2010). Resultados y perspectivas del redimensionamiento agroindustrial azucarero cubano. *Retos de la Dirección, 4*(2), 36–43.

Pérez, J. C. (2013). Inventario, biología y alternativas para el manejo de insectos asociados al garbanzo (*Cicer arietinum* L.) en el municipio Jesús Menéndez, Las Tunas. *Revista de Protección Vegetal, 28*(1), 79.

Pérez, T., Vázquez, T., & Mollineda, M. (1991). Mass rearing of *Heterorhabditis heliothidis* on the alternate host *Corcyra cephalonica. Proteccion de Plantas, 1*(1), 7–14.

Pérez, Y., Márquez, M. E., & Gómez, M. (2006). Caracterización de la fase primaria de las bacterias simbiontes de los nematodos entomopatógenos *Steinernema cubanum, Heterorhabditis indica y H. bacteriophora. Fitosanidad, 10*(3), 193–195.

Pozo, E., López, D., & Martínez, Y. (2003). Nuevos aislados de nematodos entomopatógenos en la región central de Cuba. *Centro Agrícola, 30*(4), 94–95.

Pozo, E., Valdés, R., Mora, E., & Cárdenas, M. (2004). *Consumo de área foliar del pepino y umbral económico y señalización de* Diaphania hyalinata *(Linn.) (Lepidoptera; Pyralidae) y control con nematodos entomopatógenos en organopónicos.* V Seminario Internacional de Sanidad Vegetal. La Habana, Cuba. Mayo del 2004. Cd ISBN: 959–246–137–6.

Richards, M. (2002). *Alternativas Agroecológicas como medidas de lucha de las plagas claves del cultivo de la col de repollo* (Brassicae oleracea *var.* capitata) (80 p.) [MSc Tesis]. La Habana, Cuba: Universidad Agraria de La Habana.

Rivas, A. (2012). Lepidópteros en cultivares de tabaco: elementos ecológicos y alternativas biológicas para su manejo en Las Tunas. *Revista de Protección Vegetal, 27*(3), 216.

Rodríguez, I., Martínez, M. A., Sánchez, L., & Rodríguez, M. G. (1998). Comprobación en campo de la efectividad de *Heterorhabditis bacteriophora* cepa HC1 en el control de chinches harinosas (Homoptera: Pseudococcidae) del cafeto. *Revista de Protección Vegetal, 13*(3), 195–198.

Rodríguez, M. G., Enrique, R., Gómez, L., Hernández-Ochandía, D., Miranda, I., Hernández, A., et al. (2013). Impact of entomopathogenic nematode applications on diamond back moth population. *Revista de Protección Vegetal, 28*(2), 158–160.

Rodríguez, M. G., Enrique, R., González, E., Gómez, L., Bertolí, M., Montano, R., et al. (2008). *Desarrollo y uso racional de nematodos entomopatógenos en el manejo de plagas.* In: VI Seminario Científico Internacional de Sanidad Vegetal. Cuidad de la Habana, Cuba. ISBN: 978–959–282–080–7.

Rodríguez, M. G., García, M., Gómez, G., Rodríguez, Y., Enrique, E., Alemán, J., et al. (2008). *Producción y utilización de nematodos entomopatógenos en el manejo de la broca del café* (Hypothenemus hampei) (177 p.). Informe final de Proyecto de Investigación. Programa Ramal de Entidades Exóticas. La Habana, Cuba: Ministerio de la Agricultura.

Rodríguez, M. G., Hernández, M., Borrero, Y., Gómez, L., & Enrique, R. (2011). Recycle of entomopathogenic nematodes in soils grower coffee crop (*Coffea* spp.) in Buey Arriba. *Revista de Protección Vegetal, 26*(1), 67.

Rodríguez, M. G., Hernández-Ochandía, D., & Gómez, L. (2012). Nematodos entomopatógenos: elementos del desarrollo histórico y retos para su consolidación como biorreguladores en la agricultura en Cuba. *Revista de Protección Vegetal, 27*(3), 137–146.

Rodríguez, M. G., Hernández-Sabourin, I., Enrique, R., Gómez, L., Hernández-Ochandía, D., López, L., et al. (2013). Eficacia de nematodos entomopatógenos en el manejo de *Dysmicoccus brevipes* en el cultivo de *Hedychium coronarium. Nematropica, 43*(2), 321–322.

Roig, J. T. (1965). *Diccionario botánico de nombres vulgares cubanos.* Tomos I–II (1142 p.), La Habana, Cuba: Editorial Consejo Nacional de Universidades.

Rojas, J., Gómez-Sousa, J. R., & Barreada, A. (2001). Control de *Spodoptera frugiperda* (J. E. Smith) con diferentes medios biológicos. *Centro Agrícola, año 28*(2), 19–22.

Sánchez, L. (2002). Heterorhabditis bacteriophora *HC1. Estrategia de desarrollo como agente de control biológico de plagas insectiles* (100 p.) [PhD Tesis]. La Habana, Cuba: Universidad Agraria de La Habana. (Documento depositado en el Centro Nacional de Derecho de Autor (CENDA) Número 9613–2002.

Sánchez, L., Christopher, J. E., Rodríguez, M. G., Gómez, L., & Soler, D. M. (2003). Alteraciones en el comportamiento de una cepa de *Heterorhabditis bacteriophora. Revista de Protección Vegetal, 18*(1), 49–52.

Sánchez, L., Rodríguez, M. G., Gómez, L., Soler, D. M., Hernández, M. A., Castellanos, L., et al. (2001). *Desarrollo de una Metodología para la reproducción artificial de nematodos entomopatógenos para el control de plagas en cafeto*. PNCT: Desarrollo Sostenible de la Montaña. CODIGO: 0703023. Informe final proyecto– CENSA. (Metodologías Depositadas en Centro de Derechos de Autor (www.cenda.cult.cu), Cuba, número 09613/2002).

Sánchez, L., Soler, D. M., Gómez, L., & Martín D. (2006). *Medios de cultivo para la cría in vitro de* Heterorhabditis *spp.* Oficina Cubana de Propiedad Industrial: Patente OCPI 882/2006.

Sánchez, S. (1997). El CREE de Margarita, algo más que un sueño. *Agricultura Orgánica, 3*(2–3), 5–8.

Sierra, A., Pozo, E., González, E., & Pérez, B. (2014). *Lucha biológica de* Keiferia lycopersicella *(Walsg.) en tomate en producciones protegidas de Ciego de Ávila*. VII Congreso Internacional de Sanidad Vegetal. La Habana, Cuba. http://ahorainternet.net/mail/Norlen/Multimedia/MultimediaVIICongSV/Ponencias/P12/SIERRA.%20A.pdf [revised 12/12/2014]

Stack, M., Easwaramoorthy, S. G., Metha, U. K., Downes, M. J., Griffin, C. T., & Burnell, A. M. (2000). Molecular characterisation of *Heterorhabditis indica* isolates from India, Kenya, Indonesia and Cuba. *Nematology, 2*(5), 477–487.

Valdés, Y., Lobaina, A., Márquez, M. E., Gómez, M., & Escobar, M. (2006). Reproducción de *Heterorhabditis indica* en cultivos bidimensionales elaborados con proteína animal. *Fitosanidad, 10*(1), 29–31.

Vázquez, A. (2008). *Aíslan agente de control biológico para plagas agrícolas*. Periódico "El Habanero Digital" edición del 05–02–2008. http://www.elhabanero.cubaweb.cu/ and http://www.elhabanero.cubasi.cu/ [revised 03/02/2008]

Vázquez, L. L., Moreno, S., Carr, A., Gil, J., Armas, J. L., Rodríguez, A., et al. (2010). Diagnóstico de la utilización de entomófagos y entomopatógenos para el control biológico de insectos por los agricultores en Cuba. *Fitosanidad, 14*(3), 159–169.

Vidal, M., & Llanes, G. (1999). Generalización del nematodo (*Heterorhabditis bacteriophora*). *Revista Cañaveral, 5*(2), 9–41. (ISBN 1026–0781)

White, G. F. (1927). A method for obtaining infective nematode larvae from cultures. *Science, 66*, 302–303.

Woodring, J. L., & Kaya, H. K. (1988). *Steinernematid and heterorhabditid nematodes: A handbook of techniques* (Southern cooperative series bulletin, Vol. 331). Fayetteville, AR: Arkansas Agricultural Experiment Station. 30 pp.

Chapter 15
Entomopathogenic Nematodes in Tropical Agriculture: Current Uses and Their Future in Venezuela

Ernesto San-Blas, Carolina Rosales, and Ángel Torres

15.1 Introduction

In Venezuela, agricultural pest control is very frequently done using chemical products in indiscriminate manners, with adverse effects to human health and the environment. Thank to technological advances, today it is possible to identify chemical residues in food which were undetected in the past (Mol et al., 2008). For that reason, we can assume that the ingestion of fresh and processed vegetables and fruits which were considered innocuous in the past, could be affecting the consumers health. In fact, more than 1,500 chemical poisoning cases due to pesticides were reported in 2011 in the country (MPPS, 2011). Cholinesterase levels in blood have been correlated with poisoning due to organophosphate pesticides; in 2003, a study with a farmer community at Falcón State, indicated 60 % of poisoning in the whole population (Zamora, 2003).

For decades, many scientists and academic institutions have been concerned with the dangers of abusing chemical pesticides and despite it, no permanent strategies or policies to reduce the input of those compounds in the Venezuelan fields have been developed, while some advances and attempts of introducing biological control

E. San-Blas (✉)
Instituto Venezolano de Investigaciones Científicas, Centro de Estudios Botánicos y Agroforestales, Laboratorio de Protección Vegetal, Maracaibo, Estado Zulia, Venezuela
e-mail: esanblas@ivic.gov.ve; esanblas@yahoo.com

C. Rosales
Laboratorio de Nematología, Instituto Nacional de Investigaciones Agrícolas, Maracay, Estado Aragua, Venezuela
e-mail: carolinar@yahoo.com

Á. Torres
Campo Experimental Bramon, Instituto Nacional de Investigaciones Agrícolas, Bramón, Estado Táchira, Venezuela
e-mail: atorres@inia.gob.ve

© Springer International Publishing Switzerland 2015

R. Campos-Herrera (ed.), *Nematode Pathogenesis of Insects and Other Pests*, Sustainability in Plant and Crop Protection 1, DOI 10.1007/978-3-319-18266-7_15

agents have been done. Nowadays many laboratories in the country are dedicated to the study of those organisms and their commercialization with an emphasis in bacteria (*Bacillus thuringiensis* Berliner [Bacillales: Bacillaceae]), fungi (*Beauveria bassiana* Vuill. [Hypocreales: Clavicipitaceae], *Metarhizium anisopliae* Sorokin [Hypocreales: Clavicipitaceae] and *Trichoderma harzianum* Rifai [Hypocreales: Hypocreaceae],), and insects, both predators and parasitoids (*Crisoperla* spp. [Neuroptera: Chrysopidae] and *Trichogramma* spp. [Hymenoptera: Trichogrammatidae).

The first registered experience in biological control in the country was done by Dr. Juan Iturbe and the bacteriologist Eudoro González in 1913. An outbreak of locusts *Shistocerca gregaria* Forsskal (Orthroptera: Acrididae) from Colombia occurred in the country in 1912 causing an immense devastation in the fields. The government contacted Dr. Felix D'Herelle (French–Canadian microbiologist) who discovered and used *Coccobacillus acridiorum* (now *Enterobacter aerogenes* Kruse [Enterobacteriales: Enterobactereaceae]) for controlling locusts and after some negotiations, he gave samples of the bacterium to the Venezuelan scientists. By using this organism, some experiments were performed and between the end of 1913 and 1914, the national government launched the first biological control program for locusts applying thousands of litres of a broth containing the bacteria (Quiróz-Mireles, 2005).

Other experiences in biological control programs in Venezuela were achieved by the American entomologist Charles Ballou. In 1941, he applied *Rodolia cardinalis* Rulsant (Coleoptera: Coccinellidae] to control *Icerya purchase* Maskell (Hemiptera: Monophlebidae) (cottony cushion scale) in citrus orchards (Ballou, 1941) and the woolly apple aphid (*Eriosoma lanigerum* Hausmann [Homoptera, Aphididae]). By the end of the decade (1947), the British entomologist Harold Box, bred and released the Amazonian fly (*Lydella minense* Townsed [Diptera: Tachinidae]) to control the sugarcane borer (*Diatrea saccharalis* Fabricius [Lepidoptera: Crambidae]). In the 60s an Italian entomologist, Pietro Gaugliumi, established biological control programs of froghoppers (*Aeneolamia varia* Fabricius [Hemiptera: Cercopidae]) in sugar cane fields, palm weevils (*Rhynchophorus palmarum* L. [Coleoptera: Curculionidae]) and locusts (*Shistocerca* sp. [Orthoptera: Acrididae]) (Ferrer, 2001; Guédez, Castillo, Cañizales, & Olivar, 2008).

Between 1987 and 1994 a research program was conducted in the Norwestern region of the country to evaluate the incidence of phytophagous arthropods, their population fluctuations, regulation by natural enemies and the crop damage in free–pesticide commercial plots (Chirinos & Geraud-Pouey, 2011). The results allowed the implementation of new concepts in integrated pest management programs (IPM) and the use of chemical pesticides dropped from 17 applications per cycle (in 4 months) to just two applications per cycle. The authors observed that when farmers applied chemical pesticides to tomato seedlings, for controlling leaf miners (*Liriomyza* spp. [Dipetra: Agromyzidae]), the natural populations of entomoparasitic wasps were affected, limiting their control of the miners and worsening the problem (Chirinos & Geraud-Pouey, 1996; Geraud-Pouey, Chirinos, & Rivero, 1997). Sadly, this program was abandoned by lack of financial support, and shortly after, the farmers came back to using the chemical products in high and frequent doses.

Despite the lack of official data on the numbers regarding biological control means, an increase in the interest of both government and farmers in substituting chemical control has been observed (Guédez et al., 2008), though to this day, more than 92 % of vegetable farmers still use high doses of chemical pesticides in their crops (Chirinos & Geraud-Pouey, 2011). Conversely, one of the few, most renown and durable experience in integrated pest management (IPM) and biological control programmes in the country is represented by "La Alianza", a cooperative located at the community of Las Lajitas (Lara state). For more than 20 years, this community has been producing crops without the use of chemical pesticides, which they substituted for *Trichogramma* sp., *Chrysoperla carnea* Stephen (Neuroptera: Chrysopidae) and *B. thuringiensis* (Guillén, Alcalá de, Fernández, Pire, & Álvarez, 2008). Despite introducing biological control programmes in the country has taken many year, it is probably that in the near future, many other organisms will be incorporated in the fields. The positive perception in the use of biological control agents among the population has been increased in the last few years and there are many ongoing initiatives for biological control. Entomopathogenic nematodes (EPNs) represent a new alternative in Venezuela and research stills in early stages; however the efforts made by some researchers, students and farmers have contributed to improve the visibility of EPNs in some parts of the country and in some more years the results of incorporating them in Venezuelan fields will be tangible

15.2 First Steps on Entomopathogenic Nematode Research in Venezuela

In Venezuela studies on EPNs have been limited. The initial steps in their research were achieved between 1980 and 1990 at the Instituto de Zoología Agrícola, Facultad de Agronomía, Universidad Central de Venezuela (http://www.ucv.ve/estructura/facultades/facultad-de-agronomia.html) by Prof. Giovanni Martínez and the technician H. Wassink. Their main research was focused on the development of culture media for mass production of EPNs. Before discontinuing the research on the field, they contributed in the publication of a review about the situation of EPNs in Latin America (Wassink & Poinar, 1984).

Late in the 90s, the first attempts to control insect pests with EPNs were done; in the 8th Latin–American Congress of Phytopathology, 1995, Fan and Maggiorani presented their first results in laboratory trails, testing a *Heterorhabditis* spp. (Rhabditida: Heterorhabditidae) isolated in the southern region of Maracaibo Lake, against larvae of the Guatemalan potato moth (*Tecia solanivora* Povolni [Lepidoptera: Gelechiidae]) (Fan, Maggiorani, & Gudiño, 2000). Also, Dr. Carolina Rosales from the Instituto Nacional de Investigaciones Agrícolas (INIA) (http://www.inia.gov.ve/) prepared a sampling prospection of EPNs in Aragua State (at the centre of the country) between 1996 and 1997. In that work, some *Heterorhabditis* species/populations were isolated and evaluated in laboratory conditions as potential biocontrol agents of the banana weevils (*Cosmopolites sordidus* Germar (Coleoptera: Curculionidae]) (Rosales & Súarez, 1998).

Ferrer et al. (2004) performed the first field trials against the sugar froghopper (*A. varia*) in commercial sugar cane fields, while Angel Torres started his work on the susceptibility of coffee berry borer (*Hypothenemus hampei* Ferrary [Coleoptera: Scolytidae]) to *Steinernema* (Rhabditida: Steienrnematidae) and *Heterorhabditis* spp. (Pacheco & Torres, 2005) in the Andean INIA station located in Táchira State.

In 2007, Dr. Rosales and member of her team established an international cooperation agreement with the Cuban "Centro Nacional de Sanidad Agropecuaria (CENSA)" to develop the technological basis for mass production and application of EPNs in Venezuela. From that project, some interesting results have been applied on the ongoing research in different parts of the country, such as the traditional rearing of *Galleria mellonella* L. (Lepidoptera: Pyralidae), sampling methods for native populations of EPNs, and socio–economic studies on the implementation of EPNs in pest management programs (Rosales, Suárez, Navas, & Tellechea, 1999a, 1999b; Rosales et al., unpublished; Briceño et al., unpublished). In the same year, Ernesto San–Blas founded the Laboratorio de Protección Vegetal at the Instituto Venezolano de Investigaciones Científicas (Zulia State branch) (www.ivic.gob.ve) which initiated his research focused in the ecology of EPNs as well as sampling in regional surveys in the north–western part of the country.

From 2000 on, some public and private laboratories have had an interest in mass production of EPNs for field application, but their sustainability through time has been narrow. The most limiting factor has been the producing system based on *G. mellonella* larvae instead on industrial processes. As a consequence, the number of non–formulated products offered still low, and their impact in the national crops is completely unknown.

Nowadays, three laboratories are dedicated to EPNs studies: (a) Carolina Rosales' lab, located in the central zone of the country, belongs to the Instituto Nacional de Investigaciones Agrícolas (www.inia.gov.ve) and focuses its research on the use of EPNs to control target pests such as *Spodoptera frugiperda* Smith (Lepidoptera: Noctuidae), mangoes fruit flies *Anastrepha obliqua* Macquart (Diptera: Thephritidae) and the sweet potato weevil *Euscepes postfasciatus* Fairmaire (Coleptera: Curculionidae); (b) Angel Torres' lab, at the Andean branch of the same institution, works in the control of insect pests related with highland crops such as coffee, potato, strawberry, among others. Most of his research is devoted to evaluate native populations of EPNs for controlling insect pests of agricultural importance in the Andes; and (c) Ernesto San-Blas' lab from the North Western Zulia state branch of the Instituto Venezolano de Investigaciones Científicas (www.ivic.gob.ve), dedicated to the collection of native species/populations for biological control purposes, as well as the study of ecological aspects of EPNs, development of new tools for quality control of EPNs through infrared technology and use of their symbiotic bacteria in protected crops.

Even though studies and published works with EPNs have been scarce, the gathered results are promising, many more people are being involved in the area and the founding bodies (public and private) have become engaged in this line of research. These facts will stimulate more research in order to provide technological innovations in biological control, taking advantage on the increasing demand of biocontrol agents by virtue of the growing consciousness of farmers and consumers to produce food by more ecological (or less chemical) means.

15.3 Feasibility of Using Entomopathogenic Nematodes in Venezuela

According to official data, the Venezuelan annual harvested area is equivalent to 2,500,000 ha (MPPAT, 2007). Valuable crops are vegetables (1.4 %), fruits (6.4 %) [especially plantains and bananas (*Musa* sp. [Zingiberales: Musaceae]), citrus (*Citrus* sp. [Sapindales: Rutaceae]), pineapples (*Ananas* sp. [Poales: Bromeliaceae]), avocados (*Persea* sp. [Laurales: Lauraceae]), papayas (*Carica* sp. [Brassicales: Caricaceae]) and mangoes (*Mangifera* sp. [Sapindales: Anacardiaceae])] and 19.2 % of so called "tropical crops" which include coffee (*Coffea* sp. [Gentianales: Rubiaceae]), cocoa (*Theobroma* sp. [Malvales: Malvaceae]), sugar cane (*Saccharum* sp. [Poales: Poaceae]) and tobacco (*Nicotiana* sp. [Solanales: Solanaceae]). They occupy 27 % of the harvested area, equivalent to 675,000 ha, which could be treated by biological control methods in a short term with profitable possibilities for private companies.

Despite the experience on biological control programs in Venezuela, the use of EPNs has been limited to experimental conditions, and only in few cases have been reported for commercial purposes. Ferrer et al. (2004) tested a Cuban population of *Heterorhabditis bacteriophora* Poinar (Rhabditis: Heterorhabditidae) against the sugar froghopper (*A. varia*) in commercial sugar cane fields, applying 50–100 million of infective juveniles (IJs) per ha resulted in 75 % of control (Table 15.1); however, no other reports (at commercial level) have been published after this experience. Due to political, economic, and lack of awareness of each other, for many years commercial companies and academic bodies represented by public universities and public scientific organizations have been divorced of achieving common goals in the use of EPNs. The former because of their determination to import and apply foreign species/population of nematodes with few previous experimentation; and the latter by focusing their efforts in the search for native nematodes which are subsequently tested in laboratory assays, with few field experiments (Table 15.1). Fortunately, in the last few years, a better relationship between public and private sector in terms of research, nematode production and field applications seems to be emerging.

In the last decade, politics have been designed to increase the use of biological controls in order to reduce the input of chemical pesticides into the environment (nevertheless, no official data has been released to know the current status of biological control programmes). The first mention of biological agents in the Venezuelan legislation appeared in 2002 under the "seed, animal reproductive

Table 15.1 Target insects and their control using entomopathogenic nematodes tested in Venezuela

Common name	Scientific name	Crop	Used species/strain	Control (%)	Usage level	Source
Fall armyworm	*Spodoptera frugiperda*	Corn *Zea mays*	Different native isolates	42–100	Laboratory	Rosales' Lab
West Indian sweet potato weevil	*Eucespes postfasciatus*	Sweetpotato *Ipomea batatas*	Different native isolates	36–94	Laboratory	Rosales' Lab
Coffee berry borer (larvae)	*Hypothenemus hampei*	Coffee *Coffea arabica*	*Heterorhabditis* sp.	29,73	Field experiment	Torres' Lab
Coffee berry borer (adults)				44,44		
Pinneaple weevil	*Metamasius dimidiatipennis*	Pineapple *Ananas comosus*	Different native isolates	50–95	Laboratory	Torres' Lab
White grubs	*Phyllophaga* spp.	Grass	Different native isolates	3–16	Laboratory	Torres' Lab
Mangoes fruit fly	*Anastrepha obliqua*	Mangoes *Mangifera indica*	Different native isolates	44–96	Laboratory	Rosales' Lab
Guava fruit fly	*Anastrepha striata*	Guava *Psidium guajava*	Different native isolates	4–40	Laboratory	Torres' Lab
			Different native isolates	24–100	Laboratory	San-Blas' Lab
			Heterorhabditis indica	75–79	Field experiment	
Sapodilla fruit fly	*Anastrepha serpentina*	Sapodilla *Manilkara zapota*	Different native isolates	33–100	Laboratory	San-Blas' Lab
			Heterorhabditis indica	50–78	Field experiment	
Sugar froghopper	*Anaeolamia varia*	Sugar cane *Saccharum officinarum*	*Heterorhabditis bacteriophora*	71–75	Commercial	Ferrer et al., 2004

Fungus gnat	*Bradysia* spp.	Sweet pepper seedlings *Capsicum annum*	Different native isolates	80–85	Laboratory	Rosales' Lab
		Mushroom *Agaricus bicolor*	*Heterorhabditis amazonensisSteinernema* sp.	65–85 45–62	Field experiment	San-Blas' Lab
		Tomatoes (Cherry) *Solanum lycopersicum*	*Heterorhabditis amazonensis*	75	Field experiment	San-Blas' Lab
		Gerbera flowers *Gerbera* sp.	*Heterorhabditis amazonensis*	79	Field experiment	San-Blas' Lab
Wireworms	*Agriotes* sp.	Sweet corn *Zea mays*	*Heterorhabditis amazonensis*	80–100	Laboratory	San-Blas' Lab
Western flower thrips	*Frankliniella occidentalis*	Roses *Rosa* sp.	*Heterorhabditis amazonensis*	18–85	Laboratory	San-Blas' Lab

material and biological inputs law" (Gaceta Oficial de la República Bolivariana de Venezuela, 2002), which regulates the commercialization, research, development and certification of any kind of biological control agent. Then, a number of different legal regulations promote the correct use of biological means in the agricultural activity and the creation of a laboratories network which guarantees all the processes involved in biological control activities (Gaceta Oficial de la República Bolivariana de Venezuela, 2008).

For that reason, many farmers have been changing their minds and nowadays the willingness to use biological control products has exploded; yet the country's infrastructure (public and private) is not in full capacity to supply all the biocontrol products required in the country. It is probable than in the next few years, EPNs would be a common product in the shelves. Nevertheless, some steps have to be reached before it happens. Native populations with proven efficient in controlling specific insect pests in laboratory conditions must be tested in the fields; mass production has to be adapted to an appropriate technology based in local raw materials and controlled in tropical conditions. At the same time, field demonstrations are necessary to motivate growers to introduce such biocontrol agents into IPM programs, and educational strategies for including EPNs curricula in schools and universities need to be achieved.

15.4 Factors Affecting the Advance of Entomopathogenic Nematodes Development: Ecological, Economical and Social Implications

15.4.1 Agro-Ecological

Venezuela is considered one of the 17 megadiverse countries in the world (Mittermeier, Gil, & Mittermeier, 1997). This recognition implies that apart from all cosmopolitan pests, plant diseases and weeds, the amount of endemic organisms affecting crops could be extremely high. The same principle, then might be applied to biological control agents, so, the amounts of potential biocontrol organisms should be immense. Venezuela, as a tropical country, is characterized by abundant rainfalls and high (more or less constant) temperatures. Those factors promote a continuous competition of weeds, microorganisms and insect pests in the crops. In temperate climates, winter offers a natural barrier and native populations tend to be reduced due to low temperatures. For instance, many thrip species (*Frankliniella occidentalis* Pergande, *Thrips tabaci* Lindeman, *Thrips palmi* Karny [Thysanoptera: Thripidae]) are able to complete their life cycles up to 8 times in temperate zones, whereas in tropical areas can complete their cycle every 15 days, producing more than 20 generations in a single year (Salas, 2003). Commercially available nematodes come from temperate regions and their virulence or percentage of pest mortality is reduced when compared with native nematodes, mainly because of tropical temperatures. Because of legal limitations for the incorporation of alien species into the country, sampling programmes to build native EPNs collections are currently ongoing and many trials are being established and evaluated to selected nematodes with high performance levels both in the lab and the field.

15.4.2 Economic Aspects

Mass production of EPNs in Venezuela shares some of the problems present in other parts of the world. Their potential incorporation in integrated pest programs will be limited (at least in the beginning) only to value crops (vegetables, fruits and flowers) because of their production costs. Currently, the concern of Venezuelan scientists is to reduce costs using industrial by–products to elaborate media for mass production. Some experience includes milk whey (from the cheese industry), sugar cane molasses and palm oil (San-Blas, unpublished). Another aspect to be considered is to build cottage industries using insects to produce nematodes in isolated areas with difficult access, to provide EPNs to farmers. The use of *G. mellonella* is ruled out due to the high cost of honey, required to feed larvae. For that reason, the use of *Tenebrio molitor* L. [Coleoptera: Tenebrionidae] or other alternatives are being evaluated.

15.4.3 Technological Aspects

Incorporating and developing EPNs technologies is a conflicting matter nowadays in Venezuela. One of the most important limitations is the lack of highly trained staff. Currently, no more than 20 people (including students) constitute the whole scientific body engaged in EPN research in the country, all of them working in the public sector. As a consequence, the path from isolating a nematode to finally delivering it into the field as a formulated product will take many years. Additionally, adapting foreign technology to tropical conditions will require additional time. Quality control methods are usually performed by determining mortality levels in live hosts, a time and infrastructure consuming task that is driving Venezuelan researchers to look for indirect methods to measure the quality of nematodes by means of infrared technologies. By using these new cutting–edge tools, it is possible to assess the amount of certain compounds [i.e. trehalose (Jagdale & Grewal, 2003), glycogen (Patel & Wright, 1997; Qiu & Bedding, 2000) and reserve lipids (Selvan, Gaugler, & Lewis, 1993)] which have been correlated with virulence or survival of EPNs. Molecular technology is also being incorporated to understand the function of genes involved in the differential EPNs performance in the tropics and temperate regions.

Large scale production of EPNs involves *in vitro* techniques as the only economically reasonable way to provide nematodes at low costs (Ehlers & Shapiro-Ilan, 2005; Salma & Shahina, 2012). However, there is no possibility whatsoever for Venezuelan scientists to mass produce nematodes in that way, because of high costs and local limitations related with such technology (i.e. raw materials which compete with human food, few people involved in research activities). *In vivo* techniques for cottage production as an alternative for supplying EPNs to farmers in Venezuela may be a short term solution to the introduction of indigenous EPNs (Gaugler, Brown, Shapiro-Ilan, & Atwa, 2002). While in other countries, the labour costs in terms of employees could be a limiting factor, in Venezuela, the minimum monthly wage revolves around $100, an affordable amount for a starting biocontrol company. In addition, insect breeding for nematode production must be redesigned to either reduce the cost of diets for *G. mellonella* or change the insect species to *T. molitor* or the like.

At the same time, the research needs to focuses in setting an *in vitro* mass production in liquid media. Currently, 2 up to 5 l bioreactors (Bioflo C30, New Brunswick Scientific, USA) and 1 up to 40 l bioreactor (Magnaferm, New Brunswick Scientific, USA) have been restored at IVIC to breed a *Heterorhabditis amazonensis* Andalo, Nguyen & Moino (Rhabditda. Heterorhabditidae) population (San-Blas et al., unpublished) using the recommended (and sometimes expensive) media reported in the literature (Ehlers, Lunau, Krasomil-Osterfeld, & Osterfeld, 1998; Yoo, Brown, & Gaugler, 2000). As we have gained experience, new local raw materials (referred before) are being used in order to reduce costs without affecting the nematodes quality.

Milk production in Venezuela is around 1,266. Millions L/year (MLPY); 63 % is destined to cheese production (798.7 MLPY). As a consequence, 698.8 MLPY of whey are produced, of which 75 % (524.1 MLPY) is discarded to water bodies (the remaining 25 % is used for the production of ricotta cheese and for animal feed) (González, 2012). The sugar cane industry processes 9 million tons of biomass to generate 1 million tons of sugar, and as a by–product, 300,000 t of molasses is destined mainly to animal feed. Also, the growth of African oil palm (*Elaeis guineensis* Jacq. [Arecales: Arecaceae]) is raising in the last 20 years in the country, to reach more than 80,000 t of oil refined, and the production of a residual slurry with high contents of oil in the process of clarification after pressing the fruits (Reinoso, 2009). All mentioned by–products should be considered to be part of the culture media for mass production because they are abundant, cheap and are environmental passives in our country. All these possibilities will contribute to reduce costs, with the additional environmentally friendly benefit of reduction of contaminant effluents in our water bodies.

15.4.4 Formulation

The acceptance of EPNs–formulated products depends heavily on their durability, shelf life and transportation (Georgis, 2002), areas that represent different challenges when producing tropical EPNs. Manufacturing a formulated nematode–based product, should be directed to use local inert components and develop technology to make commercial formulations capable to bear temperatures ranging from 15 to 20 °C or no refrigerating conditions at all.

One of the most important limitations in the production and commercialization of EPNs in Venezuela is the very short shelf life of the nematodes. This happens basically because the nematodes die when stored at low temperatures that prevent their metabolism from going on. One of the key aspects to develop in the future must be the introduction of new techniques capable of inducing EPN anhydrobiosis in the absence of low temperatures. Some experiments are being done to incorporate vegetal compounds that are economical and easily found in the country. Some polysaccharides extracted from *Aloe vera* L. (Asparagales: Xanthorrhoeaceae) as protectants and cassava (*Manihot esculenta* Crantz [Malpighiales: Euphorbiaceae]) starch as carrier are currently being tested but the research still is in its initial stage.

15.4.5 Educational Aspects

Adopting new agricultural technologies is a very complex process, composed with a series of variables and factors that need attendance to ensure a successful change from old to new technologies. In Venezuela, the progress of implementing integrated pests management (IPM) practices has been slow and according to studies done with tomatoes farmers (in Lara state), the better known and employed IPM techniques are (a) sticky traps, (b) pheromone traps, (c) biological insecticides (specially *B. thuringiensis*) and (d) releasing of the entomophagous insects *Chrisoperla* spp. and the parasitoid *Trichogramma* spp. (Guillén et al., 2008). At the same time, the authors pointed that the farmers' refusal to use biological insecticides laid in ignorance, disinterest and distrust of the benefits of new techniques, factors connected to years of use and abuse of chemical pesticides and resistance to learn new ways of production.

The majority of farmers can identify just fungi or bacteria as biological control agents but have no idea of the existence of EPNs. It is mandatory to make educational programs for farmers, school teachers and consumers while releasing EPNs in the fields, in order to create awareness of their advantages and promote their use in Venezuelan fields.

At the same time, the next generation of agronomists, biologists and scientists should be exposed in their formative years to programs on nematology, of which there are currently none at undergraduate levels and only two in equal number of universities as postgraduate courses, mainly in Phytoparasitic Nematology. Few courses on biological control or integrated pest management, provide limited information on EPNs. Three laboratories fully dedicated to research on EPNs belong to scientific institutes that are independent from universities. Therefore, the only chance to get students interested in nematology is as thesis supervisors or whenever invited to the universities to give sporadic lectures and conferences. Nevertheless, some advances in that direction have been done. In 2008, the Venezuelan Society of Nematology (www.sovenem.org.ve) was founded; one of its main activities is to organize courses in different nematological topics. Since then, 3 courses on EPNs have been organized and more than 50 professionals (including scientists, university teachers and farmers) and many more students have attended. Due to an international cooperation with Cuba, 15 workshops with farmers have been done with more than 450 participants and 6 more are planned for March 2015. It is worth to mention that Dr. Parwinder Grewal, a renowned professor from the University of Tennessee (U.S.), visited the country when the Venezuelan Entomological Society met in 2007 in San Cristóbal (Táchira state), delivering a conference to an audience of more than 100 people, and holding a EPNs course for 35 students.

15.4.6 Social and Cultural Issues

Agricultural communities in Venezuela face poverty, as well as high levels of scholar desertion (Pérez, Rincón, Huerta, & Urdaneta, 2001), with additional lack of public investment in infrastructure, roads, land property; therefore, the efficiency and efficacy for crop production is influenced by limitations in transport, storage and marketing (Pérez, 2005). Nevertheless, some important agricultural zones have moderate to good infrastructure and farmers in those areas have formal education, even to the graduate level in national universities. For those reasons, implementing integrated pest management programs, should be made with two different approaches: (1) a first scheme for isolated and less developed communities, to be implemented through the participation of extensionists, sociologists and local community leaders, who would organize a more intensive monitoring system in order to ensure the proper use of the biological control agents as well as setting demonstrative plots, and (2) organizing technical workshops with the more influential farmers, to raise data collection and stimulate discussion of results in order to demonstrate the benefits of changing from chemical to biological control *in situ*.

15.5 State of Art of Entomopathogenic Nematodes Research in Venezuela

15.5.1 Sampling in Venezuela

In recent years, many countries (especially those from less developed areas) have started research on Entomopathogenic Nematology and significant effort in regional surveys (San-Blas, 2013). Latin America is not the exception, and since 2000, some EPNs species/populations from the subcontinent have been reported (Edgington et al., 2010; López-Nuñez, Cano, Góngora-Botero, & Stock, 2007; Molina-Ochoa et al., 2009; Powers et al., 2009) including the description of new species. Sampling for native EPNs is considered as an obligate activity for all Venezuelan laboratories, due to agro–ecological, economical, technological and legal reasons (explained above); hence, a number of sampling efforts have been done and nowadays, many species/populations of both *Steinernema* and *Heterorhabditis* have been isolated from many ecosystems and are currently kept in all research laboratories devoted to the EPN studies (Table 15.2). The vast majority of the isolated nematodes come from the coastal line and the Andean region of the country (Fig. 15.1) because those are the most populated and developed areas, with better value crop fields and laboratories dedicated to EPNs research.

Today, 29 *Heterorhabditis* isolates (69 %), 10 *Steinernema* (24 %), and 3 unidentified isolates (7 %) have been isolated in Venezuela (Table 15.2). Many of them are under study for proper identification; so far, molecular and morphological studies have been completed (in San-Blas' lab), resulting in three isolates of *H. amazonensis* and one *Heterohabditis indica* Poinar, Karunakar & David (Rhabditida: Heterorhabditidae), and three new species of *Steinernema* (unpublished). Among those new *Steinernema* there are two different species from the "*bicornutum*–group", two populations (LPV023 and LPV687) from the same species found in both sides of the Maracaibo lake and separated by 170 Km, a single isolate (LPV 474) found in a natural grassland in the eastern side of Maracaibo lake, and a new *Steinernema* from the *glaseri*–group, isolated in a cultivated *Cynodon* sp. (Poales: Poaceae).

15.5.2 Molecular and Biochemical Studies

Infrared technology (FTIR) is a tool recently added to EPN research programs. The principle in the use of FTIR is to measure the absorption of electromagnetic radiation in the infrared spectral region, which permits chemical bonds and groups to be identified by their vibrational modes for different compounds. Therefore, a unique spectral fingerprint can be recorded for any chemical compound, tissue or the whole organism. The applications of such technique are vast and include the characterization of samples according to their composition, detecting differences between samples, tracking changes in the metabolism of an organism, etc. The first experience using FTIR was done by San-Blas et al. (2011), where the spectra generated by 2 EPNs (*Steinernema glaseri* (Steiner) Wouts, Mráček, Gerdin & Bedding [Rhabditida: Steinernematidae] and *H. indica*) and *Caenorhabditis elegans* Maupas (Rahbitida: Rahbditidae) were characterized and clearly separated from each other using mathematical and statistical means. The nematode spectra also showed a series of bands representative of lipids, proteins and sugars that were species–dependent. Also, the symbiotic bacteria of the above mentioned nematodes were compared with *Escherichia coli* Migula (Enterobacteriales: Enterobacteriaceae) and again the separation of spectra was possible between them (San-Blas, Cubillán, Guerra, Portillo, & Esteves, 2012). Ongoing experiments are currently elucidating differences on the lipid composition of EPNs reared in different hosts and studies on

Table 15.2 Entomopathogenic nematodes isolated from Venezuela

Lab. ID	Species	Locality	Associated crop/vegetation	Laboratory collection
INIA-1	*Heterorhabditis* sp.	Colonia Tovar, Tovar municipality, Aragua State	Peach (*Prunus persica*)	Rosales' laboratory
INIA-2	*Heterorhabditis* sp.	Timotes, Miranda municipality, Mérida State	Garlic (*Allium sativa*)	
INIA-3	*Heterorhabditis* sp.	Santo Domingo, Cardenal Quintero municipality, Mérida State	Potatoes (*Solanum tuberosum*)	
INIA-4	*Steinernema* sp.	Mucuchíes, INIA experimental station, Rangel municipality, Mérida State	Potatoes (*Solanum tuberosum*)	
INIA-5	*Heterorhabditis* sp.	Bramon experimental station, Acevedo municipality, Táchira State	Coffee (*Coffea arabica*)	
INIA-6	*Heterorhabditis* sp.	Pueblo Hondo station, Jaureguimunicipality, Táchira State	Potatoes (*Solanum tuberosum*)	
INIA-7	*Heterorhabditis* sp.	El Limón, Mario Briceño Iragorry municipality, Aragua State	Natural dry forest	
INIA-8	*Heterorhabditis* sp.	NUDE José Félix Rivas, Revenga municipality, Aragua State	Tomatoes (*Solanum lycopersicum*)	
INIA-9	*Heterorhabditis* sp.	NUDE José Félix Rivas, Revenga municipality, Aragua State	Pepper (*Capsicum annum*)	
INIA-10	*Heterorhabditis* sp.	Tapipa, INIA experimental station, Acevedo municipality, Miranda State	Citrus (*Citrus nobilis*, *C. sinensis*)	
INIA-11	*Heterorhabditis* sp.	Puerto Ayacucho, INIA experimental station, Atures municipality, Amazonas State	Natural grassland	
INIA-12	*Heterorhabditis* sp.	Macagua, Veroes municipality, Yaracuy State	Sweet pepper (*Capsicum annum*)	
INIA-13	*Heterorhabditis* sp.	INIA experimental station, Peña municipality, Yaracuy State	Sugar cane (*Saccharum* spp.)	
INIA-14	*Heterorhabditis* sp.	El Jarillo, Guaicaipuro municipality, Miranda State	Tree tomato (*Solanum betaceum*)	
INIA-15	*Heterorhabditis* sp.	FincaInfante, Atures municipality, Amazonas State	Cocona (*Solanum sessiliforum*)	
LPV-001	*H. indica*	CORPOZULIA experimental station (fruticulture), El Moján, Mara municipality, Zulia State	Sapodilla (*Manilkara sapota*)	San-Blas's laboratory
LPV-012	*Heterorhabditis* sp.	INZIT, La Cañada de Urdaneta municipality, Zulia State	Cultivated grassland (*Cynodon* sp.)	
LPV-023	*Steinernema* sp. (*bicornutum* group)	Finca el Tigre, Cahirí, Mara municipality, Zulia State	Cultivated grassland (*Megathyrsus* sp.)	

LPV-030	*Heterorhabditis* sp.	Los jobitos, Miranda municipality, Zulia State	Natural coastal xerophytic forest	
LPV-035	*Heterorhabditis* sp.	Sur del Lago de Maracaibo, Colón Municipality, Zulia State	Plantain (*Musa paradisiaca*)	
LPV-061	*Heterorhabditis* sp.	Sur del Lago de Maracaibo, Colón Municipality, Zulia State	Cultivated grassland (*Cynodon* sp.)	
LPV-081	*H. amazonensis*	Sur del Lago de Maracaibo, Colón Municipality, Zulia State	Plantain (*Musa paradisiaca*) – Corn (*Zea mays*)	
LPV-156	*H. amazonensis*	Bobures, Sucre municipality, Zulia State	Cultivated grassland (*Cynodon* sp.)	
LPV-299	*Heterorhabditis* sp.	Isla de patos, Valdéz municipality, Sucre State	Natural coastal dry forest	
LPV-474	*Steinernema* sp. (*bicornutum* group)	El Venado, Baralt municipality, Zulia State	Natural grassland	
LPV-498	*H. amazonensis*	Carretera Barinas-San Cristobal, Barinas municipality, Barinas State	Natural grassland	
LPV-565	*Heterorhabditis* sp.	La Danta, Brión municipality, Miranda State	Cocoa (*Theobroma cacao*)	
LPV-687	*Steinernema* sp. (same LPV-023)	Cerro Socopó, Baralt municipality, Zulia State	Natural grassland	
LPV-723	*Steinernema* sp. (*glaseri* group)	Mene de Mauroa, Mauroa municipality, falcon State	Cultivated grassland (*Cynodon* sp.)	
6SVR-E1	*Steinernema* sp.	San Vicente de la Revancha, Junín municipality, TáchiraState	Strawberry (*Fragaria* sp.) and horticultural crops area	Torres' laboratory
8SVR-A1	*Heterorhabditis* sp.			
1SVR-E1	*Steinernema* sp.			
COR-C4	*Steinernema* sp.	Montecarmelo, Andres Bello municipality, Táchira State	Coffee (*Coffea arabica*)	

(continued)

Table 15.2 (continued)

COR-E5	*Steinernema* sp.			
BRA1	*Heterorhabditis* sp.	Bramón, Junín municipality, Táchira State	Coffee (*Coffea arabica*)	
1PEN-D2	Unidentified	Pueblo Hondo, Jauregui municipality, Táchira State	Potato and horticultural crops area	
10PAL-D1	*Steinernema* sp.	Palenque, Jauregui municipality, Táchira State	Potato, strawberry, and horticultural crops area	
1MES-A21	Unidentified	Las Mesas, Rómulo Costa municipality, Táchira State	Pineaple (*Ananas comosus*)	
1CAS-D2	Unidentified	Hato de la Virgen, Libertad municipality, Táchira State	Pineaple (*Ananas comosus*)	
PRE6-B2	*Heterorhabditis* sp.	Paramito, Uribante municipality, Táchira State	Coffee (*Coffea arabica*)	
DEL1-D1	*Heterorhabditis* sp.	Delicias, Rafael Urdaneta municipality, Táchira State	Coffee (*Coffea arabica*)	
TRO16-A3	*Heterorhabditis* sp.	Bramon, Juninmunicipaliti, Táchira State	Coffee (*Coffea arabica*)	

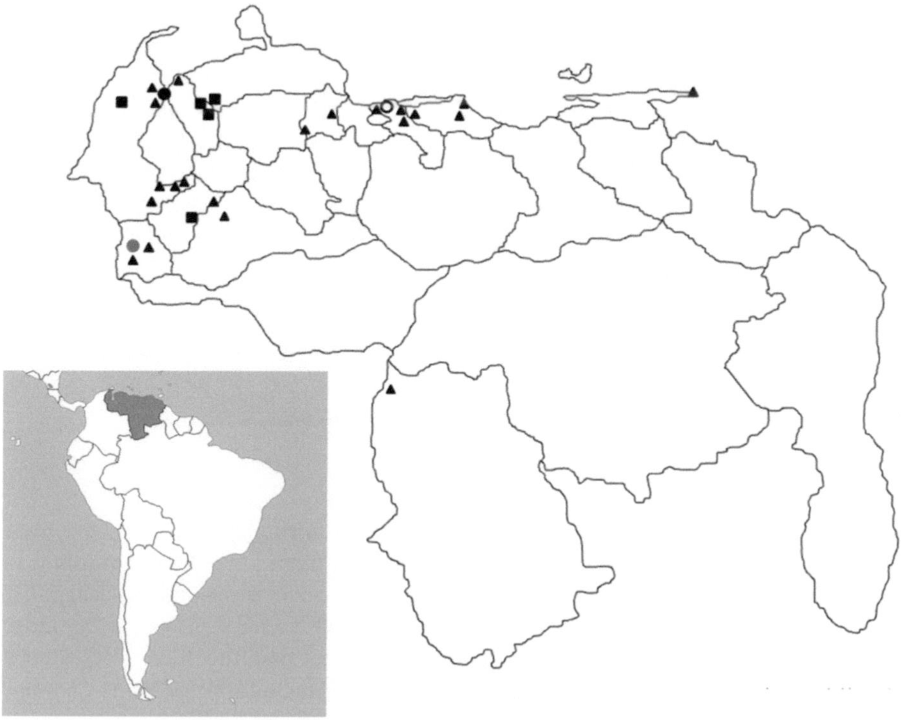

Fig. 15.1 Relative location of the isolated entomopathogenic nematodes in Venezuela and laboratories. *Steinernema* spp. (*black squares*) and *Heterorhabditis* spp. (*black triangles*); Rosales' Lab (*white circle*), San-Blas' Lab (*black circle*) and Torres' Lab (*grey circle*). Venezuela relative position in South America (insert)

the effect of the concentration of trehalose produced by the EPNs in stress conditions on their virulence and pathogenicity.

The role of nitrogenous compounds in the EPNs life cycle due to nematodes defecation has been considered in the last years as a matter of study. The IJ formation and its emergence form the insect cadavers are supposed to be related with the quality rather than quantity of the food resource, for that reason a series of regarding the effect of ammonia concentration in the emergence and recovery processes, survival and its kinetic through the *Steinernema* life cycle have been successfully accomplished (San-Blas, Gowen, & Pembroke, 2008; San-Blas, Pirela, García, & Portillo, 2014). The results indicated that emergence, recovery, and survival percentage of *Steinernema* IJ depends on the level of accumulated ammonia into the cadavers and the nematode species. Similar experiments to measure urea and ammonia in *Heterorhabditis* are currently ongoing.

Molecular biology studies have been scarce and focused on taxonomy. Random Amplification of Polymorphic DNA (RAPD) was used to analyse polymorphism of 15 different populations of EPNs and their symbiotic bacteria (Rosales, 2013).

Total amplified bands for nematodes and bacteria accounted for 493 and 496, respectively, with 99.2 and 100 % polymorphism, respectively. Nematodes clusters were related to altitude above the sea level at which they were collected, while for bacteria, clustering was in general related to the type of soil from which they came (Peteira et al., 2014). However, no further research has been done so far to locate those differences and interpret their meaning in the evolutionary path. Analysis of the Internal Transcribed Spacer (ITS) and D2D3 region of the Large Subunit rDNA sequences have been completed for some *Heterorhabditis* and *Steinernema* isolates (San-Blas, unpublished data, with the collaboration of Vladimir Puza, Czech Academy of Sciences, Czech Republic).

15.5.3 Laboratory Experiences Using Entomopathogenic Nematodes

Until 2014, around 13 insect pest of economic importance in the country have been tested in laboratory conditions and 5 insects among them have been controlled in field trials (Table 15.1). According to the data, all insects were highly susceptible to some EPNs species/populations, and the native EPNs have proven to be more pathogenic in laboratory conditions against their target than the alien species used as references. The research progresses to field experimentation for some of those insects, and every day new insect targets are included in our studies.

Fruit flies from the genus *Anastrepha* (Diptera: Thephritidae) comprises more than 200 species; between 40 and 50 species have been reported in Venezuela (Carballo, 2001; Hernández & Morales, 2004) and from them, 4 species (*Anastrepha frateculus* Wiedemann, *Anastrepha striata* Shiner, *Anastrepha obliqua* Macquart and *Anastrepha serpentina* Wiedemann) are considered serious fruit pests in the country, affecting more than 50 crops (Boscán, 1987). The control of these flies is achieved exclusively with chemical insecticides, even though, several lists of parasitoids with potential to become biological control agents have been published but no further studies have been done (Boscán & Godoy, 1995; Katiyar, Camacho, Geraud, & Matheus, 1995). The adults lay eggs in immature fruits and the newly hatched larvae eat and burrow into the fruit pulp; after few weeks, the mature larvae abandon the fruit and pupate into the soil which make them an ideal target for EPNs applications. Due to the economic importance of *Anastrepha* in the country, in the last 4 years, an intense program of testing EPNs against different fruit fly species has been conducted.

Sapodilla (*Manilkara zapota* Royen [Ericales: Sapotaceae]) is a very valuable tropical fruit which is produced almost exclusively in Zulia state. Depending on the year, fruit losses by *A. serpentina* can be variable. In 2007, yield losses reached up to 35–50 % of the total fruit production in different farms at Mara Municipality (Zulia state), a similar situation as that present in guava (*Psidium guajava* L. [Myrtales: Myrtaceae]) plantations infected with *A. striata*. Zulia state produces almost 90 %

of the national fruits (Terán, Meléndez, García-Aguilar, Acuña, & Urdaneta, 1996), and in 2000, the losses of guava fruits reached up to 60 % in some farms (Corzo, 2000). *Anastrepha* species pupate in the soil, and laboratory tests proved their high susceptibility to both *Steinernema* and *Heterorhabditis*. In laboratory conditions, a local population of *H. indica* has proven to control *A. serpentina* (with a 100 % mortality rate). The same population and other heterorhabditids have been worked successfully in *A. striata* with mortality rates of 100 % (*H. indica*) and 75–100 % (different populations of *Heterorhabditis amazonensis*). On the other hand, applying different populations of *Heterorhabditis* against *A. obliqua* (a mangoes pest), led to a mortality range between 44 and 96 %, depending on the population.

Other crop damaging dipterans are the fungus gnats from the genus *Bradysia* (Diptera: Sciaridae). Their larvae, attack the roots of many crops in the tropics and their control have been done in Venezuela exclusively with chemical pesticides with variable results. Many laboratory trials have been done to evaluate the potential of Venezuelan EPNs to control those flies in flowers, mushrooms, tomatoes and sweet pepper seedling. In the first three mentioned crops, the use of EPNs is currently in field experimentation (detailed in the next section) and the commercial applications should start in the first months of 2015. In sweet pepper seedling, some native isolates were evaluated against *Bradysia* larvae and found 2 *Heterorhabditis* capable of producing 80–85 % of mortality in Petri dishes (Rosales, 2013).

Among coleopterans, West Indian sweet potato weevil (*E. postfasciatus*) was tested against different isolates of *Heterorhabditis* sp. and one *Steinernema* sp. resulting in a range of control between 36 and 94 % (Rosales, 2013). Coleopteran are key insect pest in the Andean crops. The area produces more than 20 % of the pineapples harvested in Venezuela (around 40,000 t per year), in recent years, damage produced by the pineapple weevil (*Metamasius dimidiatipennis* Jekel [Coleoptera: Curculionidae]) has increased, and national funded efforts are contributing to its study (Torres, personal communication). *Heterorhabditis* sp. and *Steinernema* sp. were tested under laboratory conditions with promising results (between 50 and 95 % of mortality) (García-Caicedo, Torres, & Ochoa, 2013). Larvae from the genus *Phyllophaga* (Coleoptera: Scarabaeidae) are important pest in higher areas in Venezuela, damaging many crops. In the Andes, *Phyllophaga* affects the grasslands dedicated to milk production. So far, tests for EPNs evaluation have not been successful; a native group of *Steinernema* and *Heterorhabditis* were assessed in the context of these studies, but the mortality rates were very low (3–16 %). Nevertheless, nematode sampling is still ongoing in those soils affected by *Phyllophaga* in order to find a suitable species/populations.

Biological control of wireworms using EPNs has been controversial as a consequence of many unsuccessful applications in the fields (Barsics, Haubruge, & Verheggen, 2013). Moreover, different species can avoid the EPNs attacks thank to morphological barriers and their behavior (Eidt & Thurston, 1995). However, few attempts to control those insects have been successful. Ansari, Evans, and Butt (2009) achieved 67 % mortality rates, manipulating the time of EPNs application and using EPNs isolated directly from wireworm cadavers in laboratory conditions. Larvae from the genus *Agriotes* (Coleopters: Elateridae) were tested against 7 different nematodes: 3 *Steinernema* (LPV023, LPV474, and LPV723), one *H. indica* (LPV001) and 3 *H. amazonensis* populations (LPV498, LPV156 and LPV081); the results indicated that 2 *H. amazonensis* populations (LPV156 and LPV081) were very effective against the wireworms, killing between 68 and 100 % of them after 4–7 days post application.

Spodoptera frugiperda is the most important pest of maize and sorghum in Venezuela. Their larvae attack the plant whorls especially in their early stages and wreak havoc on the crop (Vélez, 1997). Biological control has been done traditionally by applications of *B. thuringiensis* and more recently by the parasitoids *Trichogramma* sp. and *Telenomas remus* Nixon (Hymenoptera: Scelionidae) (Hernández, Ferrer, & Linares, 1989). Alternatively, laboratory trials with EPNs have been done because *S. frugiperda* pupates in the soil and should be perfect target for nematodes and foliar applications to control larvae in the whorls are also suitable. Using different *Heterorhabditis* and *Steinernema* isolated in Petri dishes, a population of *Heterorhabditis* spp. led to 100 % mortality using a nematode dose of 80 IJs/larva, a promising biological control against this insect in the field.

The most important thrips species in Venezuela are the Western Flower Thrips (WFT) *F. occidentalis*, *T. palmi* and *T. tabaci* (Salas, 2003). Control of these insects should consider some variables; their cryptic behavior and resistance to chemical pesticides has become the most important limitations in their control (Immaraju, Paine, Bethke, Robb, & Newman, 1992), also, despite thrips pupate in the soil making them theoretically easy targets for EPNs, many of them remain confined in the flower buds and pupate aerially (Helyer, Brobyn, Richardson, & Edmondson, 1995). On the other hand, due to their small size, EPNs applications should be more frequent because the insects are not suitable for recycling the nematodes (Bastidas, Portillo, & San-Blas, 2014). As in many parts of the world, WFT represents the most damaging pest in flower production in Venezuela and many laboratories are undertaking experimental trials for alternative control measurements of WFT. In the case of EPNs, several populations of *Steinernema* and *Heterorhabditis* were tested against these thrips species, resulting in a control between 18 and 85 %. The most promising isolates were 2 *H. amazonensis* (LPV498 and LPV156) whereas *Steinernema* infected poorly the WFT larvae.

15.5.4 Biological Control Experience in Fields

At present, the only EPN biological control program at commercial level in Venezuela is done for sugar cane for controlling *Aeneolamia* sp. using a Cuban population of *H. bacteriophora* which is mass produced *in vivo* at CAAEZ (governmental sugar mill facilities) in Barinas state. However, no official data on how many treated hectares or nematodes production are available. Besides this specific program, recently, a series of field experiments are being conducted in a number of crops to incorporate EPNs to the normal crop management in the country. For example, in a commercial mushroom farm, field trials have been performed to control fungus gnats with the aim of substituting the input of chemicals on that facilities. The Venezuelan mushroom industry almost entirely consists on the production of white and some Portobello mushrooms (*Agaricus bisporus* Lange [Agaricales: Agaricaceae]), cultivated in plastics bags rather than boxes or trays. The complete cycle takes 8 week from the arrival of the mycelium inoculum, and the chemical control consists in one application of an ovicide, and eight further applications of insecticides (once a week). A population of *Steinernema* spp. (from the *bicornutum*–group) and a population of *H. amazonensis* (LPV498) were applied in ten production bags (randomly chosen) using 1.5×10^6 nematodes/m^2 in a mushroom farm located at El Junquito (close to Caracas). *H. amazonensis* reduced the population of fungus gnats in similar fashion as the chemical control treatment (49 %); the production of mushrooms was enhanced in 20 % (2,320 g per bag) over the chemical control (1,887 g per bag) and the control treatment (water) (963 g per bag). Remarkably, such results were obtained with a single application *H. amazonensis,* in contrasts with the chemical procedure that required eight applications per production cycle. Currently, a treatment with two nematode applications is being evaluated in the same facilities; the first application at the arrival of the inoculum (to avoid the use of the ovicides) and the second one when the caster layer is added (2 weeks after arrival). So far, encouraging results indicate that it will be feasible to substitute the use of chemical insecticides by EPNs in the Venezuelan mushroom industries in the very short term.

As mentioned before, fungus gnats (*Bradysia* spp.) are presents in many crops in Venezuela. The same *H. amazonensis* population was applied to cherry tomatoes in a commercial greenhouse (located at El Jarillo, Miranda state), severely affected by those insects. A total of 50,000 nematodes per plant were applied into the drip system, and the control was evaluated with the use of yellow sticky traps (normally used in this farm). The results confirmed a reduction of insects in the traps by 36 % the first week and by the second, the counting of adults was reduced by 68 % comparing with the traps before the application. Between the third and fourth weeks, the reduction was similar and the total control registered was 72 % of adults. In another farm at the same location, 30,000 EPNs per plant were applied to *Gerbera* spp. flowers pots, through the irrigation system for fungus gnat control in greenhouses. Sticky traps were used to weekly to monitor the procedure. As before, the effect of EPNs to control fungus gnats was very effective, after a month post–application, the number of adults stuck in the traps was reduced by 76 %.

Coffee is one of the most important crops in Venezuela due to the quality of its beans which are very appreciated in the international market. In 1995, the coffee borer (*H. hampei*) was detected (Montilla, Camacho, Quintero, & Cardozo, 2006) which presumably was introduced from the Colombian fields. The borer has moved from the Andean region towards the centre of the country. Some local sanitary barriers, educational and technical programmes for borer eradication have been implemented, to avoid its introduction to other fields (especially to the eastern part of the country). To this aim, different alternatives for controlling the coffee borers such as *B. bassiana*, adult alcoholic attractants (Fernández & Cordero, 2005) have been applied in fields with variable results. A native *Heterorhabditis* population has been tested against the borers resulting in 23.7 % mortality in larvae and 44.4 % in adults (Torres, unpublished data).

A native population of *H. indica* was applied in experimental fields to control *Anastrepha* in a 5.5 ha of different varieties of sapodilla trees and 3 ha of red guavas, using a dose of 2,500,000 IJs/m^2 under the tree canopies in February 2013. The application was repeated once a month until May 2013. The results indicated a significant inhibition in the emergence of fruit flies from the soil, both in sapodillas (50–78 %) and guavas (75–79 %) (San-Blas, unpublished data). The huge differences in the control of *A. serpentina* could be attributed to the variety of the sapodilla tree, because their latex production and composition is different for each variety. It is possible that larva flies when fed in a specific varietal tree could be accumulating secondary metabolites which make them more resistant to EPNs attack. Currently some studies to evaluate differences in the fruit and larvae composition of those different sapodilla varieties are under way to elucidate which metabolite is accumulated (Barbercheck, Wang, & Hirsh, 1995).

15.5.5 Laboratory Experiences Using Symbiotic Bacteria for the Control of Fungal Diseases in Plants

The symbiotic bacteria of EPNs (*Xenorhabdus* and *Photorhabdus* [Enterobacteriales: Enterobactereaceae]) produce antimicrobial compounds which have been investigated around the world for developing new biocontrol protocols against plant pathogenic diseases produced by fungi and bacteria (Akhurst, 1982; Böszörményi et al., 2009; Chen, Dunphy, & Webster, 1994; Yang et al., 2011). Fungal diseases are a major concern in Venezuela due to their climatic characteristics favouring the presence of these microorganisms (Beets, 1990). At present, the EPNs symbiotic bacteria have been tested in laboratory conditions against few fungi. The first experience was done in the cocoa frosty pod rot (*Moniliophthora roreri* Cif. [Agaricales: Marasmiaceae]) which is the most destructive disease in cocoa plantations from Central and South America (Aime & Phillips-Mora, 2005). *Xenorhabdus* sp. (undescribed Venezuelan population), *Xenorhabdus innexi* and *P. luminescens* (from *H. bacteriophora*) demonstrated antifungal action reaching up to 97 % after 13 days of exposure (San-Blas, Carrillo, & Parra, 2012). Currently, the effects of the most promising bacteria are being tested in different populations of Venezuelan populations of *M. roreri* and *M. perniciosa* (agents of the witch broom disease in cocoa).

The production of guava fruits (*Psidium guajava*) has been strongly affected by a disease called "apical rotten fruit" produced by *Dothiorela* sp. (Botryosphaeriales: Botryosphaeriaceae). Laboratory experiments using *Xenorhabdus bovienii* Akhurst & Boemare (Akhurst) (Enterobacteriales: Enterobacteriaceae), *Xenorhabdus innexi* Lengyel, Lang, Fodor, Szállás, Schumann & Stackenbrandt (Enterobacteriales: Enterobacteriaceae) and *Photorhabdus luminesces* Fisher le Saux, Villard, Brunel, Normand & Boemare (Thomas & Poinar) *(Enterobacteriales: Enterobacteriaceae* (from *Heterorhabditis bacteriophora*) showed good antifungal activity (reaching up to 100 % of antifungal index) (San-Blas, Parra, & Carrillo, 2013). Currently, metabolites of the most promising bacteria are being characterized and field trials to control *M. roreri* will be established by 2015.

15.6 Conclusion and Overcoming Challenges

Delivering nematodes in the field is the most important leap for all laboratories in the country. Our experience in laboratory trials is vast, but we still face limitations. Lack of funding for research, public policies and the inexistence of continuous investigation programs on EPNs makes it difficult to progress. It is important to start a campaign with policy makers and farmers to increase their consciousness on the benefits of biological control methods, including EPNs, for a more sustainable agriculture. In building a critical mass around the subject, we will be in a better position to have scientific programs formally established. Also, international cooperation in different areas must be achieved in order to amplify our frontiers on EPN research, to increase the number of PhD students and to share experiences with other laboratories in a bidirectional and symbiotic relationship, in which both partners learn and teach from each other. From 2006 onwards we have participated in few international projects, especially from a Cuba–Venezuela cooperation agreement, which has been successful in terms of introducing EPNs in the field and *in vivo* production of nematodes, but limited in its impact because of the short duration of the projects.

As EPNs are introduced in different crops, combining them with other biological control agents should be addressed in the short term. So far Venezuela (both in public institutions and private companies) produces *B. bassiana*, *T. harziarum* and entomophagous and parasitoid insects; but not a single experiment has been addressed in that direction. Many courses and workshops regarding the use of biofumigants, IPM in different crops, plant extracts and biological control agents (except EPNs or at least not as a formal subject) are being delivered to farmers, students and teachers nowadays; so, we have the obligation to be more proactive in evaluating the effects of combining EPNs with all the new alternatives currently used in the country and to make a serious effort in providing to the people involved in the agricultural activities our knowledge and the nematodes of course.

All the overcoming challenges and plans are not possible unless an approach between the private sector and governmental laboratories is reached. As far as we know, there are few companies trying to sell nematodes and few more are being created. However, the process of mass producing nematodes *in vivo* (currently the only technology achievable in the country) is not an easy task and requires experience, highly qualified staff, strict quality control protocols and a deep understanding of the nature of EPN, their symbiotic bacteria and the insect host rearing techniques. If private companies wish to become *in vitro* mass producers, they should join the governmental laboratories in order to invest in such technology. At present, there are no private facilities which can carry on the research and development of *in vitro* mass production in the country with our local raw materials, only public laboratories (including universities and research institution) have the equipment and the expertise to complete that task; but without the support of private organizations to rise funds, discuss results and proposals and participate as reliable suppliers and farmer advisors to ensure the good usage and performance of EPNs in Venezuelan fields, the goal will not be reached. On the other hand, governmental laboratories must support the companies not only in mass production techniques but also advising in the building of production facilities to ensure the availability of EPNs in the markets as they become recognised by farmers, improving their shelf life and cooperating with them in training their human resources.

The outlook seems promising for the implementation of EPNs in Venezuela in the short and medium term. Numerous questions are still waiting for an explanation to be discovered. For this, researchers are required. Students are coming to our laboratories to do undergraduate, master and PhD thesis, particularly from the areas of agronomy, chemistry and biology. The amount of new species/populations will also be on the rise due to the efforts of all laboratories in finding better candidates for biological control programmes. Finally, in the next 10 years we believe that EPNs will be in the stores as reliable biological control agents for different crops, and farmers will use them not as an experimental organism but as a regular product.

Acknowledgments The authors thank Prof. Gioconda San-Blas (InstitutoVenezolano de Investigaciones Científicas, IVIC, Caracas, Venezuela) for her valuable comments and suggestions. The authors also acknowledge the Fondo de Desarrollo Nacional FONDEN S.A. for financial support through "Convenio de Cooperación Integral Cuba-Venezuela", project "MPPF–FONDEN–CJ–CCATR–XIII–14".

References

Aime, M. C., & Phillips-Mora, W. (2005). The causal agents of witches' broom and frosty pod rot of cacao (chocolate, *Theobroma cacao*) form a new lineage of Marasmiaceae. *Mycologia, 97*, 1012–1022.

Akhurst, R. J. (1982). Antibiotic activity of *Xenorhabdus* spp., bacteria symbiotically associated with insect pathogenic nematodes of the families Heterorhabditidae and Steinernematidae. *Journal of General Microbiology, 128*, 3061–3065.

Ansari, M. A., Evans, M., & Butt, T. M. (2009). Identification of pathogenic strains of entomopathogenic nematodes and fungi for wireworm control. *Crop Protection, 28*, 269–272.

Ballou, C. H. (1941). Notas sobre la introducción de un predator para el control de la escama algodonosa del naranjo. *Boletín de la Sociedad Venezolana de Ciencias Naturales, 7*, 121–122.

Barbercheck, M. E., Wang, J., & Hirsh, I. S. (1995). Host plant effect on entomopathogenic nematodes. *Journal of Invertebrate Pathology, 66*, 169–177.

Barsics, F., Haubruge, E., & Verheggen, F. J. (2013). Wireworms' management: An overview of the existing methods, with particular regards to *Agriotes* spp. (Coleoptera: Elateridae). *Insects, 4*, 117–152.

Bastidas, B., Portillo, P., & San-Blas, E. (2014). Size does matter: The life cycle of *Steinernema* spp. in micro–insect hosts. *Journal of Invertebrate Pathology, 121*, 46–55.

Beets, W. C. (1990). *Raising and sustaining productivity of smallholder farming systems in the tropics*. Alkmaar, Holland: AgBé Publishing.

Boscán, N. (1987). Las moscas de las frutas del género *Anastrepha* en Venezuela. *FONAIAP Divulga, 23*, 22–24.

Boscán, N., & Godoy, F. (1995). Nuevos parasitoides de moscas de las frutas de los géneros *Anastrepha* y *Ceratitis* en Venezuela. *Agronomía Tropical, 46*, 465–471.

Böszörményi, E., Érsek, T., Fodor, A., Fodor, A. M., Földes, L. S., Hevesi, M., et al. (2009). Isolation and activity of *Xenorhabdus* antimicrobial compounds against the plant pathogens *Erwinia amylovora* and *Phytophthora nicotianae. Journal of Applied Microbiology, 107*, 746–759.

Carballo, J. (2001). Diagnosis y clave pictórica para las especies del género *Anastrepha* Schiner, 1868 (Diptera: Tephritidae) de importancia económica en Venezuela. *Entomotropica, 16*, 157–164.

Chen, G., Dunphy, G. B., & Webster, J. M. (1994). Antifungal activity of two *Xenorhabdus* species and *Photorhabdus luminescens*, bacteria associated with the nematodes Steinernema species and *Heterorhabditis megidis. Biological Control, 4*, 157–162.

Chirinos, D., & Geraud-Pouey, F. (1996). Efecto de algunos insecticidas sobre entomofauna del cultivo del tomate, en el noreste del estado Zulia, Venezuela. *Interciencia, 21*, 31–36.

Chirinos, D., & Geraud-Pouey, F. (2011). El manejo de plagas agrícolas en Venezuela. análisis y reflexiones sobre algunos casos. *Interciencia, 36*, 192–199.

Corzo, P. (2000). *Situación de la fruticultura a nivel regional. 1er encuentro regional sobre fruticultura (Proceedings)* (pp. 8–11). FONAIAL-LUZ-CORPOZULIA, Santa Cruz de Mara, Estado Zulia.

Edgington, S., Buddie, A. G., Moore, D., France, A., Merino, L., Tymo, L. M., et al. (2010). Diversity and distribution of entomopathogenic nematodes in Chile. *Nematology, 12*, 915–928.

Ehlers, R. U., Lunau, S., Krasomil-Osterfeld, K., & Osterfeld, K. H. (1998). Liquid culture of the entomopathogenic nematode bacterium–complex *Heterorhabditis megidis/Photorhabdus luminescens. BioControl, 43*, 77–86.

Ehlers, R. U., & Shapiro-Ilan, D. I. (2005). Mass production. In P. S. Grewal, R. U. Ehler, & D. I. Shapiro-Ilan (Eds.), *Nematode as biological control agent* (pp. 65–78). Wallingford, Oxon, UK: CAB International.

Eidt, D. C., & Thurston, G. S. (1995). Physical deterrents to infection by entomopathogenic nematodes in wireworms (Coleoptera: Elateridae) and other soil insects. *Canadian Entomologist, 127*, 423–429.

Fan, X., Maggiorani, A., & Gudiño, S. (2000). Uso de nemátodos entomopatógenos como una alternativa en el control de polilla (*Tecia solanivora*), importante plaga de la papa (*Solanum tuberosum*). Mérida–Venezuela. *Revista Forestal Venezolana, 44*, 115–118.

Fernández, S., & Cordero, J. (2005). Evaluación de atrayentes alcohólicos en trampas artesanales para el monitoreo y control de la broca del café, *Hypothenemus hampei* (Ferrari). *Bioagro, 17*, 143–148.

Ferrer, F. (2001). Biological control of agricultural insect pests in Venezuela; advances, achievements, and future perspectives. *Biocontrol News and Information, 22*, 67N–74N.

Ferrer, F., Arias, M., Trelles, A., Palencia, G., Navarro, J. M., & Colmenárez, R. (2004). Posibilidades del uso de nematodos entomopatógenos para el control de *Aeneolamia varia* en caña de azúcar. *Manejo Integrado de Plagas y Agroecología, 72*, 39–43.

Gaceta Oficial de la República Bolivariana de Venezuela. (2002). *N° 37.552*. Retrieved from: http://www.tsj.gov.ve/gaceta/octubre/181002/181002\T1\textendash37552\T1\textendash01.html.

Gaceta Oficial de la República Bolivariana de Venezuela. (2008). *N° 5.890 Extraordinario*. Retrieved from: http://www.mat.gob.ve/legislacion/5890.pdf.

García-Caicedo, M., Torres, A., & Ochoa, A. (2013). Evaluación de nematodos entomopatógenos para el control del picudo de la piña en el estado Táchira, Venezuela. *Agronomía Tropical, 63*, 1–2.

Gaugler, R., Brown, I., Shapiro-Ilan, D., & Atwa, A. (2002). Automated technology for *in vivo* mass production of entomopathogenic nematodes. *Biological Control, 24*, 199–206.

Georgis, R. (2002). The Biosys experiment: An insider's perspective. In R. Gaugler (Ed.), *Entomopathogenic nematology* (pp. 357–372). Wallingford, Oxon, UK: CAB International.

Geraud-Pouey, F., Chirinos, D., & Rivero, G. (1997). Dinámica poblacional del pasador de la hoja *Liriomyza* spp., Diptera: Agromyzidae en tomate, en la región noroccidental del estado Zulia, Venezuela. *Revista de la Facultad de Agronomía LUZ, 14*, 475–485.

González, M. (2012). Aspectos medio ambientales asociados a los procesos de la industria láctea. *Mundo Pecuario, 8*, 16–32.

Guédez, C., Castillo, C., Cañizales, L., & Olivar, R. (2008). Control biológico: una herramienta para el desarrollo sustentable y sostenible. *Academia, 13*, 50–74.

Guillén, L., Alcalá de, M. D., Fernández, S., Pire, A., & Álvarez, C. (2008). Farmer's perception on the integrated pest management in tomato (*Lycopersicum esculentum* Mill.) crop. *Revista de la Facultad de Agronomía LUZ, 25*, 223–242.

Helyer, N. L., Brobyn, P. J., Richardson, P. N., & Edmondson, R. N. (1995). Control of Western flower thrips (*Frankliniella occidentalis* Pergande) pupae in compost. *Annals of Applied Biology, 127*, 405–412.

Hernández, D., Ferrer, F., & Linares, B. (1989). Introducción de *Telenomas remus* Nixon (Hym.: Scelionidae) para controlar *Spodoptera frugiperda* (Lep.: Noctuidae) en Yaritagua, Venezuela. *Agronomía Tropical, 39*, 45–61.

Hernández, O. V., & Morales, V. P. (2004). Distribución geográfica y plantas hospederas de *Anastrepha fraterculus* (Diptera: Tephritidae). *Folia Entomologica Mexicana, 43*, 181–189.

Immaraju, J. A., Paine, T. D., Bethke, J. A., Robb, K. L., & Newman, J. P. (1992). Western flower thrips (Thysanoptera: Thripidae) resistance to insecticides in coastal California greenhouses. *Journal of Economic Entomology, 85*, 9–14.

Jagdale, G. B., & Grewal, P. S. (2003). Acclimation of entomopathogenic nematodes to novel temperatures: Trehalose accumulation and the acquisition of thermotolerance. *International Journal for Parasitology, 33*, 145–152.

Katiyar, K. P., Camacho, J., Geraud, F., & Matheus, R. (1995). Hymenopterous parasitoids of fruit flies (Diptera: Tephritidae) in western Venezuela. *Revista de la Facultad de Agronomía LUZ, 12*, 303–312.

López-Nuñez, J. C., Cano, L., Góngora-Botero, C. E., & Stock, S. P. (2007). Diversity and evolutionary relationships of entomopathogenic nematodes (Steinernematidae and Heterorhabditidae) from the Central Andean region of Colombia. *Nematology, 9*, 333–341.

Mittermeier, R. A., Gil, P. R., & Mittermeier, C. G. (1997). *Megadiversity: Earth's biologically wealthiest nations*. Mexico: Conservation International, Cemex.

Mol, H., Plaza, P., Zorrás, P., de Rij, T., Stoker, A., & Mulder, P. (2008). Toward a generic extraction method for simultaneous determination of pesticides, mycotoxines, plant toxins and veterinary drugs in feed and food matrices. *Analytical Chemistry, 80*, 9450–9459.

Molina-Ochoa, J., Nguyen, K. B., González-Ramírez, M., Quintana-Moreno, M. G., Lezama-Gutiérrez, R., & Foster, J. E. (2009). *Steinernema diaprepesi* (Nematoda: Steinernematidae): Its occurrence in western Mexico and susceptibility of engorded cattle ticks *Boophilus microplus* (Acari: Ixodidae). *Florida Entomologist, 92*, 661–663.

Montilla, R., Camacho, B., Quintero, A., & Cardozo, G. (2006). Parasitismo por *Beauveria bassiana* sobre la broca del café, en el estado Trujillo, Venezuela. *Agronomía Tropical, 56*, 183–198.

MPPAT. (2007). *VII CensoAgrícola*. Retrieved from: http://censo.mat.gob.ve.

MPPS. (2011). *Anuario de morbilidad del Ministerio del Poder Popular para la Salud de la República Bolivariana de Venezuela.* Retrieved from: http://www.mpps.gob.ve/index.php?option=com_phocadownloadandview=categoryanddownload=720:anuario2011andid=15:anuarios\T1\textendashde\T1\textendashmorbilidadandItemid=915.

Pacheco, M., & Torres, A. (2005). Evaluación de nematodos entomopatógenos en el control de la broca del café en el estado Táchira. *Entomotropica, 20*, 127.

Patel, M. N., & Wright, D. J. (1997). Glycogen: Its importance in the infectivity of aged juveniles of *Steinernema carpocapsae. Parasitology, 114*, 591–596.

Pérez, J. J. (2005). Dimensión ética del desarrollo sostenible de la agricultura. *Revista de Ciencias Sociales LUZ, 11*, 246–255.

Pérez, J. J., Rincón, N., Huerta, I., & Urdaneta, F. (2001). Diagnóstico socioeconómico de la comunidad agrícola La Estrella, estado Zulia. *Revista de Ciencias Sociales LUZ, 7*, 23–33.

Peteira, B., Rodríguez, M. G., Rosales, C., Maselli, A., Casado, R., Castro, L., et al. (2014). Variabilidad molecular de aislamientos venezolanos de nematodos entomopatógenos y sus bacterias simbiontes. *Revista de Protección Vegetal, 29*, 112–121.

Powers, T. O., Neher, D. A., Mullin, P., Esquivel, A., Giblin-Davis, R. M., Kanzaki, N., et al. (2009). Tropical nematode diversity: Vertical stratification of nematode communities in a Costa Rican humid lowland rainforest. *Molecular Ecology, 18*, 985–996.

Qiu, L., & Bedding, R. (2000). Energy metabolism and its relation to survival and infectivity of infective juveniles of *Steinernema carpocapsae* under aerobic conditions. *Nematology, 2*, 551–559.

Quiróz-Mireles, Y. (2005). Los métodos aplicados para combatir y erradicar la plaga de la langosta en la región noroccidental de Venezuela (1912–1914). In*:* G. Yépez-Colmenares (Ed.), *Temas de historia contemporánea de Venezuela* (1st ed., pp. 99–126.). Venezuela: Fondo editorial de Humanidades y Educación. Universidad Central de Venezuela.

Reinoso, D. (2009). Costos ambientales en el proceso de extracción del aceite de palma. *Estudio de un caso. Revista Venezolana de Gerencia, 14*, 228–247.

Rosales, L. C. (2013). *Evaluación de nematodos entomopatógenos como biorreguladores de algunas plagas agrícolas*. PhD Thesis, Universidad Central de Venezuela, Maracay, Venezuela.

Rosales, L. C., & Súarez, H. Z. (1998). Nematodos entomopatógenos como posibles agentes de control del gorgojo negro del plátano *Cosmopolites sordidus* (Germar 1824) (Coleoptera: Curculionidae). *Boletín Entomológico Venezolano, 13*, 123–140.

Rosales, L. C., Suárez, Z., Navas, R., & Tellechea, V. (1999a). Nematodos entomopatógenos: Generalidades. *Fonaiap Divulga, 63*, 6–9.

Rosales, L. C., Suárez, Z., Navas, R., & Tellechea, V. (1999b). Nematodos entomopatógenos: Su uso en control biológico. *Fonaiap Divulga, 64*, 9–16.

Salas, J. (2003). Plantas cultivadas y silvestres hospederas de *Thrips tabaci* y *Thrips palmi* (Thysanoptera:Thripidae) en Quíbor, estado Lara, Venezuela. *Bioagro, 15*, 47–54.

Salma, J., & Shahina, F. (2012). Mass production of eight Pakistani strains of entomopathogenic nematodes (Steinernematidae and Heterorhabditidae). *Pakistan Journal of Nematology, 30*, 1–20.

San-Blas, E. (2013). Progress on entomopathogenic nematology research: A bibliometric study of the last three decades: 1980–2010. *Biological Control, 66*, 102–124.

San-Blas, E., Carrillo, Z., & Parra, Y. (2012). Effect of *Xenorhabdus* and *Phothorhabdus* bacteria and their exudates on *Monilliophthora roreri. Archives of Phytopathology and Plant Protection, 45*, 1950–1967.

San-Blas, E., Cubillán, N., Guerra, M., Portillo, E., & Esteves, I. (2012). Characterization of *Xenorhabdus* and *Photorhabdus* bacteria by Fourier transform mid–infrared spectroscopy with attenuated total reflection (FT–IR/ATR). *Spectrochimica Acta Part A, 93*, 58–62.

San-Blas, E., Gowen, S. R., & Pembroke, B. (2008). *Steinernema feltiae*: Ammonia triggers the emergence of their infective juveniles. *Experimental Parasitology, 119*, 180–185.

San-Blas, E., Guerra, M., Portillo, E., Esteves, I., Cubillán, N., & Alvarado, Y. (2011). ATR/FTIR characterization of *Steinernema glaseri* and *Heterorhabditis indica. Vibrational Spectroscopy, 57*, 220–228.

San-Blas, E., Parra, Y., & Carrillo, Z. (2013). Effect of *Xenorhabdus* and *Photorhabdus* bacteria (Enterobacteriales: Enterobactereaceae) and their exudates on the apical rotten fruit disease caused by *Dothiorella* sp. in guava (*Psidium guajava*). *Archives of Phytopathology and Plant Protection, 46*, 2294–2303.

San-Blas, E., Pirela, D., García, D., & Portillo, E. (2014). Ammonia concentration at emergence and its effects on the recovery of different species of entomopathogenic nematodes. *Experimental Parasitology, 144*, 1–5.

Selvan, S., Gaugler, R., & Lewis, E. E. (1993). Biochemical energy reserves of entomopathogenic nematodes. *Journal of Parasitology, 79*, 167–172.

Terán, L., Meléndez, I., García-Aguilar, L., Acuña, J. G., & Urdaneta, M. (1996). Efecto de la aplicación de nitrógeno y potasio en el rendimiento del cultivo de la guayaba (*Psidium guayaba* L.). *Revista de la Facultad de Agronomía LUZ, 13*, 363–370.

Vélez, R. (1997). *Plagas agrícolas de impacto económico en Colombia: bionomía y manejo integrado*. Medellín, Colombia: Universidad de Antioquia.

Wassink, H., & Poinar, G. O., Jr. (1984). Use of *Neoaplectana carpocapsae* Weiser (Steinernematidae: Rahbditida), in Latin America. *Nematropica, 14*, 97–109.

Yang, X., Qiu, D., Yang, H., Liu, Z., Zeng, H., & Yuan, J. (2011). Antifungal activity of xenocoumacin 1 from *Xenorhabdus nematophilus* var. Pekingensis against *Phytophthora infestans. World Journal of Microbiology and Biotechnology, 27*, 523–528.

Yoo, S. K., Brown, I., & Gaugler, R. (2000). Liquid media development for *Heterorhabditis bacteriophora*: Lipid source and concentration. *Applied Microbiology and Biotechnology, 54*, 759–763.

Zamora, F. (2003). *Identificación de indicadores de sosteniblilidad en los sistemas de producción hortícola de Barrio Nuevo, Municipio Federación, estado Falcón*. Degree Thesis, Universidad Nacional Experimental Francisco de Miranda. Coro, Falcón. Venezuela, pp. 95.

Chapter 16
Orchard Applications of Entomopathogenic Nematodes in Spain

Fernando García del Pino and Ana Morton

16.1 Introduction

Entomopathogenic nematodes (EPNs) are important biological control agents for a variety of pests, especially for soil–dwelling insect pests. Worldwide, EPNs have been used in strawberry plantations, mushroom production, orchards, production nurseries, greenhouses and turfgrass (Grewal, Ehlers, & Shapiro-Ilan, 2005). Currently, over 42 EPN products from five different species (*Steinernema feltiae* (Filipjev) Wouts, Mráček, Gerdin & Bedding (Rhabditida: Steinernematidae), *Steinernema carpocapsae* (Weiser) Wouts, Mráček, Gerdin & Bedding [Rhabditida: Steinernematidae], *Steinernema kraussei* (Steiner) Travassos [Rhabditida: Steinernematidae], *Heterorhabditis bacteriophora* Poinar [Rhabditida: Heterorhabditidae] and *Heterorhabditis megidis* Poinar, Jackson & Klein [Rhabditida: Heterorhabditidae]) can be commercialized in Spain (MAGRAMA, 2014). These EPNs are supplied by ten different producers, six international and four Spanish companies. However, in Spain, EPNs have been applied against few selected pests in some crops, frequently in organic farming or in conventional management plantations as the last choice when chemical pesticides have failed. An example is the use of EPNs against the red palm weevil, *Rhynchophorus ferrugineus* (Olivier) (Coleoptera: Curculionidae). Due to the high efficacy of EPNs observed in the control of *R. ferrugineus* and the low efficacy of most of the chemical pesticides, EPNs have been one of the principal control methods of this weevil pest in Spain (Dembilio, Llácer, Martínez de Altube, & Jacas, 2010; Llácer, Martínez de Altube, & Jacas, 2009).

F. García del Pino (✉) • A. Morton
Department of Animal Biology, Universidad Autónoma de Barcelona, Barcelona, Spain
e-mail: fernando.garcia@uab.cat; ana.morton@uab.cat

© Springer International Publishing Switzerland 2015

R. Campos-Herrera (ed.), *Nematode Pathogenesis of Insects and Other Pests*,
Sustainability in Plant and Crop Protection 1, DOI 10.1007/978-3-319-18266-7_16

In other crops, as EPN products are generally more expensive than the standard chemical pesticides, the research on the use of this biocontrol agent has focused on economically important pests. In this context, pests in orchards are economically important, especially in Spain, one of the principal fruit and nut producers in Europe with approximately 1.17 million ha under cultivation (MAGRAMA, 2013a). For this reason, this chapter reviews the research on EPNs in Spain with respect to their efficacy and use against two important orchard pests: the Mediterranean flat–headed rootborer, *Capnodis tenebrionis* (L.) (Coleoptera: Buprestidae), and the hazelnut weevil, *Curculio nucum* L. (Coleoptera: Curculionidae).

16.2 The Mediterranean Flatheaded Root Borer, *Capnodis tenebrionis* (L.) (Coleoptera: Buprestidae)

16.2.1 Biology, Distribution and Damage

The Mediterranean flatheaded rootborer, *C. tenebrionis*, is a serious pest of stone fruit, *Prunus* spp., and in pome fruit (apple and pear) over Southern European and Mediterranean areas (Garrido, 1984). In Spain, this species mainly attacks cherry, apricot, peach, almond and plum fruit, with high incidence in drier regions and during times of drought. The production area of these fruits in Spain covers ca. 200,000 ha (MAGRAMA, 2013a), and data of crop losses caused by this pest is estimated at 25 % of the national production (Del Moral, Rosado, Casdomet, De la Cruz, & Sereno, 2013).

The presence of this pest is determined by high temperature and low humidity, with an optimum of 28–34 °C and soil humidity below 6 % (Malagon, 1989). The life cycle is completed in 1–2 years, and the insects overwinter in the adult stage or in different larval instars in trees (Balachowski, 1962). Adults of *C. tenebrionis* (Fig. 16.1a) emerge from tree trunks or soil in spring and summer when the temperature rises and feed on twigs and young branches sometimes damaging young trees in nurseries. Females oviposit on the ground, usually in cracks of dry soil or under stones within 50 cm of trees. Females can lay more than 1,000 eggs (Rivnay, 1944, 1946). After 7–16 days, depending on temperature (Malagon, 1989), eggs hatch and the neonate larvae move to the plant, penetrate into the root and start feeding on the cortex. Although they can survive for up to 6 days outside the trunk, the penetration occurs mainly during the first 24 h (Balachowski, 1962). During the first 1–2 years larvae burrow into the tree progressing from small to large roots and trunk as larvae mature (Fig. 16.1b). Pupation occurs in the wood of the tree trunk (Mendel, Assael, & Ben-Yehuda, 2003). Larval tunnels can reach 8 cm in length (Mfarrej & Sharaff, 2010) so that a single larva can kill 1 year old seedlings and a few larvae can lead to the death of an adult tree within 1 or 2 years (Ben-Yehuda, Assael, & Mendel, 2000).

Fig. 16.1 Infectivity of entomopathogenic nematodes (EPNs) to the Mediterranean flatheaded rootborer, *Capnodis tenebrionis*. (**a**) Healthy adult. (**b**) Healthy larva into the trunk. (**c**) Infected adult with *Steinernema carpocapsae*. (**d**) Infected larva with *Steinernema feltiae*

16.2.2 Control Methods

For many decades, chemical insecticides have been the most practical method used to control *C. tenebrionis.* Chemical products based on imidacloprid and chlorpyriphos target adults feeding on trees (Ben-Yehuda & Mendel, 1997) and on neonate larvae moving to the roots (Sanna-Passino & Delrio, 2001). Because of the difficulty to control larvae when penetrating into the roots, cultural practices have been used to reduce the pest population, such as watering the ground, using plastic bands around the trunks to prevent egg–laying or removing and burning infected trees. These methods seem to be effective but costly. Increased efforts in recent years have been focused on the potential use of different *Prunus* rootstocks resistant to *C. tenebrionis* larvae (Gindin et al., 2014; Mendel et al., 2003).

Biological control with entomopathogenic fungi has been studied as an alternative to pesticides, and their efficacy has been tested in the laboratory against neonate larvae and adults of C. *tenebrionis* (Marannino, Santiago-Álvarez, de Lillo, & Quesada-Moragas, 2006, 2008). In addition to the control provided by fungal applications, EPNs in the genus *Steinernema* have been shown effective for controlling different *C. tenebrionis* stages (Garcia-del-Pino & Morton, 2005; Morton & Garcia-del-Pino, 2009a), as we will describe in detail in the following sections.

16.2.3 Susceptibility of the Mediterranean Flatheaded Rootborer to Entomopathogenic Nematodes: Native Populations and Laboratory Assessment

Capnodis tenebrionis is susceptible to different species of EPNs (Fig. 16.1c, d). At least two populations of *S. feltiae* and one of *S. carpocapsae* have been reported to occur naturally in flatheaded rootborer larvae found inside tree trunks (Garcia-del-Pino, 1994; Morton & Garcia-del-Pino, 2008a; Santos Lobatón, García Vela, Lara López, & Canales Roca, 1998). Moreover, in a study conducted to determine the natural presence of EPNs in stone fruit orchards attacked by *C. tenebrionis* in two regions of Spain, Catalonia and Murcia, 14 populations of *S. feltiae* and 3 populations of *H. bacteriophora* were isolated from soil samples (Morton & Garcia-del-Pino, 2008a). The characterization of these populations showed interspecific and intraspecific differences in environmental tolerance, and could determine their potential to control this pest. For example, the characterization of these native populations evidenced that most of them were adapted to warm temperatures, being able to infect an insect at 35 °C, and to reproduce at 30 °C. This temperature range concurs with climatic conditions when *C. tenebrionis* is present as a pest.

The susceptibility of *C. tenebrionis* to different EPN populations/species varies depending on the insect's life stage and the EPN isolate. In laboratory trials, neonate larvae showed high susceptibility to infective juveniles (IJs) of *H. bacteriophora*, *S. carpocapsae* (Garcia-del-Pino & Morton, 2005; Marannino, Tarasco, & de Lillo, 2003), *S. feltiae* and *Steinernema arenarium* (Artykhovsky) Wouts, Mráček, Gerdin & Bedding (Rhabditida: Steinernematidae) (Garcia-del-Pino & Morton, 2005). All populations caused 96–100 % mortality after 5 days at a dose of 150 IJs/larva. However, at 10 IJs/larva significant differences were observed among the EPN populations tested, showing between 59 % (*S. carpocapsae*) and 91 % (*S. arenarium*) larval mortality (Garcia-del-Pino & Morton, 2005).

A subsequent laboratory study evaluated the virulence of different EPN species/populations against the last larval stage of *C. tenebrionis,* pupae and adults (Morton & Garcia-del-Pino, 2009a). At a concentration of 50 IJs/cm^2, *H. bacteriophora, S. carpocapsae* and *S. feltiae* caused 100 % mortality of last instar larvae (Fig. 16.2a), although some *S. feltiae* isolates worked faster than the other species. The pupae of *C. tenebrionis* were not especially susceptible to *S. feltiae* (up to 40 % mortality) and *S. carpocapsae* (20 % mortality), although one *H. bacteriophora* population reached 70 % mortality (Fig. 16.2b).

Adults of *C. tenebrionis* were less susceptible than at their larval stages, although some isolates of S. *feltiae* and *S. carpocapsae* reached 100 and 87 % mortality, respectively, at a dose of 100 IJs/cm^2 (Fig. 16.2c). Additional studies found males to be more susceptible than females (Morton & Garcia-del-Pino, 2013). These authors carried out an experiment to study the route of entry of nematodes into adults of *C. tenebrionis*, exposing males and females to *S. carpocapsae* for 36 h at a concentration of 50 IJs/cm^2. Nematodes were found in the genitalia in males (6.9 $\pm$ 1.2 nematodes) while no nematodes were detected in females, corroborating

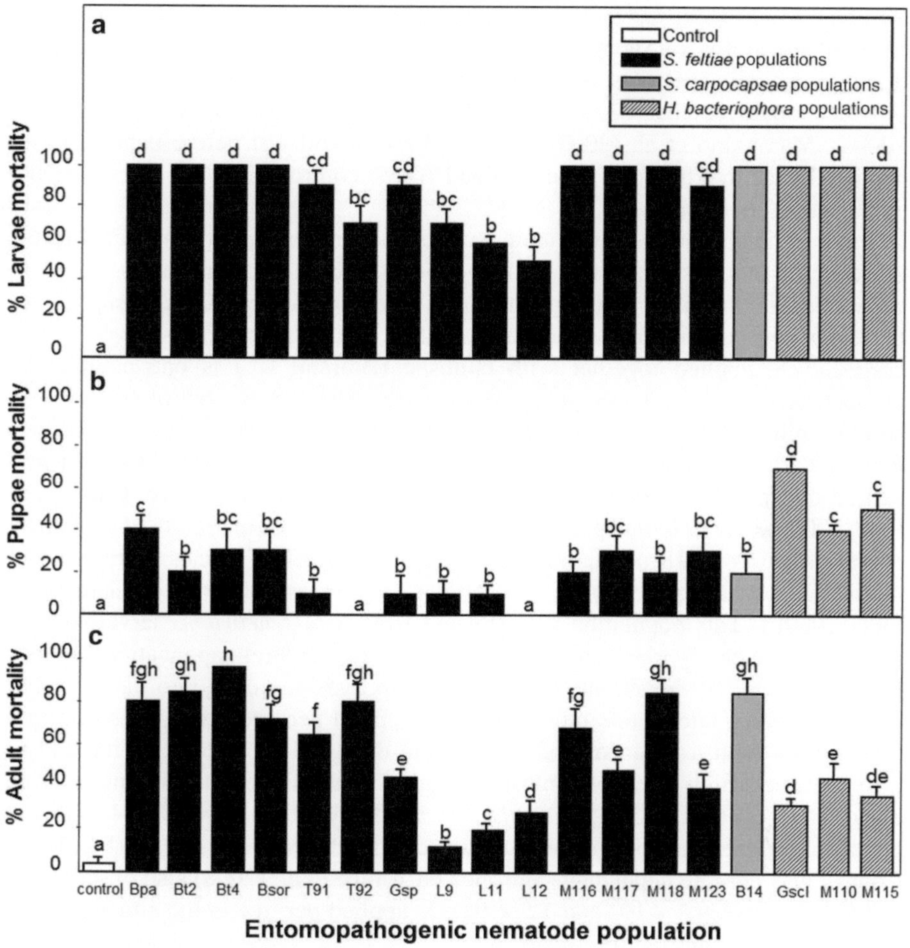

Fig. 16.2 Mortality (%) of different stages of *Capnodis tenebrionis* in Petri dishes to different *Steinernema feltiae*, *S. carpocapsae* and *Heterorhabditis bacteriophora* populations: (**a**) mortality of last instar larvae after 5 days of exposure to 50 IJ/cm^2; (**b**) mortality of pupae after 6 days of exposure to 100 IJ/cm^2; (**c**) mortality of adults after 8 days of exposure to 100 IJ/cm^2. Means with the same letter do not differ significantly (P = 0.05) (Reprinted with permission from Morton and Garcia-del-Pino (2009a))

that differences of susceptibility between sexes are related to penetration of IJs through genital apertures (Morton & Garcia-del-Pino, 2013).

16.2.4 Field Trials and Application Methods

Field studies on *C. tenebrionis* control using EPNs have focused on the larvae, when hatched and moving to the roots as well as when burrowing galleries into the tree. Semi–field tests using potted peach trees and testing 13 populations of *S. feltiae,*

S. affine, S. carpocapsae and *H. bacteriophora* for control of early larval stages revealed the ability of IJs to locate and kill *C. tenebrionis* larvae in the galleries that they make in the roots. *Steinernema feltiae* and *H.* bacteriophora populations provided up to 88 and 77 % control, respectively (Morton & Garcia-del-Pino, 2008a). Marannino et al. (2003) applied EPNs on soil just before penetration of neonate larvae into the roots, and obtained 100 % control with *S. carpocapsae* and 99 % with *H. bacteriophora*.

Environmental factors (high temperatures and low precipitation) related to the presence of this pest do not seem to provide the best environment for EPNs to be applied in the field. However, the use of *S. feltiae* and *S. carpocapsae* for *C. tenebrionis* control has been successful in the different crops. *Steinernema carpocapsae* applied together with chitosan (Biorend R®) is one commercial product used to control this pest and has given over 85 % control of larvae in apricot orchards (Martínez de Altube, Strauch, Fernández de Castro, & Martínez Peña, 2008). In another study, one *S. feltiae* population was tested in a field trial carried out in a cherry orchard attacked by *C. tenebrionis* (Morton & Garcia-del-Pino, 2008b). The *S. feltiae* population tested was chosen from a study of ecological characterization of 18 populations of three different EPN species carried out to select the most adequate isolate to control *C. tenebrionis* (Morton & Garcia-del-Pino, 2009b). This population was isolated from a *C. tenebrionis* larval cadaver found inside a cherry tree trunk. In the field trial, two different application methods were evaluated: (*i*) drench and (*ii*) injection (up to 50 cm deep into the soil). For both methods, a rate of one million IJs was applied per tree every week during 4 or 8 weeks, with a total dose of 4×10^6 and 8×10^6 IJs per tree. At the end of the experiment high levels of control resulted with all the treatments, ranging from 88 to 97 % control of *C. tenebrionis*, and mortality of *C. tenebrionis* larvae and pupae for different sections of the plant did not differ between treatments (Fig. 16.3). Control was not rate dependant. Martínez de Altube et al. (2008) also observed no difference between 1×10^6 and 1.5×10^6 IJs applied per tree in the efficacy of the *S. carpocapsae*–chitosan formulate.

Persistence of nematodes in soil is important to improve the efficacy of nematodes, and to design the best timing and frequency of nematode applications. In the cherry tree orchard assay, S. *feltiae* was recovered from soil for 6 weeks after the last application, and with no differences between the two application methods (Morton & Garcia-del-Pino, 2008b). From all of the above results it is clear that EPNs can control *C. tenebrionis* in the field, and improving the timing of EPN application could increase control efficacy. Future studies should investigate applications made during the cold weather months, when larvae and pupae remain in roots and adults are hibernating in the soil, but also during the oviposition period to infect neonate larvae in soil or when they enter into the roots.

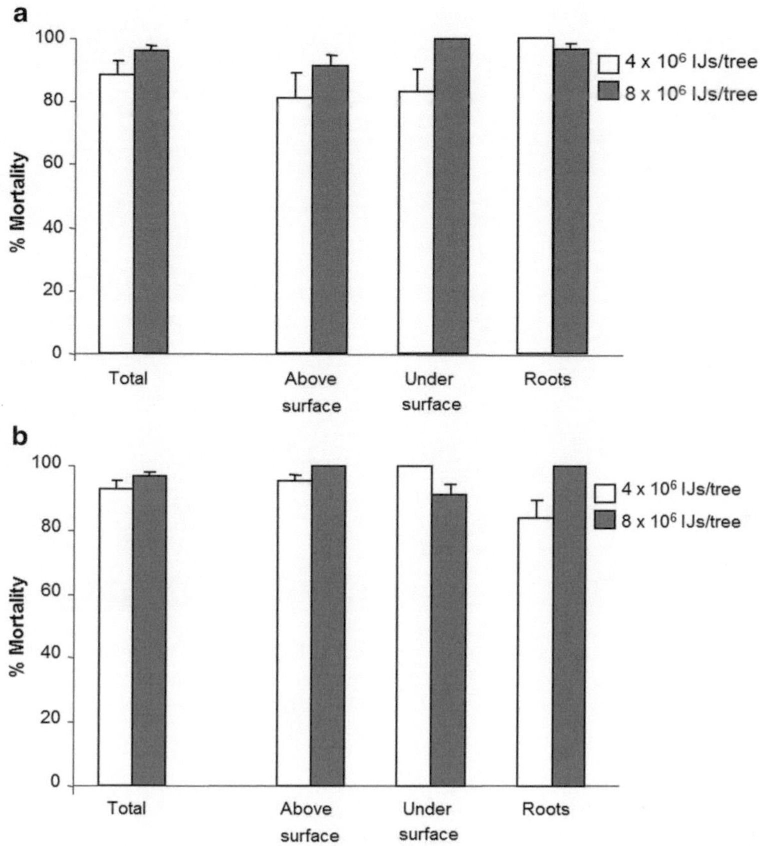

Fig. 16.3 Control–corrected mortality of *Capnodis tenebrionis* (larvae and pupae) collected from tree trunks (n = 8) above or below the soil surface and from the root system 4 weeks after (**a**) drench application or (**b**) soil injection of weekly treatments with 1×10^6 IJs/tree of *Steinernema feltiae* (population Bpa) over a period of 4 or 8 weeks (Reprinted with permission from Morton and Garcia-del-Pino (2008b)

16.3 The Hazelnut Weevil, *Curculio nucum* L. (Coleoptera, Curculionidae)

16.3.1 Biology, Distribution and Damage

The hazelnut weevil, *Curculio nucum*, is a major pest of cultivated hazelnuts (*Corylus avellana* L.) and wild hazel trees. The hazelnut is widely distributed throughout Europe, Western Asia, North Africa and the Caucasus and was introduced to the United States from Europe during the late 1800s (AliNiazee, 1998). Spain is the eighth largest hazelnut producer in the world, with 13,900 t produced during 2012

Fig. 16.4 Infectivity of entomopathogenic nematodes (EPNs) to the hazelnut weevil, *Curculio nucum*. (**a**) Healthy adult. (**b**) Healthy larva emerging from the nut to burrow into the ground. (**c**) Infected adult with *Steinernema* sp. D122 (glaseri–group). (**d**) Infected larva with *Steinernema feltiae*

(FAO, 2013). The main Spanish hazelnut–producing region is Catalonia, containing more than 95 % of the hazelnut growing areas in Spain (FAO, 2009).

Curculio nucum is widely distributed throughout Europe and Asia where it is considered the most destructive pest of hazelnuts (AliNiazee, 1998). In the Mediterranean region adults of *C. nucum* emerge from the soil in April and feed during May–June on the immature fruits (Fig. 16.4a). Oviposition takes place from June to July in the hazelnut fruit and the larvae develop inside the nuts. At the beginning of August the larvae emerge from the nuts and burrow into the ground (Fig. 16.4b), within the first 25 cm, where they spend a wintering diapause (Akça & Tuncer, 2005).

The hazelnut weevil has a 2–year life cycle, including one adult wintering in addition to larval wintering (AliNiazee, 1998; Bel-Venner et al., 2009; Coutin, 1992).

While adults feed on tender leaves, buds, female flowers and young fruits, causing in this last case aborted nuts and the drop of fruits before time (Akça & Tuncer, 2005), the principal damage is associated with larval feeding and development inside the nut. Larval damage is the most visible and well–known for the farmers, but occasionally damage caused by feeding of adults early in the season can also be serious (Akça & Tuncer, 2005). In unprotected orchards, larvae and adults are capable of causing production losses of up to 80 % (AliNiazee, 1998).

The hazelnut orchards at slightly higher altitudes are attacked more by this pest than those situated at lower elevations (Tuncer & Ecevit, 1997).

16.3.2 Control Methods

The use of chemical insecticides has been the most important method to *C. nucum* control and, due to the larvae's cryptic habitat, chemical control is directed only at emerging adults, limiting its success (Akça & Tuncer, 2005). Carbaryl and endosulfan have been widely used in the past against this pest with some success. The withdrawal of these active ingredients at the European level resulted in the use of newer insecticides such as methiocarb or chlorpyrifos with low success rates.

Cultural practices have been used for more than 50 years as a method of control. Hazelnut weevil populations were reduced by sanitation and clean orchard practices. As more of the nuts infested with larvae of this insect drop before the main crop, collecting and destroying all of the prematurely dropped nuts is a cultural practice frequently used against this insect (Leske, 1973). In addition to those practices, choosing varieties of *C. avellana* resistant or more tolerant to the hazelnut weevil is one of the modern methods of crop protection. Resistant varieties are based on some component in the endocarp tissues or on greater shell thickness (Caramiello, Me, & Radicati, 2000; Piskornik, 1992, 1994; Wojciechowicz-Zytko, 2005). Guidone, Valentini, Rolle, Me, and Tavella (2007) proposed the use of early nut development varieties as a resistance factor against the attacks of *C. nucum*.

Due to the difficulty of controlling this insect with cultural practices and chemical insecticides and the environmental issues associated with these agrochemical products, alternative biocontrol methods are needed. Biological control of this insect has been the focus through the use of entomopathogenic fungi and nematodes. Entomopathogenic fungi belonging to the order Hypocreales (Ascomycota) that inhabit the soil such as *Metarhizium anisopliae* (Metschnikoff) Sorokin and *Beauveria bassiana* (Balsamo) Vuillemin have been considered promising biological control agents of different weevils. A commercial product made of *B. bassiana* was tested by Paparatti and Speranza (2005) against *C. nucum* with a 35 % increase in larval mortality. Batalla-Carrera, Morton, Santamaria, and Garcia-del-Pino (2013b) tested in laboratory conditions the virulence towards hazelnut weevil larvae of combining *M. anisopliae*, with four EPN species: *S. carpocapsae*, *S. feltiae*, *Steinernema* sp. D122 and *H. bacteriophora*. The combination of fungi and EPN is not likely to improve suppression of *C. nucum* larvae beyond what is expected from the single application of either pathogen.

The presence of native EPN populations in Spanish orchards and the possible use against *C. nucum* has been investigated during the last few years. Garcia-Lopez, Martinez, Morton, and Garcia-del-Pino (2013) showed that from a total of 396 soil samples collected in different hazelnut areas (cultivated and wild) with presence of *C. nucum,* distributed throughout Catalonia and Asturias (Spain), 8 % were positive to EPNs. The major occurrence of EPNs was in wild hazelnut areas

(19 % of positive samples). The nematodes found in cultivated areas (3 %) were associated with organic orchards, integrated management and abandoned hazelnut orchards. No EPNs were isolated from conventionally managed hazelnut orchards. The percentage of EPN–positive samples in hazelnut areas (19 %) is similar to values obtained from croplands, woodlands and pastures in the same region (23 %) (Garcia-del-Pino & Palomo, 1996). However, the presence of EPNs in hazelnut orchards (3 %) is lower than that obtained in stone–fruit orchards in the same region (5 %) (Morton & Garcia-del-Pino, 2009b). Conventional management of hazelnut orchards requires intensive agricultural soil practices to avoid vegetation coverage that would interfere with the collection of hazelnuts, and these practices could affect EPN occurrence (Barbercheck, 1992; Campos-Herrera et al., 2008; Hummel et al., 2002). In organic orchards or wild hazelnuts the agricultural practices are less aggressive, allowing for the presence of vegetation coverage and the EPN host populations necessary for EPN establishment and persistence (Strong, 2002).

The EPNs most frequently present in hazelnut areas are steinernematids (96 % of positive soils) while heterorhabditids (only *H. bacteriophora*) are present in 4 % of positive soils. Among Steinernematidae, the predominant species were *S. feltiae* (52 % of the nematodes found) followed by *S. affine* (11 %), *S. intermedium* (7 %), and *Steinernema* sp. (26 %) (Batalla-Carrera, 2013; Garcia-Lopez et al., 2013). In another survey carried out in chestnut and oak soils in the presence of other weevils (*Curculio elephas* L. [Coleoptera: Curculionidae] and *Curculio glandium* Marsham [Coleoptera: Curculionidae]), the EPNs found are also mainly *S. feltiae* followed by *S. kraussei* and *S. carpocapsae* (Garcia-Lopez et al., 2013).

16.3.3 Susceptibility of Hazelnut Weevil to Entomopathogenic Nematodes: Laboratory Assessment

The susceptibility of the larvae and the adults of hazelnut weevil can influence the effectiveness of EPNs. Recent studies show that the last instar larvae and adults of *C. nucum* are susceptible to the EPN species found in the soil of hazelnut orchards (Batalla-Carrera, Morton, Shapiro-Ilan, Strand, & Garcia-del-Pino, 2014) (Fig. 16.4c, d). Nevertheless, differences in susceptibility among larvae and adults to these nematodes were observed. Although in general weevil larvae are more susceptible than adults to EPNs (Giblin-Davis, Pena, & Duncan, 1996; Mannion & Jansson, 1992; Shapiro-Ilan, Mizell, & Campbell, 2002), for *S. carpocapsae* larvae of *C. nucum* are less susceptible than adults (Batalla-Carrera et al., 2014). These authors observed a 5 % mortality and 0.3 % penetration rate of IJs in larvae, and a 92.5 % mortality and 4 % penetration rate in adults after a 12 h exposure to 500 IJs of *S. carpocapsae*. In the same experiment, *S. feltiae* reached a 55 % mortality and 1.4 % penetration rate in larvae, and a 2.5 % mortality and 0.02 % penetration rate in adults. The potential of *S. carpocapsae* to control *C. nucum* adults is enhanced by the short time needed by this species to cause weevil mortality. Nematode exposure for 15 min was sufficient to infect 17 % of the weevils, and adult mortality reached 100 % in 120–240 min (Batalla-Carrera, 2013).

The virulence of the symbiotic bacteria could also be an important factor in defining the effectiveness of the entomopathogenic nematode–bacteria complex. Batalla-Carrera (2013) determined a high mortality when symbiotic bacteria of EPN populations with low virulence were directly injected in larvae and adults of *C. nucum*, suggesting that it is the ability of the nematode to locate the insect, get into the host and release the symbiotic bacteria that is responsible for the effectiveness of EPNs in controlling this insect.

16.3.4 Method of Application and Nematode Persistence in the Field

Efficacy of the EPN application against hazelnut weevil depends, among other factors, on the ability of the nematodes to persist in the pest's environment. Knowing the EPN persistence after application in hazelnut orchards could help to develop appropriate patterns of EPN application timing. Batalla-Carrera, Morton et al. (2013) showed that EPNs could be present in hazelnut orchard up to 9 weeks after application following a fluctuating pattern. The fluctuations in EPN presence are related to insect population dynamics and EPN recycling in these insects (Fenton et al., 2002). A high number of insect species has been identified as being associated with hazelnuts (AliNiazee, 1998, Gantner & Jaskiewicz, 2002). These insect populations could be potential EPN reservoirs, acting as hosts for EPN recycling, thus increasing the EPN's persistence in the soil of hazelnut orchards.

One approach for the control of *C. nucum* may be to apply EPNs to the soil surface under a tree's canopy. In this situation the ability of EPNs to locate the insect in the soil could be an important factor affecting EPN efficacy. Recent field trials concerning the vertical distribution of EPNs applied in the soil of hazelnut orchards, showed that all EPN species tested (*S. feltiae*, *S. carpocapsae*, *Steinernema* sp. D122 and *H. bacteriophora*) were present down to a 40 cm depth, although 50 % of the recovered individuals were found in the first 20 cm (Batalla-Carrera, Morton et al., 2013). This study also found nearly 90 % of overwintering *C. nucum* larvae in the first 20 cm of soil, suggesting that the vertical distribution in soil of the wintering *C. nucum* is not a factor affecting EPN efficacy after application to the soil surface.

Considering the life cycle of *C. nucum*, the persistence of applied EPNs, their vertical distribution in the soil of hazelnut orchards, and the susceptibility of larval and adult stages to each EPN species, the following approach for *C. nucm* control with EPNs may be feasible: summer and spring applications. Summer applications would be a barrier strategy, attacking the insect when the larvae are burying in the soil, while spring applications would control the overwintering *C. nucum* when they are buried in the soil. Because *C. nucum* overwinter in the soil as larvae in their first winter and as adults in their second winter (Bel-Venner et al., 2009), spring

Table 16.1 Efficacy of different semi–field trials with four EPN species against *C. nucum* larvae in two different application periods (summer and spring). Efficacy is expressed as corrected mortality using Abbott's formula (Abbott, 1925). Ref: References: (1) Kuske et al., 2005, (2) Peters et al., 2007, (3) Batalla-Carrera et al., 2013a

Application period	Dosage (IJs m^{-2})	*S. feltiae* (%)	*Steinernema* sp (*glaseri*–group)	*H. bacteriophora* (%)	*H. indica*	Ref
Summer	2.2×10^6	15	–	63	48 %	(1)
Summer	5×10^5	20	–	0	32 %	(2)
Summer	5×10^5	34	44 %	52	–	(3)
Summer	5×10^5	52	61 %	61	–	(3)
Spring	5×10^5	52	64 %	61	–	(3)
Spring	5×10^5	88	50 %	47	–	(3)

EPN applications would target both stages. Due to adults and larvae differing with respect to what EPN species are most effective against them (Batalla-Carrera et al., 2014), the spring EPN application should include *S. feltiae* to target larvae and *S. carpocapsae* to target adults.

The spring application of *S. carpocapsae* under the tree's canopy covering the entire area of adult emergence could also serve to infect the emerging *C. nucum* adults due to the quick infection of adults (Batalla-Carrera, 2013). Similar strategies have been proposed by Shapiro-Ilan (2003) against the pecan weevil, *Curculio caryae* Horn (Coleoptera: Curculionidae). This author determined that field trials applying *S. carpocapsae* in a band around each pecan tree can provide 60–80 % control of emerging *C. caryae* adults, although this level of control lasts only 1 week.

16.3.5 Semi–field Trials

The efficacy of EPN applications against *C. nucum* larvae in semi–field trials has been documented in different works (Table 16.1). All of these semi–field trials were conducted in plastic tubes or buckets buried into the soil where a known number of larvae of *C. nucum* were introduced. While Kuske et al. (2005) and Peters, Sarraquigne, Blum, and Kuske (2007) achieved efficacies between 0 and 63 % using commercial EPNs, Batalla-Carrera et al. (2013a) obtained efficacies between 34 and 88 % using EPN species isolated in the soil of hazelnut orchards. This could be highlighting the importance of the use of native species, as these isolates may be better adapted or prepared to infect a particular host that cohabits in the same location.

Batalla-Carrera et al. (2013a) did not detect differences in efficacy between the summer and spring applications (Table 16.1). All EPNs showed the capacity of controlling larvae both in spring, when *C. nucum* is overwintering, and during the summer when they are burying in the soil, because EPNs have the capacity to find and invade overwintering stages of this insect at any depth. Thus, Batalla-Carrera et al. (2013a) concluded that EPNs can effectively reduce *C. nucum* populations and suggested that a spring application could be an alternative to summer application in order to minimize negative abiotic factors and improve EPN persistence.

16.4 Conclusions and Future Directions

The proven efficacy of EPNs against the Mediterranean flat–headed rootborer and the hazelnut weevil shows these nematodes as promising biocontrol agents to be used against other orchard pests in soil environments. Future research on aboveground EPN applications could extend the target insects to pests located on or in tree branches or trunks. An example is the codling moth, *Cydia pomonella* (L.) (Lepidoptera: Tortricidae), which is a serious worldwide pest of the apple, pear and walnut. Overwintering mature codling moth larvae within the bark of trees, cracks in wooden supports, leaf litter and other cryptic habitats are good candidates to use EPNs for their control. EPN formulation to retard desiccation and the use of cold–tolerant nematode populations for these overwintering larvae could improve the efficacy of EPNs against this insect (Lacey, Arthurs, Unruh, Headrick, & Fritts, 2006).

The promotion of organic farming could also create opportunities for EPN products in orchard production. In the last decade, the favorable socio–political atmosphere for organic farming production has resulted in an increase in organic orchards in Spain from 41,513 ha in 2002 to 109,416 ha in 2012 (MAGRAMA, 2013b). This growth in organic production combined with the UE restrictions on chemical pesticides in conventional management (Directive 2009/128/EC to achieve the sustainable use of pesticides) could lead to expanded EPN use in orchards. Nevertheless, the use of EPNs in orchard production in Spain is still limited. Factors limiting EPN use against such orchard pests include efficacies less predictable than are necessary for the successful market penetration of these biocontrol products (Georgis et al., 2006). Suboptimum or unpredictable efficacy may be related to the use of inappropriate nematode species and/or their application under suboptimal conditions such as high temperature, low moisture and exposure to sunlight which are climatic characteristics more typical in Mediterranean orchards. For this reason, to expand the commercial use of EPNs, research should be focused on improving application and formulation technology and on discovering new, more virulent as well as the best–adapted EPN species and populations. These beneficial traits of EPNs could also be obtained by the hybridization and genetic selection of currently available populations to improve their virulence against each orchard pest and their persistence in the adverse climatic conditions of orchards in the Mediterranean area.

References

Abbott, W. S. (1925). A method of computing effectiveness of an insecticide. *Journal of Economic Entomology, 18*, 265–267.

Akça, I., & Tuncer, C. (2005). Biological and morphological studies on nut weevil (*Curculio nucum* L., Col., Curculionidae). *Acta Horticulturae, 668*, 413–420.

AliNiazee, M. (1998). Ecology and management of hazelnut pests. *Annual Review of Entomology, 43*, 395–419.

Balachowski, A. S. (1962). Famille des Buprestidae. In J. P. Masson (Ed.), *Entomologie applique a l'Agriculture. Coleopteres* (Vol. 1, pp. 235–300). Paris: Masson et cie.

Barbercheck, M. E. (1992). Effect of soil physical factors on biological control agents of soil insect pests. *Florida Entomologist, 75*, 539–548.

Batalla-Carrera, L. (2013). Biological control of the hazelnut weevil *Curculio nucum* L. (Coleopterea: Curculionidae), with entomopathogenic nematodes (Rhabditida: Steinernematidae and Heterorhabditidae). PhD Thesis, Universitat Autònoma de Barcelona, Barcelona, Spain.

Batalla-Carrera, L., Morton, A., & Garcia-del-Pino, F. (2013a). Field efficacy against the hazelnut weevil, *Curculio nucum* and short–term persistence of entomopathogenic nematodes. *Spanish Journal of Agricultural Research, 11*(4), 1112–1119.

Batalla-Carrera, L., Morton, A., Santamaria, S., & Garcia-del-Pino, F. (2013b). Isolation and virulence of entomopathogenic fungi against larvae of hazelnut weevil *Curculio nucum* (Coleoptera, Curculionidae) and the effects of combining *Metarhizium anisopliae* with entomopathogenic nematodes in the laboratory. *Biocontrol Science and Technology, 23*, 101–125.

Batalla-Carrera, L., Morton, A., Shapiro-Ilan, D., Strand, M. R., & Garcia-del-Pino, F. (2014). Infectivity of *Steinernema carpocapsae* and *S. feltiae* to larvae and adults of the hazelnut weevil, *Curculio nucum*: Diferential virulence and entry routes. *Journal of Nematology, 46*(3), 281–286.

Bel-Venner, M. C., Mondy, N., Arthaud, F., Marandet, J., Giron, D., Venner, S., et al. (2009). Ecophysiological attributes of adult overwintering in insects: Insight from a field study of the nut weevil, *Curculio nucum. Physiological Entomology, 34*, 61–70.

Ben-Yehuda, S., Assael, F., & Mendel, Z. (2000). Improved chemical control of *Capnodis tenebrionis* and *C. carbonaria* in stone–fruit plantations in Israel. *Phytoparasitica, 28*, 1–16.

Ben-Yehuda, S., & Mendel, Z. (1997). Control of wood–boring beetles (*Capnodis* spp and *Scolytus* spp) in deciduous orchards in Israel. *Alon–Hanoteca, 51*, 170–172.

Campos-Herrera, R., Gómez-Ros, J. M., Escuer, M., Cuadra, L., Barrios, L., & Gutiérrez, C. (2008). Diversity, occurrence, and life characteristics of natural entomopathogenic nematode populations from La Rioja (Northern Spain) under different agricultural management and their relationships with soil factors. *Soil Biology and Biochemistry, 40*, 1474–1484.

Caramiello, R., Me, G., & Radicati, L. (2000). Structure and characteristics of the hazelnut shell in different cultivars and their agronomic and industrial influence. *Acta Horticulturae, 517*, 195–205.

Coutin, R. (1992). Caractères originaux des cycles évolutifs de plusieurs espèces de balanins européens: *Curculio elephas* Gyll., *C. nucum* L., *C. glandium* Marsh., *C. venosus* Grav. et *C. villosus* F. *Mémoires de la Société Royale Belge d'Entomologie, 53*, 259–266.

Del Moral, J., Rosado, P. E., Casdomet, E., De la Cruz, J., & Sereno, M. (2013). Control del gusano cabezudo en los frutales de hueso. *Vida Rural, 631*, 12–16.

Dembilio, O., Llácer, E., Martínez de Altube, M. M., & Jacas, J. A. (2010). Field efficacy of imidacloprid and *Steinernema carpocapsae* in a chitosan formulation against the red palm weevil *Rhynchophorus ferrugineus* (Coleoptera: Curculionidae) in *Phoenix canariensis. Pest Management Science, 66*, 365–370.

FAO. (2009). *Food and agricultural commodities production*. Available in: http://faostat.fao.org/faostat. Revised 28 July 2014.

FAO. (2013). *Food and agricultural commodities production*. Available in: http://faostat.fao.org/faostat. Revised 28 July 2014.

Fenton, A., Gwynn, R. L., Gupta, A., Norman, R., Fairbairn, J. P., & Hudson, P. J. (2002). Optimal application strategies for entomopathogenic nematodes: Integrating theoretical and empirical approaches. *Journal of Applied Ecology, 39*, 481–492.

Gantner, M., & Jaskiewicz, B. (2002). Beetles (Coleoptera) occurring in hezel (*Corylus* L.) in different habitat conditions. *Acta Scientiarum Polonorum – Hortorum Cultus, 1*(1), 55–56.

Garcia-del-Pino, F. (1994). Los nematodos entomopatógenos (Rhabditida: Steinernematidae y Heterorhabditidae) presentes en Cataluña y su utilización para el control biológico de insectos. PhD Thesis, Departamento de Biología Animal, Vegetal y Ecología. Facultad de Ciencias, Universidad Autónoma de Barcelona, Barcelona, Spain.

Garcia-del-Pino, F., & Morton, A. (2005). Efficacy of entomopathogenic nematodes against neonate larvae of *Capnodis tenebrionis* (L.) (Coleoptera: Buprestidae) in laboratory trials. *BioControl, 50*, 307–316.

Garcia-del-Pino, F., & Palomo, A. (1996). Natural occurrence of entomopathogenic nematodes (Rhabditida: Steinernematidae and Heterorhabditidae) in Spanish soils. *Journal of Invertebrate Pathololgy, 68*, 84–90.

García-López, D., Martinez, M., Morton, A., & García-del-Pino, F. (2013, October 21–25). Presencia natural de nematodos entomopatógenos en avellanares y castañares de Asturias y Cataluña. *Proceedings of the VIII Congreso Nacional de Entomología Aplicada* (p. 237). Mataró.

Garrido, A. (1984). Bioecologia de *Capnodis tenebrionis* L. (Col. Buprestidae) y orientaciones para su control. *Boletín de Sanidad Vegetal. Plagas, 10*, 205–221.

Georgis, R., Koppenhöfer, A. M., Lacey, L. A., Bélair, G., Duncan, L. W., Grewal, P. S., et al. (2006). Successes and failures in the use of parasitic nematodes for pest control. *Biological Control, 38*, 103–123.

Giblin-Davis, R. M., Pena, J. E., & Duncan, R. E. (1996). Evaluation of an entomopathogenic nematode and chemical insecticides for control of *Metamasius hemipterus sericeus* (Coleoptera: Curculionidae). *Journal of Entomological Science, 31*, 240–251.

Gindin, G., Mendel, Z., Levitin, B., Kumar, P., Levi, T., Shahi, P., et al. (2014). The basis for rootstock resilient to *Capnodis* species: Screening for genes encoding δ–endotoxins from *Bacillus thuringiensis*. *Pest Management Science, 70*, 1283–1290.

Grewal, P. S., Ehlers, R. U., & Shapiro-Ilan, I. (2005). *Nematodes as biocontrol agents*. New York: CABI Publishing.

Guidone, L., Valentini, N., Rolle, L., Me, G., & Tavella, L. (2007). Early nut development as a resistance factor to the attacks of *Curculio nucum* (Coleoptera: Curculionidae). *Annals of Applied Biology, 150*, 323–329.

Hummel, R. L., Walgenbach, J. F., Barbercheck, M. E., Kennedy, G. G., Hoyt, G. D., & Arellano, C. (2002). Effects of production practices on soil borne entomopathogens in western North Carolina vegetable systems. *Environmental Entomology, 31*, 84–91.

Kuske, S., Daniel, C., Wyss, E., Sarraquigne, J. P., Jermini, M., Conedera, M., et al. (2005). Biocontrol potential of entomopathogenic nematodes against nut and orchard pests. *Insect pathogens and insect parasitic nematodes: Melolontha. OIBC/OILB, West Palaeartic Regional Section (WPRS/SROP) Bulletin, 28*, 163–167.

Lacey, L. A., Arthurs, S. P., Unruh, T. R., Headrick, H., & Fritts, R., Jr. (2006). Entomopathogenic nematodes for control of codling moth (Lepidoptera: Tortricidae) in apple and pear orchards: effect of nematode species and seasonal temperatures, adjuvants, application equipment and post–application irrigation. *Biological Control, 37*, 214–223.

Leske, W. (1973). Badania nad biologia slonkowcaorzechowca, *Curculio* (*Balaninus*) *nucum* L. (Col. Curculionidae) i jego niszczeniem. *Polish Journal of Entomology, 43*, 861–873.

Llácer, E., Martínez de Altube, M. M., & Jacas, J. A. (2009). Evaluation of the efficacy of *Steinernema carpocapsae* in a chitosan formulation against the red palm weevil, *Rhynchophorus ferrugineus*, in *Phoenix canariensis*. *BioControl, 54*, 559–565.

MAGRAMA. (2013a). *Superficie y producción de frutas y hortalizas en España (años 2008–2012)*. Madrid, Spain: Ministerio de Agricultura, Alimentación y Medio Ambiente. http://www.magrama.gob.es/es/agricultura/temas/producciones\T1\textendashagricolas/tablawebspffhh_tcm7\T1\textendash318715.pdf

MAGRAMA. (2013b). *Agricultura Ecológica. Estadísticas 2012*. Madrid, Spain: Ministerio de Agricultura, Alimentación y Medio Ambiente. http://www.magrama.gob.es/es/alimentacion/temas/la\T1\textendashagriculturaecologica/Estadisticas_AE_2012_ok_tcm7\T1\textendash297880.pdf

MAGRAMA. (2014). *Registro de productos y materiales fitosanitario: otros medios de defensa Fitosanitaria (OCB)*. Madrid, Spain: Ministerio de Agricultura, Alimentación y Medio Ambiente. http://www.magrama.gob.es/agricultura/pags/fitos/registro/fichas/pdf/Registro+OCB+web+octubre+14.pdf

Malagon, J. (1989). Bioecologia de *Capnodis tenebrionis* L. (Col. Buprestidae) e influencia de ciertos factores abioticos sobre sus estados inmaduros, en el momento de la eclosion del huevo y su penetracion en huespedes de interés agricola. PhD Thesis, Universidad Politecnica de Valencia, Valencia, Spain.

Mannion, C. M., & Jansson, R. K. (1992). Comparison of ten entomopathogenic nematodes for control of sweetpotato weevil (Coleoptera: Apionidae). *Journal of Economic Entomology, 85*, 1642–1650.

Marannino, P., Santiago-Álvarez, C., de Lillo, E., & Quesada-Moragas, E. (2006). A new bioassay method reveals pathogenicity of *Metarhizium anisopliae* and *Beauveria bassiana* against early stages of *Capnodis tenebrionis* (Coleoptera; Buprestidae). *Journal of Invertebrate Pathology, 93*, 210–213.

Marannino, P., Santiago-Álvarez, C., de Lillo, E., & Quesada-Moraga, E. (2008). Evaluation of *Metarhizium anisopliae* (Metsch) Sorok to target larvae and adults of *Capnodis tenebrionis* (L.) (Coleoptera: Buprestidae) in soil and fiber band applications. *Journal of Invertebrate Pathology, 97*, 237–244.

Marannino, P., Tarasco, E., & de Lillo, E. (2003). Biological notes on larval hatching in *Capnodis tenebrionis* (L.) (Coleoptera: Buprestidae) and evaluation of entomopathogenic nematodes in controlling neonate larvae. *Redia, 86*, 101–105.

Martínez de Altube, M. M., Strauch, O., Fernández de Castro, G., & Martínez Peña, A. (2008). Control of the flat–headed root borer *Capnodis tenebrionis* (Linné) (Coleoptera: Buprestidae) with the entomopathogenic nematode *Steinernema carpocapsae* (Weiser) (Nematoda: Steinernematidae) in a chitosan formulation in apricot orchards. *BioControl, 53*, 531–539.

Mendel, Z., Assael, F., & Ben-Yehuda, S. (2003). Host selection and root colonization of cyanogenic stonefruit species by *Capnodis* spp. (Coleoptera: Buprestidae). *Annals of the Entomological Society of America, 96*(2), 127–134.

Mfarrej, M. F. B., & Sharaff, N. S. (2010). Peach rootborer *Capnodis tenebrionis* L. (Coleoptera: Buprestidae) on stone–fruit trees. *Jordan Journal of Agricultural Science, 6*, 579–589.

Morton, A., & Garcia-del-Pino, F. (2008a). Effectiveness of different species of entomopathogenic nematodes for biocontrol of the Mediterranean flatheaded rootborer, *Capnodis tenebrionis* (Coleoptera: Buprestidae) in potted peach tree. *Journal of Invertebrate Pathology, 97*, 128–133.

Morton, A., & Garcia-del-Pino, F. (2008b). Field efficacy of the entomopathogenic nematode *Steinernema feltiae* against the Mediterranean flatheaded rootborer *Capnodis tenebrionis*. *Journal of Applied Entomology, 132*, 632–637.

Morton, A., & Garcia-del-Pino, F. (2009a). Virulence of entomopathogenic nematodes to different stages of the flatheaded rootborer, *Capnodis tenebrionis* (L.) (Coleoptera: Buprestidae). *Nematology, 11*, 365–373.

Morton, A., & Garcia-del-Pino, F. (2009b). Ecological characterization of entomopathogenic nematodes isolated in stone fruit orchard soils of Mediterranean areas. *Journal of Invertebrate Pathology, 102*, 203–213.

Morton, A., & Garcia-del-Pino, F. (2013). Sex–related differences in the susceptibility of *Periplaneta americana* and *Capnodis tenebrionis* to the entomopathogenic nematode *Steinernema carpocapsae. Journal of Invertebrate Pathology, 112*, 203–207.

Paparatti, B., & Speranza, S. (2005). Biological control of hazelnut weevil (*Curculio nucum* L., Coleoptera, Curculionidae) using the entomopathogenic fungus *Beauveria bassiana* (Balsamo) Vuill. (Deuteromycotina, Hyphomycetes). *Acta Horticulturae, 686*, 407–412.

Peters, A., Sarraquigne, J. P., Blum, B., & Kuske, S. (2007). Control of hazelnut borer, *Curculio nucum*, with entomopathogenic nematodes. *OIBC/OILB, West Palaeartic Regional Section (WPRS/SROP) Bulletin, 30*, 73–76.

Piskornik, Z. (1992). Observation on the resistance of the hazelnut (*Corylus avellana* L.) to the hazelnut weevil (*Curculio nucum* L., Coleoptera, Curculionidae). *Acta Horticulturae, 317*, 163–170.

Piskornik, Z. (1994). Relationship between the resistance of hazelnut cultivars to the hazelnut weevil and the content of sugars, amino acids, and phenols in the endocarp tissue of growing nuts. *Acta Horticulturae, 351*, 617–624.

Rivnay, E. (1944). Physiological and ecological studies on the species of *Capnodis* in Palestine (Col., Buprestidae). I. Studies on the eggs. *Bulletin of Entomological Research, 33*, 235–242.

Rivnay, E. (1946). Ecological and physiological studies on *Capnodis* spp. (Col., Buprestidae) in Palestine: III. Studies on the adult. *Bulletin of Entomological Research, 37*, 273–280.

Sanna-Passino, G., & Delrio, G. (2001). Eficacia de plaguicidas sobre larvas de *Capnodis tenebrionis* (L.). *Boletín de Sanidad Vegetal. Plagas, 27*, 59–64.

Santos Lobatón, C., García Vela, J. R., Lara López, M. P., & Canales Roca, A. (1998). Ensayos de laboratorio con una población de *Steinernema carpocapsae* (Filipjev) detectada en larvas de *Capnodis tenebrionis* L. *Boletín de Sanidad Vegetal. Plagas, 24*, 679–686.

Shapiro-Ilan, D. I. (2003). Microbial control of the pecan weevil, *Curculio caryae* (Coleoptera: Curculionidae). In J. D. Dutcher, M. K. Harris, & D. A. Dean (Eds.). *Integration of chemical and biological insect control in native, seed–ling, and improved pecan production. South–Western Entomologist, 27*(Suppl.), 101–114.

Shapiro-Ilan, D. I., Mizell, R. F., & Campbell, J. F. (2002). Susceptibility of the plum curculio, *Conotrachelus nenuphar*, to entomopathogenic nematodes. *Journal of Nematology, 34*, 246–249.

Strong, D. R. (2002). Population of entomopathogenic nematodes in foodwebs. In R. Gaugler (Ed.), *Entomopathogenic Nematology* (pp. 225–240). Wallingford, UK: CABI Publishing.

Tuncer, C., & Ecevit, O. (1997). Current status of hazelnut pests in Turkey. *Acta Horticulturae, 445*, 545–552.

Wojciechowicz-Zytko, E. (2005). Infestation of hazel nuts by hazelnut weevil (*Curculio nucum* L., Coleoptera, Curculionidae) in Poland. *Journal of Plant Protection Research, 45*, 59–61.

Chapter 17
Entomopathogenic Nematodes in the Czech Republic: Diversity, Occurrence and Habitat Preferences

Vladimír Půža and Jiří Nermuť

17.1 Introduction: History of Entomopathogenic Nematode Research in the Czech Republic

In the past 70 years, EPN research in the Czech Republic has contributed to the field with a number of important achievements, especially in the taxonomy of entomopathogenic nematodes (EPNs). Leading persons involved have been insect pathologist Dr. Jaroslav Weiser and his pupil Dr. Zdeněk Mráček. In this short overview, we will focus on the most important advances in EPN research provided by investigations in the Czech Republic.

In the early 1950s, during examination of the pathogens and parasites of the codling moth *Cydia pomonella* L. (Lepidoptera: Tortricidae), Dr. Weiser found caterpillars infected with nematodes belonging to a new species. These worms were described as *Neoaplectana* (= *Steinernema*) *carpocapsae* (Weiser, 1955). This EPN species was the first EPN found in the territory of the Czech Republic. Interestingly, a recent large scale sampling has shown that *Steinernema carpocapsae* (Weiser) Wouts, Mráček, Gerdin & Bedding (Rhabditida: Steinernematidae) is one of the rarest Czech EPNs (Table 17.1). Another steinernematid, *Steinernema weiseri* Mráček, Stuarhan & Reid (Rhabditida: Steinernematidae) was described from under an apple tree growing at the roadside near České Budějovice, Czech Republic (Mráček, Sturhan, & Reid, 2003). Recently, a species belonging to the "intermedium" group *Steinernema poinari* Mráček, Půža, & Nermuť (Rhabditida: Steinernematidae) has also been described (Mráček, Půža, & Nermuť, 2014). It

V. Půža (✉) • J. Nermuť
Department of Biosystematics and Ecology, Institute of Entomology, Biology Centre, Czech Academy of Sciences, České Budějovice, Czech Republic
e-mail: vpuza@seznam.cz; Jirka.Nermut@seznam.cz

© Springer International Publishing Switzerland 2015
R. Campos-Herrera (ed.), *Nematode Pathogenesis of Insects and Other Pests*, Sustainability in Plant and Crop Protection 1, DOI 10.1007/978-3-319-18266-7_17

Table 17.1 Entomopathogenic nematode species recovered in the Czech Republic and their prevalence in selected habitats

	Total no. of records	Fruit trees	Arable fields	Meadows	Tree habitats
S. feltiae	135	54	9	4	68
S. kraussei	116	12			102
S. affine	48	15	4	9	20
S. intermedium	26	5			21
S. silvaticum	26				26
S. bicornutum	20	7	2	2	9
S. weiseri	17	5	1		11
S. poinari	4	1			3
S. arenarium	2				2
S. carpocapsae	1	1			
H. megidis	4	1			3
H. bacteriophora	2			2	
Steinernema sp.	245	–	–	–	–

is now evident, that both species are widely distributed over the Palearctic region (Mráček et al., 2003; 2014).

A substantial contribution to the taxonomy of EPNs was achieved in studies of head cuticular structures of steinernematid species (Mráček & Weiser, 1979; Mráček, Weiser, & Gerdin, 1981). Based on these results, a genus *Neoaplectana*, which, at that time contained the majority of steinernematid species, was established as a junior synonym of the genus *Steinernema* (Wouts, Mráček, Gerdin, & Bedding, 1982). Another contribution to EPN taxonomy was the re–description of *Steinernema kraussei* (Steiner) Travassos (Rhabditida: Steinernematidae). Its original description made by Steiner (1923) was incomplete and only permanent slides with females were available. Considering the fact that it was the first steinernematid species to be described, and thus represented a type species of the family, its re–description was desirable. In 1991, Mráček sampled in the original locality, re–isolated the local *S. kraussei* population (Mráček, Weiser, Bureš, & Kahounová, 1992) and completed the original description (Mráček, 1994).

17.2 Entomopathogenic Nematode Occurrence in the Czech Republic

The Czech Republic is a country with a long tradition of EPN research and thus has a large amount of data on EPN distribution at its disposal. During a long term survey performed in the last 30 years a total of 1,350 samples have been isolated from many natural and agricultural habitats (Mráček, Bečvář, & Kindlmann, 1999; Mráček, Bečvář, Kindlmann, & Jersáková, 2005; Půža et al., unpublished) making

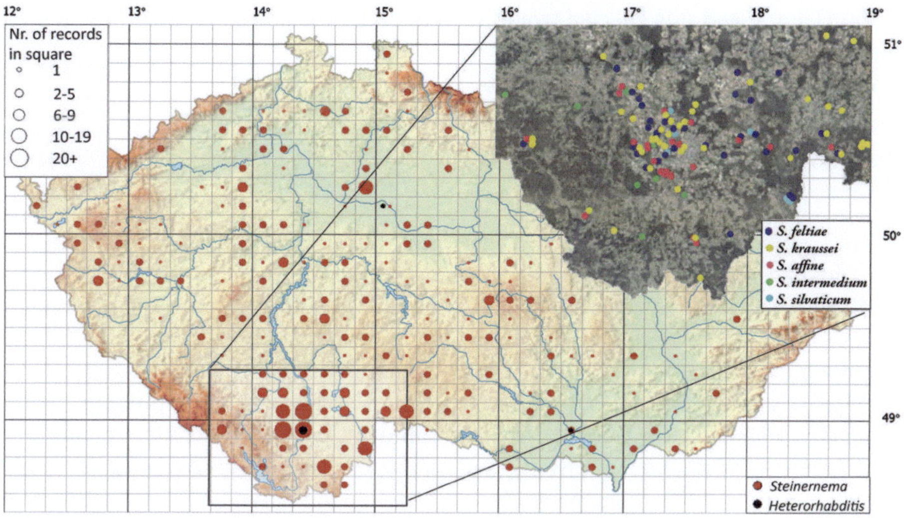

Fig. 17.1 Distribution map of the genera *Steinernema* and *Heterorhabditis* in the Czech Republic (AOPK ČR, 2014), and records of the five most common steinernematids in the area of South Bohemia

the Czech Republic one of the countries with the best explored EPN fauna. The sampling in the survey has followed the protocol described by Mráček (1980). In each sampling site, five subsamples (10–20 cm deep) are randomly taken in an area of ca. 100 m^2 and pooled together. In the laboratory, composite samples are mixed, halved, and baited with 2–4 *Galleria mellonella* L. (Lepidoptera: Pyralidae) larvae in 15 and 22 °C, respectively.

As of the year 2014, 12 EPN species have been isolated (Table 17.1), and numerous unclassified isolates belonging to the family Steinernematidae have been detected. The most prevalent species is the ubiquitous *Steinernema feltiae* (Filipjev) Wouts, Mráček, Gerdin & Bedding (Rhabditida: Steinernematidae), followed by *S. kraussei*, dominant in coniferous forests, and *Steinernema affine* (Bovien) Wouts, Mráček, Gerdin & Bedding (Rhabditida: Steinernematidae), dominating in samples from meadows (Fig. 17.1). In general, heterorhabditids have a much lower prevalence, with six records in total, representing 0.8 % of EPN positive samples. EPNs occurred in all types of soils, but light soils had a higher EPN prevalence in comparison to heavy soils (Mráček et al., 1999; 2005).

Mráček et al. (2005) showed that EPN prevalence is generally lower in agroecosystems that represent target areas for biocontrol. However, it is likely that especially in large–scale areas (fields, orchards), the standard sampling procedure often gives false negative results as EPN density is low and their presence is hardly detectable, unless the sampling is extensive enough. Such a situation was observed by Půža, Habuštová, Hussein, and Svobodová (2014) in a large corn field, where the population of *S. affine* occurred only in some parts of the field at a low density (500–

5,000 IJs/m^2), and this species would be missed by the standard sampling method (one composite sample per site).

17.2.1 Effect of Insect Occurrence

EPNs and the target insect species have a dynamic relationship corresponding to the standard parasite–host system (Mráček & Spitzer, 1983). The distribution of EPNs thus should be affected by the presence of insect hosts. This phenomenon has received considerable attention in Czech EPN research and the findings could have an important impact on the use of EPNs in biocontrol in general.

Mráček and Spitzer (1983) revealed that an outbreak of web–spinning sawfly *Cephalcia abietis* L. (Hymenoptera: Pamphiliidae), presented excellent conditions for gathering predators, parasites, and pathogens including entomopathogenic nematodes. Besides sawflies, geometrid and noctuid moths, and bibionid and sciarid flies, also characteristically aggregate as diapausing or hibernating larvae in the soil. Thus, Mráček and Bečvář (2000) performed a survey at sites with an aggregations of various insects. The revealed incidence of steinernematids was very high in all habitats characterized by high insect host aggregation. Nematodes were recovered from 61 (70 %) out of 87 sampled localities with an insect aggregation, whereas 15 % of positive samples were detected in Czech localities with no obvious insect activity (Mráček et al., 1999). Furthermore, the authors revealed seven different steinernematid species, which is a striking number considering the relatively low number of total samples. The results showed that insect hosts aggregations are followed by rapid nematode multiplications enabling their recovery in the fields.

The nests of bibionid and sciarid flies are probably among the largest soil insect aggregations. Mráček and Sturhan (2000) examined the occurrence of EPNs within the nest of *Bibio marci* L. (Diptera: Bibionidae) and found a natural epizooty of *Steinernema intermedium* (Poinar) Mamiya (Rhabditida: Steinernematidae) with 90 % mortality of the larvae and pupae. This case illustrates how insect aggregations allow a tremendous increase in EPN abundance.

Půža, Mráček, and Holuša (2007) compared the occurrence of EPNs between two sites in a larch–spruce forest: a site with an outbreak of the larch web spinning sawfly *Cephalcia lariciphila* Wachtl (Hymenoptera: Pamphiliidae) and a control site without an outbreak. *S. kraussei* and *S. feltiae* were found at both sites; however, EPN density at the outbreak site was almost ten times higher. At the outbreak site, EPN density was significantly higher under the larch canopies, while at the control site, the EPNs density did not differ under larch and spruce canopies. The conclusion can be made that EPNs reflect the host density even at a scale of a few meters.

17.2.2 Natural Control

The fact that EPNs in their abundance follow insect gradations highlights the question of the level of natural control of insect pests by EPNs. In the Czech Republic, this phenomenon has been studied in several sawfly outbreaks. Mráček (1986) estimated 20–34 % natural parasitisation of the diapausing spruce web–spinning sawfly *C. abietis* larvae by *S. kraussei*. Půža et al. (2007) found the same nematode causing an immediate 20 % natural parasitisation of the larch web–spinning sawfly *C. lariciphila* larvae and interestingly also freshly hatched adults. The authors suggested that the mortality over the whole diapause would probably be much higher.

17.3 Entomopathogenic Nematodes Experimental Applications in the Czech Republic: Case Studies in Orchards, Forestry and Ornamental Plants

17.3.1 The Red–Belted Clearwing, **Synanthedon myopaeformis** *Borkhausen (Lepidoptera: Sesiidae)*

The red–belted clearwing, *Synanthedon myopaeformis* Borkhausen (Lepidoptera, Sesiidae) has a 2 year life cycle. Fertilized females lay their eggs singly in wounds and crevices in the bark on the trunk and branches of the host tree. The larvae feed on sap and create galleries of up to 3 cm long between the bark and the cambium layer. A heavy infestation weakens the host tree, reduces the crop yield, and makes the tree more susceptible to attack by fungal diseases. Kahounová and Mráček (1991) selected *S. feltiae*, population "Hyl" that proved to be the most virulent EPN population, causing 100 % mortality of *S. myopaeformis* in Petri dish experiments. The nematodes were sprayed onto ca. four hundred infested calluses using a concentration of 30,000 IJs per tree; after a 4 week period, pest mortality reached 76 %. Based on these observations, the authors suggested EPNs as a suitable agent for the control of *S. myopaeformis* in the Czech Republic. However, to date no further applications have been performed.

17.3.2 The Black Vine Weevil **Otiorhynchus sulcatus** *F. (Coleoptera: Curculionidae)*

The black vine weevil *Otiorhynchus sulcatus* F. (Coleoptera: Curculionidae) is a serious pest of many crops in Europe and North America (Moorhouse, Charnley, & Gillespie, 1992). The larvae destroy the root system of the host plant often causing death. Broadleaved evergreen plants are particularly prone to damage. Mráček, Jiskra, and Kahounová (1993) performed preventive treatment in a nursery of ornamentals with the *S. feltiae* population "Hyl". The nematodes were applied at a rate of 10,000 IJs per plant. A highly effective control (72–88 % of plants survived in comparison to 30 % in the untreated control) was achieved in the treated sites and adjacent sites were protected by the migrating nematodes by up to 52–77 %. This system was thus proven to be a promising method of biological protection in ornamental nurseries.

17.3.3 The Sawflies, Pamphilidae and Tenthredinidae

Sawflies (families Pamphilidae and Tenthredinidae) are serious forest pests throughout the world. Feeding larvae can cause complete defoliation of the trees leading to their death. Pupation and overwintering occurs in the soil and thus the sawflies can be easily targeted by EPNs. Mráček and David (1986) used *S. kraussei* against the spruce web–spinning sawfly *C. abietis*, a significant pest in spruce monocultures in the Czechoslovakia of the time. The application of the nematodes in the autumn caused 81–97 % mortality of the diapausing sawfly larvae and later sampling revealed that the nematode established in the treated locality. The authors concluded that EPNs are suitable for the biological control of the web–spinning sawflies.

17.4 Current Use: Species, Target Pests and Market

At present, there are only two EPN species available on the Czech market. Table 17.2 summarizes EPN based products that are registered for the use in the Czech Republic. According to the main local supplier, Biocont Laboratory spol. s.r.o. (http://www.biocont.cz/en), both *S. feltiae* and *H. bacteriophora* are recommended for ornamentals and glasshouse application, the former mainly against the maggots of sciarid flies and the latter for the control of weevils. Both nematodes are formulated in an inert carrier for the price of an equivalent of ca. 26 EUR for a package of 50 million IJs. In the past, *Heterorhabditis megidis* Poinar, Jackson & Klein (Rhabditida: Heterorhabditidae) was also available, but presently it is not on the market.

Due to the high price of EPN–based products, their use is presently limited mainly to organic or integrated farming systems with reduced tillage. In most greenhouses in the Czech Republic, bioagents are often used, but for EPNs, this holds true only for ornamentals. In the vegetable production that occurs mostly in hydroponic systems with mineral substrates in plastic tunnels, nematodes cannot be successfully used.

In contrast to conventional agriculture EPNs are firmly established in the Czech hobby market and are widely used against many pests in gardens, greenhouses and also in decorative house plants. Naturally this market segment is not dependent on economic profit and therefore the higher price is accepted by the customers.

17.5 Limitations on the Use of Entomopathogenic Nematodes and Potentials

Based on our present knowledge, EPN applications seem in general to be suitable for the Czech Republic. Most of the local soil types enable good EPN movement and survival. Also the mild climate with sufficient moisture is favourable. It can be assumed, that some exotic populations adapted for high temperature would be negatively affected (e.g. *Heterorhabditis indica* Poinar, Karunakar & David [Rhabditida: Heterorhabditidae]), however, such nematodes are not on the Czech market.

There are many pest species sensitive to nematode infection in the Czech Republic, e.g. *Cydia pomonella* L. (Lepidoptera), Sesiidae (Lepidoptera), *Ostrinia nubilalis* Hübner (Lepidoptera: Pyrallidae), *Leptinotarsa decemlineata* Say (Coleoptera: Chrysomelidae), *Otiorhynchus* sp. (Coleoptera), *C. abietis* (Hymenoptera), and Sciaridae (Diptera). Nevertheless, despite these facts the use of EPNs in biological pest control is still marginal in the commercial sphere.

Several important limitations of using EPNs can be identified, e.g. price, crop, farming and tillage system, and available information. Probably the most limiting factor is the price of the commercial bioagents. At the present time, the price of nematode *S. feltiae* for 1 ha of orchard in the Czech Republic is about 2,000 EUR, while the equivalent agrochemicals are approximately fifty times cheaper. Treatment of 1 ha of apple orchard with a product based on thiaclopride actually costs ca. 65 EUR and treatment with acetamipride costs about 45 euros. Czech government programmes financially support the use of bioagens, nematodes included, and the subsidy can represent 25–70 % of the price of the bioagents, but despite this fact the price remains very high and therefore EPNs can be used only in the most profitable crops or in horticulture.

Farming and tillage system also strongly influence the use of EPNs. Especially intensive tillage systems that lead to the fast drying of the surface soil layer cause the death of EPNs. Also, the use of pesticides in conventional agriculture contributes

Table 17.2 EPN based products registered for the use in Czech Republic

Nematode species	Trademark	Producer	Registration due
S. feltiae	Entonem	Koppert B.V.	03/31/2019
S. feltiae	Nemaplus	E–Nema gmbH	06/18/2023
S. feltiae	Steinernema–System	Biobest N.V.	03/01/2019
H. bacteriophora	Larvanem	Koppert B.V.	03/31/2019
H. bacteriophora	Nematop	E–Nema gmbH	06/18/2023

to a decrease of nematode populations in fields (Nermuť & Mráček, 2010). An absence of information about the option of EPNs use and technologies for their application may also figure among important limitations. During the last 20 years the awareness of bioagents has substantially increased among farmers, and some growers have implemented them. However, many farmers still avoid biocontrol probably because of a lack of information and the skills to introduce this system of pest control.

Luckily, information on the topic has recently begun to spread more quickly thanks to the many popular articles on the use of EPNs published in Czech papers and web–based hobby magazines in the last few years (e.g. Nermuť, Půža, & Mráček, 2012; Nermuť & Půža, 2014). Thus, the demand for nematodes among gardeners and farmers is growing.

17.6 Conclusions and Future Directions

As illustrated above, EPNs in the Czech Republic can be used against a very wide spectrum of different insect pest. However, the use of these organisms in conventional agriculture on a large scale is improbable in the near future. It is likely that in the commercial sphere the use of EPN based products will be restricted to organic or integrated farming systems targeted at high value crops, e.g. wine grapes, fruits, berries, vegetables, ornamentals, or herbs for the cosmetic and pharmaceutical industry. In this area, and also in the hobby market and home plant production, the role of EPN based products will certainly grow in the future.

Acknowledgements This study was completed with the support of Czech Science Foundation, project: GAČR P504/12/2352.

References

AOPK ČR. (2014). *Map server of the agency for nature conservation and landscape protection of the Czech Republic*. Available at: http://mapy.nature.cz/ [revised 10/11/2014].

Kahounová, L., & Mráček, Z. (1991). Larval mortality of *Synanthedon myopaeformis* (Lepidoptera, Sesiidae) in apple trees sprayed with *Steinernema* strain Hylobius (Nematoda, Steinernematidae). *Acta Entomologica Bohemoslovaca, 88*, 205–210.

Mráček, Z. (1980). The use of "*Galleria* traps" for obtaining nematode parasites of insect in Czechoslovakia (Lepidoptera: Nematoda, Steinernematidae). *Acta Entomologica Bohemoslovaca, 77*, 378–382.

Mráček, Z. (1986). Nematodes and other factors controlling the sawfly, *Cephaleia abietis* (Pamphilidae: Hymenoptera), in Czechoslovakia. *Forest Ecology and Management, 15*(1), 75–79.

Mráček, Z. (1994). *Steinernema kraussei* (Steiner, 1923) (Nematoda: Rhabditida: Steinernematidae): redescription of its topotype from Westphalia. *Folia Parasitologica, 41*(1), 59–64.

Mráček, Z., & Bečvář, S. (2000). Insect aggregations and entomopathogenic nematode occurrence. *Nematology, 2*(3), 297–301.

Mráček, Z., Bečvář, S., & Kindlmann, P. (1999). Survey of entomopathogenic nematodes from the families Steinernematidae and Heterorhabditidae (Nematoda: Rhabditida) in the Czech Republic. *Folia Parasitologica, 46*(2), 145–148.

Mráček, Z., Bečvář, S., Kindlmann, P., & Jersáková, J. (2005). Habitat preference for entomopathogenic nematodes, their insect hosts and new faunistic records for the Czech Republic. *Biological Control, 34*(1), 27–37.

Mráček, Z., & David, L. (1986). Preliminary field control of *Cephalcia abietis* L. (Hymenoptera, Pamphilidae) larvae with steinernematid nematodes in Czechoslovakia. *Journal of Applied Entomology, 102*(1–5), 260–263.

Mráček, Z., Jiskra, K., & Kahounová, L. (1993). Efficiency of steinernematid nematodes (Nematoda: Steinernematidae) in controlling larvae of the black vine weevil, *Otiorhynchus sulcatus* (Coleoptera: Curculionidae) in laboratory and field experiments. *European Journal of Entomology, 90*(1), 71–76.

Mráček, Z., Půža, V., & Nermuť, J. (2014). *Steinernema poinari* sp. n. (Nematoda: Steinernematidae) a new entomopathogenic nematode from the Czech Republic. *Zootaxa, 3760*(3), 336–350.

Mráček, Z., & Spitzer, K. (1983). Interaction of the predators and parasitoids of the sawfly, *Cephalcia abietis* (Pamphilidae: Hymenoptera) with its nematode *Steinernema kraussei*. *Journal of Invertebrate Pathology, 42*(3), 397–399.

Mráček, Z., & Sturhan, D. (2000). Epizootic of the Entomopathogenic nematode *Steinernema intermedium* (Poinar) in an aggregation of the bibionid fly, *Bibio marci* L. *Journal of Invertebrate Pathology, 76*(2), 149–150.

Mráček, Z., Sturhan, D., & Reid, A. (2003). *Steinernema weiseri* n. sp. (Rhabditida, Steinernematidae), a new entomopathogenic nematode from Europe. *Systematic Parasitology, 56*(1), 37–47.

Mráček, Z., & Weiser, J. (1979). The head papillae of the Steinernematidae. *Journal of Invertebrate Pathology, 34*(3), 310–311.

Mráček, Z., Weiser, J., Bureš, M., & Kahounová, L. (1992). *Steinernema kraussei* (Steiner, 1923)(Nematoda – Steinernematidae) rediscovery of its type locality in Germany. *Folia Parasitologica, 39*, 181–182.

Mráček, Z., Weiser, J., & Gerdin, S. V. E. N. (1981). Head and cuticular structures of some species in the family Steinernematidae (Nematoda). *Nematologica, 27*(4), 443–448.

Moorhouse, E. R., Charnley, A. K., & Gillespie, A. T. (1992). A review of the biology and control of the vine weevil, *Otiorhynchus sulcatus* (Coleoptera: Curculionidae). *Annals of Applied Biology, 121*(2), 431–454.

Nermuť, J., & Mráček, Z. (2010). The influence of pesticides on the viability and infectivity of entomopathogenic nematodes (Nematoda: Steinernematidae). *Russian Journal of Nematology, 18*(2), 141–148.

Nermuť, J., & Půža, V. (2014). Parazitické hlístice hmyzu (Insect parasitic nematodes). *Živa, 4*,176–179.

Nermuť, J., Půža, V., & Mráček, Z. (2012). Entomopatogenní a moluskoparazitické hlístice – neviditelní půdní zabijáci (Entomopathogenic and mollusc–parasitic nematodes – The invisible soil killers). *Živa, 1*, 10–13.

Půža, V., Mráček, Z., & Holuša, J. (2007). Entomopathogenic nematodes and target soil insect pests in tree habitats in the Czech Republic, with focus on sawflies and cockchafers. *Comptes Rendus de la Reunion, 30*(7), 105–107.

Steiner, G. (1923). *Aplectana kraussei* n. sp., eine in der Blattwespe *Lyda* sp. parasitierende Nematodenform, nebst Bemerkungen über das Seitenorgan der parasitischen Nematoden. *Zentralblatt für Bakteriologie Parasitenkunde Infektions krankheiten und Hygiene, 2*(59), 14–18.

Weiser, J. (1955). *Neoaplectana carpocapsae* n. sp, (Anguillulata, Steinernematidae), nový cizopasník housenek obaleče jablečného *Carpocapsa pomonella* L. *Věstník České Společnosti Zoologické, 19*, 44–52. (in Czech)

Wouts, W. M., Mráček, Z., Gerdin, S., & Bedding, R. A. (1982). *Neoaplectana* Steiner, 1929 a junior synonym of *Steinernema* Travassos, 1927 (Nematoda; Rhabditida). *Systematic Parasitology, 4*(2), 147–154.

Chapter 18
Entomopathogenic Nematodes in Italy: Occurrence and Use in Microbial Control Strategies

Francesca De Luca, Mirella Clausi, Alberto Troccoli, Giovanna Curto, Giancarlo Rappazzo, and Eustachio Tarasco

18.1 Introduction

The first data of entomopathogenic nematodes (EPNs) population isolation in Italy goes back to the end of the Second World War with the discovery of some specimens of *Temnorhinus mendicus* Gyll. (Coleoptera: Curculionidae), infested with nematodes, which were described as *Neoaplectana menozii* Travassos, currently an *inquirenda* species. The first samplings, performed in Emilia Romagna, in order to ensure the dissemination of entomopathogenic nematodes in the soil were carried out in 1983 by Deseö, Grassi, Foschi, and Rovesti (1984) and some years later they published the first data on the presence of Heterorhabditidae and Steinernematidae in Italian agricultural land (Deseö, Fantoni, & Lazzari, 1988). Subsequent surveys conducted in the recent decades (Clausi, Longo, Rappazzo, Tarasco, & Vinciguerra, 2011; Clausi & Vinciguerra, 2005, 2007, 2008; Rappazzo et al., 2011; Rappazzo, Clausi, & Vinciguerra, 2005; Susurluk, Tarasco, Triggiani, & Ehlers, 2007; Tarasco

F. De Luca (✉) • A. Troccoli
Istituto per la Protezione Sostenibile delle Piante, Consiglio Nazionale delle Ricerche, Bari, Italy
e-mail: francesca.deluca@ipsp.cnr.it; a.troccoli@ba.ipp.cnr.it

M. Clausi • G. Rappazzo
Dipartimento di Scienze Biologiche, Geologiche e Ambientali, Sez. Biologia Animale "M. La Greca", University of Catania, Catania, Italy
e-mail: mclausi@unict.it; rappazzo@unict.it

G. Curto
Servizio Fitosanitario, Regione Emilia–Romagna, Laboratorio di Nematologia, Bologna, Italy
e-mail: GCurto@regione.emilia-romagna.it

E. Tarasco
Dipartimento di Scienze del Suolo, della Pianta e degli Alimenti (DiSSPA), Sez. Entomologia e Zoologia, University of Bari "Aldo Moro", Bari, Italy
e-mail: eustachio.tarasco@uniba.it

© Springer International Publishing Switzerland 2015

R. Campos-Herrera (ed.), *Nematode Pathogenesis of Insects and Other Pests*, Sustainability in Plant and Crop Protection 1, DOI 10.1007/978-3-319-18266-7_18

et al., 2009; Tarasco, Mráček, Nguyen, & Triggiani, 2008; Tarasco & Triggiani, 1997, 2005, 2007; Triggiani, Mráček, & Reid, 2004; Triggiani & Tarasco, 2000) have contributed to the discovery of new species. The discovery and knowledge of indigenous EPN populations in Italy has gone on with the increase for their use in agricultural and forest biological practices. Surveys on the occurrence and distribution of EPNs have been carried out in several Italian regions following a large–scale biogeographic approach with respect to different vegetation levels and habitats (Tarasco et al., 2015) in order to create a comprehensive map of the Italian EPN biodiversity. Currently the EPNs isolated from different habitats are almost 140 indigenous populations belonging to 12 species: *Heterorhabditis bacteriophora* Poinar (Rhabditida: Heterorhabditidae), *Heterorhabditis downesi* Stock, Griffin & Burnell (Rhabditida: Heterorhabditidae), *Heterorhabditis megidis* Poinar, Jackson & Klein (Rhabditida: Heterorhabditidae), *Steinernema feltiae* (Filipjev) (Rhabditida: Steinernematidae), *Steinernema affine* (Bovien) (Rhabditida: Steinernematidae), *Steinernema kraussei* (Steiner) (Rhabditida: Steinernematidae), *Steinernema apuliae* Triggiani, Mràcek & Reid (Rhabditida: Steinernematidae), *Steinernema ichnusae* Tarasco, Mràcek, Nguyen & Triggiani (Rhabditida: Steinernematidae), *Steinernema carpocapsae* (Weiser) (Rhabditida: Steinernematidae), *Steinernema vulcanicum* Clausi, Longo, Rappazzo, Tarasco & Vinciguerra (Rhabditida: Steinernematidae), *Steinernema* "isolate S.sp.MY7" of "*S. intermedium* group" and *Steinernema arenarium* (Artyukhovsky) (Rhabditida: Steinernematidae). Steinernematids are more widespread and biodiverse than heterorhabditids and *S. feltiae* and *H. bacteriophora* are the most commonly encountered species (Tarasco et al., 2015). EPN populations and species are mainly found in natural sites rather than in cultivated soil (Tarasco et al., 2015; Tarasco & Triggiani, 1997). *Steinernema kraussei*, *H. downesi* and *H. megidis* were collected only in Sicily, whereas two of the species recently described – *S. apuliae* and *S. vulcanicum* – are known only from Italy and seem to be endemic (Tarasco et al., 2015). The high EPN biodiversity found in Italy could be related to the unique geographic diversity of the country, which provides different vegetation habitats plus the unique island fauna. In addition to the native populations, there are several commercial EPN products used in Italian agro–forest ecosystems which will be discussed later. Herein, we will describe the peculiarities of our native EPN fauna and their contribution to the biological control of insect pest for major crops.

18.2 Native Distribution of Entomopathogenic Nematode Populations: Biodiversity Studies

18.2.1 Occurrence and Distribution in Italy: Sampling and Habitat

Surveys were carried out mostly during the humid season from November to June in ten Italian regions (Apulia with Tremiti islands, Basilicata, Calabria, Campania,

Emilia Romagna, Molise, Sardinia, Sicily with Aeolian islands, Pantelleria island, Tuscany and Veneto) collecting over a period of 24 years (1990–2014). In order to sample ecologically different habitats, several samples were taken in natural and cultivated ecosystems, at altitudes ranging from 0 to 2,000 m above sea level and from 11 different habitats, namely uncultivated land (wild vegetation, shrubland, tamarisk), orchard (olive, almond, vineyard, several kinds of fruit plants), field (several kinds of crops), sea coast (sandy beach, gravelly beach, sand dunes), pinewood (*Pinus pinea*, *P. halepensis*, *P. nigra*, *P. sylvestris*, *P. pinaster*), broadleaf wood (oak, chestnut, mulberry, walnut, hazelnut), grassland (meadows, pastures, irrigated gardens, herbs, turfgrass), river and lake borders (riparian vegetation), cave (caverns, quarries), salt pan borders and wetland (swamp, lagoons, canebrake). These different habitats have been identified based on ecological similarities and for each sampling location, the soil texture, altitude, time and type of vegetation or habitat were recorded. Within each site, composite soil samples were collected and *Galleria mellonella* L. (Lepidoptera, Pyralidae) larvae used as bait–insects.

A mapping of EPN distribution in Italy displayed 137 indigenous EPN populations belonging to 12 species: 43 isolates of *H. bacteriophora* (30 %), 1 of *H. downesi* (1 %), 1 of *H. megidis* (1 %), 52 *S. feltiae* (40 %), 12 *S. affine* (9 %), 4 *S. kraussei* (3 %), 8 *S. apuliae* (6 %), 6 *S. ichnusae* (3 %), 4 *S. carpocapsae* (2 %), 1 *S. vulcanicum* (1 %), 3 *Steinernema* "isolate S.sp.MY7" of "*S. intermedium* group" (2 %) and 1 *S. arenarium* (1 %) (Fig. 18.1) (Tarasco et al., 2015). EPNs were found in all 11 sampled habitats, thus showing a wide distribution of species in different ecosystems. More species of steinernematids were found than heterorhabditids, and *S. feltiae* and *H. bacteriophora* were the most commonly found species.

Steinernema feltiae was present in most habitats, with a preference for sandy soils, thus confirming studies by Campos-Herrera et al. (2007; 2008). *Heterorhabditis bacteriophora* was also quite widespread, and was found even in volcanic soils (on Pantelleria volcanic island), but never in broadleaf woodland; it showed a preference for sand/sandy loam soils, with 30 % of populations detected in silt/silty loam soils and 12 % in clay loam soils. Except for the two dominant species *S. feltiae* and *H. bacteriophora*, EPNs tend to be correlated with a specific vegetation habitat; for example, *S. kraussei* and *S. affine* were found in forests at quite high altitudes (Tarasco et al., 2015). Italian *S. affine* in particular was isolated from different soil types, but almost exclusively in broadleaf woodland, thus confirming its preference for a forest habitat as found by Mráček, Bečvář, Kindlmann, and Jersáková (2005). *Steinernema kraussei* was collected four times in chestnut groves with sandy loam soils on Mount Etna in Sicily (Tarasco et al., 2015). Previously this species had been isolated only once in Italy by Ricci et al. (2004) in the Alps, although indication of its habitat was not provided. Because it has been found in woodland elsewhere (Hominick, 2002; Torr, Heritage, & Wilson, 2007), it is reasonable to suppose that this species prefers forest habitats. Regarding the preferences of other species, *S. apuliae* was isolated from different habitats, but always close to coastal areas, showing a clear preference for sandy soils. *Steinernema carpocapsae* was isolated in the northern part of Apulia and also in Tuscany; this species had already been reported in northern Italy by Ehlers,

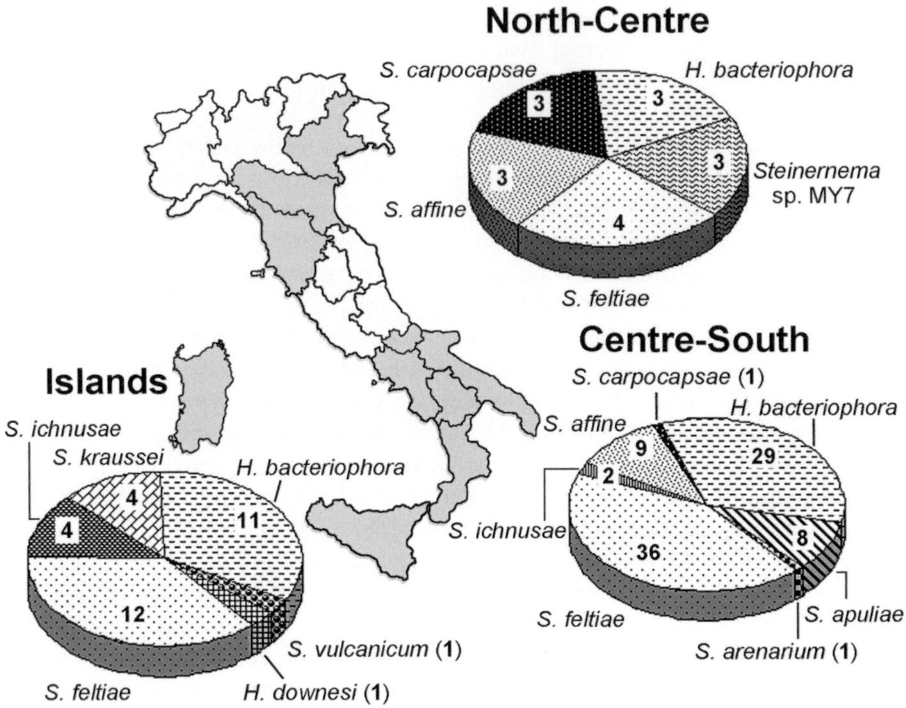

Fig. 18.1 Distribution of EPN species/populations found in Italy. Surveyed regions are in *grey* colour. *Numbers* in pie charts refer to the number of populations found for each species

Deseö, and Stackerbrandt (1991). Two species recently described – *S. apuliae* and *S. vulcanicum* – have been identified only for Italy and seem to be endemic, while *S. ichnusae* firstly isolated from Sardina only, seems to be more widespread in the Mediterranean area (Tarasco et al.). *Steinernema apuliae* and *S. vulcanicum* are 'long nematodes' and belong to *glaseri*–group. Heterorhabditids in Italy were frequently found in coastal sites with sandy soils, but were also isolated from many other inland locations including hilly agricultural and uncultivated lands, even at 800 m.a.s.l. with clay loam and silty loam soils, which is somewhat unusual as heterorhabditids are often reported as coastal and sandy EPNs (Glazer, Liran, & Steinberger, 1991; Griffin, Moore, & Downes, 1991; Stock, Pryor, & Kaya, 1999). Sampling carried out in Sicily revealed the presence of *H. bacteriophora* and the first Italian populations of *H. megidis* and *H. downesi*; this means that three of the four *Heterorhabditis* species reported for Europe are present in Sicily, although *H. bacteriophora* was also found on the two small islands of Pantelleria and Salina (close to Sicily), where *S. feltiae* was also found. Some of Sicily's small islands hosted just the most widespread EPN species, while others had no EPNs at all. The persistence of EPNs at a particular site is probably due to the presence of suitable hosts (Mráček & Bečvář, 2000; Mráček & Webster, 1993; Peters, 1996), although some authors have reported an independent correlation between EPN density and insect hosts (Campbell, Lewis, Yoder, & Gaugler, 1995; Půža & Mráček, 2005).

The distribution of Italian EPN populations agrees with studies which found that EPN populations are both spatially and temporally extremely patchy, within and among sites (Cabanillas & Raulston, 1994; Campbell, Orza, Yoder, Lewis, & Gaugler, 1998; Garcia del Pino & Palomo, 1996; Glazer, Kozodoi, Salame, & Nestel, 1996; Koppenhöfer & Kaya, 1996; Stuart & Gaugler, 1994). Soil characteristics are an additional factor influencing the presence of EPNs, and our surveys showed a clear correlation between the EPN presence and soil texture, with a preference for sand/sandy loam and silt/silty loam soils. Only a few populations were recovered from clay loam soil and none from clay soil. This is probably because soils with a high sand/silt content favour EPN mobility and survival, whereas soils with high clay content restrict nematode movements (Kung, Gaugler, & Kaya, 1990; Barbercheck & Kaya, 1991). This study on EPN distribution among habitats suggests the importance of two natural habitats: pinewoods and broadleaf woodland. These two forest environments present the highest values of parameters (biodiversity indexes) describing community structures (Tarasco et al., 2015); in particular, only three out of the seven species found in broadleaf woodland are shared with other habitats. Two out of the three new species were found in the broadleaf woodland, and *S. vulcanicum* was found only in this habitat. The EPN biodiversity data, collected so far, does not exhaustively cover all geographical areas and habitats in Italy, but still makes a significant contribution to knowledge of EPN occurrence and geographical distribution in relation to the country's great variety of habitats (Table 18.1).

18.2.2 Molecular Analysis

For Italian EPN, the ITS–containing regions and the mitochondrial COI of *Steinernema* spp., and *Heterorhabditis* spp. were characterized. Among Italian *Steinernema* species, four groups were found: *feltiae–kraussei–oregonense* group, *carpocapsae* group, *affine–intermedium* group and *glaseri*–group. The sequence analyses of the ITS containing region showed little sequence variation between individual nematodes and populations of *S. feltiae*. The intra–specific variability of the ITS sequences of *S. feltiae* from Italy ranged from 0 to 0.33 % and reached 3.3–3.5 % among European isolates (26 bp, JX544069) and other isolates coming from all over the world, respectively. The alignment of the Italian ITS sequences of *S. feltiae* with the corresponding sequences of *S. feltiae* from the database revealed that all Italian isolates, at position 150–170, contained ITS sequences of the first type without deletions, as reported by Spiridonov, Reid, Podrucka, Subbotin, and Moens (2004), and can be distinguished from the third type by the presence of *Dde* I restriction site. Low intra–specific variability was observed between *S. ichnusae* isolates. The alignment of ITS1 sequences of Italian *S. feltiae*, *S. ichnusae* and

Table 18.1 The occurrence, with accession numbers (na = not available), of isolates of EPNs from sampling sites in Italy (1990–2014)

Species	No. of populations	Italian regions	Altitude	Vegetation	Habitat	Soil texture	Accession number
S. feltiae	52	SIC, APU, BAS, CAM, CAL, TUS	5–1,200	Pine, Cherry, Wild vegetation, Tomato, Olive, Apple, Pear, Kaki, Chestnut, Meadows, Oak, Artichoke, Cave, Swamp, Pasture, Wheat, Mulberry, Eucalyptus	Pinewood, Orchard, Uncultivated land, Field, Broadleaf wood, Grassland, Cave, Wetland	Sand, sandy loam, silt, silty loam, clay loam	HQ412811.1 to HQ412832.1; HQ412833.1 to HQ412838.1; HQ416966.1, HQ41416967.1; GU599904.1 to GU599910.1; GU599912.1, GU599913.1
H. bacteriophora	43	SIC, APU, BAS, CAM, TUS	0–650	Wild vegetation, Oak, Vineyard, Sandy beach, Salt pan border, Pine, Pasture, Artichoke, Wheat, Eucalyptus, Olive, Mulberry	Pinewood, Orchard, Uncultivated land, Field, Grassland, River border, Sea coast	Sand, sandy loam, silt, silty loam, clay loam	n.a.
S. affine	12	APU, BAS, CAM, TUS	150–650	Oak, Vineyard	Broadleaf wood, Orchard	Sandy loam, clay loam, silt, silty loam	HQ412839.1
S. apuliae	8	APU, BAS	0–50	Sandy beach, Tamarisk, Wild vegetation, Pine, Salt pan border	Sea coast, Salt pan, Pinewood, Uncultivated land	Sand, silt, silty loam	GU599901.1; HQ416968.1
S. ichnusae	6	SAR, SIC	0–950	Sandy beach, Oak, Chestnut	Sea coast, Broadleaf wood	Sand, sandy loam	HQ412833
S. kraussei	4	SIC	700–1,400?	Chestnut?	Broadleaf wood?	Sandy loam?	GU599903.1; GU599916.1 to GU599918.1

S. carpocapsae	4	APU, TUS	50–600	Artichoke, Olive, Maize	Field, Orchard	Silt, silt loam	n.a.
Steiner. sp. MY7	3	TUS	280	Oak	Broadleaf wood	Sandy loam	n.a.
S. arenarium	1	APU	50	Pine	Pinewood	Sand	n.a.
H. downesi	1	SIC	70	Eucalyptus	Pinewood	Sand	n.a.
S. vulcanicum	1	SIC	1,300	Chestnut	Broadleaf wood	Sandy loam	GU929442.1; GU929443.1
H. megidis	1	SIC	20	Pine	Pinewood	Sandy loam	n.a.

Acronyms of Italian regions: *APU* Apulia, *BAS* Basilicata, *CAL* Calabria, *CAM* Campania, *SAR* Sardinia, *SIC* Sicily, *TUS* Tuscany

S. kraussei isolates revealed, in positions 150–170, the presence of ten nucleotide deletions in *S. ichnusae* and eight nucleotide deletions in *S. kraussei* compared to *S. feltiae*. The deletion observed in *S. ichnusae* and *S. kraussei* determined the absence of *Dde* I site.

The sequence divergence of *S. affine* among Italian isolates ranged from 0 to 0.67 % (0–6 bp) but reached 2.4 % (20–22 bp) compared to UK and Belgium isolates (AY230159 and AY171289). The sequence divergence between *S. carpocapsae* isolates from Italy and all corresponding sequences from Europe and USA varied between 0.6 and 1.9 % (5–16 bp). Italian isolates of *S. vulcanicum* showed 83 % similarity with *S. apuliae* (68 bp; 37 gaps).

A second marker largely used for diagnostics of EPN is the *mt*COI gene. The nucleotide sequences of *mt*DNA COI gene from Italian *S. feltiae* isolates showed a 7–9 % dissimilarity (25–30 bp, respectively) with the corresponding *mt*DNA COI sequences present in the database, whereas at amino acid level they showed a 95 % similarity (86/91 identities; 88/91 positives). The nucleotide sequences of *mt*DNA COI gene from the two Italian *S. apuliae* isolates showed a 99 % similarity (5 bp different), revealing two transitions (C/T) and three transversions (A/T). Most of nucleotide changes were synonymous, only the transversion at position 185 (A/T) resulted in a change in the COI amino acid sequence. The nucleotide sequences of *mt*DNA COI gene from Italian *S. affine* isolates showed 99 % similarity, at nucleotide and amino acid level, with the corresponding COI sequences of *S. affine* present in the database. The nucleotide sequences of *mt*DNA COI gene from the two Italian *S. carpocapsae* isolates showed a 99 % similarity (2 bp different), whereas 98 % similarity with those present in the database (16 bp different). Most of nucleotide changes are 13 transitions (9 C/T and 4 A/G) and 2 transversions (A/T). As the rapid rate of evolution is a key requirement for a prospecting marker, thus *mt*DNA is preferable to ITS to prospect for cryptic species of EPN and to read: to reveal species with recent ancestry.

18.3 Entomopathogenic Nematodes Application in Sustainable Plant/Crop Protection: Some Case Studies in Italy

Commercial products containing *H. bacteriophora*, *H. megidis*, *S. feltiae* and *S. carpocapsae* are successfully used for the control of several species of Heteroptera, Lepidoptera, Coleoptera, Diptera and Hymenoptera (Table 18.2). In Italy, EPN formulations have been effectively tested and applied to control *Tomicus piniperda* L. (Coleoptera: Scolitydae) (Triggiani, 1983) and *Thaumetopoea pityo-campa* Denis & Schiffermüller (Lepidoptera: Thaumetopoeidae) on pine (Triggiani & Tarasco, 2002) (Fig. 18.2a); *Curculio elephas* Gyll. and *Curculio glandium* Mars. (Coleoptera: Curculionidae), *Pammene fasciana* L., *Cydia splendana* Hüb. and *Cydia fagiglandana* (Zeller) (Lepidoptera: Tortricidae) on chestnut (Vinciguerra & Clausi, 2006; Curto, Reggiani, Dallavalle, & Bariselli, 2009); *Curculio nucum*

Table 18.2 Species of target insects for nematodes available on the market

	Species	Susceptible stage to EPN	Application site	Cultivation	EPN species recommended
Class insect	Order and family				
Coleoptera					
Buprestidae	*Capnodis tenebrionis*	Larva	Soil	Stone fruits, mainly Apricot	*H. bacterio-phora*, *S. carpocapsae*, *S. feltiae*
Cerambycidae	*Saperda carcharias*	Larva	Wood tunnels	Poplar	*S. carpocapsae*, *S. feltiae*
Chrysomelidae	*Chaetocnema* spp., *altica* spp.	Larva	Soil	Potato, Beet, Vegetables	*S. carpocapsae*
	Diabrotica virgifera virgifera	Larva	Soil	Maize	*H. bacterio-phora*, *S. carpocapsae*
	Xanthogaleruca luteola	Larva	Leaves	Elm	*S. carpocapsae*
Curculionidae	*Otiorhynchus* spp.,	Larva, pupa	Soil	Ornamental, Strawberry, Small fruits	*H. bacterio-phora*, *S. carpocapsae*, *S. feltiae*
	Rhynchophorus ferrugineus	Larva, pupa	Plant	Palm	*S. carpocapsae*, *S. feltiae*
	Temnorhinus mendicus	Larva, pupa	Soil	Sugar beet	*H. bacterio-phora*, *S. carpocapsae*
	Curculio elephas, *C. glandium*, *C. nucum*	Larva, pupa	Soil	Chestnut, Hazelnut	*H. bacterio-phora*, *S. carpocapsae*, *S. feltiae*
	Tomicus piniperda	Larva, pupa	Bark, stumps	Pine	*H. bacterio-phora*, *S. carpocapsae*, *S. feltiae*
Diptera					
Agromyzidae	*Liriomyza* spp.	Larva	Leaves	Ornamental, Vegetables	*S. feltiae*
Sciaridae	*Bradysia* spp., *Lycoriella* spp.	Larva	Soil	Ornamental, Mushrooms	*S. feltiae*
Hemiptera					
Tingidae	*Corythuca ciliata*	Adult over-wintering	Bark	Plane tree	*H. bacterio-phora*, *S. carpocapsae*

(continued)

Table 18.2 (continued)

Hymenoptera					
Tenthredinidae	*Hoplocampa brevis*	Larva	Soil, leaves	Pear	*H. bacteriophora*, *S. feltiae*
	Caliroa varipes	Larva	Soil, leaves	Oak	*H. bacteriophora*, *S. feltiae*
Lepidoptera					
Castniidae	*Paysandisia archon*	Larva	Crown, leaves insertion	Palm	*S. carpocapsae*, *S. feltiae*
Cossidae	*Cossus cossus*, *Zeuzera pyrina*	Larva	Wood tunnels	Orchards	*S. carpocapsae*, *S. feltiae*
Gelechiidae	*Tuta absoluta*	Larva	Leaves	Tomato, Potato, Pepper, Aubergine	*H. bacteriophora*, *S. carpocapsae*, *S. feltiae*
Noctuidae	*Agrotis* spp. *Spodoptera* spp.	Larva	Soil, leaves	Ornamental, Vegetables	*S. carpocapsae*, *S. feltiae*
Pyralidae	*Euzophera bigella*	Larva	Trunk, branches	Orchards	*S. carpocapsae*, *S. feltiae*
Sesiidae	*Synanthedon* spp.	Larva	Trunk, bark	Orchards, Small fruits	*S. carpocapsae*, *S. feltiae*
Thaumetopoeidae	*Thaumetopoea pityocampa*	Larva	Insect nest	Pine	*S. feltiae*
Tortricidae	*Pammene fasciana*, *Cydia splendana*, *C. fagiglandana*	Larva	Soil	Chestnut	*H. bacteriophora*, *S. carpocapsae*, *S. feltiae*
	C. pomonella, *C. molesta*, *C. funebrana*	Larva	Trunk, branches	Apple, Pear, Peach, Apricot, Plum	*S. carpocapsae*, *S. feltiae*
Thysanoptera					
Thripidae	*Frankliniella occidentalis*	Juveniles	Leaves	Ornamental, Vegetables	*S. feltiae*
Class Gasteropoda	Order and family				
Pulmonata					
Milacidae		Juveniles	Soil	Various	*Phasmarhabditis hermaphrodita*
Limacidae		Juveniles	Soil	Various	*P. hermaph rodita*
Arionidae		Juveniles	Soil	Various	*P. hermaph rodita*

Fig. 18.2 Application of EPNs against target insect: (**a**) native *S. feltiae* inside *T. pityocampa* larva; (**b**) native *S. feltiae* IJ entering the mouth of *Capnodis tenebrionis* larva; (**c**) commercial *H. bacteriophora* infecting *C. varipes* larvae; (**d**) native *H. bacteriophora* infecting *R. plicatus* larvae (Pictures: (**a**, **d**) made by E. Tarasco; (**b**) by P. Marannino; (**c**) by G. Curto

L. (Coleoptera: Curculionidae) on hazel (Blum, Kron Morelli, Vinotti, & Ragni, 2009); *Cossus cossus* L. and *Zeuzera pyrina* L. (Lepidoptera: Cossidae) on orchards (Deseö, 1982; Deseö et al., 1984), *Synanthedon tipuliformis* (Clerck) (Lepidoptera: Sesiidae) on persimmon (Caruso, Vergnani, Reggiani, & Curto, 2014); *Hoplocampa brevis* Klug. (Hymenoptera: Tenthredinidae) (Curto, Boselli, Vergnani, & Reggiani, 2007) and *Cydia pomonella* L. (Lepidoptera: Tortricidae) (Curto, Caruso, Reggiani, & Vergnani, 2009; Reggiani, Curto, Vergnani, Caruso, & Boselli, 2008) on pear; *Capnodis tenebrionis* L. (Coleoptera: Buprestidae) on apricot (Marannino, Tarasco, & De Lillo, 2003) (Fig. 18.2b); *Saperda carcharias* L. (Coleoptera: Cerambycidae) on poplar (Curto et al., unpublished); *Xanthogaleruca luteola* Müller (Coleoptera: Chrysomelidae) on elm (Triggiani & Tarasco, 2007); *Corythucha ciliata* Say (Hemiptera: Tingidae) on plane tree (Tarasco & Triggiani, 2006); *Caliroa varipes* Klug. (Hymenoptera Tenthredinidae) on oak (Curto, Vai, & Dallavalle, 2008) (Fig. 18.2c); *Paysandisia archon* Burm. (Lepidoptera: Castniidae) (Nardi et al., 2009) and *Rhynchophorus ferrugineus* Oliver (Coleoptera: Curculionidae) (Clausi & Vinciguerra, 2008; Nardi et al., 2011; Triggiani & Tarasco, 2011) on palm; *Otiorhynchus sulcatus* F. (Coleoptera: Curculionidae) on ornamental plants and strawberry (Curto, Boselli, & Ricci, 2001); *Temnorhinus mendicus* Gyll. (Coleoptera: Curculionidae) on sugarbeet (Boselli, Curto, & Tacconi, 1997);

Rhytidoderes plicatus Oliv. (Coleoptera: Curculionidae) on savoy cabbage (Tarasco & Triggiani, 2002) (Fig. 18.2d), Lepidoptera Noctuids on artichokes (Ippolito & Triggiani, 1995); *Liriomyza trifolii* (Burgess) (Diptera; Agromyzidae) on flowering plants (Colombo & Locatelli, 1985); slugs and snails in horticulture with the specific nematode *Phasmarhabditis hermaphrodita* Sch. (Rhabditida: Rhabditidae) (Gengotti, Censi, & Curto, 2006).

18.3.1 Vine Weevil **(Otiorhynchus sulcatus)** *Control*

The most suitable EPNs for vine weevil control belong to the genus *Heterorhabditis*, although satisfactory results have been achieved with formulations based on *S. feltiae* and, to a lesser extent, *S. carpocapsae* (Curto et al., 2001). Formulation based on *H. bacteriophora* are marketed in Italy in packs containing 50 million IJs. The concentrations correspond to 200,000–400,000 IJs per m^2 (one pack for 125–250 m^2) or 25,000–40,000 IJs per plant (one package for 1,250–2,000 m^2).

The late summer and an *optimum* soil temperatures around 18–22 °C represent the best conditions for the EPN application against the first larval stages of *O. sulcatus*, the most susceptible as at the start of their feeding action. Spring treatments on overwintering mature larvae and newly formed pupae could be effective only with soil temperatures >15 °C and applying the concentration of 500,000 nematodes per m^2. The EPN treatment can be applied to the soil either by drip irrigation or a watering can *i.e.* for small surfaces, on ornamental nurseries and strawberry crops.

18.3.2 Codling Moth **(Cydia pomonella)** *Control on Pear and Apple Orchards*

Another interesting use of EPN, widespread in recent years on more than 1,500 ha throughout northern Italy, is represented by autumnal treatments against overwintering larvae of codling moth (*C. pomonella*) on pear and apple (Curto, Caruso, Reggiani, & Vergnani, 2010). *Steinernema feltiae* and *S. carpocapsae* are the suitable species in order of efficacy, EPNs are sprayed on the trunks and lower branches, where the codling moth larvae overwinter in bark cracks, protected in a slight cocoon that EPNs are able to enter. The application dose is 1.5×10^9 IJs in 15 hl of water per hectare, distributed by an atomizer, closing the higher nozzles and removing the filters. The application timing takes into account the weekly forecasts of rainfall and temperature in September–October, because the EPNs are completely effective with temperatures ≥10–12 °C and a prolonged wetting of the substrate. The treatment is more effective when carried out at the beginning of a rain, when the logs get wet, because, at the rain end, the plant can dry out quickly and nematodes die; in rain absence, it is crucial to spray on the plants high volumes of water before and after the EPN treatment, in cloudy, not windy, foggy weather and with high relative humidity. *Steinernema feltiae* is more resistant than *S. carpocapsae* at low temperatures and therefore acts even at temperatures <12 °C; the EPNs dragged to the ground by the rain remain active for 2 weeks from their application. The spring applications of EPNs do not show efficacy on overwintering larvae of codling moth due mainly to the progressive pupation, stage wherein the nematodes are not able to penetrate. The treatments with EPNs can be effectively inserted within an Integrated Pest Management (IPM) strategy, with the aim of reducing the populations of *C. pomonella* in the orchard and in a territory, for a more effective control strategy in the following spring–summer.

18.3.3 The Experience of an Italian Central–Eastern Region in the Control of Red Palm Weevil (Rhynchophorus ferrugineus)

An IPM strategy for the control of red palm weevil (*R. ferrugineus*) was applied in the Marche region (Central–Eastern Italy) during 2008–10, aimed to limit the spread of this pest in a demarcated area according to the Commission Decision 2010/467/EU (Nardi et al., 2011). The integrated management of RPW was carried out on infested plants and consisted in: the pest survey with visual inspections of susceptible hosts, the mass trapping of weevil adults in the buffer zone by commercial traps baited with the aggregation pheromone of *R. ferrugineus*, the spherical pruning of the top trees, the washing of the crown (head of palm) with water at high pressure, and finally the application of either chemical insecticides or EPNs. All tested methods proved to be promising tools for IPM of *R. ferrugineus*. Regarding the EPN application, 15–20 L of *S. carpocapsae* suspension, corresponding to 50×10^6 IJs per plant, were sprayed at low pressure over the palm top 30, 60 and 180 days after spherical pruning to prevent new weevil infestations. *Steinernema carpocapsae* was applied alone or in combination with chitosan or azadirachtin and compared with chlorpyrifos + pyrethroids or acetamiprid + bifentrin treatments. The mean efficacy of the integrated approach (spherical pruning plus treatment application) in the control of *R. ferrugineus* exceeded 70 % in all treatments: the effectiveness of IPM with *S. carpocapsae* application ranged between 71 and 79 %, with chemicals between 86 and 93 %. The addition of chitosan to EPN treatments seemed not increase their efficacy (59–81 %). *Steinernema carpocapsae* was able to parasitize all the weevil stages (larva, pupa and adult) inside or outside their cocoon, as already observed by Triggiani and Tarasco (2011) in laboratory tests, even if the mortality of adults was lower than larvae. The IPM of RPW was demonstrated to be possible and effective in controlling its spreading of *ferrugineus* when applied as sanitation measures.

18.4 Regulation of Entomopathogenic Nematodes Use in Italy

Before their use, entomopathogenic organisms should generally be registered, with the exception of the EPNs for their multicellular structure and recognized specificity against insects (Akhurst & Smith, 2002). Regulation (EC) no 1107/2009 of the European Parliament provides registration for three "categories" of formulated: chemical biocidal products, micro–organisms and viruses. Nematodes and macro–organisms (auxiliary insects and mites) are not mentioned by pursuing the principle of avoiding registration procedure for products with a low environmental impact, uniquely interpreted by European Union Countries. However, at European level there are Countries that do not require any registration as Denmark, Finland, France, Germany, Greece, Italy, Portugal and Spain, while there are others that request some form of registration, such as Austria (a registration like chemicals is requested), Belgium and the Netherlands (registration is required only for new formulates), Poland, Czech Republic and Hungary (experimental screening in field tests), Ireland, Switzerland, Norway and Sweden (all biological control agents require a registration), and the United Kingdom (the indigenous EPNs do not require any registration, but the introduction of non–indigenous agents is controlled). In Italy, in relation to the introduction of exotic species of EPNs, it should be taken into consideration that is still in force the "DPR 12/3/2003, n. 120" on the conservation of natural habitats, flora and fauna, that, in article 12 paragraph 3, "prohibits the reintroduction, restocking and nature of non–native species and populations". For indigenous EPNs there are no restrictions. Italian legislation tends to follow the recommendations of REBECA Action (Regulation of Biological Control Agent, http://www.rebeca-net.de), financed by the European Union for a correct use and commercialization of the EPNs: knowledge of the exact identity of EPNs, any recommendation when using native EPN species, information about origin, distribution and target hosts should be requested for non–autochthonous EPNs before their use. The conclusion of the REBECA action reiterates that for the EPNs there are the same *criteria* of "Environmental Risk Assessment" (ERA: Environmental Risk Assessment) used for auxiliary insects and mites; in addition, information about their presence, dispersal, target hosts and direct and indirect effects may not even be necessary given the limited potential of EPNs in dispersion and persistence in the environment in which they are applied (more information, Chap. 10).

18.5 Conclusions and Future Directions

In Italy the use of EPNs in IPM strategies has been included not only in the guidelines for organic farming, but also in those for integrated management agriculture. Inclusion of EPNs in IPM strategies is conditioned by the proved effectiveness in multiannual field trials. Treatments should be applied in the rainy periods of autumn or spring, on susceptible host stages (larvae and in case of Coleoptera also pupae) and when climatic conditions of temperature and substrate wetting can promote the nematode mobility towards the host. EPNs have been therefore inserted in the IPM guidelines for the control of Lepidoptera such as codling moth in orchards of apple, pear and walnut, currant clearwing in persimmon, acorn moth and beech moth in chestnut, and for the control of Coleoptera as vine weevil on strawberry fields.

Farmers know the EPN based products and how to store and applied them. The acceptance of farmers is generally very positive but requires a network of technical assistance for the implementation of integrated agriculture (i.e. use of mathematical models, days of the highest pest presence in a susceptible stage, installation of pheromone traps, and an efficient network of meteorological forecast) useful for effective treatment by EPNs. Future research plans involve surveys in the Italian regions not yet explored in order to identify EPN species present and to estimate their geographical occurrence and biodiversity. Molecular studies on the response of EPN to abiotic stresses (temperature, water, etc.) will be also carried out. All these data will prove useful to design new IPM strategies that include EPNs as stable tools for biological control.

References

Akhurst, R., & Smith, K. (2002). Regulation and safety. In R. Gaugler (Ed.), *Entomopathogenic nematology* (pp. 311–332). Wallingford, UK: CABI Publishing.

Barbercheck, M. E., & Kaya, H. K. (1991). Effect of host condition and soil texture on host finding by the entomogenous nematodes *Heterorhabditis bacteriophora* (Rhabditida: Heterorhabditidae) and *Steinernema carpocapsae* (Rhabditida: Steinernematidae). *Environmental Entomology, 20*, 582–589.

Blum, B., Kron Morelli, R., Vinotti, V., & Ragni, A. (2009). Control of *Curculio nucum*, the hazelnut borer, by entomopathogenic nematodes. *Acta Horticulturae, 845*, 567–570.

Boselli, M., Curto, G. M., & Tacconi, R. (1997). Field efficacy of entomopathogenic nematodes against the sugar–beet weevil *Temnorhinus* (=*Conorrhynchus*) *mendicus* Gyll. (Coleoptera: Curculionidae). *Biocontrol Science and Technology, 7*, 231–238.

Cabanillas, H. E., & Raulston, J. R. (1994). Evaluation of the spatial pattern of *Steinernema riobravis* in corn plots. *Journal of Nematology, 26*, 25–31.

Campbell, J. F., Lewis, E., Yoder, F., & Gaugler, R. (1995). Entomopathogenic nematode (Heterorhabditidae and Steinernematidae) seasonal population dynamics and impact on insect populations in turfgrass. *Biological Control, 5*, 598–606.

Campbell, J. F., Orza, G., Yoder, F., Lewis, E., & Gaugler, R. (1998). Spatial and temporal distribution of endemic and released entomopathogenic nematode in turfgrass. *Entomologia Experimentalis et Applicata, 86*, 1–11.

Campos-Herrera, R., Escuer, M., Cuadra, L., Labrador, S., Robertson, L., & Gutierrez, C. (2007). Distribution of entomopathogenic nematodes from La Rioja (Northern Spain). *Journal of Invertebrate Pathology, 95*, 125–139.

Campos-Herrera, R., Gómez-Ros, J. M., Escuer, M., Cuadra, L., Barrios, L., & Gutiérrez, C. (2008). Diversity, occurrence, and life characteristics of natural EPN population from La Rioja (Northern Spain) under different levels of disturbance associated with agricultural management and their relationships with soil factors. *Soil Biology and Biochemistry, 40*, 1474–1484.

Caruso, S., Vergnani, S., Reggiani, A., & Curto, G. (2014, March 18–21). Valutazione di nematodi entomopatogeni per il controllo della sesia del kaki in Emilia–Romagna [Evaluation of entomopathogenic nematodes for the control of currant clearwing (*Synanthedon tipuliformis*) on persimmon] (Vol. 1, pp. 133–136). In *Proceedings of "Giornate Fitopatologiche"*, Chianciano Terme, Clueb Eds. Bologna.

Clausi, M., Longo, A., Rappazzo, G., Tarasco, E., & Vinciguerra, M. T. (2011). *Steinernema vulcanicum* n. sp. (Rhabditida: Steinernematidae), a new entomopathogenic nematode species from Sicily (Italy). *Nematology, 13*, 409–423.

Clausi, M., & Vinciguerra, M. T. (2005). I nematodi entomopatogeni in un progetto per lo sviluppo sostenibile dei castagneti. *Nematologia Mediterranea, 33*, 91–94.

Clausi, M., & Vinciguerra, M. T. (2007). Efficacy and environmental impact of entomopathogenic nematodes used against nut insect pests in some chestnut woods on Etna (Italy). *IOBC WPRS Bulletin, 30*, 33–37.

Clausi, M., & Vinciguerra, M. T. (2008). Prove di lotta biologica contro il Punteruolo rosso delle palme per mezzo di Nematodi entomopatogeni. In: Servizi allo Sviluppo (Ed.), *La ricerca scientifica sul punteruolo rosso e gli altri fitofagi delle palme in Sicilia* (Vol. 1, pp. 165–168). Regione Siciliana Assessorato Agricoltura E Foreste Dipartimento Interventi Infrastrutturali Servizio XI, Servizi allo Sviluppo Eds. Palermo.

Colombo, M., & Locatelli, D. P. (1985). Laboratory evaluation of the activity of *Steinernema feltiae* Filip. and *Heterorhabditis* spp. on *Liriomyza trifolii* (Burgess) and *Opogona sacchari* (Bojer) infesting cultivated flowering plants. *La Difesa delle Piante, 8*, 263–269.

Curto, G., Boselli, M., & Ricci, M. (2001). Winter field trial against blackvine weevils *Otiorhynchus sulcatus* in potted *Euonymus japonica* using the cold active nematode *Steinernema kraussei* N0093 strain. In: C. T. Griffin, A. M. Burnell, M. J. Downes, & R. Mulder (Eds.), *Developments in entomopathogenic nematode/bacterial research* (pp. 285–290). Luxembourg: COST Action 819–European Commission EUR 1969, Office for official publications of the European Communities.

Curto, G., Boselli, M., Vergnani, S., & Reggiani, A. (2007). Effectiveness of entomopathogenic nematodes in the control of sawflies (*Hoplocampa brevis*) in pear orchards. *IOBC WPRS Bulletin, 30*, 13–17.

Curto, G., Caruso, S., Reggiani, A., & Vergnani, S. (2009). Efficacia dei nematodi entomopatogeni nel contenimento delle larve di *Cydia pomonella* L. nell'Italia settentrionale. *Frutticoltura, 3*, 45–49.

Curto, G., Caruso, S., Reggiani, A., & Vergnani, S. (2010). Difesa del melo dalla carpocapsa con nematodi. *L'Informatore Agrario, 39*, 48–50.

Curto, G., Reggiani, A., Dallavalle, E., & Bariselli, M. (2009). Control of carpophagous Lepidoptera in chestnut by means of entomopathogenic nematodes. *IOBC WPRS Bulletin, 45*, 345–348.

Curto, G., Vai, N., & Dallavalle, E. (2008). Control of *Caliroa varipes* (Klug, 1814) (Hymenoptera, Tenthredinidae) with entomopathogenic nematodes, in an urban park of Bologna suburbs. *Redia, 91*, 173–176.

Deseö, K. V. (1982, March 16–19). Prove di lotta col nematode entomopatogeno *Neoaplectana carpocapsae* Weiser contro i rodilegno *Cossus cossus* L. e *Zeuzera pyrina* L. (Lepidoptera; Cossidae). [Trials with *Neoaplectana carpocapsae* Weiser (Nematoda) in the control of two treeborers *Cossus cossus* L. and *Zeuzera pyrina* L. (Lepidoptera; Cossidae)] (Vol. 3, pp. 3–10). In *Proceedings of "Giornate Fitopatologiche"*, Sanremo, Clueb Eds. Bologna.

Deseö, K. V., Fantoni, P., & Lazzari, G. L. (1988, May 16–20). Presenza di nematodi entomopatogeni (*Steinernema* spp., *Heterorhabditis* spp.) nei terreni agricoli d'Italia [Natural occurrence of Enthomopathogenic Nematodes (*Steinernema* spp. and *Heterorhabditis* spp.) in farmland soil in Italy] (Vol. 2, pp. 269–280). In *Proceedings of "Giornate Fitopatologiche"*, Lecce, Clueb Eds. Bologna.

Deseö, K. V., Grassi, S., Foschi, F., & Rovesti, L. (1984, March 26–29). Un sistema di lotta biologica contro il Rodilegno giallo (*Zeuzera pyrina* L., Lepidoptera Cossidae) [Biological methods for the control of the Leopard Moth (*Zeuzera pyrina* L., Lepidoptera Cossidae)] (Vol. 2, pp. 403–414). In *Proceedings of "Giornate Fitopatologiche"*, Sorrento, Clueb Eds. Bologna.

Ehlers, R.-U., Deseö, K. V., & Stackerbrandt, E. (1991). Identification of *Steinernema* spp. and their symbiotic bacteria *Xenorhabdus* spp. from Italian and German soil. *Nematologica, 37*, 360–366.

Garcia del Pino, F., & Palomo, A. (1996). Natural occurrence of entomopathogenic nematodes (Rhabditida: Steinernematidae and Heterorhabditidae) in Spanish soil. *Journal of Invertebrate Pathology, 68*, 84–90.

Gengotti, S., Censi, D., & Curto, G. (2006). Alternative alla metaldeide per il controllo dei gasteropodi. *L'Informatore Agrario, 12*, 56–59.

Glazer, I., Kozodoi, E., Salame, L., & Nestel, D. (1996). Spatial and temporal occurrence of natural populations of *Heterorhabditis* sp. (Nematode: Rhabditida) in a semiarid region. *Biological Control, 6*, 130–136.

Glazer, I., Liran, N., & Steinberger, Y. (1991). A survey of entomopathogenic nematodes (Rhabditida) in the Negev desert. *Phytoparasitica, 19*, 291–300.

Griffin, C. T., Moore, J. F., & Downes, M. J. (1991). Occurrence of insect–parasitic nematodes (Steinernematidae, Heterorhabditidae) in the Republic of Ireland. *Nematologica, 37*, 92–100.

Hominick, W. M. (2002). Biogeography. In R. Gaugler (Ed.), *Entomopathogenic nematology* (pp. 115–143). New York: CABI Publishing.

Ippolito, R., & Triggiani, O. (1995). Efficacia di *Heterorhabditis* e *Steinernema feltiae* Fili. (=*bibionis*) (Nematoda) nei riguardi di larve di Lepidotteri Nottuidi su carciofo. *Informatore Fitopatologico, 9*, 55–60.

Koppenhöfer, A. M., & Kaya, H. K. (1996). Coexistence of entomopathogenic nematode species (Steinernematidae and Heterorhabditidae) with different foraging behaviour. *Fundamental and Applied Nematology, 19*, 175–183.

Kung, S. P., Gaugler, R., & Kaya, H. K. (1990). Soil type and entomopathogenic nematode persistence. *Journal of Invertebrate Pathology, 55*, 401–406.

Marannino, P., Tarasco, E., & De Lillo, E. (2003). Biological notes on larval hatching in *Capnodis tenebrionis* (L.) (Coleoptera Buprestidae) and evaluation of entomopathogenic nematodes in controlling neonate larvae. *Redia, 86*, 101–105.

Mráček, Z., & Bečvàř, S. (2000). Insect aggregation and entomopathogenic nematodes occurrence. *Nematology, 2*, 297–301.

Mráček, Z., Bečvář, S., Kindlmann, P., & Jersáková, J. (2005). Habitat preference for entomopathogenic nematodes, their insect hosts and new faunistic records for the Czech Republic. *Biological Control, 34*, 27–37.

Mráček, Z., & Webster, J. M. (1993). Survey of Heterorhabditidae and Steinernematidae (Rhabditida: Nematoda) in Western Canada. *Journal of Nematology, 25*, 710–717.

Nardi, S., Ricci, E., Lozzi, R., Marozzi, F., Ladurner, E., Chiabrando, F., et al. (2011). Control of *Rhynchophorus ferrugineus* (Olivier, 1790) according to EU Decision 2007/365/CE in the Marche region (Central–Eastern Italy). *OEPP/EPPO Bulletin, 41*, 103–115.

Nardi, S., Ricci, E., Lozzi, R., Marozzi, F., Ladurner, E., Chiabrando, F., et al. (2009). Use of entomopathogenic nematodes for the control of *Paysandisia archon* Burmeister. *IOBC WPRS Bulletin, 45*, 375–378.

Peters, A. (1996). The natural host range of *Steinernema* and *Heterorhabditis* spp. and their impact on insect populations. *Biocontrol Science and Technology, 6*, 389–402.

Půža, V., & Mráček, Z. (2005). Seasonal dynamics of entomopathogenic nematodes of the genera *Steinernema* and *Heterorhabditis* as a response to abiotic factors and abundance of insect hosts. *Journal of Invertebrate Pathology, 89*, 116–122.

Rappazzo, G., Avanzato, C., Petronio, G., Clausi, M., Tarasco, E., Buccheri, M. A., et al. (2011). Bacterial symbionts of entomopathogenic nematodes of the genus *Steinernema* from South Italy: Phenotypic features and evolutionary implications. *IOBC WPRS Bulletin, 66*, 321–325.

Rappazzo, G., Clausi, M., & Vinciguerra, M. T. (2005). Molecular identification of entomopathogenic nematodes found in pristine chestnut growing areas on the foothills of Etna (Sicily). *Nematologia Mediterranea, 33*, 101–104.

Reggiani, A., Curto, G., Vergnani, S., Caruso, S., & Boselli, M. (2008). Effectiveness of entomopathogenic nematodes in the control of *Cydia pomonella* overwintering larvae in Northern Italy. *IOBC WPRS Bulletin, 31*, 287–293.

Ricci, M., Fifi, A. P., De Bernardis, M., Colli, M., Barcarotti, R., & Ragni, A. (2004). Development of a bio–insecticide based on a cold–active entomopathogenic nematode. *IOBC WPRS Bulletin, 27*, 139.

Spiridonov, S. E., Reid, A., Podrucka, K., Subbotin, S. A., & Moens, M. (2004). Phylogenetic relationships within the genus *Steinernema* (Nematoda: Rhabditida) as inferred from analyses of sequences of the ITS1–5.8S–ITS2 region of rDNA and morphological features. *Nematology, 6*, 547–566.

Stock, S. P., Pryor, B. M., & Kaya, H. K. (1999). Distribution of entomopathogenic nematodes (Steinernematidae and Heterorhabditidae) in natural habitats in California. *Biodiversity and Conservation, 8*, 535–549.

Stuart, R. J., & Gaugler, R. (1994). Patchiness in populations of entomopathogenic nematodes. *Journal of Invertebrate Pathology, 64*, 39–45.

Susurluk, I. A., Tarasco, E., Triggiani, O., & Ehlers, R. U. (2007). Molecular characterisation of entomopathogenic nematodes isolated in Italy by PCR–RFLP analysis of the ITS region of the ribosomal DNA repeat unit. *Nematologia Mediterranea, 35*, 23–28.

Tarasco, E., Clausi, M., Rappazzo, G., Panzavolta, T., Curto, G., Sorino, R., et al. (2015). Biodiversity of entomopathogenic nematodes in Italy. *Journal of Helminthology, 89*, 359–366.

Tarasco, E., Clausi, M., Rappazzo, G., Vinciguerra, M. T., Longo, A., & Triggiani, O. (2009). Could be Italy considered a favorite place in Europe for EPN biodiversity? *IOBC WPRS Bulletin, 45*, 387–389.

Tarasco, E., Mráček, Z., Nguyen, K. B., & Triggiani, O. (2008). *Steinernema ichnusae* sp. n. (Nematoda: Steinernematidae) a new entomopathogenic nematode from Sardinia Island (Italy). *Journal of Invertebrate Pathology, 99*, 173–185.

Tarasco, E., & Triggiani, O. (1997). Survey of *Steinernema* and *Heterorhabditis* (Rhabditida: Nematoda) in Southern Italian soils. *Entomologica, 31*, 117–123.

Tarasco, E., & Triggiani, O. (2002). Biocontrol of *Rhytidoderes plicatus* Oliv. (Coleoptera, Curculionidae) in potted savoy cabbage with entomopathogenic nematodes. *Entomologica, 36*, 157–164.

Tarasco, E., & Triggiani, O. (2005). Entomopathogenic nematodes of forest habitats in Southern Italy and their pathogenicity for overwintering larvae of *Thaumetopoea pityocampa* (Den. et Schiff.) (Lepidoptera, Thaumetopoeidae). In F. Lieutier & D. Ghaioule (Eds.), *Entomological research in Mediterranean forest ecosystems* (pp. 257–263). Paris, France: INRA.

Tarasco, E., & Triggiani, O. (2006). Evaluation and comparison of entomopathogenic nematodes and fungi to control *Corythucha ciliata* Say (Rhynchota Tingidae). *Redia, 89*, 51–54.

Tarasco, E., & Triggiani, O. (2007). Entomopathogenic nematodes and forest insects in Italy. *IOBC WPRS Bulletin, 30*, 63–65.

Torr, P., Heritage, S., & Wilson, M. J. (2007). *Steinernema kraussei*, an indigenous nematode found in coniferous forest: Efficacy and filed persistence against *Hylobius abietis. Agricultural and Forest Entomology, 9*, 181–188.

Triggiani, O. (1983). Sensibilità del *Tomicus* (*Blastophagus*) *piniperda* L. (Coleoptera: Scolytidae) a Nematodi della famiglia Steinernematidae e Heterorhabditidae. *Entomologica, 18*, 215–223.

Triggiani, O., Mráček, Z., & Reid, A. (2004). *Steinernema apuliae* sp. n. (Rhabditida: Steinernematidae): A new entomopathogenic nematode from southern Italy. *Zootaxa, 460*, 1–12.

Triggiani, O., & Tarasco, E. (2000). Indagini sui nematodi entomopatogeni (Rhabditida: Steinernematidae e Heterorhabditidae) in pinete e querceti dell'Italia meridionale. *Entomologica, 34*, 23–32.

Triggiani, O., & Tarasco, E. (2002). Efficacy and persistence of entomopathogenic nematodes in controlling larval populations of *Thaumetopoea pityocampa* (Lepidoptera: Thaumetopoeidae). *Biocontrol Science and Technology, 12*, 747–752.

Triggiani, O., & Tarasco, E. (2007). Applying entomopathogenic nematode to *Xanthogaleruca luteola* (Coleoptera Chrysomelidae) infested foliage. *Redia, 90*, 29–31.

Triggiani, O., & Tarasco, E. (2011). Evaluation of the effects of autochthonous and commercial isolates of Steinernematidae and Heterorhabditidae on *Rhynchophorus ferrugineus. Bulletin of Insectology, 64*, 175–180.

Vinciguerra, M. T., & Clausi, M. (2006). Biological control of chestnut insect pests by means of entomopathogenic nematodes. *Advances in Horticultural Science, 20*, 40–44.

Chapter 19
Entomopathogenic Nematodes in Iran: Research and Applied Aspects

Javad Karimi and Elham Salari

19.1 Introduction

Entomopathogenic nematodes (EPNs) have been used successfully to control a variety of economically important insect pests. Their pathogenic effect is mostly due to the symbiotic bacteria of the genera *Xenorhabdus* Thomas & Poinar and *Photorhabdus* Boemare, Akhurst & Mourant (Enterobacteriales: Enterobacteriaceae), located in the intestine of the infective juveniles (IJs) (Poinar, 1990; Lacey & Shapiro-Ilan, 2008). The bacto–helminthic complexes of *Heterorhabditis*–*Photorhabdus* and *Steinernema*–*Xenorhabdus* have high potential as effective biological control agents against many soil dwelling insect pests or those in cryptic habitats such as galleries in plants (Burnell & Stock, 2000; Koppenhöfer, 2007). Within 30–60 min after entering the host's haemocoel through natural openings (mouth, anus, and spiracles) or directly through the cuticle, nematodes release their bacterial symbionts, which multiply and cause septicaemia by producing toxins and kill the insect host within 24–48 h (Grewal, Ehlers, & Shapiro-Ilan, 2005).

Recently, control of some economically important pests become very difficult due to the development of insect resistance to insecticides (Bughio & Wilkins, 2004). Another problem raised during last few years is to alter the population of insect pests as well their geographical distribution and pest incidence. One main reason for such changes could be resulted of climate changes, which reflect the effects on the pests themselves, their host plants and the interactions between them (Franklin, 2009; Ulrichs & Hopper, 2008). Moreover, the insect physiology, behavior, development and species distribution may also be affected in a changing climate (Merrill et al., 2008; Thomson, Macfadyen, & Hoffmann, 2010).

J. Karimi (✉) • E. Salari
Faculty of Agricultura, Department of Plant Protection, College of Agriculture, Ferdowsi University of Mashhad, Mashhad, Iran
e-mail: jkb@um.ac.ir; salari.elham@stu-mail.um.ac.ir

© Springer International Publishing Switzerland 2015

R. Campos-Herrera (ed.), *Nematode Pathogenesis of Insects and Other Pests*, Sustainability in Plant and Crop Protection 1, DOI 10.1007/978-3-319-18266-7_19

Climate change could have positive, negative or no impact on occurrence and distribution of some insect pests (Franklin, 2009; Torchin, Lafferty, Dobson, McKenzie, & Kuris, 2003). For example, central parts of Iran including salt lakes, deserts, and sand dune areas with very high temperatures and windy weather were predicted to be less suitable for habitat distribution of some heteropteran pests such as *Apodiphus amygdali* (Germar) and *Nezara viridula* (L.) (Hemiptera: Pentatomidae); While other regions, mainly northern parts, were more suitable (Solhjouy-Fard, Sarafrazi, Moeini, & Ahadiyat, 2013). It was probably due to the impact of hot, dry, and windy weather on the insect embryo development (Grichanov & Ovsyannikova, 2009).

Climate change related factors like rising global temperature, changes in precipitation patterns, milder and shorter winters, and rise of sea levels as well as lake drying may have strong influence on the development and survival of insect pests. Research showed that such changes in climatic conditions may have significant impacts on the population dynamics and status of insect pests of crops (Woiwod, 1997). In addition, global climate is expected to increase 1.4–5.8 °C during the next century (Meehl, 2007). One of such climate change that have been occurred in Iran was drying of the Urmia lake which will be encountered the region with irreparable environmental and agricultural damages. Urmia Lake is one of the largest permanent hyper–saline lakes in the world located in northwest Iran. Recently, more than 95 % of the Urmia Lake has become dry for some reasons and the water level of this lake decreased nearly three meters in comparison to the past 20 years water level due to continuous dry years in the past 10 years (Eimanifar & Mohebbi, 2007). One of the adverse effects of such changes that observed in this region during these years was population growth and establishment of some insect pests. For example, regional studies on poplar nurseries and plantations of the Urmia indicated that the density of poplar pests was numerically higher in 2012 than in 2011. Also, it was showed that the emergence of some important pests [i.e., *Lithocolletis populifoliella* (Treitschke) (Lepidoptera: Gracillariidae)] started approximately a week or 10 days earlier in 2012 than the previous year Zargaran, Akbarian, & Pourfarhadi (2013, unpublished). Overall, climate change could profoundly affect the status of insect pests, but the precise impact of climate change on insects is somewhat uncertain because some climate changes may favour insects while others may inhibit a few insect species.

Meanwhile, Iran has various types of climates from arid or semiarid, to subtropical along the Caspian coast and the northern forests. Iranian climate is divided into hyper-arid (35 %), arid (29 %), semi-arid (20 %), Mediterranean (5 %) and wet (10 %) which is susceptible to extreme temperature differentiation from –20 °C in the high mountainous land to 50 °C in the dessert regions (Amiri & Eslami, 2010). Therefore, this country has a rich fauna and flora biodiversity resource due to its rather diverse geographic, climatic, and weather conditions which is resulted in high undisturbed habitats with great diversity of herbivore insects and their natural enemies.

Hence, there is also a tremendous opportunity for discovery new nematode species and populations with higher tolerance adapted to local environmental conditions. There are several target pests in Iran that can be controlled with EONs. Currently, due to less successful of some chemical applications, effects of climate changes on altering the population dynamics of some insect pests as well as rich fauna resource of EPNs in the country, several universities and research institutes have contributed to the development of EPN research for the control of such insect pests in Iran. This chapter provides an overview on the status of EPN research and applied aspects in Iran. Also, activities of research on identification and characterization of EPNs from various locations of this country are reported.

19.2 Basic Research on Entomopathogenic Nematodes

Several surveys have been conducted for the isolation of EPNs, which led to reporting of several known or new species (Fig. 19.1, Table 19.1). Initially, Parvizi, Barooti, and Adldoost (1988, unpublished) recovered *Steinernema arenarium* (Artyukhovsky) Wouts, Mráček, Gerdin & Bedding (syn. *S. anomalae*, Kozodoi) (Rhabditida: Steinernematidae) and *Heterorhabditis bacteriophora* Poinar (Rhabditida: Heterorhabditidae) as natural pathogens of *Agrotis ipsilon* (Hufnagel) (Lepidoptera: Noctuidae) from irrigated cultivation of West Azerbaijan province, Iran. Subsequently, Parvizi (2001) isolated an unnamed isolate of *Steinernema* sp. as well as *H. bacteriophora* from this province. Karimi, Kharazi-Pakdel, and Robert (2003) found a number of severely nematode–infected slugs, *Parmacella*

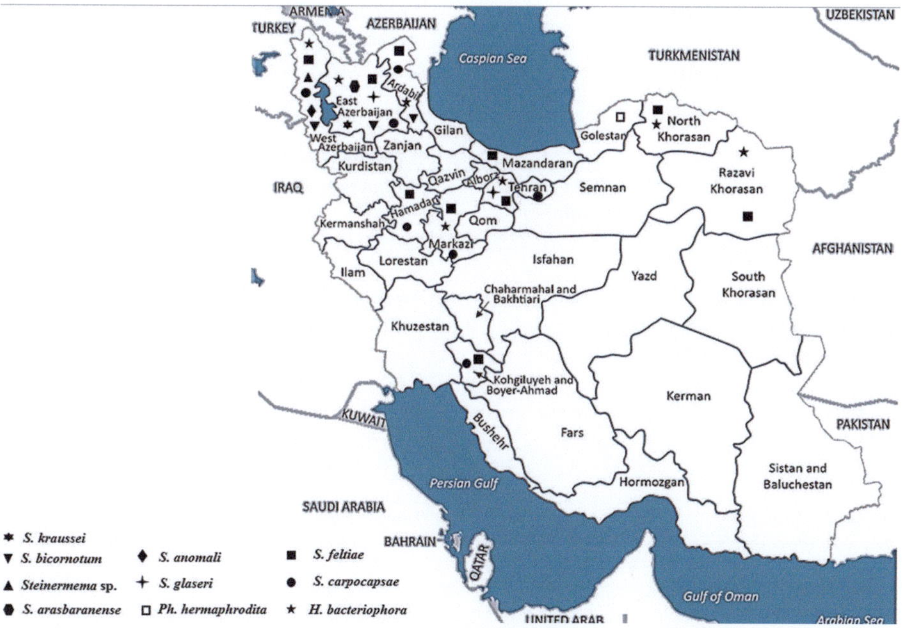

Fig. 19.1 Map of Iran showing distribution of entomopathogenic nematodes

Table 19.1 List of valid species of entomopathogenic nematodes with their original localities and sources of isolation in Iran

Nematode species	Original source	Original locality	Reference
S. arenarium and *H. bacteriophora*	*Agrotis ipsilon*	West Azerbaijan province	Parvizi et al. (1988, unpublished)
Steinernema sp. and *H. bacteriophora*	Soil	West Azerbaijan province	Parvizi (2001)
Ph. hermaphrodita	*Parmacella iberica*	Citrus trees in Gorgan, Iran	Karimi et al. (2003)
S. feltiae	Soil	Mazandaran and Tehran provinces	Tanha Ma'afi et al. (2006, unpublished)
S. glaseri, *S. carpocapsae* and *H. bacteriophora*	*Polyphylla olivieri*	Tehran province	Karimi and Kharazi-pakdel (2007)
S. carpocapsae and *S. feltiae*. *H. bacteriophora*	Soil	Arasbaran forests, North West Iran	Nikdel, Niknam, Shojaee, Askary, and Mohammadi (2008)
S. carpocapsae, *S. feltiae*, *S. bicornutum* and *H. bacteriophora*	Orchards, grasslands and alfalfa fields	Ardabil, East Azerbaijan and West Azerbaijan provinces	Eivazian Kary, Niknam, Griffin, Mohammadi, and Moghaddam (2009)
S. glaseri	*P. olivieri*	Tehran province	Karimi et al. (2009)
S. feltiae	Soil	Kohgiluyeh and Boyerahmad province	Abdollahi (2010, unpublished)
H. bacteriophora, *S. carpocapsae* and *S. feltiae*	Potato fields	East and West Azerbaijan provinces	Agazadeh, Mohammadi, and Eivazian Kary (2010)
H. bacteriophora NIR1	Soil	North West	Eivazian Kary, Niknam, et al. (2010)
S. bicornutum	Soil	Marand, East Azerbaijan province	Eivazian Kary, Rafiee et al. (2010)
S. carpocapsae	*Polyphylla olivieri*	Tehran province	Karimi, Kharazi-Pakdel, Yoshiga, and Koohi-Habibi (2010)
H. bacteriophora, *S. carpocapsae*, *S. bicornutum*, *S. feltiae*, *S. glaseri* and *S. kraussei*	Soil	Arasbaran forests, north-west Iran	Nikdel et al. (2010)
H. bacteriophora, *S. feltiae* and *S. carpocapsae*	Soil	Arak, Markazi province	Ashtari, Karimi, Rezapanah, and Hassani-kakhki (2011)
S. feltiae and *H. bacteriophora*	Soil	Tabriz, East Azerbaijan province	Ebrahimi and Niknam (2011)

(continued)

Table 19.1 (continued)

Nematode species	Original source	Original locality	Reference
S. kraussei	Soil	Chichakloo, Varzeghan, East Azerbaijan	Nikdel, Niknam, & Eivazian Kary et al. (2011)
S. arasbaranense sp. n.	Soil	Kerengan village, East Azerbaijan province	Nikdel et al. (2011)
S. carpocapsae and *S. feltiae*	Soil	Hamedan province	Saffari et al. (2012)
S. feltiae and *S. carpocapsae*	Soil	Kohgiluyeh and Boyer Ahmad province	Roodaki, Haghani, Falahi, and Abdollahi (2012), Roodaki, Haghani and Abdollahi (2012)
S. feltiae and *H. bacteriophora*	Soil	Mashhad, North east Iran	Hassani-Kakhki et al. (2013)
H. bacteriophora and *S. feltiae*	Soil	Bojnourd, North Khorasan Province	Kamali, Karimi, Hosseini, Campos-Herrera, and Duncan (2013)
H. bacteriophora and *S. feltiae*	potato fields	Farooj, North Khorasan province	Rahatkhah, Karimi, Ghadamyari and Brivio (2015)

ibera Eichwald (Heterobranchia: Parmacellidae), from Citrus trees in Gorgan, Iran and it was identified as *Phasmarhabditis hermaphrodita* Schneider (Rhabditida: Rhabditidae). This is the first record of the genus from Iran. Tanha Ma'afi, Ebrahimi, Abootorabi, & Spiridonov (2006) isolated two *Steinernema* species from Mazandaran and Tehran provinces which identified as *Steinernema feltiae* (Filipjev) Wouts, Mráček, Gerdin & Bedding (Rhabditida: Steinernematidae) and a member from "*affine–intermedium*" species group. Karimi and Kharazi-pakdel (2007) collected and identified eight isolates from three EPN species as natural pathogens of the white grub, *Polyphylla olivieri* Castelnau (Coleoptera: Scarabaeidae). Those species were identified as *Steinernema glaseri* (Steiner) Wouts, Mráček, Gerdin & Bedding (Rhabditida: Steinernematidae), *Steinernema carpocapsae* (Weiser) Wouts, Mráček, Gerdin & Bedding (Rhabditida: Steinernematidae) and *H. bacteriophora*. Subsequently, five EPN species were collected and identified from Arasbaran forests in North West Iran. The most commonly found species were reported were *S. carpocapsae* and *S. feltiae*. The *Heterorhabditis bacteriophora* was also isolated from different regions (Nikdel, Niknam, Shojaee, Askary, & Mohammadi, 2008). Eivazian Kary, Niknam, Griffin, Mohammadi, and Moghaddam (2009) reported occurrence of several EPNs from natural areas of the three provinces in the north-west (Ardabil, East Azerbaijan and West Azerbaijan provinces). Among the 833 soil samples, 27 samples (3 %) were positive for EPN. They extracted and identified *S. carpocapsae*, *S. feltiae*, *Steinernema bicornutum* Tallosi, Peters & Ehlers (Rhabditida: Steinernematidae) and *H. bacteriophora*. Again *H. bacteriophora* was shown to be the most common heterorhabditid and *S. feltiae* was the most common steinernematid species. *Steinernema* spp. were isolated mainly from orchards and grasslands whereas *Heterorhabditis* was isolated mainly from grasslands and alfalfa fields.

Steinernema glaseri was isolated from larval stages of the white grub, *P. olivieri* from different sites in Tehran province of Iran in 2005–2006 (Karimi, Kharazi-pakdel, Yoshiga, & Koohi-habibi, 2009). It was reported as natural pathogen of local populations of the white grub larvae. After that, Abdollahi (2010, unpublished) collected and reported *S. feltiae* in Kohgiluyeh and Boyer Ahmad province, Iran. Agazadeh et al. (2010) collected EPNs from potato fields in two provinces in the north–west of Iran during 2009. Heterorhabditid isolates were identified as *H. bacteriophora* and one new undescribed species whereas the steinernematid isolates were identified as *S. carpocapsae* and *S. feltiae*. Eivazian Kary, Niknam, Mohammadi, Moghaddam, and Nikdel (2010) isolated *S. bicornutum* from soil samples collected near Marand, East Azerbaijan province, North–west of Iran. Also, Eivazian Kary Rafiee, Mohammadi, and Afghahi (2010) recovered a geographical isolate of *Heterorhabditis* from soil in the North West and characterized as *H. bacteriophora* NIR1. Moreover, out of a total 11,800 soil samples studied from natural pathogens in larval populations of the white grub in the Tehran province of Iran, two isolates of *Steinernema* spp. were isolated and morphological characters identified those as members of "*carpocapsae*" group Karimi, Kharazi-Pakdel, & Hasani-Kakhki (2010, unpublished).

Nikdel, Niknam, Griffin, and Eivazian Kary (2010) conducted a survey on diversity of EPNs in the Arasbaran forests and rangelands, north–west Iran, during 2006–2008. From out of 691 soil samples from 62 localities, 21 samples (3 %) were positive for EPN. Seven isolates (four *Steinernema* and three *Heterorhabditis*) were recovered from rangelands and 14 isolates (eight *Steinernema* and six *Heterorhabditis*) from forest soil samples. The *Heterorhabditis* isolates were identified as *H. bacteriophora* and the *Steinernema* isolates as *S. carpocapsae*, *S. bicornutum*, *S. feltiae*, *S. glaseri*, *Steinernema kraussei* (Steiner) Travassos (Rhabditida: Steinernematidae) and three undescribed species referred to here as *Steinernema* spp.

In another survey, three native isolates of *Steinernema* and *Heterorhabditis* were isolated from soil orchards of walnut trees in Arak, Markazi province, Iran, which were identified as *H. bacteriophora*, *S. feltiae* and *S. carpocapsae* (Ashtari et al., 2011). Also, native isolate of *S. feltiae* and *H. bacteriophora* from Tabriz and its suburb soils were reported (Ebrahimi & Niknam, 2011). Nikdel, Niknam, and Eivazian Kary (2011) reported the first record of *S. kraussei* from soil samples collected from rangelands, near Chichakloo, Varzeghan, and East Azerbaijan. Furthermore, Nikdel, Niknam, and Ye (2011) collected a new species of EPN in 2007 from Arasbaran forests, the Kerengan village, and East Azerbaijan province. The new species is described as *Steinernema arasbaranense* Nikdel, Niknam & Ye (Rhabditida: Steinernematidae). This is the first record of a new EPN species from the Iran.

In another survey, of the 100 soil samples studied from Hamedan, 20 samples were positive for EPNs from *Steinernema*. They were identified as three species–groups "*carpocapsae*", "*feltiae*" and "*intermedium*" and most of the identified isolate were belonged to the "*feltiae*" species–group Saffari, Karimi, & Madadi (2012, unpublished). In addition, Roodaki, Haghani, and Abdollahi (2012, unpublished) and Roodaki, Haghani, Falahi, and Abdollahi (2012) found native isolates of EPNs from different districts of Kohgiluyeh and Boyer Ahmad province. Those EPN species were *S. feltiae* and *S. carpocapsae*.

Hassani-Kakhki, Karimi, and Hosseini (2013) collected native isolates of EPNs from Mashhad, Razavi Khorasan province characterized as *S. feltiae* and *H. bacteriophora*. In a regional survey, two native isolate of EPNs, *H. bacteriophora* and *S. feltiae* were isolated from soil in Bojnourd, North Khorasan Province (Kamali et al., 2013). Moreover, during postharvest season of potato tuber, native EPNs were collected from potato fields in Farooj, North Khorasan province, Iran and characterized as *H. bacteriophora* and *S. feltiae* (Rahatkhah et al., 2015). These results indicated that the country has a rich fauna biodiversity resource of EPNs and therefore, it could be an opportunity for discovery of new nematode species and populations from other provinces.

19.3 Isolates of Symbiotic Bacteria

In addition to native EPNs, their symbiotic bacteria were identified and characterized based on phenetic characters and 16S ribosomal RNA gene sequence. Characterization of symbiotic bacteria associated with EPN is a main step to study on different aspects of this complex whereas this part of EPN researches is still in its early stages in the country.

The first of such study was developed by Karimi, Kharazi-Pakdel and Yoshiga (2009) who isolated and identified the symbiotic bacteria associated with native isolates of *S. feltiae*, as *Xenorhabdus* spp. using 16S rRNA sequence and phenotypic characterization. DNA sequence had sharing high similarity related to *Xenorhabdus bovienii* (Akhurst) Akhurst & Boemare (Enterobacteriales: Enterobacteriaceae), the symbiont of *S. feltiae*. After that, Agazadeh et al. (2010) did an identification and molecular characterization of six isolates of *Xenorhabdus* and five bacterial isolates of *Photorhabdus* and identified them as *X. bovienii*, *Xenorhabdus nematophila* (Poinar & Thomas) Thomas & Poinar (Enterobacteriales: Enterobacteriaceae), and *Xenorhabdus budapestensis* Lengyel, Lang, Fodor, Szállás, Schumann & Stackebrandt (Enterobacteriales: Enterobacteriaceae), as well as *Photorhabdus luminescens* subsp. *laumondii* Fischer–Le Saux, Viallard, Vrunel, Normand & Boemare (Enterobacteriales: Enterobacteriaceae). Moreover, Karimi, Kharazi-Pakdel, Yoshiga, Koohi Habibi, and Hasani-Kakhki (2011) have determined the taxonomic affiliation of three bacterial isolates isolated from *S. glaseri* and *S. carpocapsae*, natural pathogens of the white grubs, *P. olivieri*. In this study, a polyphasic approach led to the identification of two species as *X. nematophila* and *Xenorhabdus poinarii* (Akhurst) Akhurst & Boemare (Enterobacteriales: Enterobacteriaceae).

In another survey, the subspecies of *P. luminescens* subsp. *laumondii*, were described from Iranian isolates of *H. bacteriophora* using 16S ribosomal RNA gene sequence and phenetic characters, which were collected from soil in Mashhad, Razavi Khorasan province, Iran (Karimi, Mokarm, & Hasani-Kakhki, 2012). Also, symbiotic bacteria of native *H. bacteriophora* from potato fields in Farooj, North Khorasan province, Iran and characterized as *Photorhabdus luminescens* based on main phenotypic and molecular characteristics (Rahatkhah et al., 2015). These results provide new insights for the biodiversity of bacto–helminthic complex in Iran.

19.4 Application of Entomopathogenic Nematodes for Pest Control

Several studies have been conducted on efficacy of EPNs as biocontrol agents for control of a number of important pests (Table 19.2). Current research on potential of EPNs is focused on the infectivity, pathogenicity and virulence of native and commercial species/populations of *Steinernema* spp. and/or *Heterorhabditis* spp. against different developmental stages of target pests under laboratory and field conditions. In general, these studies have shown promising results indicating that these nematodes could be developed for biological control and incorporated into integrated pest management programs for some economic important pests in the country.

19.4.1 Entomopathogenic Nematodes Application Against Vegetable and Greenhouse Pests

In 2000, Parvizi initiated research on possibility of biocontrol potential of *H. bacteriophora* and *S. carpocapsae* against Colorado potato beetle (CPB), *Leptinotarsa decemlineata* (Say) (Coleoptera: Chrysomelidae) which is the most economically damaging pest to potatoes in most areas of Iran. He found that at the concentration of 160 IJ cm^{-2}, these EPNs could cause a mean mortality of 84 and 90 % in CPB larvae, respectively. Subsequently, Eivazian, Rafiee, et al. (2010) investigated the biocontrol potential of four geographical isolate of *H. bacteriophora* and three species of *Steinernema* include *S. bicornutum*, *S. carpocapsae* and *S. feltiae* against *L. decemlineata* with three arenas: filter paper assay, leaf assay and soil assay. In filter paper assay and leaf assay, *H. bacteriophora* IRA10 had the highest virulence and *S. bicornutum* IRA7 the least, as a dose of 1,000 IJ per larva of *H. bacteriophora* IRA10 after 120 h of exposure time caused 60 % and 83 % mortality, whereas *S. bicornutum* IRA7 caused 9 % and 17 % mortality, respectively. In soil assay, similar results were found and *H. bacteriophora* IRA10 caused the highest mortality percentage (93 %) and *S. bicornutum* IRA7 (6 %) showed the lowest. Their results showed that the most effective EPN among the studied isolates might be *H. bacteriophora* IRA10.

Table 19.2 Target pests used for research on entomopathogenic nematodes in Iran

Order	*Family*	Scientific name	Commodity	Nematode Sp.	Region	Reference
Coleoptera	Curculionidae	*Conorrhynchus brevirostris*	Sugar Cane	Hb, S sp.	West Azerbaijan	Parvizi et al. (1988, unpublished)
	Curculionidae	*Curculio glandium*	oak trees	HbSb	East Azerbaijan, Tabriz	Nikdel et al. (2008)
	Curculionidae	*Hypera postica* (Gyllenhal)	alfalfa	Sc, Sf	Kohgiluyeh and Boyer Ahmad province	Falahi et al. (2011) Roodaki et al. (2011)
	Chrysomelidae	*Leptinotarsa decemlineata* (Say)	Potato	Hb, S sp.Sb, Sc, Sf	West and East Azerbaijan	Parvizi (2000, unpublished) Eivazian Kary, Rafiee, et al. (2010)Ebrahimi et al. (2011)
	Cerambyciae	*Osphranteria coerulescens* Redtenbacher	Apricot	Hb, Sc	Mashhad, Razavi Khorasan province	Sharifi, Karimi, Hosseini, and Rezapanah (2014)
	Melolonthidae	*Polyphylla olivieri* Cast.	Turf	Hb, S sp.Sc, Sg	West AzerbaijanTehran province	Parvizi (2001) Karimi and Kharazi-pakdel (2007)
	Melolonthidae	*Polyphylla adspersa*	Turf	Hb, Sc, Sf	Mashhad, Razavi Khorasan province	Karimi et al. (2010)
	Curculionidae	*Sitophilus oryzae* (L.)	Stored Products	Sf	Kohgiluyeh and Boyer Ahmad province	Roodaki et al. (2012)
	Tenebrionidae	*Tribolium confusum* (Duval)	Stored Products	Sf	Kohgiluyeh and Boyer Ahmad province	Roodaki, Haghani and Abdollahi (2012)
Dermaptera	Forficulidae	*Forficula auricularia L.*		Hb	Mashhad, Razavi Khorasan province	Kordestani, Karimi, Modarres Awal, Hasani Kakhaki, and Hoseini (2013)
Diptera	Tephritidae	*Dacus ciliatus* Loew	Cucurbit	Hb, Sc	Mashhad, Razavi Khorasan province	Kamali et al. (2013)
	Agromyzidae	*Liriomyza trifolii*	Vegetable Crops	Sf	East Azerbaijan, Tabriz	Ebrahimi et al. (2012)

(continued)

Table 19.2 (continued)

Order	*Family*	Scientific name	Commodity	Nematode Sp.	Region	Reference
Hemiptera	Aleyrodidae	*Trialeurodes vaporariorum*	sweet pepper cucumber	Hb, Sf	Mashhad, Razavi Khorasan province	Rezaei et al. (2014)
Hymenoptera	Argidae	*Arge ochropus*	Rose	Hb, Sc		Sheykhnejad, Ghadamyari, Ghasemi, Jamali, &and Karimi (2014)
Lepidoptera	Noctuidae	*Agrotis ipsilon*	Field Crops	Hb	West Azerbaijan	Parvizi (2004)
	Pyralidae	*Ephestia kuehniella*	Stored Products	Hb		
	Lymantriidae	*Euproctis chrysorrhoea*	Forest Trees	Hb isolate IRAZ5, Sc isolate IRAZ9	East Azerbaijan, Tabriz	Nikdel et al. (2010a)
	Galleridae	*Galleria mellonella*		Hb	West Azerbaijan	Saghaei et al. (2004)
	Noctuidae	*Helicoverpa armigera*	Cotton	Hb, SfSc	East and West Azerbaijan	Ebrahimi et al. (2012) Eivazian Kary et al. (2012)
	Noctuidae	*Heliothis viriplaca*	Pea	Hb, S sp.	West Azerbaijan	Parvizi et al. (1988)
	Pyralidae	*Ostrinia nubilalis*	Corn	Sf	Tehran province	Abootorabi (2011, 2012)
	Pieridae	*Pieris rapae*	Cabbage	Hb, S sp.	West Azerbaijan	Parvizi et al. (1988)
	Pyralidae	*Plodia interpunctella*	Stored Products	Sc	Kohgiluyeh and Boyer Ahmad province	Roodaki et al. (2012)
	Gelechiidae	*Phthorimaea operculella*	Potato	Sc, Sf, Sg Hb	Mashhad, Razavi Khorasan province	Hassani-Kakhki et al. (2013)

	Noctuidae	*Spodoptera exigua* (Hübner)	Sugar beet Cane	Hb, S sp.Sc, Sf	West AzerbaijanMashhad, Razavi Khorasan province	Parvizi et al. (1988)Aramideh, Safarali Zadeh, Purmirza, and Parvizi (2004) Karimi, Kharazi-Pakdel, and Yoshiga 2009)
	Sesiidae	*Synanthedon myopaeformis*	Apple	Hb, S sp.	West Azerbaijan	Parvizi (2001, 2003)
	Gelechiidae	*Tuta absoluta*	Tomato	SfHb, Sc	Tehran province, Mashhad, Razavi Khorasan province	Abootorabi (2014) Kamali et al. (2014)
	Cossidae	*Zeuzera pyrina*	Walnut	Hb, Sc	Arak province, Kerman province	Ashtari et al. (2011) Salari, Karimi, Sadeghi-Nameghi, and Hosseini (2014)
Thysanoptera	Thripidae	*Thrips tabaci*	Oniongreen bean	Hb, Sc, Sf, Sf isolate H1		Kashkouli et al. (2014) Saffari, Madadi, and Karimi (2013)

In another study, Ebrahimi, Niknam, and Lewis (2011) assessed the lethal and sub–lethal effects of two isolates of *S. feltiae* and *H. bacteriophora* against the prepupae of CPB in soil at two different temperatures. Their results revealed that both isolates were effective against *L. decemlineata* although *H. bacteriophora* was more effective at lower concentrations than *S. feltiae*. LC_{50} values of *H. bacteriophora* against progeny of field–collected adults and laboratory–reared adults were estimated as 8.5 and 7.6 IJ per prepupa, respectively. For *S. feltiae* the value was calculated as 51.2 IJ per prepupa against offspring of laboratory–reared adults of *L. decemlineata* only. Also, they reported that sub–lethal nematode concentrations had adverse effects on CPB adult fitness manifesting as wing and elytra deformation and delayed metamorphosis.

Ebrahimi, Niknam, & Askari Saryazdi (2012, unpublished) assessed biocontrol potential of *S. feltiae* against the vegetable serpentine leafminer, *Liriomyza trifolii* (Burgess) (Diptera: Agromyzidae), a serious and important pest of vegetable crops. Their result indicated 19–63 % mortality of the leafminer. Development and reproduction of the nematodes were also observed inside the insect cadavers. Hassani-Kakhki et al. (2013) studied the susceptibility of potato tuber moth (PTM), *Phthorimaea operculella* (Zeller) (Lepidoptera: Gelechiidae) to five EPN isolates including *S. carpocapsae*, *S. feltiae*, *S. glaseri* and *H. bacteriophora* (FUM7 and commercial isolates). The initial assessment (on filter paper) showed that *S. carpocapsae* and *H. bacteriophora* isolates caused the highest mortality on both larval and prepupal stages of PTM. The calculate LC_{50} for *H. bacteriophora* and *S. carpocapsae* were lowest values with 81 and 84 IJs per early PTM larvae, while LC_{50} *H. bacteriophora* FUM 7, *S. feltiae* and *S. glaseri* were 323, 392 and 427 IJs per early larvae, respectively. In general, prepupa was the most susceptible stage (more than 90 % mortality from *H. bacteriophora* and *S. carpocapsae*). In complementary test in soil columns, the results indicated that the larval mortality induced by *S. carpocapsae* (88 %) was higher than those caused either by the commercial population of *H. bacteriophora* (79 %) or by *H. bacteriophora* FUM 7 (78 %). The high pathogenicity of *S. carpocapsae* and *H. bacteriophora* against immature stages of PTM, suggest that both species have potential as biocontrol agents for management of *P. operculella*. Also, a study was conducted by Kamali et al. (2013) to determine the efficacy of *S. carpocapsae* and *H. bacteriophora* against the cucurbit fly, *Dacus ciliatus* Loew (Diptera: Tephritidae) in laboratory and greenhouse experiments. Larvae and adult flies were susceptible to nematode infection, but both nematode species induced low mortality of pupae with mortality percentages ranging from 9 % (*H. bacteriophora*) to 12 % (*S. carpocapsae*). *Steinernema carpocapsae* had a significantly lower LC_{50} value (28 IJs cm^{-2}) against larvae than *H. bacteriophora* (326 IJs cm^{-2}) in filter paper assays. Both species of EPNs were effective against adult flies but *S. carpocapsae* caused higher adult mortality (56 %) than did *H. bacteriophora* (45 %). In greenhouse experiments, when EPN species were applied to naturally infested fruit (150 and 300 IJs cm^{-2}), the mortality rates of *D. ciliates* larvae were 28 % for *S. carpocapsae* and 12 % for *H. bacteriophora*. Their findings provided the first insight into the biocontrol efficacy of *S. carpocapsae* against *D. ciliatus*. Moreover, pathogenicity and reproductive potential of *H. bacteriophora* on the European earwig, *Forficula auricularia* L. (Dermaptera: Forficulidae) was investigated under laboratory conditions and it was showed that common earwig could be suitable or alternative host for *H. bacteriophora* (Kordestani et al., 2013).

Saffari et al. (2013) evaluated the pathogenicity of a native isolate of *S. feltiae* (H1) and two exotic isolates, *H. bacteriophora* and *S. carpocapsae* against second instar larvae, prepupa and pupa of the onion thrips, *Thrips tabaci* Lindeman (Thysanoptera: Thripidae) under laboratory conditions. Their results indicated that prepupa was the most susceptible stage, while the second instar larvae showed the least susceptibility to EPNs. After 24 h of treatment, *H. bacteriophora* caused the highest mortality in prepupae (54 %) while its effect was significantly reduced on second instar larvae (25 %) at 10,000 nematode/mL. Similarly, *S. carpocapsae* was most effective against prepupae (49 %) and significantly ineffective against second instar larvae (20 %). In contrast, native isolate of *S. feltiae* (H1) had same effect on all developmental stages. Also, pathogenicity of commercial EPNs isolates, *S. feltiae*, *S. carpocapsae* and *H. bacteriophora*, as well as native isolate of *S. feltiae* were evaluated on second larvae, prepupae and pupae of *T. tabaci* under laboratory and semi field conditions by Kashkouli, Khajeali, & Poorjavad (2014). The results revealed that all EPNs have pathogenic effect on onion thrips. The native population of *S. feltiae* had more efficiency against the immature stages of *T. tabaci* and caused 92 and 81 % mortality at concentration of 1,000 IJs cm^{-2} against thrips prepupae and pupae, respectively. Moreover, prepupae and pupae of the thrips are more sensitive to the tested EPN species than second instar larvae.

Kamali, Karimi, and Kordestani (2014) assessed the pathogenecity of *S. carpocapsae* and *H. bacteriophora* against tomato leafminer *Tuta absoluta* (Meyrick) (Lepidoptera: Gelechiidae) under laboratory condition. Results showed that *H. bacteriophora* with the LC_{50} 1 IJs/cm^2 were more effective compare with *S. carpocapsae* with the LC_{50} 20 IJs/cm^2 and the last instar larvae was more susceptible compare with other stages.

Pathogenicity of *S. feltiae* and *H. bacteriophora* were studied against adult and second instar nymphs of greenhouse whitefly, *Trialeurodes vaporariorum* (Westwood) (Hemiptera: Aleyrodidae) in greenhouse on two different host plants, sweet pepper and cucumber, under laboratory and greenhouse conditions by Rezaei, Karimi, Hosseini, & Goldani (2014). Their results indicated that the highest mortality rate of second instar nymph instars of whitefly were obtained from the application of *S. feltiae* with the concentration of 250 IJs on cucumber (49 %). The lower mortality was reached from the application of *H. bacteriophora* with the concentration of 25 IJs on pepper (4 %). The application of *S. feltiae* against the second instar nymph on both host plants resulted significant higher mortality of the pest compared with *H. bacteriophora*. Both EPN species showed higher virulence on cucumber host plant than sweet pepper.

Recently, the susceptibility of larvae of the rose sawfly, *Arge ochropus* (Gmelin) (Hymenoptera: Argidae), an important pest of roses and wild rose bushes in northern Iran, to infection by *H. bacteriophora* and *S. carpocapsae* was reported in laboratory bioassays (Sheykhnejad, Ghadamyari, Koppenhöfer, & Karimi, 2014). The results indicated that *H. bacteriophora* (LC_{50}: 32 IJs/larva) had low infectivity to *A. ochropus* than *S. carpocapsae* (LC_{50}: 21 IJs/larva); however, their efficacy were not at a level that would make foliar applications of EPNs alone feasible.

19.4.2 Entomopathogenic Nematodes Application Against Fruit and Forest Tree Pests

White grubs, which is a serious pest of fruit and other trees in most parts of Iran, was one of the first test insect used for determining its susceptibility to EPNs (Parvizi, 2001). During a preliminary study, two EPNs species, *H. bacteriophora* and *Steinernema* sp. which were isolated from soils of West Azerbaijan, were conducted against *P. olivieri* and IJs with concentration 5×10^5 per m^2 caused a mean mortality of 34 %, and 46 % in third larval stage of this scarabaeid. Also, effectiveness of *Steinernema* sp. and *H. bacteriophora* was investigated on trunk borer red–belted clearwing moth, *Synanthedon myopaeformis* (Borkhausen) (Lepidoptera: Sesiidae), on apple trees (Parvizi, 2003).

Nikdel et al. (2008) evaluated the efficiency of two native EPNs, *H. bacteriophora* and *S. bicornutum* against the last instar larvae of acorn weevil, *Curculio glandium* Marsham (Coleoptera: Curculionidae) under laboratory condition. This Curculionid is an important forest pest of oak trees in Iran. Experiments were conducted at two temperature ranges (21–24 and 25–28 °C) and *H. bacteriophora* and *S. bicornutum* were applied at different concentrations. The penetration rate of *H. bacteriophora* (%1.6) was higher than *S. bicornutum* (0.55 %). Maximum mortality caused by *H. bacteriophora* and *S. bicornutum* were 58.3 %, and 25 % (at 21–24 °C) and 63.5 % and 30.5 % (at 25–28 °C), respectively. The results of this research indicated that the highest penetration in larvae and the highest mortality of fifth instar larvae of *C. glandium* was observed for *H. bacteriophora* under the both temperature ranges. Karimi, Rezapanah, Monfared, and Mirsaeidi (2010) examined biocontrol potential of a commercial formulation of *H. bacteriophora* (LARVANEM®) on second and third larval instars and pupae of the white grub *Polyphylla adspersa* Motschulsky (Coleoptera: Melolonthidae) which is one of the most important scarabaeid pest of trees in the north east of Iran. Their finding indicated that second instar larvae were significantly more susceptible than third instar to *H. bacteriophora*. The mean mortality in larval stages was about 42 %. Pupal stage of the white grub had high susceptibility to this entomopathogen.

In laboratory studies, the pathogenicity of *H. bacteriophora* isolate IRAZ5 and *S. carpocapsae* isolate IRAZ9 was assessed against third, fourth and fifth instar larvae of brown tail moth, *Euproctis chrysorrhoea* (L.) (Lepidoptera: Lymantriidae) (Nikdel, Niknam, & Dordaei, 2010). They observed that both nematode species were highly effective on last instar larvae of the pest. The greatest mortality of *E. chrysorrhoea*, by both nematodes, was achieved with last instar larvae at the rate of 5,000 IJs of the nematodes per ml suspension. Mean mortalities for *H. bacteriophora* and *S. carpocapsae* on the third, fourth and fifth larval stages of the pest at over all rates were 45, 44 and 51 and 42, 58 and 69, respectively. They concluded that *S. carpocapsae* caused significantly greater mortality of the fourth and fifth instar larvae of the insect compared with *H. bacteriophora*. In another work, it was observed how the larvae of the leopard moth, *Zeuzera pyrina* L. (Lepidoptera: Cossidae), the most destructive pest of walnut trees in Iran, were highly susceptible to infection by EPNs. Native isolate of *H. bacteriophora* and commercial products of *S. carpocapsae* and *H. bacteriophora* showed high pathogenicity against the larvae in laboratory bioassays and field applications at 2,000 IJ per larva and 2,000 IJ per active hole, respectively (Ashtari et al., 2011). In laboratory tests, *S. carpocapsae* caused 100 % mortality in 2nd, 3rd, and 4th instars larvae 54, 30 and 36 h after treatment, respectively. *H. bacteriophora* caused 100 % mortality in 2nd, 3rd, and 4th instars larvae 44, 40 and 52 h after treatment, respectively. In field experiment, mortality in treatments with *S. carpocapsae* under plastic cover and without cover was 100 % and 63 %, respectively.

Ghaffarpoor, Niknam, and Toorchi (2013) evaluated the efficacy of live and dead IJs of *H. bacteriophora* on egg hatching inhibition and second stage juveniles (J2) mortality of the root–knot nematode, *Meloidogyne javanica* (Treub) Chitwood (Tylenchida: Heteroderidae), under laboratory conditions. The data recorded after 48, 72, 96 and 120 h and indicated that both live and dead *H. bacteriophora* have potential for biocontrol of root–knot nematode with significant difference on J2 mortality and egg hatching inhibition at different concentrations and times post application. The highest mortality was achieved in 100 IJ per ml and after 120 h and the most inhibition of hatching in concentration of 100 IJ per ml and after 48 h. Concerning the mixture of 50 eggs and larvae, the results indicated that the effect of live juveniles of *H. bacteriophora* on second stage juveniles (J2) mortality have no significant difference among the various time treatments.

Salari, Karimi, Sadeghi-Nameghi, and Hosseini (2014) assessed the biological traits of *S. carpocapsae* and *H. bacteriophora* in a comprehensive study including the pathogenicity assay in plate and branch (branch assays included two separate experiments using the same procedures and different prepared experimental units as healthy fresh walnut branches by exposing healthy larvae into the healthy branches and infested branches with active hole), reproduction and penetration potential as well as foraging behavior of EPNs versus *Z. pyrina*. In plate and branch assays, the larvae were susceptible to both EPN species. Significantly higher mortality rates occurred in the larger larvae (97 and 53 %) after exposure to *S. carpocapsae* at two concentrations: 20 IJs/larva and 6 IJs/larva (6 IJs/larva calculated as the corresponded LC_{50} value for *S. carpocapsae*), respectively. Both EPN species successfully penetrated and reproduced in the *Z. pyrina* larvae. Also, the proportional response of *H. bacteriophora* to the host–associated cues was strongly higher than *S. carpocapsae* in petri dishes containing agar 1, 12 and 24 h after EPN application. These results highlight the efficiency of EPNs for the control of *Z. pyrina*. Overall, due to the cryptic habitat of larvae in their tree galleries which is close to the natural habitats of EPNs, field trails need to be conducted to further evaluate this potential.

Sharifi et al. (2014) evaluated the efficacy of *H. bacteriophora* and *S. carpocapsae*, against the larvae of the rosaceae longhorned beetle, *Osphranteria coerulescens* Redtenbacher (Coleoptera: Cerambycidae), the serious and economically important pest of fruit trees in cold regions of Iran. The plate assay showed that the larvae were susceptible to both EPN species but were more susceptible to *S. carpocapsae* (65–97 % mortality) than *H. bacteriophora* (42–88 %). The EPN species located and killed the larvae in branch experiments and were able to penetrate and reproduce within *O. coerulescens* larvae, with higher reproduction for *H. bacteriophora* than for *S. carpocapsae*. In a migration test, *H. bacteriophora* was strongly attracted to the sector of Petri dishes containing larvae. These findings highlighted the potential of EPNs as potential biocontrol agent of the larvae and warrant further field experiments to evaluate their efficacy under the wide environmental conditions in which rosaceae longhorned beetle larvae are found.

19.4.3 Entomopathogenic Nematodes Application Against Field Crop (Cereals) and Stored Product Pests

Initially, Parvizi et al. (1988) evaluated the infectivity of some EPNs against a few field crop pests such as *Heliothis viriplaca* (Hufnagel) (Lepidoptera: Noctuidae), *Pieris rapae* (L.) (Lepidoptera: Pieridae) and *Spodoptera exigua* (Hübner) (Lepidoptera: Noctuidae) under laboratory condition. After that, Parvizi (2004) reported the efficacy of *H. bacteriophora* on *A. ipsilon* under field and laboratory experiments. The results indicated that *H. bacteriophora* at the rates of 4×10^5 and 8×10^5 IJs/m^2 was able to parasitize 76.6 % and 75.2 % of the pest larvae, respectively. Aramideh, Safaralizadeh, Pourmirza, & Parvizi (2004) evaluated efficiency of native isolate of *S. carpocapsae* from apple orchards soil of West Azerbaijan against different larval, prepupal and pupal stages of beet armyworm, *S. exigua* under laboratory condition as well on sugar cane plant. In laboratory conditions, this EPN could cause a mean mortality of 80 % at 4×10^4 IJs/L in the pre–pupa. Based on this, the pre–pupa was highly susceptible to EPN and they recommended possible potential on using this pathogen against the larval and prepupal stages of this pest in the fields.

In another study, Ebrahimi, Niknam, Nikdel, and Hassanpour (2008) studied on efficiency of *H. bacteriophora* and *S. feltiae* at various concentrations against cotton bollworm, *Helicoverpa armigera* (Hubner) (Lepidoptera: Noctuidae) under laboratory conditions. Results indicated a mean mortality percentage of 83 % with *S. feltiae* and 67 % with *H. bacteriophora*. The high susceptibility of adults of alfalfa weevil, *Hypera postica* (Gyllenhal) (Coleoptera: Curculionidae) to infection by native isolates of *S. carpocapsa* (97 % mortality) and *S. feltiae* (90 % mortality) (Falahi, Adollahi, Roodaki, & Haghani, 2011; Roodaki, Haghani, Fallahi, & Abdollahi, 2011) were observed under laboratory conditions.

Furthermore, Eivazian Kary et al. (2012) examined the insecticidal effect of three species of EPNs, *H. bacteriophora*, *S. carpocapsae* and *S. feltiae* against

H. armigera using three methods of filter paper assay, food assay and soil assay under laboratory conditions. In all trials, *H. bacteriophora* IRA10 had the highest infectivity and *S. carpocapsae* IRA18 had the least. In food and soil assay, similar results were found and *H. bacteriophora* IRA10 showed higher infectivity against cotton bollworm compared to *S. feltiae* and *S. carpocapsae*. Roodaki et al. (2012) performed a study to determine the effectiveness of *S. carpocapsae* (the native isolated from Kohgiluyeh and Boyer Ahmad province) on larval and pupal stages of Indian moth, *Plodia interpunctella* (Hübner) (Lepidoptera: Pyralidae). Their result indicated that pupal stage was more sensitive than larval stage to lower concentration of EPNs. The highest mortality was recorded after 48 h which was 99 % pupal stage at the concentration 1,000 IJs, whereas it was 2,000 IJs for the larval stage which caused a total mortality of the tested insect. In this study, pupal stage was more sensitive than larval stage to lower concentration of EPNs.

These studies have shown promising results indicating that EPNs could be developed for further evaluation as potential agents for biological control and incorporated into integrated pest management programs of some economic important pests in Iran.

19.5 Ecological Characterization Studies

Ecological characterization of EPNs in the laboratory such as the effects of temperature, host plant or soil texture on some EPN isolates has been conducted. Due to climate change that have been accrued in Iran during these years, it is necessary to find adapted isolates to higher temperature and limited water. Temperature may have effect survival and pathogenicity of EPN (Chen, Li, Han, & Moens, 2003). In addition, soil texture may affect host finding behaviour of EPNs (Kaspi et al., 2010). Karimi and Kharazi-pakdel (2007) studied heat tolerance as well as life cycle and natality/mortality of the three Iranian EPNs (IRAN1 of *H. bacteriophora*, IRAN2 of *S. glaseri* and IRAN3 of *Steinernema* sp.) in *G. mellonella*, at a range of temperatures from 5 to 30 °C. Heat tolerance study showed that Iranian isolates of EPN were more tolerant than European ones. *H. bacteriophora* isolate was the most tolerant nematode at 32 °C, but no nematodes could survive at 36 °C after a 4–5 h exposure. All isolates developed and produced progeny between 10 and 25 °C. At 28 °C, mortality rate of *Galleria* larvae was 100 %, and no progeny was produced. Also, they reported that the highest IJs production was observed at 15 °C for all isolates. After that, in order to determine the thermal optimal range of EPNs, Ebrahimi and Niknam (2011) studied a native isolate of two common EPN species (*S. feltiae* and *H. bacteriophora*) collected from suburb soils of Tabriz. Results of this study showed that the preferred temperature for *S. feltiae* activity is lower than *H. bacteriophora* and the optimum thermal ranges should be considered as they are used for biological control of insect pests.

In another work, laboratory experiments were conducted to determine the effect of two host plants (green bean pods and onion leaves) on the mortality of onion thrips, *T. tabaci*, by *S. carpocapsae* and *H. bacteriophora* Kashkouli, Khajeali, & Poorjavad (2014). This study indicated that mortality from *H. bacteriophora* at higher concentration (1,000 IJs/cm^2) was 12 % for thrips larvae that had been fed on pods of green bean, while it was 30 % for thrips that had been fed on onion leaves. The rate of onion thrips mortality from *S. carpocapsae* at this concentration was 42 % for thrips that had been fed on pods of green bean, while it was higher for thrips that had been fed on onion leaves. These finding showed that the host plant of thrips could affect susceptibility of insect hosts to entomopathogenic nematodes.

Hassani-Kakhki et al. (2013) determined the effects of soil type (loamy, loamy–sandy and sandy) on virulence of three EPN isolates, *S. carpocapsae*, commercial and native isolate of *H. bacteriophora* against the second and fourth instar larvae as well as prepupa of potato tuber moth, *P. operculella*. Their results indicated that *S. carpocapsae* and both isolates of *H. bacteriophora* caused high mortality on larval and prepupal stages of *P. operculella* in all tested soil types. The mortality of the second larval stage was not significantly influenced by the effect of nematode isolates or soil types alone; Although higher mortality rate was observed in second larval stages of *P. operculella* in sandy soil type after exposure to S. *carpocapsae* (100 %), *H. bacteriophora* (98 %) and FUM 7 isolates (98 %) with 2,000 IJs (160 IJs/cm^2). The results also showed that *S. carpocapsae* and *H. bacteriophora* isolates have higher efficiency in lighter soils (sandy and sandy–loamy soils), therefore caused higher mortality than loamy soil. Kamali et al. (2013) investigated the influence of *H. bacteriophora* (native isolate) and *S. carpocapsae* (commercial isolate) against third instars larvae of cucurbit fly, *D. ciliatus* in sand, sandy loam and clay loam soils. Higher rates of larval mortality observed in sandy loam and sand (more than 60 %) than clay loam. Also the optimal temperature for infectivity of *S. carpocapsae* and *H. bacteriophora* were 30 °C and 25 °C, respectively. Recently, Kamali et al. (2014) assessed the effects of soil type (loamy–sandy, sandy–loamy and cocopeat), temperatures (19, 25 and 31 °C) and exposure time (65, 240 and 480 min) on the susceptibility of the last instar larvae of tomato leafminer, *T. absoluta*, to *S. carpocapsae* and *H. bacteriophora* under laboratory condition. Their results indicated that both EPNs have higher efficiency in loamy–sandy than sandy–loamy soil. The highest mortality rate was recorded at 480 min and the optimal temperature for nematode infection was 25 °C.

19.6 Compatibility of Entomopathogenic Nematodes with Other Insecticidal Agents

Another aspect of insect nematology studies of the country emphasized on the compatibility of nematodes with insecticidal agents. EPNs may have a good potential to control different pests but this agents effect slowly. Studies on the compatibility of nematodes with insecticidal agents could be a promising new window to increasing the efficiency of other insecticidal agents to lowering population density of pests (Koppenhöfer, Grewal, & Kaya, 2000; Morales-Rodriguez & Peck, 2009). Hence it is important to test various methods for application these agents to select the best way for application. Also, it is necessary to investigate the time intervals between applications of EPNs and other control methods in the field.

Recently, few basic studies have been initiated about this subject. The effect of *H. bacteriophora* (Isolate Iran 3) and native isolate of *Metarhizium anisopliae* (Metch) Sorokin (Ascomycota: Hypocreales) simultaneously was studied on second instar larvae of white grub, *P. adspersa* Karimi et al. (2010). Resulted data as mean corrected mortality and pathogenicity rate revealed that application of both pathogens had an additive effect and high compatibility with ecological niche of the pest habitat. Moreover, Sheykhnejad, Ghadamyari, Ghasemi, Jamali, & Karimi (2014) investigated the interactions between two rates of the insecticide imidacloprid (LC_{30} and LC_{50}) and four concentrations of two EPN species, *H. bacteriophora* and *S. carpocapsae* (LC_{25} to LC_{75}) as control agents of fifth instar larvae of rose sawfly, *A. ochropus* in the laboratory. The LD_{30} and LD_{50} values for imidacloprid 48 h after topical application were 5.6 and 9.6 ng, respectively. They found that interactions were generally stronger at the lower imidacloprid rate and were stronger for *S. carpocapsae* than for *H. bacteriophora*. In combinations with the higher imidacloprid rate, only one combination with *H. bacteriophora* (LC_{50}) and two combinations with *S. carpocapsae* (LC_{25}, LC_{30}) caused higher mortality than both respective single agent treatments. They suggested that synergistic imidacloprid/*S. carpocapsae* combinations could be a useful tool for the control of *A. ochropus* larvae that would simultaneously control other common pests susceptible to imidacloprid.

19.7 Immunological Aspect in Insect Nematodes Research

Another basic criterion which was considered in studies of insect pathogenic nematodes in the country was insect defence mechanisms against EPNs. Despite the efficiency of many EPN species in biological control, insects have evolved different kinds of defence mechanisms against them (Dunphy & Thurston, 1990; Feldhaar & Gross, 2008). Innate immune systems of insects play important roles in defence mechanisms against pathogens which were divided into cellular and humoral systems. Cellular responses involve haemocytes, which participated in phagocytosis, nodulation, and metazoan encapsulation (Schmidt, Theopold, & Strand, 2001). Therefore knowledge about cellular immune response against microorganisms is a trend in insect pathology that it may contribute to effectively selecting the best biocontrol agent against economic important pests.

Ebrahimi, Niknam, and Dunphy (2011) conducted the first study on the immune response of insect host against EPNs. They examined cellular encapsulation of two Iranian isolates of EPNs, *S. feltiae* and *H. bacteriophora*, against the prepupae of *L. decemlineata*. Encapsulation of nematodes in *L. decemlineata* was more frequently observed for *S. feltiae* than for *H. bacteriophora*. They stated that despite the frequent encapsulation of *S. feltiae* the number of cadavers producing offspring was the same for both nematode species. The results showed that *L. decemlineata* was a more responsive host than *G. mellonella* and the haemocyte responses occurred sooner and more extensively. In *G. mellonella* there were no encapsulation or melanisation responses against *S. feltiae*, whereas *H. bacteriophora* showed encapsulated and melanised (17 %), the encapsulation level being lower than in *L. decemlineata*. Weak hemocyte reaction to nematodes in *G. mellonella*, were reported previously (Dunphy & Thurston; Milstead, 1979). These results indicated that cellular encapsulation of nematodes in CPB at the dosage used is not an effective defensive mechanism, and despite substantial encapsulation of the nematodes, insect mortality occurred for *S. feltiae* and *H. bacteriophora* 72 h after injection.

In another survey, phenoloxidase (PO) and protease activity of *G. mellonella* haemolymph was determined against *S. carpocapsae* Ebrahimi & Niknam (2012). Maximum activity of PO in nematode–injected and phosphate buffer–injected insects occurred at 1 h and 4 h post injection, respectively. Maximum protease activity occurred 8 h post injection in both experiments, while in nematode–injected insects it was found 6.2 fold greater than buffer–injected insects. In overall, both enzyme activities in nematode–injected insects were higher than buffer injected insects, in all time intervals.

In order to determine the possible role of haemocytes in cellular defences, Alvandi, Karimi, and Dunphy (2014) investigated cytology of the white grub, *P. adspersa* and its cellular reactions against *H. bacteriophora* and *S. glaseri*. Six haemocyte types in the haemolymph of second instar larvae of white grub were identified as prohemocytes, granulocytes, plasmatocytes, oenocytoids, coagulocytes and spherulocytes. The granulocytes were the dominant haemocyte type followed by the plasmatocytes and both haemocyte types encapsulate EPNs. The maximum total haemocyte counts (THC) of the white grub larvae when challenged with *S. glaseri* occurred at 12 h post–injection. The cell reactions of the grubs against *H. bacteriophora* in terms of THC and differential haemocyte counts and the encapsulation rate started earlier and were more pronounced than those against *S. glaseri*. EPN–triggered encapsulation in *P. adspersa* larvae was more extensive than in *G. mellonella* larvae. Overall, their results indicated that the cellular immune system of *P. adspersa* to be weak for dealing with the EPN *S. glaseri* and *H. bacteriophora*. The weak reaction may also be related to variability in the insect species.

In another work regarding immunology of fifth instar larvae of beet armyworm *S. exigua* against EPNs, it was indicated high cellular responses of *S. exigua* larvae against *H. bacteriophora*, while these reactions were weaken for *S. carpocapsae* Darsouei, Karimi & Rahatkhah (2014). Ebrahimi, Niknam, Dunphy, and Toorchi (2014), and Ebrahimi, Niknam, Toorchi, and Dunphy (2014) investigated lethal and sub–lethal effects of *S. carpocapsae* on CPB, surviving adults and PO activity in haemolymph of nematode–injected last instar larvae. Sub–lethal effects of *S. carpocapsae* on surviving adult CPB include discoloration and thinning of the cuticle in colour defective adults, anatomical deformation in wings, antenna and legs, and decreasing fertilized eggs production in females when infected with low concentration of the nematodes in the prepupal stage. Discoloration of the infected adults was the most prominent sub–lethal effect. Increasing nematode concentration increased PO activity.

Sheykhnejad et al. (2014) assessed cellular reactions of larvae of rose sawfly, *A. ochropus* against *S. carpocapsae* and *H. bacteriophora*. Their results showed strong immune responses of the rose sawfly larvae against *H. bacteriophora*, while these reactions were weaken for *S. carpocapsae*. It was showed that the rate of encapsulation and melanization of EPN depend on the host and EPN species. Higher encapsulation rate and melanization was observed in the sawfly larvae treated with *H. bacteriophora* and melanization ability of *H. bacteriophora* by *A. ochropus* increased over time of injection.

Recently, immune reactions of *Agriotes lineatus* (L.) (Coleoptera: Elateridae) and *G. mellonella* larvae (as susceptible host) against native isolates of EPN species, *H. bacteriophora* and *S. feltiae* is being investigated (Rahatkhah, 2015). Their study provided the first insight into survey on immune system of wireworms against EPNs in Iran. Encapsulation efficacy was significantly different against two EPN species; *S. feltiae* was almost unrecognized by host haemocytes (6 % of encapsulated parasites), instead, assays with *H. bacteriophora* showed 24 % of encapsulated nematodes. The higher PO activity was detected at 8 and 12 h post injection in response to *H. bacteriophora*. Both *H. bacteriophora* and *S. feltiae* were melanized in the *A. lineatus* haemocoel cavity, though with different percentages, 23 % and 5 %, respectively.

19.8 Conclusions and Future Directions

In Iran, research about EPNs has started since 1990. A number of new species/populations of EPNs have been isolated and some of these species are being evaluated against various soil and cryptic–habitat pests. There is no any formal record about application of EPNs by users in the country. All information are restricted to those resulted from research.

However, it is important to note that native biocontrol agents are often preferable in biological control programs, since they are adapted to local conditions. Further, due to the availability of a range of undisturbed habitats with high diversity of insect species in the country, it is expected that the diversity of EPN will be more than those reported. Therefore, opportunity exists for the discovery of novel EPN species and isolates with higher tolerance to stressing environmental conditions.

In addition, most studies are restricted to laboratory condition and few to field survey application. Due to the fact that EPNs effectiveness and persistence may vary in the field, it is necessary to study the biology, ecology and infectivity of these agents in the field conditions. Also, in order to select the best application method, it is indispensable to assessment various methods under the wide environmental conditions in selected regions to evaluate the efficiency of these biocontrol agents.

Due to the biological control potential of EPNs, it is useful that scientists in this area provide training for other researchers and students through workshops. Awareness amongst farmers and users must be created on the safety of EPNs, their usage, advantages and limitations. In addition to scientific studies, regulatory strategies of the government should also aim at supporting the easy introduction of EPNs based products as a part of control performance. The industry requirements for future research includes fundamental research on the characterisation of available EPNs isolates, screening for virulent isolates and mass production, formulation and application techniques. Recently, a research on mass production of EPN has started in Tehran University. This research would provide initial and preliminary information toward applied aspects of EPNs in biocontrol programs.

References

Abootorabi, E. (2011, August 27–28). Evaluation of susceptibility of *Ectomyelois ceratoniae* to entomopathogenic nematode (p. 146). *Proceedings of the National Conference of Pomegranate.* Mashhad.

Abootorabi, E. (2012, March 8). Biological control of insect pests by entomopathogenic nematodes (EPNs). *Proceedings of First National Conference on Sustainable Agricultural Development and Healthy Environment*. Hamadan: Islamic Azad University of Hamadan.

Abootorabi, E. (2014). Report of native isolate pathogenicity of *Steinernema feltiae* on tomato leafminer, *Tuta absoluta. Biocontrol in Plant Protection, 1*(2), 107–109.

Agazadeh, M., Mohammadi, D., & Eivazian Kary, N. (2010). Molecular identification of Iranian isolates of the genus *Photorhabdus* and *Xenorhabdus* (Enterobacteriaceae) based on 16S rRNA. *Munis Entomology and Zoology, 5*(2), 772–779.

Alvandi, J., Karimi, J., & Dunphy, G. B. (2014). Cellular reactions of the white grub larvae, *Polyphylla adspersa*, against entomopathogenic nematodes. *Nematology, 16*(4), 1–12.

Amiri, M. J., & Eslami, S. S. (2010). Investigation of climate change in Iran. *Journal of Environmental Science and Technology, 4*, 208–216.

Ashtari, M., Karimi, J., Rezapanah, M. R., & Hassani-kakhki, M. (2011). Biocontrol of leopard moth, *Zeuzera pyrina* L. (Lep.: Cossidae) using entomopathogenic nematodes in Iran. *IOBC/wprs Bulletin, 66*, 333–335.

Aramideh, S. H., Safarali Zadeh, M. A., Purmirza, A. A., & Parvizi, R. (2004). Study on sensitivity of larvae, pupa and pre–pupa stages of *Spodoptera exigua* H. to *Steinernema carpocapsae* under laboratory condition and on sugar cane plant. *Journal of Agricultural Science and Natural Sources, 12*(5), 159–166.

Armer, C. A., Rao, S., & Berry, R. E. (2004). Insect cellular and chemical limitations to pathogen development: The Colorado potato beetle, the nematode *Heterorhabditis marelatus*, and its symbiotic bacteria. *Journal of Invertebrate Pathology, 87*, 114–122.

Bughio, F. M., & Wilkins, R. M. (2004). Influence of Malathion resistance status on survival and growth of *Tribolium castaneum* (Coleoptera: Tenebrionidae), when fed on flour from insect resistant and susceptible grain rice cultivars. *Journal of Stored Products Research, 40*, 65–75.

Burnell, A. M., & Stock, S. P. (2000). *Heterorhabditis*, *Steinernema* and their bacterial symbionts—lethal pathogens of insects. *Nematology, 2*, 31–42.

Chen, S., Li, J., Han, X., & Moens, M. (2003). Effect of temperature on the pathogenicity of entomopathogenic nematodes (*Steinernema* and *Heterorhabditis* spp.) to *Delia radicum. Biological Control, 48*, 713–724.

Dunphy, G. B., & Thurston, S. (1990). Insect immunity. In R. Gaugler & H. K. Kaya (Eds.), *Entomopathogenic nematodes in biological control* (pp. 301–323). Boca Raton, FL: CRC Press.

Ebrahimi, L., & Niknam, G. H. (2011). Detection of thermal preference range of two endemic isolates of entomopathogenic nematodes, *Steinernema feltiae* (Steinernematidae, Tylenchina) and *Heterorhabditis bacteriophora* (Heterorhabditidae, Rhabditina) for application in biological control of insect pests. *Journal of Sustainable Agriculture and Production Science, 21*(2), 77–86.

Ebrahimi, L., Niknam, G. H., & Dunphy, G. B. (2011). Hemocyte responses of *Leptinotarsa decemlineata* and *Galleria mellonella* to the entomopathogenic nematodes, *Steinernema feltiae* and *Heterorhabditis bacteriophora. Journal of Insect Science, 11*, 76 available online: insectscience.org/11.76.

Ebrahimi, L., Niknam, G., Dunphy, G. B., & Toorchi, M. (2014). Side effects of immune response of Colorado potato beetle, *Leptinotarsa decemlineata* against the entomopathogenic nematode, *Steinernema carpocapsae* infection. *Invertebrate Survival Journal, 11*, 132–142.

Ebrahimi, L., Niknam, G. H., & Lewis, E. E. (2011). Lethal and sublethal effects of Iranian isolates of *Steinernema feltiae* and *Heterorhabditis bacteriophora* on the Colorado potato beetle, *Leptinotarsa decemlineata. BioControl, 56*, 781–788.

Eimanifar, A., & Mohebbi, F. (2007). Review: Urmia Lake (Northwest Iran): A brief review. *Saline Systems, 3*:5. http://www.salinesystems.org/content/3/1/5

Eivazian Kary, N., Golizadeh, A., Rafiee, H. D., Mohammadi, D., Afghahi, S., Omrani, M., et al. (2012). A laboratory study of susceptibility of *Helicoverpa armigera* (Hübner) to three species of entomopathogenic nematodes. *Munis Entomology and Zoology, 7*(1), 372–379.

Eivazian Kary, N., Niknam, G. H., Griffin, C. T., Mohammadi, S. A., & Moghaddam, S. A. (2009). A survey of entomopathogenic nematodes of the families Steinernematidae and Heterorhabditidae (Nematoda: Rhabditida) from north–west of Iran. *Nematology, 11*(1), 107–116.

Eivazian Kary, N., Niknam, G. H., Mohammadi, S. A., Moghaddam, M., & Nikdel, M. (2010). Morphology and molecular study of an entomopathogenic nematode, *Steinernema bicornutum* (Nematoda, Rhabditida, Steinernematidae) from Iran. *Journal of Entomological Society of Iran, 29*(2), 25–34.

Eivazian Kary, N., Rafiee, H. D., Mohammadi, D., & Afghahi, S. (2010). Efficacy of some geographical isolates of entomopathogenic nematodes against *Leptinotarsa decemlineata* (Say) (Col.: Chrysomelidae). *Munis Entomology and Zoology, 5*, 1066–1074.

Falahi, M., Abdollahi, M., Roodaki, M., & Haghani, M. (2011, September 10–12). Efficacy of *Steinernema carpocapsae* for control of the adults of alfalfa weevil, *Hypera postica* (pp. 518–521).*Proceedings of National Conference on Modern Agricultural Sciences and Technologies (MAST)*. Zanjan.

Feldhaar, H., & Gross, R. (2008). Immune reactions of insects on bacterial pathogens and mutualists. *Microbes and Infection, 10*, 1082–1088.

Franklin, J. (2009). *Mapping species distributions: spatial inference and prediction*. New York: Cambridge University Press.

Ghaffarpoor, T., Niknam, G., & Toorchi, M. (2013). Efficacy of entomopathogenic Nematode *Heterorhabditis bacteriophora* as biological control of root knot nematode *Meloidogyne javanica* under laboratory conditions. *Journal of Applied Researches in Plant Protection, 2*(1), 101–114.

Grewal, P. S., Ehlers, R.-U., & Shapiro-Ilan, D. I. (2005). *Nematodes as biocontrol agents* (p. 528). Wallingford, UK: CABI Publishing.

Grichanov, I. Y., & Ovsyannikova, E. I. (2009). *Interactive agricultural ecological atlas of Russia and neighboring countries*. Available online: http://www.agroatlas.ru/

Hassani-Kakhki, M., Karimi, J., & Hosseini, M. (2013). Efficacy of entomopathogenic nematodes against potato tuber moth, *Phthorimaea operculella* (Lepidoptera: Gelechiidae) under laboratory conditions. *Biocontrol Science and Technology, 23*(2), 146–159.

Kamali, S. H., Karimi, J., Hosseini, M., Campos-Herrera, R., & Duncan, L. W. (2013). Biocontrol potential of the entomopathogenic nematodes *Heterorhabditis bacteriophora* and *Steinernema carpocapsae* on cucurbit fly, *Dacus ciliatus* (Diptera: Tephritidae). *Biocontrol Science and Technology, 23*(11), 1307–1323.

Kamali, S. H., Karimi, J., & Kordestani, M. (2014, August 23–26). Influence of abiotic factors on pathogenicity of entomopathogenic nematodes, *H. bacteriophora* and *S. carpocapsae* against tomato leafminer *Tuta absoluta* (Meyrick) (Lepidoptera: Gelechiidae) (p. 638). In *Proceedings of the 21th Iranian Plant Protection Congress*. Urmia: Urmia University.

Kaspi, R., Ross, A., Hodson, A. K., Stevens, G. N., Kaya, H. K., & Lewis, E. E. (2010). Foraging efficacy of the entomopathogenic nematode *Steinernema riobrave* in different soil types from California citrus groves. *Applied Soil Ecology, 45*, 243–253.

Karimi, J., & Kharazi-pakdel, A. (2007). Incidence of natural infection of the white grub *Polyphylla olivieri* (Coleoptera: Scarabaeidae) with entomopathogenic nematodes in Iran. *OILB/srop Bulletin, 30*(1), 35–39.

Karimi, J., Kharazi-Pakdel, A., & Robert, S. J. (2003). Report of pathogenic nematode of slugs, *Phasmarhabditis hermaphrodita* (Nematoda: Rhabditida) in Iran. *Journal of Entomological Society of Iran, 22*(2), 77–78.

Karimi, J., Kharazi-pakdel, A., Yoshiga, T., & Koohi-habibi, M. (2009). First Report of *Steinernema glaseri* (Rhabditida: Steinernematidae) from Iran. *Russian Journal of Nematology, 17*, 83–85.

Karimi, J., Kharazi-Pakdel, A., & Yoshiga, T. (2009). Insect pathogenic nematode, *Steinernema feltiae* from Iran. *IOBC/wprs Bulletin, 45*, 409–412.

Karimi, J., Kharazi-Pakdel, A., Yoshiga, T., Koohi Habibi, M., & Hasani-Kakhki, M. (2011). Characterization of *Xenorhabdus* (ɣ–Proteobacteria) strains associated bacteria with the *Steinernema* (Nematoda: Steinernematidae) isolates from Iran. *Journal of Entomological Society of Iran, 31*(1), 1–13.

Karimi, J., Kharazi-Pakdel, A., Yoshiga, T., & Koohi-Habibi, M. (2010). Introduction of *Steinernema carpocapsae* Weiser, 1955 (Rhabditida: Steinernematidae) From Natural Population of White Grub, *Polyphylla Olivieri* (Coleoptera: Melolonthidae) From Iran. *Journal of Agricultural Faculty of Uludag University, 1*, 47–54.

Karimi, J., Mokarm, A. H., & Hasani-Kakhki, M. (2012). A study of a native isolate of an entomopathogenic nematode, *Heterorhabditis bacteriophora* and its symbiont, Photorhabdus luminecens subsp. laumondii. *Iranian Journal of Plant Protection Science, 43*(2), 353–363.

Kashkouli, M., Khajeali, J., & Poorjavad, N. (2014, August 23–26). Effect of entomopathogenic nematodes for controlling the onion thrips, *Thrips tabaci* (Thys.: Thripidae) under semi-field condition (p. 457). In *Proceedings of the 21th Iranian Plant Protection Congress*. Urmia: Urmia University.

Koppenhöfer, A. M. (2007). Nematodes. In L. A. Lacey & H. K. Kaya (Eds.), *Field manual of techniques in invertebrate pathology application and evaluation of pathogens for control of insects and other invertebrate pests* (pp. 249–266). New York City: Springer.

Koppenhöfer, A. M., Grewal, P., & Kaya, H. K. (2000). Synergism of imidacloprid and entomopathogenic nematodes against white grubs: The mechanism. *Entomologia Experimentalis et Applicata, 94*, 283–293.

Kordestani, M., Karimi, J., Modarres Awal, M., Hasani Kakhaki, M., & Hoseini, M. (2013). Pathogenecity of entomopathogenic nematode, *Heterorhabditis bacteriophora* Poinar, 1975 on the indigenous population of common earwig, *Forficula auricularia* L. *Biological Control of Pests and Plant Disease, 2*, 35–41.

Lacey, L. A., & Shapiro-Ilan, D. I. (2008). Microbial control of insect pests in temperate orchard systems: Potential for incorporation into IPM. *Annual Review of Entomology, 53*, 121–144.

Meehl, G. A. (2007). Global climate projections. In S. Solomon, D. Qin, M. Manning, Z. Chen, M. Marquis, K. B. Averyt, M. Tignor, & H. L. Miller (Eds.), *Climate change 2007: The physical science basis. Contribution of Working Group I to the Fourth Assessment Report of the Intergovernmental Panel on Climate Change*. Cambridge, U.K/New York: Cambridge University Press.

Merrill, R., Gutieŕrez, D., Lewis, O., Gutieŕrez, J., Diez, S., & Wilson, R. (2008). Combined effects of climate and biotic interactions on the elevational range of a phytophagous insect. *Journal of Animal Ecology, 77*, 145–155.

Milstead, J. E. (1979). Pathophysiological influences of *Heterorhabditis bacteriophora* complex on seventh–instar larvae of the greater wax moth, *Galleria mellonella*: Changes in the hemolymph refractive index. *Journal of Invertebrate Pathology, 33*, 274–277.

Morales-Rodriguez, A., & Peck, D. C. (2009). Synergies between biological and neonicotinoid insecticides for the curative control of the white grubs *Amphimallon majale* and *Popillia japonica. Biological Control, 51*, 169–180.

Nikdel, M., Niknam, G. H., & Dordaei, A. A. (2010). Evaluation of susceptibility of the brown tail moth, *Euproctis chrysorrhoea* (L.) to entomopathogenic nematode under laboratory condition. *Nematologia Meditaranea, 38*, 3–6.

Nikdel, M., Niknam, G. H., & Eivazian Kary, N. (2011). First report of an entomopathogenic nematode, *Steinernema kraussei* (Rhabditida, Steinernematidae) from Iran. *Journal of Entomological Society of Iran, 30*(2), 39–51.

Nikdel, M., Niknam, G. H., Griffin, C. T., & Eivazian Kary, N. E. (2010). Diversity of entomopathogenic nematodes (Nematoda: Steinernematidae, Heterorhabditidae) from Arasbaran forests and rangelands in north–west Iran. *Nematology, 12*, 767–773.

Nikdel, M., Niknam, G., Shojaee, M., Askary, H., & Mohammadi, S. A. (2008). A survey on the response of the last instar larvae of acorn weevil, *Curculio glandium* (Col.: Curculionidae), to entomopathogenic nematodes *Steinernema bicornutum* and *Heterorhabditis bacteriophora* in the laboratory. *Journal of Entomological Society of Iran, 28*, 45–60.

Nikdel, M., Niknam, G. H., & Ye, W. (2011). *Steinernema arasbaranense* sp. n. (Nematoda: Steinernematidae), a new entomopathogenic nematode from Arasbaran forests, Iran. *Nematologia Mediterranea, 39*, 17–28.

Parvizi, R. (2001). Survey on pathogenicity of entomopathogenic nematodes, *Steinernema sp.* and *Heterorhabditis bacteriophora* infesting larval *Polyphylla olivieri. Journal of Entomological Society of Iran, 21*, 63–72.

Parvizi, R. (2003). Efficacy of insect pathogenic nematodes *Steinernema* sp on control of trunk borer butterfly larvae in apple trees *Synanthedon myopaeformis. Iranian Journal of Agricultural Science, 34*(2), 303–311.

Parvizi, R. (2004). Investigation on efficacy of entomopathogenic nematodes, *Heterorhabditis bacteriophora* against black cutworm *Agrotis ipsilon. Journal of Agricultural Science, 14*, 35–42.

Poinar, G. O. J. (1990). Biology and taxonomy of Steinernematidae and Heterorhabditidae. In R. Gaugler & H. K. Kaya (Eds.), *Entomopathogenic nematodes in biological control*(pp. 23–62). Boca Raton, FL: CRC Press.

Rahatkhah, Z., Karimi, J., Ghadamyari, M., & Brivio, M. F. (2015). Immune defenses of Agriotes lineatus larvae against entomopathogenic nematodes. *BioControl.*

Rezaei, N., Karimi, J., Hosseini, M., & Goldani, M. (2014, August 23–26). Pathogenecity of entomopathogenic nematode, *Steinernema feltiae* and *Heterorhabditis bacteriophora* against greenhouse whitefly, *Trialeurodes vaporariorum* in laboratory and greenhouse conditions (p. 530). In *Proceedings of the 21th Iranian Plant Protection Congress*. Urmia: Urmia University.

Roodaki, M., Haghani, M., Falahi, M., & Abdollahi, M. (2011, September 10–12). The response of the alfalfa weevil, *Hypera postica*, adults to *Steinernema feltiae* in vitro (pp. 514–517). *Proceedings of National Conference on Modern Agricultural Sciences and Technologies (MAST)*. Zanjan.

Roodaki, M., Haghani, M., & Abdollahi, M. (2012). Isolation of *Steinernema feltiae* and its pathogenicity on rice weevil, *Sitophilus oryzae* (Col.: Curculionidae). *Journal of Entomological Research, 5*(2), 83–93.

Roodaki, M., Haghani, M., Falahi, M., & Abdollahi, M. (2012). The effect of *Steinernema carpocapsae* (Weiser) on mortality of the last instar larvae and pupae of *Plodia interpunctella* (Huebner) under laboratory conditions. *Plant Pest Research, 2*(2), 33–40.

Saghaei, N., Pourmirza, A. A., Sadeghi Nasab, F., Hashemi Khabir, Z., Javie, E., & Hossein Zadeh, A. (2004, August 29–September 2). Studies on the susceptibility of *Galleria mellonella* L. to *Heterorhabditis bacteriophora* P. under laboratory conditions. *16th Iranian Plant Protection Congress*. Tabriz: Tabriz University, 60 P.

Salari, E., Karimi, J., Sadeghi-Nameghi, H., & Hosseini, M. (2014). Efficacy of two entomopathogenic nematodes *Heterorhabditis bacteriophora* and *Steinernema carpocapsae* for control of the leopard moth borer *Zeuzera pyrina* (Lepidoptera: Cossidae) larvae under laboratory conditions. *Biocontrol Science and Technology, 25*(3), 260–275.

Saffari, T., Karimi, J., & Madadi, H. (2012, February 12). Characterization some entomopathogenic nematodes from Hamedan using ITS and 28S gene sequences (pp. 199–200). *Proceedings of First Conference on DNA Barcoding in the Taxonomy*. Mashhad.

Saffari, T., Madadi, H., & Karimi, J. (2013). Pathogenicity of three entomopathogenic nematodes against the onion thrips, *Thrips tabaci* Lind. (Thys.; Thripidae). *Archives of Phytopathology and Plant Protection, 46*(20), 2459–2468.

Schmidt, O., Theopold, U., & Strand, M. R. (2001). Innate immunity and evasion by insect parasitoids. *BioEssays, 23*, 344–351.

Sharifi, S. H., Karimi, J., Hosseini, M., & Rezapanah, M. R. (2014). Efficacy of two entomopathogenic nematode species as potential biocontrol agents against the rosaceae longhorned beetle, *Osphranteria coerulescens*, under laboratory conditions. *Nematology, 16*, 729–737.

Sheykhnejad, H., Ghadamyari, M., Ghasemi, V., Jamali, S., & Karimi, J. (2014). Hemocytes immunity of rose sawfly, *Arge ochropus* (Hym.: Argidae) against entomopathogenic nematodes, *Steinernema carpocapsae* and *Heterorhabditis bacteriophora. Journal of Asia-Pacific Entomology, 17*, 879–883.

Sheykhnejad, H., Ghadamyari, M., Koppenhöfer, A. M., & Karimi, J. (2014). Interactions between entomopathogenic nematodes and imidacloprid for rose sawfly control. *Biocontrol Science and Technology, 24*(12), 1481–1486.

Solhjouy-Fard, S., Sarafrazi, A., Moeini, M. M., & Ahadiyat, A. (2013). Predicting habitat distribution of five heteropteran pest species in Iran. *Journal of Insect Science, 13*,116. Available online: www.insectscience.org/13.116

Thomson, L. J., Macfadyen, S., & Hoffmann, A. A. (2010). Predicting the effects of climate change on natural enemies of agricultural pests. *Biological Control, 52*, 296–306.

Torchin, M. E., Lafferty, K. D., Dobson, A. P., McKenzie, V. J., & Kuris, A. M. (2003). Introduced species and their missing parasites. *Nature, 421*, 628–629.

Ulrichs, C., & Hopper, K. R. (2008). Predicting insect distributions from climate and habitat data. *BioControl, 53*, 881–894.

Wang, Y., Gaugler, R., & Cui, L. (1994). Variations in immune response of *Popillia japonica* and *Acheta domesticus* to *Heterorhabditis bacteriophora* and *Steinernema* species. *Journal of Nematology, 26*, 11–18.

Woiwod, I. (1997). Detecting the effects of climate change on Lepidoptera. *Journal of Insect Conservation, 1*, 149–158.

Chapter 20
Entomopathogenic Nematode Exploitation: Case Studies in Laboratory and Field Applications from South Africa

Antoinette P. Malan and Justin L. Hatting

20.1 Introduction

The first record of an entomopathogenic nematode (EPN) (belonging to the order Rhabditidae, and to the families Steinernematidae and Heterorhabditidae) in South Africa was that of Harington (1953), who reported nematodes from the larval, pupal, and adult stages of the black maize beetle, *Heteronychus arator* Fabricius (Coleoptera: Scarabaeoidea), which were collected from a maize field near Grahamstown in the Eastern Cape province. After an elapse of 35 years, the first attempt was made to use EPN for the control of the sugarcane stalk borer, *Eldana saccharina* Walker (Lepidoptera: Pyralidae), and three local EPN isolates were evaluated in laboratory and field trials by the South African Sugarcane Research Institute (SASRI) in KwaZulu-Natal (Spaull, 1988, 1990, 1991). From 1993 to 1994, soil samples were collected from deciduous fruit orchards in the Western Cape province. *Heterorhabditis* were then isolated from the soil samples, and used for the control of the banded fruit weevil, *Phlyctinus callosus* (Schönerr) (Coleoptera: Curculionidae) (Basson, 1993). The specimens were sent to France, where they were the first to be identified as *Heterorhabditis bacteriophora* Poinar (Rhabditida: Heterorhabditidae), using species–specific satellite DNA as diagnostic probes (Grenier, Bonifassi, Abad, & Laumond, 1996; Grenier, Laumond, & Abad, 1996). Ten years later, the first new species to be described for South Africa was *Steinernema khoisanae* Nguyen,

A.P. Malan (✉)
Department of Conservation Ecology and Entomology, Faculty of AgriSciences, Stellenbosch University, Stellenbosch, South Africa
e-mail: apm@sun.ac.za

J.L. Hatting
South African Agricultural Research Council (ARC), Small Grain Institute, Bethlehem, South Africa
e-mail: hattingJ@arc.agric.za

© Springer International Publishing Switzerland 2015
R. Campos-Herrera (ed.), *Nematode Pathogenesis of Insects and Other Pests*, Sustainability in Plant and Crop Protection 1, DOI 10.1007/978-3-319-18266-7_20

Table 20.1 Target pests research using entomopathogenic nematodes in South Africa

Order	Family	Scientific name	Common name	Nematode	Reference
Coleoptera	Curculionidae	*Phlyctinus callosus*	Banded fruit weevil	Hz	Ferreira, Addison and Malan (2014)
Diptera	Tephritidae	*Ceratitis capitata*; *C. rosa*	Fruit fly	Hb, Hz, Sk	Malan and Manrakhan (2009)
Hemiptera	Pseudococcidae	*Planococcus ficus*	Vine mealybug	Sy	Le Vieux and Malan (2013a, 2013b)
Hemiptera	Pseudococcidae	*Planococcus citri*	Citrus mealybug	Sy	Van Niekerk and Malan (2012, 2013, 2014a, 2014b)
Hemiptera	Pseudococcidae	*Pseudococcus viburni*	–	Hz	Stokwe (2009)
Lepidoptera	Tortricidae	*Thaumatotibia leucotreta*	False codling moth	Hz	Malan, Knoetze, and Moore (2011), Manrakhan et al. (2013)
Lepidoptera	Tortricidae	*Cydia pomonella*	Codling moth	Hz, Sk, Sy, S sp	De Waal et al. (2010, 2011a, 2011b, 2013)
Lepidoptera	Noctuidae	*Helicoverpa armigera*	Bollworm	Hb, St	Jankielsohn and Hatting (2005), ARC-SGI (unpublished)
Lepidoptera	Pyralidae	*Eldana saccharina*	Sugarcane borer	Hb, St, Sk, Si	ARC-SGI (unpublished)
Lepidoptera	Noctuidae	*Busseola fusca*	Maize stalk borer	Hb, Hz, Si, Sy	Ramakuwela Erasmus, & Hatting (2011), Steenkamp, Erasmus, and Malan (2011)

Hb *Heterorhabditis bacteriophora,* Hz *H. zealandica,* Sk *Steinernema khoisanae*, Si *S. innovationi,* St *S. tophus*, Sy *S. yirgalemense,* S sp *Steinernema sp*

Malan & Gozel (Rhabditida: Steinernematidae) (Nguyen, Malan, & Gozel, 2006). A revived interest in applied research on EPN ensued during early 2000 (Hatting & Kaya, 2001) with research starting in earnest in 2003 at the South African Agricultural Research Council–Small Grain Institute (ARC–SGI) near Bethlehem, Free State province, continuing a year later at Stellenbosch University, in the Western Cape province (Table 20.1).

20.2 Occurrence and Distribution of Entomopathogenic Nematodes in Africa

South Africa has a diverse climate where summer rainfall, mostly in the form of thundershowers, dominates with a gradient of increasing rainfall towards the east, reaching a maximum along the eastern escarpment and south eastern coastal areas. Much of the interior is classified as semi–arid, but arid to hyper–arid towards the western interior and west coast while dry sub–humid and humid over the eastern high lying areas and coastal regions. Total annual rainfall ranges from less than 150 mm in the west to more than 500 mm over much of the eastern parts, exceeding 1,000 mm over parts of the escarpment and along the south eastern coastal belt. Maximum temperatures during summer can occasionally exceed 40 °C especially over the north western and north eastern low–lying interior, while winter night–time temperatures can drop below freezing over much of the plateau. The extreme south western part of the country has a Mediterranean climate, where precipitation is mainly associated with cold fronts during winter and summers are warm to hot and dry. Total rainfall is closely related to topography, ranging between 200 mm in low–lying areas to more than 1,000 mm in the mountainous terrain in the southwest. Towards the east of this, the coastal belt in the south has a dry sub–humid to humid climate and receives rainfall of between 350 and 1,000 mm throughout the year, also associated strongly with topography. These climatic extremes are likely to impact the distribution of EPN in South Africa, underscoring the need for country–wide surveys across the nine provinces.

Only in three previous surveys that were conducted in South Africa have EPN been identified to species level. The identification included that of two non–targeted surveys, in an effort to establish the occurrence, and the distribution, of EPN in South Africa (Hatting, Stock, & Hazir, 2009; Malan, Nguyen, & Addison, 2006). From 2009 to 2010, surveys targeting citrus orchards were conducted, to determine the diversity, and frequency, of native EPN in the Western and Eastern Cape, and Mpumalanga, provinces of South Africa (Malan et al., 2011). The main aim of the surveys was to obtain nematodes to use as outdoor biological control agents in subsequent research against key South African insect pests. From the results of the surveys undertaken, it can be concluded that *H. bacteriophora* was the most frequently found species. The occurrence of EPN species in the different provinces of South Africa is indicated in Fig. 20.1.

In the previous century, only two species, namely *Heterorhabditis taysearae* Shamseldean, El-Sooud, Abd-Elgawad & Saleh (Rhabditida: Heterorhabditidae), in 1996 from Egypt (Shamseldean, Abou-El-Sooud, Abd-Elgawad, & Saleh, 1996), and *Steinernema karii* Waturu, Hunt & Reid (Rhabditida: Steinernematidae) in 1997 from Kenya (Waturu, Hunt, & Reid, 1997), were described as being from the African continent. Other reports of EPN from Africa before the twentieth century include those of *H. bacteriophora* from both South Africa (Grenier, Bonifassi et al., 1996) and Kenya (Waturu, 1998), and of *Heterorhabditis indica* Poinar, Karunakar

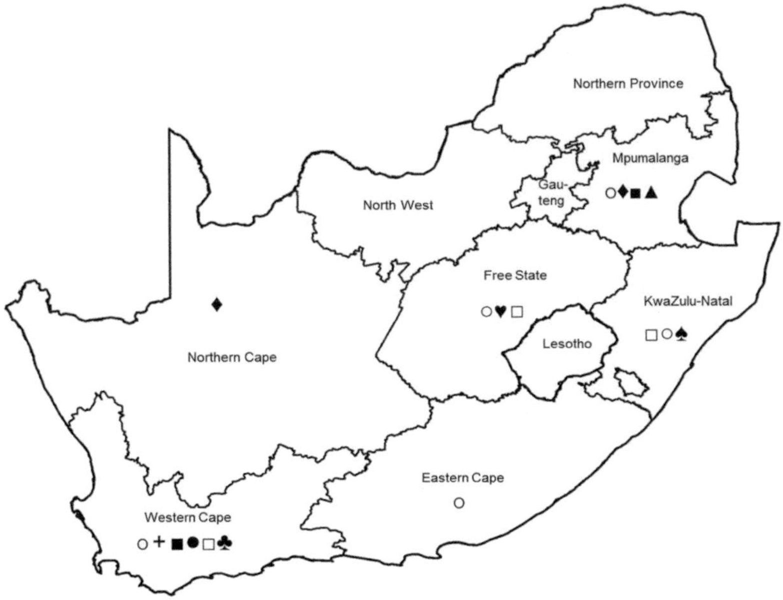

Fig. 20.1 Occurrence and distribution of entomopathogenic nematodes in South Africa. *Key*: o *H. bacteriophora*, ♦ *H. noenieputensis*, ♣ *H. safricana*, ■ *H. zealandica*, + *S. khoisanae*, ▲ *S. yirgalemense*, • *S. citri*, □ *S. tophus*, ♥ *S. innovationi*, ♠ *S. sacchari*

& David (Rhabditida: Heterorhabditidae) in 1992, from Egypt (Shamseldean, Adb-Elgawad, & Atwa, 1998) and Kenya (Shamseldean et al., 1996; Waturu, 1998).

A total of 24 species are currently described as being from Africa, of which eight represent *Heterorhabditis*, and 16 *Steinernema* (Table 20.2). Of these, five *Steinernema* and two *Heterorhabditis* were described from South Africa, indicating the strong potential for new EPN species and isolates from the African continent, and highlighting the necessity of bioprospecting. New isolates reported from this century include *H. indica* and *H. bacteriophora* from Kenya and Egypt (Hominick, 2002; Stack et al., 2000). New isolates of *Steinernema yirgalemense* Nguyen, Tesmafariam, Gozel, Gaugler & Adams (Rhabditida: Steinernematidae) have also been reported from South Africa (Malan et al., 2011) and Ethiopia (Mekete et al., 2005). *Steinernema karii* and *Steinernema weiseri* Mráček, Sturhan & Reid (Rhabditida: Steinernematidae) have been reported from the Central Rift Valley Region of Kenya (Mwaniki et al., 2008). Tarasco et al. (2009) reported 13 isolates of *Steinernema feltiae* (Filipjev) Wouts, Mráček, Gerdin & Bedding (Rhabditida: Steinernematidae) and two of *H. bacteriophora*, during a survey that was undertaken of EPN in Algeria. This report is the first record of *S. feltiae* on the African continent. In Ethiopia, the dominant species was found to be *S. yirgalemense*, which was reported, together with two isolates of *H. bacteriophora* (Mekete et al., 2005). Kanga, Waeyenberge, Hauser, and Moens (2012) reported *Heterorhabditis*

Table 20.2 Occurrence of entomopathogenic nematodes in Africa

Country	Nematode species	Report of occurrence
Algeria	*H. bacteriophora*	Tarasco, Triggiani, Sai, and Zamoum (2009)
Algeria	*S. feltiae*	Tarasco et al. (2009)
Benin	*H. indica*	Zadji et al. (2013)
Benin	*H. sonorensis*	Zadji et al. (2013)
Cameroon	*H. amazonensis*	Kanga et al. (2012)
Cameroon	*H. baujardi*	Kanga et al. (2012)
Cameroon[a]	*S. cameroonense*[b]	Kanga et al. (2012)
Cameroon[a]	*S. nyetense*[b]	Kanga et al. (2012)
Egypt	*H. bacteriophora*	El-Rahman, El-Razzik, Osman, and Mangoud (2012)
Egypt	*H. baujardi*	El-Rahman et al. (2012)
Egypt	*H. indica*	El-Rahman et al. (2012)
Egypt*[a]	*H. taysearae*[b]	Shamseldean et al. (1996)
Egypt[a]	*S. abbasi*[b]	Elawad, Ahmad, and Reid (1997)
Egypt	*S. kushidai*	Mamiya (2008)
Egypt	*S. carpocapsae*	El-Rahman et al. (2012)
Ethiopia	*H. bacteriophora*	Mekete, Gaugler, Nguyen, Mandefro, and Tessera (2005)
Ethiopia[a]	*S. ethiopiense*[b]	Tamiru et al. (2012)
Ethiopia[a]	*S. yirgalemense*[b]	Nguyen, Tesfamariam, Gozel, Gaugler, and Adams (2004), Mekete et al. (2005)
Kenya	*H. bacteriophora*	Waturu (1998)
Kenya	*S. arenarium*	Waturu (1998)
Kenya	*S. glaseri*	Waturu (1998)
Kenya[a]	*S. karii*[b]	Waturu et al. (1997)
Kenya	*S. weiseri*	Mwaniki, Nderitu, Olubayo, Kimenju, and Nguyen (2008)
Kenya	*S. yirgalemense*	Mwaniki et al. (2008)
South Africa	*H. bacteriophora*	Malan et al. (2006)
South Africa[a]	*H. noenieputensis*[b]	Malan, Knoetze, and Tiedt (2014)
South Africa[a]	*H. safricana*[b]	Malan, Nguyen, De Waal, and Tiedt (2008)
South Africa	*H. zealandica*	Malan et al. (2006)
South Africa[a]	*S. citrae*[b]	Stokwe, Malan, Nguyen, Knoetze, and Tiedt (2011)
South Africa[a]	*S. innovationi*[b]	Çimen, Lee, Hatting, Hazir, and Stock (2014a)
South Africa[a]	*S. khoisanae*[b]	Nguyen et al. (2006)
South Africa[a]	*S. tophus*[b]	Çimen et al. (2014b)
South Africa[a]	*S. sacchari*[b]	Nthenga et al. (2014)
South Africa	*S. yirgalemense*	Nguyen et al. (2004)

[a]Locality type
[b]Type specimen

baujardi Phan, Subbotin, Nguyen & Moens (Rhabditida: Heterorhabditidae) from Cameroon, which is a species that was originally described from Vietnam, and which was later also recorded from Brazil (Dolinski, Del Valle, Burla, & Machado,

2007). Surveys in the Guinean zone of Southern Benin reported two species, *Heterorhabditis sonorensis* Stock, Rivera-Orduño & Flores-Lara (Rhabditida: Heterorhabditidae) (Stock, Rivera-Orduño, & Flores-Lara, 2009), and *H. indica* (Zadji et al., 2013), described from the Sonoran Desert in Mexico. *Heterorhabditis amazonensis* Andaló, Nguyen & Moinohas (Rhabditida: Heterorhabditidae) has been described from the Amazonas in Brazil, and it was also recently found during a survey in Cameroon (Kanga, Waeyenberge et al., 2012).

A total of five species described from Africa belong to Clade V (Spiridonov, Reid, Podrucka, Subbotin, & Moens, 2004), and, morphologically, to the *glaseri*-group (Nguyen, Hunt, & Mráček, 2007). A new group, called the Cameroonian Clade VI (Nthenga, Knoetze, Berry, Tiedt, & Malan 2014), is formed by *Steinernema cameroonense* Kanga, Trinh, Waeyenberge, Spiridonov, Hauser & Moens (Rhabditida. Steinernematidae), *Steinernema nyetense* Kanga, Trinh, Waeyenberge, Spiridonov, Hauser & Moens (Rhabditida: Steinernematidae), and *Steinernema sacchari* Nthenga, Knoetze, Berry, Tiedt & Malan (Rhabditida: Steinernematidae) (Fig. 20.2a). The *Heterorhabditis* spp. described from Africa belong to both of the two broad clades, the *indica* group, with *Heterorhabditis noenieputensis* Malan, Knoetze & Tiedt (Rhabditida: Heterorhabditidae) and *Heterorhabditis baujardi* Phan, Subbotin, Nguyen & Moens (Rhabditida: Heterorhabditidae), and the *megidis* group, with *Heterorhabditis safricana* Malan, Nguyen, De Waal & Tiedt (Rhabditida: Heterorhabditidae), *Heterorhabditis zealandica* Poinar (Rhabditida: Heterorhabditidae), and *H. bacteriophora* (Nguyen et al., 2007) (Table 20.1).

Four symbiotic bacteria (Fig. 20.2), of which two were *Xenorhabdus*, and two *Photorhabdus*, were identified from endemic South African EPN, with three being described as new species, including *Xenorhabdus khoisanae* Ferreira, Van Reenen, Gozel, Malan & Dicks (Enterobacteriales: Enterobacteriaceae), associated with *S. khoisanae* (Ferreira et al., 2013b), *Photorhabdus zealandica* Ferreira, Van Reenen, Endo, Tailiez, Pagès, Spröer, Malan & Dicks (Enterobacteriales: Enterobacteriaceae) associated with *H. zealandica* (Ferreira et al., 2014a), and *Photorhabdus luminescence* subsp. *noenieputensis* Ferreira, Van Reenen, Pagès, Tailiez, Malan & Dicks (Enterobacteriales: Enterobacteriaceae), associated with *H. noenieputensis* (Ferreira et al., 2013). The bacteria associated with *S. yirgalemense* was identified as being *Xenorhabdus indica* (Ferreira et al., 2014b), previously described from *Steinernema abbasi* (syn. *S. termophylum*). The bacteria of *H. zealandica* found in South Africa differed from those of *H. zealandica* that were originally found in New Zealand and Florida. The associated bacterium from *H. zealandica* from New Zealand was identified as being *Photorhabdus temperata* Fischer-Le Saux, Viallard, Brunel, Normand & Boemare (Enterobacteriales: Enterobacteriaceae), while those that were from the South African *H. zealandica* were identified as being *P. zealandica* (Ferreira et al., 2014a).

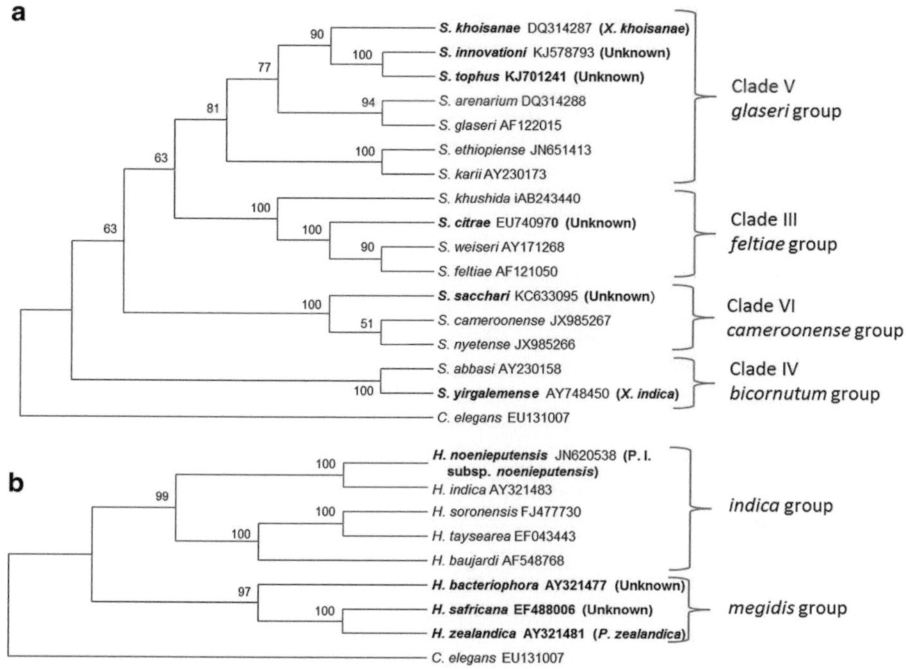

Fig. 20.2 The evolutionary analysis of *Steinernema* (**a**) and *Heterorhabditis* (**b**) (including the associated bacteria) reported from South Africa, as inferred using the maximum parsimony method. The percentage of replicate trees in which the associated taxa clustered together in the bootstrap test (1,000 replicates) is shown next to the branches. All positions containing gaps and missing data were eliminated. Evolutionary analyses were conducted in MEGA6. Species in *bold* = present in South Africa

20.3 Entomopathogenic Nematode Biological Control of Major Insect Pests in South Africa

20.3.1 The Codling Moth, Cydia pomonella *(L.) (Lepidoptera: Tortricidae) in Apples and Pears*

The codling moth, *Cydia pomonella* (L.) (Lepidoptera: Tortricidae), is the key pest of apples and pears in South Africa (Barnes, 1991). Apples are mainly produced in the Western and Eastern Cape provinces. In South Africa, infestation rates in certain areas can be as high as 80 %, if no control measures are taken (Pringle, Eyles, & Brown, 2003). A key factor in the biology of codling moth is that the total population is represented as a diapausing overwintering population during the winter months of June to August. During early spring, when the temperature increases, the larvae again turn into pupae, from which the moths emerge in late spring. They lay their eggs on the young fruit, and on the adjacent leaves, with feeding larvae creating frass–filled tunnels that are equipped with an exit hole (with, usually, one codling moth larva per fruit), rendering the fruit unmarketable (Welter, 2008). In early, and late, autumn, when the fruit are ready for harvest, the last instar of the codling moth larvae move from the fruit to such cryptic habitats as pruning wounds, main spurs, and the trunk of the tree, close to the soil, as well as to debris around the tree, especially when the tree in question is a smooth–barked young apple tree (Cossentine, Sholberg, Jensen, Bedford, & Sheperd, 2004; Riedl, Blomefield, & Giliomee, 1998).

Codling moth is mostly chemically controlled throughout the growing season; however, integrated pest management (IPM) options are currently being employed in commercial orchards (Addison, 2005). Some of the tactics that are currently being used include mating disruption and 'attract and kill', as standard practice. The sterile insect technique (SIT) is being employed on a semi–commercial basis in specific regions (Pringle et al., 2003), with it being low density dependent (Judd & Gardiner, 2005). To use EPN successfully for the control of codling moth, the nematode isolate used should be: highly virulent; able to infect codling moth at low temperatures; and effective at low–water activity levels. The window of opportunity for aerial applications of EPN in a low humidity, water–scarce region is only a period of approximately 24 h, during which it is essential to maintain humidity of above 80 %, with a few hours of temperatures exceeding 20 °C. The nematode isolate used should also be able to locate cocooned larvae hiding on the tree. Different aspects involved in efficacy have been investigated in studies that have been undertaken into the potential of using EPN for the control of codling moth in South Africa (De Waal, Malan, & Addison, 2011a, 2011b; De Waal, Addison, & Malan, 2013; De Waal, Malan, Levings, & Addison, 2010).

IPM measures are currently hampered by infested wooden fruit bins, acting as a potential source of re–infestation. The investigation evaluated mini wooden fruit bins (built from the planks taken from old bins) that were artificially infested with last–instar diapausing codling moth larvae, which were inoculated with 25 infective juveniles (IJ)/mL (De Waal et al., 2010). Maximum mortality was achieved when the bins were pre–wet for at least 1 min, and then maintained at maximum humidity post–treatment for at least 3 days (De Waal et al., 2010). Tarping of the bin was the method used to obtain the desired high level of humidity that was required for effective insect control. By adding an adjuvant, increased mortality of the codling moth larvae was obtained. Moreover, the study revealed that, by using the correct concentrations of *H. zealandica* and high humidity, the addition of adjuvants to the nematode suspension has the potential to disinfest wooden fruit bins of codling moth successfully (De Waal et al., 2010). More research is required to evaluate the logistics of handling the wooden fruit bins, and their successful treatment with nematodes, in terms of commercial orchards.

The concept of mulching in orchards has been investigated in a further study, especially in the case of smooth–barked apple trees, where codling moth can be tempted to hide, and to overwinter, in the mulch. De Waal et al. (2011b) evaluated the potential of using *H. zealandica* in combination with mulches (pine chips, wheat straw, pine wood shavings, blackwood and apple wood chips) to control diapausing codling moth. Mesh cages filled with the different mulches, used as a larval confinement method, showed high levels of codling moth mortality (88 %), with pine wood shavings as mulch. Again, it was imperative that a high humidity of above 95 % was maintained for at least 3 days, to ensure nematode efficacy. A noteworthy point in this regard was that nematodes were found to have the ability to move 10 cm upwards into moist mulch, so as to infect codling moth larvae. Low temperatures (<15 °C) recorded during the first field trial resulted in low levels of control (<48 %), as opposed to the higher mortality recorded during the second field trial, with temperatures between 20 and 25 °C (De Waal et al., 2011b).

The biocontrol potential of six isolates, namely *H. zealandica*, *S. citrae*, *S. khoisanae* (J96, SF87), *S. yirgalemense*, and *Steinernema* sp., was evaluated (De Waal et al., 2011a). At optimum conditions in the laboratory, codling moth was found to be highly susceptible to all nematode isolates, at a low concentration of 50 IJ/insect, with mortalities between 78 and 100 %. A laboratory study at a suboptimal low temperature cycle, starting with 10 h at 17 °C, and 14 h at 12 °C, negatively affected the efficacy of all isolates to below 3 % codling moth mortality. The levels of free water in which a nematode is able to move, in a form of movement that is called water activity (a_w), were investigated for the above–mentioned nematode isolates, with the average a_{w50}–values for all isolates tested found to be 0.94, except for *S. khoisanae*, which had a higher a_w level of 0.97 (De Waal et al., 2011b).

Laboratory conditions, and the containment method used for evaluating the field mortality levels of codling moth, were found to be not necessarily representative of the related field performance. In most of the previous studies with codling moth in field trials, cardboard strips (Lacey & Unruh, 1998) were used as a containment method, with high codling moth mortality. Three isolates, *H. zealandica*, *S. khoisanae*, and *Steinernema* sp., were used for field testing, with the latter being proven to be more effective, with a mortality of 70 %, compared to *H. zealandica*, with 59 % mortality. Insect containment methods used during field trials were shown to influence efficacy against codling moth, as different levels of mortality were obtained with the use of various containment methods (wooden planks vs. pear tree logs *versus* mesh cages) (De Waal et al., 2011a). Predictive equations were subsequently developed, enabling future trials to be conducted using either planks or cages (with pear tree logs proving impractical), and enabling the prediction of the expected level of control on the tree logs. All tested isolates showed a certain degree of biological control potential, although none of the experiments showed clear efficacy differences among the isolates. As the study showed that higher levels of control were obtained using the containment methods mentioned, the factor in question should be taken into consideration, when reporting the actual level of control during normal field applications (De Waal et al.).

All laboratory and field trials indicated that the main problem with the control of codling moth by means of EPN is the maintenance of adequate moisture levels that are required for nematode survival, and for their efficacy as biocontrol agents. De Waal et al. (2013) investigated the addition of a superabsorbent polymer, Zeba®, on the performance of *H. zealandica*, which was able to infect codling moth larvae only at $a_w \geq 0.92$, with $aw_{50} = 0.94$ and $aw_{90} = 0.96$. Laboratory experiments showed the highest level of mortality recorded to take place at 80 IJs/codling moth larva, which required at least 4 h of optimum conditions to ensure infectivity, and subsequent efficacy. Further studies showed that the addition of Zeba® to nematode suspensions improved the level of control obtained at 60 and 80 % RH in the laboratory, as well as enhancing the survival, and the infection ability, of the nematodes in the field.

20.3.2 The False Codling Moth Thaumatotibia leucotreta *Meyrick (Lepidoptera: Tortricidae) in Citrus*

False codling moth (FCM), *Thaumatotibia leucotreta* Meyrick (Lepidoptera: Tortricidae), is a key pest of citrus in South Africa. It is indigenous to South Africa, while also occurring elsewhere south of the Sahara, as well as on the Indian Ocean islands (CIBC, 1984), and Israel. The Eastern Cape produces the most citrus in South Africa, followed by Limpopo, Mpumalanga, the Western Cape, and KwaZulu–Natal. Current control of false codling moth in South Africa consists of orchard sanitation, chemical application, mating disruption, 'attract and kill', and SIT, combined with other biological control methods (Carpenter, Bloem, & Hofmeyer, 2007; Moore, 2002; Moore & Hattingh, 2012; Moore, Kirkman, & Stephen, 2004). The harvesting season is usually from May to October. Production in South Africa is confined to areas with mild, virtually frost–free winters. The average minimum temperature in the coldest month should not be below 3 °C to achieve ongoing production. Where rainfall is poor, the use of drip, or sprinkler, irrigation should be used to ensure good growth and production (CABI, 2011).

The FCM moth lays eggs on fruit, or leaves, with the larvae burrowing into the fruit, where they develop into the final instar (Daiber, 1979b), which drop, on a silken thread, to the soil, where they burrow a few mm into dry soil, and spin themselves into a cocoon (Daiber, 1980, 1989). After a few days, the prepupae turn into pupae (Daiber, 1979a), remaining, as such, in the soil for 8–10 days, depending on the prevailing temperature, after which they emerge from the soil as adult moths. False codling moth is multivoltine, producing up to six generations per year (Newton, 1998). The soil stages that are targeted by nematodes include the final–instar larvae, the prepupae, and the pupae, and the emerging moth. The soil stage of FCM, spanning approximately 14–18 days (Daiber, 1980, 1989), depending on the prevailing temperature, offers a long window period for the use of EPN.

Laboratory bioassays have shown isolates of six local EPN species to be highly virulent against the last instar of false codling moth larvae (Malan et al., 2011). This was the first research to be undertaken on the potential use of EPN to control the soilborne life stages of false codling moth, including larvae, pupae, and emerging moths. *Steinernema yirgalemense*, at a concentration as low as 50 IJ/insect, caused 100 % mortality of codling moth larvae, while, in most cases, the pupae concerned were at least half as sensitive to infection as were larvae using higher concentrations of nematodes. An important finding that was made during this study was that the emerging moths were infected with nematodes, thus potentially facilitating control, and their long–distance dispersal (Malan et al.).

Semi–field trials were conducted with contained FCM larvae in soil mesh cages. Six days after field nematode application, no significant differences were found in FCM mortality between three concentrations (5, 10, and 20 IJs/cm^2) of *H. zealandica* applied, which caused >80 % control. In a field trial using three nematode species (*H. bacteriophora*, *H. zealandica* and *S. khoisanae*), treatment with *H. zealandica* resulted in significant persistence for each evaluation day, up to day 49.

As soil is the natural habitat for nematodes, they are especially suited to control the soil stages of FCM. All life stages, including the prepupae, the pupae, and the emerging moth, were found to be susceptible to nematodes (Malan et al., 2011). Results from these studies showed local EPN isolates to hold major potential for the control of the soil stages of FCM, with the added advantage of good persistence. Currently, large–scale efficacy trials are under way, with imported formulated *H. bacteriophora* in the different production areas, with promising results for future commercial use. However, more research into the ecology of nematodes, with regard to persistence in citrus orchards in different production areas in South Africa is required.

20.3.3 Mealybugs (Pseudococcidae) in Deciduous Fruit, Citrus and Grapevine

Mealybugs (Pseudococcidae) are severe agricultural pests that pose major problems for farmers in South Africa. The obscure mealybug, *Pseudococcus viburni* (Signoret) (Hemiptera: Pseudococcidae), is one of the most common, and serious, pests of apples and pears in South Africa (Wakgari & Giliomee, 2004), while the citrus mealybug, *Planococcus citri* (Risso) (Hemiptera: Pseudococcidae), is a highly destructive pest of citrus (Hattingh & Moore, 2003), with both occurring only in the aerial parts of trees. In the case of grapevine, the vine mealybug, *Planococcus ficus* (Signoret) (Hemiptera: Pseudococcidae), has been shown to be the dominant mealybug species in South African vineyards. Although they remain predominantly above ground, they can also occur up to 30 cm deep, as colonies in the soil, on grapevine roots (Walton, 2003).

Mealybugs are difficult to control with chemicals, due to their cryptic lifestyles of hiding in crevices, under bark, and below ground on roots, where they are protected from insecticidal sprays. Their hydrophobic waxy secretions repel water–based insecticides, and they have the ability to rapidly develop resistance (Walton & Pringle, 2004). In citrus orchards, mealybug populations are usually suppressed by a complex of natural enemies (Hattingh & Moore, 2003), which is disrupted by the application of chemicals. However, there is a need for new and improved, *P. ficus* control options, potentially including EPN (Le Vieux & Malan, 2013a).

Laboratory bioassays were conducted to identify isolates of EPN that could cause high percentage mortality against *P. viburni* (Stokwe, 2009). Notable variation was found in the mortality caused by the different nematode isolates, leading to *H. zealandica* being selected as the most promising isolate for use in further studies. The biological development of a steinernematid and a heterorhabditid in adult *P. viburni*, *P. ficus*, and *P. citri* females was investigated, with *H. zealandica* and *S. yirgalemense* both being found to reproduce successfully in *P. viburni* (Le Vieux & Malan, 2013b; Stokwe, 2009; Van Niekerk & Malan, 2012).

The effect that mealybug size has on EPN infectivity was assessed. Adult and intermediate *P. viburni* were found to be more susceptible to nematode infection than were crawlers, because of the latter's small size (Bastidas, Edgar, & San-Blas, 2014; Stokwe, 2009). Nematodes were tested for their ability to locate, and to infect, mealybugs on the surface, and in the ovary and calyx, of *P. viburni* field–infested apples. Results from the study indicated that the nematodes are capable of locating, and of infecting mealybugs, even when they are in the cores of infested apples. The LC_{50} and LC_{90} values were 54 and 330 nematodes per insect, respectively, with the LT_{50} and LT_{90} values being 30 h and 62 h, respectively. The study showed good potential for the use of EPN to control *P. vibruni*.

To determine the potential of local isolates of EPN to control *P. citri*, various laboratory bioassays were conducted (Van Niekerk & Malan, 2012). Adult female *P. citri* were found to be most susceptible to *S. yirgalemense* and *H. zealandica*, causing >90 % mortality. Further bioassays illustrated a linear relationship between mealybug mortality, and the concentration of nematodes applied. If nematodes are to be used as an above–ground application to control *P. citri* in citrus orchards, the amount of water that is available can be a major limiting factor. Insecticidal activity proved to be dependent on the available surface moisture after nematode application. An aw–bioassay indicated *S. yirgalemense* to be twice as tolerant to relatively low levels of free water. After application, nematodes have a limited time frame in which to locate, and infect, hosts, as the level of available free water gradually decreases, as trees dry out. *Steinernema yirgalemense* proved able to locate, and to infect, *P. citri* more quickly than were *H. zealandica*. An interesting result in this study was that *S. yirgalemense* were able to infect *P. citri* after an exposure time as short as 30 min. The results also showed the first 2–4 h post–application to be the most decisive time for establishing successful infection of mealybugs. The report was the first on the potential use of nematodes for the control of *P. citri* (Van Niekerk & Malan).

Humidity is one of the key factors to consider when using EPN as biological control agents. The addition of adjuvants to suspensions of EPN, to improve control in a foliar application, was investigated (Van Niekerk & Malan, 2013). An aqueous suspension, containing *H. zealandica* and 0.3 % Zeba®, significantly increased *P. citri* mortality at 80 % relative humidity (RH), with a temperature cycle starting at 22 °C for 14 h, and continuing at 11 °C for 11 h. The same polymer formulation was tested for *S. yirgalemense*, with the mortality of *P. citri* increasing by 21 % at 60 % RH, and by 27 % at 80 % RH. The addition of Nu–Film–P® and Zeba® to *H. zealandica* suspensions did not significantly retard application runoff from citrus leaves. The combination of Nu–Film–P® and Zeba®, however, was able to retard sedimentation significantly, increasing the average number of nematodes deposited on 2–cm^2 leaf discs by 10 nematodes.

The compatibility of two endemic EPN with biological control agents and agrochemicals, which were likely to be used in an IPM programme for citrus in South Africa, was investigated (Van Niekerk & Malan, 2014a). This is the first report to have been produced on the possible negative effect of EPN against *Cryptolaemus montrouzieri* Mulsant (Coleoptera: Coccinellidae), a commercially produced biocontrol predatory insect, which is used against mealybugs. Results from bioassays in the laboratory showed the beetle larvae to be highly susceptible to *H. zealandica* and *S. yirgalemense*. Adult beetles were found to be twice as susceptible to *S. yirgalemense* as they were to *H. zealandica*. Tolerance of both species of IJ to aqueous solutions of Cyperfos 500 EC® (Chlorpyrifos and cypermethrin), Cryptogran™ (*Cryptophlebia leucotreta* granulovirus), Helicovir™ (Nucleopolyhedrovirus), Nu–Film–P® (Poly–1–P–menthene), and Zeba® (starch–g–polypotassium salt) for infectivity, and survival, was evaluated. *Heterorhabditis zealandica* proved to be highly compatible with all products tested, with no significant increase occurring in terms of nematode mortality. The products concerned also did not affect the ability of *H. zealandica* to infect mealworm larvae after exposure to products over a 24–h period.

Laboratory bioassays were conducted to establish the potential of EPN as biocontrol agents of *P. ficus* (Le Vieux & Malan, 2013b). Screening of local EPN isolates showed promising results for *H. zealandica* and *S. yirgalemense*. Bioassays indicated a concentration–dependent susceptibility of *P. ficus* to *H. zealandica*, to *S. yirgalemense*, and to commercially produced *H. bacteriophora*, with LC_{50} and LC_{90} values of 19, 82; 13, 80; and 36, 555, respectively. In soil column bioassays, both *H. zealandica* and *S. yirgalemense* were able to move 15 cm vertically downwards, so as to infect *P. ficus*, with respective mortalities of 82 and 95 %.

EPN can potentially be used within an IPM scheme to control *P. ficus*, which also occurs on grapevine roots. When *S. yirgalemense* was applied to the soil of two vineyards together with *P. ficus*, contained in pierced Eppendorf tubes, and buried at a depth of 15 cm in the soil, mortalities of up to 50 % were obtained after 48 h (Le Vieux & Malan, 2014). The persistence of *S. yirgalemense*, measured using codling moth larval mortality, was found to be zero in one vineyard, whereas, in another vineyard, it was 70 %, 12 weeks after application. Tests were conducted to establish the production of scavenger–deterrent factors (Le Vieux & Malan, 2015) by *H. zealandica* and *S. yirgalemense*. Of the cadavers that were presented 6 days after nematode infection, 49 % of the *H. zealandica*, and 60 % of the *S. yirgalemense* infected cadavers were left intact. Olfactometry tests indicated a significant difference concerning the number of *S. yirgalemense* IJ that were attracted to damaged *Vitis vinifera* L. (Vitales: Vitaceae) roots, and to *P. ficus*, indicating the active movement of the IJ, and the attractive ability of organic compounds produced by the roots. These studies showed that EPN, and specifically *S. yirgalemense*, have promising potential as biological control agents for the control of *P. ficus* soil populations (Le Vieux & Malan, 2015).

20.3.4 *The Banded Fruit Weevil* Phlyctinus callosus *Schönherr (Coleoptera: Curculionidae)*

Phlyctinus callosus Schönherr (Coleoptera: Curculionidae) was first reported from New Zealand in 1899, from where it spread to Australia (Kuschel, 1972). *Phlyctinus callosus* is indigenous to South Africa and is described for the first time in 1834 (Barnes, 1987). In deciduous fruit orchards, it is the main weevil pest, amongst others, as well as being a serious pest in grapevine (Allsopp, Barnes, Blomefield, & Pringle, 2015; Annecke & Moran, 1982; Myburgh, 1980), and in blueberries (Bredenhand, Van Hoorn, May, Ferreira, & Johnson, 2010). In South Africa, most damage occurs during November and December, when grape bunches are actively developing. In apple and plum orchards, most of the damage is inflicted on the lower parts of trees. In the Western Cape province, with a Mediterranean climate, it is estimated that *P. callosus* has the ability to cause up to 40 % damage on apples (Witt, Little, & Crowe, 1995).

In South Africa, *P. callosus* has one, or two, generations per year. The eggs are laid either just below the soil surface, or in organic matter, with the first–instar larvae then feeding on the roots of the host plant (Barnes, 1987, 1989; Barnes & Pringle, 1989). The majority of the larval stages of *P. callosus* tend to stay in the top 10 cm of soil during the winter months (Barnes, 1989). *Phlyctinus callosus* passed through up to 11 instars, with pupation lasting approximately 14 days. Emerging during late spring and early summer, they migrate, as flightless adults, up the tree trunks to reach the available fruits (Barnes; Barnes & Giliomee, 1992).

Since *P. callosus* has developed a high tolerance to pyrethroids, with an indication of cross–tolerance to acephate, chemical control is not successful against this pest (Barnes, Knipe, & Calitz, 1994, 1996). Trunk barriers are only used for monitoring purposes, as such use is very labour–intensive. The larvae, pupae, and emerging adults remain in the soil throughout the winter months, offering a window of opportunity for the use of EPN.

Research undertaken by Ferreira and Malan (2014a) showed that higher concentrations, and longer exposure times, were required to obtain satisfactory control of *P. callosus* larvae, in bioassay trials using local EPN. The trials in question involved three isolates, two *H. bacteriophora* and one *H. zealandica*, at a concentration of 400 IJ/ insect, with a 4–day exposure time for the adults and larvae. The percentage mortality was found to range between 41 and 73 % for the larvae, and between 13 and 35 % for the adults, with *H. zealandica* causing the highest mortality.

Optimum control is, however, obtainable by means of applying nematodes during winter and early spring. However, during the mentioned period, the temperature is generally low, with all local South African isolates being inactive at low temperatures. More local isolates still need to be screened, as only three isolates have been tested so far, with the current isolate giving only 43 % control after 2 days. Superior isolates should be selected. The best time for application in South Africa would be when the soil temperature is relatively low, with a low–temperature active nematode being selected.

20.3.5 *The Fruit Flies* **Ceratitis capitata** *(Wiedemann) (Diptera: Tephritidae) and* **Ceratitis rosa** *Karsch (Diptera: Tephritidae)*

In South Africa, two species of fruit flies of economic importance occur in the Western Cape province, namely the Mediterranean fruit fly (Medfly), *Ceratitis capitata* (Wiedemann); and the Natal fruit fly, *Ceratitis rosa* Karsch (Diptera: Tephritidae), which are important pests of many fruits (Annecke & Moran, 1982; Prinsloo & Uys, 2015). Not only are the fruit flies responsible for economic crop losses, and for the cost of control, they are also international quarantine pests, causing restrictions on the international trade in fruit. Current control strategies for fruit fly mainly use the application of baits, mixed with insecticides while, in some areas of South Africa, medfly is commercially controlled through the use of the Sterile Insect Technique or SIT (Barnes, Eyles, & Franz, 2002).

Adult fruit fly tend to lay their eggs on the fruit, where the larvae go through several instars before leaving the infested fruit, dropping to the ground, and burrowing a few mm into the soil. After only a few hours, the pre–pupae turn into pupae in the soil (Annecke & Moran, 1982; Prinsloo & Uys, 2015).

The potential of three local isolates of *H. bacteriophora*, *H. zealandica*, and *S. khoisanae* to infect pupariating larvae, pupae, and adults of *C. capitata* and *C. rosa* was investigated, using 24–well bioassay plates in the laboratory (Malan & Manrakhan, 2009). Results from the study showed that pupariating larvae and adult flies were susceptible to nematode infection, with no infection being recorded for the pupae. However, some pupariating larvae infected with nematodes still managed to pupate, giving rise to malformed puparia, while trapping the nematodes inside the puparium. Pupariating larvae of *C. capitata* were generally more susceptible to infection than were those of *C. rosa*. Significantly, more larvae of *C. capitata* were infected with *H. bacteriophora*, and, in the case of *C. rosa*, the highest infectivity of larvae was obtained with *H. zealandica*. In contrast, adults of both species were highly susceptible to infection with *S. khoisanae*.

20.3.6 *Noctuids, the African (Old World) Bollworm*

Members of the Noctuidae family (Order: Lepidoptera) are agricultural pests of worldwide significance, of which *Helicoverpa armigera* (Hübner), *Helicoverpa zea* (Boddie), and *Heliothis virescens* (Fabricius) have achieved major pest status (Fitt, 1989). In South Africa, at least 38 commodities have chemical insecticide registrations listed against *H. armigera*, underscoring the importance of this ubiquitous pest (CropLife South Africa, [s.d.]). Biological control of *H. armigera* has gained global attention, given the development of resistance against all the major chemical groups, i.e. synthetic pyrethroids, organophosphates, organochlorines and carbamates (Regupathy, Kranthi, Singh, Iqbal, & Russell, 2003). According to the

Arthropod Pesticide Resistance Database (http://www.pesticideresistance.com), this insect has shown resistance to at least 48 insecticidal active ingredients, including DDT. The South African scenario raises particular concern, as a large proportion of registered insecticides belong to the synthetic pyrethroid group.

The use of EPN has been attempted against above–ground noctuid pests (Bong & Sikorowski, 1983; Richter & Fuxa, 1990; Vyas, Patel, Yadav, Ghelani, & Patel, 2003), but the application of EPN to plant foliage is challenged by the general intolerance of IJ to desiccation and/or to UV radiation. For this reason, the mixing of EPN with surfactants, gels, polymers, and/or other adjuvants remains an area that is actively explored (see review by Shapiro-Ilan, Han, & Dolinksi, 2012). In contrast, the use of EPN against the soilborne stages of noctuid pests (Bell, 1995; Cabanillas & Raulston, 1995; Feaster & Steinkraus, 1996; Hussain, Ahmad, & Ahmad, 2014) is a more reasonable approach. Rather than applying the IJ directly onto the soil, a more 'natural' approach would be to apply the EPN inside their nematode–killed (carrier) hosts (Jansson, Lecrone, & Gaugler, 1993). Doing so has, in the past, demonstrated improved nematode dispersal (Shapiro & Glazer, 1996), infectivity (Shapiro & Lewis, 1999), survival (Perez, Lewis, & Shapiro-Ilan, 2003), and efficacy (Shapiro-Ilan, Lewis, Tedders, & Son, 2003). In an attempt to explore this approach, the Agricultural Research Council–Small Grain Institute evaluated two South African populations of *H. bacteriophora* (populations SGI 22 and SGI 173) (Hatting et al., 2009), as well as population SGI 148, *S. tophus* (Çimen, Lee, Hatting, Hazir, & Stock 2014b), against the pre–pupal and pupal stages of *H. armigera*, using final instar *Tenebrio molitor* L. (Coleoptera: Tenebrionidae) as carrier host. A glasshouse trial was conducted to measure not only the percentage mortality caused, but also to improve the understanding of the bionomics of the approach in terms of (1) the survival and infective capacity of IJ over a 16–week period post emergence (IJ age); (2) the total number of IJ emerging from the host; (3) the duration of IJ emergence (*i.e.* the 'release period'); and (4) the number of IJ emerging from *H. armigera*, following infection over the above–mentioned period (*i.e.* in terms of EPN fitness). Briefly, the methodology entailed: the establishment of bean seedlings of *Phaseolus vulgaris* L. (Fabales: Fabaceae) in pots, the once–off inoculation of soil by means of EPN–infected *T. molitor* larva, and the release of two *H. armigera* final instar larvae per pot after 4, 8, 12, and 16 weeks (*i.e.* approximate IJ ages). Eight days after each release, all insects were removed, mortality noted, and both IJ yield, and day of IJ emergence, recorded (Jankielsohn A & Hatting, J.L. 2005).

Highest mortality (88 %) was noted after 2 and 4 weeks, with populations SGI 173 and SGI 148, respectively. No statistical differences were noted among any of the populations tested over the 16–week period. However, compared to the control, SGI 22 showed insignificant mortality from week 8 onwards. In general, mortalities decreased significantly over time, with an average mortality of only 8 % being recorded among all populations 16 weeks post inoculation. This finding was clearly reflected by the negative correlation coefficient values of −0.957, −0.926, and −0.977 for populations SGI 22, SGI148 and SGI 173, respectively. Compared

Table 20.3 Average number of IJ produced per *Helicoverpa armigera* cadaver by three EPN isolates, after varying lengths of time (infective juvenile age) spent in the soil without a host

IJ age (weeks)	SGI 22	SGI 148	SGI 173	
2	58,772a[a]	26,616b	114,636c	F = 21.70; $P < 0.05$
4	43,419a	8,658b	194,679c	F = 13.26; $P < 0.05$
8	14,442a	736b	25,527a	F = 18.14; $P < 0.05$
12	1,702a	0	16,317b	F = 13.89; $P < 0.05$
16	0	0	1,664	–
Mean ± SEM	23,667 ± 11,705	7,202 ± 3,215	70,564 ± 36,721	–

[a]Means within rows, followed by different letters, differ significantly at the 5 % test level (Tukey's HSD test)

Table 20.4 Average duration (days) of infective juvenile emergence from two different insect hosts (IJ ≤2 weeks old)

EPN isolate	*Tenebrio*[a]	*Helicoverpa*[b]
H. bacteriophora (SGI 22)	19.6 ± 5.5a[c]	18.4 ± 5.5a
S. tophus (SGI 148)	28.8 ± 4.3b	25.4 ± 2.3b
H. bacteriophora (SGI 173)	12.8 ± 4.5a	21.6 ± 1.3a
	F = 13.81 ($P < 0.05$)	F = 10.16 ($P < 0.05$)

[a]Artificial infection with 100 IJ per larva in laboratory
[b]Natural infection during glasshouse pot trial (IJ emerging from a single *T. molitor* cadaver)
[c]Means within columns, followed by different letters, differ significantly at the 5 % test level (Tukey's HSD test)

with week 2, a significant decrease in percentage mortality among all populations was observed from week 8 onwards.

IJ production in *H. armigera* was generally higher for SGI 173, with a pooled average of 70,564 IJs produced over the entire duration of the trial, compared with the 23,669 and 7,202 produced by SGI 22 and SGI 148, respectively (Table 20.3). Again, measured against IJ age, negative correlation coefficient values of –0.960, –0.855, and –0.777 for SGI 22, SGI 148 and SGI173, respectively, were apparent.

For IJ ≤2 weeks old, significant differences were observed within host species, with the longest duration being 28 days for SGI 148 from *T. molitor* (Table 20.4). Considering the impact of the time spent in the soil without a host on the duration of IJ emergence, the bollworm data generally showed a negative correlation in this regard, with correlation coefficients of –0.548, –0.742, and –0.366 for SGI populations 22, 148 and 173, respectively. The average durations are presented in Table 20.5.

In vivo production and application of EPN via *T. molitor* proved successful against *H. armigera*, with mean pooled mortalities of 80 ± 10 %, and 78 ± 11 %, recorded with 2– and 4–week–old IJ, respectively. A noticeable decline was, however, evident from the eighth week onwards, with only 8 ± 2 % (pooled) mortality being recorded with 16–week–old IJ. Whether this decline was due to a loss of symbiotic bacterial load associated with IJ aging (Flores-Lara, Renneckar,

Table 20.5 Average duration (days) of IJ emergence from *Helicoverpa armigera* following infection by IJ, having spent varying lengths of time in the soil without a host

Time (weeks)	SGI 22	SGI 148	SGI 173
2	18.4	25.4	21.6
4	24.2	9.5	22.8
8	17.2	13.0	19.4
12	15.2	0	21.4
16	0	0	4.0

Forst, Goodrich-Blair, & Stock, 2007), to decreasing IJ survival (Kung, Gaugler, & Kaya, 1990; Molyneux, 1985; Perez et al., 2003; Shapiro-Ilan, Stuart, & McCoy, 2006), and/or to a change in IJ foraging behaviour/ability (Grewal, Selvan, & Gaugler, 1994; Lewis, Campbell, & Gaugler, 1997; O'Leary, Stack, Chubb, & Burnell, 1998) affecting the eventual 'dose', is unknown. In any event, rapid dispersal/contact between the 'young' IJ and its host is critical during the initial stages of emergence, as has been pointed out by Stuart, Lewis, and Gaugler (1996). Coherently, this trait was found to be positively supported by the application of EPN by means of infected host cadavers, compared to aqueous suspension (Shapiro & Glazer, 1996). The ability of EPN to produce offspring was another fitness trait that was found to be negatively correlated with IJ age. In all three isolates, this ability deteriorated, with only SGI 173 producing some offspring (1, 664 IJ/cadaver), following infection with 16–week–old IJ.

An equally important aspect relates to the susceptibility of the pest at the time (life stage) of exposure to the EPN. In a study with *S. feltiae*, neonate larvae of the noctuid *Spodoptera exigua* (Hübner) (Lepidoptera: Noctuidae) were found to be significantly less susceptible to EPN infection than were 3– or 8–day old larvae (Kaya, 1985). Likewise, the high susceptibility of final instar *H. armigera* to *S. riobrave*, *S. carpocapse*, and *Heterorhabditis* sp. was reported by Tahir, Otto, and Hague (1995), with a similar trend also being noted with *S. glaseri*, *S. feltiae*, and *H. indica* (Karunakar, Easwaramoorhy, & David, 1999). In contrast, Glazer and Navon (1990) reported a negative relationship between the larval age of *H. armigera* and susceptibility to a population of *S. feltiae*, a phenomenon that was also observed with several EPN isolates tested against pecan weevil larvae (Shapiro-Ilan, 2001). These observations seem to support the general notion of employing EPN against latter larval stages of noctuid pests, even under relatively high temperature conditions (Ali, Pervez, Abid Hussain, & Ahmad, 2007; Cabanillas, Poinar, & Raulston, 1994; Grewal et al., 1994), to which species such as *H. armigera* have shown good adaptability (highest intrinsic rate of increase measured at 27.5 °C [Mironidis & Savopoulou-Soultani, 2008]).

The three EPN isolates tested here originate from the Free State province of South Africa, an area in a climatic zone defined as "humid subtropical with summer rainfall and cool (warmest month <22 °C)" (Hatting et al., 2009), and where the soil is typically expected to harbour *H. armigera*, given major crops, such as soya beans, wheat, maize, sunflower, apples, and selected vegetables, typically cultivated in the region. Although surveys to quantify the (natural) level of pre–pupal and pupal stage parasitism have not yet been conducted locally, the phenomenon has been explored elsewhere. Surveys over a 5–year period in the Lower Rio Grande Valley, Texas, found infection of 9 and 12 % in fall armyworm and corn armyworm, respectively (Raulston, Pair, Loera, & Cabanillas, 1992). Optimism gained from observing such natural levels of parasitism has led to augmentative attempts against noctuid species such as *H. virescens* (Bell, 1995; Bell & Hardee, 1994) and *H. zea* (Cabanillas & Raulston, 1995, 1996; Feaster & Steinkraus, 1996). Recently, a trial under seemingly challenging environmental conditions, and against the soil stages of *H. armigera* on chickpea, found up to 70 % moth suppression with *Steinernema masoodi* Alie, Shaheen, Pervez & Hussain (Rhabditida: Steinernematidae) at a rate of 6×10^9 IJ/ha (Hussain et al., 2014). The authors involved proposed that further research should be undertaken to optimise the timing of EPN applications, so as to coincide with irrigation during critical stages of the crop. To further underscore the importance of the correct timing of application, the data presented here suggest time–mediated fitness among IJ, as has been noted for mortality and IJ production (Table 20.2), as well as for the duration of IJ emergence (Table 20.4). Although a similar 'worst case scenario' (*i.e.*, no alternative insect host in the soil) is unlikely to occur under natural field conditions, the potential deterioration in IJ fitness over time should be taken into account when considering EPN applications.

20.3.7 The Sugarcane Stalk Borer **Eldana saccharina** *Walker (Lepidoptera: Pyralidae)*

The sugarcane stalk borer, *Eldana saccharina* Walker (Lepidoptera: Pyralidae), is the most injurious insect pest of sugarcane, *Saccharum* spp., in Southern Africa (Goebel & Sallam, 2011; Leslie, 2004). With its local identification dating back to the 1940s (Dick, 1945), it is today known to infest numerous host plants that are of economic significance throughout Africa (Polaszek & Khan, 1998). Larval feeding is associated with stalk tissue damage, with reduced sucrose levels, and with compromised plant vigour (Goebel & Way, 2003). Control of *E. saccharina* in South Africa is based on the adoption of an IPM approach encompassing cultural, genetic, chemical, and biological strategies (Carnegie, 1981; Conlong & Rutherford, 2009; Conlong & Way, 2015; Keeping, 2006; Leslie, 2009; Rutherford & Conlong, 2010).

Some of the earliest attempts at pest suppression using EPN in South Africa, have been against the larval stages of *E. saccharina* (Spaull, 1988, 1990, 1991). During the first trial, EPN were applied at midday to the foliage, at concentrations ranging from 100,000 to 1,000,000 IJ, in 200 ml suspension per infested stalk, with up to 56 % larval mortality being recorded (Spaull, 1988). By reducing both the number of IJ (87,000), and the water volume (57 mL) per stalk, late afternoon applications realised control levels ranging from 40 to 45 % (Spaull, 1990). In a subsequent trial, using a more cold–tolerant population at 100,000 IJ, borer mortality was

Fig. 20.3 Cane stalks with drilled holes positioned in vermiculate (Photo: JL Hatting, ARC-SGI)

6 %, suggesting a typical dose response rather than a temperature–linked response (Spaull, 1991).

In addition to the EPN concentration, another aspect to be considered is the potential impact of sugarcane sap (sucrose), as osmolyte, on the surviving IJ after entry into the borer tunnel. Delivery of IJ directly to the stubble and/or root stool shortly after harvesting might serve as another strategy for targeting *E. saccharina*, especially in older fields where infestation increases progressively. As follow–up to the earlier research by Spaull, bioassays were conducted by ARC–SGI, in collaboration with the South African Sugar Research Institute (SASRI; Dr Des Conlong), to (1) verify the pathogenicity of several newly collected indigenous EPN (Hatting et al., 2009) against *E. saccharina*; (2) measure the survival of IJ in sugarcane sap; and (3) investigate the movement and ability of IJ to infect *E. saccharina* inside infested sugarcane stalks. Briefly, the methodologies entailed the use of a piece of filter paper in a Petri dish assay, exposing final instar *E. saccharina* to five *H. bacteriophora* populations (SGI 32, SGI 43, SGI 180, INF 61, SASRI 75), three *S. tophus* (R 343, SASRI 356, SASRI 426), two *S. innovationi* (SGI 35, SASRI 198), and one *S. khoisanae* (R 293); the exposure of five EPN populations (SGI 32, SGI 35, SGI 43, SASRI 75, and SASRI 426) to sugarcane (cultivar N12) sap concentrations of 50 % (diluted with sterile distilled water) and 100 % (undiluted sap), with survival checks being undertaken after 24, 48, and 72 h; and the artificial infestation of sugarcane stalks with mid–instar *E. saccharina* larvae by way of vertically drilled holes, and topical inoculation with 1 ml EPN (SGI 35) suspension per stalk (Fig. 20.3).

Six EPN populations (SGI 35, SGI 43, R 293, R 343, SASRI 75, SASRI 356) caused 100 % mortality, with positive recycling in the host. Three populations (SGI 32, SGI 180, INF 61) killed only 33 % of larvae, with no recycling being recorded for SGI 32 and INF 61. Population SGI 35 was selected for

Table 20.6 Effect of pure sugarcane sap on % survival of five EPN populations after 72 h

Species/population	Control (water)[a]	Treatment[a]
H. bacteriophora (SGI 32)	100a	48bc
S. innovationi (SGI 35)	90ab	81a
H. bacteriophora (SGI 43)	99a	67ab
H. bacteriophora (SASRI 75)	81b	28c
S. tophus (SASRI 426)	100a	79a

[a]Values within columns, followed by different letters, differ significantly at the 0.5 % (0.05/10) test level (chi^2 value >7.9)

a concentration–response assay with an LC_{90} of 44 (fiducial limits: 26–617) IJ per larva. The apparent susceptibility of *E. saccharina* to EPN infection is also supported by the findings of Pillay, Martin, Rutherford, and Berry (2009). The authors concerned tested ten indigenous EPN, recording 100 % mortality after 48 h with two *Steinernema* populations, EST3D and GING13G.

Of the five populations tested, only SGI 35 showed >80 % survival after 72 h in 100 % (undiluted) sap (Table 20.6). Survival in 50 % sap after 72 h decreased markedly among all isolates tested, with the highest survival being only 64 % noted with SGI 32. In a study by Glazer and Salame (2000), the effect of different osmolytes on the viability of *Steinernema carpocapsae* (Weiser) Wouts, Mráček, Gerdin & Bedding (Rhabditida. Steinernematidae) 'All' was evaluated. Viability was found not to have been affected by sucrose concentrations ranging from 1.2 to 3.7 mol/L after 24 h, but declined to only 27 %, in the higher concentration after 72 h. Similarly, differences in EPN tolerance towards a 20 % sucrose solution were also noted by Shamseldean, El-Sadawy, and Allam (2004). These authors found *S. carpocapsae* 'All' to be the most tolerant, while *H. taysearae* was found to survive for only 31 h. Superior osmotic tolerance to a mixture of fortified artificial seawater and glycerol at 15 °C was also noted for *S. carpocapsae* 'All' by Yan et al. (2010). Seemingly, the selection of EPN species/populations, based on their ability to tolerate the sucrose–rich environment within a sugarcane plant, should be considered when targeting *E. saccharina* and other borer species such as *Diatraea saccharalis* (Fabricius) (Lepidoptera: Crambidae). The latter species had previously been targeted with *H. baujardi* LPP7 and *S. carpocapsae* NCAll (Bellini & Dolinski, 2012).

Three of the 40 treated stalks had missing larvae on day 7, resulting in their omission from further calculations. Of the 37 remaining stalks, 24 (65 %) harboured dead *E. saccharina* (Fig. 20.4), of which 14 (58 %) larvae showed positive EPN recycling. Control mortality was 5 %. The data concerned support the notion of using EPN to target *E. saccharina* larvae inside the stubble, and directly after harvest (Fig. 20.4), while the cut wounds are still 'fresh' and relatively uncontaminated. Moreover, the low water volume of only 1 ml per stalk supports the practicality of adopting such a strategy under field conditions. Additional research aimed at optimising the dose (IJ/mL), the formulation and the application method, is warranted.

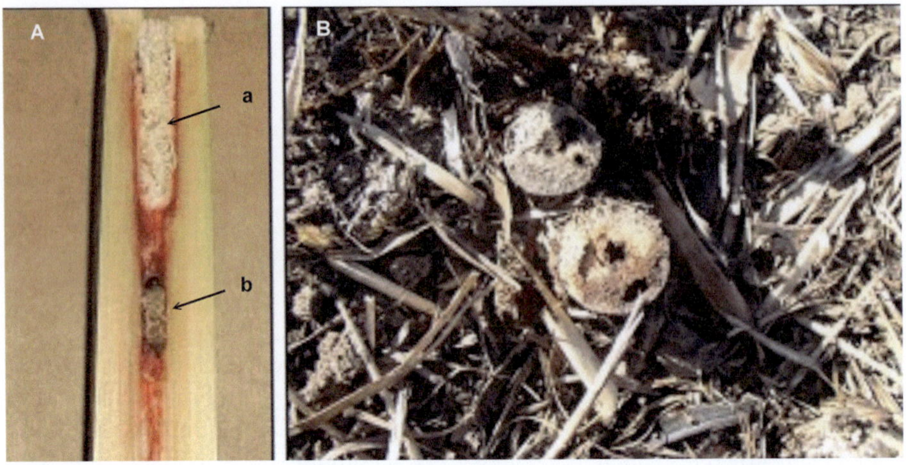

Fig. 20.4 (**a**) Split stalk, showing EPN–infected *Eldana saccharina* larva (*a*: entrance filled with frass, and *b*: EPN–infected cadaver); (**b**) cane stubble after harvest, revealing tunnel damage by *Eldana saccharina* as potential entry point for EPN (Photos: JL Hatting, ARC–SGI)

20.4 Current Legislation with Regard to Entomopathogenic Nematodes in South Africa

In South Africa, EPN–based products constitute an 'agricultural remedy', and, as such, are governed by the Fertilizers, Farm Feeds, Agricultural Remedies and Stock Remedies Act 36 of 1947. According to this Act, an 'agricultural remedy' means any chemical substance, or biological remedy, or any mixture, or combination, of any substance, or remedy, that is intended, or offered, to be used (*a*) for the destruction, control, repelling, attraction, or prevention of any undesired microbe, alga, nematode, fungus, insect, plant, vertebrate, or invertebrate, or any product thereof. The use of EPNs, however, is still in its infancy in this country, largely impeded by the lack of locally–produced/formulated (indigenous) products. Although many chemical insecticides are available on the local market, the South African government has, through legislation, banned, or limited, the use/sale of several insecticides since the late 1970s. Recent interventions include the withdrawal of monocrotophos (2005), chlorpyrifos (products for home use, 2010), endosulfan (2012), and aldicarb (2012), fuelling the need for alternative remedies, including EPN–based products.

20.5 Conclusions and Future Directions

South Africa relies heavily on chemical pesticides with several hundred registered pesticides available on the local market (CropLife South Africa, [s.d.]). Not surprisingly, South Africa is also one of the four largest importers of pesticides in sub–Saharan Africa (Osibanjo et al., 2002). The economic implications and potential environmental impact thereof was reviewed by Quinn et al. (2011). Over the past 10 years much research input has been directed towards the base–line characterization of indigenous EPN species as well as the verification of target pest suitability for biocontrol with EPN. Ideally, such endeavours should be supported by the commercialization of indigenous species/populations, thereby negating the need for importation and release of exotic organisms. Interest in mass production of indigenous species/populations of EPN is evident from recent work by Fasemore (2012), Van Zyl (2012), Ferreira and Addison & Malan (2014), Ferreira and Malan 2014b), Ramakuwela, Hatting, Laing, and Hazir (2014), and Van Zyl and Malan (2014a, 2014b). According to the USA–based company 'MarketsandMarkets' the global market for biopesticides was valued at $1,796.56 Million in 2013 and is expected to reach $4,369.88 Million by 2019, growing at a CAGR of 16.0 % from 2014 to 2019. The need, in South Africa, for safer, environmentally sound, alternatives to chemical pesticides is expected to contribute to the abovementioned market expansion.

References

Addison, M. F. (2005). Suppression of codling moth *Cydia pomonella* L. (Lepidoptera: Tortricidae) populations in South African apple and pear orchards using sterile insect release. *Acta Horticulturae, 671*, 313–327.

Ali, S. S., Pervez, R., Abid Hussain, M., & Ahmad, A. (2007). Effect of temperature on survival of *Steinernema seemae*, *S. masoodi* and *S. carpocapsae* (Rhabditida: Steinernematidae) and their subsequent infectivity to prepupa of *Helicoverpa armigera* (Hubner). *Archives of Phytopathology and Plant Protection, 40*(3), 183–187.

Allsopp, E., Barnes, B. N., Blomefield, T. L., & Pringle, K. L. (2015). Grapevine. In G. L. Prinsloo & V. M. Uys (Eds.), *Insects of cultivated plants and natural pastures in Southern Africa* (pp. 420–437). Petroria, South Africa: Entomological Society of Southern Africa.

Annecke, D. P., & Moran, V. C. (1982). *Insects and mites of cultivated plants in South Africa*. Durban/Pretoria, South Africa: Butterworths.

Barnes, B. M., Eyles, D. K., & Franz, G. (2002). South Africa's fruit fly SIT programme – the Hex River Valley pilot project and beyond. In B. M. Barnes (Ed.), *Proceedings of the 6th international symposium on fruit flies of economic importance* (pp. 131–141). Isteg Scientific Publications, Irene, South Africa.

Barnes, B. N. (1987). *Bionomics, behaviour and monitoring of the vine snoutbeetle,* Phlyctinus callosus *Boh., indeciduous fruit orchards, with proposals for an improved control strategy*. PhD thesis, Stellenbosch University, Stellenbosch.

Barnes, B. N. (1989). Embryonic and immature stages of *Phlyctinus callosus* Boh. (Coleoptera, Curculionidae) – Aspects of biology and behavior with respect to control in deciduous fruit orchards. *Journal of the Entomological Society of Southern Africa, 52*, 165–178.

Barnes, B. N., & Giliomee, J. H. (1992). Fruit–feeding behavior of banded fruit weevil, *Phlyctinus callosus* (Schönherr) (Col, Curculionidae), in apple orchards. *Journal of Applied Entomology Zeitschrift für Angewandte Entomologie, 113*, 407–415.

Barnes, B. N., Knipe, M. C., & Calitz, F. J. (1994). Trunk barriers provide effective control of banded fruit weevil on apples and nectarines. *Deciduous Fruit Grower, 44*, 327–340.

Barnes, B. N., Knipe, M. C., & Calitz, F. J. (1996). Latest results with trunk exclusion barriers for weevil control on apples (Jongste resultate met stamsperbande vir kalanderbeheer op appels). *Deciduous Fruit Grower, 46*, 284–287.

Barnes, B. N., & Pringle, K. L. (1989). Oviposition by the banded fruit weevil, *Phlyctinus callosus* (Schoenherr) (Coleoptera, Curculionidae), in deciduous fruit orchards in South Africa. *Bulletin of Entomological Research, 79*, 31–40.

Barnes, M. M. (1991). Codling moth occurrence, host race formation, and damage. In L. P. S. Van der Geest & H. H. Evenhuis (Eds.), *Tortricid pests: Their biology, natural enemies and control* (pp. 313–327). Amsterdam: Elsevier.

Bastidas, B., Edgar, P., & San-Blas, E. (2014). Size does matter: The life cycle of *Steinernema* spp. in micro–insect hosts. *Journal of Invertebrate Pathology, 121*, 46–55.

Basson, S. (1993) Biological control of *Phlyctinus callosus* by entomogenous nematodes. Proceedings of the 11th Symposium of the Nematological Society of Southern Africa, Rustenburg, South Africa. *Phytophylactica, 25*, 303.

Bell, M. R. (1995). Effects of an entomopathogenic nematode and nuclear polyhedrosis virus on emergence of *Heliothis virescens* (Lep: Noctuidae). *Journal of Entomological Science, 30*, 243–250.

Bell, M. R., & Hardee, D. D. (1994, January 5–8). Tobacco budworm: Possible use of various entomopathogens in large area pest management. *Proceedings of the Beltwide Cotton Production conference* (pp. 1168–1171), San Diego, CA: Beltwide Cotton Production.

Bellini, L. L., & Dolinski, C. (2012). Foliar application of entomopathogenic nematodes (Rhabditida: Steinernematidae and Heterorhabditidae) for the control of *Diatraea saccharalis* in greenhouse. *Ciências Agrárias Londrina, 33*, 997–1004.

Bong, C. F. J., & Sikorowski, P. P. (1983). Use of DD136 strain of *Neoaplectana carpocapsae* Weiser (Rhabditidae: Steinernematidae) for control of corn earworm (Lepidoptera: Noctuidae). *Journal of Economic Entomology, 76*, 590–593.

Bredenhand, E., Van Hoorn, A., May, F., Ferreira, T., & Johnson, S. (2010). Evaluation of techniques for monitoring banded fruit weevil, *Phlyctinus callosus* (Schoenherr) (Coleoptera: Curculionidae), infestation in blueberry orchards. *African Entomology, 18*, 205–209.

Cabanillas, H. E., Poinar, G. O., & Raulston, J. R. (1994). *Steinernema riobravis* n.sp. (Rhabditida: Steinernematidae) from Texas. *Fundamental and Applied Nematology, 17*, 123–131.

Cabanillas, H. E., & Raulston, J. R. (1995). Impact of *Steinernema riobravis* (Rhabditida: Steinernematidae) on the control of *Helicoverpa zea* (Lepidoptera: Noctuidae) in corn. *Journal of Economic Entomology, 88*, 58–64.

Cabanillas, H. E., & Raulston, J. R. (1996). Effects of furrow irrigation on the distribution and infectivity of *Steinernema riobravis* against corn earworm in corn. *Fundamental and Applied Nematology, 19*, 273–281.

CABI. (2011). *Crop protection compendium. Thaumatotibia leucotreta*. Wallingford, CT: CABI. http://www.cabi.org/cpc

Carnegie, A. J. M. (1981). Combatting *Eldana saccharina* Walker: A progress report. *Proceedings of the International Society of Sugar Cane Technologists, 55*, 116–119.

Carpenter, J., Bloem, S., & Hofmeyer, H. (2007). Area–wide control tactics for the false codling moth *Thaumatotibia leucotreta* in South Africa: A potential invasive species. In: M. J. B. Vreysen, A. S. Robertson, & J. Hendrichs (Eds.), *Area–wide control of pest insects* (pp. 351–359). Springer Science & Business Media, The Netherlands.

CIBC. (1984). Possibilities for the control of false codling moth *Cryptophlebia leuctreta* (Lepidoptera: Tortricidae). *Biocontrol News and Information, 5*, 217–220.

Çimen, H. C., Lee, M. M., Hatting, J., Hazir, S., & Stock, S. P. (2014a). *Steinernema innovationi* n. sp. (Panagrolaimomorpha: Steinernematidae), a new entomopathogenic nematode species from South Africa. *Journal of Helminthology*.

Çimen, H. C., Lee, M.-M., Hatting, J., Hazir, S., & Stock, P. S. (2014b). *Steinernema tophus* sp. n. (Nematoda: Steinernematidae), a new entomopathogenic nematode from South Africa. *Zootaxa, 3821*, 337–353.

Conlong, D. E., & Rutherford, R. S. (2009). Conventional and new biological and habitat interventions for integrated pest management systems: Review and case studies using *Eldana saccharina* Walker (Lepidoptera: Pyralidae). In R. Peshin & A. K. Dhawan (Eds.), *Integrated pest management: Innovation–development process* (pp. 241–260). Dordrecht, The Netherlands: Springer Science and Business Media B.V.

Conlong, D. E., & Way, M. J. (2015). Sugarcane. In G. L. Prinsloo & V. M. Uys (Eds.), *Insects of cultivated plants and natural pastures in Southern Africa* (pp. 156–176). Petroria, South Africa: Entomological Society of Southern Africa.

Cossentine, J. E., Sholberg, P. L., Jensen, L. B. J., Bedford, K. E., & Sheperd, T. C. (2004). Fumigation of empty fruit bins with carbon dioxide to control diapausing codling moth larvae and *Penicillium expansum* Link. ex Tom spores. *HortScience, 39*, 429–432.

CropLife South Africa. [s.d.]. *Agricultural remedies database*. http://www.croplife.co.za/

Daiber, C. C. (1979a). A study of the biology of the false codling moth [(*Cryptophlebia leucotreta* (Meyr.)]: The cocoon. *Phytophylactica, 11*, 151–157.

Daiber, C. C. (1979b). A study of the biology of the false codling moth [(*Cryptophlebia leucotreta* (Meyr.)]: The larva. *Phytophylactica, 11*, 141–144.

Daiber, C. C. (1980). A study of the biology of the false codling moth *Cryptophlebia leucotreta* (Meyr.): The adult and generations during the year. *Phytophylactica, 12*, 187–193.

Daiber, C. C. (1989). The false codling moth, *Cryptophlebia leucotreta* (Meyr) (Lepidoptera, Tortricidae), in Southern Africa. *Journal of Plant Diseases and Protection, 96*, 71–80.

De Waal, J. Y., Addison, M. F., & Malan, A. P. (2013). A superabsorbent polymer formulation for improved efficacy of *Heterorhabditis zealandica* (Rhabditida: Heterorhabditidae) control of codling moth larvae, *Cydia pomonella* (L.) (Lepidoptera: Tortricidae). *Biocontrol Science & Technology, 23*, 62–78.

De Waal, J. Y., Malan, A. P., & Addison, M. F. (2011a). Efficacy of entomopathogenic nematodes (Rhabditida: Heterorhabditidae and Steinernematidae) against codling moth, *Cydia pomonella* (Lepidoptera: Tortricidae) in temperate regions. *Biocontrol Science & Technology, 20*, 489–502.

De Waal, J. Y., Malan, A. P., & Addison, M. F. (2011b). Evaluating mulches together with *Heterorhabditis zealandica* (Rhabditida: Heterorhabditidae) for the control of diapausing codling moth larvae, *Cydia pomonella* (L.) (Lepidoptera: Tortricidae). *Biocontrol Science & Technology, 21*, 225–270.

De Waal, J. Y., Malan, A. P., Levings, J., & Addison, M. F. (2010). Key elements in the successful control of diapausing codling moth, *Cydia pomonella* (L.) (Lepidoptera: Tortricidae) in fruit bins with a South African isolate of *Heterorhabditis zealandica* (Rhabditida: Heterorhabditidae). *Biocontrol Science and Technology, 20*, 489–502.

Dick, J. (1945). Some data on the biology of the sugarcane borer *Eldana saccharina* Walker. *Proceedings of the South African Sugar Technologists Association, 19*, 75–79.

Dolinski, C., Del Valle, E. E., Burla, R. S., & Machado, I. R. (2007). Biological traits of two native Brazilian entomopathogenic nematodes (Rhabditida: Heterorhabditidae). *Nematologia Brasileira, 31*(32), 180–185.

Elawad, S., Ahmad, W., & Reid, A. P. (1997). *Steinernema abbasi* sp. n. (Nematoda: Steinernematidae) from the Sultanate of Oman. *Fundamental and Applied Nematology, 20*, 435–442.

El-Rahman, R. M. A., El-Razzik, M. I. A., Osman, E. A., & Mangoud, A. A. H. (2012). Efficacy of the entomopathogenic nematodes and their based–product on some species of mealybugs (Hemiptera: Pseudococcidae) in Egypt. *Egyptian Academic Journal of Biological Sciences Entomology, 5*, 193–196.

Fasemore, O. A. (2012). *An investigation of the possiblities of scaling–up and mass–producing entomopathogenic nematodes (EPNs)* (154 p.). PhD thesis, University of the Witwatersrand, Johannesburg, South Africa.

Feaster, M. A., & Steinkraus, D. C. (1996). Inundative biological control of *Helicoverpa zea* (Lepidoptera: Noctuidae) with the entomopathogenic nematode *Steinernema riobravis* (Rhabditida: Steinernematidae). *Biological Control, 7*, 38–43.

Ferreira, T., Van Reenen, C. A., Endo, A., Spröer, C., Malan, A. P., & Dicks, L. M. T. (2013). Description of *Xenorhabdus khoisanae* sp. nov., a symbiont of the entomopathogenic nematode *Steinernema khoisanae*. *International Journal of Systematic and Evolutionary Microbiology, 63*, 3220–3224.

Ferreira, T., Van Reenen, C. A., Pagès, S., Patrick Tailliez, P., Malan, A. P., & Dicks, L. M. T. (2013). Description of *Photorhabdus luminescens* subsp. *noenieputensis* subsp. nov., a symbiotic bacterium associated with a new *Heterorhabditis* species related to *Heterorhabditis indica*. *International Journal of Systematic and Evolutionary Microbiology, 63*, 1853–1858.

Ferreira, T., & Malan, A. P. (2014a). Potential of entomopathogenic nematodes for the control of the banded fruit weevil, *Phlyctinus callosus* (Schönherr) (Coleoptera: Curculionidae). *Journal of Helminthology, 88*, 293–301.

Ferreira, T., & Malan, A. P. (2014b). *Xenorhabdus* and *Photorhabdus*, bacterial symbionts of the entomopathogenic nematodes *Steinernema* and *Heterorhabditis* and their *in vitro* liquid mass culture: a review. *African Entomology, 22*(1), 1–14.

Ferreira, T., Addison, M. F., & Malan, A. P. (2014). *In vitro* liquid culture of a South African isolate of *Heterorhabditis zealandica* for the control of insect pests. *African Entomology, 22*, 80–92.

Ferreira, T., Van Reenen, C. A., Endo, A., Spröer, C., Malan, A. P., & Dicks, L. M. T. (2014). Description of *Photorhabdus zealandica* sp. nov., a symbiont of the entomopathogenic nematode *Heterorhabditis zealandica*. *International Journal of Systematic and Evolutionary Microbiology, 64*(Pt 5), 1540–1545.

Ferreira, T., Van Reenen, C. A., Tailliez, P., Pagès, S., Malan, A. P., & Dicks, L. M. T. (2014). First report of the symbiotic bacterium, *Xenorhabdus indica*, associated with the entomopathogenic nematode *Steinernema yirgalemense*. *Journal of Helminthology*.

Fitt, G. P. (1989). The ecology of *Heliothis* species in relation to agroecosystems. *Annual Review of Entomology, 34*, 17–52.

Flores-Lara, Y., Renneckar, D., Forst, S., Goodrich-Blair, H., & Stock, P. (2007). Influence of nematode age and culture conditions on morphological and physiological parameters in the bacterial vesicle of *Steinernema carpocapsae* (Nematoda: Steinernematidae). *Journal of Invertebrate Pathology, 95*, 110–118.

Glazer, I., & Navon, A. (1990). Activity and persistence of entomoparasitic nematodes tested against *Heliothis armigera* (Lepidoptera: Noctuidae). *Journal of Economic Entomology, 83*(5), 1795–1800.

Glazer, I., & Salame, L. (2000). Osmotic survival of the entomopathogenic nematode *Steinernema carpocapsae*. *Biological Control, 18*, 251–257.

Goebel, F. R., & Sallam, N. (2011). New pest threats for sugarcane in the new bioeconomy and how to manage them. *Current Opinion in Environmental Sustainability, 3*, 81–89.

Goebel, F. R., & Way, M. J. (2003). Investigation of the impact of *Eldana saccharina* (Lepidoptera: Pyralidae) on sugarcane yield in field trials in Zululand. *Proceedings of the South African Sugar Technologists Association, 77*, 256–265.

Grenier, E., Bonifassi, E., Abad, P., & Laumond, D. (1996). Use of species specific satellite DNAs as diagnostic probes in the identification of Steinernematidae and Heterorhabditidae entomopathogenic nematodes. *Parasitology, 113*, 483–489.

Grenier, E., Laumond, C., & Abad, P. (1996). Molecular characterization of two species–specific tandemly repeated DNAs from entomopathogenic nematodes Steinernema and Heterorhabditis (Nematoda: Rhabditida). *Molecular and Biochemical Parasitology, 83*, 47–56.

Grewal, P. S., Selvan, S., & Gaugler, R. (1994). Thermal adaptation of entomopathogenic nematodes: Niche breadth for infection, establishment, and reproduction. *Journal of Thermal Biology, 19*, 245–253.

Harington, J. S. (1953). Observation on the biology, the parasites and the taxonomic position of the maize beetle – *Heteronychus san–helenae* Blanch. *South African Journal of Science, 50*, 11–14.

Hatting, J., Stock, P. S., & Hazir, S. (2009). Diversity and distribution of entomopathogenic nematodes (Steinernematidae, Heterorhabditidae) in South Africa. *Journal of Invertebrate Pathology, 102*, 120–128.

Hatting, J. L., & Kaya, H. K. (2001, July 2–5). Entomopathogenic nematodes: prospects for biological control in South Africa. In: *Proceedings of the 13th Entomological Congress, Entomological Society of Southern Africa* (pp 28–29). Pietermaritzburg, South Africa: Entomological Society of Southern Africa. ISBN 0–620–27806–4.

Hattingh, V., & Moore, S. D. (2003). Mealybugs. In T. G. Grout (Ed.), *Integrated production guidelines for export citrus. Integrated pest and disease management* (pp. 65–69). Nelspruit, : Citrus Research International.

Hominick, W. (2002). Biogeography. In R. Gaugler (Ed.), *Entomopathogenic nematology* (pp. 115–143). Wallingford, UK: CABI.

Hussain, M. A., Ahmad, R., & Ahmad, W. (2014). Evaluation of *Steinernema masoodi* (Rhabditida: Steinernematidae) against soil–dwelling life stage of *Helicoverpa armigera* (Lepidoptera: Noctuidae) in laboratory and microplot study. *Canadian Journal of Plant Protection, 2*(1), 4–8.

Jankielsohn, A., & Hatting, J. L. (2005, August 7–11). Efficacy of entomopathogenic nematodes, applied in an insect cadaver, as biological control agent against soil-dwelling stages of bollworm (*Helicoverpa armigera* Hübner). *Proceedings of the 38th Annual Meeting of the Society for Invertebrate Pathology*. Anchorage, Alaska.

Jansson, R. K., Lecrone, S. H., & Gaugler, R. (1993). Field efficacy and persistence of entomopathogenic nematodes (Rhabditida: Steinernematidae, Heterorhabditidae) for control of sweet potato weevil (Coleoptera: Apionidae) in southern Florida. *Journal of Economic Entomology, 86*, 1055–1063.

Judd, G. J. R., & Gardiner, M. G. T. (2005). Towards eradication of codling moth in British Columbia by complimentary actions of mating disruption, tree banding and sterile insect technique: Five–year study in organic orchards. *Crop Protection, 24*, 718–733.

Kanga, F. N., Trinh, P. Q., Waeyenberge, L., Spiridonov, S. E., Hauser, S., & Moens, M. (2012). Two new species of *Steinernema* Travassos, 1972 from the humid forest of southerm Cameroon. *Russian Journal of Nematology, 20*, 15–36.

Kanga, F. N., Waeyenberge, L., Hauser, S., & Moens, M. (2012). Distribution of entomopathogenic nematodes in Southern Cameroon. *Journal of Invertebrate Pathology, 109*, 41–51.

Karunakar, G., Easwaramoorhy, S., & David, H. (1999). Susceptibility of nine lepidopteran insects to *Steinernema glaseri S. feltiae and Heterorhabditis indicus* infection. *International Journal of Nematology, 9*(1), 68–71.

Kaya, H. K. (1985). Susceptibility of early larval stages of *Pseudaletia unipuncta* and *Spodoptera exigua* (Lepidoptera: Noctuidae) to the entomogenous nematode *Steinernema feltiae* (Rhabditida: Steinernematidae). *Journal of Invertebrate Pathology, 46*, 58–62.

Keeping, M. G. (2006). Screening of South African sugarcane cultivars for resistance to stalk borer, *Eldana saccharina* Walker (Lepidoptera: Pyralidae). *African Entomology, 14*, 277–288.

Kung, S.-P., Gaugler, R., & Kaya, H. K. (1990). Soil type and entomopathogenic nematode persistence. *Journal of Invertebrate Pathology, 55*, 401–406.

Kuschel, G. (1972). The foreign Curculionidae established in New Zealand (Insecta: Coleoptera). *New Zealand Journal of Science, 15*, 273–289.

Lacey, L. A., & Unruh, T. R. (1998). Entomopathogenic nematodes for control of codling moth, *Cydia pomonella* (Lepidoptera: Tortricidae): Effect of nematode species, concentration, temperature, and humidity. *Biological Control, 13*, 190–197.

Le Vieux, P. D., & Malan, A. P. (2013a). An overview of the vine mealybug (*Planococcus ficus*) in South African vineyards and the use of entomopathogenic nematodes as potential biocontrol agents. *South African Journal of Enology and Viticulture, 34*, 34–45.

Le Vieux, P. D., & Malan, A. P. (2013b). The potential use of entomopathogenic nematodes to control *Planococcus ficus* (Signoret) (Hemiptera: Pseudococcidae). *South African Journal of Entomology and Viticulture, 34*, 296–306.

Le Vieux, P. D., & Malan, A. P. (2014). Prospects for using entomopathogenic nematodes to control the vine mealybug, *Planococcus ficus*, in South African vineyards. *South African Journal of Enology and Viticulture, 36*, 59–70.

Le Vieux, P. D., & Malan, A. P. (2015). Prospects for using entomopathogenic nematodes to control the vine mealybug, *Planococcus ficus*, in South African vineyards. *South African Journal of Enology and Viticulture, 36*(1), 59–70.

Leslie, G. (2004). Pests of sugarcane. In G. James (Ed.), *Sugarcane* (pp. 78–100). Oxford, U.K.: Blackwell Science.

Leslie, G. W. (2009). Estimating the economic injury level and the economic threshold for the use of alpha–cypermethrin against the sugarcane borer, *Eldana saccharina* Walker (Lepidoptera: Pyralidae). *International Journal of Pest Management, 55*, 37–44.

Lewis, E. E., Campbell, J. F., & Gaugler, R. (1997). The effects of aging on the foraging behaviour of *Steinernema carpocapsae* (Rhabdita: Steinernematidae). *Nematologica, 43*, 355–362.

Malan, A. P., Knoetze, R., & Moore, S. D. (2011). Isolation and identification of entomopathogenic nematodes from citrus orchards and their biocontrol potential against false codling moth. *Journal of Invertebrate Pathololog*y, *108*, 115–125.

Malan, A. P., Knoetze, R., & Tiedt, L. R. (2014). *Heterorhabditis noenieputensis* n. sp. (Rhabditida: Heterorhabditidae), a new entomopathogenic nematode from South Africa. *Journal of Helminthology, 88*, 138–151.

Malan, A. P., & Manrakhan, A. (2009). Susceptibility of the Mediterranean fruit fly (*Ceratitis capitata*) and the Natal fruit fly (*Ceratitis rosa*) to entomopathogenic nematodes. *Journal of Invertebrate Pathology, 100*, 47–49.

Malan, A. P., Nguyen, K. B., & Addison, M. F. (2006). Entomopathogenic nematodes (Steinernematidae and Heterorhabditidae) from the southwestern parts of South Africa. *African Plant Protection, 12*, 65–69.

Malan, A. P., Nguyen, K. B., De Waal, J. Y., & Tiedt, L. (2008). *Heterorhabditis safricana* n. sp. (Rhabditida: Heterorhabditidae), a new entomopathogenic nematode from South Africa. *Nematology, 10*, 381–396.

Mamiya, Y. (2008). *Steinernema kushidai* n. sp. (Nematoda: Steinernematodae) associated with scarabaeid beetle larvae from Shizuoca, Japan. *Applied Entomology and Zoology, 23*, 313–320.

Manrakhan, A., Daneel, J.-H., & Moore, S. D. (2013). The impact of naturally occurring entomopathogenic nematodes on false codling moth, *Thaumatotibia leucotreta* (Lepidoptera: Tortricidae), in citrus orchards. *Biocontrol Science and Technology, 24*, 241–245.

Mekete, T., Gaugler, R., Nguyen, K. B., Mandefro, W., & Tessera, T. (2005). Biogeography of entomopathogenic nematodes in Ethiopia. *Nematropica, 35*, 31–35.

Mironidis, G. K., & Savopoulou-Soultani, M. (2008). Development, survivorship, and reproduction of *Helicoverpa armigera* (Lepidoptera: Noctuidae) under constant and alternating temperatures. *Environmental Entomology, 37*(1), 16–28.

Molyneux, A. S. (1985). Survival of infective juveniles of *Heterorhabditis* spp., and *Steinernema* spp. (Nematoda: Rhabditida) at various temperatures and their subsequent infectivity for insects. *Revue de Nematologie, 8*(2), 165–170.

Moore, S. B. (2002). Entomopathogens and microbial control of citrus pests in South Africa: A review. *South African Fruit Journal, 1*, 30–32.

Moore, S. D., & Hattingh, V. (2012). A review of current pre–harvest control options for false codling moth in citrus in southern Africa. *South African Fruit Journal, 11*, 82–85.

Moore, S. D., Kirkman, W., & Stephen, P. (2004). Cryptogran: A virus for the biological control of false codling moth. *South African Fruit Journal, 3*, 35–39.

Mwaniki, S. W., Nderitu, J. H., Olubayo, F., Kimenju, J. W., & Nguyen, K. (2008). Factore influencing the occurrence of entomopathogenic nematodes in the Central Rift Valley Region of Kenya. *African Journal of Ecology, 46*(Suppl. 1), 79–84.

Myburgh, A. C. (1980). Infestation potential of the codling moth. *The Deciduous Fruit Grower, 30*, 368–377.

Newton, P. J. (1998). False codling moth *Cryptophlebia leuctreta* (Meyrick). In E. C. G. Bedford, M. A. Van den Berg, & E. A. De Villiers (Eds.), *Citrus pests in the Republic of South Africa* (pp. 194–200). Nelspruit, South Africa: Dynamic Ad.

Nguyen, K. B., Hunt, D. J., & Mráček, Z. (2007). Steinernematidae: Species descriptions. In K. B. Nguyen & D. J. Hunt (Eds.), *Entomopathogenic nematodes: Systematics, phylogeny and bacterial simbionts* (Nematology Monographs and Perspectives, pp. 121–609). Leiden, The Netherlands/Boston: Brill.

Nguyen, K. B., Malan, A. P., & Gozel, U. (2006). *Steinernema khoisanae* n. sp. (Rhabditida: Steinernematidae), a new entomopathogenic nematode from South Africa. *Nematology, 8*, 157–175.

Nguyen, K. B., Tesfamariam, M., Gozel, U., Gaugler, R., & Adams, B. J. (2004). *Steinernema yirgalemense* n. sp. (Rhabditida: Steinernematidae) from Ethiopia. *Nematology, 6*, 839–856.

Nthenga, I., Knoetze, R., Berry, S., Tiedt, L. R., & Malan, A. P. (2014). *Steinernema sacchari* n. sp. (Rhabditida: Steinernematidae), a new entomopathogenic nematode from South Africa. *Nematology, 16*, 475–494.

O'Leary, S. A., Stack, C. M., Chubb, M. A., & Burnell, A. M. (1998). The effect of day of emergence from the insect cadaver on the behavior and environmental tolerances of infective juveniles of the entomopathogenic nematode *Heterorhabditis megidis* (strain UK211). *Journal of Parasitology, 84*(4), 665–672.

Osibanjo, O., Bouwman, H., Bashir, N. H. H., Okond'Ahoka, J., Choong Kwet Yve, R., & Onyoyo, H. A. (2002). *Regionally based assessment of persistent toxic substances* (Sub–Saharan Africa Regional Report). Geneva, Switzerland: UNEP Chemicals.

Perez, E. E., Lewis, E. E., & Shapiro-Ilan, D. I. (2003). Impact of host cadaver on survival and infectivity of entomopathogenic nematodes (Rhabditida: Steinernematidae and Heterorhabditidae) under desiccating conditions. *Journal of Invertebrate Pathology, 82*, 111–118.

Pillay, U., Martin, L. A., Rutherford, R. S., & Berry, S. D. (2009). Entomopathogenic nematodes in sugarcane in South Africa. *Proceedings of the South African Sugar Technologists Association, 82*, 538–541.

Polaszek, A., & Khan, Z. R. (1998). Host plants. In: A. Polaszek (Ed.), *African cereal stem borers: Economic importance, taxonomy, natural enemies and control* (pp. 3–10). Wallington, UK: CAB International, in association with ACP–EU Technical Centre for Agricultural and Rural Co–operation (CTA).

Pringle, K. L., Eyles, D. K., & Brown, L. (2003). Trends in codling moth activity in apple orchards under mating disruption using pheromones in the Elgin area, Western Cape province of South Africa. *African Entomology, 11*, 65–75.

Prinsloo, G. L., & Uys, V. M. (2015). *Insects of cultivated plants and natural pastures in Southern Africa*. Pretoria, Southern Africa: Entomological Society of Southern Africa.

Quinn, L. P., De Vos, B. J., Fernandes-Whaley, M., Roos, C., Bouwman, H., Kylin, H., et al. (2011). Pesticide use in South Africa: One of the largest importers of pesticides in Africa, In M. Stoytcheva (Ed.), *Pesticides in the modern world – Pesticides use and management*. InTech ISBN: 978–953–307–459–7. Available from: http://www.intechopen.com/books/pesticides–in–the–modern–world–pesticidesuse–and–management/pesticide–use–in–south–africa–one–of–the–largest–importers–of–pesticides–in–africa

Ramakuwela, T., Erasmus, A., & Hatting, J. (2011). Maize stalk borer, *Busseola fusca*, as potential target for control with entomopathogenic nematodes. *South African Journal of Plant and Soil, 28*(4), 255–256.

Ramakuwela, T., Hatting, J., Laing, M.D., Hazir, S. (2014, May 4–9). Cost effective solid-state production of an entomopathogenic nematode (Steinernematidae). *Proceedings, 6th International Congress of Nematology* (pp. 32). Cape Town.

Raulston, J. R., Pair, S. D., Loera, J., & Cabanillas, H. E. (1992). Prepupal and pupal parasitism of *Helicoverpa zea* and *Spodoptera frugiperda* (Lepidoptera: Noctuidae) by *Steinernema* sp. in cornfields in the lower Rio Grande Valley. *Journal of Economic Entomology, 85*(5), 1666–1670.

Regupathy, A., Kranthi, K., Singh, J., Iqbal, A. & Russell, D. (2003, March 9–13). Patterns and magnitude of insecticide resistance levels in India, Pakistan and China. *Proceedings of World Cotton Research Conference*. Cape Town. Abst. PS30.9.

Richter, A. R., & Fuxa, J. R. (1990). Effect of *Steinernema feltiae* on *Spodoptera frugiperda* and *Heliothis zea* (Lepidoptera: Noctuidae) in corn. *Journal of Economic Entomology, 83*, 1286–1291.

Riedl, H., Blomefield, T. L., & Giliomee, J. H. (1998). A century of codling moth control in South Africa: II. Current and future status of codling moth management. *Journal of South African Society of Horticultural Science, 8*, 32–54.

Rutherford, R. S., & Conlong, D. E. (2010). Combating sugarcane pests in South Africa: From researching biotic interactions to bio–intensive integrated pest management in the field. *Proceedings of the International Society of Sugar Cane Technologists, 27*, 1–17.

Shamseldean, M. M., Abou-El-Sooud, A. B., Abd-Elgawad, M. M., & Saleh, M. M. (1996). Identification of a new heterorhabditid species from Egypt, *Heterorhabditis tayserae* n. sp. (Rhabditidae: Heterorhabditidae). *Egypt Journal of Biological Control, 6*, 129–138.

Shamseldean, M. M., Adb-Elgawad, M. M., & Atwa, A. A. (1998). Effects of soil temperature, exposure time and host introduction on pathogenicity and recycling of an Egyptian strain of *Heterhabditis bacteriophora* strain EASD98 and its comparison with *H. bacteriophora* strain HP88 and *Steinernema riobravae* from USA. *Journal of Nematology, 8*, 71–76.

Shamseldean, M. M., El-Sadawy, H., & Allam, S. F. M. (2004). Comparative safety of entomopathogenic nematodes on honeybee workers. *Egyptian Journal of Biological Pest Control, 14*(1), 147–153.

Shapiro, D. I., & Glazer, I. (1996). Comparison of entomopathogenic nematode dispersal from infected hosts versus aqueous suspension. *Environmental Entomology, 25*, 1455–1461.

Shapiro, D. I., & Lewis, E. E. (1999). Comparison of entomopathogenic nematode infectivity from infected hosts versus aqueous suspension. *Environmental Entomology, 28*, 907–911.

Shapiro-Ilan, D. I. (2001). Virulence of entomopathogenic nematodes to pecan weevil larvae, *Curculio caryae* (Coleoptera: Curculionidae), in the laboratory. *Journal of Economic Entomology, 94*(1), 7–13.

Shapiro-Ilan, D. I., Han, R., & Dolinksi, C. (2012). Entomopathogenic nematode production and application technology. *Journal of Nematology, 44*(2), 206–217.

Shapiro-Ilan, D. I., Lewis, E. E., Tedders, W., & Son, Y. (2003). Superior efficacy observed in entomopathogenic nematodes applied in infected–host cadavers compared with application in aqueous suspension. *Journal of Invertebrate Pathology, 83*, 270–272.

Shapiro-Ilan, D. I., Stuart, R. J., & McCoy, C. W. (2006). A comparison of entomopathogenic nematode longevity in soil under laboratory conditions. *Journal of Nematology, 38*(1), 119–129.

Spaull, V. W. (1988). A preliminary evaluation of the entomogenous nematodes to control the African sugarcane stalk borer *Eldana Saccharina* (Lepidoptera: Pyralidae). *Proceedings of the South African Sugar Technologists Association, 62*, 120–123.

Spaull, V. W. (1990). Field tests to control the phylarid, *Eldana saccharina*, with an entomogenous nematode, *Heterorhabditis* sp. *Proceedings of the South African Sugar Technologists Association, 64*, 103–106.

Spaull, V. W. (1991). *Heterorhabditis* and *Steinernema* species (Nematoda: Rhabditida) for the control of a sugarcane stalk borer in South Africa. *Phytophylactica, 23*, 213–215.

Spiridonov, S. E., Reid, A. P., Podrucka, K., Subbotin, S. A., & Moens, M. (2004). Phylogenetic relationships within the genus *Steinernema* (Nematoda: Rhabditida) as inferred from analyses of sequences of the ITSI–5.8S–ITS2 region of rDNA and morphological features. *Nematology, 6*, 547–566.

Stack, C. M., Easwaramoorthy, S. G., Metha, U. K., Downes, M. J., Griffin, C. T., & Burnell, A. M. (2000). Molecular characterisation of *Heterorhabditis indica* isolates from India, Kenya, Indonesia and Cuba. *Nematology, 2*, 477–487.

Steenkamp, S., Erasmus, A., & Malan, A. (2011). The effect of South African entomopathogenic nematodes on Bt-resistant and non-resistant lifestages of *Busseola fusca. South African Journal of Plant & Soil, 28*(4), 268–269.

Stock, S. P., Rivera-Orduño, B., & Flores-Lara, Y. (2009). *Heterorhabditis sonorensis* n. sp. (Nematoda: Heterorhabditidae), a natural pathogen of the seasonal cicada *Diceroprocta ornea* (Walker) (Homoptera: Cicadidae) in the Sonoran Desert. *Journal of Invertebrate Pathology, 100*, 175–184.

Stokwe, N. F. (2009). *Entomopathogenic nematodes: Characterization of a new species, long–term storage and control of obscure mealybug,* Pseudococcus viburni *(Hemiptera: Pseudococcidae) under laboratory conditions*. MScAgric thesis, Department of Conservation Ecology and Entomology, Stellenbosch University, Stellenbosch.

Stokwe, N. F., Malan, A. P., Nguyen, K. B., Knoetze, R., & Tiedt, L. (2011). *Steinernema citrae* n. sp. (Rhabditida: Steinernematidae), a new entomopathogenic nematode from South Africa. *Nematology, 13*, 567–587.

Stuart, R. J., Lewis, E. E., & Gaugler, R. (1996). Selection alters the pattern of emergence from the host cadaver in the entomopathogenic nematode, *Steinernema glaseri. Parasitology, 113*(02), 183–189.

Tahir, H. I., Otto, A. A., & Hague, N. G. M. (1995). The susceptibility of *Helicoverpa* (*Heliothis*) *armigera* and *Erias vitella* larvae to entomopathogenic nematodes. *Afro Asian Journal of Nematology, 5*(2), 161–165.

Tamiru, T., Waeyenberge, L., Tesfaye, H., Ehlers, R. U., Puza, V., & Mráček, Z. (2012). *Steinernema ethiopiense* sp. n. (Rhabditida: Steinernematidae), a new entomopathogenic nematode from Ethiopia. *Nematology, 14*, 741–757.

Tarasco, E., Triggiani, O., Sai, K., & Zamoum, M. (2009). Survey on entomopathogenic nematodes in Algerian soils and their activity at different temperatures. *Frustula Entomologica, 32*(45), 31–42.

Van Niekerk, S., & Malan, A. P. (2012). Potential of South African entomopathogenic nematodes (Heterorhabditidae and Steinernematidae) for control of the citrus mealybug, *Planococcus citri* (Pseudococcidae). *Journal of Invertebrate Pathology, 111*, 166–176.

Van Niekerk, S., & Malan, A. P. (2013). Adjuvants to improve control of *Planococcus citri* (Hemiptera: Pseudococcidae) using entomopathogenic nematodes (Heterorhabditidae and Steinernematidae). *Journal of Helminthology, 89*, 189–195.

Van Niekerk, S., & Malan, A. P. (2014a). Compatibility of *Heterorhabditis zealandica* and *Steinernema yirgalemense* with agrochemicals and biological control agents. *African Entomology, 22*, 49–56.

Van Niekerk, S., & Malan, A. P. (2014b). Evaluating the efficacy of a polymer–surfactant formulation to improve control of *Planococcus citri* (Risso) (Hemiptera: Pseudococcidae) under simulated natural conditions. *African Plant Protection, 17*, 1–8.

Van Zyl, C. (2012). *The in vivo production of* Heterorhabditis zealandica *and* Heterorhabditis bacteriophora (99 p). MSc thesis, University of Stellenbosch.

Van Zyl, C., & Malan, A. P. (2014a). The role of entomopathogenic nematodes as biological control agents of insect pests, with emphasis on the history of their mass culturing and *in vivo* production. *African Entomology, 22*(2), 235–249.

Van Zyl, C., & Malan, A. P. (2014b). Optimisation of inoculation techniques for *in vivo* mass culture of entomopathogenic nematodes through nematode and insect host manipulation. *African Entomology, 24*, 405–416.

Vyas, R. V., Patel, N. B., Yadav, P., Ghelani, Y. H., & Patel, D. J. (2003). Performance of entomopathogenic nematodes for management of gram pod borer, *Helicoverpa armigera. Annals of Plant Protection Sciences, 11*, 107–109.

Wakgari, W. M., & Giliomee, J. H. (2004). Insect pests of astringent persimmon in the Western and Eastern Cape provinces of South Africa. *African Plant Protection, 10*, 1–5.

Walton, V. M. (2003). *Development of an integrated pest management system for vine mealybug,* Planococcus ficus *(Signoret), in vineyards in the Western Cape province, South Africa.* PhD thesis, Department of Conservation Ecology and Entomology, Stellenbosch University, Stellenbosch, South Africa.

Walton, V. M., & Pringle, K. L. (2004). Vine mealybug, *Planococcus ficus* (Signoret) (Hemiptera: Pseudococcidae), a key pest in South African vineyards: A review. *South African Journal of Enology and Viticulture, 25*, 54–62.

Waturu, C. N. (1998). *Entomopathogenic nematodes (Steinernmatidae and Heterorhabditidae) from Kenya*. PhD thesis, University of Reading, Reading, U.K.

Waturu, C. N., Hunt, D. J., & Reid, A. P. (1997). *Steinernema karii* sp. n. (Nematoda: Steinernematidae), a new entomopathogenic nematode from Kenya. *International Journal of Nematology, 7*, 65–75.

Welter, S. C. (2008). Codling moth. In H. C. R. T. Resh (Ed.), *Encyclopedia of insects* (pp. 189–199). Amsterdam: Elsevier.
Witt, A. B. R., Little, R. M., & Crowe, T. M. (1995). The effectiveness of helmeted guineafowl *Numida meleagris* (Linnaeus 1766) in controlling the banded fruit weevil *Phlyctinus callosus* (Schönherr, 1826), and their impact on other invertebrates in apple orchards in the Western Cape province, South Africa. *Agriculture, Ecosystems & Environment, 55*, 169–179.
Yan, X., Liu, X., Han, R., Chen, S., De Clerc, P., & Moens, M. (2010). Osmotic induction of anhydrobiosis in entomopathogenic nematodes of the genera *Heterorhabditis* and *Steinernema. Biological Control, 53*, 325–330.
Zadji, L., Baimey, H., Afouda, L., Houssou, F. G., Waeyenberge, L., De Sutter, N., et al. (2013). First record on the distribution of entomopathogenic nematodes (Rhabditida: Steinernematidae and Heterorhabditidae) in Southern Benin. *Russian Journal of Nematology, 21*, 117–130.

Chapter 21
Phasmarhabditis hermaphrodita as a Control Agent for Slugs

Michael Wilson and Robert Rae

21.1 Introduction

While there has been intense study on entomopathogenic nematodes (EPNs) by numerous research groups throughout the world, there has been much less work on nematodes that can control slugs. A survey of academic research papers listed in the Scopus database in December 2014 using the search terms "*Steinernema* or *Heterorhabditis*" revealed over 2,100 hits compared with a mere 82 hits for "*Phasmarhabditis*". Conversely, while the EPNs have a relatively recent history, with the first *Steinernema* spp. being described in 1923 (Steiner, 1923), *Phasmarhabditis hermaphrodita* Schneider (Rhabditida: Rhabditidae) was first described as a parasite of slugs in 1859 (Schneider, 1859). The slug parasitic nematode, *P. hermaphrodita* featured prominently in Maupas' classic paper on nematode reproduction (Maupas, 1900). This paper is best known for including the original description of *Caenorhabditis elegans* Maupas (Rhabditida: Rhabditidae) but the paper also contained drawings, measurements, and experimental observation on the reproduction of *P. hermaphrodita* which Maupas called *Rhabditis caussaneli.* However, there was virtually no more work done on this nematode until its potential for commercialisation as a bio–pesticide was first realised, and published as a patent (Wilson, Glen, & Pearce, 1993). The lifecycle of *P. hermpahrodita* is similar in many ways to EPNs (Table 21.1). The infective stage is a dauer larva that penetrates slugs through the dorsal integumental pouch (Wilson, Glen, & George, 1993; Tan & Grewal, 2001a). Larvae feeding within the slug develop into adults and eventually

M. Wilson (✉)
Innovative Farm Systems, AgResearch, Ruakura, New Zealand
e-mail: michael.wilson@agresearch.co.nz

R. Rae
School of Natural Sciences & Psychology, Liverpool John Moores University, Liverpool, UK
e-mail: R.G.Rae@ljmu.ac.uk

© Springer International Publishing Switzerland 2015

R. Campos-Herrera (ed.), *Nematode Pathogenesis of Insects and Other Pests*, Sustainability in Plant and Crop Protection 1, DOI 10.1007/978-3-319-18266-7_21

kill the host within 4–21 days. However, there are certain key differences from EPNs, most notably that *P. hermaphrodita* is a facultative parasite that can reproduce on a wide range of substrates including slug faeces, dead earthworms, dead insects, compost and leaf litter (Tan & Grewal, 2001a; MacMillan et al., 2009; Nermut', Půža, & Mráček, 2014). But apart from this basic understanding, we have very little knowledge of the nematode's biology and ecology.

In spite of the low level of research interest by the academic community, *P. hermaphrodita* has been commercialised and used successfully as a biological molluscicide since 1994. The nematode was first commercialised by MicroBio Ltd, a company which was acquired by Becker Underwood in 2000, which in turn was acquired by BASF in 2012. Currently, production of *P. hermaphrodita* is the responsibility of BASF Agricultural Specialities, Littlehampton, UK, who are the sole producers, even though the original patent has expired. The nematodes are mass produced in fermenters and formulated using similar technologies to those used for EPNs.

21.2 The Relationship Between Bacteria and *Phasmarhabditis* spp. Nematodes

It can be seen from above that *P. hermaphrodita* and EPNs share some lifestyle characteristics and that products containing either EPNs or *P. hermaphrodita* are produced and used in a similar manner. However, while there are many similarities there are also some key differences (see Table 21.1). One area which has been

Table 21.1 Similarities and differences in biological attributes and commercial availability of *P. hermaphrodita* and entomopathogenic nematodes (Steinernematidae and Heterorhabditidae families)

Characteristic	Entomopathogenic nematodes	*Phasmarhabditis hermaphrodita*
Parasitic association with hosts	Obligate	Facultative
Association with bacteria in nature	Obligate association with *Xenorhabdus* or *Photorhabdus* spp.	None known
Association with bacteria in commercial products	*Xenorhabdus* or *Photorhabdus*	*Moraxella osloensis*
Time to kill	Rapid, 1–5 days	Slower, 5–21 days
Host range	Broad	Broad
Geographical range	Cosmopolitan	Not well studied, but appears less widespread than for EPN
Commercial Producers	Many	Only BASF Agricultural Specialties
Commercial Availability	Most inhabited continents	Only in Europe

the subject of much confusion and misinformation is the association between *P. hermaphrodita* and the bacterium *Moraxella osloensis* Jebasingh, Lakshmikandan, Rajesh, & Raja (Pseudomonadales: Moraxellaceae). It has been assumed that the association between *P. hermaphrodita* and *M. osloensis* is a natural symbiosis, directly analogous to EPNs and *Xenorhabdus* or *Photorhabdus* spp. For example, it has been stated that "*The bacterium* Moraxella osloensis *is a mutualistic symbiont of the slug–parasitic nematode, P. hermaphrodita. In nature, P. hermaphrodita vectors M. osloensis into the shell cavity of the slug host* Deroceras reticulatum *in which the bacterium multiply and kill the slug*" (An, Sreevatsan, & Grewal, 2008). There are numerous other published examples of this widespread belief. In fact, the association between *P. hermaphrodita* and *M. osloensis* is entirely a manmade convenience.

Early investigations concentrated on isolating bacteria from *P. hermaphrodita* dauer larvae, xenic foam chip cultures and cadavers of infected *Deroceras reticulatum* (Muller) (Gastropoda: Agriolimacidae) (Wilson, Glen, Pearce, & Rodgers, 1995). This approach yielded more than 100 isolates, distinguished solely by bacterial colony morphology. From this collection, 16 isolates comprising 13 species were selected for identification and further work. *P. hermaphrodita* was able to grow prolifically on the majority of isolates which included such diverse species as *Aeromonas* sp., *Bacillus cereus* Frankland & Frankland (Bacillales: Bacillaceae), *Flavobacterium* sp., and *Pseudomonas fluorescens* Migula (Pseudomonadales: Pseudomonadaceae). Clearly *P. hermaphrodita* did not exhibit the growth specificity of EPNs for their own, or closely related bacteria. Further studies tested whether *P. hermaphrodita* grown on numerous bacterial species would differ in the numbers of dauers produced but also the quality of dauers used to kill *D. reticulatum* (Wilson, Glen, George, & Pearce, 1995). *P. hermaphrodita* was shown to retain many different bacterial species, again in contrast to the high specificity association of bacteria and EPNs. The bacterium *M. osloensis* was selected for commercial production as it consistently produced high yields of *P. hermaphrodita* dauers that were pathogenic to *D. reticulatum* (Wilson, Glen, George et al., 1995). Since this work, commercial production of *P. hermaphrodita* has always been in monoxenic association with *M. osloensis* and this bacterium is present in, and easily isolated from commercial product.

Much confusion arose from the work of Tan and Grewal and subsequent work form the same laboratory (Tan & Grewal, 2001b, 2002). These authors studied interactions between the bacterium, nematodes and slugs. Their work conclusively showed that aged *M. osloensis* cultures and lipopolysaccharide (endotoxin) from the cell walls of *M. osloensis* caused mortality in slugs. However, their work did not conclusively show a role for this type of interaction during the natural infection process.

While there is little evidence that *M. osloensis* or any other bacterium is naturally associated with *P. hermaphrodita*, in the absence of detailed studies of bacteria associated with *P. hermaphrodita* in nature, the possibility cannot be completely ruled out. However, there are several factors involved in *P. hermaphrodita*'s life cycle that differ from EPNs that would not favour a highly specific bacterial mutualism similar to that seen in EPNs. Firstly, *P. hermaphrodita* is a facultative, rather than obligate parasite. The nematode can reproduce on a wide range of substrates including, slug faeces, dead earthworms, dead insects, compost and leaf litter (Tan & Grewal, 2001a; MacMillan et al., 2009; Nermuť et al., 2014). The ability to grow on such a wide range of substrates, with complex bacterial flora associated with them, is not consistent with selective growth and transport of a single mutualistic bacterium. In a similar vein, slugs lack a rigid cuticle and most reproduction takes place on the surface of the cadaver (Wilson, Glen & George, 1993). When slugs are killed by *P. hermaphrodita* they die in the soil matrix (Pechova & Foltan, 2008) and are exposed to a vast diversity of bacteria. Conversely, when EPNs kill host, the host cuticle tends to stay intact, acting as a barrier to influx of soil bacteria. This, combined with the known antibiotic production of *Xenorhabdus* and *Photorhabdus* allows the nematodes to feed on almost pure cultures of their symbionts, although occasional opportunistic bacteria can also reproduce in the cadaver (Enright & Griffin, 2004). *Xenorhabdus* spp. and *Photorhabdus* spp. have very limited ability to survive outside hosts, (Poinar, Thomas, Haygood, & Nealson, 1980; Morgan, Kuntzelmann, Tavernor, Ousley, & Winstanley, 1997). Conversely, *M. osloensis* has been isolated from numerous environments such as sink traps in hospitals (Rosenthal, 1978), the nasopharynx of healthy adults (Berger & Felsen, 1976), dairy farm drains (Muramatsu & Kikuchi, 2005) and from ear, nose and throats of outpatients (Bovre, 1970). However, again, this evidence is largely circumstantial, and it is now acknowledged that the EPN/bacterium symbiosis is not quite as rigid as earlier believed (see Chap. 1, this volume).

It is always difficult to prove a negative – e.g. that there is no one bacterium symbiotically associated with *P. hermaphrodita.* The most compelling data to date was a study that used PCR–DGGE profiling to investigate bacteria associated with *P. hermpahrodita* after growth and development in different slug species (Rae, Tourna, & Wilson, 2010). These authors found diverse and variable bacterial associates on nematodes reared on different slugs, but all nematodes were equally pathogenic to the host slug *D. reticulatum.* While these authors found *M. osloensis* in commercial product, they found no evidence of retention after passage through slugs. These authors used two species of EPNs, *Heterorhabditis megidis* Poinar, Jackson & Klein (Rhabditida: Heterorhabditidae) and *Steinernema carpocapsae* (Weiser) (Rhabditida: Steinernematidae) as positive controls for their methods. Bacterial retention was as predicted with these species, with only a monoxenic culture of the symbiont being found within dauer larvae after passage through insects. While for reasons stated above, we believe it is unlikely that a mutualistically associated bacteria for *P. hermaphrodita* will be found, it would be interesting to study bacteria associated with natural populations of *P. hermpahrodita.* A similar approach has been used to study bacteria naturally associated with *Pristionchus pacificus* Sommer, Carta, Kim & Sternberg, (Diplogasterida: Neodiplogasteridae), a nematode that associates with beetles (Rae et al., 2008).

21.3 Slug Pests and Problems with Chemical Control of Terrestrial Molluscs

Terrestrial gastropods (slugs and snails) are among the most successful land invertebrates with more than 35,000 species being described (Barker, 2001). Their importance as crop pests has been largely neglected but in areas with moist climates, they are becoming increasingly common for a number of reasons. Many of the practices associated with more sustainable agriculture lead to increased slug populations e.g. direct drilling or minimal tillage, and incorporation of crop residues, and thus molluscs pest status is increasing globally (Barker, 2002). Furthermore, in addition to their status as crop pests, slugs and snails can serve as intermediate hosts to medical and veteriniary parasites (South, 1992).

There are numerous advantages to developing biological control, as opposed to chemical pesticides. Benefits attributed include greater specificity (although this can sometimes be problematic, especially if the target pest exists as a species complex). Other benefits include reduction of toxic chemicals in the environment. From a commercial point of view, biopesticides are more rapid and cheaper to develop to market than chemical pesticides (Glare et al., 2012). The existence of "organic" or "biological" growers who don't use synthetic pesticides also offers an additional market for products. More recently, strict residue limits imposed by regulatory bodies as well as supermarket buyers have also driven interest in biological pesticides as has widespread development of resistance to multiple chemical pesticides. However, in the case of mollusc pests, a key driver to develop biopesticides is the lack of efficacy and currently available products, and the apathy of the major agrochemical producers for developing new active ingredients for a fragmented market. Currently all mainstream chemical molluscicides are sold as bait pellets typically containing metaldehyde, chelated iron phosphate or methiocarb. These are typically surface applied and molluscs have to find and ingest the pellets. The active ingredients are repellent and slugs frequently cease feeding before ingesting a lethal dose (Bailey, 2002). Thus the development of *P. hermaphrodita* as a biological molluscicide was a welcome addition to the market.

21.4 Use in Glasshouse Crops and Field Vegetables

Most use of *P. hermaphrodita* relates to application in glasshouse crops and field vegetables. Slugs feed on a huge range of field vegetables, including lettuce, cabbage, Brussels sprouts and asparagus. Such vegetables command high prices enabling growers to spend more on biological pesticides. Furthermore, unlike arable crops where slugs are only major pests during crop establishment, slugs can be pests throughout the cropping cycle of most field vegetables. For example, in lettuce, in addition to killing seedlings shortly after planting, slug grazing on the outer leaves late in the growing cycle causes cosmetic damage that can substantially reduce the marketable value of the crop. Usually, feeding damage is accompanied by slug faecal contamination. Furthermore, slugs can rest in between the leaves of the lettuce head, contaminating the crop and making it unacceptable to consumers. Growers are reluctant to apply slug pellets at this late stage of the cropping cycle because of the risk of contaminating the crop with pellets, again making the product unacceptable to buyers and potentially posing a health hazard to humans. Thus, biological control with *P. hermaphrodita* represents the only viable slug control method at this stage of cropping.

BASF Agricultural Specialities sells *P. hermaphrodita* as a single product, Nemaslug®, for use in all glasshouse and vegetable crops. As will be discussed in the following examples, best slug control is often achieved by repeated applications of low concentration rather than a single high concentration. This reflects the fact that many vegetable crops need protection throughout the growing cycle and not just at establishment. This fits in with standard practice for growers who are used to making repeat applications of slug pellets. The nematode is sold in packs of 30 million or 250 million that treat between 100 and 5,000 m^2 depending on dose rate and application strategy. The product can be applied as a single high concentration of 300,000 m^2 or repeated lower concentrations of either 50,000 or 150,000 m^2. Two to three repeat applications are recommended at intervals of 2–6 weeks depending on crop and pest pressure. It is recommended that product is applied in a large amount of water e.g. 1 l/m^2. Other application recommendations are as for EPN products (i.e. remove fine filters, avoid application in direct sunlight).

For larger areas of field crops, e.g. potatoes, it is possible to apply *P. hermaphrodita* through irrigation lines. Brown, Barker, Hopkins, and Nelson (2011) described nematode–specific application equipment that allows *P. hermaphrodita* to be mixed with irrigation water. The Wroot water Nemaslug Xtra applicator injects nematodes onto the irrigation water to be applied via boom or gun irrigators. The applicator agitates and aerates the nematodes during injection so that the nematodes can be used to treat large areas over prolonged periods of time. Ester and Wilson (2005) documented how *P. hermaphrodita* can be used to protect high value orchids from slug damage in glasshouses. In glasshouses the slug *Lehmania valentiana* Férussac (Stylommatophora: Limacidae) can be a particular problem. The slugs rest in and under plant pots during the day, but at night they become active and climb the plants. The slugs feed on the flower reducing the marketable value considerably. The slug is susceptible to *P. hermaphrodita* and an application of the 300,000 nematode per plant pot virtually eliminated slug damage and performed as well as metaldehyde pellets (Ester & Wilson, 2005).

Lettuce can be grown either as a field vegetable or in polythene tunnels. In two polythene tunnel experiments, *P. hermaphrodita* was tested at a range of doses between 1×10^8 and 1×10^{10}/ha for its ability to protect lettuce from slug damage (Wilson, Glen, George, & Hughes, 1995). These authors found that the recommended application rate of 3×10^9/ha could reduce both slug damage and slug numbers. Two main types of lettuce are grown commercially under field conditions, Butterhead lettuce (*Lactuca sativa* var. *captitata*) and Iceberg lettuce

(*L. sativa*). Both are damaged and contaminated by slugs and because many growers irrigate the crop, at all stages, slug problems are almost ubiquitous. There have been several accounts of using *P. hermaphrodita* to control lettuce in field crops (Ester & Wilson, 2005; Speiser & Andermatt, 1996). In Switzerland, an initial trial using *P. hermaphrodita* to protect outdoor lettuce was extremely successful, with *P. hermaphrodita* applied at 1×10^6/ha significantly reducing slug damage, whereas metaldehyde pellets did not (Speiser & Andermatt, 1996). However, in a second experiment testing the same concentration, and a tenfold dilution of this concentration, neither concentration reduced slug damage. The authors suspected this to be a result of large Arionid slugs causing damage, which are more resistant to *P. hermaphrodita* (Grimm, 2002). In the Netherlands, an experiment testing the ability of *P. hermaphrodita* applied at a range of rates and timings to control slug damage to Iceberg lettuce showed repeated low rate applications gave protection at much lower nematode rates than the standard single application rate (Ester & Wilson, 2005).

Green asparagus (*Asparagus officinalis*) is often grown on clay soils that favour slugs as they retain moisture. Slugs damage the crop largely by feeding on growing tips, causing substantial damage and rendering the crop unmarketable. Ester and Wilson (2005) reported on the successful use of *P. hermaphrodita* to control slugs in asparagus. A range of concentrations (10,000–300,000/m^2) with differing numbers of applications (1–4) were tested. For early asparagus harvests, all but the lowest dose of nematodes significantly reduced slug damage. At later harvest, the benefits of multiple low–concentration applications over a single high concentration application could be seen (Ester & Wilson, 2005).

Slugs are a serious pest of Brussels sprouts in most growing regions. As with asparagus, they tend to be grown on clay soils that favour slugs. Slugs can kill plants at the seedling stage at crop establishment, but they also damage the mature crop. In this case, slugs graze on the outer leaves of the developing sprouts. The damage is cosmetic and does not lower yields. However, market value of the crop is reduced considerably. Ester and Wilson (2005) reported on two large scale field experiments in which multiple low concentration nematode applications were compared with metaldehyde pellets. In addition, band applications were used to further lower the concentration. In one experiment, three band applications at a rate of 50,000/m^2 gave highly significant protection to the sprouts, at 10 % of the recommended single high application rate. Thus, Brussels sprouts represents a high value target market for using *P. hermaphrodita* and there are several papers published on this topic (Ester, Van Rozen, & Molendijk, 2003; Glen et al., 2000).

Potatoes are severely damaged by slugs. Subterranean slugs, particularly *D. reticulatum* but also keeled slug species, particularly *Tandonia budapestensis* Hazay (Stylomatophora: Milacidae) feed on the maturing tubers. In addition to the feeding damage, slug grazing makes the potato susceptible to attack by many soil fungi. Because the slugs cause damage underground toward the end of the growing cycle, application of bait pellets is ineffective, leaving biological control using nematodes as the only option available. There is little published data on efficacy of *P. hermaphrodita* in field potatoes, but BASF Agricultural Specialities report widespread use of the nematode in this crop and have designed special irrigation equipment that can be used to apply *P. hermpahrodita* to large areas of potatoes (Brown et al., 2011).

21.5 Use in Arable Crops

Slugs cause damage in a wide range of arable crop throughout the world. In Europe, wheat and oilseed rape are the main crops damaged but also sugar beet can be important in certain locales (Ester & Wilson, 2005)

In wheat, slugs feed directly on seeds, causing characteristic seed–hollowing that kills the plants. This in turn leads to bare patches in the established crops that may need to be re–drilled. Slugs also graze on the emerged seedlings, but it is thought that this is less likely to reduce yield. There is no doubt that *P. hermaphrodita* has potential to be used as a control agent in wheat. The original work leading to the discovery of *P. hermaphrodita* was done at the UK's Institute of Arable Crops Research (now Rothamsted Research) with winter wheat as the main target. When a range of application rates of *P. hermaphrodita* was tested in winter wheat, rates of 3×10^9/ha gave plant protection equivalent to standard molluscicidal baits applied at the recommended rate (Wilson, Glen, George, Pearce, & Wiltshire, 1994). This finding was confirmed in another wheat trial that also showed the benefits of incorporating *P. hermaphrodita* into soil following application (Wilson, Hughes, Hamacher, Barahona, & Glen, 1996).

In oilseed rape, slugs do not damage the seeds, but damage the newly emerged seedlings. Frequently slugs consume the apical meristem, thus killing the plants. As with wheat, this may result in large bare patches as the crop establishes. However, in rape there is much less potential for re–drilling the crop because rape has a much narrower time period during which successful establishment can be achieved (Moens & Glen, 2002). The only published work showing successful control of slugs by *P. hermpahrodita* is that of Ester and Wilson (2005) who reported a significant increase of plant stand by treatment with 3×10^9/ha. While the increase in plant stand was significant compared to untreated plots, the plant stand in *P. hermaphrodita* treated plots was significantly less than in plots treated with methiocarb baits.

Another arable crop that can be damaged by slug is sugar beet. In this crop, most economic damage is not done to seeds, or newly emerged seedlings which are fairly resistant to slug grazing, but when the seedlings reach the 4–6 leaf stage. It is common practice to grow sugarbeets in a cover crop that provides dense cover and a layer of organic matter on the soil that favours slugs. The damage at the 4–6 leaf stage can be sufficiently severe to kill plants. Again there is little published data on use of *P. hermaphrodita* to control slugs in sugar beet, but Ester and Wilson (2005) reported that the standard dose of 3×10^9/ha provided plant protection equivalent to bait pellets at the recommended rate.

Clearly there is no doubt that it is feasible to control slugs in arable crops using *P. hermpahrodita* but at present, the limited available production capacity and high production costs mean that *P. hermaphrodita* is not likely to be used widely in arable crops in the foreseeable future. However, the recent interest in using entomopathogenic nematodes to control *Diabrotica* in corn shows that there is potential for using nematodes as biocontrol agents in such broadacre crops (Pilz et al., 2014). Furthermore, novel approaches such as repeated low application concentrations as shown for Brussels sprouts, and band application may further lower application costs.

21.6 Future Opportunities

There is much potential for using the *P. hermaphrodita* in geographical ranges outside Europe as slugs are a pest throughout the world (Barker, 2002). Many countries have strict regulations for using biopesticides that allow only indigenous species to be used (Ehlers, 2005). Such regulations have slowed the uptake of use of *P. hermaphrodita* in several countries outside Europe. For example the mild, moist climate of New Zealand is very suitable for slugs and a range of European invasive slugs, particularly *D. reticulatum* damage a wide range of agricultural and horticultural crops. In New Zealand use of biological pesticides is in part regulated by the 1998 Hazardous Substances and New Organisms (HSNO) Act. Any organisms that was not known to be present prior to the 1998 HSNO act is classified as a 'new organisms' and subject to strict regulation precluding use as a biocontrol agent. Recently, *P. hermaphrodita* has been found parasitizing slugs in New Zealand and thus has potential to be developed as a biological molluscicide (Wilson, Burch, Tourna, Aalders, & Barker, 2012). However, because this discovery was after 1998 *P. hermaphrodita* is still classified and regulated as a "new organism". To have this classification removed, we need to demonstrate that the nematode is widely distributed in New Zealand and a survey is currently being undertaken on nematodes associated with slugs in New Zealand. This represents a fairly sizeable task, as there is no equivalent to the *Galleria* baiting method to isolate *P. hermaphrodita*. The easiest way to detect *P. hermpahrodita* is to decapitate field collected slugs and incubate at 15 °C for approximately ten days. Nematodes can then be seen reproducing on the cadaver. Isolates need to be checked to ensure they are *Phasmarhabditis* spp. and not free–living nematodes which use the slug for phoretic transport (e.g. *C. elegans*).

Similarly in the USA there are many markets is which invasive European slugs damage a range of crops that represent target markets for *P. hermaphrodita*. For example, maize and soybeans are severely damaged by slugs (Hammond & Byers, 2002), and the cool moist climate of the pacific North West is particularly favourable to slugs (Gavin, Mueller–Warrant, Griffith, & Banowetz, 2012). However, the use of exotic nematode species as biocontrol agents is regulated by the Animal and Plant Health Inspection Service (APHIS) and other federal organisations (Ehlers, 2005).

In order to exempt *P. hermaphrodita* from regulation as an exotic organism, Becker Underwood commissioned a survey of nematodes parasitizing slugs in the USA (Ross, Ivanova, Severns, & Wilson, 2010). In this survey, levels of parasitism by all nematode species in European slugs invasive in the USA were significantly lower than in their home range. Not a single *Phasmarhabditis* spp. isolate was recorded from 2,126 slugs collected from 70 US sites, compared with 26 slugs being infected with *Phasmarhabditis* spp. isolates among 1,469 slug samples in the UK. While the data of Ross et al. (2010) suggest that *P. hermaphrodita* is much less prevalent in the USA than in the UK, they do not preclude its presence there. Recently *P. hermaphrodita* has been found associated with three species of European slug pests in California (Tandingan De Ley, McDonnell, Lopez, Paine, & De Ley, 2014).

Outside Europe, USA and New Zealand, *P. hermaphrodita* has been isolated in Chile (France & Gerding, 2000), Iran (Karimi, Kharazi–Pakadel, & Roberts, 2003) and Egypt (Genena, Mostafa, Fouly, & Yousef, 2011) and thus these countries represent more potential markets for exploitation as a biological molluscicide.

In addition to the potential to sell *P. hermaphrodita* into new geographical ranges, there are other opportunities to lower application costs and capture a larger share of the molluscicide market.

One potential strategy is band application of nematodes to row crops. Advances in precision agriculture machinery now mean it is possible to apply pesticides as narrow bands either side of the crop rows. This may be a particularly suitable strategy for applying *P. hermaphrodita* because in addition to its lethal parasitic activity, slugs are known to avoid crawling on soil treated with the nematode in laboratory studies (Wilson, Hughes, Jefferies, & Glen, 1999). As a result, two outdoor plot trials investigating the potential for reducing the area treated with *P. hermaphrodita* were conducted and both studies showed potential (Hass, Glen, Brain, & Hughes, 1999; Hass, Hughes, & Glen, 1999). Another novel approach to reducing numbers of applied *P. hermpahrodita* and hence application cost is to put out shelters for slugs to rest in, and only treat the soil directly under the shelters (Grewal, Grewal, Taylor, & Hammond, 2001). This approach achieved a 63 % reduction in numbers of nematodes applied without any loss of efficacy in slug control.

In conclusion, *P. hermaphrodita* has been sold successfully as a biological control agent in Europe for 20 years. There is much potential for expansion into new crops, and particularly new regions of the world.

References

An, R., Sreevatsan, S., & Grewal, P. S. (2008). *Moraxella osloensis* gene expression in the slug host *Deroceras reticulatum*. *BMC Microbiology, 8*, 19.

Bailey, S. E. R. (2002). Molluscicidal baits for control of terrestrial gastropods. In G. M. Barker (Ed.), *Molluscs as crop pests* (pp. 33–54). Wallingford, UK: CABI Publishing.

Barker, G. M. (2001). *The biology of terrestrial molluscs*. Wallingford, UK: CABI Publishing.

Barker, G. M. (2002). *Molluscs as crop pests*. Wallingford, UK: CABI Publishing.

Berger, U., & Felsen, E. (1976). Distribution of species of *Moraxella* and *Moraxella*–like organisms in the nasopharynx of healthy human adults. *Medical Microbiology and Immunology, 162*, 239–240.

Bovre, K. (1970). Oxidase positive bacteria in the human nose, incidence and species distribution, as diagnosed by genetic transformation. *Acta Pathologica et Microbiologica Scandinavica Section B: Microbiology, 78*, 780–784.

Brown, A. P., Barker, A., Hopkins, A., & Nelson, D. (2011). Application of *Phasmarhabditis hermaphrodita* (Nemaslug) to commercial broad acre crops. *IOBC/WPRS Bulletin, 64*, 99–104.

Ehlers, R. U. (2005). Forum on safety and regulation. In P. S. Grewal, R. U. Ehlers, & D. Shapiro–Ilan (Eds.), *Nematodes as biocontrol agents* (pp. 107–114). Wallingford, UK: CABI Publishing.

Enright, M. R., & Griffin, C. T. (2004). Specificity of association between *Paenibacillus* spp. and the entomopathogenic nematodes, *Heterorhabditis* spp. *Microbial Ecology, 48*(3), 414–423.

Ester, A., Van Rozen, K., & Molendijk, L. P. G. (2003). Field experiments using the rhabditid nematode *Phasmarhabditis hermaphrodita* or salt as control measures against slugs in green asparagus. *Crop Protection, 22*(5), 689–695.

Ester, A., & Wilson, M. J. (2005). Application of slug parasitic nematodes. In P. S. Grewal, R. U. Ehlers, & D. Shapiro–Ilan (Eds.), *Nematodes as biocontrol agents* (pp. 421–429). Wallingford, UK: CABI Publishing.

France, A., & Gerding, M. (2000). Discovery of *Phasmarhabditis hermaphrodita* in Chile and its pathological differences with the U.K. isolate in slug control. *Journal of Nematology, 32*, 430.

Gavin, W. E., Mueller–Warrant, G. W., Griffith, S. M., & Banowetz, G. M. (2012). Removal of molluscicidal bait pellets by earthworms and its impact on control of the gray field slug (*Derocerus reticulatum* Mueller) in western Oregon grass seed fields. *Crop Protection, 42*, 94–101.

Genena, M. A. M., Mostafa, F. A. M., Fouly, A. H., & Yousef, A. A. (2011). First record for the slug parasitic nematode, *Phasmarhabditis hermaphrodita* (Schneider) in Egypt. *Archives of Phytopathology and Plant Protection, 44*(4), 340–345.

Glare, T., Caradus, J., Gelernter, W., Jackson, T., Keyhani, N., Köhl, J., et al. (2012). Have biopesticides come of age? *Trends in Biotechnology, 30*(5), 250–258.

Glen, D. M., Wiltshire, C. W., Hughes, L. A., Ester, A., Van Rozen, K., Castillejo, J., Iglesias, J., Speiser, B., Coupland, J., & Gwynn, R. (2000). *The use of slug–parasitic nematodes and other techniques for control of slug and snail damage in horticultural crops*. Paper presented at the BCPC Conference, Pests and Disease, Brighton.

Grewal, P. S., Grewal, S. K., Taylor, R. A. J., & Hammond, R. B. (2001). Application of molluscicidal nematodes to slug shelters: A novel approach to economic biological control of slugs. *Biological Control, 22*(1), 72–80.

Grimm, B. (2002). Effect of the nematode *Phasmarhabditis hermaphrodita* on young stages of the pest slug *Arion lusitanicus*. *Journal of Molluscan Studies, 68*(1), 25–28.

Hammond, R. B., & Byers, R. A. (2002). Ariolimacidae and Arionidae as pests in conservation tillage soybean and maize cropping in North America. In G. M. Barker (Ed.), *Molluscs as crop pests* (pp. 301–314). Wallingford, UK: CABI Publishing.

Hass, B., Glen, D. M., Brain, P., & Hughes, L. A. (1999). Targeting biocontrol with the slug–parasitic nematode *Phasmarhabditis hermaphrodita* in slug feeding areas: A model study. *Biocontrol Science and Technology, 9*(4), 587–598.

Hass, B., Hughes, L. A., & Glen, D. M. (1999). Overall versus band application of the nematode *Phasmarhabditis hermaphrodita* with and without incorporation into soil, for biological control of slugs in winter wheat. *Biocontrol Science and Technology, 9*(4), 579–586.

Karimi, J., Kharazi–Pakadel, A., & Roberts, S. J. (2003). Report of pathogenic nematodes of slugs, *Phasmarhabditis hermaphrodita* (Nematoda: Rhabditida) in Iran. *Journal of Entomological Society of Iran, 22*, 77–78.

MacMillan, K., Haukeland, S., Rae, R., Young, I., Crawford, J., Hapca, S., et al. (2009). Dispersal patterns and behaviour of the nematode *Phasmarhabditis hermaphrodita* in mineral soils and organic media. *Soil Biology and Biochemistry, 41*(7), 1483–1490.

Maupas, E. (1900). Modes et formes de reproduction des nématodes. *Archives de Zoologie Expérimentale et Générale, 8*, 463–624.

Moens, M., & Glen, D. M. (2002). Agriolimacidae, Arionidae and Milacidae as pests in West European oilseed rape. In G. M. Barker (Ed.), *Molluscs as crop pests* (pp. 425–439). Wallingford, UK: CABI Publishing.

Morgan, J. A. W., Kuntzelmann, V., Tavernor, S., Ousley, M. A., & Winstanley, C. (1997). Survival of *Xenorhabdus nematophilus* and *Photorhabdus luminescens* in water and soil. *Journal of Applied Microbiology, 83*(6), 665–670.

Muramatsu, K., & Kikuchi, M. (2005). Screening and isolation of trehalase–producing bacteria from a dairy farm drain. *Journal of Rakuno Gakuen University Natural Science, 30*, 15–19.

Nermut', J., Půža, V., & Mráček, Z. (2014). The effect of different growing substrates on the development and quality of *Phasmarhabditis hermaphrodita* (Nematoda: Rhabditidae). *Biocontrol Science and Technology, 24*(9), 1026–1038.

Pechova, H., & Foltan, P. (2008). The parasitic nematode *Phasmarhabditis hermaphrodita* defends its slug host from being predated or scavenged by manipulating host spatial behaviour. *Behavioural Processes, 78*(3), 416–420.

Pilz, C., Toepfer, S., Knuth, P., Strimitzer, T., Heimbach, U., & Grabenweger, G. (2014). Persistence of the entomoparasitic nematode *Heterorhabditis bacteriophora* in maize fields. *Journal of Applied Entomology, 138*(3), 202–212.

Poinar, G. O., Thomas, G. M., Haygood, M., & Nealson, K. H. (1980). Growth and luminescens of the symbiotic bacteria associated with the terrestrial nematode *Heterorhabditis bacteriophora. Soil Biology and Biochemistry, 12*, 5–10.

Rae, R., Riebeseil, M., Dinkelacker, I., Wang, Q., Herrmann, M., Weller, A. M., Dieterich, C., & Sommer, R. J. (2008). Isolation of naturally associated bacteria of necromenic *Pristionchus* nematodes and fitness consequences. *Journal of Experimental Biology, 211*(12), 1927–1936.

Rae, R. G., Tourna, M., & Wilson, M. J. (2010). The slug parasitic nematode *Phasmarhabditis hermaphrodita* associates with complex and variable bacterial assemblages that do not affect its virulence. *Journal of Invertebrate Pathology, 104*(3), 222–226.

Rosenthal, S. L. (1978). Clinical role of *Actinetobacter* and *Moraxella*. In G. L. Gilardi (Ed.), *Glucose non–fermenting Gram–negative bacteria in clinical microbiology* (pp. 105–107). Boca Raton, FL: CRC Press.

Ross, J. L., Ivanova, E. S., Severns, P. M., & Wilson, M. J. (2010). The role of parasite release in invasion of the USA by European slugs. *Biological Invasions, 12*(3), 603–610.

Schneider. (1859). Ueber eine Nematodenlarve und gewisse Verschiedenheiten in den Geschlechtsorganen der Nematoden. *Zeitschrift für wissenshaftliche Zoologie, 10*, 176–178.

South, A. (1992). *Terrestrial slugs, biology, ecology and control*. London: Chapman & Hall.

Speiser, B., & Andermatt, M. (1996). *Field trials with* Phasmarhabditis hermaphrodita *in Switzerland*. Paper presented at the Slug and Snail Pests in Agriculture, Canterbury, UK.

Steiner, G. (1923). Aplectana kraussei n. sp., eine in der Blattsespe *Lyda* sp. Parasitierende Nematodenform, nebst Bomerkungen uber das Seitenorgan der parasitishen Nematoden. *Zendralblatt für Bakteriologie Parasitenkunde Infecktionskankheiten und Hygien Abteilung I Original, 59*, 14–18.

Tan, L., & Grewal, P. S. (2001a). Infection behavior of the rhabditid nematode *Phasmarhabditis hermaphrodita* to the grey garden slug *Deroceras reticulatum. Journal of Parasitology, 87*(6), 1349–1354.

Tan, L., & Grewal, P. S. (2001b). Pathogenicity of *Moraxella osloensis*, a bacterium associated with the nematode *Phasmarhabditis hermaphrodita*, to the slug *Deroceras reticulatum. Applied and Environmental Microbiology, 67*(3–12), 5010–5016.

Tan, L., & Grewal, P. S. (2002). Endotoxin activity of *Moraxella osloensis* against the grey garden slug, *Deroceras reticulatum. Applied and Environmental Microbiology, 68*(8), 3943–3947.

Tandingan De Ley, I., McDonnell, R. D., Lopez, S., Paine, T. D., & De Ley, P. (2014). First record of *Phasmarhabditis hermaphrodita* (Nematoda: Rhabditidae) in North America. *Nematology, 16*, 1129–1138.

Wilson, M. J., Burch, G., Tourna, M., Aalders, L. T., & Barker, G. M. (2012). The potential of a New Zealand strain of *Phasmarhabditis hermaphrodita* for biological control of slugs. *New Zealand Plant Protection, 65*, 161–165.

Wilson, M. J., Glen, D. M., & George, S. K. (1993). The rhabditid nematode *Phasmarhabditis hermaphrodita* as a potential biological control agent for slugs. *Biocontrol Science and Technology, 3*(4), 503–511.

Wilson, M. J., Glen, D. M., George, S. K., & Hughes, L. A. (1995). Biocontrol of slugs in protected lettuce using the rhabditid nematode *Phasmarhabditis hermaphrodita. Biocontrol Science and Technology, 5*(2), 233–242.

Wilson, M. J., Glen, D. M., George, S. K., & Pearce, J. D. (1995). Selection of a bacterium for the mass production of *Phasmarhabditis hermaphrodita* (Nematoda Rhabditidae) as a biocontrol agent for slugs. *Fundamental and Applied Nematology, 18*, 419–425.

Wilson, M. J., Glen, D. M., George, S. K., Pearce, J. D., & Wiltshire, C. W. (1994). Biological control of slugs in winter wheat using the rhabditid nematode *Phasmarhabditis hermaphrodita. Annals of Applied Biology, 125*(2), 377–390.

Wilson, M. J., Glen, D. M., & Pearce, J. D. (1993). Biological control of molluscs. *World Intellectual Property Organisation, International Patent Publication Number:* WO/93/00816.

Wilson, M. J., Glen, D. M., Pearce, J. D., & Rodgers, P. B. (1995). Monoxenic culture of the slug parasite, *Phasmarhabditis hermaphrodita* with different bacteria in liquid and solid phase. *Fundamental and Applied Nematology, 18*, 159–166.

Wilson, M. J., Hughes, L. A., Hamacher, G. M., Barahona, L. D., & Glen, D. M. (1996). Effects of soil incorporation on the efficacy of the rhabditid nematode, *Phasmarhabditis hermaphrodita*, as a biological control agent for slugs. *Annals of Applied Biology, 128*(1), 117–126.

Wilson, M. J., Hughes, L. A., Jefferies, D., & Glen, D. M. (1999). Slugs (*Deroceras reticulatum* and *Arion ater* agg.) avoid soil treated with the rhabditid nematode *Phasmarhabditis hermaphrodita. Biological Control, 16*(2), 170–176.

Index

© Springer International Publishing Switzerland 2015

R. Campos-Herrera (ed.), *Nematode Pathogenesis of Insects and Other Pests*, Sustainability in Plant and Crop Protection 1, DOI 10.1007/978-3-319-18266-7

I

J

K

L

M

MIX
Papier aus verantwortungsvollen Quellen
Paper from responsible sources
FSC® C105338

If you have any concerns about our products,
you can contact us on
ProductSafety@springernature.com

In case Publisher is established outside the EU,
the EU authorized representative is:
Springer Nature Customer Service Center GmbH
Europaplatz 3, 69115 Heidelberg, Germany

Printed by Libri Plureos GmbH
in Hamburg, Germany